IMMUNOLOGY
of
INSECTS
and
OTHER ARTHROPODS

Ayodhya P. Gupta, Ph.D., Editor
Rutgers University

CRC Press
Boca Raton Ann Arbor London

Library of Congress Cataloging-in-Publication Data

Immunology of insects and other arthropods / Ayodhya P. Gupta, editor.
p. cm.
Includes bibliographical references and index.
ISBN 0-8493-7381-6
1. Insects—Immunology. 2. Arthropoda—Immunology. I. Gupta, A. P., 1928-
QL492.5.I45 1991
595.7′029—dc20 91-17675
CIP

Direct all inquiries to CRC Press, Inc., 2000 Corporate Blvd., N.W., Boca Raton, Florida 33431.

International Standard Book Number 0-8493-7381-6

Library of Congress Card Number 91-17675

Printed in the United States of America 1 2 3 4 5 6 7 8 9 0

Comparative Arthropod Morphology, Physiology, and Development

Ayodhya P. Gupta, Ph.D., Series Founder and Editor-in-Chief
Rutgers University, New Brunswick, NJ

CONTENTS

SERIES PREFACE

The series, *Comparative Arthropod Morphology, Physiology, and Development,* has been established to publish critical reviews and syntheses of topics that may have matured to a critical stage and call for organization of the available recent information into free-standing books. Comparative aspects of arthropod morphology (including gross morphology, paleomorphology, ultrastructure, and molecular morphology), physiology of organ systems, and developmental biology will receive priority, because such accounts on major arthropod groups (Aquatic and Terrestrial Chelicerata, Crustacea, Myriapoda, and Insecta) are often lacking. The series editor-in-chief, the Advisory Editorial Board and the Life Sciences editor of CRC Press will choose subjects for the series and invite researchers to serve as editors and contributors. Proposals for volumes from prospective authors will also be considered. The decision to accept proposals will rest with the series editor-in-chief and CRC Press, the advice and recommendation of the Advisory Editorial Board, and outside referees, when appropriate. The series will not publish proceedings of conferences or symposia. Volume authors and editors will be responsible for organizing, editing, and overseeing the publication of the volume under the overall supervision of the series editor-in-chief.

The first volume of the series, *Immunology of Insects and Other Arthropods,* organized and edited by the series editor-in-chief, fulfills the objectives of the series.

Ayodhya P. Gupta, Ph.D.
Series Founder and Editor-in-Chief
Professor
Department of Entomology
Rutgers University, New Brunswick
New Brunswick, NJ

PREFACE

As new evidence unfolds, the notion that the arthropod immune system, compared with that of vertebrates, is primitive or unsophisticated, is becoming increasingly untenable. We now know that at least six of the monosaccharides or their derivatives found on the plasma membranes of vertebrate cells are present on the immunocyte plasma membrane of one of the most primitive arthropods, the horseshoe crab, *Limulus polyphemus*. They most likely participate in foreign tissue recognition, phagocytosis, and encapsulation, and at least one (sialic acid or NANA) is involved in signaling and cell-cell recognition and activation of the complement pathway. Furthermore, specific inhibitors of the classical and alternative complement (C) pathway of mammalian C are present in the insect hemolymph, and it is now thought that the key components of the vertebrate C exist in insects. Proteins similar to the human α_2-macroglobulin have been found in the hemolymph of *L. polyphemus* and the American lobster, *Homarus americanus*. Evidence is accumulating that in some insects, a humoral adaptive response that possesses specificity and memory comparable to the functional attributes of the mammalian immunoglobulins, exists. These and other findings regarding the various components of the arthropod immune system strongly suggest that we should more vigorously explore other functions analogous to those of the vertebrate immune system. Indeed, we should not hesitate to propose plausible and provocative concepts even at the risk of appearing to venture into what might be considered as purely theoretical immunology at the moment. My chapter dealing with the molecular basis of encapsulation is an attempt in that direction. Some five years ago, I had proposed some functional analogies between the arthropod immunocytes and vertebrate B- and T-lymphocytes, a concept that has now been independently confirmed by two groups of workers on either side of the Atlantic. I hope the book as a whole will prompt such bold ventures.

The book is organized into four sections: Immunocytes and their Immune Reactions, Coagulation, Humoral Factors, and Biomedical Applications and Techniques. It includes the most comprehensive review to date on insect immunocytes and other hemocytes, including computer image analysis of immunocyte serial sections. The two basic cellular immune reactions (phagocytosis and encapsulation) are covered in detail, including the molecular basis of and roles of gap junctions in encapsulation, and how the latter is affected by polydnavirus and encapsulation-promoting factors. One chapter discusses why insect cells are immune to HIV. Chapters on coagulation, opsonins, and many of the humoral factors (such as α_2-macroglobulins, antibacterial immune factors in several insect orders, hemolymph proteins interacting with the mammalian complement cascade, adaptive humoral response in the American cockroach, and antigenic stimulation of hemagglutinin production in insects) provide new insights in insect immunology. Finally, two chapters discuss the applications of LAL (*Limulus* Amebocyte Lysate) in detecting endotoxin contamination in pharmaceuticals, medical devices, clinical diagnoses, and hygienic control.

In any multiauthored book some overlap is inevitable, and this one is no exception. Such overlaps, however, are retained only where necessary for understanding the discussion at hand. Both similarities and differences of opinions of authors discussing similar topics are cross-referenced. Each chapter presents an in-depth review of the topic it covers, and most point out gaps and areas for future research on those topics. Each contributor has had complete freedom to develop, interpret, and present his/her views. In the chapters of non-English-speaking contributors, editing has been confined to removing obvious infelicities in order to retain the original style and content.

An endeavor such as this book could not have been successfully completed without the cooperation and assistance of all the authors who responded to my invitation to contribute, and thus made thc book possible. To all of them, I am most grateful. Many of the contributors and I appreciate the permissions of various publishers and authors to reproduce copyrighted material in some chapters. For more than a decade, I have benefited from the understanding, ungrudging support, and enthusiasm of my wife and children on nine previous book projects and the present one. I know I have usurped much time that was rightfully theirs. To them I am forever indebted.

Ayodhya P. Gupta
New Brunswick, New Jersey
June 15, 1991

THE EDITOR

Ayodhya P. Gupta, Ph.D., is a Professor in the Department of Entomology at Rutgers University, New Brunswick, New Jersey.

Professor Gupta obtained a B.Sc. degree (Biology) from St. Andrew's College, India; and M.Sc. (Zoology) (standing first in order of merit among all M.Sc. students) from Banaras Hindu University, India; another M.Sc. (Zoology/Entomology) from the University of British Columbia, Canada; and the Ph.D. (Entomology) from the University of Idaho (UI) in 1964. He commenced his graduate studies at the University of Idaho after being awarded one of the five prestigious University fellowships. Professor Gupta joined the Department of Entomology at Rutgers in 1964 as a Research Fellow and became Assistant Professor in 1966, Associate Professor in 1968, and full Professor in 1973. He has also served as Director of the Graduate Program (1981 to 1985) and chairman of the Zoology section (1983 to 1984) at Rutgers.

Professor Gupta is a Life Fellow of the Entomological Society of India, Fellow of the Royal Entomological Society of London, England, member of the Entomological Society of America, American Association for the Advancement of Science, Society of Sigma Xi, Electron Microscopy Society of America, was charter member (1978 to 1981) of the Society of Scholarly Publications, and several other professional societies. He was made the honorary member of the Arthropodan Embryological Society of Japan in 1982. A clerid beetle (*Colyphus guptai*) is named in his honor (*Syst. Entomol.*, 2, 217, 1977).

Professor Gupta founded the *International Journal of Insect Morphology and Embryology (IJIME)* in 1970, and has nurtured it into one of the most prestigious world-class journals in his discipline, as its Editor-in-Chief since its inception. He also founded the *Distinguished International Award in Insect Morphology and Embryology* ($1500 and a gold medal) in 1984, under the sponsorship of the *IJIME,* and its publisher, Pergamon Press, to be awarded at each International Congress of Entomology, held at 4-year intervals. The third award will be presented by him at the 19th International Congress in China in 1992. In 1986, he founded a new international series titled *Recent Advances in Comparative Arthropod Morphology, Physiology, and Development,* (previously published by Rutgers University Press), to be published now by CRC Press.

Professor Gupta was twice honored by the Food and Agriculture Organization of the United Nations and United Nations Development Program by serving as their consultant and representative in India (1974) and Afghanistan (1982). He was selected by the President of the 15th International Congress of Entomology to serve as Chairman of the National Brochure Committee to organize the book, *Entomology in the U.S.A.* (1974 to 1976). He was also appointed by the President of the Council of Biology Editors to serve on its Editorial Policy Committee (1973 to 1976) that organized and published the book, *The Scientific Journal: Editoral Policies and Practices* (Mosby, 1976). Since 1976, he has been continuously invited by the international congresses of entomology to chair symposia and meetings, and has been awarded travel grants by the Entomological Society of America to attend these congresses. The Organizing Committee of the 19th Congress, to be held in China in 1992, has appointed him as Convenor of the program section on Morphology and Ultrastructure. He was invited (February 12, 1990) to serve as a member of the International Scientific Advisory Council for the *International Conference on Design and Function of the Nervous System,* to be held in Goa, India in September, 1992. Professor Gupta has served as chairperson and member of many committees of the Entomological Society of America (Eastern

Branch). He has served as external examiner of Ph.D. theses of foreign universities for many years, as referee for academic promotion at the request of the Academic Council of Riyadh University, Kingdom of Saudi Arabia (1980), and as reviewer for the MacArthur Fellowship (''genius awards'') of the MacArthur Foundation (1985).

Professor Gupta has expertise in four major areas of entomology: Insect Immune System, Endocrinology, Morphology, and Urban Pest (cockroaches) Management, and has published extensively in these areas over the past 29 years. His current research program deals with arthropod immunity, especially cellular immunity. He is a renowned world authority on insect hemocytes and their immunologic functions, and has contributed several invited review chapters, many research papers, and two books, *Insect Hemocytes* (1979) and *Hemocytic and Humoral Immunity in Arthropods* (1986). In another of his books, *Arthropod Phylogeny* (1979), he has discussed the significance of the major hemocyte types on the phylogeny of arthropods. During 1975 to 1985, Dr. Gupta conducted research on the morphogenetic effects of insect growth regulators. In addition to many research publications in this area, his three volumes on *Morphogenetic Hormones of Arthropods* were recently published (1989 to 1991) by Rutgers University Press. Professor Gupta is a leading world authority in insect morphology and has published two books, *Neurohemal Organs in Arthropods* (1983) and *Arthropod Brain* (1987) (judged to be ''beyond any doubt...a classic volume in its field,'' by *Crustaceana*) and numerous research papers. During 1968 to 1974, he conducted research on the population management of German cockroaches in hospitals and nursing homes in the inner-city areas in New Jersey. The objectives and the results of this study constitute a part of the Urban Pest Management Report, sponsored by the National Research Council (1980). His pioneering work on the significance of good and/or bad sanitation on the effectiveness of chemical control of roaches is often cited in works dealing with the management of these pests.

In recognition of his many contributions to his discipline, Professor Gupta was elected and inducted into the prestigious *Alumni Hall of Fame* of his alma mater (University of Idaho) on May 17, 1991, and on May 18, 1991, received the *Recognition and Honor* award of the College of Agriculture, University of Idaho.

CONTRIBUTORS

Peter B. Armstrong, Ph.D.
Department of Zoology
University of California
Davis, California

W. Peter Aston, Ph.D.
Department of Microbiology and Immunology
Queen's University
Kingston, Ontario

Ronald N. Berzofsky, Ph.D.
Research and Development — Biotechnology
Whittaker Bioproducts, Inc.
Walkersville, Maryland

June S. Chadwick, Ph.D.
Department of Microbiology and Immunology
Queen's University
Kingston, Ontario

D. Huw Davies, Ph.D.
Division of Biomolecular Sciences
King's College
London, England

Osmond J. D'Cruz, Ph.D.
Department of Obstetrics & Gynecology
College of Medicine
University of Oklahoma
Oklahoma City, Oklahoma

Stanley M. Dunn, Ph.D.
Department of Bioengineering
College of Engineering
Rutgers University
Piscataway, New Jersey

Michèle Durliat, Ph.D.
Biologie du Développement des Vertébrés Inférieurs
Université Paris-Sud — Orsay
Orsay, France

Laura E. Duwel-Eby
Department of Biological Sciences
University of Cincinnati
Cincinnati, Ohio

Peter L. Ey, Ph.D.
Department of Microbiology and Immunology
University of Adelaide
Adelaide, South Australia

Lenore M. Faulhaber
Department of Biological Sciences
University of Cincinnati
Cincinnati, Ohio

Petter Friberger, Dr.Med.Sci.
Biochemistry Research and Development
Kabi Diagnostica
Mölndal, Sweden

Peter Götz, Ph.D.
Division of Biology
Institute for General Zoology
Free University of Berlin
Berlin, Germany

Ayodhya P. Gupta, Ph.D.
Department of Entomology
Rutgers University
New Brunswick, New Jersey

Richard D. Karp, Ph.D.
Department of Biological Sciences
University of Cincinnati
Cincinnati, Ohio

Hiroto Komano, Ph.D.
Faculty of Pharmaceutical Sciences
University of Tokyo
Tokyo, Japan

Karen Zink McCullough
MMI Associates
Whitehouse Station, New Jersey

Michael F. Minnick, Ph.D.
Division of Biological Sciences
University of Montana
Missoula, Montana

Claude E. Monken, Ph.D.
The Wistar Institute
Philadelphia, Pennsylvania

Robert H. Painter, Ph.D.
Department of Biochemistry
University of Toronto
Ontario, Canada

James P. Quigley, Ph.D.
Department of Pathology
Health Sciences Center
State University of New York
Stony Brook, New York

Mike T. Siva-Jothy, Ph.D.
Department of Animal and Plant Sciences
University of Sheffield
Sheffield, United Kingdom

Kemet D. Spence, Ph.D.
Department of Microbiology
Washington State University
Pullman, Washington

Stefan Spycher, Ph.D.
Institute for Biochemistry and Molecular Biology
University of Bern
Bern, Switzerland

Alagarsamy Srinivasan, Ph.D.
The Wistar Institute
Philadelphia, Pennsylvania

Tina Trenczek, Ph.D.
Division of Biology
Institute for General Zoology
Free University of Berlin
Berlin, Germany

Ayyavoo Velpandi, Ph.D.
The Wistar Institute
Philadelphia, Pennsylvania

Haruhisa Wago, D.Sc.
Department of Medical Technology
Saitama Medical School Junior College
Saitama, Japan

Immunocytes and Their Immune Reactions

Chapter 1

CELLULAR AND HUMORAL IMMUNITY IN THE HORSESHOE CRAB, *LIMULUS POLYPHEMUS*

Peter B. Armstrong

TABLE OF CONTENTS

I. INTRODUCTION

Individual metazoans are threatened by a nearly continual challenge by microbial and metazoan pathogens. The life cycle time for most microbial pathogens is significantly shorter than that of the metazoan host, making it essentially impossible for the host population to outgrow populations of the pathogens. Instead, metazoans have evolved a varied array of defense strategies that limit their susceptibility to attack by potential pathogens. These include the presence of surface tissues that restrict penetration of pathogens into the interior, and a variety of specialized cells and humoral factors that fight pathogens which have succeeded in gaining access into the body. In higher animals, a majority of the internal defense systems involve cells and soluble factors of the blood. This chapter will review the defense systems of the arthropod, *Limulus,* with particular emphasis on the involvement of the blood cells and blood-borne soluble factors in defense.

II. PATHOGENIC CHALLENGE IN *LIMULUS*

Several pathogens of *Limulus* have been identified. Strains of the Gram-negative marine bacterium *Vibrio* have been isolated from moribund animals.[1] The animal is also subject to cutaneous infection by a variety of organisms, including filamentous and unicellular cyanobacteria[2] that eventually penetrate the cuticle and provide a pathway for invasion by other bacteria. The triclad turbellarid worm, *Bdelloura candida,* which is often abundant in the gills, causes lesions in the cuticle that are portals for entry of microbial pathogens into the body.[3-5] Although *Limulus* lives in a septic environment and is subject to frequent breach of the surface integument, the hemolymph of a majority of recently collected animals is sterile.[6] Presumably, invading pathogens are effectively controlled by the internal defenses. These defense systems will be the subject of this review.

III. CHARACTERIZATION OF THE HEMOCYTES OF *LIMULUS*

Limulus is one of the largest, easily obtainable arthropods. The blood occupies approximately one-third of its body mass. It is relatively easy to obtain 100 or more milliliters of blood from an average-sized adult.[7] *Limulus* is, thus, an ideal subject for hematologic studies requiring large volumes of material. Based on cell morphology, there appears to be only one blood cell type in the systemic circulation of the adult intermolt animal — the granulocyte or amebocyte.[8-12] This cell is an oval, plate-shaped structure, 15 to 20 μm in its longest dimension (Figure 1). The cytoplasm is packed with large, highly refractile exocytotic granules. The discoid shape is stabilized by a circumferential bundle of microtubules.[13] The amebocyte is stimulated by extravasation to become motile[14-18] and adhesive[15,19,20] and to undergo degranulation.[10,21-24] The alterations in morphology attendant with these responses result in a pleiomorphic population of cells that are very different in appearance from the unstimulated circulating cells, but nevertheless are derived from them[16-18,25] (Figure 2). This variation in cell morphology led some investigators to postulate the existence of multiple cell types in the blood.[26-31] However, careful study of both living and quickly fixed samples has convinced us that the peripheral blood of the adult contains but one cell type, with the more overt morphological heterogeneity ascribable to alterations resultant from cellular activation. Some degree of heterogeneity, such as a variable content of autofluorescent cytoplasmic granules[11] and electron-dense inclusions,[12] is real, but of unknown significance. *Limulus* does have a second blood cell, the cyanocyte,[32,33] but in the adult this cell is restricted to the vascular spaces of the ocular ganglion, and is apparently absent from the general circulation of the adult, intermolt animal.[11,34] The cyanoblast does appear, however, to be systemically distributed in the early larval stages.[35]

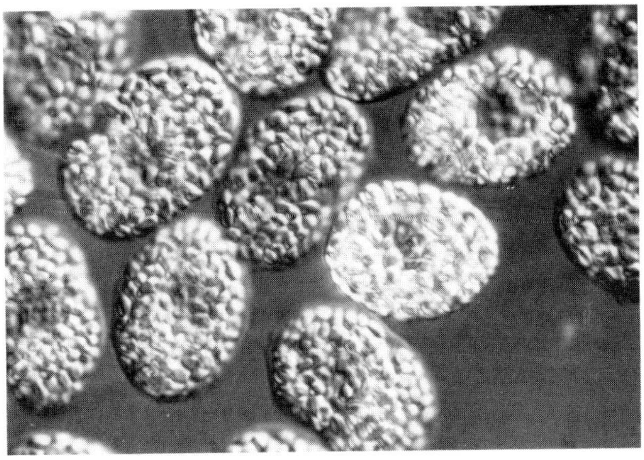

FIGURE 1. *Limulus* blood cells in the oval configuration characteristic of unextravasated blood. The cells are oval, nonadhesive, immobile, and fully granular. Cell morphology and behavior change dramatically upon the transformation that follows removal of the cells from the animal. (× 1243) (From Armstrong, P. B., *Exp. Cell Res.*, 107, 127, 1977. With permission.)

IV. THE ACTIVATION REACTION

As indicated above, the circulating amebocyte is an ovoid, granular, nonadhesive, and nonmotile cell. At the morphological level, the cell activation that follows extravasation establishes an alternative phenotype, in which the cell becomes pleiomorphic, adhesive, and motile, and from which the contents of the cytoplasmic granules are released into the external milieu by exocytosis. *Limulus* lacks the explosive blood cells of crustaceans, which respond to extravasation by cytolysis (reviewed in Hose et al.[36]). Instead, the extravasated *Limulus* blood cells survive,[16,17,37,38] but lose the cytoplasmic granules by exocytosis. The amebocyte is responsive to bacterial products, such as lipopolysaccharide (LPS, endotoxin — LPS is the principal toxin of Gram-negative bacteria).[1,21-23,39-42] The cells also activate, albeit more slowly, if prepared under LPS-free conditions,[17,23,25] indicating the existence of LPS-independent activation pathways.

Relatively little is known of the pathways that couple initial external stimulation with the family of intracellular responses that characterize the activated state. Activation is significantly retarded in the absence of extracellular divalent cations,[23,43] which suggests that movement of divalent cations into the cell may be involved. In this regard, Mg^{2+} appears to be more effective than Ca^{2+} in stimulating isolated cells to degranulate.[44] Presumably second messengers, such as cyclic nucleotides and/or the metabolic products of phosphoinositide breakdown, are involved, possibly acting through membrane receptors for LPS.[45,46] Cyclic-AMP appears to be involved, although, whether in the potentiation or in the inhibition of exocytosis is controversial. Agents that elevate c-AMP levels have been variously reported to retard[23] or potentiate[47] activation. The acquisition of locomotory capabilities and adhesiveness, and the extrusion of the cytoplasmic granules by exocytosis all imply that activation is accompanied by reorganizations of the cytoskeleton and alteration of the composition of the cell surface. Bundles of aligned actin filaments appear in the filopodial extensions of motile cells.[11,48] Surprisingly, a circumferential bundle of microtubules, which apparently stabilizes the discoidal shape of the unactivated cell, remains intact, at least through the early stages of activation, although it may be displaced from its original sub-plasmalemmal location.[48] It has been proposed that the aggregation of the amebocytes is mediated by an

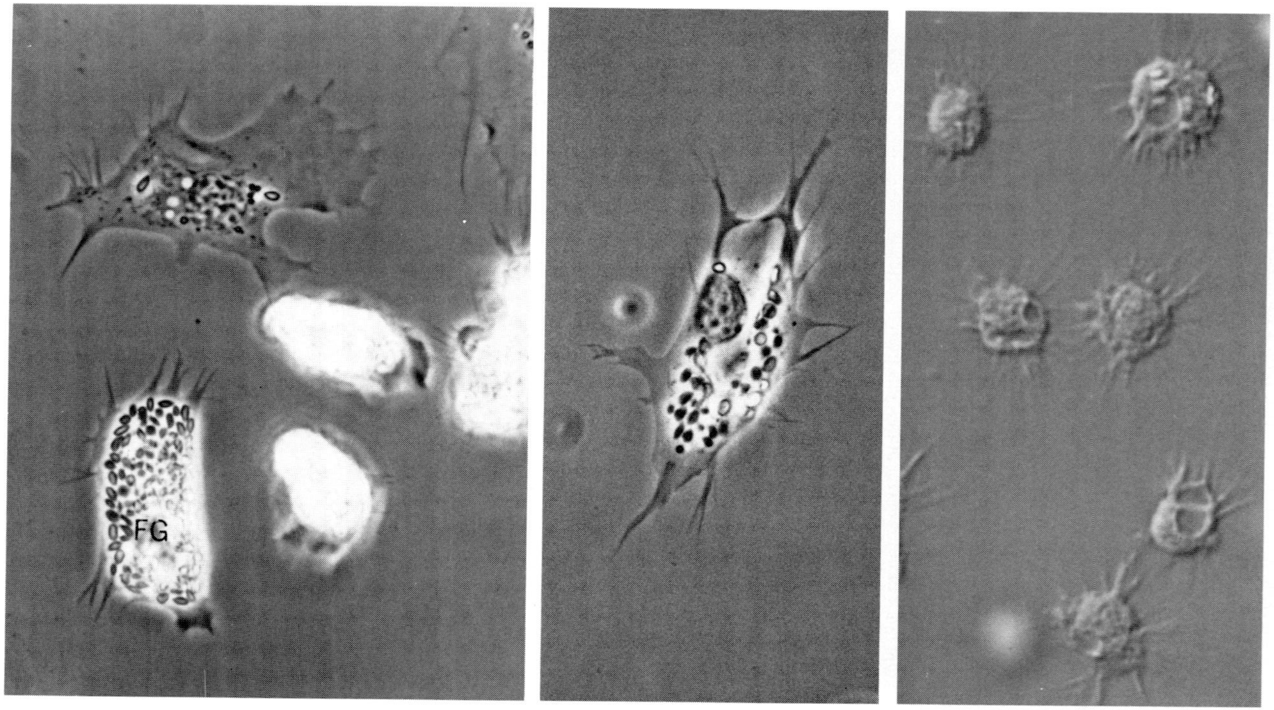

FIGURE 2. Extravasated blood cells adherent to a microscopic coverglass showing the range of phenotypes adopted by cells under these conditions. In the left panel, the two cells in the lower right are compact and highly motile, with hyaline pseudopods at the anterior ends. The cell in the lower left is partially flattened and fully granular. Such cells are motile and may revert to the highly motile phenotype. The cell in the upper center is highly flattened on the substratus and has suffered exocytosis of all but two of its complement of cytoplasmic granules. (From Armstrong, P. B., *Exp. Cell Res.*, 107, 127, 1977. With permission.) In the middle panel, the flattened cell is in the process of degranulation, with the large clear region in the center being the site of recent exocytosis. (From Armstrong, P. B., *Blood Cells of Marine Invertebrates: Experimental Systems in Cell Biology*, W. D. Cohen, Ed., Alan R. Liss, p. 77. With permission.) The cells of the right panel have been stimulated to degranulate by exposure to bacterial lipopolysaccharide. From Armstrong, P. B., *Blood Cells of Marine Invertebrates: Experimental Systems in Cell Biology*, W. D. Cohen, Ed., Alan R. Liss, p. 77. With permission.) (× 1046)

interaction between a cell surface hemagglutinin and coagulogen bound to the cell surface,[43,49,50] but this proposal has not been rigorously proved. (Coagulogen is the structural protein of the extracellular blood clot in *Limulus*.)

The functional significance of activation is more clear. One of the major functions of the granular amebocyte is the regulation of hemostasis. Two of the principal requirements of any hemostatic system are that it be inactive in the intact animal and that it be responsive to bleeding by a rapid activation that is spatially restricted to the wound site. The activation response of the granular amebocyte clearly fits these requirements. The acquisition of cellular adhesiveness allows the formerly nonadhesive amebocytes to attach to the borders of the wound and to one another to form a cellular hemostatic plug that arrests bleeding.[14,19,26,51,52] In addition to the cellular clot, the hemostatic response includes the formation of an extracellular clot,[42,53] in which the amebocytes play a dominant role. The extracellular clot is composed of fibers of the protein coagulin.[54-64] In the intact animal, the soluble precursor of this protein, coagulogen, is absent from the plasma, but is instead sequestered in the cytoplasmic granules of the amebocytes.[65-67] Granule exocytosis releases coagulogen into the external milieu. Coagulogen is converted to coagulin by a proteolytic cleavage administered by a particular serine protease, known as clotting enzyme.[68-73] The best characterized exogenous activator of the clotting pathway is the bacterial toxin, lipopolysaccharide, which activates a serine protease, factor C, which then initiates a proteolytic cascade that culminates in the activation of proclotting enzyme and the cleavage of coagulogen.[74-77] After proteolysis, the coagulin monomers polymerize into the fibrillar elements of the clot.[55-58,60,78,79] Like the coagulogen, clotting enzyme appears to be present within the cytoplasmic granules prior to blood cell activation.[66] Thus, unlike the familiar clotting system of mammals, in which major elements of the clotting pathway are free at all times in the plasma, the entire system in *Limulus* is sequestered within the secretory granules of the blood cells until granule release and expression of the clotting pathway.

When bacteria are injected into the hemocoel, the hemocytes are activated within the confines of the circulatory system. Exocytosis occurs and the released coagulogen undergoes polymerization into the extracellular clot.[1,42,80] Large doses of bacteria are invariably fatal, but surprisingly, if the bacterial challenge is not excessive, or when moderate doses of purified LPS are used as the challenging agent, a fraction of the animals survive this episode of disseminated intravascular coagulation, with its attendant loss of most or all of the hemocytes.[1,34,42,81] These observations have suggested that a second function of the coagulogen-based extracellular clotting system to hemostasis is the prevention of systemic dissemination of Gram-negative bacteria that have gained access to the hemocoel.[1,9,40,65,82] Indeed, the coagulin clot does immobilize bacteria,[1,42] supporting the suggestion that it functions *in vivo* to restrict their movement beyond the initial site of entry.

The post-challenge physiology of injected animals has not been much studied. It would be interesting, for example, to investigate the processes of clot resolution in surviving animals. To my knowledge, proteases analogous to the plasminogen activator-plasmin system of mammalian clot resolution have not been identified in *Limulus*. Also of interest is the recovery of the normal abundance of blood cells. Restoration of cell number after bleeding requires several weeks.[83] The relative contributions of mitosis of a stem cell population and of release of cells from sequestered sites in the animal have not been clarified. The hemopoietic stem tissue has recently been identified as tissue associated with pillars that support the hematologic space of the gill leaves. Isolated gill tissue has been demonstrated to produce mature amebocytes in organ culture.[84]

V. PHAGOCYTOSIS

One of the most important antimicrobial reactions of the blood cells of higher animals is phagocytosis of small particles and encapsulation of foreign bodies too large to be phago-

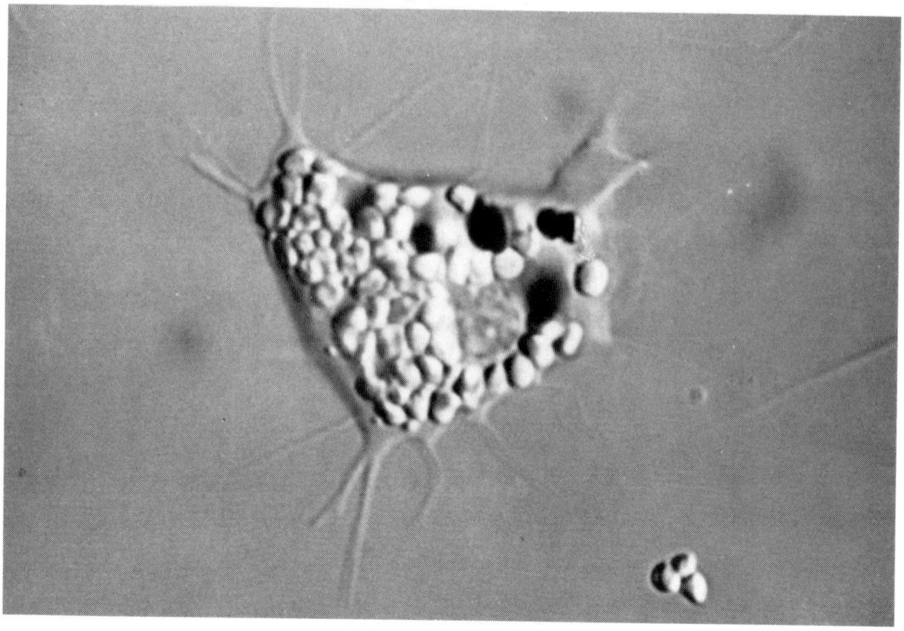

FIGURE 3. Flattened, granulated amebocyte in culture that has phagocytosed particles of carbonyl iron (black inclusions). It has been established that the particles are within the cells, rather than on the cell surface, by careful focusing with high numerical aperture microscope objectives and because the particles move with the flow of the cytoplasm of the locomoting cell. (× 2394) (From Armstrong, P. B., *Blood Cells of Marine Invertebrates: Experimental Systems in Cell Biology,* W. D. Cohen, Ed., Alan R. Liss, p. 77.With permission.)

cytosed.[85,86] In the latter reaction, a number of blood cells attach to the foreign body and establish a many-layered cellular capsule around it.[87-89] In arthropods, the responsible blood cells for encapsulation and phagocytosis are, variously, the granulocyte,[88,90-92] the plasmotocyte,[84,93-102] or both cell types.[89] Both phagocytosis and encapsulation require the ability to discriminate between self and nonself (see Chapter 4). Surprisingly, for such a well-studied organism as *Limulus*, little is known of the possible import of these reactions for defense. The amebocytes are definitely capable of phagocytosis,[103] but their phagocytic capabilities appear to be relatively feeble. Although many of the reports of phagocytosis by the blood cells of arthropods do not comment on the avidity of the response, it appears that arthropod hemocytes are not very active phagocytes, in comparison to the activities of vertebrate and molluscan blood cells.[85,86,104-106] In studies of the phagocytic processing of carbonyl iron particles by *Limulus* amebocytes *in vitro,* we seldom found more than two to five particles in any one cell (Figure 3), and most cells even in our best preparations contained no particles. It is possible that in this and the few other studies of phagocytosis in this organism,[1,8,65,80,82,107-109] optimal conditions were not utlized, but present evidence does not suggest that phagocytic attack by blood cells is very important.

Other arthropods have been shown to possess fixed pinocytic and phagocytic tissues and participate in the clearance of foreign particles from the circulation.[110-115] To my knowledge, analogous systems have not been described for *Limulus*. Likewise, I am not aware of reports of an encapsulation reaction in *Limulus*. One feature of *Limulus* that may impede the demonstration of a phagocytic system based either on activities of the blood cells or fixed phagocytic tissues is the above-mentioned clotting system. My attempts to search for a fixed phagocytic tissue *in vivo* were confounded by the rapid development of a clot at the site of injection that entrapped most of the injected particles. This was observed even when

LPS-free particles (carbonyl iron rendered free of LPS by prolonged dry heat sterilization at 180°C) were utilized. Presumably, foreign bodies enmeshed in a coagulin gel would be insulated from contact with phagocytic cells. This response to the introduction of foreign particles is an expression of the exocytosis/clotting reaction. Similar responses, in which foreign particles in the circulation are entrapped in a clot of blood cells and a matrix of polymerized extracellular clot, may be common in arthropods, and may serve to prevent systemic dissemination of the foreign particles, similar to the proposed containment function of the *Limulus* clotting system.[8,105,116-119]

VI. OTHER CELL-MEDIATED DEFENSE REACTIONS

One defense reaction present in most other arthropods that appears to be absent in *Limulus* is the prophenoloxidase system. In this system, the enzyme phenoloxidase, when activated at the surfaces of bacterial and fungal invaders, generates toxic quinones that participate in killing of the target cells.[120] The prophenoloxidase system of crustaceans is contained within the granulocytes.[121,122] Its release and activation is triggered by the presence of LPS or fungal polysaccharides (the β-1,3-glucans).[123] A thorough study of the blood and blood cells of *Limulus* failed to produce evidence for the presence of this system in that organism.[34,124] The coagulogen-based clotting system is apparently inactive in the face of challenge by naturally occurring fungal β-1,3-glucans.[124] Thus, it appears that *Limulus*, unlike crustaceans and insects, lacks this antifungal defense system. The prophenoloxidase system of crustaceans appears to participate in the recognition of foreign particles and to facilitate their uptake by phagocytes.[125] Its absence in *Limulus* may contribute to the relatively weak phagocytic response shown by the blood cells.

There is evidence for the presence of antimicrobial systems associated with serum and the blood cells.[82,126-132] *Limulus* serum has hemolytic and opsonic activities that resemble vertebrate complement and that may contribute to antimicrobial activity[133,134] (see also Chapter 14). The biochemical identification of the cellular and plasma components with antimicrobial activity has revealed several interesting elements. Important cellular contributions to antimicrobial defense include small peptide antibiotics, and the proteins α_2-macroglobulin and anti-LPS factor.

The family of peptide antibiotics isolated from the blood cells, termed the tachyplesins and the polyphemusins, have potent activity against Gram-positive and Gram-negative bacteria and fungi.[135-138] These peptides contain 17 or 18 residues with a preponderance of cationic and hydrophobic amino acids giving them amphipathic characteristics.[139]

The protease inhibitors that are released from the amebocytes during exocytosis of the cytoplasmic granules are a second class of factors of possible importance in antimicrobial defense.[38] Included in the inhibitors released from the cells are a collection of active-site inhibitors, with both low molecular weight, acid-stable[38,140] and larger, acid-labile[38] members. In addition, the blood cells release the protease inhibitor, α_2-macroglobulin.[38,141] The cells from a given volume of blood contain significantly less α_2-macroglobulin than is present in solution in the plasma, but the cell-associated form of α_2-macroglobulin may be significant in the inactivation of proteases in the interiors of the densely cellular hemostatic plugs that form at sites of vascular injury. As reviewed in Chapters 10 and 11, α_2-macroglobulin appears to play a role in the clearance of proteases of both endogenous and exogenous origin, and in this role, serves as an important element in immunity.

The pathogenicity of Gram-negative bacteria is largely due to the release of lipopolysaccharide into the internal milieu of the host organism.[142] As was described above, LPS is a physiological activator of the amebocyte and of the clotting system released from the activated cells. Detoxification and clearance of LPS in *Limulus* appears to involve a low molecular mass protein, anti-LPS factor, which is released from the activated cells, and which binds LPS.[143-145] Anti-LPS factor shows antibacterial activity for rough strains of Gram-negative bacteria[146] and is hemolytic for LPS-sensitized mammalian erythrocytes.[147] The first attribute may contribute to the antimicrobial activities of serum noted above[132] and the second to the complement-like hemolytic activity described for *Limulus* serum.[133,134] In addition to anti-LPS factor, factor C, the serine protease that initiates the clotting pathway, also binds LPS.[75,77]

VII. HUMORAL DEFENSE SYSTEMS

The humoral α_2-macroglobulin system of *Limulus* will be reviewed elsewhere in this volume.[148] In addition, the plasma contains several lectins (e.g., carbohydrate-binding proteins) with differing spectra of reactivity to oligosaccharides. Limulin is a Ca^{++}-dependent lectin[149,150] that binds the sugars, 2-keto-3-deoxyoctonate (KDO),[151,152] and sialic acid,[153-160] with high affinity for homopolymers of *N*-acetylneuraminic acid.[161] Reactivity with KDO enables the lectin to bind to KDO residues on bacterial lipopolysaccharide[162-164] and to agglutinate Gram-negative bacteria.[132] The binding of limulin to polymers of *N*-acetylneuraminic acid also contributes to agglutination of Gram-negative organisms, since these oligosaccharides also occur in the bacterial capsules.[161] In addition to limulin, the plasma contains other lectins with sialic acid recognition specificities.[165,166] Polyphemin is another serum lectin with affinity for galactosyl[167] and teichoic acid.[168] This latter activity enables polyphemin to agglutinate certain Gram-positive bacteria.[168,169] Polyphemin appears to contain bound copper and may associate with hemocyanin or even be one of the subunits of hemocyanin.[170] A third hemagglutinin that is associated with the hemocytes rather than the plasma has been described by Srimal et al.[50] The hemocytes lack limulin.[50] The cell-associated hemagglutinin has not been well characterized, but it appears to bind coagulogen and may participate in the aggregation of activated hemocytes.[50]

One of the most abundant plasma proteins is a homologue to mammalian C-reactive protein.[171-173] C-reactive protein is present at approxiamtely 2 mg/ml of plasma and is distinct from the plasma lectins, limulin and polyphemin.[166] The molecular masses of the subunits of C-reactive protein (22, 28, and 30 kDa) are similar to those of limulin (30 kDa), which appears to have resulted in some confusion in the literature. Limulin is present in smaller quantities than C-reactive protein (e.g., 2 μg/ml).[166] Mammalian C-reactive protein is an acute-phase protein that reacts directly with bacterial coat oligosaccharides in a Ca^{++}-dependent manner to initiate several humoral defense systems including agglutination of the microbes and the activation of complement.[174] *Limulus* C-reactive protein is a doubly-stacked hexagonal aggregate of 24-kDa subunits[175] with a native molecular mass of approximately 5×10^5. Although the function of *Limulus* C-reactive protein has not been characterized, its abundance in the plasma suggests that it is important in antimicrobial defense.

VIII. SUMMARY

Limulus, like many large arthropods, is relatively long lived, requiring 9 to 12 years to reach maturity,[176-177] and with a maximum life span estimated to be in excess of 14 to 19 years.[178] One requirement for such longevity is the ability to defend against potentially pathogenic microbes. Although our understanding of the cellular and humoral defense systems in *Limulus* is obviously incomplete, a number of cellular- and humoral-based systems have been identified. The coagulogen-based clotting system is reasonably well studied and is clearly important in processes of hemostasis and containment of invading Gram-negative organisms. In contrast, the several possible systems of defense against fungal, protozoan, and metazoan pathogens are less well characterized. Whether, and under what circumstance, the organism possesses a phagocytic/encapsulation system of defense is not well understood.

Clearly *Limulus* differs from members of other classes or arthropods in the absence of a prophenoloxidase system. The defense functions of the humoral lectins, α_2-macroglobulin and C-reactive protein, are incompletely understood. In many respects, *Limulus* is an ideal subject for the characterization of the internal defense systems. The animal is large, reasonably hardy, and readily available. An average-sized adult yields 100 to 150 ml of blood and it is relatively easy to collect one or more liters of blood in an hour or so. It is probable that no other readily available arthropod offers so much blood for so little effort. Furthermore, the presence of a single cell type in the blood is a simplification not seen in many other arthropods (which typically have several more or less distinct cells in the blood;[36,97,179-184] see Chapter 2) or, for that matter, in other higher animals. Thus, one can suggest that relatively rapid progress is possible in defining and characterizing the processes responsible for immunity in *Limulus*.

ACKNOWLEDGMENTS

Original research reported in this review was supported by Grant No. GM35185 from the NIH. I thank Dr. Frederick R. Rickles for helpful suggestions for improving the manuscript.

REFERENCES

1. **Bang, F. B.,** A bacterial disease of *Limulus polyphemus, Bull. Johns Hopkins Hosp.*, 98, 325, 1956.
2. **Leibovitz, L.,** Cyanobacterial diseases of the horseshoe crab *(Limulus polyphemus), Biol. Bull. (Woods Hole, Mass.)*, 171 (Abstr.) 482, 1986.
3. **Goff, J. M. and Leibovitz, L.,** A gill disease of *Limulus polyphemus* associated with triclad trubellarid worm infection, *Biol. Bull. (Woods Hole, Mass.)*, 163 (Abstr.) 392, 1982.
4. **Landy, R. B. and Leibovitz, L.,** Preliminary study on the toxicity and therapeutic efficacy of formalin in the treatment of triclad turbellarid worm infestations in *Limulus polyphemus, Proc. 16th Annu. Meet. Soc. Invertebr. Pathol.*, Cornell University (Abstr.), 1983.
5. **Leibovitz, L. and Lewbart, G. A.,** A green algal (Chlorophycophytal) infection of the dorsal surface of the exoskeleton, and associated organ structures in the horseshoe crab, *Limulus polyphemus, Biol. Bull. (Woods Hole, Mass.)*, 1973 (Abstr.) 430, 1987.
6. **Brandin, E. R. and Pistole, T. G.,** Presence of microorganisms in the hemolymph of the horseshoe crab, *Limulus polyphemus, Appl. Environ. Microbiol.*, 49, 718, 1985.
7. **Armstrong, P. B.,** Amebocytes of the American "horseshoe crab", *Limulus polyphemus*, in *Blood Cells of Marine Invertebrates: Experimental Systems in Cell Biology and Comparative Physiology*, Cohen, W. D., Ed., Alan R. Liss, New York, 1985, 253.
8. **Loeb, L.,** On the blood lymph cells and inflammatory processes of *Limulus, J. Med. Res.*, 7, 145, 1902.
9. **Levin, J. and Bang, F. B.,** The role of endotoxin in the extracellular coagulation of *Limulus* blood, *Bull. Johns Hopkins Hosp.*, 115, 265, 1964.
10. **Dumont, J. N., Anderson, E., and Winner, G.,** Some cytologic characteristics of the hemocytes of *Limulus* during clotting, *J. Morphol.*, 119, 181, 1966.
11. **Armstrong, P. B.,** Adhesion and motility of the blood cells of *Limulus*, in *Blood Cells of Marine Invertebrates: Experimental Systems in Cell Biology and Comparative Physiology*, Cohen, W. D., Ed., Alan R. Liss, New York, 1985, 77.
12. **Copeland, D. E. and Levin, J.,** The fine structure of the amebocyte in the blood of *Limulus polyphemus*. I. Morphology of the normal cell, *Biol. Bull. (Woods Hole, Mass.)*, 169, 449, 1985.
13. **Cohen, W. D. and Nemhauser, I.,** Marginal bands and the cytoskeleton in blood cells of marine invertebrates, in *Blood Cells of Marine Invertebrates: Experimental Systems in Cell Biology and Comparative Physiology*, Cohen, W. D., Ed., Alan R. Liss, New York, 1985, 3.
14. **Loeb, L.,** Studies on cell granula and amoeboid movement of the blood cells of *Limulus, Univ. Penn. Med. Bull.*, May 1905, 1, 1905.
15. **Loeb, L.,** The movements of the amebocytes and the experimental production of amebocyte (cell fibrin) tissue, *Washington University Studies*, 8, 3, 1920.

16. **Armstrong, P. B.,** Interaction of the motile blood cells of the horseshoe crab, *Limulus*. Studies on contact paralysis of pseudopodial activity and cellular overlapping *in vitro*, *Exp. Cell Res.*, 107, 127, 1977.
17. **Armstrong, P. B.,** Motility of the *Limulus* blood cell, *J. Cell Sci.*, 37, 169, 1979.
18. **Armstrong, P. B.,** Motility of the *Limulus* amebocyte, *Prog. Clin. Biol. Res.*, 29, 73, 1979.
19. **Loeb, L.,** Amoeboid movement, tissue formation and consistency of protoplasm, *Am. J. Physiol.*, 56, 140, 1921.
20. **Loeb, L.,** Amoeboid movement and agglutination in amoebocytes of *Limulus* and the relation of these processes to phagocytosis, tissue formation and thrombosis, in *Colloid Chemistry*, Vol. 2, J. Alexander, Ed., Chemical Catalogue Co., New York, 1928, 487.
21. **Ornberg, R. L. and Reese, T. S.,** Beginning of exocytosis captured by rapid-freezing of *Limulus* amebocytes, *J. Cell Biol.*, 90, 40, 1981.
22. **Ornberg, R. L. and Reese, T. S.,** Comparison of compound with plasmalemmal exocytosis in *Limulus* amebocytes, *Meth. Cell. Biol.*, 23, 301, 1981.
23. **Armstrong, P. B. and Rickles, F. R.,** Endotoxin-induced degranulation of the *Limulus* amebocyte, *Exp. Cell Res.*, 140, 15, 1982.
24. **Ornberg, R. L.,** Exocytosis in *Limulus* amebocytes, in *Blood Cells of Marine Invertebrates: Experimental Systems in Cell Biology and Comparative Physiology*, W. D. Cohen, Ed., Alan R. Liss, New York, 1985, 127.
25. **Armstrong, P. B.,** Adhesion and spreading of *Limulus* blood cells on artificial surfaces, *J. Cell Sci.*, 44, 243, 1980.
26. **Loeb, L.,** On the coagulation of the blood of some arthropods and on the influence of pressure and tractions on the protoplasm of the blood cells of arthropods, *Biol. Bull.* (*Woods Hole, Mass.*), 4, 301, 1903.
27. **Grégoire, C.,** Sur la coagulation du sang de *Limulus polyphemus*, *Arch. Int. Physiol.*, 60, 97, 1952.
28. **Grégoire, C.,** Blood coagulation in arthropods. VI. A study by phase contrast microscopy of blood reactions *in vitro* in Onychophora and in various groups of arthropods, *Arch. Biol.* (Liège), 66, 489, 1955.
29. **Grégoire, C.,** Haemolymph coagulation in arthropods, *Symp. Zool. Soc. London*, 27, 45, 1970.
30. **Shur-Jessen, P., Baek, L., and Jakobsen, P. P.,** Microscopical, biochemical and immunological studies of the immune defense system of the horseshoe crab, *Limulus polyphemus*, *Biol. Bull.* (*Woods Hole, Mass.*), 176, 290, 1989.
31. **Jakobsen, P. P. and Shur-Jessen, P.,** The horseshoe crab *Tachypleus tridentatus* has two kinds of hemocytes: granulocytes and plasmatocytes, *Biol. Bull.* (*Woods Hole, Mass.*), 178, 55, 1990.
32. **Fahrenbach, W. H.,** The cyanoblast, hemocyanin formation in *Limulus polyphemus*, *J. Cell Biol.*, 44, 445, 1970.
33. **Fahrenbach, W. H.,** The visual system of the horseshoe crab *Limulus polyphemus*, *Int. Rev. Cytol.*, 41, 285, 1975.
34. **Srimal, S.,** The role of lipopolysaccharide binding proteins in the coagulation system of *Carcinoscoprius rotundicauda*, Ph.D. thesis, University of Calcutta, India, 1984.
35. **Sherman, R. G.,** Chelicerates, in *Invertebrate Blood Cells*, Vol. 2, Ratcliffe, N. A. and Rowley, A. F., Eds., Academic Press, New York, 1981, 355.
36. **Hose, J. E., Martin, G. G., and Gerard, A. S.,** A decapod hemocyte classification scheme integrating morphology, cytochemistry, and function, *Biol. Bull.* (*Woods Hole, Mass.*), 178, 33, 1990.
37. **Belamarich, F. A., Stiller, R. A., and Shepro, D.,** The release reaction: is it a common mechanism in all types of hemostatic cells? *Ser. Hematol.*, 6, 418, 1973.
38. **Armstrong, P. B. and Quigley, J. P.,** Proteinase inhibitory activity released from the horseshoe crab blood cell during exocytosis, *Biochim. Biophys. Acta*, 827, 453, 1985.
39. **Levin, J. and Bang, F. B.,** A description of cellular coagulation in the *Limulus*, *Bull. Johns Hopkins Hosp.*, 115, 337, 1964.
40. **Levin, J.,** Blood coagulation and endotoxin in invertebrates, *Fed. Proc.*, 26, 1707, 1967.
41. **Levin, J.,** Blood coagulation in the horseshoe crab *(Limulus polyphemus)*: a model for mammalian coagulation and hemostasis, in *Animal Models of Thrombosis and Hemorrhagic Disease* [DHEW Publication No. (NIH) 76-982], Washington, D.C., 1976, 87.
42. **Bang, F. B.,** Ontogeny and phylogeny of response to Gram-negative endotoxins among the marine invertebrates, *Prog. Clin. Biol. Res.*, 29, 109, 1979.
43. **Kenney, D. M., Belamarich, F. A., and Shepro, D.,** Aggregation of horseshoe crab *(Limulus polyphemus)* amebocytes and reversible inhibition of aggregation by EDTA, *Biol. Bull.* (*Woods Hole, Mass.*), 143, 548, 1972.
44. **Cannon, G. W., Tsuchiya, M., Rittschof, D., and Bonaventura, J.,** Magnesium dependence of endotoxin-induced degranulation of *Limulus* amebocytes, *Biol. Bull.* (*Woods Hole, Mass.*), 171, 330, 1986.
45. **Liang, S.-M., Sakmar, T. P., and Liu, T.-Y.,** Studies on *Limulus* amebocyte lysate. III. Purification of an endotoxin-binding protein from *Limulus* amoebocyte membranes, *J. Biol. Chem.*, 255, 5586, 1980.
46. **Liang, S.-M., Liang, C.-M., and Liu, T.-Y.,** Studies on *Limulus* amoebocyte. Isolation and identification of a membrane-bound protein activator of cyclic nucleotide phosphodiesterase from *Limulus* amoebocyte, *J. Biol. Chem.*, 256, 4968, 1981.

47. **Liang, S.-M., Hascall, G., Liu, T., and Liu, T.-Y.,** *Limulus* amebocyte clotting cascade: roles of endotoxin and adenylate cyclase, *J. Protein Chem.*, 4, 151, 1985.
48. **Tablin, F. and Levin, J.,** The fine structure of the amebocyte of *Limulus polyphemus*. II. The amebocyte cytoskeleton: a morphological analysis of native, activated, or endotoxin-stimulated amebocytes, *Biol. Bull. (Woods Hole, Mass.)*, 175, 417, 1988.
49. **Belamarich, F. A.,** Nature of the contents of large granules of *Limulus* amebocytes, in *Animal Models of Thrombosis and Hemorrhagic Disease* [DHEW Publication No. (NIH) 76-982], Washington, D.C., 1976, 101.
50. **Srimal, S., Dorai, D. T., Somasundaran, M., Bachhawat, B. K., and Miyata, T.,** A new hemagglutinin from the amoebocytes of the horseshoe crab *Carcinoscorpius rotundicauda*. Purification and role in cell aggregation, *Biochem. J.*, 230, 321, 1985.
51. **Howell, W. H.,** Observation on the chemical composition and coagulation of the blood of *Limulus polyphemus, Callinectes hastatus,* and *Cucumaria* sp., *Johns Hopkins Univ. Circulars,* 5, 4, 1885.
52. **Maluf, N. S.,** The blood of arthropods, *Q. Rev. Biol.*, 14, 149, 1939.
53. **Bursey, C. R.,** Histological response to injury in the horseshoe crab, *Limulus polyphemus, Can. J. Zool.*, 55, 1158, 1977.
54. **Holme, R. and Solum, N. O.,** Electron microscopy of the gel protein formed by clotting of *Limulus polyphemus* hemocyte extracts, *J. Ultrastruct. Res.*, 44, 329, 1973.
55. **Nakamura, S., Iwanaga, S., Harada, T., and Niwa, M.,** A clottable protein (coagulogen) from amoebocyte lysate of Japanese horseshoe crab *(Tachypleus tridentatus)*. Its isolation and biochemical properties, *J. Biochem. (Tokyo)*, 80, 1011, 1976.
56. **Nakamura, S., Takagi, T., Iwanaga, S., Niwa, M., and Takahaski, K.,** A clottable protein (coagulogen) of horseshoe crab hemocytes. Structural change of its polypeptide chain during gel formation, *J. Biochem. (Tokyo)*, 80, 649, 1976.
57. **Nakamura, S., Iwanaga, S., Niwa, M., and Harada, T.,** A clottable protein (coagulogen) from hemocyte lysate of Japanese horseshoe crab *(Tachypleus tridentatus):* isolation, properties and gel formation, *Jpn. J. Med. Sci. Biol.*, 30, 74, 1977.
58. **Tai, J. Y., Seid, R. C., Huhn, R. D., and Liu, T.-Y.,** Studies on *Limulus* amoebocyte lysate. II. Purification of the coagulogen and the mechanism of clotting, *J. Biol. Chem.*, 252, 4773, 1977.
59. **Mosesson, M. W., Wolfenstein-Todel, C., Levin, J., and Bertrand, O.,** Characterization of amebocyte coagulogen from the horseshoe crab *Limulus polyphemus, Thromb. Res.*, 14, 765, 1979.
60. **Takagi, T., Hokama, Y., Morita, T., Iwanaga, S., Nakamura, S., and Niwa, M.,** Amino acid sequence studies on horseshoe crab *(Tachypleus tridentatus)* coagulogen and the mechanism of gel formation, *Prog. Clin. Biol. Res.*, 29, 169, 1979.
61. **Morita, T., Nakamura, T., Ohtsubo, S., Tanaka, S., Miyata, T., Hiranaga, M., and Iwanaga, S.,** Intracellular proteinases and inhibitors associated with the hemolymph coagulation system of the horseshoe crabs *(Tachypleus tridentatus Limulus polyphemus),* in *Proteinase Inhibitors: Medical and Biological Aspects,* Katunuma, N., Umezawa, H., and Holzer, H., Eds., Springer-Verlag, Berlin, 1983, 229.
62. **Miyata, T., Hiranaga, M., Umezu, M., and Iwanaga, S.,** Amino acid sequence of the coagulogen from *Limulus polyphemus* hemocytes, *J. Biol. Chem.*, 259, 8924, 1984.
63. **Takagi, T., Hokama, Y., Miyata, T., Morita, T., and Iwanaga, S.,** Amino acid sequence of Japanese horseshoe crab *(Tachypleus tridentatus)* coagulogen B chain: completion of the coagulogen sequence, *J. Biochem. (Tokyo)*, 95, 1445, 1984.
64. **Srimal, S., Miyata, T., Kawabata, S., Miyata, T., and Iwanaga, S.,** The complete amino acid sequence of coagulogen isolated from Southeast Asian horseshoe crab, *Carcinoscorpius rotundicauda, J. Biochem. (Tokyo)*, 98, 305, 1985.
65. **Levin, J. and Bang, F. B.,** Clottable protein in *Limulus:* its localization and kinetics of its coagulation by endotoxin, *Thromb. Diathes. Haemmorrh.*, 19, 186, 1968.
66. **Mürer, E. H., Levin, J., and Holme, R.,** Isolation and studies of the granules of the amebocytes of *Limulus polyphemus,* the horseshoe crab, *J. Cell. Physiol.*, 86, 533, 1975.
67. **Shishikura, F. and Sekiguchi, K.,** Comparative studies on hemocytes and coagulogens of the Asian and American horseshoe crabs, *Prog. Clin. Biol. Res.*, 29, 185, 1979.
68. **Sullivan, J. D. and Watson, S. W.,** Purification and properties of the clotting enzyme from *Limulus* lysate, *Biochem. Biophys. Res. Commun.*, 66, 848, 1975.
69. **Tai, J. Y. and Liu, T.-Y.,** Studies on *Limulus* amoebocyte lysate. Isolation of pro-clotting enzyme, *J. Biol. Chem.*, 252, 2178, 1977.
70. **Liu, T.-Y., Seid, R. S., Tai, J. Y., Liang, S.-M., Sakmar, T. P., and Robbins, J. B.,** Studies on *Limulus* lysate coagulating system, *Prog. Clin. Biol. Res.*, 29, 147, 1979.
71. **Nakamura, S., Morita, Y., Harada-Suzuke, T., Iwanaga, S., Takahashi, K., and Niwa, M.,** A clotting enzyme associated with the hemolymph coagulation system of a horseshoe crab *Tachypleus tridentatus):* its purification and characterization, *J. Biochem. (Tokyo)*, 92, 781, 1982.

72. **Torano, A. E., Nakamura, S., and Levin, J.,** Properties of the clotting enzyme responsible for endotoxin-mediated *Limulus* coagulation, *Thromb. Res.*, 34, 407, 1984.
73. **Nakamura, T., Morita, T., and Iwanaga, S.,** Intracellular proclotting enzyme in Limulus *(Tachypleus tridentatus)* hemocytes: its purification and properties, *J. Biochem.* (Tokyo), 97, 1561, 1985.
74. **Ohki, M., Nakamura, T., Morita, T., and Iwanaga, S.,** A new endotoxin sensitive factor associated with hemolymph coagulation system of horseshoe crab (Limulidae), *FEBS Lett.*, 120, 217, 1980.
75. **Nakamura, T., Morita, T., and Iwanaga, S.,** Lipopolysaccharide-sensitive serine-protease zymogen (factor C) found in *Limulus* lysates, *Eur. J. Biochem.*, 154, 511, 1986.
76. **Shishikura, F.,** Hemocytes and coagulogen, a coagulation factor, in *Biology of Horseshoe Crabs*, Sekiguchi, K., Ed., Science House, Ltd., Tokyo, 1988, 334.
77. **Tokunaga, F., Miyata, T., Nakamura, T., Morita, T., Kuma, K., Miyata, T., and Iwanaga, S.,** Lipopolysaccharide-sensitive serine-protease zymogen (factor C) of horseshoe crab hemocytes. Identification and alignment of proteolytic fragments produced during the activation show that it is a novel type of serine protease, *Eur. J. Biochem.*, 167, 405, 1987.
78. **Young, N. S., Levin, J., and Prendergast, R. A.,** An invertebrate coagulation system activated by endotoxin: evidence for enzymatic mediation, *J. Clin. Invest.*, 51, 1790, 1972.
79. **Shishikura, F., Nakamura, S., Takahashi, K., and Sekiguchi, K.,** Horseshoe crab phylogeny based on amino acid sequences of the fibrino-peptide-like peptide C, *J. Exp. Zool.*, 223, 89, 1982.
80. **Stagner, J. I. and Redmond, J. R.,** The immunological mechanisms of the horseshoe crab, *Limulus polyphemus, Marine Fish. Rev.*, 37, 11, 1975.
81. **Smith, W. R.,** Interactions between *Limulus polyphemus* and two species of marine bacteria, *Biol. Bull.* (*Woods Hole, Mass.*), 127 (Abstr.), 390, 1964.
82. **Shirodkar, M. V., Warwick, A., and Bang, F. B.,** The *in vitro* reaction of *Limulus* amebocytes to bacteria, *Biol. Bull.* (*Woods Hole, Mass.*), 118, 324, 1960.
83. **Levin, J., Quigley, J., and Stracker, K.,** unpublished data.
84. **Hilly, J. B. and Gibson, D. G.,** Culture of amebocytes on opened gill lamellae of the horseshoe crab, *Limulus polyphemus, Am. Zool.*, 29 (Abstr.), 112A, 1989.
85. **Ratcliffe, N. A. and Rowley, A. F.,** A comparative synopsis of the structure and function of the blood cells of insects and other invertebrates, *Dev. Comp. Immunol.*, 3, 189, 1979.
86. **Ratcliffe, N. A., White, K. N., Rowley, A. F., and Walters, J. B.,** Cellular defense systems of the arthropoda, in *The Reticuloendothelial System, A Comprehensive Treatise*, Vol. 3, Cohen, N. and Segel, M. M., Eds., Plenum Press, New York, 1982, 167.
87. **Salt, G.,** *The Cellular Defense Reactions of Insects*, Cambridge University Press, Cambridge, 1970.
88. **Akai, H. and S. Sato, S.,** Ultrastructure of the larval hemocytes of the silkworm, *Bombyx mori* L. (Lepidoptera, Bombycidae), *Int. J. Insect Morphol. Embryol.*, 2, 207, 1975.
89. **Sato, S., Akai, H., and Sawada, H.,** An ultrastructural study of capsule formation by Bombyx hemocytes, *Annot. Zool. Jpn.*, 49, 177, 1976.
90. **Lai-Fook, J.,** The structure of the haemocytes of *Calpodes ethlius* (Lepidoptera), *J. Morphol.*, 139, 79, 1973.
91. **Neuwirth, M.,** Granular hemocytes, the main phagocytic blood cells in *Calpodes ethlius* (Lepidoptera, Hesperiidae), *Can. J. Zool.*, 52, 783, 1974.
92. **Dunphy, G. B. and Nolan, R. A.,** Response of eastern hemlock looper hemocytes to selected stages of *Entomopththora egressa* and other foreign particles, *J. Invertebr. Pathol.*, 36, 71, 1980.
93. **Hardy, W. B.,** The blood corpuscles of the crustacea, together with a suggestion as to the origin of the crustacean fibrin-ferment, *J. Physiol.*, 13, 165, 1892.
94. **Tait, J. and Gunn, J. D.,** The blood of *Astacus fluvirtis:* a study in crustacean blood, with special reference to coagulation and phagocytosis, *Q. J. Exp. Physiol.*, 12, 35, 1918.
95. **Wigglesworth, V. B.,** The haemocytes and connective tissue formation in an insect, *Rhodnius prolixus* (Hemiptera), *Q. J. Microsc. Sci.*, 97, 89, 1956.
96. **Anderson, R. S., Holmes, B., and Good, R. A.,** *In vitro* bactericidal capacity of *Blaberus craniifer* hemocytes, *J. Invertebr. Pathol.*, 22, 127, 1973.
97. **Arnold, J. W.,** The hemocytes of insects, in *The Physiology of Insects*, Rockstein, M., Ed., Vol. 5, 2nd ed., Academic Press, New York, 1974, 201.
98. **Schmit, A. R. and Ratcliffe, N. A.,** The encapsulation of foreign tissue implants in *Galleria mellonella* larvae, *J. Insect Pathol.*, 23, 175, 1977.
99. **Ourth, D. D. and Smalley, D. L.,** Phagocytic and humoral immunity in the adult cotton boll weevil, *Anthonomus grandis* (Coleoptera: Curculionidae), to *Serratia marcescens, J. Invertebr. Pathol.*, 36, 104, 1980.
100. **El Shoura, S. M.,** Fine structure of the hemocytes and nephrocytes of *Argas (Persicargas) arboreus* (Ixodoidea, Argasidae), *J. Morphol.*, 189, 17, 1986.
101. **Dannhorn, D. R. and Seitz, K. A.,** Hemocytes of *Leiobunum limbatum* and two other species of harvestmen (Arachnida, Opitiones): morphological classification and functional aspects, *J. Morphol.*, 193, 185, 1987.

102. **Han, S. S. and Gupta, A. P.,** Arthropod immune system. II. Encapsulation of implanted nerve cord and "plain gut" surgical suture by granulocytes of *Blattella germanica* (L.) (Dictyoptera: Blattellidae), *Zool. Sci.*, 6, 303, 1989.
103. **Armstrong, P. B. and Levin, J.,** *In vitro* phagocytosis by *Limulus* blood cells, *J. Invertebr. Pathol.*, 34, 145, 1979.
104. **Rabinovitch, M. and De Stefano, M.-J.,** Interactions of red cells with phagocytes of the wax-moth (Galleria mellonella, L.) and mouse, *Exp. Cell Res.*, 59, 272, 1970.
105. **Johnson, P. T.,** Bacterial infection in the blue crab *Callinectes papidus:* course of infection and histopathology, *J. Invertebr. Pathol.*, 28, 25, 1976.
106. **Paterson, W. D. and Stewart, J. E.,** Rate and duration of phagocytic increase in lobsters induced by *Pseudomonas perolens* endotoxin, *Dev. Comp. Immunol.*, 3, 353, 1979.
107. **Ruediger, R. F. and Davis, D. J.,** Phagocytosis and opsonins in lower animals, *J. Infect. Dis.*, 4, 333, 1907.
108. **Chorney, M. J.,** *Abstract MBL Course on Invertebr. Pathol.*, Woods Hole, 1978.
109. **Richardson, E. C., Banerji, B., Seid, R. C., Levin, J., and Alving, C. R.,** Interactions of lipid A and liposome-associated lipid A with *Limulus polyphemus* amoebocytes, *Infect. Immunol.*, 39, 1385, 1983.
110. **Nutting, W. L.,** A comparative anatomical study of the heart and accessory structure of the orthopteroid insects, *J. Morphol.*, 89, 501, 1951.
111. **Bowers, B.,** Coated vesicles in the pericardial cells of the aphid (*Muzus persicae* Sulz), *Protoplasma,* 59, 351, 1964.
112. **Mills, R. P. and King, R. C.,** The pericardial cells of *Drosophila melanogaster, Q. J. Microsc. Sci.*, 106, 261, 1965.
113. **Aggarwal, S. K. and King, R. C.,** The ultrastructure of the wreath cells of *Drosophila melanogaster* larvae, *Protoplasma,* 63, 343, 1967.
114. **Wigglesworth, V. B.,** The pericardial cells of insects, analogue of the reticuloendothelial system, *J. Reticuloendoth. Soc.*, 7, 208, 1970.
115. **Johnson, P. T.,** A review of fixed phagocytic and pinocytic cells of decapod crustaceans, with remarks on hemocytes, *Dev. Corp. Immunol.*, 11, 679, 1987.
116. **Gagen, S. J. and Ratcliffe, N. A.,** Studies on the *in vivo* cellular reactions and fate of injected bacteria in *Galleria mellonella* and *Pieris brassicae* larvae, *J. Invertebr. Pathol.*, 28, 17, 1976.
117. **Neuwirth, M. and Lai-Fook, J.,** Cytochemistry of the early stages of *in situ* coagulation in *Calpodes ethlius* (lepidoptera), *Can. J. Zool.*, 55, 1767, 1977.
118. **Smith, V. J. and Ratcliffe, N. A.,** Host defense reactions of the shore crab, *Carcinus maenas* (L): clearance and distribution of injected test particles, *J. Mar. Biol. Assoc. U.K.*, 60, 89, 1980.
119. **Smith, V. J. and Ratcliffe, N. A.,** Cellular defense reaction of the shore crab, *Carcinus maenas: in vivo* hemocytic and histopathological reactions to injected bacteria, *J. Invertebr. Pathol.*, 35, 63, 1980.
120. **Johansson, M. W. and Söderhäll, K.,** Cellular immunity in crustaceans and the proPo system, *Parasitol. Today,* 5, 171, 1989.
121. **Smith, V. J. and Söderhäll, K.,** Induction of degranulation and lysis of haemocytes in the freshwater crayfish *Astacus astacus* by components of the prophenoloxidase activating system *in vitro, Cell Tissue Res.*, 233, 295, 1983.
122. **Söderhäll, D. and Smith, V. J.,** Separation of the haemocyte populations of *Carcinus maenas* and other marine decapods, and prophenoloxidase distribution, *Dev. Comp. Immunol.*, 7, 229, 1983.
123. **Söderhäll, K. and Smith, V. J.,** Prophenoloxidase-activating cascade as a recognition and defense system in arthropods, in *Hemocytic and Humoral Immunity in Arthropods,* Gupta, A. P., Ed., John Wiley & Sons, New York, 1986, 251.
124. **Söderhäll, K., Levin, J., and Armstrong, P. B.,** The effect of β-1,3-glucans on blood coagulation and amoebocyte release in the horseshoe crab, *Limulus polyphemus, Biol. Bull. (Woods Hole, Mass.)*, 169, 661, 1985.
125. **Söderhäll, K., Smith, B. J., and Johansson, M.,** Exocytosis and uptake of bacteria by isolated haemocyte populations of two crustaceans: evidence for cell cooperation in the host defense reactions of arthropods, *Cell Tissue Res.*, 345, 43, 1985.
126. **Johannsen, R., Anderson, R. S., Good, R. A., and Day, N. K.,** A comparative study of the bactericidal activity of horseshoe crab *(Limulus polyphemus)* hemolymph and vertebrate serum, *J. Invertebr. Pathol.*, 22, 372, 1973.
127. **Furman, R. M. and Pistole, T. G.,** Bactericidal activity of hemolymph from the horseshoe crab, *Limulus polyphemus, J. Invertebr. Pathol.*, 28, 239, 1976.
128. **Pistole, T. G. and Britko, J. L.,** Bactericidal activity of amebocytes from the horseshoe crab, *Limulus polyphemus, J. Invertebr. Pathol.*, 31, 376, 1978.
129. **Nachum, R.,** Antimicrobial defense mechanisms in *Limulus polyphemus, Prog. Clin. Biol. Res.*, 29, 513, 1979.

130. **Nachum, R., Watson, S. W., and Siegel, S. E.,** Antimicrobial defense mechanisms in the horseshoe crab, *Limulus polyphemus:* effect of sodium chloride on bactericidal activity, *J. Invertebr. Pathol.*, 36, 382, 1980.
131. **Smith, R. H. and Pistole, T. G.,** Bactericidal activity of granules isolated from amebocytes of the horseshoe crab, *Limulus polyphemus, J. Invertebr. Pathol.*, 45, 272, 1985.
132. **Pistole, T. G. and Graf, S. A.,** Antibacterial activity in *Limulus,* in *Hemocytic and Humoral Immunity in Arthropods,* Gupta, A. P., Ed., John Wiley & Sons, New York, 1986, 331.
133. **Noguchi, H.,** A study of immunization — haemolysins, agglutinins, precipitins, and coagulins in cold-blooded animals, *Zentralb. Bakteriol. Parasitenk. Infekt.*, 33, 353, 1903.
134. **Day, N. K. B., Gewurz, H., Johannsen, R., Finstad, J., and Good, R. A.,** Complement and complement-like activity in lower vertebrates and invertebrates, *J. Exp. Med.*, 132, 941, 1970.
135. **Miyata, T., Tokunaga, F., Yoneya, T., Yoshikawa, K., Iwanaga, S., Niwa, M., Takao, T., and Shimonishi, Y.,** Antimicrobial peptides isolated from horseshoe crab hemocytes, Tachyplesin II, and Polyphemusins I and II: chemical structures and biological activity, *J. Biochem.* (*Tokyo*), 106, 663, 1989.
136. **Muta, T., Miyata, T., Tokunaga, F., Nakamura, T., and Iwanaga, S.,** Primary structure of anti-lipopolysaccharide factor from American horseshoe crab, *Limulus polyphemus, J. Biochem.* (*Tokyo*), 101, 1321, 1987.
137. **Nakamura, T., Furunaka, H., Miyata, T., Tokunaga, F., Muta, T., Iwanaga, S., Niwa, M., Takao, T., and Shimonishi, T.,** Tachyplesin, a class of antimicrobial peptide from the hemocytes of the horseshoe crab, *(Tachypleus tridentatus), J. Biol. Chem.*, 263, 16709, 1988.
138. **Muta, T., Fujimoto, T., Nakajima, H., and Iwanaga, S.,** Tachyplesins isolated from hemocytes of Southeast Asian horseshoe crabs *(Carcinoscorpius rotundicauda* and *Tachypleus gigas):* identification of a new Tacheplesin, Tachyplesin III, and a processing intermediate of its precursor, *J. Biochem.* (*Tokyo*), 108, 261, 1990.
139. **Kawano, K., Yoneya, T., Miyata, T., Yoshikawa, K., Tokunaga, F., Terada, Y., and Iwanaga, S.,** Antimicrobial peptide, Tachyplesin I, isolated from hemocytes of the horseshoe crab *(Tachypleus tridentatus), J. Biol. Chem.*, 265, 15365, 1990.
140. **Nakamura, T., Hirai, T., Tokunaga, F., Kawabata, S., and Iwanaga, S.,** Purification and amino acid sequence of Kunitz-type protease inhibitor found in the hemocytes of the horseshoe crab *(Tachypleus tridentatus), J. Biochem.* (*Tokyo*), 101, 1297, 1987.
141. **Armstrong, P. B., Quigley, J. P., and Rickles, F. R.,** The *Limulus* amebocyte releases α_2-macroglobulin by exocytosis, *Biol. Bull.* (*Woods Hole, Mass.*), 178, 137, 1990.
142. **Morrison, D. C. and Ryan, J. L.,** Endotoxins and disease mechanisms, *Annu. Rev. Med.*, 38, 417, 1987.
143. **Tanaka, S., Nakamura, T., Morita, T., and Iwanaga, S.,** *Limulus* anti-LPS factor, an anticoagulant which inhibits the endotoxin-mediated activation of *Limulus* coagulation system, *Biochem. Biophys. Res. Commun.*, 105, 717, 1982.
144. **Aketagawa, J., Miyata, T., Ohtsubo, S., Nakamura, T., Morita, T., Hayashi, H., Miyata, T., Iwanatga, S., Takao, T., and Shimonishi, Y.,** Primary structure of *Limulus* anticoagulant anti-lipopolysaccharide factor, *J. Biol. Chem.*, 261, 7357, 1986.
145. **Wainright, N. R., Miller, R. J., Pares, E., Novitsky, T. J., Gletcher, M. A., McKenna, T. M., and Williams, T.,** Endotoxin binding and neutralizing activity by a protein from *Limulus polyphemus,* submitted, 1990.
146. **Morita, T., Ohtsubo, S., Nakamura, T., Tanaka, S., Iwanaga, S., Ohashi, K., and Niwa, M.,** Isolation and biological activities of *Limulus* anticoagulant (anti-LPS factor) which interacts with lipopolysaccharide (LPS), *J. Biochem.* (*Tokyo*), 97, 1611, 1985.
147. **Ohashi, K., Niwa, M., Nakamura, T., Morita, T., and Iwanaga, S.,** Anti-LPS factor in the horseshoe crab, *Tachypleus tridentatus.* Its hemolytic activity on the red blood cell sensitized with lipopolysaccharide, *FEBS Lett.*, 176, 207, 1984.
148. **Armstrong, P. B. and Quigley, J. P.,** α_2-Macroglobulin: a recently-discovered defense system in arthropods, in *Immunology of Insects and Other Arthropods,* Gupta, A. P., Ed., Boca Raton, FL, 1991.
149. **Marchalonis, J. J. and Edelman, G. M.,** Isolation and characterization of a hemagglutinin from *Limulus polyphemus, J. Mol. Biol.*, 32, 453, 1968.
150. **Finstad, C. L., Good, R. A., and Litman, G. W.,** The erythrocyte agglutinin from *Limulus polyphemus* hemolymph: molecular structure and biological function, *Ann. N.Y. Acad. Sci.*, 234, 170, 1974.
151. **Vaith, P., Uhlenbruck, G., Müller, W. E. G., and Cohen, E.,** Reactivity of *Limulus polyphemus* hemolymph with D-glucuronic acid containing glycosubstances, *Prog. Clin. Biol. Res.*, 29, 579, 1979.
152. **Stebbins, M. R. and Harper, K. D.,** Isolation, characterization and inhibition of arthropod agglutinins, in *Hemocytic and Humoral Immunity in Arthropods,* Gupta, A. P., Ed., John Wiley & Sons, New York, 1986, 463.
153. **Oppenheim, J. D., Nachbar, M. S., Salton, M. R. J., and Aull, F.,** Purification of a hemagglutinin from *Limulus polyphemus* by affinity chromatography, *Biochem. Biophys. Res. Commun.*, 58, 1127, 1974.

154. **Roche, A.-C. and Monsigny, M.**, Purification and properties of limulin: a lectin (agglutinin) from hemolymph of *Limulus polyphemus, Biochim. Biophys. Acta,* 371, 242, 1974.
155. **Nowak, T. P. and Barondes, S. H.**, Agglutinin from *Limulus polyphemus*. Purification with formalinized horse erythrocytes as the affinity absorbent, *Biochim. Biophys. Acta,* 393, 115, 1975.
156. **Roche, A.-C., Schauer, R., and Monsigny, M.**, Protein-sugar interactions. Purification by affinity chromatography of limulin on *N*-acylneuraminyl-binding proteins, *FEBS Lett.*, 57, 245, 1975.
157. **Kaplan, R., Li, S. S. L., and Kehoe, J. M.**, Molecular characterization of Limulin, a sialic acid binding lectin from the hemolymph of the horseshoe crab, *Limulus polyphemus, Biochemistry,* 16, 4297, 1977.
158. **Shimizu, S., Ito, M., and Niwa, M.**, Lectins in the hemolymph of Japanese horseshoe crab, *Tachypleus tridentatus, Biochim. Biophys. Acta,* 500, 71, 1977.
159. **Mureson, V., Iwanij, V., Smith, Z. D. J., and Jamieson, J. D.**, Purification and use of limulin: a sialic acid-specific lectin, *J. Histochem. Cytochem.*, 30, 938, 1982.
160. **Cohen, E., Vasta, G. R., Korytnyk, W., Petrie, C. R., and Sharma, M.**, Lectins of the Limulidae and hemagglutination-inhibition by sialic acid analogs and derivatives, *Prog. Clin. Biol. Res.*, 157, 55-69, 1984.
161. **McSweegan, E. F. and Pistole, T. G.**, Interaction of the lectin limulin with capsular polysaccharides from *Neisseria meningitidis* and *Escherichia coli, Biochem. Biophys. Res. Commun.*, 106, 1390, 1982.
162. **Rostam-Abadi, H. and Pistole, T. G.**, Studies on the lipopolysaccharide molecule reactive with *Limulus* agglutinins, *Prog. Clin. Biol. Res.*, 29, 537, 1979.
163. **Pistole, T. G.**, Limulus lectins: analogues of vertebrate immunoglobulins, *Prog. Clin. Biol. Res.*, 81, 283, 1982.
164. **Rostam-Abadi, H. and Pistole, T. G.**, Lipopolysaccharide-binding lectin from the horseshoe crab, *Limulus polyphemus,* with specificity for 2-keto-3-deoxyoctonate (KDO), *Dev. Comp. Immunol.*, 6, 209, 1982.
165. **Shishikura, F.**, Agglutinins in hemolymph and in embryos, in *Biology of Horseshoe Crabs,* Sekiguchi, K., Science House, Ltd., Tokyo, 1988, 353.
166. **White, M., Wainwright, N., and Novitsky, T. J.**, Isolation of three lectins with sialic acid specificity from the hemolymph of the horseshoe crab, *Limulus polyphemus,* in preparation, 1990.
167. **Gilbride, K. J. and Pistole, T. G.**, Isolation and characterization of a bacterial agglutinin in the serum of *Limulus polyphemus, Prog. Clin. Biol. Res.*, 29, 525, 1979.
168. **Brandin, E. R. and Pistole, T. G.**, Polyphemin: a teichoic acid-binding lectin from the horseshoe crab, *Limulus polyphemus, Biochem. Biophys. Res. Commun.*, 113, 611, 1983.
169. **Pistole, T. G.**, Bacterial agglutinins from *Limulus polyphemus* — an overview, *Prog. Clin. Biol. Res.*, 29, 547, 1979.
170. **Gilbride, K. J. and Pistole, T. G.**, The presence of copper in a purified lectin from *Limulus polyphemus:* possible new role for hemocyanin, *Dev. Comp. Immunol.*, 5, 347, 1981.
171. **Robey, F. A. and Liu, T.-Y.**, Limulin: a C-reactive protein from *Limulus polyphemus, J. Biol. Chem.*, 261, 969, 1981.
172. **Nguyen, N. Y., Suzuki, A., Cheng, S.-M., Zon, G., and Liu, T.-Y.**, Isolation and characterization of *Limulus* C-reactive protein genes, *J. Biol. Chem.*, 261, 10450, 1986.
173. **Nguyen, N. Y., Suzuki, A., Boykins, R. A., and Liu, T.-Y.**, The amino acid sequence of *Limulus* C-reactive protein, *J. Biol. Chem.*, 261, 10456, 1986.
174. **Kushner, I.**, Erythrocyte sedimentation rate and the acute phase reactants, in *Textbook of Rheumatology,* 3rd ed., Kelly, W. N., DuHarris, E., Ruddy, S., and Sledge, C. B., Eds., W. B. Saunders, Philadelphia, PA, 1989, 719.
175. **Fernandez-Moran, H., Marchalonis, J. J., and Edelman, G. M.**, Electron microscopy of a hemagglutinin from *Limulus polyphemus, J. Mol. Biol.*, 32, 467, 1968.
176. **Shuster, C. N.**, Observations on the natural history of the American horseshoe crab, *Limulus polyphemus, Third Report on Investigations of Methods of Improving Shellfish Resources of Massachusetts* (Woods Hole Oceanogr. Inst. Contr. No. 564), 1950, 18.
177. **Shuster, C. N.**, A horseshoe "crab" grows up, *Ward's Nat. Sci. Bull.* (Rochester, N.Y.), 28, 3, 1954.
178. **Ropes, J. W.**, Longevity of the horseshoe crab, *Limulus polyphemus, Trans. Am. Fish. Soc.*, 90, 79, 1961.
179. **Jones, J. C.**, Current concepts concerning insect hemocytes, *Am. Zool.*, 2, 209, 1962.
180. **Gupta, A. P.**, Arthropod hemocytes and phylogeny, in *Arthropod Phylogeny,* Gupta, A. P., Ed., Van Nostrand Reinhold Co., New York, 1979, 669.
181. **Bauchau, A. G.**, Crustaceans, in *Invertebrate Blood Cells,* Ratcliffe, N. A. and Rowley, A. F., Eds., Vol. 2, Academic Press, London, 1981, 385.
182. **Ravindranath, M. H.**, Onychophorans and myriapods, in *Invertebrate Blood Cells,* Vol. 2, Ratcliffe, N. A. and Rowley, A. F., Eds., Academic Press, London, 1981, 327.
183. **Rowely, A. F. and Ratcliffe, N. A.**, Insects, in *Invertebrate Blood Cells,* Vol. 2, Ratcliffe, N. A. and Rowley, A. F., Eds., Academic Press, London, 1981, 421.
184. **Amirante, G. A.**, Cellular immune responses in crustaceans, in *Hemocytic and Humoral Immunity in Arthropods,* Gupta, A. P., Ed., John Wiley & Sons, New York, 1986, 61.

Chapter 2

INSECT IMMUNOCYTES AND OTHER HEMOCYTES: ROLES IN CELLULAR AND HUMORAL IMMUNITY*

Ayodhya P. Gupta

TABLE OF CONTENTS

* This is the ninth paper in a series on arthropod immune systems.

I. INTRODUCTION

Insects and other arthropods have several kinds of hemocytes, two of which (the *granulocytes* [GRs] and *plasmatocytes* [PLs]) provide cellular immunity against foreign antigens. Because of their commonly known capacity to distinguish among self, altered self, and nonself tissues and to synthesize and secrete several humoral factors, I first designated them as *immunocytes,*[1] and we have consistently used this term in our subsequent works;[2-11] its usage is gaining acceptance by other authors[12] (see also Chapter 7).

Other hemocyte types, such as *prohemocytes, spherulocytes, oenocytoids, and adipohemocytes* are thought to be nonimmunocompetent.[13] Note that in this and another review,[10] I have merged the category of *coagulocytes* with that of the GR. Although insect hemocytes have been studied for nearly 150 years, there are still controversies regarding their classification, largely owing to problems of identifying various types in a uniformly consistent manner. Consequently, many authors are reluctant to assign any specific function(s) to a particular hemocyte type; instead, they prefer to use the generic term "hemocyte", which, in most cases, can be avoided, because the major hemocyte types can be morphologically identified.

The purpose of this chapter is to present a detailed ultrastructural account of those organelles involved in the immunologic functions of immunocytes, and to update our most recent review.[14]

II. MAJOR IMMUNOCYTES

The most important measure of the competence of an immune system is its ability to distinguish between self and nonself tissues (foreign antigens). In insects and other arthropods, this is accomplished by the combined actions of the two immunocytes (GRs and PLs), several humoral (hemolymph) factors (hemagglutinins, complement-like factors, antimicrobial factors, and prophenoloxidase systems), and coagulation.

A. GRANULOCYTE/COAGULOCYTE

1. Synonymy with Coagulocyte (CO)

In my previous works and reviews, I have recognized the category of *coagulocyte,* despite a lingering suspicion spanning more than two decades that these cells (in terms of their ultrastructure and functional role in coagulation) are in fact GRs. This suspicion has been consistently sustained by our observations that hemolymph samples that would not apparently contain any so-called COs, would exhibit these cells, following localized coagulation brought about by lysing GRs (Figures 1A and 2C—E). In fact, as far back as 1966, we had suggested[15] that the CO is the *result* rather than the *cause* of coagulation, and we restated this opinion in our recent work,[2] although we still recognized the CO as a separate category. It is time, however, that, instead of perpetuating this situation, the category of COs be merged with that of the GRs, and that the COs be regarded as GRs. In fact, many authors have reported that GRs cause coagulation in many arthropods, owing to coagulation factors contained in the granules.[16-21] Thus, features, such as "nucleus with chromatin arranged in a cartwheel-like fashion, scant hyaline cytoplasm with some spherical or elongate granular inclusions, and cystlike blebs in the process of exocytosis"[14,22,23] (Figure 2A, B), observed in light microscopy (LM), are artifacts caused by degranulation of the GRs. This is substantiated by the ultrastructural features of a degranulating GR, in which very few cytoplasmic granules are present. Two ultrastructural features (the microruptures in the plasma membrane and the perinuclear cisterna) have been suggested by Grégoire and Goffinet[24] as diagnostic features of the COs. The microruptures have not been confirmed by others, even in the most recent works,[2,25] and the latter is also present in PRs and PLs as well as GRs (Figure 1B—D).[2]

2. Distinguishing Features

The GRs are easily recognizable owing to the presence of numerous intracytoplasmic granules. The GRs are very labile and therefore easily degranulate or lyse after bleeding, rapidly losing the cytoplasmic contents, especially the granules; this in turn makes them appear hyaline when viewed by LM and causes localized coagulation in many instances (Figures 1A and 2C—E). The GRs are often discoid (Figure 3C) or oval (Figures 3A, B and 4A), vary in size from small to large (4 to 32 μm wide and 10 to 45 μm long, rarely larger), and appear nodulose in scanning electron microscopy (SEM) (Figure 3C). The nucleus may be relatively small compared with that of the PL, is round or elongate (Figure 4A, C), and generally is centrally located. The nucleus size is variable (2 to 8 μm wide and 2 to 8 μm long).

a. The Granules

The membrane-bounded granules have been described in detail in the GRs of various insects and some Crustacea.

Structureless, electron-dense granules — They are the unstructured inclusion (Type 1) bodies of Baerwald and Boush,[26] melanosome-like granules of Hagopian,[27] opaque bodies of Moran,[28] Type 2 bodies of Scharrer,[29] and electron-dense granules of Raina,[30] Chiang et al.,[2] Barracco and Loch,[25] and others (Figure 4A—D).

Structureless, electron-lucent granules — These are the Type 3 bodies of Scharrer[29] and electron-lucent granules of Raina[30] and Barracco and Loch[25] (Figure 4 D). The length (or the diameter) of the structureless granules varies from 0.5 to 2 μm, and the shape of the granules may be spherical, ovoid, elongate, or irregularly polygonal.

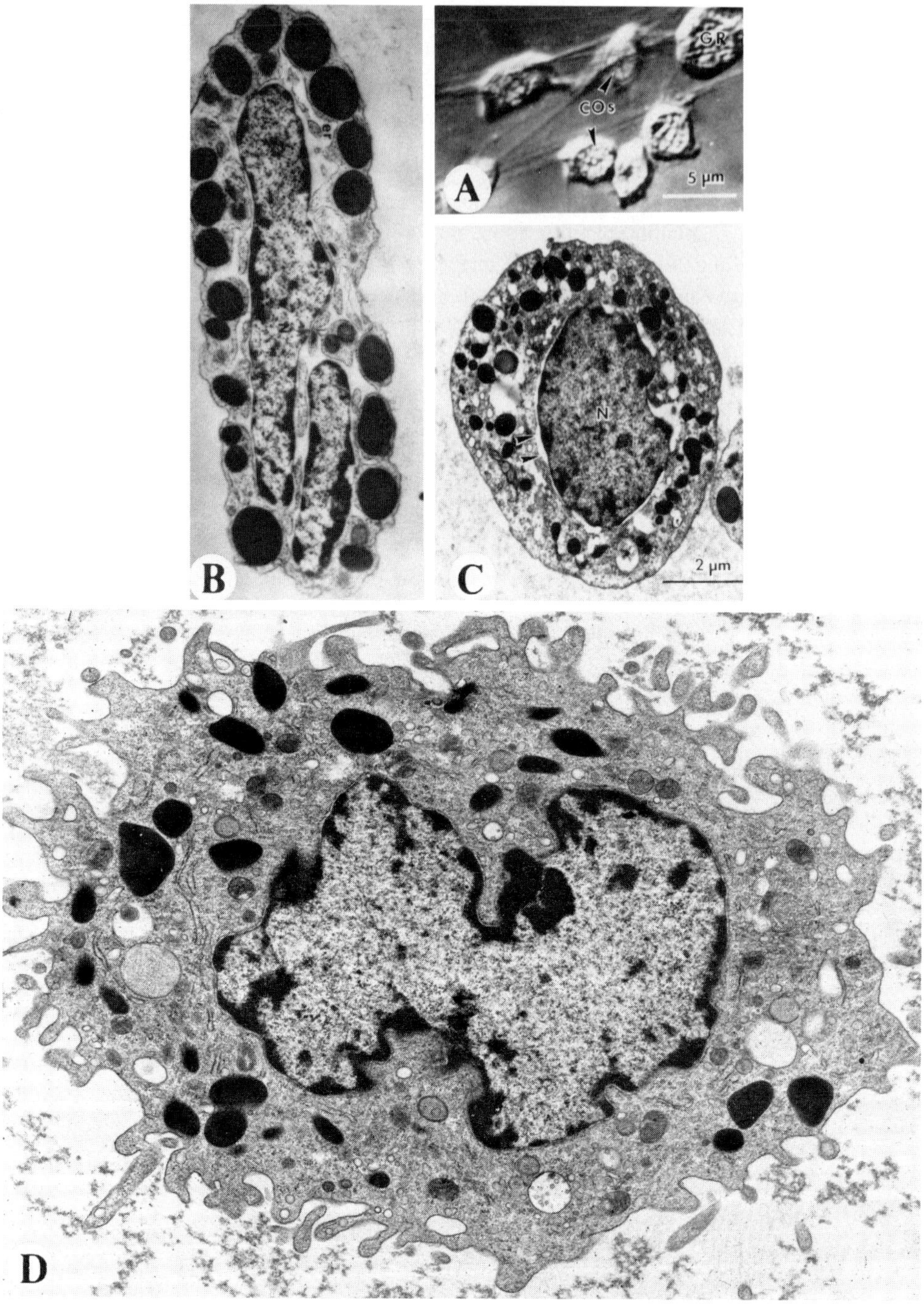

FIGURE 1. (A) Coagulation spot (Nomarski) from *B. germanica,* showing degranulating (lysing) granulocytes (previously called coagulocytes). Note that one granulocyte (GR) (upper right corner) has not degranulated. (B) A so-called coagulocyte (= GR) (TEM) of *Melolontha melolontha,* showing perinuclear space (cisterna). Note electron-dense (structureless) granules, characteristic of GRs (× 13,400). (C) A so-called coagulocyte (= GR) (TEM) of *B. germanica,* showing perinuclear space (arrowheads). N = nucleus. (D) Granulocyte (TEM) (of unknown insect), showing perinuclear space as in B and C, suggesting that it is not characteristic of the so-called coagulocytes. (× 12,600) (Parts A and C from Chiang, A. S., Gupta, A. P., and Han S. S., *J. Morphol.,* 198, 257, 1988; part B from Gupta, A. P., *Comprehensive Insect Physiology, Biochemistry and Pharmacology,* Vol. 3, Kerkut, G. A. and Gilbert, L. I., Eds., Pergamon Press, Oxford, 1985, 401. With permission. Part D courtesy of Drs. Goffinet and Grégoire.)

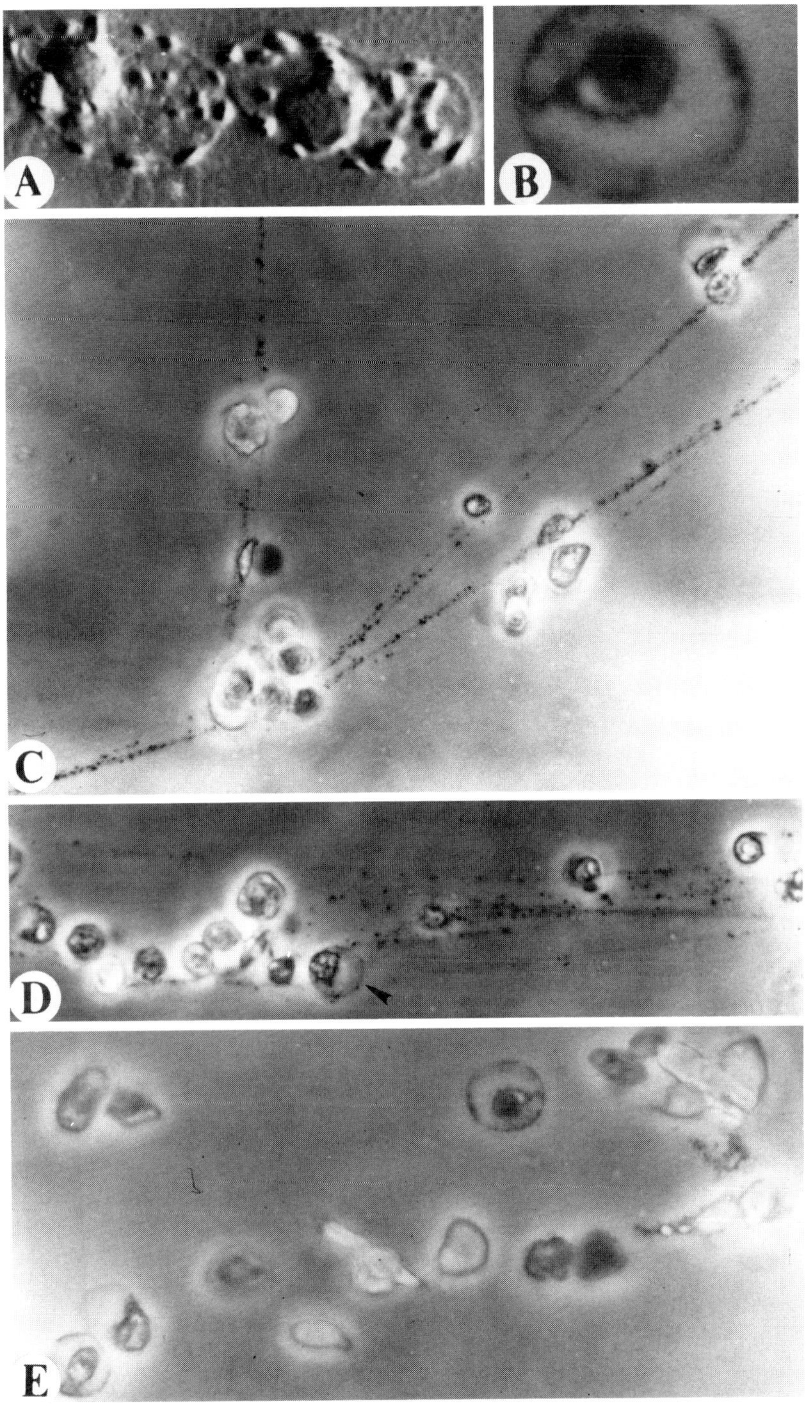

FIGURE 2. (A) A degranulating (lysing) granulocyte previously called coagulocyte) of *G. portentosa* (Nomarski) (× 5600). (B) Same as A of *Epicauta cinerea* (× ca. 28,480). (C) Coagulation pattern II of *E. cinerea* (phase contrast), showing long, straight, thread-like cytoplasmic strands, and some lysed granulocytes (so-called coagulocytes) (× 7,700). (D) Same as C, with one (center) lysed granulocyte, showing a typical cartwheel-like arrangement of the nuclear chromatin material (× 7,700). (E) Same as C and D, but without cytoplasmic threads. Note several lysed, hyaline granulocytes (× 7,700). (Parts A and C from Gupta, A. P., *Comprehensive Insect Physiology, Biochemistry and Pharmacology,* Vol. 3, Kerkut, G. A. and Gilbert, L. I., Eds., Pergamon Press, Oxford, 1985, 401. With permission.)

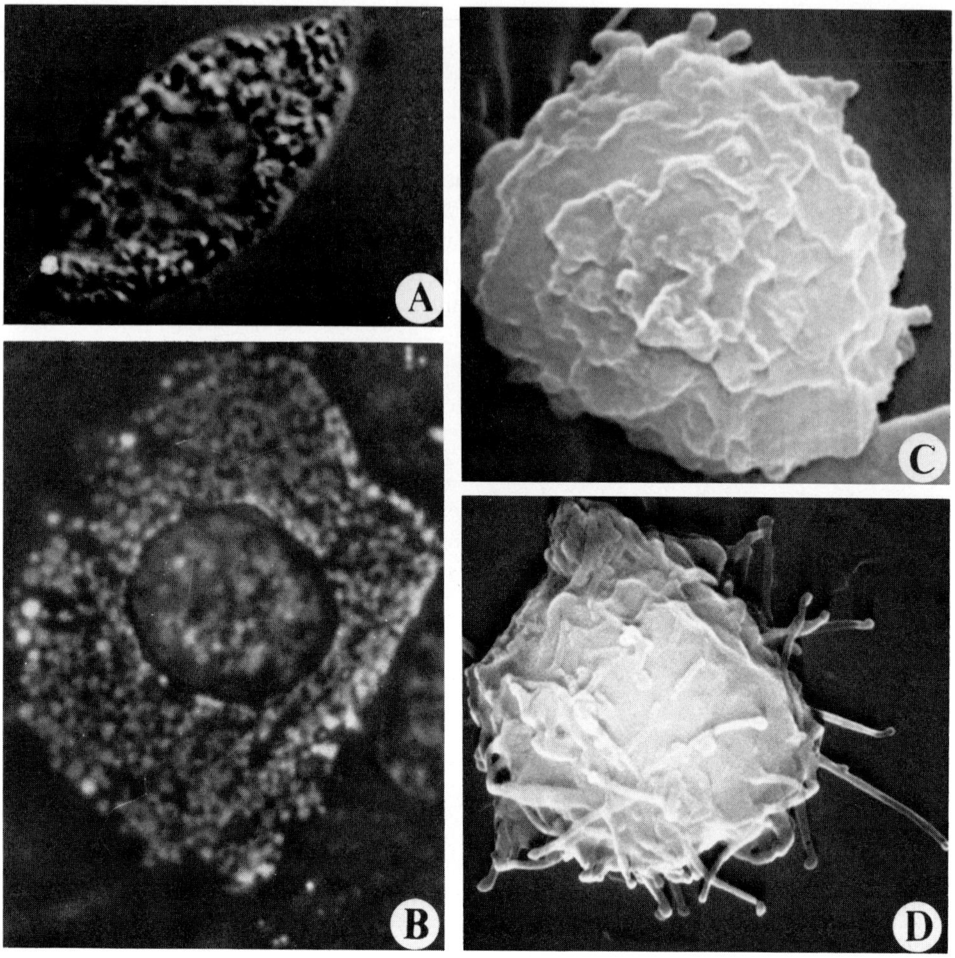

FIGURE 3. (A) Granulocyte (Nomarski) of *G. portentosa* (× 8300). (B) Granulocyte (phase contrast) of *Locusta migratoria* (× ca. 8300). (C) Granulocytes (SEM) of *B. germanica* (× ca. 8300). (D) An activated granulocyte (SEM) of *B. germanica,* showing long extensions of the plasma membrane (× ca. 7470). (Part B from Gupta, A. P., *Comprehensive Insect Physiology, Biochemistry and Pharmacology,* Vol. 3, Kerkut, G. A. and Gilbert, L. I., Eds., Pergamon Press, Oxford, 1985, 401. With permission.)

Structured granules — These are the *grains denses structures* (in the adipohemocyte) of Devauchelle[31] and Landureau and Grellet,[32] *corpus fibrillaires* of Hoffmann et al.,[33,34] cylinder inclusions (Type 4) of Baerwald and Boush,[26] *Granula mit tubularer Binnenstruktur* of Stang-Voss,[35] premelanosome-like granules of Hagopian,[27] tubule-containing bodies (TCB) of Moran,[28] Type 1 bodies of Scharrer,[29] and granules with a microtubular structure of Ratcliffe and Price,[36] Barracco and Loch,[25] and others (Figures 4B and 5A—C). The diameter of the microtubules within the structured granules (Figure 5A, B) varies from 15 to 80 nm in the GRs of various insects. Internally, the microtubules may show micro-microtubules (Figure 5C) about 5 nm in diameter.[27] Akai and Sato[37] also have described "subunits of fibrils" in what they call secretory vesicles. The number of microtubules per granule may vary from 9 to 80. The granules are derived from the Golgi bodies, the microtubules developing during the later stages of morphogenesis, although it has been suggested that they form during the earlier stages of their formation.[38]

Unfortunately, the biochemical nature of the various kinds of granules has not been thoroughly studied. Furthermore, it is not known if GRs in insects and other arthropods possess biochemically different kinds of granules or those comparable to α-granules and dense bodies of human platelets. Han and Gupta[6] reported some ultrastructural (e.g., the peripheral microtubule bands) and functional (e.g., coagulation) similarities between human platelets and *Blattella germanica* GRs. It is conceivable that both the structured and/or structureless, electron-dense granules contain sulfated, periodate-reactive sialomucin and other glycoproteins or neutral mucopolysaccharides.[25,39-43] That the granules contain the factors that cause coagulation has been reported by several authors[16-21] (see also Section VII. F on coagulation). Stang-Voss[44] has suggested that they are the sites of hemagglutinin storage in *Astacus astacus*. Amirante[45] and Yeaton[46] have reported that GRs synthesize hemagglutinins (see also Section VII.A on hemagglutinins). Occasionally, lipid droplets may be present, especially in older GRs.

In addition to the structureless and structured granules, the cytoplasm of the GR is rich in free ribosomes (polysomes), lysosomes (membrane-bounded, electron-dense bodies, 0.1 to 1.30 μm in size), numerous Golgi complexes, microtubules, multivesicular bodies, and both smooth (ER) and rough endoplasmic reticulum (RER), the latter with well-developed cisternae. Mitochondria are generally few in number. Another organelle of significance in cellular immunity is the microtubule band, located near the margin or periphery of the cell in the plane of flattening in some insects[2,3,6,7] (Figure 4D, *B. germanica*) (see also discussion in Section II.A.4.c on microtubules). Because of their peripheral or marginal location, the microtubules appear as apical bands at the two poles in elliptically sectioned cells (Figure 4E). In sections passing along the plane of flattening, the peripheral and circumferential orientation of the microtubules becomes evident (Figure 4D). At least in this insect, the GR constitutes 91 to 96% of the hemocyte population, and is the major immunocyte that participates in encapsulation[7] and phagocytosis (unpublished observation).

According to Arnold,[47] the GRs of hemimetabolous insects are generally larger and have more prominent granules than those of holometabolous insects. I have proposed elsewhere[48] that the arthropod GRs perform some of the analogous functions of vertebrate B-lymphocytes as well as the helper and suppressor functions of T-lymphocytes. In this respect, the GRs are functional hybrids of both B- and T-lymphocytes as well as of the macrophages in some primitive arthropods (e.g., the horseshoe crab, *Limulus polyphemus*) (see Sections VIII and IX).

Brehélin and Zachary[49] categorized insect granulocytes into four categories (GH1—GH4), primarily on the basis of their granules. They claimed that GH1 cells possess three types of granules, including the "heterogeneous, multivesicular bodies as a 'granular' inclusion" (erroneously, I might add, because multivesicular bodies are not granules) and that GH2 cells possess only one kind of electron-dense granules. These authors claimed to be able to distinguish these two types of GRs on the basis of such features as the rapidity with which they transformed following bleeding (GH1 transforming rapidly, but GH2 being more stable), endocytic activity (considerable in GH1 cells, but limited in GH2), and capsule formation (only by GH2). GH3 cells supposedly contain only multivesicular or myelin-like bodies with phosphatase activity and, like the GH1 cells, are phagocytic. GH4 cells (no micrographs provided) possess electron-dense granules with an inner electron-dense core and an outer granular zone and a swollen perinuclear cisterna. Note that Rowley's[50] coagulocytes have precisely these features. The GH4 as well as GH1 cells take part in coagulation.

According to Brehélin and Zachary,[49] GH1 cells are present in all orders, except Diptera, which also lacks GH2 cells; GH2 and GH3 cells are absent in Lepidoptera. GH3 cells are numerous in Orthoptera, Dictyoptera, and Diptera, and the GH4 cells are found in *Carausius morosus* and *Clitumnus extradentatus*.

Unfortunately, the well-intentioned attempt of Brehélin and Zachary[49] to "rule out controversy" in insect hemocyte classification does not fulfill that goal, primarily because the criteria selected to separate the various subtypes are not diagnostic. In our own work

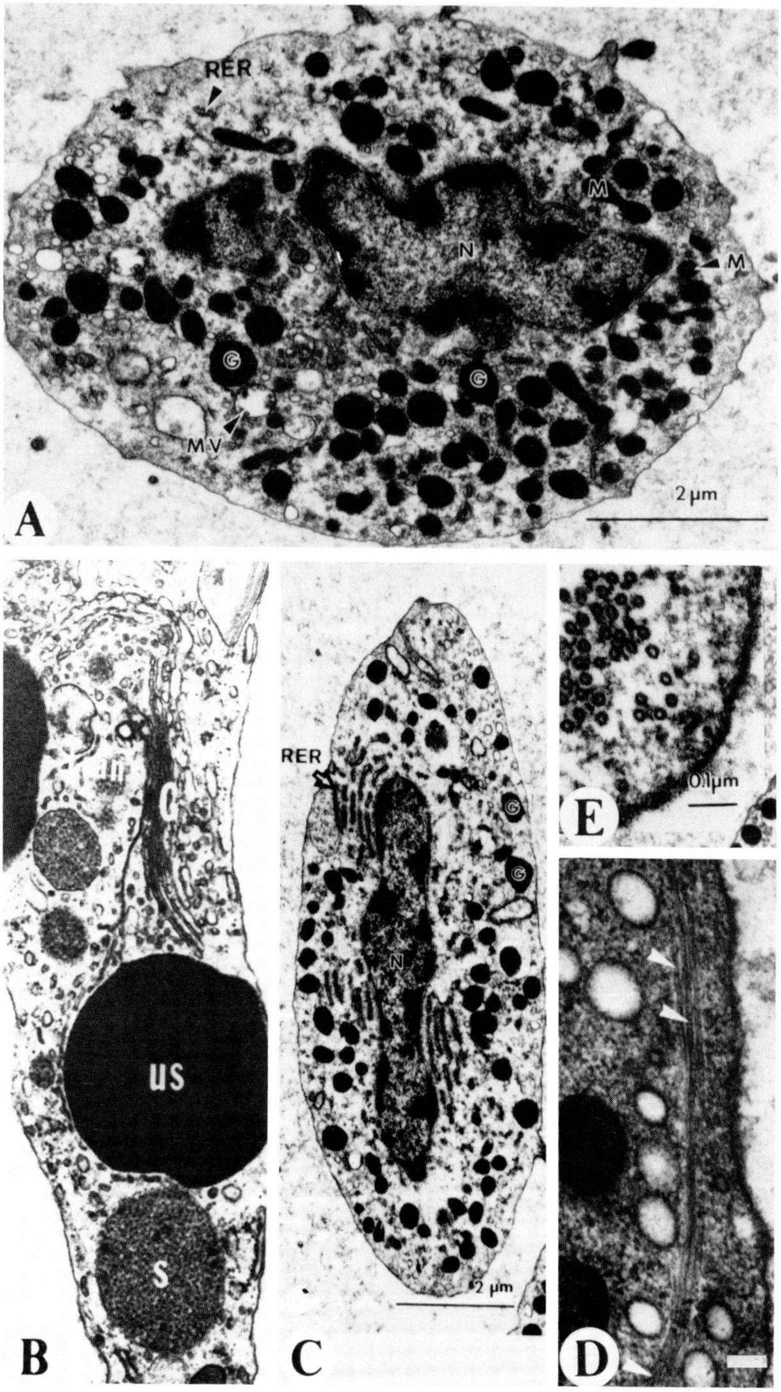

FIGURE 4. (A) Granulocyte (TEM) of *B. germanica,* showing granules (G), free ribosomes, some mitochondria (M), RER, and multivesicular bodies (MV). N = nucleus. (B) Portion of a granulocyte (TEM) of *Leucophaea maderae,* showing unstructured (US) granules (= melanosome-like) and structured (S) (= premelanosome-like). Note also the *cis* and *trans* faces of the Golgi apparatus (G) (× ca. 40,000). (C) Granulocyte (elliptical TEM section) of *B. germanica,* showing abundant granules and ribosomes, and RER. N = nucleus. Note some microtubules in cross section at the two poles. (D) Portion of a granulocyte (TEM) of *B. germanica* through the peripheral region in the plane of flattening, showing marginal band of microtubules (arrowheads). (Bar = 0.1 µm.) (E) Portion of a granulocyte (TEM) of *B. germanica* in elliptical section, showing microtubules in cross section. (Parts A, C, D, and E from Chiang, A. S., Gupta, A. P., and Han, S. S., *J. Morphol.,* 198, 257, 1988; part B from Gupta, A. P., *Comprehensive Insect Physiology, Biochemistry and Pharmacology,* Vol. 3, Kerkut, G. A. and Gilbert, L. I., Eds., Pergamon Press, Oxford, 1985, 401. With permission.)

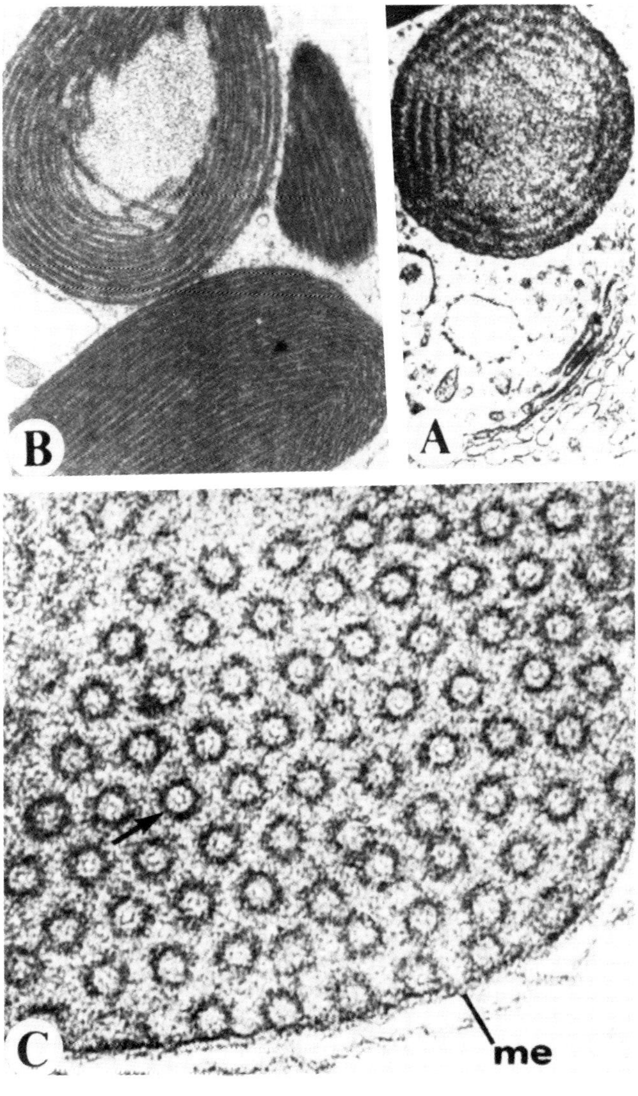

FIGURE 5. (A) Portion of granulocyte (TEM) of *L. maderae*, showing a structured granule and an earlier stage of development of internal microtubules (× ca. 30,000). (B) Same as A, showing a later stage of concentric arrangement of internal microtubules (× ca. 30,000). (C) Highly magnified view of the microtubules, showing micro-microtubules (5 nm in diameter) within microtubules. me = limiting membrane of the granules (× 230,000). (From Gupta, A. P., *Comprehensive Insect Physiology, Biochemistry and Pharmacology,* Vol. 3, Kerkut, G. A. and Gilbert, L. I., Eds., Pergamon Press, Oxford, 1985, 401. With permission.)

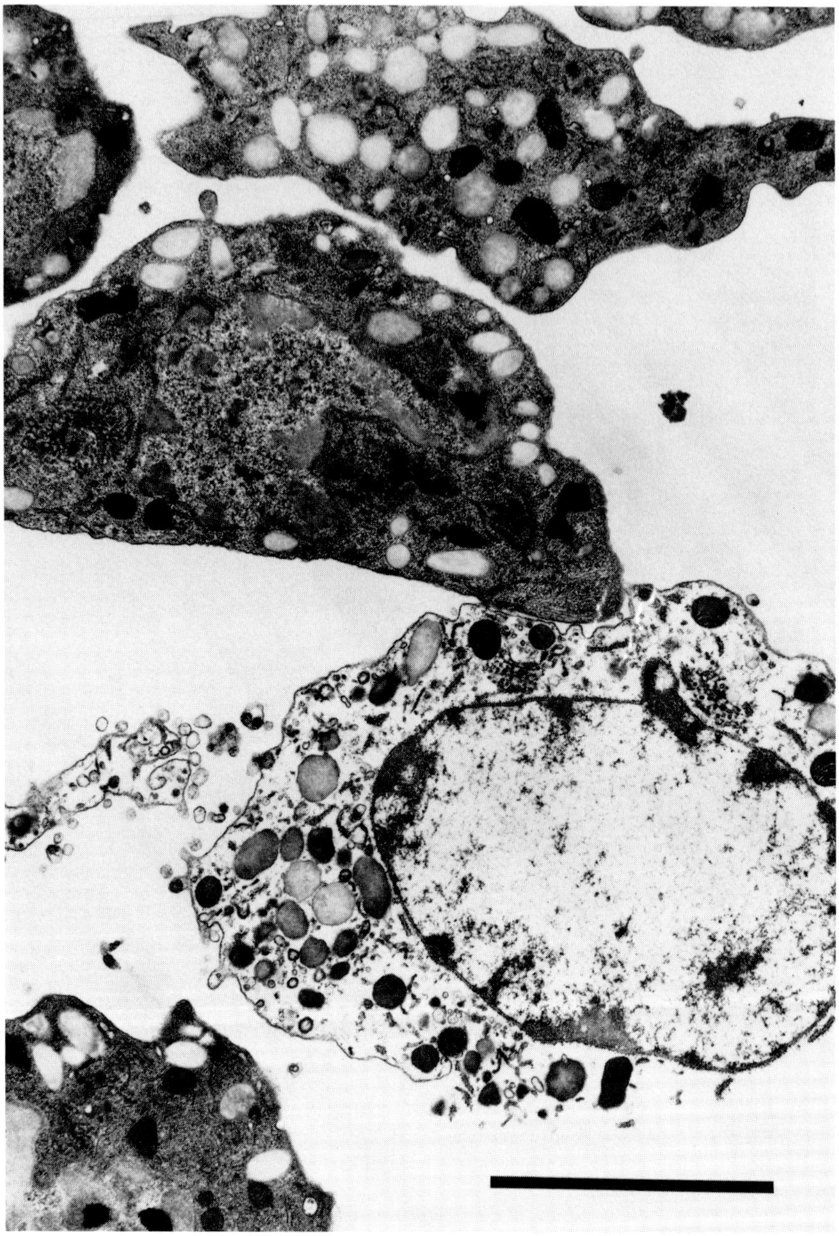

FIGURE 6. Section of a partially formed capsule (40 min postimplantation of a piece of surgical suture) (TEM) of *B. germanica,* showing two kinds (electron-dense and electron-lucent, latter showing partial lysis) of granulocytes. (Bar = 2.0 μm.) (From Han, S. S. and Gupta, A. P., *Zool. Sci.,* 6, 303, 1989. With permission.)

on the *B. germanica,* we suspected the existence of three subpopulations of GRs (electron-dense, electron-lucent, and normal)[2,7] (Figure 6), but were unable to find any diagnostic ultrastructural features to separate them, and could not apply any of the criteria suggested by Brehélin and Zachary[49] for that purpose. Furthermore, the attributed functions of some of the four (GH1—GH4) subtypes (e.g., of GH1 in phagocytosis and coagulation) and the assignment of these subtypes to taxa (e.g., absence of GH1 cells in Diptera) are unwarranted, because no detailed comparative studies of these subtypes are currently available.

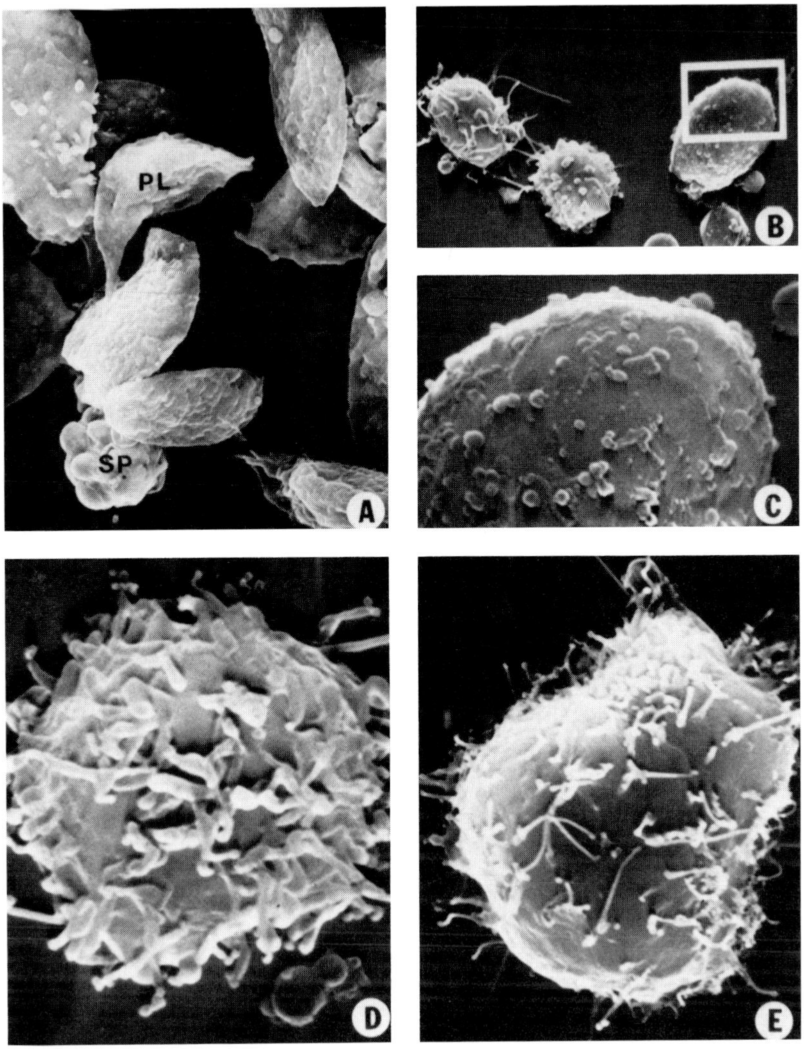

FIGURE 7. Granulocyte (SEM) of *B. germanica* in various stages of activation. (A) Cluster of hemocytes (plasmatocyte [PL], spherulocyte [SP], and granulocytes [unlabeled]) (× 3,000). (B) Three granulocytes (× 3,000). (C) Magnified view of portion of one granulocyte (Figure B, right) (× 12,000). (D) Activated granulocyte (× 10,000). Note that this immunocyte, when reacting to foreign materials, including chemicals in saline solution, gives rise to numerous extensions (pseudopods) of its plasma membrane. This type of cellular reaction of the granulocyte is analogous to similar reaction of the human T-lymphocyte. (E) Activated granulocyte of *G. portentosa* (× 9,500). (From Gupta, A. P., Das, Y. T., and Rao, B. R., *Cockroaches as Model Systems for Neurobiology: Applications in Biomedical Research,* Huber, I., Masler, E. P., and Rao, B. R., Eds., CRC Press, Boca Raton, FL, 1990, 11. With permission.)

In *B. germanica,* the normal or resting (unactivated) GR (Figure 4A) is a flattened, discoid cell, and contains, among other organelles, microtubules that are arranged in the form of bundles or bands in its peripheral region in the plane of flattening (Figure 4D). The activated (e.g., following implantation of surgical suture in the animal) GRs (Figures 3D and 7A—E) show different ultrastructural details; they are characterized by an irregular outline, presence of numerous extensions of the plasma membrane (Figure 7A—E), and a pronounced development of the peripheral microtubules.

From the standpoint of cellular immunity, the GR plasma membrane (its receptors and cytoskeleton) and the cell junctions among encapsulating GRs play significant roles. The following account includes a general discussion of these organelles in other eukaryotic cells, and provides specific information, wherever possible, on the German cockroach immunocyte, based on our work on this insect. It also points out the gaps in our knowledge on several aspects of cellular immunity in insects, and thus the need for further investigations.

3. The Plasma Membrane

In transmission electron microscopy (TEM), the plasma membrane may appear smooth (Figure 4A—C) or with a few filopodia, depending on the resting or activated state of the cell. Although the plasma membrane of the immunocyte and its receptors actively participate in nonself tissue recognition (phagocytosis and encapsulation) in insects and other arthropods, virtually nothing is known of its molecular structure.[11,48] Specifically, the biochemical nature of the plasma membrane, its intramembranous particles (IPs) (Figures 8B, C and 9C), markers or receptors, and the cytoskeleton are unknown. The basic components of the plasma membrane of a eukaryotic cell are largely phospholipids, some glycolipids, carbohydrates, phosphoproteins, and glycoproteins. Some of the glycoproteins are large transmembranous (intrinsic) molecules, and it is these that appear as IPs. The molecular homology of the IPs with receptor molecules is also unknown.

a. Biochemistry of the Plasma Membrane and Intramembranous Particles (IPs)

Study of the biochemistry of the insect immunocyte plasma membrane is important for three reasons: (1) to determine its various proteins/glycoproteins (e.g., G-protein, adenylate cyclase, gap junction proteins), (2) to find out which of its several glycoproteins serve as receptors or markers, and (3) if it is these markers or their copies that appear as IPs of the membrane in freeze-fractured specimens (Figures 8B, C and 9C). In other words, do the same glycoprotein molecules serve as both plasma membrane receptors and IPs? Note that the plasma membrane of the GR in *B. germanica* is 7.5 nm thick, and is distinguishable as two electron-dense layers, separated by one electron-lucent region, each being 2.5 nm thick.

The IPs of the hemocyte plasma membrane have been previously demonstrated by Baerwald,[51,52] and those of the hemocytic gap junctions by Han and Gupta[3,6,7] (Figure 8B, C). However, their homology with the receptor molecules is unknown. For ultrastructural and molecular analogy of the IPs of the plasma membrane, we may turn to the human red blood cell (RBC), the molecular structure of whose plasma membrane is well known.[53-56] Although no homology exists between the insect immunocyte and the RBC, it should be noted that certain components of the plasma membranes (e.g., the IPs, actin, and tubulin) are well conserved even in cells of phylogenetically unrelated taxa. Furthermore, of the two RBC plasma membrane glycoproteins (glycophorin and band 3), the latter is a transmembranous molecule and appears as IPs on both the P (cytoplasmic) and E (extracytoplasmic) fracture faces of the membrane. From the standpoint of their role in cellular immunity in insects, however, the important question to address is whether the IPs are biochemically similar to the receptor molecules; in other words, do the same glycoproteins serve both as surface receptors and IPs?

Although it has not yet been demonstrated that the IPs of the plasma membrane serve as receptors, if one compares the molecular structure of a known receptor with those of the IPs of the gap junction, one is struck with the remarkable similarity between the two organelles. A receptor with a well-known molecular structure is the acetylcholine receptor,[54,57] which appears as a doughnut-like ring in electron micrographs. It is composed of four different polypeptide subunits (alpha, beta, gamma, and delta), each with a different molecular weight, and arranged so as to form a channel with a diameter of 0.65 to 0.8 nm, sufficient to allow passage of both Na and K ions.[54] The receptor, according to these authors,

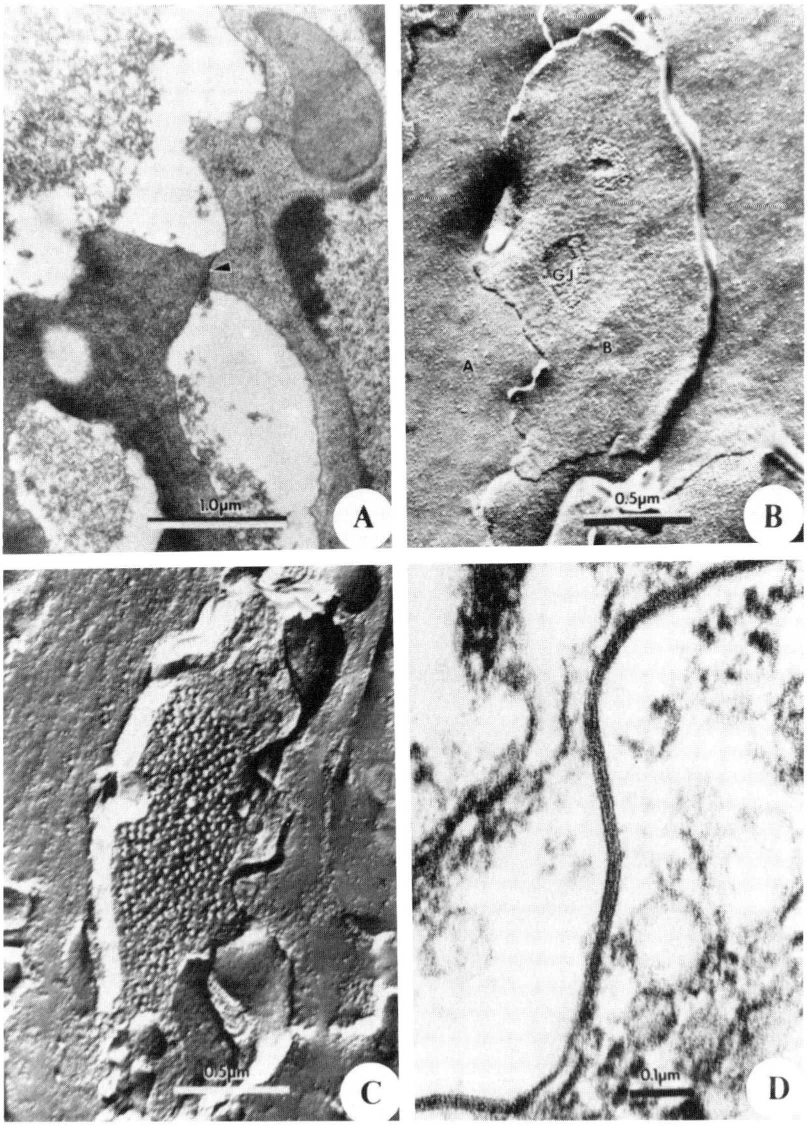

FIGURE 8. (A) Section of a 3 day old capsule (TEM) of *B. germanica,* showing gap junction (arrowhead) in the inner region. Note that whenever granulocytes first contact each other during encapsulation, they form junctions. (B) Freeze-fracture and etching replica of the inner region of a 3-day-old capsule of *B. germanica,* showing gap junction (G). Note that of the two E-type junctions, one shows greater dispersion of intramembranous particles (IPs). A = P-fracture face; B = E-fracture face. (C) Freeze-fracture and etching replica of the outer region of a 3-day-old capsule of *B. germanica,* showing gap junction. The IPs on the E-fracture face are 20 nm in diameter, and their number is ca 640/μm^2. (D) Section of a 3-day-old capsule (TEM) of *B. germanica,* showing gap junction in the outer region. The intercellular space between the two junction-forming granulocytes is 2.5 nm. (From Han, S. S. and Gupta, A. P., *Zool. Sci.,* 6, 303, 1989. With permission.)

projects 7 nm on the extracellular (E) side and 2 nm on the cytoplasmic (P) side of the plasma membrane, which indicates that it is a transmembranous protein.

This molecular structure of the acetylcholine receptor seems remarkably similar to the IP in the gap junction, in terms of the arrangement of its polypeptide subunits to form a channel for the passage of ions.[56,58] Each IP of a gap junction serves as a "connexon" and

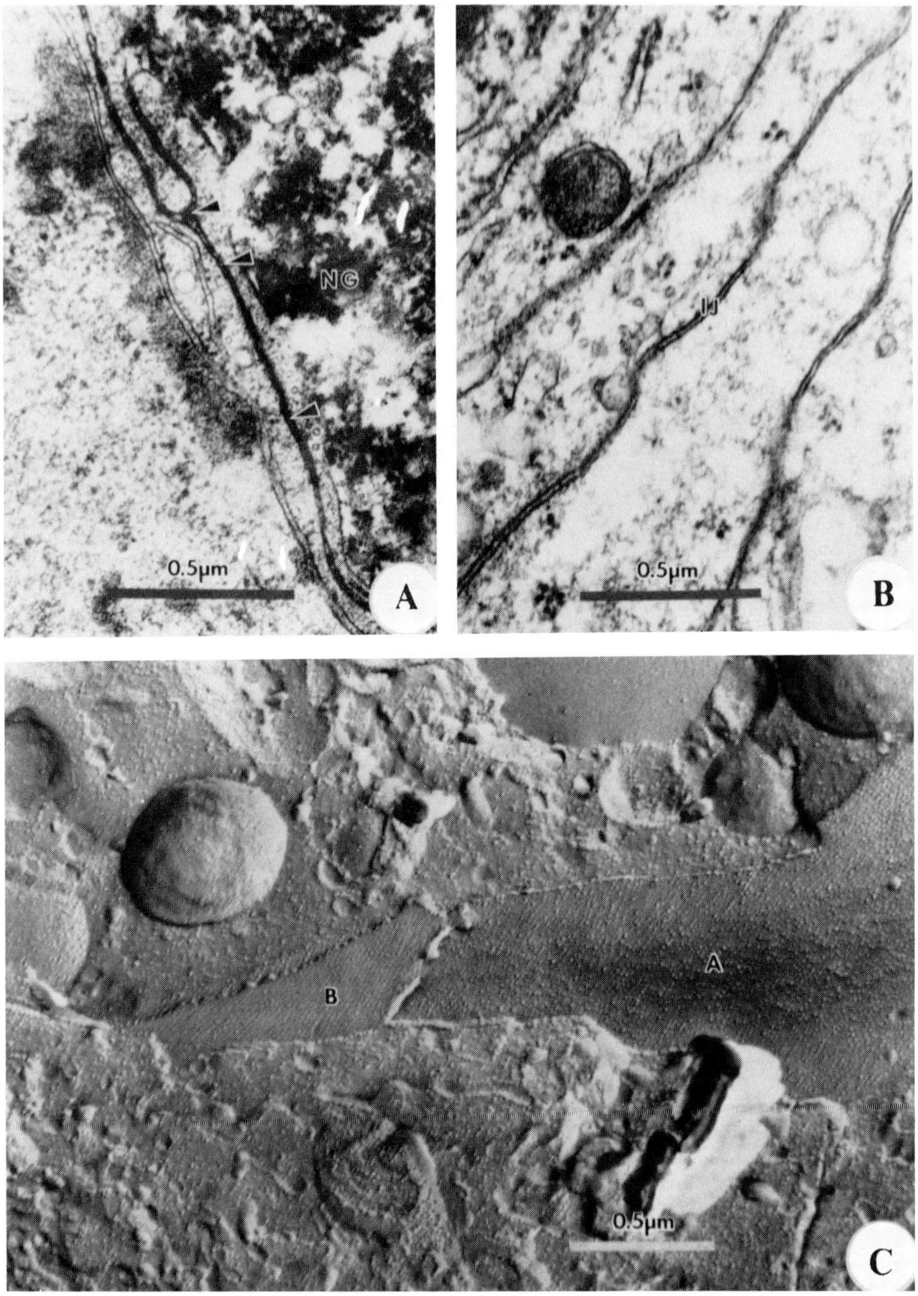

FIGURE 9. (A) Section of the inner region of a 3-day-old capsule of *B. germanica,* showing septate junctions (arrowheads). The space between the two plasma membranes in the junctional region is ca. 3.5 nm and the septa are often about 15 nm wide, but may vary in width. NG = necrotic granulocyte. (B) Section of the inner region of a 3-day-old capsule of *B. germanica,* showing intermediate junction (IJ). The two junction-forming plasma membranes run parallel to each other, the electron-lucent space between the two being always 9 nm. (C) Freeze-fracture and etching replica of the inner region of a 3-day-old capsule (TEM) of *B. germanica,* showing septate junction. The septate junctions reveal rows of IPs (8 to 14 nm in diameter) on the P-fracture face and corresponding furrows (6 nm wide, 16 nm center to center) on the E-fracture face. (From Han, S. S. and Gupta, A. P., *Zool. Sci.*, 6, 303, 1989. With permission.)

is thought to be composed of six protein subunits arranged such that a channel is formed. For details of junction-forming connexons, see Chapter 4. Since disassembly of MTs also changes cell shape (see discussion in Section II. A.4.c.iii on cell shape maintenance), it is possible that change in cell shape and formation of gap junction might be interrelated events, and the former might trigger the formation of the latter. Hanna et al.[59] suggested that the IPs of the gap junction may be involved in the flow of ions from the presynaptic to postsynaptic axons in the crayfish, *Procambarus,* during neural transmission. Finally, although the biochemical homology between the IPs and the receptor molecules remains to be demonstrated in an insect immunocyte, Yahara and Edelman[60] have indeed suggested that the transmembranous protein molecules (which appear as IPs) serve as receptors for lectins or antibodies.

b. Receptors

The study of the plasma membrane receptors is important for four reasons.

Elucidation — We can elucidate the role of surface receptors in foreign antigen recognition (e.g., in phagocytosis and encapsulation) (see Section VI. A and B).

Characterization of insect immunocytes — In order to specifically assign immunologic functions to the insect immunocytes (and not mistakenly to any other hemocyte type) we must be able to more precisely characterize them. In fact, various agglutination studies of lectin receptors (meaning glycoconjugates on the plasma membranes to which lectins bind) have clearly demonstrated not only that specific lectin-binding receptors exist on the plasma membranes of some cells, but also that these receptors can be used to discriminate among morphologically and functionally distinct cell populations.[61] These authors concluded that the wheat germ agglutinin (WGA)-binding lectin receptor serves as a marker of a particular developmental stage (epimastigote) of the protozoan, *Trypanosoma cruzi.* Several studies on other animals have also shown that surface receptors of hemocytes are characteristic of certain stages of cells or cell populations. For example, concanavalin A (Con A) receptors are present on the circulating hemocytes of the earthworm, *Eisenia foetida,* but are absent on sessile ones.[62] In the snail, *Biomphalaria glabrata,* sessile granular hemocytes react with Con A, but sessile agranular ones do not.[63]

Monosaccharide determination — We can determine, from the evolutionary and phylogenetic standpoints, how many of the known monosaccharides or their derivatives (e.g., D-mannose, D-glucose, D-galactose, *N*-acetyl-galactosamine, *N*-acetylglucosamine, L-fucose, and *N*-acetylneuraminic acid (NANA) [sialic acid]), commonly found on the surface membranes of eukaryotic cells,[56] are present on the insect immunocyte surface.

Determination of the presence of lectin receptors — We can determine whether any of the lectin receptors known to be of medical significance (e.g., sialic acid-, wheat germ agglutinin (WGA)-, and peanut agglutinin (PNA)-binding receptors), are present on the immunocyte plasma membrane. For example, *N*-acetylneuraminic acid (sialic acid)-binding receptor is known to have differential agglutinability between normal and neoplastic cells.[64] Burger and Goldberg[65] reported a similar property of the WGA-binding receptor, and Klein et al.[66] showed that the PNA-binding receptor could differentiate between hormone-sensitive and hormone-insensitive breast carcinoma cells. Thus, an insect immunocyte with sialic acid (NANA)-, WGA-, and PNA-binding receptors could be used as a probe to identify neoplastic conditions in human beings. Several commercially available lectins are known to have specific binding affinities for the seven above-mentioned monosaccharides found on the plasma membranes of vertebrate erythrocytes and other eukaryotic cells. For example, Con A-, peanut (PNA), soybean (SBA), wheat germ (WGA), and winged pea agglutinin (WPA), and limulin (a major lectin in the hemolymph of *Limulus polyphemus*) are known to bind to D-mannose, D-galactose, *N*-acetylgalactose, *N*-acetylglucosamine, L-fucose, and *N*-acetylneuraminic acid, respectively.[11,67-69] Note that according to Pendland and Boucias,[69] Con

A also binds to D-glucose, and it has been demonstrated[61] that WGA binds to sialic acid residues, and, more importantly, that it does not compete with limulin, which indicates that these two lectins (WGA and limulin) interact with different oligosaccharide moieties (see also Section VI). Furthermore, multivalent lectins, upon binding with surface receptors of cells at 37°C, are known to cause redistribution of these receptors into "patches", followed by "capping"[56,70] (see also Section II. A.4.c.iii). In our study of labeling, mapping, isolation, and molecular characterization of the GR surface receptors of *L. polyphemus* (Figure 10A—D), we found that Con A and, to a lesser extent, soybean agglutinin produced both "patching" and "capping" of the surface receptors[11] (Figure 10E, F) (see Section VI on phagocytosis and encapsulation).

4. Cytoskeleton

The plasma membrane in eukaryotic cells is anchored to an underlying filamentous cytoskeleton, which, in the human RBC, platelets, and other cells, is known to determine the shape and mobility of the cell.[54,56,71,72]

Studying the ultrastructural and molecular details of the cytoskeleton is important because of its role in cell movement, "capping" of lectin-binding receptors, and changes in cell shape (e.g., differences in cytoskeletal components of GR and PLs; the GR in *B. germanica,* for example, is flat and discoid, but the PL is spherical). Furthermore, the study of the assembly and disassembly phenomena of the cytoskeleton, especially those of the peripheral microtubule bands (MBs), will shed light on the roles of these organelles in the maintenance of the GR shape (see discussion in Section II.A.4.c).

The cytoskeleton of eukaryotic cells is composed of a complex network of protein filaments, which generally are of three types: actin microfilaments (diameter ca. 5 nm), intermediate filaments (diameter 10 nm), and microtubules (MTs) (diameter 25 nm).[56]

a. Actin Microfilaments

Actin microfilaments are often grouped in bundles, located close to the plasma membrane in association with MTs, and participate in cell shape changes.[71] The cytoskeleton of the human RBC is attached to the cytoplasmic (P) face of the plasma membrane at many points, and is composed of a network of filamentous proteins, which include spectrin (= bands 1, 2), ankyrin (band 4.1), and actin (= band 5).[53,55] Spectrin, actin (both of which determine the shape and mobility of RBC), and band 4.1 have also been detected in the cytoskeletons of various nucleated cells.[73] Boyles et al.[74] reported the filamentous nature of the cytoskeleton of the discoid human platelet.

Like tubulin (which is the main constituent of MTs), actin is thought to be a ubiquitous protein component of the cytoskeleton and is responsible for changes in cell shape, "capping" of receptors on lymphocytes, and phagocytosis (endocytosis).[71,75] Indeed, in a preliminary SDS-PAGE analysis of the *L. polyphemus* GR plasma membrane, we found protein fractions that have the same electrophoretic mobility as actin and tubulin (Figure 11). Actin microfilaments are known to be associated with actin-binding proteins, such as α-actinin, filamin, myosin, tropomyosin, and vinculin.[76,77]

b. Intermediate Filaments (IFs)

Intermediate filaments (IFs) are the most stable of the three cytoskeletal filaments, and consist of at least five different tissue-specific classes, such as *neurofilaments* in neurons, *desmin* or *skeletin* in muscles, *keratin* in epithelial cells, *vimentin* in cells of mesenchymal origin, and *glial acidic fibrillary protein* in astrocytes (see reviews in Goldman and Steinert[78]). Because of their insoluble nature in cells treated with detergents (e.g., Triton X-100), the term cytoskeleton was at first coined for them.[56] We have demonstrated their presence in Triton X-100-treated GRs of *B. germanica.*[79] They play a tension-bearing role in cells, and

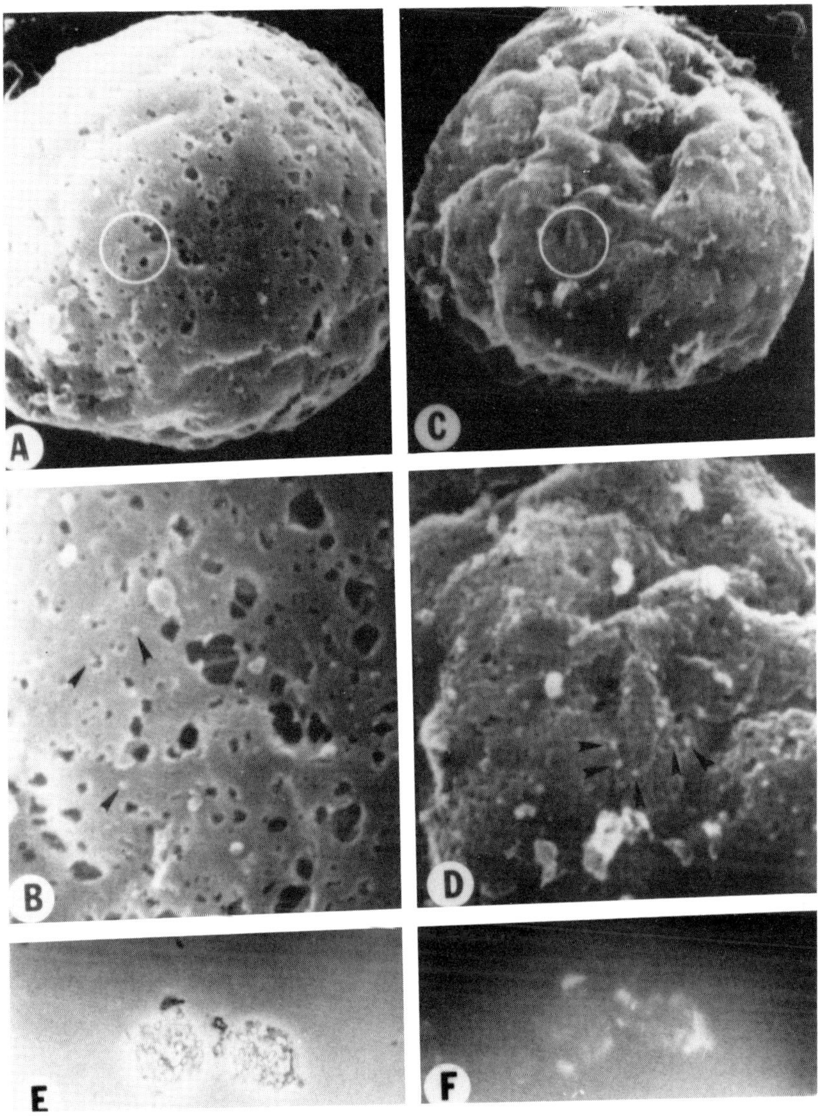

FIGURE 10. Labeling of lectin-binding plasma membrane receptors of the granulocyte of *Limulus polyphemus*. (A) Con A (concanavalin A) labeling (specific for D-mannose and D-glucose binding) (× 8,000). (B) Magnified view of the area within circle in (A), showing labeled receptors (three arrowheads) (× 20,000). (C) WGA (wheat germ agglutinin) labeling (specific for *N*-acetylglucosamine and NANA binding) (× 8,000). (D) Magnified view of the area within circle in C, showing labeled receptors (five arrowheads) (× 20,000). (E) Phase contrast of granulocyte, treated with Con A-FITC (fluorescein isothiocyanate) (× 1,330). (F) Immunofluorescence microscopy of the granulocyte treated with Con A-FITC, showing "patching" (left granulocyte) and "capping" (right granulocyte) of receptors (× 1,330). Hemocyanin (35 nm in diameter) was used as a marker to label the lectin-binding receptors for SEM. Positive labeling was confirmed by inhibition of the lectins by specific sugar inhibitors for the lectins. (From Gupta, A. P., Orenberg, S. D., Das, Y. T., and Chattopadhyay, S. K., *Exp. Cell Res.*, 194, 83, 1991. With permission.)

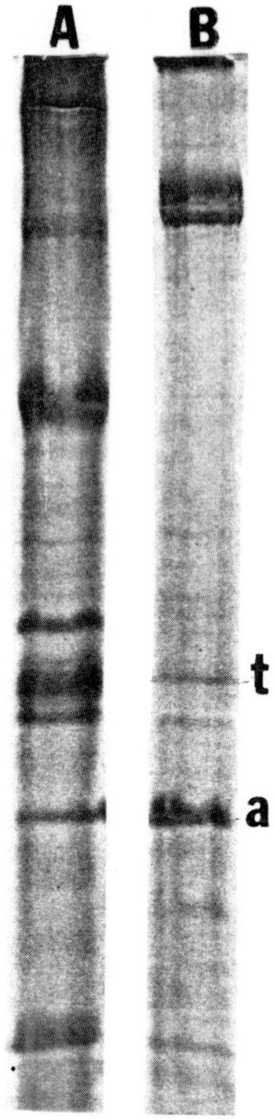

FIGURE 11. SDS-PAGE of human erythrocyte (A) and *Limulus* granulocyte (B) ghosts. Molecular weights were calculated with reference to BRL standards. Note the presence of bands comparable to actin (a) and tubulin (t) in the *Limulus* granulocyte. Other bands in the *Limulus* immunocyte might represent microtubule-associated proteins of the cytoskeleton, receptor and/or intramembranous particle molecules, and gap junction polypeptides and/or their precursors. (Courtesy of Drs. E. V. Oaks and Y. T. Das.)

also have associated proteins (IFAPs), whose functions are unknown. Unlike those of actin and tubulin, assembly of IFs is irreversible. Their association with MTs and actin microfilaments is poorly understood.

c. *Microtubules (MTs)*

MTs (25 nm outer diameter and 15 nm inner diameter) are composed of helical assemblies of α and β subunits (M.W. ca 55 kDa) of tubulin, which is highly conserved in nature. Immunohistochemical methods indicate that this protein from various vertebrate taxa and insects[80] has similar antigenic determinants.[81] During assembly, tubulin polypeptides (α and β units) become aligned and form protofilaments, which are arranged side-by-side around a central core to form the hollow MT. Isolated MTs, when negatively stained, clearly show these protofilaments, which vary in number in various taxa, the most frequent number being 13 in some vertebrate cells.[71] Most MTs are associated with accessory proteins (MAPs), such as MAP1 and *tau* that are thought to regulate MT assembly. Some MAPs form cross-bridges among MTs;[82] such cross-bridges, however, are absent in the GRs of *B. germanica.*[6] Finally, a protein identical to tubulin may be present in the plasma membrane;[83] in some cases, is firmly attached to it; and copolymerizes with the cytoplasmic tubulin in *in vitro* studies.[84]

Microtubule bands (MBs) are found in both vertebrate and invertebrate erythrocytes,[82] which are often flat and elliptical. Among arthropods, they have been reported in *L. polyphemus, Homarus americanus, Cancer borealis,* and *Libinia emarginata;*[82] *Periplaneta americana;*[26] *Thermobia domestica;*[41,85] *Locusta migratoria* and *Melolontha melolontha;*[86] and *B. germanica.*[2,6,7,79]

i. Assembly of MTs from Tubulin Pool

Weisenberg[87] was the first to discover that MTs are assembled from tubulin *in vitro.* Assembly and disassembly of MTs depend on an exchange of tubulin molecules between the MTs and a soluble cytoplasmic pool.[56] Tubulin molecules rapidly aggregate in a helical fashion into long tubular structures, and the process is reversible because of a constant balance between tubulin pool and MT in relation to cell activities.[71]

Measurements of the total amount of cell tubulin and the amount used in MT assembly are complex and have been discussed by Fulton and Simpson.[88] According to them, in addition to the MTs, a tubulin pool exists in most cells. For example, radioimmunoassays have shown that in neuroblastoma cells undergoing neurite formation *in vitro,* both differentiated and undifferentiated cells contained about 4 pg per cell of tubulin, 70% of which was assembled in MT in the differentiated cells.[89] The cytoplasmic tubulin pool is regulated by the assembly of MTs,[90] and tubulin mRNA decreases when MTs are disassembled by colchicine. The pioneering work of these authors has provided some insight into the regulatory mechanism of tubulin synthesis. Furthermore, intracytoplasmic injection of tubulin depresses tubulin synthesis.[91,92] The short half-life (1 to 2 h) of tubulin mRNA might also regulate tubulin mRNA transcription.[93] Note that other factors, such as MAPs, Ca ions, calmodulin, nerve growth factor, cAMP, and proper ionic conditions are all known to determine the extent of MT assembly in various types of cells.[71]

In recent studies, we found that in *B. germanica* there was a considerable increase in the number of the MTs (Figure 12A—C) as well as of the nuclear pores (Figure 13A, B) in the encapsulating (activated) GRs. That nuclear pores play a role in maintaining the cytoplasmic network of MTs has been reported earlier in mast cells.[94,95] It was obvious to us[6,7] that the encapsulating GRs were rapidly synthesizing MTs (apparently commencing within 40 to 45 min postencapsulation) from tubulin pools. This supports Dustin's[71] observation that "a fundamental property of the MT is to assemble and disassemble rapidly in relation to cell's requirements". Rapid tubulin synthesis seems to be consistent with the fact that tubulin mRNAs have a very short (1 to 2 h) half-life.[93] Furthermore, microinjection of fluorescent tubulin in dividing sea urchin cells also demonstrated a very rapid assembly of MTs, on the basis that fluorescence showed up in the asters almost immediately after injection.[96] Also, in the unicellular *Chlamydomonas reinhardii,* synthesis of tubulin starts immediately after deciliation, and between 45 and 90 min later, about 14% of the synthesized protein is tubulin.[97]

From the standpoint of the role of MTs in encapsulation,[7] two fundamental questions arise: (1) Does a rapid synthesis of tubulin indeed occur in the activated GRs? In other words, what is the turnover rate of the tubulin pool in the cytosol of the encapsulating GRs, compared with that in the normal GRs? (2) What is the mode of assembly of the MTs? Both these questions are of much interest in cellular immunity and need to be investigated in insect immunocytes that participate in encapsulation.

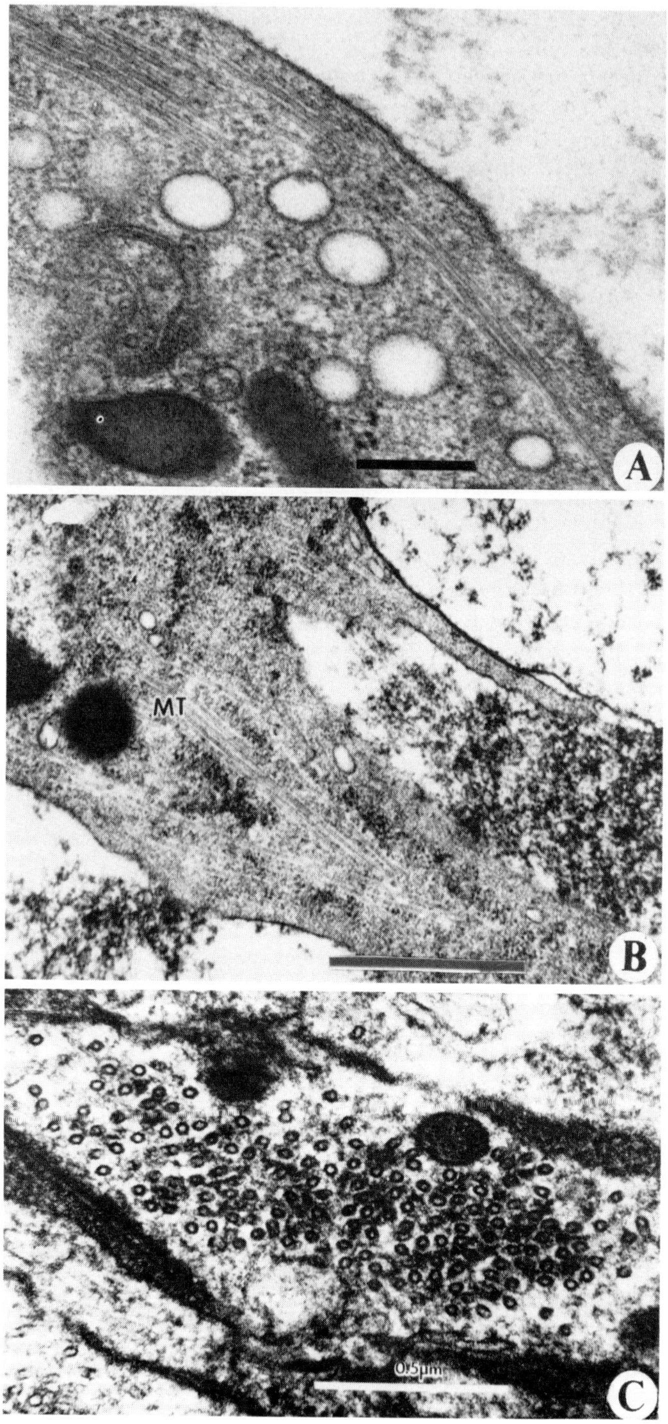

FIGURE 12. (A) Section of an unactivated granulocyte (TEM) of *B. germanica,* showing peripheral microtubule band. (Bar = 0.5 um.) (B) Section of an activated granulocyte (TEM) of *B. germanica* 8 h after implantation, showing cytoplasmic extensions of the plasma membrane and pronounced development of microtubules in longitudinal section. (C) Section of a 10-day-old, completed capsule (TEM) of *B. germanica,* showing the cytoplasm of the granulocyte with pronounced development of microtubules in cross-section. (From Han, S. S. and Gupta, A. P., *Zool. Sci.,* 6, 303, 1989. With permission.)

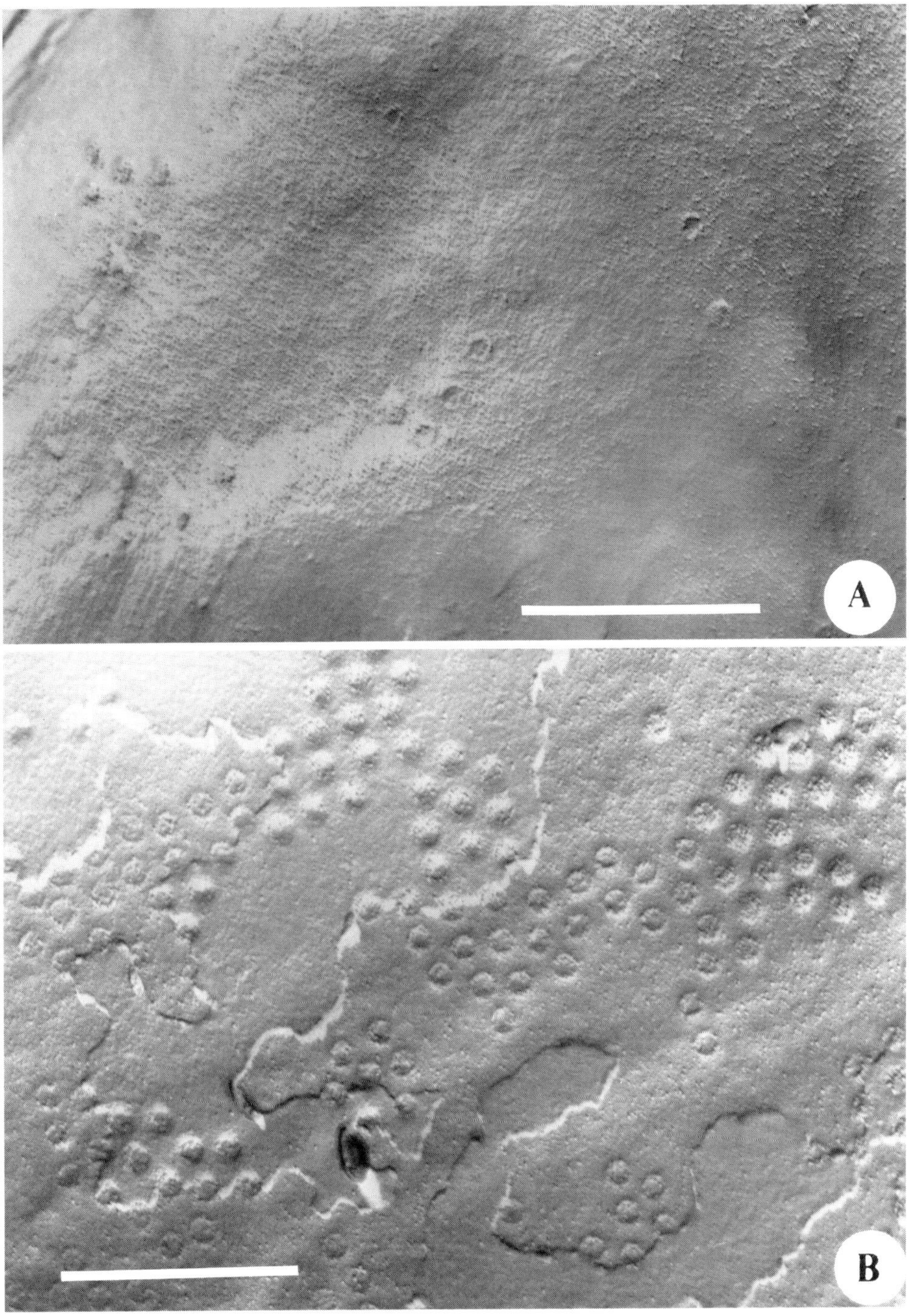

FIGURE 13. (A) Freeze-fracture and etching replica of unactivated granulocyte of *B. germanica* showing distribution ($3.75 + 1.86/\mu m^2$; mean standard error) of nuclear pores. (Bar = 0.5 μm.) (B) Freeze-fracture and etching replica of granulocyte (harvested from 10-day-old, completed capsule) of *B. germanica*. Note increase in number of nuclear pores ($26.14 + 6.49/\mu m^2$; mean standard error). (Bar = 0.5 μm.) (From Han, S. S. and Gupta, A. P., *Cell Struct. Funct.*, 13, 333, 1988. With permission.)

Pipeleers et al.[98,99] have assayed the distribution of tubulin in several kinds of tissues and found that the assembled tubulin varies from 30 to 60%, the highest being in human platelets. More relevant to the situation in the GRs of *Blattella,* Sherline and Mundy[100] showed that the percentage of formed MTs in lymphocytes activated by phytoagglutinin (PHA) paralleled the increase in total tubulin pool; the maximum fraction assembled was 30 to 40% of the total tubulin measured. Note that the activated GRs in *Blattella* also have a considerably increased number of MTs, and, by contrast, the PLs do not even have these organelles.[6]

ii. Roles of Microtubule Organizing Center (MTOC), MT Poisons, and Stabilizers in MT Assembly/Disassembly

The MTOC or other comparable organelles seem to be necessary for MT assembly; the MTOC is present in the GRs of *B. germanica.*[6,79] Nemhauser et al.,[101] in a study of the reassembly sequence of MBs in the clam *(Noetia ponderosa)* erythrocyte, showed that MTs grow from the granular material surrounding the centriole. *In vitro* studies indicate that "seeds" or rings, or MT fragments as well as appropriate ionic conditions are necessary for MT assembly, the MTOC in living cells serving the same function.[71] It is generally accepted that the growth of MT is linear and occurs in a treadmill fashion *in vitro.*[71] Note that treadmilling has not been demonstrated *in vivo.*[102] Note also that, in addition to the treadmilling mode of MT assembly/disassembly, there is a condensation equilibrium mode of MT formation, which occurs only when the MT protein concentration exceeds a critical or threshold level.[103] In fact, these authors[103] have suggested a unified or combined equilibrium-treadmilling model of MT assembly/disassembly. Reports from various sources suggest that assembly of MTs is indeed polarized. For example, MT poisons (e.g., colchicine) block assembly at one end but allow disassembly to contine at the other end. Studies of labeled GTP also suggest a polarized assembly, and this is further confirmed by regenerative growth of cilia.[104] Bergen et al.[105] and Borisy and Bergen,[106] using flagellar axonemes, defined (+) and (−) ends of doublets, the (+) end being distal, the direction in which cilia grow. Jordan and Farrell[107] have distinguished the opposite MT ends by differentially labeling the MTs with [^{3}H]- and [^{14}C]guanine nucleotides. Whether the MBs in insect immunocytes (e.g., GRs) are constructed of MTs of opposite polarity, as reported by Cohen and Nemhauser[82] in the clam erythrocyte, or grow by continuous lengthening of a single strand, remains to be determined.

A number of chemicals, such as colchicine (and some of its derivatives) and benzimidazole, bind specifically to tubulin and thus interfere with MT functions.[58,71,102] Dustin[71] has extensively reviewed the literature on MT poisons. Colchicine binds to tubulin with high specificity to form a 1:1 complex,[108] and this property has made it a useful tool for determining the amount of tubulin (colchicine-binding protein) in various cell types. It is suggested that colchicine probably blocks the (+) end of the MT, and thus prevents elongation and assembly of tubulin subunits, the so-called tip-poisoning hypothesis of Margolis et al.[109] According to this hypothesis, if the other end is considered to be the site of disassembly, this could explain the disappearance of MT, assuming that the turnover rate of tubulin subunits is relatively fast. Note, however, that it is controversial whether colchicine poisons only the MT assembly or that colchicine-tubulin dimer interferes with assembly with no preference for (+) or (−) ends of the MT. Note also that the centriole and/or MTOC are resistant to colchicine.[71] Some of the commonly known antagonists for colchicine are podophyllotoxin, vinblastine, oncodazole, and bleomycin.

Taxol, an antitumor drug, is known to favor formation and stability of MTs. In contrast to colchicine, taxol binds to the MT polymer, rather than the subunit, and shifts the dimer-MT equilibrium in favor of the MT by decreasing its critical concentration for assembly.[110] According to Schliwa,[110] taxol can also bind to preformed MTs, which are thus stabilized against depolymerizing conditions (e.g., elevated Ca ions or depolarizing drugs). Finally, it should be remembered that in any study of MT assembly and disassembly in taxol-treated GRs, it will be necessary to be careful in assessing the role of MTOC in MT assembly, because Wright et al.[111] found that taxol-treated cells, when treated with nocodazole that disassembles MT, showed no association of MTs with the MTOC.

Although MTs are thought to be composed of tubulin, actin and spectrin (as well as tubulin) are also found in MBs.[71] Spectrin (α and β subunits), previously thought to be present in only mammalian cells, is found in many cell types in other taxa.[112] Cohen and Nemhauser[82] have reviewed the literature on the components of the cytoskeleton in many invertebrates and have stated that, while actin appears to be a ubiquitous cytoskeletal component, spectrin is not so conserved. According to them, in the clam erythrocyte, a component (240 kDa) of the cytoskeleton comigrates with the RBC α-spectrin and with α-fodrin (the latter is also known as "nonerythroid" spectrin or "brain spectrin") in nonerythroid cells. They concluded that little is known about cytoskeletal components in invertebrate clotting cells (note that the granulocytes/coagulocytes of arthropods participate in clotting). MBs, when present, are associated with the P face of the cytoskeleton and act as a flexible supporting frame across which tension is applied.[82]

Finally, although qualitative biochemical analysis of the plasma membrane could reveal the presence of the major protein components of the cytoskeleton, direct visualization of its molecular architecture (i.e., organization of the protein microfilaments of the cytoskeleton and its relationship with the MBs and the plasma membrane itself) should be done in negatively stained fragments of the immunocyte plasma membrane by high resolution electron microscopy.[53,55] Note that direct visualization of MTs in the *same* cell by both indirect immunofluorescence procedure and transmission electron microscopy (TEM) has been demonstrated by Osborn et al.[113] in PtK2 cells. This method has the advantage that one can show direct correspondence of fluorescent MTs in a light microscope with uninterrupted MTs in the TEM.

iii. *Role of MTs in Cell Shape Maintenance and "Patching" and "Capping" of Surface Receptors*

Although the exact molecular mechanism involved in capping is not well known, it is thought that displacement of surface receptors for antigens or for lectins (e.g., Con A) into "caps" involves MTs or MT-associated structures, and has been observed in lymphocytes, polymorphonuclear and some other cells.[114] More importantly, it has been shown by immunofluorescence techniques that in B-lymphocytes from mouse spleen, actin and tubulin co-cap with immunoglobulin, on the basis that the fluorescence for these proteins becomes restricted to the caps.[115] Indeed, Yahara and Edelman[60] have suggested a possible modulation of surface receptors by MTs and possibly actomyosin (their MF). In their model, MTs form a support a short distance from the cell membrane, MF becomes loosely attached at one end to this support, and at the other end is fixed to the cytoplasmic (P) ends of the transmembranous protein molecules, which, according to these authors, serve as receptors for lectins and antibodies. Note that these transmembranous protein molecules appear as intramembranous particles (IPs) on either the P or E (extracellular) or both faces of the plasma membrane in freeze-fractured replicas.[3,48]

Capping, unlike patching, is an active process and involves ATP and actin microfilaments, which form a subcap under the cap. Cap-reversal experiments have demonstrated that subcaps are an integral part of caps, and not a secondary accumulation of actin microfilaments following capping.[116] These authors[116] explained the molecular mechanism for capping (aggregation of surface receptors and concentration of the contractile actin microfilaments in a subcap) by suggesting that "*each* receptor to be capped, upon being patched by the binding of its specific multivalent ligand, becomes associated with actin and myosin in such a way that *only* that ligand-bound receptor is then collected into the cap''. Bourguignon and Singer[117] suggested previously that actin is directly or indirectly bound to a ubiquitous plasma membrane protein, called X (perhaps a transmembranous IP?), to which the aggregated receptors effectively bind (unaggregated receptors do not), thus also becoming linked to actin on the P face of the plasma membrane. This linkage ultimately collects the receptor aggregate into a cap in a manner analogous to the sliding-filament mechanism of muscle contraction.

It is generally known that MT's regulate cell morphogenesis, changing spherical cells to a flattened, elliptical, and discoid shape.[82] According to Fischer,[118] the discoid shape is suitable for flow conditions and metabolic exchanges — conditions under which the GR in *Blattella* seems to function during encapsulation. Thus, from the standpoint of immunologic function of the GR, the study of the changes in its shape and of patching and capping of surface receptors (which play an important role in the initial recognition of nonself tissue) is necessary.

The GRs in *B. germanica* are discoid in shape, and the MTs in these immunocytes are located peripherally in the form of bands under the plasma membrane in the plane of flattening. This arrangement of the MBs apparently helps maintain the discoid shape of the GRs. Indeed, Cohen and Nemhauser[82] have reported that 98% of the MB-containing erythrocytes of the blood clam, *N. ponderosa,* when intentionally subjected to mechanical stress ''by fluxing through glass capillary tubes'', maintain their flattened and elliptical shape, and many (45%) without MBs are deformed. These observations support the view that in *N. ponderosa* erythrocytes, ''MB plays a continuing role in cell shape maintenance''. Behnke[119] has also suggested that in human platelet (which, like the GR, is flat and discoid) ''MBs resist change in shape''.

Low temperature also affects cell shape. According to Behnke,[119] at 0°C, MBs in human platelets disappear rapidly, which results in an irregular or spherical shape of the cell. On rewarming, the MBs may either reassemble normally (in which case the platelet assumes its discoid shape) or in straight MTs, thus making the cell spindle-shaped. Ikeda[120] reported that chilled human platelets showed significantly decreased amounts of MTs. Also, in clam erythrocytes, at 0°C, the MBs disappear and reappear within minutes of rewarming, originating from the centriole.[82] The reassembly of the MBs even at 0°C can be induced by adding the MT-stabilizing agent taxol.[121] Indeed, because GRs *with* and *without* MBs both at 0°C (taxol-stabilized vs. nontaxol-stabilized) and at room temperature (colchicine and nocodazole inhibition of reassembly vs. no inhibition) could be made available, it should be possible to observe the MB function in cell-shape maintenance *in vitro.* I have suggested earlier (see discussion on Biochemistry of Plasma Membrane and IPs, Section II.A.3) that since disassembly of MTs is associated with an increased number of gap cell junctions, these two cellular events might somehow be interrelated and that the MT disassembly might trigger the formation of gap junctions (for details of gap junctions, see Chapter 4).

Finally, on the basis of the cell-shape maintenance function of the MBs, we can explain why these organelles are *absent* in the PLs of *B. germanica,*[2,6,7] which, unlike the GRs, are neither discoid nor participate in encapsulation.

As stated earlier, we found that in this cockroach, there is a considerable increase in the number of MTs as well as nuclear pores in the encapsulating (activated) GRs. In addition, we noticed the presence of a microtubule-organizing center (MTOC) in the periphery of the cell. Because both the MBs and the MTOC bind the antitubulin antibody antibody, they were easily seen by indirect immunofluorescence technique.[6,79] The number of MTs in the MBs ranged from 27 to 56 in resting or normal (unactivated) GRs, whereas in the encapsulating (activated) cells there was a considerable increase in MTs, ranging from 105 to 257 (Figure 12B, C). More interestingly, concomitant with this increase in the number of MTs, there was a sevenfold increase in the number of nuclear pores (Figure 13A, B).

B. PLASMATOCYTE (PL)

1. Distinguishing Features

Compared with that of the GR, the number of PLs is greater in some insects, and in such cases, these immunocytes play a major role in cellular immunity (phagocytosis and encapsulation). However, we have reported that at least in *B. germanica,* the PLs play no role in encapsulation,[7] and that the absence of the peripheral microtubule bands and the very fine and considerably smaller size of the cytoplasmic granules, distinguish this immunocyte from the GR. It should be emphasized here that at the ultrastructural level, hemocytes with granules comparable to those found in the GRs should not be designated as PLs in any insect, because such designation obscures the distinction between these two hemocyte types and inevitably creates confusion. Furthermore, it is most likely that a PL with a few granules is a transitional or intermediate form between PL and GR, and is on its way to transforming into the latter type. This latter fact may be verified by the presence of active Golgi complexes engaged in synthesizing and packaging these granules. Therefore, to avoid any confusion and misidentification, any PL with even a few granules, comparable to those of the GR, should be designated as an immature GR, not a PL. I have suggested elsewhere[48] that arthropod PLs are functionally analogous to killer T-lymphocytes and macrophages of vertebrates (see also Sections VIII and IX). Its role in phagocytosis and encapsulation will be discussed in Section VI.A,B).

2. General Features

The PLs are large polymorphic (spindle-shaped or oval, rarely spherical) (Figure 14C, D) cells of variable size (3.3 to 5 μm wide and 3.3 to 40 μm long) and may appear agranular or show a few granules, the latter being considerably smaller and finer than those in the granulocyte (GR). The nucleus of the PL is centrally located and may be round, elongate, or even lobate, varying in size (3 to 9 μm wide and 4 to 10 μm long) in various insects, and may even appear punctate. Scattered chromatin masses may be present along with a nucleolus. Occasionally, a binucleate PL may be found.

In SEM and TEM (Figure 14C, E), the plasma membrane may show micropapillae or other irregular processes, as well as pinocytotic or vesicular invaginations. The cytoplasm is often abundant, basophilic, and rich in organelles. Generally, there is a well-developed and extensive RER, which may form greatly distended cisternae or a vacuolar system. Golgi bodies (also referred to as dictyosomes, golgiosomes, or internal reticular apparatus) and lysosomes may be numerous; lysosomes are often associated with RER or the vacuolar system and can be identified by the presence in them of the reaction products of the hydrolytic marker enzymes, acid phosphatase, and thiamine pyrophosphate. Microsomes and cisternae of ER (called *ergastoplasme* by French authors) may be present. Free ribosomes (polysomes or polyribosomes) or ribosomes attached to microsomes or RER may be present.

PLs may or may not (e.g., in the cockroaches, *B. germanica* and *Gromphadorhina portentosa*) be abundant in various insects. I have proposed elsewhere (Gupta[48] and Section VIII.B in this review) that PLs are functionally analogous to the killer T-lymphocytes and macrophages of vertebrates.

3. Variant Forms

These should not be regarded as separate categories.

a. Podocyte (PO)

These hemocytes have not been recognized in ultrastructural studies, and are not ordinarily observed in hemocyte samples in LM studies. Arnold[47] commented that their presence might be due to fixation artifacts. According to Arnold,[122] they have been correctly identified only in *Prodenia eridania.*[123,124] However, I have observed them in *Periplaneta americana* nymph.[125] More often than not, radiate PLs with pseudopods are often mistaken POs.

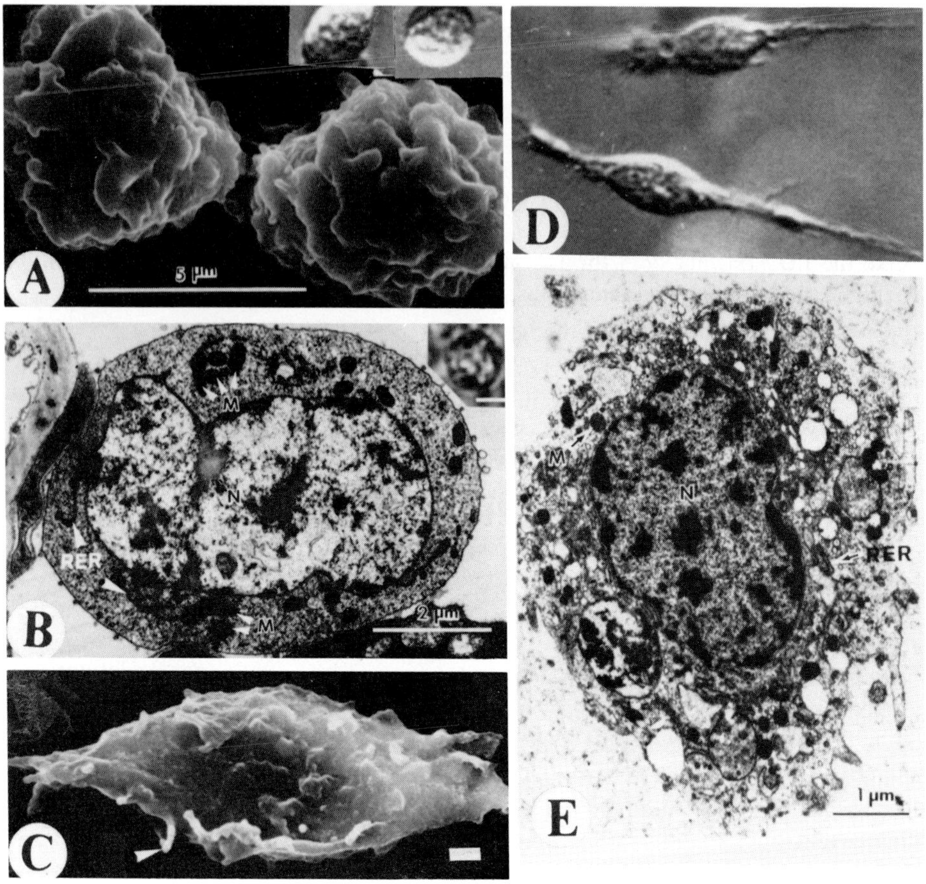

FIGURE 14. (A) Prohemocytes (SEM) of *Blattella germanica;* **inset,** prohemocytes (LM). (B) Prohemocyte (TEM) of *B. germanica,* showing large nucleus (N) almost filling the cell, numerous free ribosomes, some rough endoplasmic reticulum (RER), and mitochondria (M); **inset,** prohemocyte *in vivo.* (C) Spindle-shaped plasmatocyte (SEM) of *B. germanica* with micropapillae (arrowhead). (D) Plasmatocytes (Nomarski interference contrast) of *Gromphadorhina portentosa* (× 6500). (E) Plasmatocyte (TEM) of *B. germanica,* showing finer granules than in granulocyte, nucleus (N), abundant free ribosomes, RER, and some mitochondria (M). (Parts A, B, C, and E from Chiang, A. S., Gupta, A. P., and Han, S. S., *J. Morphol.*, 198, 257, 1988; part D from Gupta, A. P., *Comprehensive Insect Physiology, Biochemistry and Pharmacology,* Vol. 3, Kerkut, G. A. and Gilbert, L. I., Eds., Pergamon Press, Oxford, 1985, 401. With permission.)

The POs are very large, extremely flattened, PL-like cells with several cytoplasmic extensions. The nucleus is generally large and centrally located, and may appear punctate.

We suggested their transformation from PLs,[15] and this seems to be supported by Rizki's[126] observation that in *Drosophila melanogaster,* when POs increased in differential counts, PLs decreased. As PLs, they are known to ingest bacteria.[38]

b. Thrombocyte (TH)

Like the POs, these anucleate cells are variant forms or fragments of PLs (e.g., the filamentous PLs of Rowley and Ratcliffe[127]), have a tendency to fragment, and were erroneously termed "thrombocytoid" by Zachary and Hoffmann[128] on the basis of their superficial resemblence with the thrombocyte of vertebrates. Barracco and Cestari[129] did not find them in another dipteran, *Trichosia pubescens.* Like the POs, they should not be recognized as a separate category, contrary to the suggestion by Brehélin and Zachary.[49] These cells apparently take part in encapsulation in *Calliphora erythrocephala.*[130]

c. Vermicyte (VE)

A variant form of PL,[14,15,131,132] the VE is also called vermiform cell, and should not be regarded as a separate hemocyte type. According to Arnold,[122] "they seem to occur mainly just prior to pupation, but never in large number", and may be caused by faulty fixation.[47] As the name suggests, VEs are extremely elongated cells with slightly granular or agranular cytoplasm. The nucleus may be central or eccentric.

d. Lamellocyte (LA)

Like the PO, TH, and VE, the LA also is a variant form of PL, was first introduced by Rizki,[126] and has been consistently used by him and others in *D. melanogaster.*

III. CONTINUOUS LINEAR INCREASE IN IMMUNOCYTE POPULATIONS DURING ONTOGENY

Given the fact that immunocytes are important components of the immune system, one would expect a gradual increase in their population as the animal progresses from the early immature stages toward the adult stage. Such an increase in fact occurs. Several authors have reported both qualitative and quantitative changes in hemocyte populations in various insects.[133] In our studies of the effects of insect growth regulators on hemocyte population in various developmental stages of *B. germanica,* we found a progressively linear increase not only in the total hemocyte counts (THCs) (Figure 15A) in the controls, but also in the differential hemocyte counts (DHCs) of the two immunocytes (Figure 15B, C),[134] which suggests a correlation between the THCs and DHCs of the PLs and GRs. These two immunocytes represent more than 70 to 80% of the THCs in all developmental stages, thus determining the quantitative picture of the hemocyte populations in any of the seven developmental stages of this cockroach. The continuous and linear increases of the PLs and GRs seem to be commensurate with the growing need for effective cellular immunity as the insect progresses toward the adult stage. Because the effectiveness of phagocytosis, encapsulation, and other related defense mechanisms depends largely on the available circulating immunocytes (PLs and GRs),[14] it is reasonable to suggest that the larger the populations of these cells, the stronger the cellular defense. By implication, the early immature instars would be immunologically less competent than the adults and thus more susceptible to pathogen attack.

IV. OTHER HEMOCYTE TYPES

A. CONTROVERSIES ABOUT HEMOCYTE TYPES

Anyone venturing to study insect hemocytes knows only too well how discouraging it is to have to contend with numerous variant forms of hemocytes in hemolymph samples, and a bewildering array of terminologies that exists in the literature. The tendency, and perhaps irresistible urge, on the part of some workers to introduce new terms for variant forms of the morphologically distinct hemocyte types, is mainly responsible for the plethora of terms. I believe that descriptions of "new" hemocytes, based on superficial dissimilarities from the six morphologically distinct types, should be avoided. Similarly, designating subtypes of certain hemocyte types[49] on the basis of features that are not diagnostic of each subtype, serves no useful purpose, except that of further exacerbating the currently existing problems in insect hemocyte classification.

Hemocyte classifications both in insects and other arthropods have been variously based on hemocyte morphology of both immature and adult insects, various hemocytic functions, and staining or histochemical reactions of hemocytes. Thus, it is not unusual to find the same hemocyte type or its variant forms being referred to by different names in various

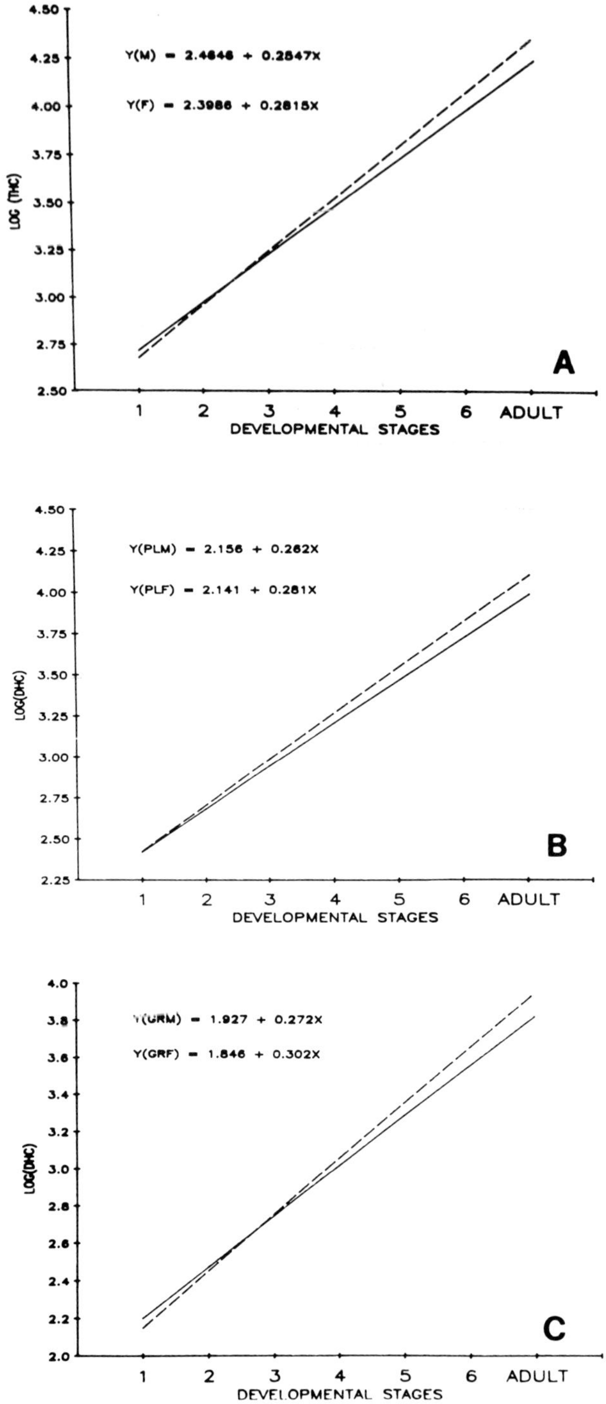

FIGURE 15. Linear increases in the total hemocyte counts and differential counts of plasmatocytes and granulocytes during various developmental stages of *B. germanica* (six nymphal stages). (A) Total hemocyte counts of male [Y (M), ——; $r^2 = 0.96$] and female [Y (F), - - - -; $r^2 = 0.95$]. (B) Differential plasmatocyte counts of male [Y (PLM), ——; $r^2 = 0.95$] and female [Y (PLF), - - - -; $r^2 = 0.96$]. (C) Differential granulocyte counts of male [Y (GRM), ——; $r^2 = 0.95$] and female [Y (GRF), - - - -; $r^2 = 0.96$]. (From Hazarika, L. K. and Gupta, A. P., *Zool. Sci.*, 4, 307, 1987. With permission.)

arthropods, by different authors — a situation that has inevitably resulted in a confusing mass of terminologies. Consequently, it becomes very difficult to compare hemocytes of one species with those of others. This has particularly hindered any phylogenetic consideration of the evolution of hemocyte types in various arthropod groups and the Onychophora, and often makes it difficult to assign specific physiologic and immunologic functions to them.

Categorizing hemocyte types on the basis of their functions *alone* can be confusing because (1) physiology and functions of blood cells is based on the knowledge of their morphology and ultrastructure; and (2) functional categories, such as phagocytes, trephocytes, or immunocytes include not just a single hemocyte type, but several. For example, the PLs, GRs, and other types are known to phagocytose foreign particles in various insects, yet they are morphologically distinct. Furthermore, a certain type (e.g., the GR) may be phagocytic in one species, but not in another. Consequently, the same cell (e.g., the GR) would have to be classified into more than one category on the basis of function. Most significantly, however, classification based on function alone incorrectly perpetuates the notion that the major insect hemocyte types are morphologically indistinguishable from one another.

Insect hematologists often disagree as to the actual number of hemocyte types in various insects. Indeed, from one to as many as nine or more types have been described, particularly by light microscopy (see reviews by Gupta[1,14,22,23,48,133,135] and Chiang et al.[2]). Ultrastructurally, however, only six types should be recognized: *prohemocyte* (PR), *plasmatocyte* (PL), *granulocyte* (GR) (including the so-called coagulocytes), *spherulocyte* (SP), *oenocytoid* (OE), and *adipohemocyte* (AD). Of these six, the AD has been reported only by Devauchelle[31] and Chiang et al.,[2] and the coagulocyte (CO) as a separate category by Goffinet and Grégoire,[136] Ratcliffe and Price,[36] and recently by us[2] and Barracco and Loch.[25] Note that of the six types, only the GR has been reported from all major arthropod groups as well as from the Onychophora. I have, therefore, regarded it as the most primitive (plesiomorphic) hemocyte in arthropods[14,22,48] from which the other types seem to have evolved (differentiated) during the course of arthropod evolution.

Note that of these six types, the identity of the so-called CO as a separate type, is doubtful; in fact, I believe that COs are degranulating and/or degranulated GRs, because ultrastructurally they look very similar to the GRs.

The hemocyte classification that is generally used has evolved over a period of more than half a century. According to Millara,[137] Cuénot[138] was the first to classify insect hemocytes into four categories, and was later followed in this attempt by Hollande[139,140] and others. Wigglesworth[141] summarized most of the earlier classifications, modified it later,[142] and presented a classification that was widely accepted. On the American side, Yeager's work (see references within work by Jones[143]) stimulated considerable interest in the study of hemocytes, and Jones[143] revised and greatly improved Yeager's classification. J. F. Yeager and C. Jack Jones of the U.S. and John W. Arnold of Canada have made many significant contributions to insect hematology.

A general description of various hemocyte types, based on light (LM), scanning (SEM), and transmission electron microscopy (TEM) follows. Synonyms of each type are listed in Table 1. Note that this classification is based on morphological features; however, for the GR and PL, I have suggested a functional name, *immunocyte* to emphasize their immunologic roles. These cells have already been described, and the descriptions of the remaining four types follow.

B. OTHER TYPES

1. Prohemocyte (PR)

The PRs are the stem cells and often the smallest of all the other types, and may appear either agranular or show a few granules, droplets, and/or vacuoles. They are often round,

TABLE 1
Synonyms for Various Hemocyte Types

Synonyms	Synonyms of	Authors
Adipocyte	AD	Wigglesworth[166,510]
Adipohemocytes	GR	Shrivastava and Richards,[511] Lea and Gilbert[132]
Adipohemocytes	SP	Harpaz et al.[146]
Adipoleucocyte	AD	Hollande[140]
Adipo-spherule	AD	Kollmann[512]
Amoebocyte	PL	Many authors
Amoebocyte	GR	Many authors
Amoebocyte with granular cytoplasm	GR	Cuénot[138]
Anucleate crescent body	OE	Ritter,[164] Scharrer[165]
"Cellules spheruleuses" or "Cellules a spherules"	SP	Paillot,[513] Paillot and Noel[514]
Coagulocyte	OE	Hoffmann and Stoekel[457]
Crescent cell	OE	Gupta[1]
Coagulocyte	GR	Many authors; Chiang et al.[2]
Crystal cell	OE	Rizki,[515] Nappi[516]
Crystalloid cell	OE	Selman[517]
Cystocyte	GR	Devauchelle[31]
Cystocyte	GR	Yeager[123] and others
Dark hyaline hemocyte	OE	Selman[517]
Eleocyte	SP	Rowley and Ratcliffe[38]
Eruptive cell	SP	Yeager[123]
Formative cell	PR	Muller[518]
Giant fusiform cell	PL	Tuzet and Manier[519]
Granulocytophagous cell	GR	Jones[520]
Hemocytoblast	PR	Bogojevlensky[521]
Hyaline cell	GR	Many authors
Hyaline cell	SP	Whitten[522]
"Jeune globule"	PR	Bruntz[523]
"Jeune leukocyte"	PR	Millara[137]
Lamellocyte	PL	Rizki,[126] other authors
Later stages of spherule	AD	Whitten[522]
Leukoblast	PR	Arvy and Gabe,[524] Arvy[525]
Leukocyte	PL	Kollmann,[512] Metalnikov[175]
Lymphocyte	PL	Many authors
Macronucleocyte	PR	Paillot[513]
Micronucleocyte	PL	Paillot[513]
Nematocyte	PL	Yeager,[123] Rizki,[515] Arnold[526]
Nongranular spindle cell	OE	Wigglesworth[510,527]
Nonphagocytic giant hemocyte	OE	Wigglesworth[510,527]
Oenocyte	OE	Metalnikov and Gaschen,[528] Muller,[518] Tateiwa,[529] Metalnikov and Chorine,[530] Bogojavlensky,[521] Cameron[176]
Oenocyte-like cell	OE	Yeager[123]
Oenocytoid	SP	Dennell[149]
Phagocyte	PL	Many authors
Phagocyte	GR	Many authors
Plasmatocyte-like cell	PR	Jones[124]
Podocyte	PL	Yeager,[123] Devauchelle,[31] other authors
Prohemocytoid	PR	Jones,[531] Jones and Tauber[532]
Proleukocyte	PR	Hollande,[140] others
Proleukocytoid	PR	Yeager[123]
"Pycnoleukocyte"	GR	Wille and Vecchi[533]
Radiate cell	PL	Lutz[534]
Rhegmatocyte	SP	Hrdy[535]
Smooth-contour chromophil cell	PR	Yeager[123]

TABLE 1 (continued)
Synonyms for Various Hemocyte Types

Synonyms	Synonyms of	Authors
Spherocyte	SP	Bogojavlensky[521]
Spheroidocyte	AD	Yeager,[123] Arnold,[536] Rizki,[515] Jones[124]
Spherulocyte	GR	François[85]
Star-shaped amoebocyte	PL	Graber[537]
Thromocytoids	GR	Zachary and Hoffmann[128]
Vermiform cell	PL	Yeager,[123] Lea and Gilbert,[132] Devauchelle[31]
Young granulocyte	PR	François[41]
Young plasmatocyte	PR	Gupta and Sutherland,[15] Gupta[125]

but may be oval or elliptical (Figure 14A, inset), and vary in size (4 to 10 μm wide and 4 to 14 μm long). The most characteristic features of these cells are their comparatively small size and a nucleus (with heterochromatin and euchromatin blocks, and occasionally with a nucleolus) of variable size (3.6 to 12 μm) that almost always nearly fills the cell (Figure 14B), and may have a perinuclear space; the nucleo-cytoplasmic ratio may vary from 0.5 to 1.9 or more. A thin, or dense, homogeneous and intensely basophilic layer of cytoplasm surrounds the nucleus. The cytoplasm generally contains a low concentration of smooth (ER) or rough endoplasmic reticulum (RER), mitochondria, Golgi bodies, and free ribosomes. Centrioles (indicating the mitotic nature of PRs) and microtubules have been observed in the cytoplasm.

In SEM and TEM (Figure 14A, B), the plasma membrane may appear smooth or may show vesiculation. The PRs may constitute 1 to 7% of the hemocyte population and may be more numerous in early instars,[144] and represent the lymphoid cells of other invertebrates.[145] They are generally found in groups, and may appear indistinguishable from young or small PLs. They may be numerous, rare, or absent, depending on the developmental and physiologic state of the insect at the time of observation. PRs are seldom seen *in vivo*.

2. Spherulocyte (SP)

These are ovoid (Figure 16A, inset) or round cells of variable size (9 to 25 μm long and 5 to 10 μm wide), usually larger than GRs. The most characteristic features of these hemocytes are the prominent membrane-bounded, electron-dense, intracytoplasmic spherules (Figure 17A, B), which are considerably larger than the GR granules, and can hardly be confused with the latter. The nucleus is generally small (2.5 μm wide and 5 to 9 μm long) (Figures 16A and 17A), central or eccentric, rich in chromatin bodies, and generally obscured by the spherules.

In SEM (Figure 16B) and TEM, the plasma membrane may or may not show micropapillae, filopodia, or other irregular processes. In addition to the spherules, the cytoplasm contains polyribosomes, moderately to well-developed Golgi bodies in close proximity to spherules, membrane-bounded vacuoles (= lysosomes) (Figure 17B), numerous, randomly distributed microtubules, elongated mitochondria, and RER (Figure 17A, B). Devauchelle[31] has also described a more or less loose network of fibrils in the cytoplasm.

The number of spherules may vary from few to many, and the diameter from 1.1 to 5 μm. On the basis of their ultrastructural details, several types of structured and unstructured spherules have been reported: with granular, fine-textured, filamentous or flocculent material (Figure 17B, inset),[30] with crystal-like lattice arranged in concentric layers,[46] with a crystalline substructure,[39,147] with cross-linked microtubules in the center,[36] and the unstructured, electron-dense spherules.[31]

Histochemically, the spherules have been reported to contain neutral or acid mucopolysaccharide[148] and glycomucoproteins by several authors (see Gupta[14,133]). Much

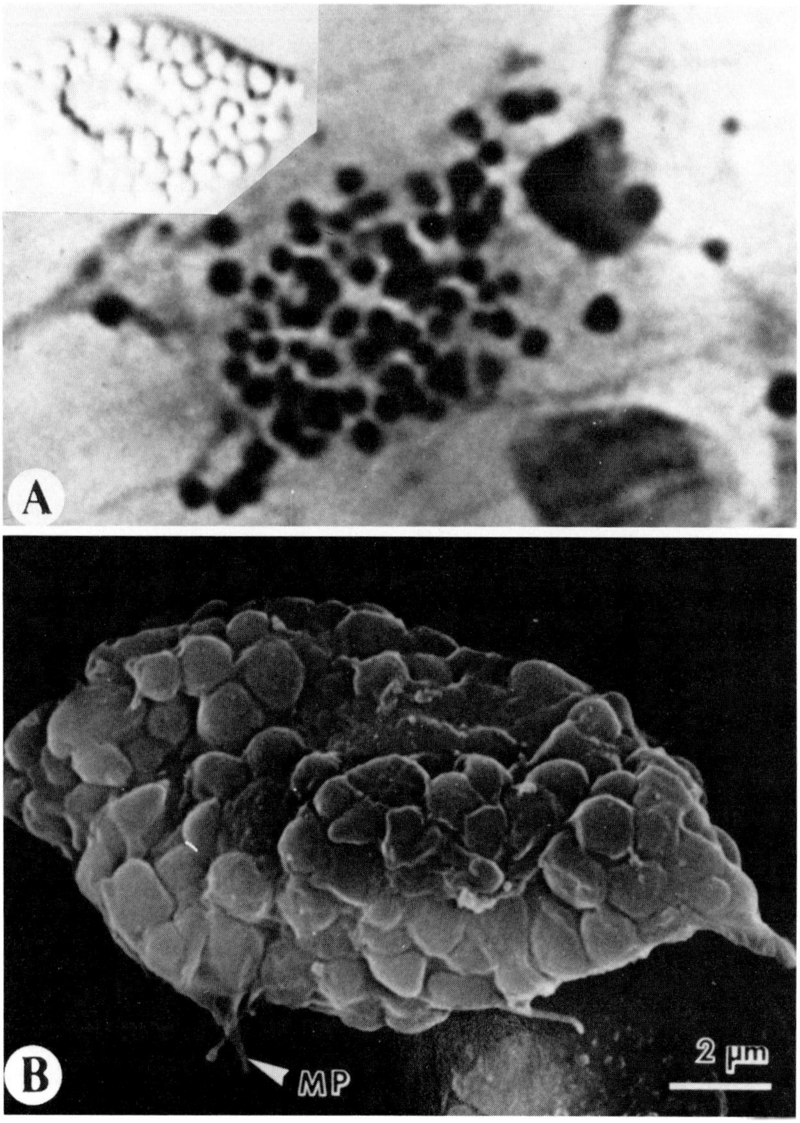

FIGURE 16. (A) A disintegrating spherulocyte (stained with Pyronin-methyl Green, specific for acid mucopolysaccharide) (LM) of 2nd instar nymph of *Nauphoeta cinerea,* showing desspherulation of the cell and release of spherules and the nucleus. **Inset,** a normal spherulocyte of the same animal (× ca. 8000). (B) Spherulocyte (SEM) of *B. germanica,* showing granules and micropapillae (MP). (Part A from Gupta, A. P. and Sutherland, D. J., *Ann. Entomol. Soc. Am.,* 60(3), 557, 1967; part B from Chiang, A. S., Gupta, A. P., and Han, S. S., *J. Morphol.,* 198, 257, 1988. With permission.)

earlier, Hollande[139] reported that the spherules contain ''lipochrome'' (a kind of carotenoid lipid). The presence of tyrosinase has been reported by several authors.[149-151] Costin[40] reported the presence of nonsulfated sialomucin, in addition to glycoproteins and neutral mucopolysaccharides, and Ashhurst[152] showed a polyanionic substance with a predominance of sulfate groups.

Possible functions of SPs include silk production,[153] melanization,[124] phagocytosis,[128,154] secretion of some hemolymph proteins,[155] regulation of clotting, control of cell adhesion and migration, and basement membrane permeability.[156] The SPs constitute 1 to 5% of the hemocyte population.

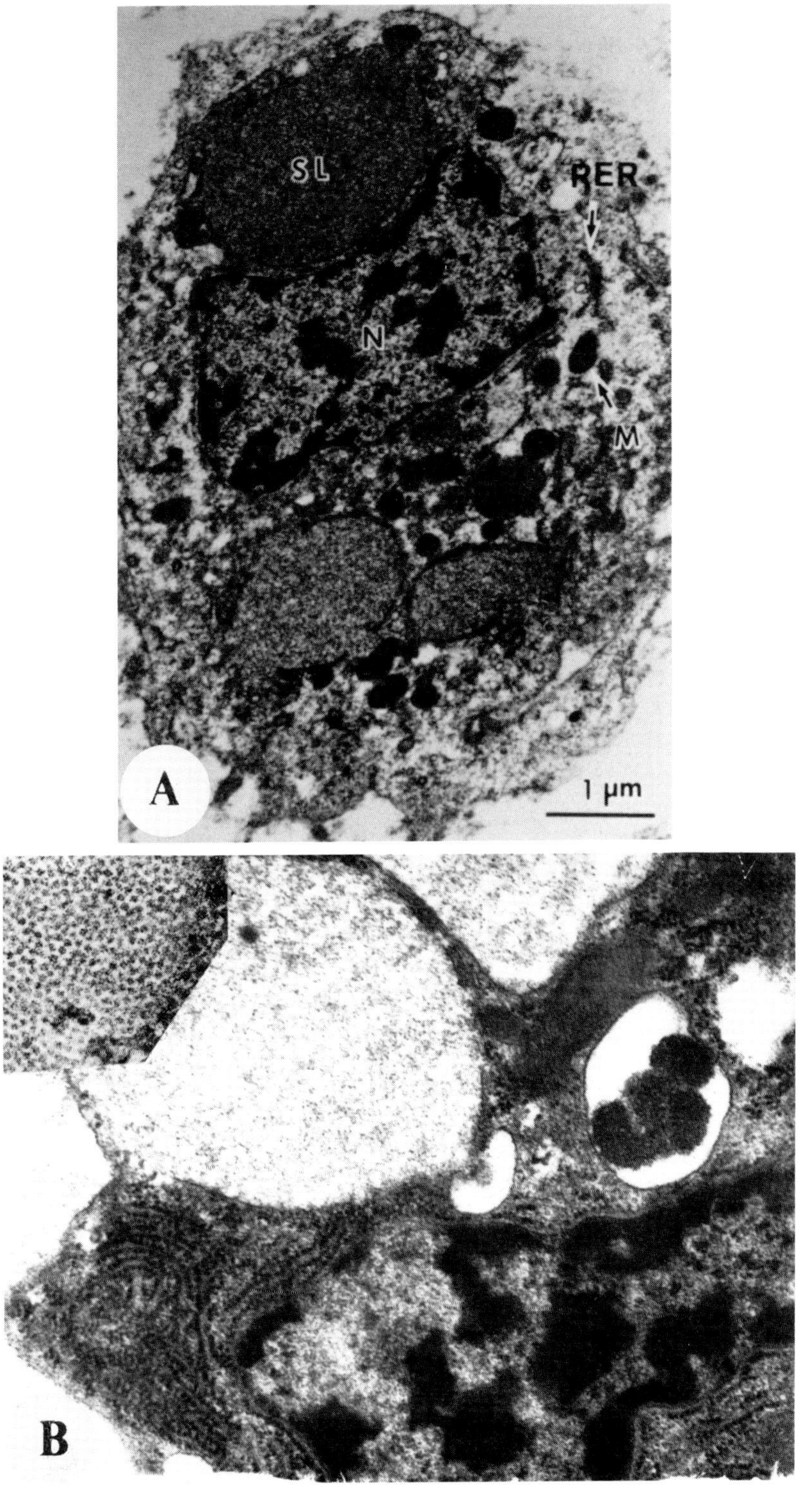

FIGURE 17. (A) Portion of a spherulocyte (TEM) of *B. germanica,* showing spherules (SL), RER, mitochondria (M), and nucleus (N). (B) Portion of spherulocyte (TEM) of *Pectinophora gossypiella,* showing spherule, RER, and lysosome (× ca. 38,400); **inset,** portion of a spherule (TEM) of *Bombyx mori,* showing fine granules (× 48,400). (Part A from Chiang, A. S., Gupta, A. P., and Han, S. S., *J. Morphol.,* 198, 257, 1988; part B from Gupta, A. P., *Comprehensive Insect Physiology, Biochemistry and Pharmacology,* Vol. 3, Kerkut, G. A. and Gilbert, L. I., Eds., Pergamon Press, Oxford, 1985, 401. With permission.)

Superficially, the insect SPs resemble vertebrate mast cells, which are connective tissue cells always found in bone marrow, lymph nodes, thymus, and spleen; and contain heparin and the biogenic amines, histamine, and serotonin.[157] Mast cells release histamine and other mediators after two exposures to the foreign antigen; the first sensitizes the IgE, which combines with the receptor on the mast cell surface, and the second triggers the allergic reactions, and causes the release of histamine. Earlier, Serafini-Fracassini[158] demonstrated that mast cell granules from ox liver capsules are composed of a network of elongated heparin-protein macromolecules embedded in a homogeneous particulate matrix. Note that Ashhurst,[152] on the basis of histochemical tests on the SPs (which also contain mucopolysaccharides) of the greater wax moth, *Galleria mellonella,* suggested that the presence of sulfated glycosaminoglycan polymer (confirmed by very long periodate oxidation) indicates the presence of heparin-like molecules in the SPs. Since histamine has been demonstrated in the coelomocyte of the echinoderm, *Mellita quinquiesperforata,*[159] attempts should be made to find out if SPs or any of the other immunocytes contain this chemical. It is well known that serotonin causes exocytosis in nerve axons;[160,161] and, although not demonstrated in any hemocyte type to date, might play this role in the release of the spherule contents, if indeed its presence could be demonstrated in the SPs. Serotinin can be demonstrated by Falk-Hillarp and immunoreactive methods.[162] Note also that in addition to causing exocytosis, serotinin has been implicated in glycoregulation in *Astacus leptodactylus,* because injection of serotonin in this animal resulted in hyperglycemia.[163] It might play some role in the release of the mucopolysaccharides that are present in these cells.[148]

3. Oenocytoid (OE)

The OEs are small to large, opaque, polymorphic (oval, spherical, crescent-shaped, or elongate) cells of widely variable size (16 to 54 μm or more). The most characteristic feature of these cells is their opaque appearance in LM (Figure 18A) due to the thick and homogeneous cytoplasm and the presence of plate-, rod-, filament-, or needle-like inclusions or crystals, hence the name "crystalloid" or "crystal" cells for these hemocytes. In some insects, the OEs are labile and thus lyse quickly, ejecting material into the hemolymph;[1,14,22,23] in others (e.g., the hissing cockroach, *G. portentosa*) they remain stable for up to 45 min, without undergoing any shape change (personal observation). They constitute only 1 to 2% of the hemocyte population and are not phagocytic.

The nucleus is generally small, round or elongate, and generally eccentrically located. Nuclear size may vary (3 to 15 μm), and occasionally two nuclei may be present. I reported a mitotic nucleus in the OE of *G. portentosa*[1] (Figure 20F). In this cockroach, the OEs occur as polymorphic (spindle- or barrel-shaped, elongated, dumbbell-shaped, oval, reniform, crescent-shaped, pyriform, and spherical) (Figure 21A—I) nucleate cells in LM, and were at one time mistakenly identified as anucleate cells.[164,165] Some 20 years later, it was confirmed[1] that these so-called anucleate, crescent cells were indeed polymorphic OEs. We also occasionally found a crescent-shaped OE in the German cockroach, *B. germanica.*[2] The most peculiar feactures of these cells in *G. portentosa* are (1) a single inclusion body — which generally fills the cell (Figure 21A) and gradually diminishes in size and density, as the cell matures (Figure 20A—E) — instead of several, as in most other insects; and (2) gradual extrusion of the nucleus as these cells mature (Figure 22A—J), thus providing the false impression that these cells are anucleate. No nuclear extrusion has been observed in any of the other five hemocyte types in insects; in *G. portentosa,* it may occur at any point along the length of the cell, thus producing the various forms of the so-called anucleate cells (Figure 22A—J). It is important to emphasize that the inclusion body is not always crescent-shaped and the nucleus not always eccentric. As a matter of fact, both these features seem to represent the later stages in the aging and disintegration processes of these cells, the nuclear extrusion being the terminal stage, followed by the disintegration of the cell. Dis-

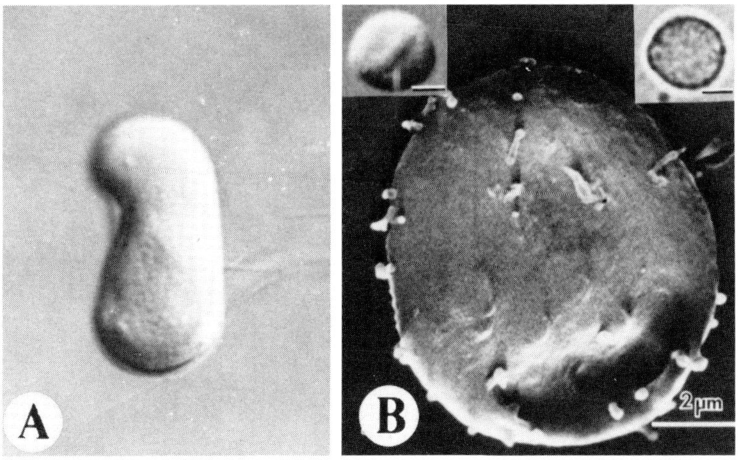

FIGURE 18. (A) Enucleate oenocytoid (Nomarski) of *G. portentosa* (× 5525). (B) Oenocytoid (SEM) of *B. germanica;* **inset** — (left) oenocytoid (Nomarski) *in vitro,* showing opaque appearance; (right) another oenocytoid (phase contrast). (Bars = 2 µm.) (Part A from Gupta, A. P., *Comprehensive Insect Physiology, Biochemistry and Pharmacology,* Vol. 3, Kerkut, G. A. and Gilbert, L. I., Eds. Pergamon Press, Oxford, 1985, 401; part B from Chiang, A. S., Gupta, A. P., and Han, S. S., *J. Morphol.,* 198, 257, 1988. With permission.)

integrating OEs, following nuclear extrusion, were more commonly found in nymphs than in adults. In terms of the nuclear extrusion, OEs in this cockroach are comparable to the maturing mammalian erythrocyte, which, during its maturation process, extrudes its nucleus (Figure 22K—M). Note that disintegrating OEs cannot be confused with disintegrating GRs/COs. Furthermore, the extruded material from the OEs produces only localized cytoplasmic strands, and does not cause plasma gelation, as does that from the GR/CO.

In SEM (Figure 18B) and TEM, the plasma membrane may show filopodia or micropapillae. The OEs may show an elaborate system of intracytoplasmic filaments or microfilaments (Figure 19A—C). In addition to the membrane-bounded intracytoplasmic inclusions, there are abundant polyribosomes (Figure 19D) and large mitochondria; other organelles, such as ER and Golgi bodies may be poorly developed. In the OEs of *G. mellonella,* microtubule bundles have been reported.[38] Supposedly, lysosomes are absent.

Histochemically, OEs in insects are reported to contain tyrosinase,[149] protein,[14,37] and PAS-positive only granules, indicating the presence of glycoproteins or glycosaminoglycans, and sulfated, periodate-reactive sialomucin.[40] The inclusion body in *G. portentosa* is reported to be intensely periodic acid-Schiff positive,[164,165] the longer oxidation producing deeper stain. Wigglesworth[166,167] reported that OEs play an indirect role in the final stages of cuticle formation in *Rhodnius prolixus,* and suggested that they are involved in the metabolism of phenols, and thus might be involved in sclerotization. Indeed, the presence of mucopolysaccharides in these hemocytes supports their possible role in cuticle formation. As a matter of fact, we suggested much earlier[148] that perhaps the mucopolysaccharides undergo hydrolysis and break into β-D-glucosamine and β-D-glucuronic acid; the former might be used in glycolysis or, after phosphorylation, might enter into the formation of new chitin molecules. The OEs also showed positive reaction with mercury-bromophenol and ninhydrin-Schiff stains, which indicates the presence of proteins. Negative reaction with DOPA (dihydroxyphenylalanine)-oxidase tests suggests the absence of tyrosinase;[1] the latter has, however, been reported in the OEs of some insects.[14,49,170] Ashida and Yoshida[168] reported that phenoloxidase, which catalyzes melanin formation, is present in its inactive form (pro-

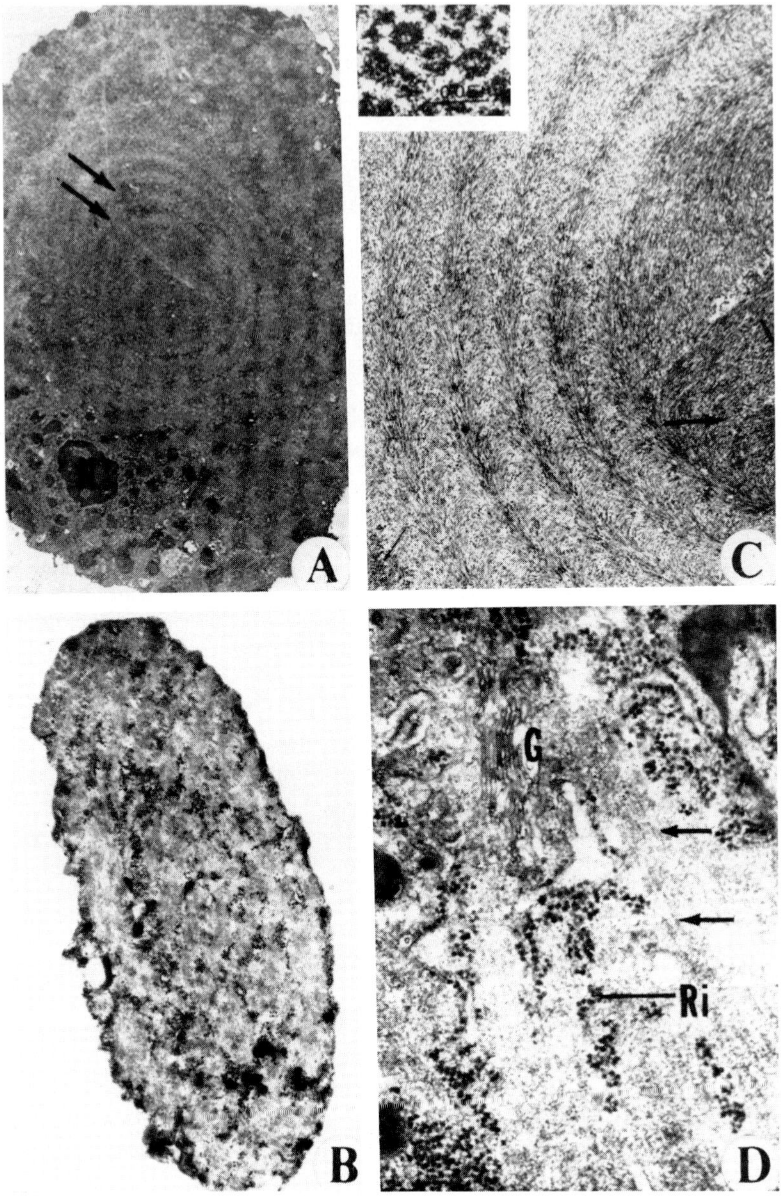

FIGURE 19. (A) Oenocytoid (TEM) of *B. mori,* showing concentric arrangement of intracytoplasmic fibrils (arrows) and eccentric nucleus (N) (× 3,000). (B) Oenocytoid (TEM) of *P. gossypiella* (× 6,600). (C) Portion of oenocytoid (TEM) of *B. mori,* showing highly magnified view of concentric rings of intracytoplasmic fibrils. Note also unoriented fibrils (arrow) (× ca. 30,000); **inset** showing fibrils in cross-section (× ca. 225,000). (D) Portion of oenocytoid (TEM) of *P. gossypiella,* showing longitudinally arranged intracytoplasmic microtubules (arrows), Golgi (G), and ribosomes (Ri) (× 35,475). (From Gupta, A. P., *Comprehensive Insect Physiology, Biochemistry and Pharmacology,* Vol. 3, Kerkut, G. A. and Gilbert, L. I., Eds., Pergamon Press, Oxford, 1985, 401. With permission.)

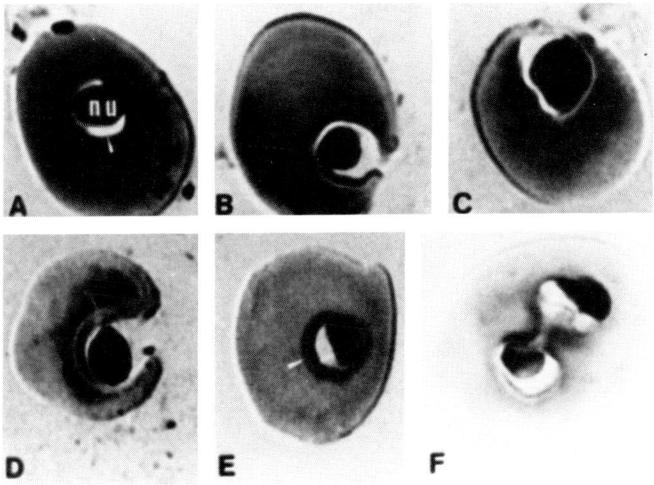

FIGURE 20. Giemsa-stained preparations of oenocytoids (LM) of *G. portentosa*. (A) Oenocytoid with inclusion body filling the cell. Note perinuclear space (arrowhead) and density of inclusion body (× 1600). (B) Oenocytoid with subapical nucleus in the process of extrusion. Note reduction in the density and size of inclusion body and increase in the perinuclear space (× 1600). (C) Later stage of B (× 1600). (D) Nuclear extrusion about to occur. Note ruptured cell wall and reduction in the size of the inclusion body (× 1600). (E) Oenocytoid in which the perinuclear inclusion body is considerably reduced prior to nuclear extrusion (× 1500). (F) Oenocytoid with mitotic nucleus. (From Gupta, A. P., *Cytologia*, 50, 739, 1985. With permission.)

phenoloxidase) in OEs, and they suggested a direct relationship between the activity of the prophenoloxidase and recognition of foreignness.

4. Adipohemocyte (AD)

In LM observations, ADs and SPs (Figure 23A, B, inset) can be easily confused with each other, except that the former are very refringent owing to the presence of lipid droplets, which obscure the nucleus both *in vivo* and *in vitro*. ADs are small to large, spherical or oval cells of variable size (7 to 45 μm in diameter). The nucleus is relatively small (as compared with that in PL or SP); round or slightly elongate; centrally or eccentrically located; and may appear concave, biconvex, punctate, or lobate. Nuclear size is variable (4 to 10 μm in diameter). The presence of ADs in blood samples is very erratic, and they constitute less than 1% of the hemocyte population. It is possible to accelerate the transformation or differentiation of these cells in the hemolymph by chilling the insect at 5°C for 20 to 25 h, as first suggested by Gupta[133] for *Tenebrio molitor* larvae.

In SEM, the surface looks very nodulose (Figure 23B), but without any micropapillae. In TEM, the plasma membrane may or may not show filopodia or other irregular processes. The cytoplasm contains the characteristic fat droplets (0.5 to 15 μm in diameter) and other nonlipid granules (0.5 to 9 μm in diameter), and vacuoles, which, according to Arnold,[122] become filled with lipids under certain conditions. In addition, there are well-developed Golgi bodies, mitochondria, RER, and polyribosomes.

Histochemically, the fat droplets are strongly Sudan black B-positive.[25] In addition, ADs are reported to contain PAS-positive substances in some granules.[132,169]

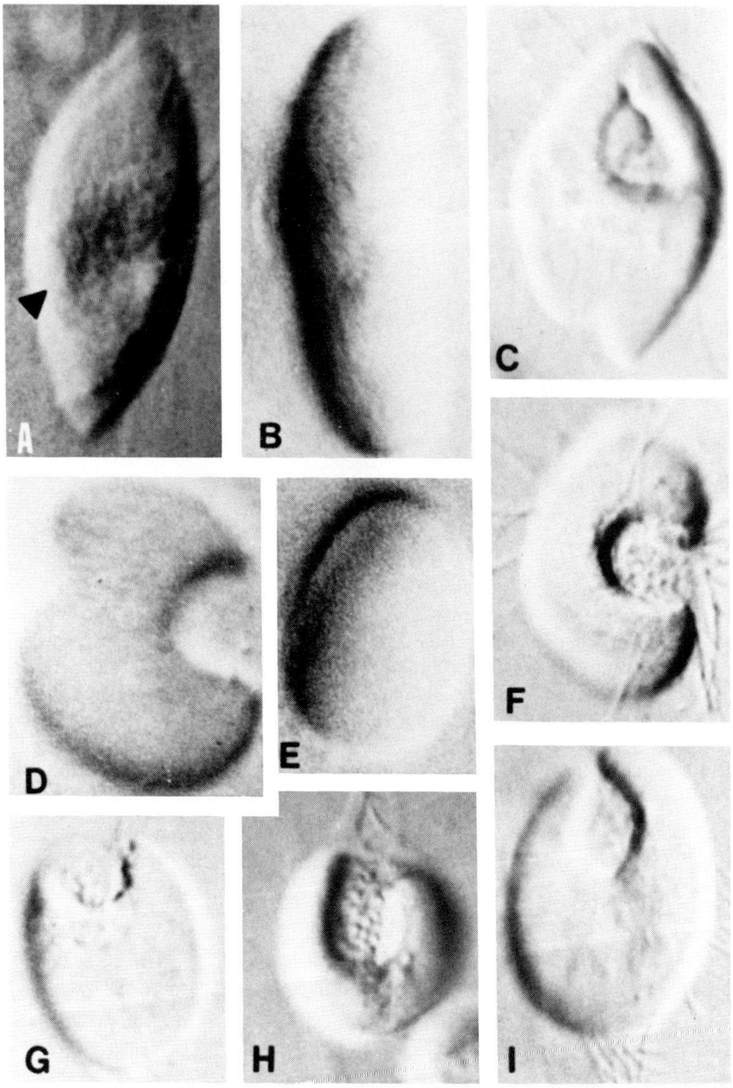

FIGURE 21. Various polymorphic oenocytoids and apical and subapical nuclear extrusions (Nomarski) of *G. portentosa.* (A) spindle-shaped oenocytoid. Note outline of inclusion body (arrowhead) (× 1800). (B) Elongated oenocytoid (× 1900). (C) Barrel-shaped oenocytoid (× 1700). (D) Dumbbell-shaped oenocytoid (× 1700). (E) Oval oenocytoid (× 1600). (F) Reniform oenocytoid (× 1600). Note cytoplasmic strands. (G) Oval oenocytoid showing apical nuclear extrusion (× 1400). (H) Spherical oenocytoid showing apical nuclear extrusion (× 1300). (I) Oval oenocytoid showing initial stage of subapical nuclear extrusion (× 1700). (From Gupta, A. P., *Cytologia,* 50, 739, 1985. With permission.)

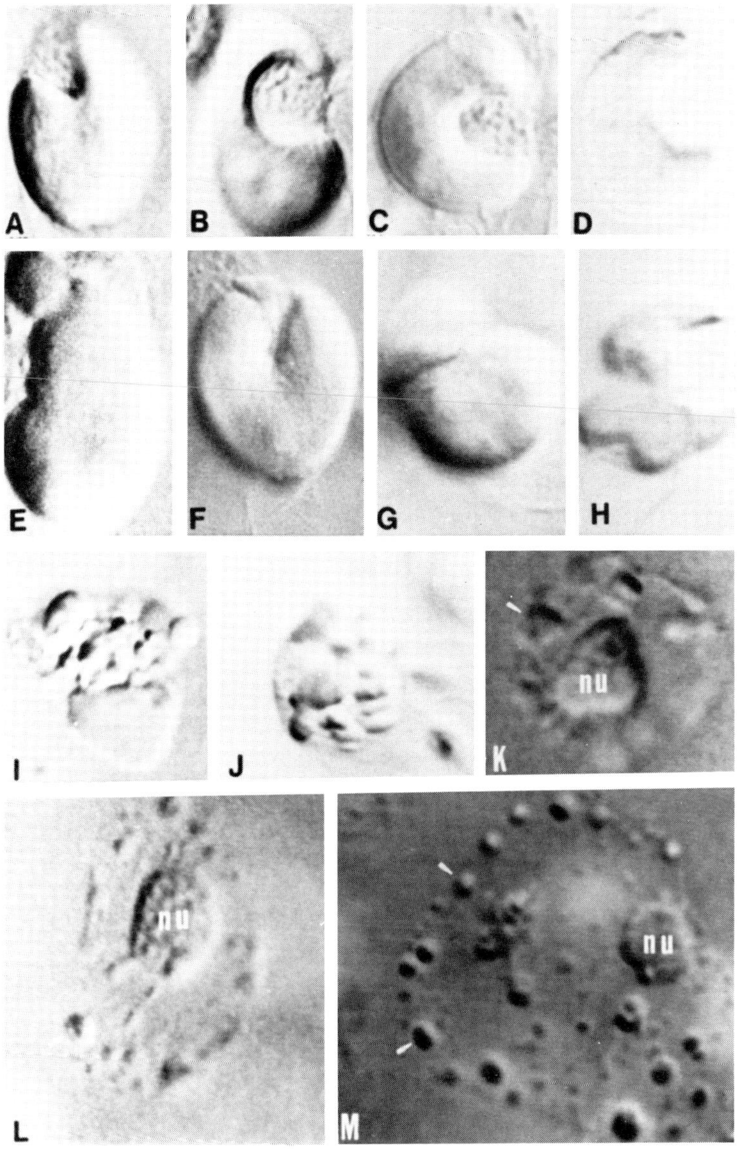

FIGURE 22. Various types of nuclear extrusions and degranulation and disintegration of oenocytoids and granulocytes (previously called coagulocytes) (Nomarski) of *G. portentosa*. (A) Oval oenocytoid showing later stage of subapical nuclear extrusion (× 1400). (B) Oval oenocytoid showing mediolateral nuclear extrusion (× 1400). (C) Spherical oenocytoid showing mediolateral nuclear extrusion (× 1400). (D) Crescent-shaped oenocytoid after nuclear extrusion (× 1800). (E—J) Stages in the disintegration of oenocytoids. (E) Apical disintegration (× 1800). (F) Apical disintegration (× 1400). (G) Oenocytoid splitting across its entire width (× 1400). (H) A crescent-shaped oenocytoid disintegrating (× 1400). (I) Advanced stage of disintegration. (J) Same as (I). (× 1300). (K) Granulocyte (previously called coagulocyte), showing initial stage of degranulation. Note presence of peripheral cysts (arrowheads) and central nucleus (nu) (× 1600). (L) Granulocyte (previously called coagulocyte) showing a later stage of degranulation and disintegration (× 1800). (M) Granulocyte showing the terminal stage of disintegration (× 1800). (From Gupta, A. P., *Cytologia*, 50, 739, 1985. With permission.)

V. IDENTIFICATION OF IMMUNOCYTES AND OTHER HEMOCYTE TYPES BY LIGHT MICROSCOPY

The following identification key is meant for the beginner, who might attempt to identify the various types in fresh hemocyte preparations by light microscopy. Detailed methods for preparation of hemocytes for light, scanning, and transmission electron microscopy have been described elsewhere.[10] Finally, note that coagulocyte is not recognized as a separate category.

Identification Key

1. Nucleus compact, large in relation to cell size, centrally located, almost filling the cell; very thin, peripheral layer of homogeneous cytoplasm around the nucleus; cells often round, but may be oval or elliptical; always small (compared with other cells in the sample) (Figure 14A, inset) . . . .*Prohemocyte (PR)*

1′. Nucleus not compact, generally small in relation to cell size, not nearly filling the cell 2

2(1′). Nucleus with chromatin arranged in cartwheel-like fashion; generally eccentric, oval, and sharply outlined; cytoplasm hyaline, generally scant, may contain some spherical or elongate granular inclusions; cell sometimes with cyst-like blebs in process of exocytosis (Figures 2A—E and 22K—M). . . . .*Lysing Granulocyte* (Note that this kind of lysing or distintegrating granulocyte is often referred to in the literature as *coagulocyte,* but in this review it has been treated as a granulocyte; see Section II.A)

2′. Nuclear chromatin not arranged in cartwheel-like fashion, nucleus eccentric or central; cytoplasm not hyaline, abundant, homogeneous, and without any plate-, rod-, and needle-like inclusions and filaments, or a single large inclusion 3

3(2′). Cytoplasm generally agranular or very finely granular; nucleus round or elongate and central, and may or may not appear punctate; cells polymorphic (spindle-shaped or oval) and variable in size in various insects (Figure 14D) . . *Plasmatocyte (PL)* (See Section II.B for variant forms of PLs.)

3′. Cytoplasm generally agranular, thick, and homogeneous often with several kinds of plate-, rod-, and needle-like inclusions and filaments, or a single large inclusion, almost filling the cell; "vacuoles" may or may not be present; nucleus generally small, round or elongate, and generally eccentric; cells variable in size and shape, generally lyse quickly *in vitro,* ejecting material into the hemolymph (Figures 18A, 20A—E, 21A—I, and 22A—I) . . . .*Oenocytoid (OE)* (Note that in some insects, e.g., the hissing cockroach, nuclear extrusion occurs in this cell type, giving rise to crescent-shaped cells that may seemingly appear anucleate; see Section IV.B.3)

4(3′). Cytoplasm distinctly granular . . . . .4′

4′. Cytoplasm prominently and characteristically granular; granules may or may not be numerous; nucleus relatively small (compared with that in plasmatocyte) and compact, round or elongate, and generally central (Figure 3A, B) . . *Granulocyte (GR)* (Note that GRs in lower orders are generally larger than in higher orders.)

5(4′). Instead of granules in the cytoplasm, considerably enlarged and distinct spherules or droplets generally fill the cell . . . . .5′

5′. Spherules nonrefringent, generally obscuring the nucleus; nucleus rather small, central or eccentric; cells ovoid or round with variable sizes, usually larger than granulocyte, and may be observed releasing spherules into the hemolymph by exocytosis (Figure 16A) . . . . *Spherulocyte (SP)*

6(5′). Spherules or droplets refringent owing to presence of lipid; nucleus relatively small (compared with that in plasmatocyte or spherulocyte), round or slightly elongate, central or eccentric, and may or may not appear concave, biconvex, punctate, or lobate; cytoplasm may contain other nonlipid granules (Figure 23A, B, inset) . . . .*Adipohemocyte (AD)*

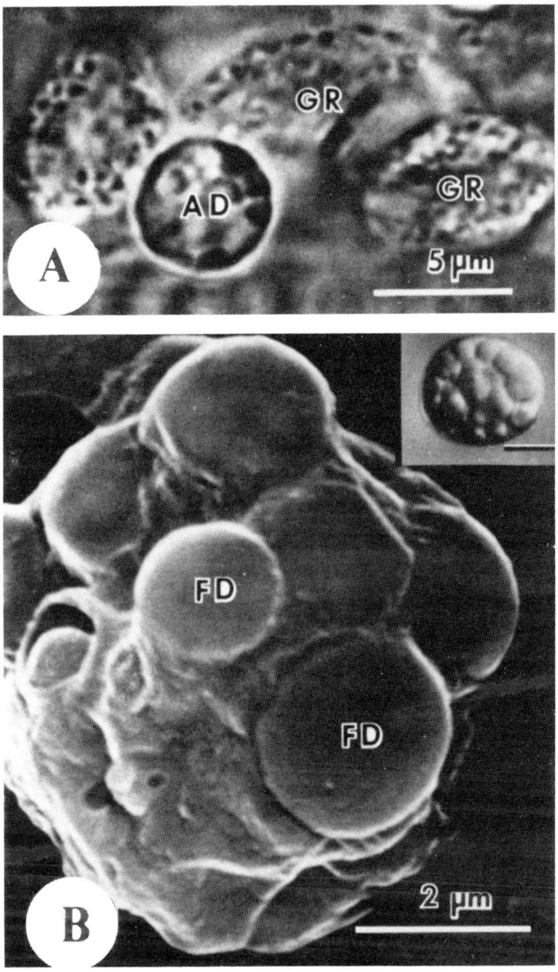

FIGURE 23. (A) Adipohemocyte (phase contrast) of *B. germanica in vivo*. Note difference in appearance from that of granulocyte (GR) (B) Adipohemocyte (SEM) of *B. germanica*, showing spherical shape and fat droplets (FD) in the cytoplasm; **inset** (Nomarski) *in vitro*, showing large, refringent fat droplets in cytoplasm. (From Chiang, A. S., Gupta, A. P., and Han, S. S., *J. Morphol.*, 198, 257, 1988. With permission.)

VI. CELLULAR IMMUNITY

Most of the following account has been adopted from my previous reviews and our recent works[6,7,10,14,22,48] and those of others. The hemocytic immunity in insects is accomplished directly by phagocytosis, encapsulation, and nodule formation, and indirectly by immunologic factors, coagulation, prophenoloxidase-activating system, and poison detoxification mechanisms; these processes are triggered as defense reactions against foreign antigens (immunogens) and toxic substances. Artifically injected animate or inanimate objects also induce some of these reactions.

A. PHAGOCYTOSIS

1. Historical

Elie Metchnikoff,[170] a Russian zoologist, first conceived the idea of both phagocytosis and encapsulation from an experiment (1892) in which he inserted a few rose thorns into a transparent starfish larva; to his utter surprise the next day, he found that the thorns were surrounded by tiny, wandering, transparent cells, which he called *phagocytes* (Greek *phagon,* meaning ''devouring or eating'' cells) and the process, *phagocytosis.* The story goes[171] that Metchnikoff performed the experiment at Christmas time at his home, and soon after (1882) he resigned his professorship at the University of Odessa (U.S.S.R.) for political and personal reasons. He was later (1888) invited by Pasteur to join the Pasteur Institute, where he spent most of the next 28 years, and later (1908) shared the Nobel Prize with Paul Ehrlich, a German scientist and one of the early critics of his theory of phagocytosis. Metchnikoff's[172] studies of cellular defense in the crustacean, *Daphnia* provided the most convincing evidence of phagocytosis. He later extended his studies to mammals, and called the mononuclear phagocytes, which were capable of ingesting large particles, *macrophages.*[170] Some of the other early works on phagocytosis in insects are those of Cuénot,[138,173,174] Metalnikov,[175] and Cameron.[176]

2. Uncertainties and Controversies

The least understood aspects of phagocytosis are the mechanisms involved in the initial recognition of the foreign antigen and the role of opsonins. In addition, we have either vague or conflicting reports about the effects of such factors as electrostatic charge on the antigen, the PL/GR population (DHCs), type of antigen (e.g., bacteria, fungi, or erythrocytes), frequency of attack, and the effect of PL stimulation by the GR on the rate of phagocytosis, especially from the standpoint of inducing immune deficiency. Although it is suspected that opsonins enhance the rate of phagocytosis, we do not know whether hemagglutinins and/or polyphenols or other molecules act as opsonins in all arthropods; equally unknown are the location and phagocytic function of the hemopoietic organs in most insects. What about immunologic memory, specific or nonspecific? Karp and his associates have demonstrated specific immunologic memory to injected proteins in *P. americana* (see Chapter 15).

Some useful information on phagocytosis has been provided by both *in vivo* (by injecting foreign materials)[177-180] and *in vitro* (by either incubating the immunocytes and the test antigens/particles in saline or cell culture medium or in monolayer preparations of the immunocytes, overlaid with test particles)[38] studies. However, much remains to be elucidated.

3. Cells Involved

In both *in vivo* and *in vitro* studies in insects and crustaceans, the GRs and/or PLs have been reported to phagocytose. The GRs in the most primitive arthropod, *L. polyphemus,* are weakly phagocytic and are known to phagocytose carbonyl iron particles[181] (see also Chapter 1). In some species of higher arthropods (e.g., Crustacea, Insecta), however, both GRs and/or PLs participate in this process. In addition, oenocytoids, spherulocytes (Figure 24C), the so-called coagulocytes (= GRs), and adipohemocytes have been occasionally reported as phagocytic, but it is likely that some of these hemocytes were incorrectly identified.[48]

4. Effectiveness

The effectiveness of phagocytosis depends on, among other factors, the phagocytic index (number of phagocytosing immunocytes), the nature of the microorganisms (antigen types), and the frequency of attack.[48] It has been reported that when the number of phagocytosing cells is high, ingestion and hence removal of the microorganisms from the hemolymph, is

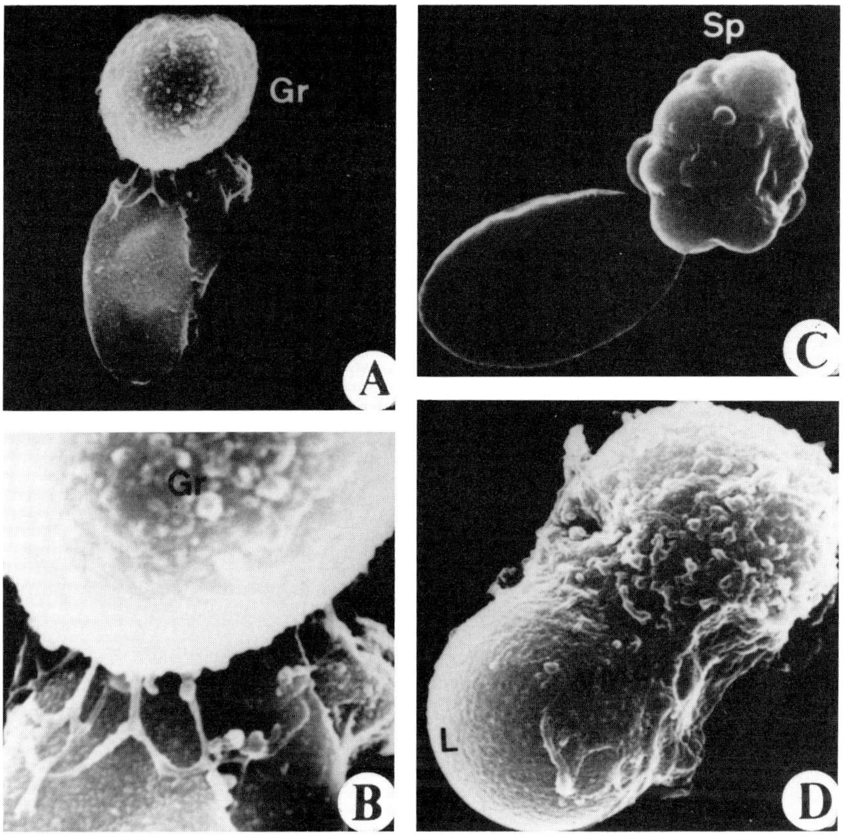

FIGURE 24. (A) Attachment of goose erythrocytes to granulocyte (Gr) of *B. mori* (× 4000). (B) Higher magnification showing details of attachment (× 7500). (C) Attachment of goose erythrocyte to spherulocyte (Sp) of *B. mori* (× 5300). (D) Granulocyte of *B. mori* phagocytosing a latex particle (× 6700). (Courtesy of Dr. H. Wago.)

hastened.[182-184] The availability of circulating hemocytes is important for the effectiveness of phagocytosis, because it is known that insects become susceptible to infection if the activity of the phagocytosing cells is blocked *in vivo*.[177,185] However, Hoffmann et al.[186] reported that in *L. migratoria,* circulating hemocytes do not play a major role in taking up bacteria; although in another acridid, *Poecilocerus pictus,* both the PLs and GRs phagocytose,[187] and Kaaya et al.[188] reported that because blocking of phagocytes with India ink had no effect on infection rate of *Glossina morsitans morsitans* by *Trypanosoma brucei,* hemocytes played little or no role in protecting the flies from the trypanosomes.

Wago and Ichikawa[189] reported that the rate of phagocytosis of injected goose erythrocytes in *Bombyx mori* larvae was highest in the fifth larval instar. This was probably owing to the increased number of circulating hemocytes in this instar. Wago[190] demonstrated that in the fifth larval instar, the capacity of GRs and PLs to adhere to sheep erythrocytes was increased in the presence of hemolymph, although hemolymph-independent binding also occurred *in vitro* (Figure 24 A—D) (see also Chapter 7). Unfortunately, little or no precise information regarding the phagocytic rate at any given time in insects is available. Furthermore, not all microorganisms invading insects are equally affected by the phagocytosing cells: phagocytosis appears to be ineffective in some insects against the bacteria, *Bacillus thuringiensis* and *B. cereus;* microsporidia, *Nosema* sp., some viruses; and certain fungi, *Sporosporella uvella* and *Coelomomyces*.[122] By contrast, a single hemocyte can ingest a number of particles.[191] The β-1,3-glucan (the carbohydrate moieties on the cell walls of fungi and algae) enhances the rate of phagocytosis of bacteria *in vitro* by causing degranulation of the GRs, which in turn causes stimulation of the PLs.[192] According to these authors,[192] prophenoloxidase (proPO) facilitates *in vitro* phagocytosis; however, its action is time dependent, i.e., the proPO should not be activated before inclusion of the bacteria in the reaction mixture. Opsonins are known to facilitate phagocytosis; however, the exact nature of these molecules remains controversial.[48] The rate of *in vitro* phagocytosis is also affected by temperature, pH, and incubation time in insects,[193] and by cationic substances in the phagocytosis of bacteria by human neutrophils.[194]

5. Role of Hemopoietic Organs

The role of the so-called phagocytic (= hemopoietic) organs or tissues in the effectiveness of phagocytosis (and perhaps also in encapsulation) deserves mention. To the extent that these organs in many insects are also hemopoietic in nature, and thus produce hemocytes (e.g., PLs, GRs, and ADs in *L. migratoria*),[186,195] they surely must play a role in phagocytosis — a fact that was first observed by Cuénot.[138,173,174] He reported[138] that the phagocytic tissue of the orthopteroid insects takes up bacteria. However, according to Hoffmann et al.,[186] at least in *L. migratoria,* it is the reticular cells of the hemopoietic organ that engulf the bacteria, not the circulating hemocytes. Furthermore, according to these authors, these reticular cells can be induced to produce large amounts of proteinaceous antibacterial substances by injecting immunizing doses of pathogens; they have suggested the possibility that, as in vertebrates, some of the reticular cells may be differentiated into immunocompetent cells, and that it is these cells that synthesize the antibacterial factor, as evidenced by marked dilations of the cisternae of the RER and the presence of crystalloid inclusions in them.

6. Stages of Phagocytosis

Phagocytosis of foreign antigens is a dynamic, energy-requiring cellular reaction and involves three stages: (1) recognition and targeting of the foreign antigen, (2) endocytosis and ingestion, and (3) clearance. These processes have been described and reviewed.[14,38,48,196] In this chapter, I will comment on only those aspects that need to be emphasized, and point out areas for future research.

a. Recognition and Targeting of the Foreign Antigen

This initial stage involves: (i) unidirectional(?) or random(?) chemotactic movement of the immunocyte toward the foreign antigen, and (ii) binding the latter by means of surface receptors, and/or (iii) opsonic molecules present in the hemolymph.

i. Chemotactic Migration of the Immunocyte toward the Foreign Antigen

Unfortunately, virtually nothing concrete is known about this *in vivo* phase in insects. *In vitro* studies provide no clues as to how the immunocyte actually first interacts with the target *in vivo;* chemotaxis/chemokinesis is often suggested.[12] However, for any chemotactic interaction to occur, at least two conditions would have to be met: the target must somehow be biochemically attractive to the immunocyte, and it must present to the latter gradients of the recognition or chemotactic factors (CFs), so that the immunocyte could orient itself and migrate toward it. Assuming that CFs are involved in this process, where do they originate? They could be produced by either the foreign antigen (target) or the immunocyte itself. Bacterial CFs have been reported from several organisms,[197] and all CFs contain a sequence of amino acids that is made up of formyl-methionine-leucine-phenylalanine (F-Met-Leu-Phe). Human neutrophils are known to possess surface receptors for the F-Met-Leu-Phe complex and react to synthetic polypeptides made up of this complex. Since certain bacteria are phagocytosed by insect immunocytes, it would be reasonable to assume that the latter also have receptors for this or similar factors secreted by these microorganisms. However, this remains to be demonstrated.

The other likely source of the CFs might be the immunocytes themselves, for it is likely that as soon as the foreign antigen enters the insect's immune system, it inevitably but randomly comes in contact with one or more of the immunocytes, which then release some sort of host factors (HFs) (''recognition'' factors? see Chapter 4), which attach to the antigen and provide the gradients and/or attraction to other immunocytes to converge to the site of action. Wago[198,199] reported that following injection of bacteria into *B. mori,* GRs became adhesive to each other and to other tissue. And Mohrig et al.[200] have reported the appearance of phagocytosis-stimulating factor in the hemolymph of the larvae of *G. mellonella,* following injection of latex beads. Whether the immunocytes secrete this factor was not ascertained. Hemagglutinins,[45] components of the prophenoloxidase system,[192] coagulins,[201] and lysozymes[202,203] are thought to function as HFs in insects and other arthropods. Indeed, several kinds of HFs (e.g., lymphokines and leukotriene B_4 [Leuk-B4] and some chemotactic peptides derived from cleavage of C3 and C5) are known in humans.[197] Leuk-B4 is known to act chemotactically, cause adhesion of the lymphocyte to the target cell, and cause aggregation and degranulation of other lymphocytes.[204] Furthermore, according to Powell,[205] phagocytic granulocytes form arachidonic acid metabolites, such as the Leuk-B4, from the phospholipids of their plasma membrane. Because Leuk-B4 is known to cause aggregation of marine sponge cells[206] and because insects have enzyme that oxidize arachidonic acid,[200] it is very like that such HFs exist in insects as well, but they remain to be investigated and confirmed. I discuss in Chapter 4 that arachidonic acid is probably generated in the GR when the plasma membrane phosphoinositides are hydrolyzed upon contact with a foreign antigen. Note that the presence of lymphokine-like substances, following grafting of foreign tissue, has been reported in the earthworm.[207]

Recently, Stefano et al.[12] attributed the chemotactic and cytoadherence properties of immunocytes (GRs) in the cockroach, *Leucophaea maderae,* and the mollusk, *Mytilus edulis,* to an opioid-like substance in the hemolymph, and suggested the presence of opiate receptors on the plasma membranes of the immunocytes (see also Section VII.E). Note that Strange[208] considered the existence[209] of opiate receptors the ultimate ''realization of Langley's receptor concept''. Langley[210] was the first to introduce the concept of receptors (receptive substance), on the basis of his studies on the effects of nicotine and curare on muscle.

Both CFs and HFs alert the insect that its homeostasis has been compromised, and trigger the cellular reactions against the intruder. HFs are thought to cause cell proliferation in hemopoietic organs, mitosis, adherence and migration behaviors of circulating cells, degranulation, and release of lysosomal factors with antimicrobial and anti-inflammatory effects[200] (for additional information, see Section VII.C on antimicrobial factors).

ii. Binding the Foreign Antigen by Surface Receptors

I have discussed the importance of surface receptors in cellular immunity in Section II.A. The existence of surface receptors on the immunocyte has been suspected and even demonstrated in some arthropods by cytoadherence or rosette formation[12,45] and labeling techniques.[11,13,211] It is known that lectins act as surface receptors[212] because they specifically bind to carbohydrates on the surface of other cells or particles. In fact, in mammals, lectin-carbohydrate interaction occurs in nonimmune phagocytosis, independent of antibody-complement interaction, and lectins participate in phagocytosis.[213] However, Bradley et al.[214] concluded that agglutinins in the grasshopper, *Melanoplus differentialis* and *M. sanguinipes* do not generally function as recognition molecules.

As stated earlier, using five different kinds of lectins — concanavalin A (Con A), wheat germ agglutinin (WGA), soybean agglutinin (SBA), peanut agglutinin (PNA), and winged pea agglutinin (WPA) — we have demonstrated respective lectin-binding receptors on the plasma membrane of the GR in the xiphosuran, *L. polyphemus*. Note, however, that Beaulaton[13] did not find any peanut agglutinin binding sites on the surface of PLs and GRs of the oak silkworm, *Antheraea pernyi*. He did, however, report that the WAG-binding sites were completely lost when these immunocytes were pretreated with neuraminidase, which splits off the sialic acid residues from sialoglycoconjugates. Beaulaton[13] concluded that the distribution of WGA-binding glycoproteins varies according to cell types and that this variation may be related to a "hierarchy in cell recognition, adhesiveness, and response to signal molecules". Although lectins are proteins/glycoproteins of nonimmune origin, they specifically bind to carbohydrates (in the form of glycoproteins, glycolipids, and polysaccharides) that occur on the surface of most cells. Furthermore, soluble lectins (hemagglutinins) may function as bridges to bind carbohydrates on two different cells (e.g., immunocyte and the target cell) and finally, lectins are involved in uptake and killing of cells and lymphocyte migration.[215] Indeed, Ofek and Sharon[216] used the term *lectinophagocytosis* for lectin-mediated nonopsonic phagocytosis (as opposed to opsonic phagocytosis — antibody and complement) of lectin-carrying bacteria that bind to the phagocytic macrophages.

Because Con A, WGA, PNA, WPA, and SBA bind specifically to D-mannose, *N*-acetylglucosamine (and NANA), D-galactose, L-fucose, and *N*-acetylgalactosamine, respectively,[61,67-69] our results confirm not only the presence of these monosaccharides on the arthropod immunocyte plasma membrane, but also provide insight into their immunologic function in cell-cell recognition.[11] Indeed, Alberts et al.[56] have suggested that the lectin molecules present on the cell surface recognize oligosaccharides on the surface of other cells, and thus guide cell-cell interaction, which is a prerequisite for the two cellular immune reactions (phagocytosis and encapsulation) against foreign antigen. Note that, according to Pendland and Boucias,[69] in addition to D-mannose, Con A also binds to D-glucose. It should be pointed out here that "an enormous number of compounds can be derived from a relatively limited number of monosaccharides" and that "a great diversity of carbohydrates associated with soluble and surface-bound glycoconjugates" have been reported.[215]

Of the five surface receptors specific for the six carbohydrates — mannose, galactose, *N*-acetylgalactosamine, L-fucose, *N*-acetylglucosamine, and *N*-acetylneuraminic acid (NANA or sialic acid) — those for the first five are known to participate in foreign tissue recognition and phagocytosis, and that for the last (NANA), in signaling and cell-cell recognition, and activation of the complement pathway.

Note that *N*-acetylglucosamine-specific lectin causes lectinophagocytosis of microorganisms that carry complementary sugars on their surfaces.[217] Petri et al.[218] have isolated lectins specific for galactose and *N*-acetylgalactosamine from the protozoan, *Entamoeba histolytica,* which causes dysentery in humans by disrupting the colonic mucosa. In fact, membrane-bound galactose- and *N*-acetylgalactosamine-specific lectin on rabbit hepatocytes provided the first evidence that this lectin was involved in clearance of glycoproteins from the circulatory system.[219,220] What is more interesting from the standpoint of its possible role as a receptor molecule is that this lectin has two subunits (M.W. 40 and 48 kDa), whose carbohydrate-binding domain is located extracellularly (on E face). Note also that peritoneal macrophages in rats attach to erythrocytes and other cells by means of a galactose-specific lectin on its surface and then lyse and phagocytose them.[221] Oda et al.[222] reported that a galactose- and *N*-acetylgalactosamine-specific lectin (with 45- and 60- kDa subunits) enables activated macrophages to discriminate between tumor and normal cells and to kill the former. In addition, peritoneal macrophages in mice with these two sugar-specific lectins recognize and phagocytose trypomastigotes of the protozoan, *T. cruzi* that causes Chagas disease.[223]

Pereira et al.[61] demonstrated that, in addition to *N*-acetylglucosamine, WGA also binds to sialic acid (*N*-acetylneuraminic acid or NANA) residues; and more importantly, that the WGA does not compete with limulin (major agglutinin in *Limulus*), which suggests that these two lectins (limulin and WGA) interact with different oligosaccharide moieties. Note also that sialic acid is widely found in nature[224] and its presence on the surface of eukaryotic cells has important biological significance.[225] Periera et al.[61] suggested that one of the roles of sialic acid residues on plasma membranes might be the activation of alternate complement pathways. Their preliminary experiment showed that neuraminidase-treated cells were much more susceptible to Mg-dependent lytic action of normal guinea pig serum than untreated cells. Furthermore, gangliosides (glycolipids containing one or more NANA residues) may serve as receptors for signaling and normal cell-cell recognition between cells.[56] More importantly, because antibodies for many gangliosides are available, it should be possible to study the specific functions of gangliosides by binding antibodies to them and studying stimulation or inhibition of specific functions (e.g., during encapsulation or phagocytosis).

iii. Binding the Foreign Antigen by Opsonins

In addition to the surface receptors, opsonins also play an important role in the recognition of the foreign antigens (hence these molecules are sometimes called ''recognition'' factors), and are responsible for binding the former to the latter. In vertebrates, the antibodies (IgG) and the complement factor C3 constitute opsonins, whose function is to facilitate phagocytosis, although they are also thought to function in activating the killer T-cells, lysis of target cells, and activation of the hexose monophosphate shunt.[200]

The presence or absence of opsonins is often determined by adding serum-treated or -untreated test particles (e.g., erythrocytes, bacteria) to immunocyte suspensions or monolayers, the presence being indicated by increased binding or uptake of particles by the immunocytes. Such experiments have shown the presence of opsonins in some crustaceans[226-229] (see also Chapter 6). The nature of opsonin molecules in insects, however, remains controversial. Hemagglutinins basically perform the same function in arthropods as antibodies (arthropods do not possess antibodies) do in vertebrates (see Section VII.A.6 on agglutinin function), and the existence of complement-like compounds in arthropods is suspected (see Section VII.B on complement-like factors). Because hemolymph hemagglutinins agglutinate vertebrate erythrocytes, they are thought to function as recognition factors.[230] Brewer and Vinson[231] suggested that tyrosine-containing proteins or polyphenols may be involved in opsonization of alien surfaces. According to Rizki and Rizki,[232] in *in vitro* phagocytosis of bacteria by *D. melanogaster* hemocytes, opsonins are not required; also, Dularay and Lackie[233] reported that phenoloxidases do not act as opsonins. However, phenoloxidase does opsonize nonself tissue in preparation for both phagocytosis and encapsulation.[234] Apparently, the prophenoloxidase-activating system (PAS) in arthropods produces a number of sticky proteins that might function as opsonins; phenoloxidase itself, although sticky, does not function as opsonin (see Section VII.D).

Whether coagulin, in fact, acts as opsonin is unknown. Cellular coagulogen (the clottable protein) is converted into an insoluble coagulin by a serine protease (clotting enzyme), both substances being located in the GR in *L. polyphemus*[235] (see also Chapter 8) and other arthropods. The serine protease is remarkably similar to the enzyme that activates the prophenoloxidase for conversion into phenoloxidase[236,237] (see Section VII.D for discussion on coagulation). The opsonic role of lysozyme also is controversial; however, its lytic action on microorganisms is well known (see Section VII.C.3).

b. Endocytosis and Digestion

Once the foreign antigen is recognized and contact with it established, ingestion (endocytosis) commences. Ingestion is carried out[238,239] by (1) forming pinocytotic or coated vesicles to engulf particles, such as viruses;[184] (2) encircling the particles by pseudopods;[240,241] and/or (3) spreading the plasma membrane around the particle[242] (see Figure 9C in Chapter 7). However, very little is known about the ultrastructural details of internalization processes in insects. Pastan and Willingham[243] described the details of receptor-mediated endocytosis in cultured vertebrate fibroblasts and suggested that similar processes occur in other types of cells. According to them, the fibroblasts have a wide variety of mobile, specific surface receptors. After a ligand (L) (e.g., virus) binds with the receptor (R), the L-R complex moves to a bristle-coated pit (150 nm in diameter) embedded in the plasma membrane, and then enters the cell in an uncoated vesicle, the receptosome (250 nm in diameter), which then shuttles it to the interior of the cell. The bristles of the coated pit are composed of the protein, clathrin.[244] Pastan and Willingham[243] chose the term receptosome to emphasize that the L-R complex appears in this organelle and not in the lysosome. Crossley[245] described flask-shaped (120 to 165 nm in diameter) coated vesicles in the plasma membrane of the hemocytes that approach cells undergoing phagocytosis in *C. erythrocephala*. Coated vesicles have also been found in insect pericardial cells,[246,247] because of the involvement of these cells in endocytosis. Crossley experimentally demonstrated that these coated vesicles in the pericardial cells seem to be involved in selective uptake of some molecules, and rejection of others.

Pastan and Willingham[243] speculated that substances that enter the cell by coated pits and receptosomes might include, among others, hormones and growth factors; and according to them, the action of many peptide hormones occurs on the cell surface.

Insect hemocytes absorb soluble, proteinaceous foreign material by coated vesicles and incorporate particulate matter, such as viruses, bacteria, and some tracer substances by "true phagocytosis".[239] In this case, the particulate matter (e.g., *Escherichia coli*), upon ingestion by PLs, in *G. mellonella* becomes enclosed in a membrane-bounded phagosome, which later fuses with the lysosome (the so-called multivesicular body, MVB).

Materials that do not bind to the specific receptors enter the cell via another type of organelle, the pinosome (ca. 0.08 to 3 μm in diameter), which fuses with the lysosome. The receptosome eventually transfers its contents to the lysosome via the Golgi endoplasmic reticulum. Although Pastan and Willingham[243] did not describe the mechanism of this transfer, it is probably akin to the transport pathway suggested by Rothman[248] on the basis of the structure of the Golgi apparatus.

The Golgi typically consists of a stack of flattened, plate-like, membrane-bounded cisternae (ca 1 μm in diameter) (comparable to a distillation tower) from which many small vesicles (ca 5 to 10 nm in diameter) bud off or fuse (or both), especially at the rims. The stack of cisternae is arranged in two compartments, the *cis* (the forming face) and the *trans* (the maturing face) (Figure 4B); the former, consisting of most of the cisternae, is nearer the nucleus and the endoplasmic reticulum (ER); the latter, consisting in many cases, of only the last one or two cisternae of the stack, is on the opposite face of the *cis* portion and acts as the receiver of the distilled product. The protein molecules synthesized by the ER enter the stack at its *cis* face and exit at the *trans* face. Thus, the newly formed secretion granules (vesicles) are always associated with the *trans* face, which ensures that only the most refined fraction is secreted for distribution in the cell.

The lysosomes are formed and their enzymes (e.g., acid phosphatase and β-glucuronidase) transferred to them in the region of the Golgi endoplasmic reticulum. The enzymes are concentrated within the *trans*-most cisternae and some of the coated vesicles. The endocytosed material is digested by these hydrolytic enzymes.

c. *Clearance*

Once transferred to the lysosome, the microorganism is broken down by the mucolytic and other enzymes and antimicrobial agents. The foreign antigens generally phagocytosed are bacteria, protozoa, fungi, erythrocytes, yeast cells, and autolysing tissue.

B. HEMOCYTIC ENCAPSULATION

As phagocytosis, encapsulation is a major defense reaction against foreign biotic or abiotic antigens that are too large to be phagocytosed by individual immunocytes.

1. Historical

As stated in Section VI.A on phagocytosis, Metchnikoff[107] was the first to observe this cellular reaction in the crustacean, *Daphnia*. Since then it has been studied in many insects.

2. Uncertainties and Controversies

The two most uncertain and controversial aspects of encapsulation are the mechanisms by which the immunocytes recognize the foreign antigen and recruit others that form the capsule, and how and when the process stops. The questions of cell types involved, sources of phenols and phenoloxidase for melanization, and the number of capsular layers are also controversial.

3. Cells Involved

As in phagocytosis, GRs and/or PLs (including its variant forms; see Section II.B.3) are involved,[1,7,48] although SPs, COs (considered GRs in this review; see Section VI.A.3), ADs and OEs also have been reported to encapsulate; these, however, could have been misidentified.

In our recent work on encapsulation in the German cockroach, *B. germanica,* we found that only the GRs are capable of encapsulation, because these blood cells alone possess a marginal band of microtubules that seems to be necessary for flattening and encapsulating the foreign antigen.[7] We also found that the number of microtubules (MTs) in the activated GRs, following implantation of pieces of surgical sutures, increased sevenfold, with a concomitant increase in the nuclear pores (Figure 13A, B). The microtubule band is absent in the PLs in this cockroach, which explains why the PLs do not participate in encapsulation. In *L. migratoria* and *M. melolontha* also only GRs participate in encapsulation.[86]

4. Effectiveness

All manner of biotic and abiotic objects, such as bacteria, fungi, protozoa, nematodes, insect eggs and larvae, latex or sephadex beads, araldite, cellophane, cotton or nylon threads, glass needles, and surgical sutures are encapsulated by insect immunocytes.[7,38,249]

Most of the same factors (as discussed in Section VI.A under phagocytosis) influence encapsulation. The number of circulating immunocytes is considered a limiting factor in Diptera.[250-252] Götz et al.[250] reported that only species with more than about 6000/mm^3 hemocytes formed hemocytic capsules. By the same token, immature stages of insects contain fewer hemocytes[134,238] and could conceivably have less effective encapsulation capability. Temperature and electrical charge on foreign objects are thought to influence encapsulation. Rizki and Rizki[253] have shown that fat body is encapsulated to form "tumors" only at 26°C in melanotic tumor mutants of *D. melanogaster*.

Although it was thought that capsules are generally formed around positively charged objects,[252,254-257] apparently both positively and negatively charged objects are encapsulated. Because the basement membrane of self tissue is negatively charged at physiologic pH,[255] it was interpreted that foreign objects lacking the host's negatively charged basement membrane are encapsulated. Indeed, Dunphy and Nolan[256] reported that the protoplasts of the fungus, *Entomophthora egressa,* in the hemolymph of the hemlock looper, *Lambdina fiscellaria,* avoid encapsulation owing to the negative charge on them. However, Lackie[258] has shown that in *P. americana* negatively charged polystyrene and sepharose beads were encapsulated. This is in contrast to Lackie and Takle's[259] report in which negatively charged beads were not encapsulated in the locust, *Schistocerca gregaria*. They further stated that "locust hemocytes are significantly more negatively charged than cockroach *(P. americana)* cells". Finally, Ratner and Vinson[252,260] have reported an encapsulation-promoting factor that regulates encapsulation (see also Chapter 3).

5. Role of Hemopoietic Organs

To the extent that these organs produce hemocytes, they play a major role in encapsulation. In *B. germanica,* we observed three subpopulations of the GRs in the hemopoietic organs (located on either side of the heart between abdominal segments one to four of this cockroach.[5] One of these subpopulations consists entirely of intensely electron-dense GRs (Figure 6), and we observed an increasing number of these cells in the hemopoietic organs and in their vicinity soon after implanting the surgical suture.[7]

6. Stages of Encapsulation

a. Foreign Antigen Recognition

This stage is essentially the same as in phagocytosis, and the reader is referred to Section VI.A.6 for additional comments. Membrane receptors, membrane-bound agglutinins, phenoloxidase, and even the differences in electrical charge between the immunocyte membrane and the foreign object are thought to be involved in foreign antigen recognition.[249] Essentially, in most cases, the GRs first contact the foreign antigen, probably randomly, and then lyse[1,7,48] and release the so-called recognition factors that attract other immunocytes. I have proposed a molecular basis of encapsulation in Chapter 4.

The discharge of material from the GRs (or COs of other authors), following lysis, apparently occurs.[239,245,252,260-262] According to Götz,[249] material released from the GRs is electron-dense, swells after release, and adheres to the foreign antigen in the vicinity of the degenerating GR. He considers this sticky substance to be the activated phenoloxidase on the basis that melanization first occurs in the innermost areas of nodules and capsules. It has been reported that during encapsulation in *G. mellonella,* sticky acid mucopolysaccharide-like substances "appear to be discharged onto the foreign surfaces by GRs, . . . and, independently of or in combination with material(s) from the foreign bodies, apparently specifically attracts PLs to the alien surface".[239] Parish[263] hypothesized that 5-glycosyltransferases secreted by hemocytes form subunits of recognition factors and polymerize, after being secreted into the hemolymph, into hexamers that in turn react with the foreign antigen. It has been suggested that a positive charge on the foreign antigen attracts GRs to them. However, capsules are formed around negatively charged objects.[252]

b. Layer Formation

Cellular capsules in insects vary in terms of the cell layers (distinguishable into two to three regions) in the completed capsule and the ultrastructural changes in the encapsulating cells.

i. Layers and Regions

Baerwald[52] reported a range of 20 to 75 layers in various insects owing to interspecific variation in the free circulating hemocytes. In *B. germanica,* we found about 20 layers, arranged into inner and outer regions.[7] Grimstone et al.[242] reported three regions in the *Ephestia kühniella* capsules, and François[85] reported four in *Thermobia domestica.* Lackie and Takle[259] stated that capsules are thicker in cockroaches than in locusts, because the former have more PLs; this is questionable, however, because *B. germanica,*[2,7] *G. portentosa,*[14] and *P. americana*[7] have more GRs than PLs.

ii. Ultrastructural Changes in Encapsulating Immunocytes (GRs and/or PLs)

These changes have been described in many insects. Essentially, the cells forming the innermost layer in contact with the foreign antigen show general elongation (Figure 25A—D), flattening, and lysis. In *B. germanica,* we recognized two regions: the inner one, composed of six to nine layers, is about 4 μm thick; and the outer, made up of 8 to 11 layers, is 2.5 μm thick; the two are separated by a narrow band (in which distinct layering is obscure) of loosely aggregated, necrotic GRs (Figure 26).

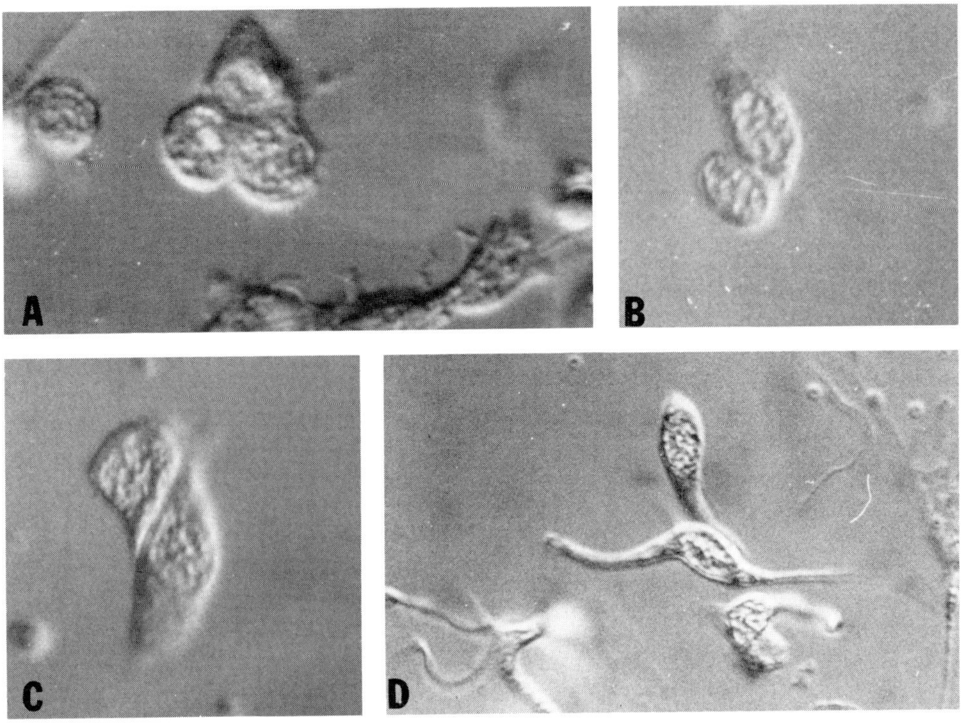

FIGURE 25. Granulocytes (Nomarski) of *B. germanica*, harvested from capsules (implants) of various ages. (A) Loosely attached granulocytes (center and upper left of micrograph) harvested from 1 × 1-mm plastic implant 30 min postimplantation. Note that most of the immunocytes are still discoid. (Capsule formation commences 40 to 45 min postimplantation.) (B) Granulocytes from a 40-min-old implant/capsule. Note that elongation of the granulocytes has begun. (C) Granulocytes from a 50-min-old implant/capsule. Note further elongation and close coupling of the cells. (D) Granulocytes from a 2-day-old capsule. Note elongation that is accompanied with flattening. (× ca. 3300)

The cells in the first or innermost layer of the inner region show well-developed Golgi bodies and reduced electron density and number of granules, which indicates that the GRs are probably engaged in synthetic and secretory activities (see Chapter 4). In addition, microvesicles exocytosed from intact cells and mitochondria were visible (Figure 27). The plasma membranes of the cells were not obscure. The next four layers (second through fifth) of the inner region show similar changes, the fifth layer being surrounded by two (18 to 23 nm-thick) laminae, which appear to be composed of amorphous coagulated hemolymph. Between the two laminae is a GR layer represented by fragments of electron-dense material. In the next three to four layers (sixth through ninth), the GRs are less compactly arranged, but ultrastructurally resemble the preceding five layers. The ninth, or both the eighth and ninth, layers show some necrotic GRs.

The first four to five layers of the outer region are visible in a 3-day-old capsule and consist of GRs that essentially show the same ultrastructural details as those of the inner region. The outer region is completed by the deposition of an additional four to five layers, which are extremely flattened and marked by pronounced development of microtubules (Figure 12B and C). At this stage of capsule formation, the innermost layer and the two amorphous laminae of the inner region become indistinguishable. Autophagic vacuoles, large lipid vacuoles, masses of ER, vacuolated mitochondria, fragmented nuclei, randomly scattered melanin deposits, and cell remnants are commonly found throughout the capsule (Figure 28), and the plasma membranes of the GRs are highly electron-dense.

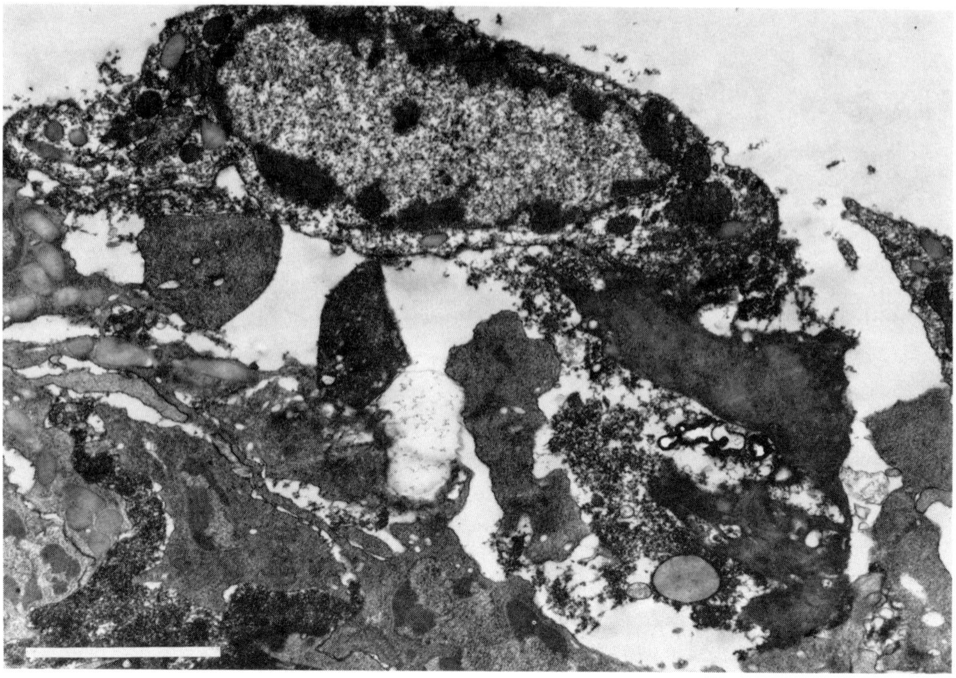

FIGURE 26. Section of a 5-day-old capsule (TEM) of *B. germanica,* showing the outer region, which is formed by continuous layering of granulocytes, and is separated from inner region by a band of loosely aggregated necrotic granulocytes. (From Han, S. S. and Gupta, A. P., *Zool. Sci.,* 6, 303, 1989. With permission.)

The completed (10-day-old) capsule as a whole (Figure 28) is very compact, consists of about 20 layers in some regions of the implant, and is about 6.5 μm thick. The general features of a finished capsule include an outer sheath, numerous microtubules, free ribosomes, few mitochondria and RER, and randomly distributed melanin deposits throughout the capsule. Golgi apparatus and granules are rarely seen. Only gap junctions are found in the outer region of the completed capsule.

iii. Melanization and Its Source

Melanization occurs in all cellular capsules to some degree. In *B. germanica,* we consistently found two sublayers of melanin in the innermost layer of the inner region during the initial stages of encapsulation, and numerous randomly distributed melanin deposits in various parts of the completed (10-day-old) capsule. The question of whether the source of melanization is the animate foreign tissue or the host's immunocytes is still debated, although it has been reported in many cases that the immunocytes provide both the phenolic substrate and the phenoloxidase,[1,48] with the GRs and/or PLs[264] supposedly providing the phenols and the OEs supplying the enzyme.[265-267]

It should be stated, however, that the hemocytic origin of both the phenols and the phenoloxidases continues to be controversial. Furthermore, in addition to GRs and/or PLs, COs (considered GRs in this review; see Section II.A), and OEs have been reported by one author or another to provide both the substrate and the enzyme,[39,268] although I found no tyrosinase in the OEs of the cockroach, *G. portentosa.*[1] Our studies in *B. germanica* suggest that both the phenols and the phenoloxidase are supplied by the GRs, because we noticed the melanin sublayers even in capsules around surgical sutures, which are the most unlikely sources of these compounds. Indeed, Schmit et al.[268] have demonstrated the presence of melanin precursors in the GRs of *G. mellonella.* However, the possibility that the hemolymph

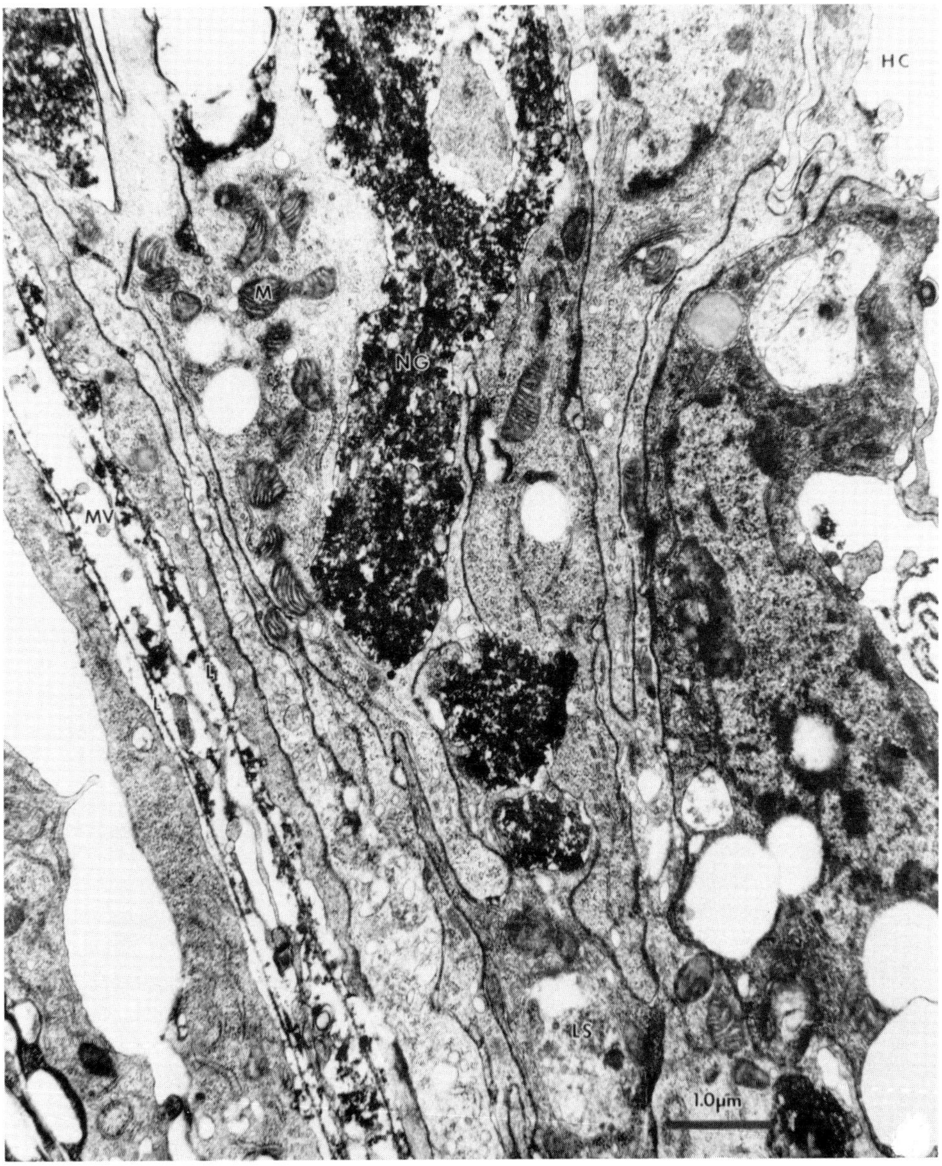

FIGURE 27. Micrograph (TEM) showing granulocytes in a 3-day-old capsule of *B. germanica*. In the region of the capsule facing the hemoceol (HC), the granulocytes are less compactly arranged. Note necrotic granulocytes (NG), several vacuoles, lysosomes (LS), well-developed Golgi apparatus, mitochondria (M), laminae (L), RER, and microvesicles (MV) exocytosed from intact cells. (From Han, S. S. and Gupta, A. P., *Zool. Sci.*, 6, 303, 1989. With permission.)

could be the source of the phenols and the phenoloxidase cannot be ignored, because melanization always occurs in humoral capsules.[249] Mazzone[269] reported that cultured gypsy moth cells produced melanin for long periods.

iv. Cell Junctions in Capsules and Their Functional Significance

The presence of cell junctions and their functional significance in capsule formation have been rarely discussed, with few exceptions.[7,51,52,270] Although desmosomes and gap junctions (Figure 8A, D) have been reported before in hemocytic capsules (see reviews by

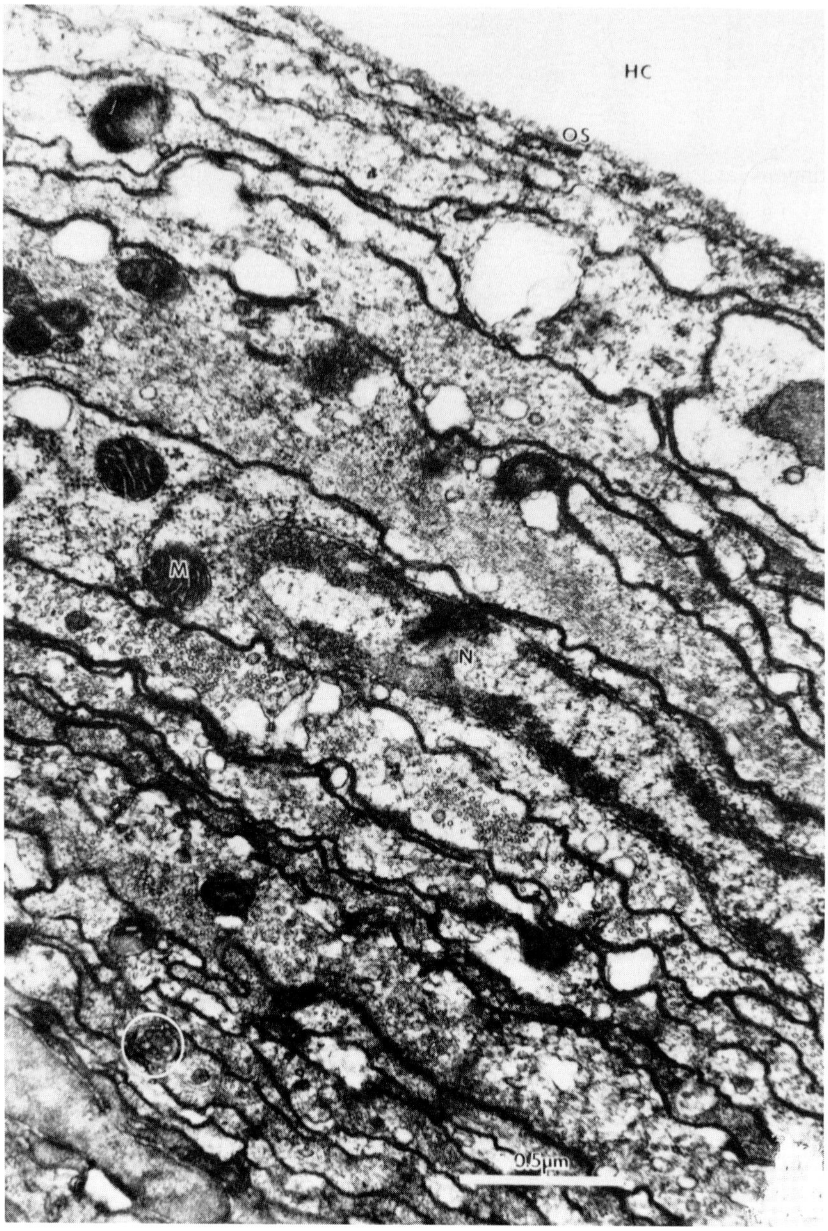

FIGURE 28. Section of a 10-day-old, completed capsule (TEM) of *B. germanica*. It is about 6.5 μm thick. Numerous microtubules, but few mitochondria (M), RER, and flattened nuclei (N) are seen. The outer sheath (OS) surrounding the capsule is 64 nm thick. HC = hemocoel; multivesicular body in white circle. (From Han, S. S. and Gupta, A. P., *Zool. Sci.*, 6, 303, 1989. With permission.)

Baerwald[52] and Gupta[48]), we reported septate junctions for the first time in the capsules of *B. germanica* by both thin section and freeze fracture methods[3] (Figure 9A, B).

I have discussed in detail the presence and functional significance of gap and other cell junctions in Chapter 4.

c. Cessation of Layer Formation

i. The Outer Sheath

The most characteristic feature of the capsule in *B. germanica* is a 64 nm-thick outer sheath (Figure 28), whose completion marks the end of encapsulation. From then on, no further attachment of GRs to the sheath occurs. The mechanism involved in the maintenance of continuous layering in the capsule and its cessation when the capsule has been completed, remain conjectural. I have postulated a molecular basis for these processes in Chapter 4. Grimstone et al.[242] suggested that as each layer is formed, the surface properties of the plasma membrane of the cells forming the capsule change, and consequently they are recognized as nonself tissue and in turn attract new cells. Amirante[45] hypothesized that as hemocytes adhere to the foreign tissue, they form "new acceptors" (some sort of lectin receptors) on their plasma membranes that attract the lectins on the plasma membranes of other hemocytes, which then form the next layer. I have suggested in Chapter 4 the possible role of gap junctions in establishing a continuous ion gradient, which might promote continuous positioning of GRs on the developing capsule. Ratcliffe et al.[193] suggested the production of a lymphokine-like substance by the encapsulating PLs as the cause of any further activation of these cells and their attachment to the capsule.

Whatever the mechanism(s), it must cease to operate when the capsule reaches a certain thickness or age, because no further recruitment and attachment of immunocytes occur. In fact, this cessation of attachment is not abrupt, but seems to occur gradually. How this mechanism is turned off in other insects is unknown. In *B. germanica,* the completed capsule (10-day-old) develops an outer sheath, which, because it is composed of the host material, seems to render the capsule as a self tissue, and therefore does not trigger any further immunoreaction among the GRs. The sheath must also act as a barrier and interrupt any further cell-cell communication through the gap junctions. Because we found this sheath only when we allowed encapsulation to proceed for at least 10 day, it is likely that the occurrence of a sheath was missed in the capsules of other insects, which were not allowed to fully mature.

ii. Killing of the Foreign Antigen

The occurrence of melanization in hemocytic and humoral capsules is an important evolutionary adaptation, and is primarily responsible for the death of the live foreign antigen. Taylor[271] apparently was the first to suggest that the activation of the phenoloxidase not only produces "disinfectants" in the form of toxic quinones, but phenoloxidase is also a prerequisite for hemocytic adhesion to nonself tissue. Beresky and Hall[272] also have reported the toxic effect of the phenoloxidase system. The quinones eventually produce melanin as a result of a series of nonenzymatic reactions. In fact, melanins and other intermediate products produced by phenoloxidase activity are known to show antimicrobial cytotoxic effects.[192,273-276] Lysozymes, β-glucuronidase and β-glucosaminidase, are also implicated in the killing of foreign objects,[193] but their roles as killing agents in hemocytic capsules are not confirmed.

7. Avoidance of Encapsulation by Foreign Organisms

Although encapsulation provides the insect with defense against many invading microorganisms, it is ineffective against certain bacteroids and parasitoids that successfully live within insects. Obviously, these organisms have evolved certain strategies that inactivate or thwart encapsulation. Such avoidance strategies may be summarized under the following major categories.

a. Secretion of Defensive Membranes

This strategy seems to be the most prevalent and the best documented. The presence of an outer membrane, which might be recognized as self by the host's immune system, seems to be the strategy adopted by bacteroids (rod-shaped, bacteria-like endocytobionts) that occur in many insects (see review by Gupta[4]); the protective membrane may be secreted by the host or the bacteroid itself. According to Pearson et al.,[277] facultative intracellular parasites are surrounded by host-provided membranes, but obligate parasites are not. Gresson and Threadgold[278] described, in the bacteroid of *Blatta orientalis,* membrane-like structures that originated from the "bacterium" and formed an outer covering around it. According to Grinyer and Musgrave,[279] the bacteroids of the rice weevil, *Sitophilus* possess one or two trilaminar unit membranes, but these are not secreted by the host. Lamb and Hinde[280] reported that in addition to the two unit membranes, the bacteroids of the cabbage aphid, *Brevicorynae brassicae* "usually appeared to be enclosed by an extra cytoplasmic membrane produced by the cytoplasm of the mycetocyte". In the cockroach, *Cryptocerus relictus,* the bacteroids in the fat body are surrounded by several host membranes.[281] Griffiths and Beck[282] described three membranes in the primary (located in the large primary mycetocyte) of the pea aphid, *Acyrthosiphon pisum.* The outer of the three membranes represents a host plasma membrane that surrounds the bacteroids soon after infection of the embryo. Furthermore, the bacteroids do not possess this outer membrane when they enter the embryonic bacteriocytes (= mycetocytes), where they acquire this membrane. These authors found that this outer membrane is connected to the rough endoplasmic reticulum of the bacteriocyte.

The presence of these membranes and their receptors is crucial for the survival of the bacteroid during its transmission phase. These membranes, secreted by the host or the bacteroids themselves, are essential to prevent recognition as a foreign antigen by the hemagglutinins and subsequent phagocytosis and/or encapsulation by the host's immunocytes. The helminth, *Moniliformis dubius* surrounds itself in an envelope of polyanionic mucin that probably does not trigger the prophenoloxidase system of its host, *P. americana,*[283] and thus avoids immunorecognition. It has been suggested that fungal pathogens may avoid immunorecognition by replacing β-1,3-glucans with α-glucans that do not trigger the phenoloxidase system.[284]

b. Molecular Mimicry of the Host's Antigens

Several insect parasites practice this particular strategy. It has been reported that in mosquitoes, the microfilariae, *Brugia pahangi* possess the host's antigens on their cuticle during midgut penetration, and thus avoiding recognition by the host's immune system.[285] According to Lafond et al.,[286] these microfilariae actually coat themselves with the material(s) from the host's midgut during migration to the hemocoel and almost totally avoid immunorecognition (no melanization occurs). Miller et al.[287] found that the infective sporozoites of the protozoan, *Plasmodium falciparum* in the mosquito's salivary gland, secrete a so-called circumsporozoite (CS) protein that coats the entire surface of the sporozoite. This protein is absent or exists only in small amounts on the noninfective sporozoites existing in the oocysts of the midgut.[288] According to Miller et al.,[287] such protective antigens (CS-proteins) have been detected in several *Plasmodium* species by monoclonal antibodies, and are essential for the parasite's survival.

c. Secretion/Injection of Anti-Immune Factors/Particles

Such factors are secreted by several bacteria.[289,290] According to these authors, *Serratia marcescens* secretes three different protease fractions that act as anti-immune enzymes, which destroy cecropins (the antibacterial proteins; see Section VII.C.1), secreted by the host in defense against bacteria. Siden et al.[291] reported that *B. thuringiensis* also secretes two inhibiting factors against the immune system of the pupae of the saturnid moth, *Hyalophora cecropia.* It has been reported that the nematode, *Neoaplectana carpocapsae,* in *G. mellonella* destroyed the immune proteins P9A, P9B, and P5.[292]

Kitano[293,294] reported that the eggs of the braconid, *Apanteles glomeratus,* oviposited in the larvae of *Pieris rapae crucivora,* caused reduced adherence of PLs. Teratocytes (giant cells produced from the trophamnions of eggs in braconid wasps[238]) are thought to produce encapsulation-inhibiting factors,[293,294] and may also entice encapsulating immunocytes away from the eggs themselves.

The eggs of *Nemeretis* (= *Venturia) canescens* avoid being perceived as foreign by the host, *E. kühniella* by means of a coating of calyx particles[295,296] that have since been identified as a dense suspension of virus-like particles,[297-299] which apparently occur widely in the reproductive tracts of other wasps.[300-304] *Cardiochiles nigriceps* and *Campoletis sonorensis* also employ calyx particles to suppress encapsulation.[305,306] These virus-like particles (VLPs) thwart encapsulation only when adherent to the eggs. In contrast, the polydnaviruses prevent encapsulation remotely from the egg. Using mammalian antisera against the VLPs, it has been shown that they share antigenic similarities with a 42-kDa protein found on the surface of tissues lining the hemocoel and on the hemocytes,[307,308] and that this protein is induced upon bacterial infection.[309]

Polydnaviruses, found in the reproductive tracts of braconids (these have a single membrane) and ichneumonids (these have two membranes),[310-312] can coexist with other viruses. These viruses, unlike the VLPs, have complex genomes, penetrate the host's tissues, and express proteins. For details of gene expression, see Chapter 3.

Finally, it has been shown that injection of podynaviruses prevents encapsulation by suppressing certain immunocytes. In the case of *Pseudeucoila bochei* that parasitizes the larvae of *D. melanogaster,*[313] suppression of the immunocyte is accomplished by a parasitoid-derived factor (lamellolysin),[314] which eliminates the so-called lamellocytes (variant forms of PLs); possibly it also interferes with the OEs (= crystal cells) that contain the components of the phenoloxidase system.[315]

d. Causing Hemocytopenia

I have stated earlier that a reduced number of hemocytes may adversely affect the cellular defense capability of the host. Several microorganisms take advantage of this vulnerability of the host by causing severe lysis of the host's hemocytes. For example, extensive (64%) lysis of hemocytes is caused by *Salmonella typhimurium,*[316] and by *B. cereus*[317] in *G. mellonella,* by *B. thuringiensis* in *H. zea,*[318] and by *Nosema apis* in *Apis mellifera.*[319] It has been suggested that hemocytic lysis may be due to the bacterial toxins, bacteremia accompanied by pH changes, and depletion of hemolymphatic nutrients.[318]

e. Defense against Lysozymes and Complement-Like Factors

Lysozymes, along with hemagglutinins and complement-like substances, play an important role in insect immunity (see discussion in Section VII.C.3 on Secretion of Immunologic Factors).

f. Existence of Acquired Immunologic Tolerance in the Host

It is very likely that permanently housed parasitoids, especially those that are present during the immature stage (when the immune system is still developing) and are later transmitted to the adult stage, benefit from the fact that owing to their presence, the host's immune system learns not to recognize them as foreign antigens, and in fact develops some kind of tolerance to them (acquired immunologic tolerance); in other words, it perceives them as self tissue. When this tolerance to self tissue breaks down under certain conditions, the immunocytes destroy even their own tissue (autoimmune reaction). Apparently, this is what happens at the time of molting and under some pathological conditions.

C. NODULE FORMATION

Nodule formation is similar to encapsulation in some respects, but may be considered to involve both phagocytosis and encapsulation,[45] and occurs in response to [illegible]mate (e.g., masses of bacteria) and inanimate particulate material that cannot be normally phagocytosed. Both encapsulation and nodule formation may occur within the same insect in response to the same parasite,[239] and it may be difficult to distinguish between the two defense reactions.[238] Nodules or granuloma can be described as multicellular sheaths formed variously by GRs and PLs.[320-323] Apparently, some of the encapsulating immunocytes detach from the nodule and enter the bacterial mass to phagocytose them.[45]

Typically, a nodule is formed around a melanized core of coagulum, produced by the degranulating GRs (including COs), and surrounds the particulate material. As in encapsulation, the PLs/GRs are soon attracted to the core and form a sheath.[239] According to these authors, the nodules in *G. mellonella* and *Pieris brassicae,* in response to certain bacteria, contained the same three distinct layers as in encapsulation: the inner layer, composed of flattened PLs; the middle layer, also composed of flattened PLs; and the outer layer, made up of normal PLs. The melanization of the nodules seems to occur in the same way as in encapsulation. Nodulation can release factors that result in the production of antibacterial products in the fat body.[276]

D. HUMORAL ENCAPSULATION

Humoral encapsulation has been recently reviewed by Götz and Vey[324] and should be consulted for details. It was first reported from mosquito and chironomid larvae parasitized by nematodes,[325-327] and has often been reported from Diptera in which it replaces hemocytic encapsulation.

VII. HUMORAL IMMUNITY: SECRETION OF IMMUNOLOGIC FACTORS

It is well known that mammals possess serum proteins (γ-globulins), called immunoglobulins (Igs) or antibodies that act as humoral recognition factors in removing the foreign antigen. Although arthropods do not possess Igs, they do produce humoral recognition factors called hemagglutinins, that are secreted by the immunocytes and fat body (Section VII.A). In addition, the hemolymph complement-like factors and antimicrobial factors (such as cecropins, attacins, and the lysozymes) are believed to originate in the GRs and/or the fat body (see Section VII.C).

A. HEMAGGLUTININS (LECTINS)

1. General Description

Since annelid hemocytes are capable of distinguishing between "self", "closely related nonself", and "distantly related nonself",[328,329] it is not surprising that those of their descendants, the arthropods, also possess this capability. This is accomplished by the surface receptors on the plasma membranes of the immunocytes and one of the humoral recognition factors, variously called hemagglutinins, heteroagglutinins (structurally different agglutinins in the same organism), or agglutinins (lectins). These humoral factors may be likened to guided missiles that are meant to deal with the invading foreign antigens. Thus, they are an important component of the humoral immune system, and provide a second line of defense.

They are called agglutinins because of their capacity to agglutinate microorganisms, such as bacteria,[330,331] protozoa,[332-334] vertebrate erythrocytes, metazoan parasites,[335] and mammalian erythrocytes,[336] because of the presence of certain polysaccharides on the surface of these foreign cells. It should be noted, however, that other substances also agglutinate erythrocytes.[337,338]

Agglutinins are not unique to arthropods, but rather ubiquitous glycoproteins (as suggested by their loss of activity after incubation with trypsin and 2-mercaptoethanol, trichloroacetic acid, and bromelain) with multiple carbohydrate-binding sites. They are found widely in plants (first observed by Renkonen[339] and Boyd and Reguera[340]), microorganisms,[341] and other animals, and are inhibited by carbohydrates and chelating agents. As a matter of fact, the carbohydrate-binding specificity of an agglutinin is defined by the chemical structure of its inhibitor(s) (e.g., monosaccharide or oligosaccharide) that reduces or obliterates its agglutinating property. Furthermore, the agglutination specificity and inhibition of agglutinins depend on both the erythrocyte type and agglutinin purity.[67,342,343]

Perhaps the most fully characterized and studied arthropod heteroagglutinins are those of the most primitive arthropod, *L. polyphemus*.[64,67,342,344] These heteroagglutinins agglutinate vertebrate erythrocytes and have electrophoretic mobility similar to those of the γ- and β-globulins.[345]

It should be noted that, according to Yeaton,[346] variation in the lectin concentration (unrelated to age and sex) occurs among individuals of the same population both in mollusks and arthropods, and that lectins may be absent temporarily. For example, in the spider crab *(Maia squinado)* no lectin appears prior to molting,[332] and in the coconut crab *(Birgus latro)* lectin is apparently absent in the young. Lectin, however, is present even in the eggs of the silkmoth, *H. cecropia;* it appears 3 day prior to hatching and steadily increases in concentration until pupation and adult development.[346]

2. Amino Acid Sequence and Composition

The α and β units of the agglutinin from *Sarcophaga bullata* have similar peptide maps, showing a common amino acid sequence[347] (see also Chapter 16). The amino acid sequence of limulin from *L. polyphemus*[348,349] shows no amino acid sequence homology with vertebrate immunoglobulins.

The amino acid compositions of agglutinins from *Melanoplus sanguinipes* and *M. differentialis* (grasshoppers), *Balanus balanoides* and *Megabalanus rosa* (crustaceans), and *L. polyphemus, Carcinoscorpius rotunda cauda,* and *Tachypleus tridentatus* (xiphosurans) have been reviewed by Hapner and Stebbins[342] and the reader is referred to their review for details. Briefly, all these agglutinins have high amounts of aspartic and glutamic acids. According to them, the two grasshopper agglutinins "have presumably identical structure".

3. Hemocytic Origin

Let us first consider where the vertebrate antibodies originate. According to Alberts et al.,[56] in vertebrates the first antibodies (known as *membrane-bound antibodies*) are made by the B lymphocytes, and instead of being secreted into the blood, are inserted into the plasma membrane and serve as receptors. Each B-lymphocyte has ca 10^5 such receptors on its plasma membrane. The *soluble* antibodies, with the same antigenic-binding sites as the membrane-bound antibodies, are produced by stimulated plasma cells (B lymphocytes) and secreted into the blood.

As a matter of fact, the various receptors or antibodies on the plasma membrane are produced by different *families* or *clones,* each consisting of B- or T-lymphocytes, descended from a common stem cell. These receptors are predestined to specifically bind to a specific foreign antigen or its determinants; it is the latter that stimulate the production of antibodies by B-lymphocytes or reactions in T-lymphocytes.

During their early development, the B- and T-lymphocytes that have receptors for "self" tissue are suppressed or eliminated. This enables the immune system, which is capable of recognizing both self and nonself, to respond to only nonself or foreign antigen. However, if a nonself antigen is continuously present, the immune system also learns not to respond to the nonself antigen; in other words, it develops an *acquired immunologic tolerance.* When the tolerance to self tissue breaks down, the B- and T-lymphocytes, or both, destroy their own tissue (a reaction called *autoimmune reaction*).

From the standpoint of the origin of hemagglutinins in insects, the most interesting works are those of Amirante and his associates,[45,350-352] who demonstrated for the first time that in the cockroach, *Leucophaea maderae,* the hemagglutinins are present in the vacuoles and the plasma membranes of the GRs. Amirante[350] also showed that in this cockroach, the hemagglutinins are in the β-globulin range. It should be mentioned here that Marek[353] reported an ''immunoglobulin'' in *G. mellonella* that is induced by cooling, and is within the ''range of vertebrate γ-globulin''. That hemagglutinins are synthesized by the GRs (and possibly also by the PLs) in many arthropods, has been reported by several authors. For example, Stang-Voss[44] reported that the granules of the so-called clotting cells in *Astacus* are the sites of hemagglutinin. Yeaton[46,346,354] reported that in the American silkmoth, *H. cecropia,* the GRs synthesize, contain, and release hemagglutinin into the culture medium, and Bradley et al.[214] concluded that only the GR (not the PL) contained agglutinin in *M. differentialis,* although it is synthesized in the fat body. In the American lobster, *H. americanus,* also the agglutinins are synthesized by hemocytes and released into the hemolymph.[355,356] Furman and Pistole[357] reported that *Limulus* plasma from which GRs were removed did not show any bactericidal action, but the serum (obtained after hemolymph clotting with GRs) did. In addition to GRs, SPs in *L. maderae*[352] and all hemocyte types in *P. americana*[358] are known to contain hemagglutinins. However, one must remember that the presence of hemagglutinins in hemocytes does not necessarily mean that they synthesize them. It is likely that these hemocytes simply sequester the agglutinins produced in some other organ (e.g., fat body in the grasshopper, *M. differentialis*).[214,359]

Nelstrop et al.[360] have cautioned that, although proteins with electrophoretic mobility comparable to vertebrate γ-globulin may be detected in insects (e.g., *Locusta*), they should not be characterized as immunoglobulins just on the basis of their electrophoretic mobility in the γ-globulin range. Considering the lower phylogenetic position of the arthropods, one would not expect their hemagglutinins to be comparable to the γ-globulins of the most highly evolved vertebrates! It is not surprising, therefore, that attempts[330,361,362] to demonstrate properties of hemagglutinins comparable to those of vertebrate immunoglobulins have been unsuccessful.

At any rate, one may draw several conclusions regarding hemagglutinins from the works of Amirante, his associates, and others.

1. Hemolymph hemagglutinins, at least in insects and crustaceans, and probably in other arthropods as well, seem to be both of hemocytic as well as fat body origin (see Section VII.A).
2. Because they have been reported to be present on the plasma membranes of GRs, hemagglutinins are thought to function as surface receptors in foreign body recognition, in addition to their opsonic role as humoral recognition factors (see Section VII.A.6). This is supported by the fact that GRs are the first to be attracted to the foreign body during phagocytosis, encapsulation, and nodule formation.
3. The hemolymph-independent attachment of vertebrate erythrocytes to hemocytes[190,239] is most likely due to these hemagglutinins. In addition, insects, like vertebrates, do possess serum-independent surface recognition factors on their immunocytes.

4. Nonhemocytic Origin

Agglutinins are also known to be secreted by the fat body.[359,363] Although the presence of agglutinins has been reported in several tissues other than hemocytes, we do not know whether these tissues actually synthesize or sequester them; the function of the agglutinins in these tissues also is unknown.

According to Hall and Rowlands,[364,365] agglutinins (lectins) are present in several tissues in *H. americanus,* such as lymphatic gland, pericardium, ovaries and testes, and hepatopancreas. Among insects, agglutinins have been reported in the alimentary canal of *R.*

prolixus and *Glossina austeri*;[333,366] pupal wing of *P. brassicae*;[367] coxal depressor muscle of the cockroach, *P. americana*;[368] fat body of *Sarcophaga peregrina*;[369] and peritrophic membrane of *C. erythrocephala*.[370]

5. Agglutinating Property

Arthropod hemagglutinins generally possess multiple binding sites that are responsible for their binding specificity with certain polysaccharides on the plasma membranes of microorganisms or vertebrate erythrocytes. This binding specificity is variable and inhibited by several mono- and oligosaccharides and glycoproteins. Ca^{2+} is necessary for carbohydrate binding.[342,371] According to Yeaton,[346] vertebrate erythrocytes contain only 7 (L-fucose, D-galactose, *N*-acetyl-galactosamine, D-glucose, *N*-acetylglucosamine, D-mannose, and *N*-acetyl [or glycolyl]-neuraminic acid) of the 100 known monosaccharides in nature.

Limulin reacts with bacterial lipopolysaccharides (LPS), and the specific site responsible for this reaction is 2-keto-3-deoxyoctonate (KDO), which is a key component of the LPS in Gram-negative bacteria.[372-374] According to Luderitz et al. (1973) (cited by Pistole[374]), the KDO directly links the polysaccharide side chain with the glucosamine backbone of lipid A. There is structural similarity between sialic acid (*N*-acetylneuraminic acid) (NANA) and KDO.[374] It is this reaction of the lectin on the NANA residues on vertebrate erythrocytes that is responsible for its agglutinating the erythrocytes. *Limulus* serum has three recognized heteroagglutinins.[345]

Pistole[374] has suggested two models to explain the dual binding of the lectin with the GR and the foreign microbe. According to the first model, the lectin molecule is supposedly not a part of the surface membrane of the GR, but rather a humoral factor; and it has two sites, one binding with the GR and the other with the microbe. According to Pistole, this is comparable to the dual binding of certain mammalian immunoglobulins that bind to specific antigens via antigen-specific sites in the *Fab* region of the molecule and to the phagocyte via the immunoglobulin-specific sites in the *Fc* region. Alternatively (the second model), the lectin is an integral part of the surface membrane of the GR (the so-called amoebocyte) and so the GR directly binds with the microbe *in vivo*. Thus, according to Pistole, the *in vitro* binding of the serum lectin to the pathogen is an artifact.

Note that this property of lectins to agglutinate cells has been found to vary with antigenic type, state of infection, stage of cell cycle, cell density, and contact during embryonic development.[375] Rabinovitch and De Stefano[178] have shown that the "attractiveness" of the sheep erythrocytes to phagocytosing cells is altered by various modifications of the surface charge by iron salt; altering the protein structure by tanning or formalin-fixation stimulated increased phagocytosis.

a. Inhibition of Agglutinating Property

The binding capacity of arthropod agglutinins is inhibited by a wide variety of compounds, including sialic acid-containing carbohydrates, glycoproteins, and glycolipids. Of the various carbohydrate inhibitors, *N*-acetylgalactosamine and *N*-acetylneuraminic acid (NANA) seem to be the most common. It seems that the effectiveness of certain inhibitors varies in different arthropod groups. For example, according to Hapner and Stebbins[342] and Stebbins and Hapner,[343] although agglutinins from aquatic Chelicerata and Crustacea are inhibited by hexuronic acid and sialic acid, the latter is not a prominent inhibitor of insect agglutinins. Agglutinin from the cricket, *Teleogryllus commodis,* is an exception[376] in that it "shows merostom-crustacean type of carbohydrate inhibition", and that of *B. mori* is inhibited by hexuronic acids and acidic oligosaccharides. Insect agglutinins are generally inhibited by galactosidic and glucosidic carbohydrates. Recently, Stynen et al.[377] reported that the most effective saccharide inhibitors for lectins in the adult female *S. bullata* were α-D (+)-melibiose (0.25 m*M*) and D-galactose (0.50 m*M*). On the whole, however, the sialic acid-binding characteristic of arthropod agglutinins seems to be widespread among various arthropod groups, based on the fact that vertebrate erythrocytes, when treated with neuraminidase, are not agglutinated by sialic acid-binding agglutinins of arthropods.

6. Functions

Lectins were originally used to characterize blood group specificities,[378] and are now commonly used in cell biology as structural probes to study cell membrane structure and function, especially to determine the number of lectin-binding receptors and their affinities. Fluorescent lectins are often used to study localization of lectin receptors by fluorescence and electron microscopy.

Because of their capacity to bind to a variety of carbohydrates (e.g., D-galactose, 2-keto-3-deoxyoctonate, glucuronic acid, *N*-acetylmuramic acid, and colominic acid) present on foreign cell membranes, agglutinins show multiple specificities, and thus may function as a carbohydrate-based recognition system for nonself tissue. From what we know of these molecules, the following functions are or can be attributed to them.

a. As Opsonins

There are conflicting reports as to whether or not agglutinins serve as opsonins. Agglutinins act as recognition molecules and opsonize (to increase ingestion by the immunocytes) the foreign antigen, and thus act as antibody-like surveillance components of the immune system.[213,347,377,379,380] These authors suggest that lectins act as opsonins. According to Weir,[381] lectins constitute the main portion of the serum opsonins in many invertebrates. Tyson and Jenkin[382,383] reported that in crayfish, hemolymph hemagglutinins are responsible for opsonic activity. Apparently, the same occurs in *H. americanus.*[228,384] It has been suggested that in mollusks, hemagglutinins may be involved in opsonization.[385,386] Rowley and Ratcliffe[38] believe that in insects agglutinins function as opsonins, and they found a non-agglutinating factor in the hemolymph of *P. americana* that causes uptake of *B. cereus* isolates, and thus, according to them, functions as an opsonin.[387] Recently, Stynen et al.[377] suggested that *S. bullata* lectins act as opsonins. It is reasonable to assume that in all these cases, the hemagglutinin first opsonizes the foreign antigens, which are then recognized and phagocytosed by the GRs and/or PLs.

Although agglutinins are present in the hemolymph, whether their secretion can be antigenically stimulated is not well documented. However, in an experiment on the giant stick insect, *Extatosoma tiaratum,* some insects showed higher titers of hemagglutinin following second immunization.[193] No such increase occurred in *Clitumnus extradentatus* and *P. americana.* Komano and associates,[347,379] however, found that in *S. bullata,* wounding stimulated production of hemagglutinin. Of the two subunits of this hemagglutinin (α and β), only the α unit (32 kDa) is present in the uninjured larva, and is converted into a β unit (30 kDa) by a wound-dependent protease. This conversion leads to the production of the active hemagglutinin $\alpha_2\beta_2$, consisting of four α units of 32 kDa and two β units of 30 kDa. Note that this agglutinin is synthesized by the fat body and upon injury, is secreted into the hemolymph, where it is sequestered by the hemocytes;[369] see also Chapter 16.

However, lack of opsonic activity of agglutinins has been reported by some authors.[214,362,387,388] Furthermore, even in the same animal one heteroagglutinin may have opsonic function, the other may not. For example, in the tunicate, *Bottrylloides leachii,* HA-1 does not have opsonic function, but HA-2 does.[389] Earlier, Schluter et al.[390] reported that HA-1 is specific for guinea pig erythrocytes and does not require Ca^{2+}, while HA-2 agglutinates erythrocytes from pigeons, rabbits, mice, and guinea pigs, and requires Ca^{2+}.

b. Causing Attraction of the GRs/PLs to the Foreign Antigen

This function is based on the assumption that the GR, upon lysis, releases granules that in fact contain hemagglutinins. Following exposure to *B. cereus,* GRs become "sticky" and contain Alcian blue-positive granules (absent in PLs) that are secreted in abundance.[391,392]

Apparently, the attraction of the PLs/GRs to the GR-foreign antigen complex occurs owing to: (1) local stickiness caused by the lysis of the GRs,[393] or (2) local clotting of the plasma,[322] and (3) release of the recognition factors into the hemolymph.[239] However, Lackie[257] commented that because insects have an open circulatory system, a chemotactic gradient (as implied by recognition factors) exists only in the tissues. Thus, according to Lackie, the question of GR releasing a chemotactic, chemokinetic, or "sticky" substance should be carefully examined. Note that phenoloxidase also is sticky.[192,394,395]

c. Causing Proliferation of GRs/PLs

Because of this function, agglutinins may be regarded as modulators of phagocytosis, encapsulation, and perhaps also of nodule formation. This is most likely accomplished by inducing mitosis. As a matter of fact, Majerus and Brodie[396] reported that lectins induce mitosis and platelet release in humans. If these compounds indeed cause the proliferation of the immunocytes (PLs and GRs), it should be possible to demonstrate the cessation of proliferation by inactivating the hemagglutinin(s) on the surface of the immunocytes or those in the hemolymph. This could have important implications, because the failure of the PLs and GRs to release these compounds could result in nonrecognition of the foreign antigen.

d. As Surface Receptors

I have stated earlier that in both vertebrates and arthropods, the antibodies or hemagglutinins, respectively, are produced by blood cells (B-lymphocytes and GRs, respectively) and act as membrane surface receptors. In insects, Amirante and Mazzalai[352] have shown that molecules (agglutinins) that agglutinate rabbit erythrocytes are associated with the hemocyte membrane. It is this agglutinating property of these compounds that enables us to pinpoint the location of surface receptors by labeling them with markers, such as hemocyanin, ferritin, or fluorescein. The latter has recently been used by Rizki and Rizki[315,397] and Nappi and Silvers[398] to demonstrate blood cell surface changes in two hemocyte populations in *Drosophila* mutants with melanotic tumors. Bradley et al.[214] attempted to show the presence of potential agglutinin receptors on the plasma membranes of the hemocytes of *M. differentialis* and *M. sanguinipes* by incubating them with FITC-PNA (because the agglutinins of these insects show binding specificity for galactose) and concluded that only GRs (all GRs were stained) bound PNA (not the PLs). However, because the GRs in monolayers did not bind pure agglutinin, they concluded that "haemocytes, *in vitro,* lack accessible agglutinin receptors on their surface". We recently demonstrated the presence of several lectin-binding receptors on the plasma membrane of the GR of *L. polyphemus.*[11]

In addition to hemagglutinins, several types of chemical surface receptors (some of which may in fact be hemagglutinins) have been reported in insects, primarily on the basis of the affinity of hemocytes to attach to vertebrate erythrocytes. For example, it has been shown that in *P. americana,* the attachment of the hemocytes to mammalian erythrocytes is due to trypsin-labile surface receptors.[330,361] Anderson[399,400] suggested the presence of nonspecific surface receptors on the hemocytes of *Spodoptera eridania* and *Estigmene acrea.* And it is known that in *P. brassicae,* the SPs (= specialized GRs in this species) have a layer of "sticky" acid mucopolysaccharide that probably is responsible for binding sheep erythrocytes and the bacteria, *E. coli* and *Staphylococcus aureus,* to these hemocytes.[239]

e. Role in Metamorphosis (Histogenesis) and Wound Healing

It has been suggested that lectins act as opsonins not only against foreign microorganisms but against disintegrating pieces of tissue in pupae; they also mediate cell-to-cell communication during wound healing.[247,280,377,379]

f. As Vitellogenins (?)

Stynen et al.[401] reported that in the adult Colorado potato beetle, *Leptinotarsa decemlineata,* hemagglutinins were identical to vitellogenins. Interestingly, they found that egg homogenates of this beetle showed considerable hemagglutinin activity. However, vitellogenin of *S. bullata* is not involved in hemagglutination.[377]

g. Other Functions

In addition to the above functions, lectins may mimic hormone action[375] and inhibit cell growth, cell movement, phagocytosis,[402] and delayed hypersensitivity, allograft rejection, and fertilization.[342,375]

Because of their presence in protozoa and many other invertebrate and vertebrate taxa, one can argue that they represent an ancient group of molecules and probably evolved several times in animals to serve some nonimmune functions, such as sugar transport (because they bind to sugars) and enzymatic activity.[403]

B. COMPLEMENT-LIKE FACTOR(S)

Like the antibodies, components of the complement system in vertebrates have opsonic function. According to Alberts et al.,[56] the vertebrate complement system consists of a complex of soluble proteins (M.W. 24 to 400 kDa), most of which are inactive unless stimulated by an immune response caused by a foreign antigen. There are about 20 interacting components in the complement system, designated as C1, C2, C3 . . . up to C9, factors B and D, and a variety of other proteins. Of these, the first four are called *early complement components,* and the remaining C5 through C9, the *late complement components* which form a large complex that participates in cell lysis by making the plasma membrane leaky, and thus causing it to swell and burst. The aggregation of C5 through C9 is brought about by a cascade of proteolytic activation reactions that sequentially cleave the early components (most of which are proenzymes) to generate a series of proteolytic enzymes, each of which cleaves the next proenzyme and so on. The reactions occur on the plasma membrane, and the C3 is the most important component of the proteolytic cascade. An antigen-antibody complex is generally necessary to activate the sequential reaction of the complement complex, although cobra venom factor (CVF) and bacterial endotoxin are also known to activate it.[404]

It is also interesting to note that arthropods possess at least some components of the vertebrate complement system necessary for the operation of the many immunologic reactions, such as antigen-antibody reaction and agglutination of cells. Day et al.[405] demonstrated the presence of a proactivator of the vertebrate complement component 3 (C3) in *L. polyphemus* and several other invertebrates; apparently a similar factor is present in *Blaberus craniifer*[404] and *Galleria.*[406] Because *Limulus* has only one (GR) of the two arthropod immunocytes (PLs and GRs),[22] it is reasonable to suggest that this complement factor is probably secreted by the GRs in this animal, and probably in higher arthropods as well. Using CVF, Chadwick et al.[407] suggested that *G. mellonella* might have a complement-like effector system with which the CVF interferes. Because immunoglobulins activate the complement system, it is likely that the hemagglutinins in *Limulus* activate the C3 proactivator in this arthropod. Note, however, that the hemocytic origin of C3 proactivator in *Limulus* has not yet been confirmed.

D'Cruz[408] reported three peptide components in the larval hemolymph of *B. mori* which reacted with a human complement-mediated lysis of rabbit erythrocytes (RaRBC). One of these hemolymph complement activators (HCA) (M.W. <1400 kDa) enhanced the lytic activity of the human serum, while another (of the three peptides) factor prevented this lysis, the activity of the HCA being modulated by a major hemolymph protein (MHP) (termed LSP by D'Cruz) (M.W. 500 kDa). The MHP is induced during pupation and on injury or immunization. The HCA peptide exists both in bound and free forms in the larval stages, complete release occurring *in vivo* during onset of pupation and *in vitro* by heat treatment (56°C/10 min). More importantly, the HCA acts as a specific opsonin for Protein A-positive *S. aureus*. D'Cruz et al.[409] isolated an MHP (M.W. 580 to 600 kDa), with two subunits (M.W. 90 kDa and 84 kDa) from the larval hemolymph of *Spodoptera*, which enhanced fourfold the lysis rate of sheep erythrocytes by human complement (C). According to them, "the interaction of *Spodoptera* LSP with mammalian C implicates a new function for the highly conserved, developmentally regulated, neuroendocrine-dependent storage protein of insects". D'Cruz (see Chapter 14) reported the presence of specific inhibitors of classical and alternative C pathways of mammalian C in the insect hemolymph. These protein inhibitors selectively interact at the C3 and C4 regulatory steps in the mammalian C cascade.

It has been suggested that the prophenoloxidase system functions like the mammalian complement system.[192] According to these authors, the sticky proteins produced during prophenoloxidase activation serve as opsonins in the same manner as does complement factor C3b, and have antimicrobial and lytic properties, as do complement factors.

C. ANTIMICROBIAL FACTORS

Various antibacterial and antiviral toxins have been reported in arthropods, but the direct role of immunocytes in the secretion of most of these toxins is documented in only a few cases. Hoffmann et al.[186] speculated that some of the reticular cells of the hemopoietic organs in *L. migratoria* perhaps differentiate into immunocompetent cells and synthesize the antibacterial factor.

Pupae of *H. cecropia* can be induced to produce three classes of antibacterial proteins — which were classified as *cecropins, attacins,* and *lysozymes* — on the basis of their activity;[276] see also Chapter 13. According to Götz and Boman,[410] both the cecropins and lyzosymes are necessary for complete destruction of the bacterial cell wall.

1. Cecropins

Cecropins (M.W. ca. 4 kDa) were discovered by Boman and associates when they separated them from cecropia lysozyme.[411] In earlier investigations, they were referred to as P9 proteins.[412] Seven different cecropins (designated A through F and a factor G) are known,[413] the major ones being A, B, and D. They have strong antibacterial activity against some Gram-negative bacteria. It has been suggested that insect cecropins may be related to mammalian cecropins,[414] and the high antibacterial activity is apparently owing to high proline content.[415] According to Boman and Hultmark,[416] cecropins have activity against prokaryotic membranes, but melittin (the main component of honey bee venom) is toxic to both pro- and eukaryotic cells. Cecropins and melittin are structurally similar, except that the polarity is reversed, i.e., the C-terminus in melittin is basic and the N-terminus hydrophobic. Both proteins have proline in the middle. According to Boman et al.,[411] five of the cecropins are basically similar with a "strongly basic N-terminal region and a long hydrophobic stretch in the C-terminal half". The high degree of homology among the five cecropins led these authors to suggest that they have evolved through gene duplication. Cecropins are known to affect differentiation, growth, and function of murine hemopoietic cells.[417]

2. Attacins

Attacins (M.W. 20 to 30 kDa) are considered immunologically similar to P5 proteins, described by Pye and Boman.[418] Hultmark et al.[419] showed that three of the basic N-terminals of the five (of the six known attacins) have similar sequences and the two acidic ones identical, but slightly different sequences from the basic. Boman et al.[411] suggested that the basic and acid or neutral forms are encoded by two different genes and that they also have arisen by gene duplication. They were unable to find any real functional differences between the three cecropins and the two attacins in the cecropia moth, and suggested that probably each of these five proteins has separate targets of microorganisms on which it acts. According to Engström et al.,[420] the attacins affect the outer membrane of *E. coli,* and are believed to facilitate the antibacterial activities of cecropins and lysozyme.

Apadecins from honey bees[421] and diptericin A from Diptera[422] also have antibacterial activities. These authors have extensively reviewed these factors in various insects,[276,411,416] and the reader is referred to them for details (see also Chapters 12 and 13).

3. Lysozymes

Lysozymes are mucolytic enzymes that are located in the lysosomes (acid phosphatase-positive organelles found in hemocytes, pericardial cells,[267] and ecdysial (prothoracic) glands[14]), and it is likely that hemolymph lysozymes originate from these three sources as well as from the fat body. In fact, Chain and Anderson[391] found that both immunocytes (PLs and GRs) are rich in lysosomes, and according to Zachary and Hoffmann,[202] lysozymes are stored in the GRs/COs. Lysozymes, along with the hemagglutinins and other complement-like substances in the hemolymph, play important roles in arthropod immunity. They are naturally occurring proteins and are found in the hemolymph of most Lepidoptera and other orders. Hemolymph lysozyme may be important in eliminating any symbionts that escape from the mycetocytes,[423] and is considered as a possible antibacterial factor in several insects.[424-426] Many studies have demonstrated that during phagocytosis, the lysosome fuses with the phagosome that contains the ingested microorganism.[239] According to Crossley,[245] lysozyme is readily identifiable in the hemolymph of injured *Galleria* and rises from 25 to 500 $\mu g/ml^{-1}$ to 9000 $\mu g/ml^{-1}$ in insects stressed by injection of Gram-positive bacteria. Powning and Davidson[426] purified the lysozymes from *G. mellonella* and *B. mori*. Lysozyme is also found in *C. erythrocephala,*[245] *Locusta,*[427] leafhoppers,[203] and in many other insects;[276,428] (see also Chapter 13).

Apparently, the lytic effect of lysozymes is due to their capacity to cleave the β-1,4-glycosidic bond between *N*-acetylmuramic acid and *N*-acetylglucosamine of the murein in the cell wall of intracellular symbionts. Lysozyme from the cecropia moth was isolated by Hultmark et al.,[412] and Engström et al.[429] elucidated its amino acid sequence.

a. Stimulation of Lysozyme Production

Anderson and Cook[430] reported that injection of *E. coli* lipopolysaccharide (LPS) into *S. eridania* increased the level of hemocytic lysozyme. Zachary and Hoffmann[202] showed the presence of lysozymes in the granules of GRs of *Locusta,* and the release of this enzyme, as expected, is associated with coagulation during which the GRs disintegrate and release the enzyme. Higher levels of lysozyme have been demonstrated in *G. mellonella* following immunization.[179,424,425,431]

4. Gallysin 1 and Gallysin 2

Chadwick and associates discovered these hemolysins in the normal and immune hemolymph of the larvae of *G. mellonella.* Gallysin 1 is active against sheep, human, guinea pig, and rabbit erythrocytes. The active material is a 70- to 75- kDa protein, and is different from cecropins, attacins, lysozyme, phospholipases, and Gallysin 2.[432]

Gallysin 2 is possibly secreted by the PL, and is highly toxic to murine and human tumor cell lines (for details, see Chapter 13).

5. Defensin and Hemolin

Defensin, discovered in Diptera,[433,434] is also found in rabbit macrophages and rabbit and human neutrophils;[435,436] (see also Chapter 12).

Hemolin, just reported in *Hyalophora cecropia,*[437] belongs to the immunoglobulin superfamily. Thus, these proteins (defensin and hemolin) suggest certain humoral components of the arthropod immune system are probably common to that of mammals. Hemolin was previously known as P4,[438] and is a 48-kDa protein. Although present in low amounts in the hemolymph of cecropia pupae, its concentration rises 18-fold after injection of live bacteria into the pupae. However, it does not exhibit direct bactericidal effects.[439] After characterizing cDNA of hemolin it was discovered that the deduced 413-amino acid sequence contained four internal repeats that are characteristic of immunoglobulin (Ig)-like domain. Hemolin apparently lacks a transmembrane sequence, although its immunocytochemical localization on fat body and certain hemocytes[439] suggests that it may attach to membranes. What is also interesting is its 20% relative sequence homology with mouse neural cell adhesion molecules (NCAM), and thus possibly its role in aggregation of GRs during encapsulation (see Chapter 4).

D. PROPHENOLOXIDASE-ACTIVATING SYSTEM (PAS)

Unestam and Nylund[440] reported that prophenoloxidase is stored in the GR granules of the crayfish, *A. astacus*. Later, Söderhäll and Smith[395] also concluded that phenoloxidase is present only in the GRs of the four decapod species that they studied. According to these authors, in the crayfishes, *A. astacus* and *Pacifastacus tenuiculus,* the PAS contains opsonin-like factors and can lyse crayfish hemocytes and produce fungitoxic compounds. In crustaceans, phagocytosis, encapsulation, and microbial killing are influenced by the PAS present within the hemocytes.[192,284,395,441,442] Phenoloxidase extracted from *A. astacus* hemocytes attaches to the fungal wall, Ca^{2+} being necessary for attachment,[394] as is the case for agglutinins.

Let us briefly look at the role of the prophenoloxidase-activating system (PAS) in the recognition of foreign antigen. It is well known that parasites that invade insects and other arthropods are melanized by the host's phenoloxidase (tyrosinase or phenolase); this enzyme also plays a role in wound healing. Pye[443] and later Unestam and Söderhäll[444] showed that β-1,3-glucans found on membranes of yeast and fungi produced melanization in arthropods by activating prophenoloxidase, the key enzyme in melanization reaction. In addition, the lipopolysaccharide (LPS) of the bacterial wall also activates the PAS. Apparently, these compounds activate the prophenoloxidase via a proteolytic activator (serine protease) in the hemolymph,[445] which itself needs to be activated by other proteolytic factor(s). The end result of this proteolytic action is the production of "sticky" substance(s), Ca^{2+} being necessary for the completion of the reaction. Note that high ionic concentration of Ca^{2+} inhibits the PAS,[445] whereas low concentration spontaneously activates it.[284,441] The PAS activation produces sticky proteins (opsonins), which adhere to foreign surfaces, and thus act as nonspecific foreign recognition molecules; phenoloxidase also is sticky. Serine protease is present in the hemocyte lysate (70,000 g supernatant of homogenized hemolymph) in an inactive form, and its inhibitors completely prevent activation of the proenzyme prophenoloxidase.[284] It should be mentioned here that prophenoloxidase may also be activated by other agents.[192] The PAS has been extensively studied in *B. mori*[446,447] and in the crayfish, *A. astacus*.[192,284,445]

E. SYNTHESIS AND RELEASE OF OPIOID-LIKE MATERIAL

Most recently and for the first time, Stefano et al.[12] have demonstrated the presence of an opioid-like substance in the immunocytes (GRs) of *L. maderae* and the mollusk, *M. edulis,* and suggested the existence of opiate receptors on these immunocytes, as found on vertebrate lymphocytes.[448] Furthermore, they stated that not only are the cell-mediated immune reactions (activation and directed migration of the immunocytes and their capacity for cellular adherence, as suggested by Van Epps and Saland[449] and Van Epps and Kutvirt[450]) of the invertebrate immunocytes common with those of vertebrates, but confirmed the involvement of endogenous neuropeptides in the activation of the immune system. The derivation of neuropeptides from immunocytes (as well as from neural and endocrine tissues) has been reported earlier (see reviews in References 451 and 452). According to Stefano et al.,[12] these neuropeptides function in both bidirectional interactions between the endocrine and immune systems and autoregulatory cell-cell communication among immunocytes.

F. HEMOSTASIS: COAGULATION

Wound or injury to insects immediately induces coagulation, which prevents any further loss of hemolymph, and thus maintains hemostasis.

In mammals, disintegration of platelets releases thromboplastin, which reacts with prothrombin and converts it into thrombin, which in turn converts soluble fibrinogen into insoluble fibrin. Thrombin also attracts more platelets, which adhere to fibrin and build up the clot.

It should be noted that substances such as thromboplastin, prothrombin, and thrombin have not yet been found in arthropods, although the possibility of their occurrence has been suggested.[453,454] Gilliam and Shimanuki[455] have claimed that the clotting system of the honey bee and man are similar.

In arthropods, coagulation essentially involves the proclotting enzyme (zymogen), the clotting enzyme (= serine protease), and the clottable protein (= coagulogen), which is converted to coagulin (= clot), as a result of polymerization. For the sake of convenience, the coagulation process will be discussed in terms of its two major aspects, cellular (cell involved and cellular changes during coagulation) and biochemical aspects. Coagulation in insects and other arthropods has been reviewed by Grégoire and Goffinet[24] and Bohn.[456]

1. Cellular Aspects

a. Cell Types Involved

The GRs undoubtedly are involved in coagulation in insects and other arthropods. However, SPs, PLs, and the so-called thrombocytes of Diptera have been reported to cause coagulation in various insects. In the coakroach, *Leucophaea maderae,* the PLs lyse as quickly as GRs and contribute material to the clot.[456]

b. Cellular Changes in GRs during Coagulation

Grégoire and Goffinet[24] have described four coagulation patterns (I through IV) in arthropods by phase contrast microscopy on the basis of the reactions of the GRs (their COs) and structural changes in them. In the so-called pattern I, the GRs, without lysing, discharge small amounts of cytoplasmic materials that later form circular islets around them, owing to plasma gelation. In Pattern II, the GRs extrude long, straight, thread-like, granular, cytoplasmic strands that eventually form a network in which other types of hemocytes become agglutinated (Figure 2C—E), as found in the blister beetle, *Epicauta cinerea.*[125] This network is probably formed by lipoprotein and shows "veils". Starved or diseased specimens show only a pattern II reaction. Pattern III is actually a combination of patterns I and II, both patterns occurring either separately in the same coagulating sample or superimposed on each other. In pattern IV, the GRs do not apparently undergo any changes. It seems that the changes in the labile GRs involved in coagulation are the effect rather than the cause of coagulation.[15]

The basic structural changes in the GRs at the light microscope level involve formation of vacuoles, radial expansions of cytoplasmic blebs, nuclear swelling, and exudation (Figure 2A) or explosive discharge of cytoplasmic and nuclear material without cell lysis. These changes produce some sort of granular "islands of coagulation" of clotted material that appears to enclose the GRs (Figure 2C—E). The cartwheel-like dispersal of the nuclear chromatin (Figure 2D) and the hyaline cytoplasm of the so-called cystocytes (= GRs and/or COs), are not the normal features, but effects of coagulation on the hemocyte.

Ultrastructural changes in GRs during coagulation have been described in many insects (see review by Grégoire and Goffinet[24]). These authors point out two major changes: appearance of microruptures in the plasma membrane and enlarged perinuclear cisterna of the GRs. In fact, they use these two features to distinguish what they call the COs from the GRs, although Hoffmann and Stoeckel[457] did not report any microruptures in the COs in *L. migratoria*. Another change includes the ejection of nuclear and cytoplasmic substances into the plasma from nonlysing GRs.

2. Biochemical Aspects

Durliat[235] (see also Chapter 8) provides thorough reviews of the clotting processes in Xiphosura and Crustacea. According to her, in the Horseshoe crab, *L. polyphemus*, the cellular coagulogen (M.W. ca. 20 kDa) is converted into an insoluble coagulin by a serine protease (the clotting enzyme) located in the GRs. The *L. polyphemus* coagulogen contains 220 amino acids with high half-cystine content of 18 residues.[236] The clotting enzyme (a serine protease) is remarkably similar to the enzyme that activates the prophenoloxidase for conversion into phenoloxidase.[192,236,237] It exists in two active forms (M.W. 78 and 40 kDa, respectively), with very similar amino acid composition.[236] It originates from an inactive proclotting enzyme that can be activated either by the lipopolysaccharide (LPS) moiety of the endotoxin of Gram-positive bacteria or by β-1,3-glucans from the cell walls of certain fungi and algae.[235] Endotoxin causes degranulation of the GRs in *Limulus*,[458] resulting in the release of their contents into the hemolymph.

The role of the crustacean GR in coagulation has been directly reported or indirectly interpreted by several authors (see reviews by Gupta[14,22]). Durliat and Vranckx[459] reported three soluble proteins in the hemocytes of *Astacus*, one of which they referred to as fibrinogen, which is the origin of the clottable hemolymph protein. In an arthropod, this is the first report of the presence of this protein, which is considered to be the main component of the clot in mammals. From the foregoing, it is obvious that in Crustacea as well the clottable protein (coagulogen) originates in the hemocytes, as in the case in *Limulus*.[21,235] It is of interest to note that Stang-Voss[44] recognized a separate category of clotting cells in *Astacus* (although these look like PLs or immature GRs). Crossley[245] reported that the clottable protein in several crustaceans is a PAS-positive mucoprotein or glycoprotein. Apparently, a multienzyme system, including both hemocytic and humoral factors, is involved in coagulation in Crustacea.[235] It has been suggested that the crustacean hemocytes seem to be more potent than vertebrate cells in clotting, and for a given number of cells, crustacean hemolymph coagulates 2 to 20 times faster than human blood.[460] Finally, Ghidalia et al.[461] reported that in the decapod crustacean, *Macropipus puber*, the coagulation process depends not only on a nonspecific, thermo-labile cellular factor, but also on several plasmatic factors (possibly comparable to Bohn's hemolytic coagulogen in some insects?).

Insects consist of two different kinds of coagulogens: (1) hemocytic coagulogen, which is released during cell lysis and forms the basic clot, and (2) plasma coagulogen (= Lipophorin), which further strengthens the basic clot.[456]

The hemocytic coagulogen has long been known as the only coagulogen that takes part in coagulation in insects as well as other arthropods (see reviews[1,22-24,48,235]). Evidence from literature suggests that hemocytic coagulogen is contained primarily in the granules of the GRs (including COs), but Bohn[456] has shown their presence in the SPs and PLs of the cockroach, *L. maderae*. Both structured and unstructured granules occur in this cockroach,[27] but it is not indicated whether both types of granules contain the coagulogen or only one of the two. It is believed that the structured granules contain the active material that forms the clot.[28,29,35,462] Moran[28] reported that the tubule-containing bodies (= structured granules) in *Blaberus* disappear as clotting progresses, and the presence of many structured granules correlates with high coagulability. Gupta and Sutherland[148] and Costin[153] have reported the presence of neutral or acid mucopolysaccharides and glycoproteins in the SPs and GRs, and it is likely that the latter constitute the coagulogen. The coagulum in *Locusta* is formed from a glycoprotein.[463] Note that periodic acid-Schiff-positive contents have been observed in the hemocytes of several insects.[29,169,464]

The hemolymph coagulogen or lipophorins are multifunctional lipoproteins, consisting of 30 to 50% lipids and 1 to 4% carbohydrates and involved in both lipid transport and coagulation.[465] They are synthesized in the fat body,[466] and not in the hemocytes, as shown by Bohn et al.[456] The hemocytic and hemolymph coagulogens do not cross-react immunologically.[456]

The clot in *Leucophaea* is formed by the cross-linking of both the hemocytic and hemolymph coagulogens. Note that the age of the hemocytic coagulogen is crucial in forming the clot with the hemolymph coagulogen; within about 30 min after lysis, there was almost total loss of clotting.[456] Bohn[456] attributed the loss of clottability beyond 30 min to either the formation of new disulfide bridges, which results in conformational changes in the hemocyte gel, which in turn prevent reaction with the hemolymph coagulation, or to blocking of the SH residues that are necessary for cross-linking the two coagulogens. He has suggested three models for cross-linking.

Ca^{2+} is necessary for clotting, as for prophenoloxidase activation. It induces both the lysis of the GRs as well as the action of the serine protease or the clotting enzyme, which does not seem to participate in clotting in *Leucophaea*.[456] Gregoire[467] found that sequestering Ca^{2+} ions by potassium oxalate (0.2%) prevented coagulation in *Periplaneta* and *Gryllotalpa*. In some insects, however, the blood can coagulate in the absence of Ca^{2+}.[468,469]

It should be mentioned here that an inducible protein, called M13 and produced by the epidermal and midgut cells of *Manduca sexta* (see details in Chapter 9), when introduced into naive hemolymph, triggers coagulation (which can be prevented by adding glucose to the hemolymph prior to introduction of the protein), activates PLs (indicated by the formation of filopodia), and induces their adhesion and aggregation. It also agglutinates sheep RBCs, and is considered a lectin.

3. Inhibition of Coagulation

Several kinds of metabolic inhibitors and anticoagulants are known to inhibit coagulation *in vivo* or *in vitro*. Sensitivity of coagulation mechanisms to metabolic inhibitors (e.g., cyanide and azide) and to antithrombins (e.g., heparin and hirudin) has been reported.[469,470] However, Crossley[245] did not observe inhibition of the hemocytic agglutination in *Calliphora* by certain concentrations of cyanide (10^{-3} *M*), azide (10^{-2} *M*), EDTA (ethylenediamine tetra-acetic acid) (5×10^{-3} *M*), thiourea (5×10^{-3} *M*), cycloheximide (10^{-4} *M*), or saturated aqueous cytochalasin B, when injected into the hemolymph 2 or 30 min before bleeding the insect for clotting assay. Acetic acid vapor,[471] sodium hydrosulphite, 0.1 *M* ascorbic acid,[469] X-irradiation,[472] and veronal buffer[473] are known to inhibit coagulation.

G. DETOXIFICATION OF POISONS

Virtually nothing is known about the direct role of immunocytes in the detoxification of poisons.[14] However, an indirect role of hemocytes in detoxification mechanisms via esterases in them might be involved in detoxification of parathion. Whitten[474] suggested an active role for hemocytes in detoxification of contact insecticides that penetrate the tarsal cuticle in the fly, *S. bullata*. Yeager and Munson[475] reported that hemocytes detoxify poisons. And we have shown that exposure to sublethal doses of the insecticide chlordane caused increases in the total and differential hemocyte counts of PLs, GRs (and COs), and SPs in *P. americana*.[476] Yeager and Munson,[477] however, demonstrated some pathological effects on the PLs and COs (= GRs) in *P. eridania* larvae after feeding them turnip-cornstarch "sandwiches" with sodium fluoride and mercuric chloride.

VIII. FUNCTIONAL ANALOGIES BETWEEN INSECT IMMUNOCYTES AND VERTEBRATE B- AND T-LYMPHOCYTES

In comparing the immune systems of major animal taxa, including arthropods, Du Pasquier[478] concluded that (1) "there may be molecules involved in the defense mechanisms of invertebrates suggested by functional tests and by amino acid composition comparison"; (2) "the lymphocyte of the primitive animals has properties of both B and T mammalian lymphocytes"; and (3) "a spectrum of antigens similar to that recognized by the mammalian immune system can be recognized by the immune system of small animals expressing no more than 10^4 antibody-forming clones". Thus, it is both possible and useful to point out at least some functional analogies between the insect immunocytes and the vertebrate B- and T-lymphocytes, macrophages, and human platelets on the basis of their known functions (see Section IX).

A. VERTEBRATE B- AND T-LYMPHOCYTES

The vertebrate white cells could be broadly categorized into polymorphonuclear leukocytes or granulocytes (consisting of eosinophils, basophils, and neutrophils) and mononuclear leukocytes (consisting of B- and T-lymphocytes and monocytes, which are the precursors of the macrophages). Of these leukocytes, the eosinophils, basophils, neutrophils, and monocytes (including the macrophages) are strongly phagocytic.[197] In man, both the B- and T-lymphocytes are produced in the lymphoid tissues (thymus, lymph nodes, and spleen) and can be distinguished from each other by the presence (on B-cells) and absence (on T-cells) of the immunoglobulin (Ig) molecules on their plasma membranes.[197]

B-lymphocytes or B-cells — These cells have several kinds of Igs (IgM, IgD, IgG), Fc (macrophages also have this molecule), C3 (CR1, CR2 receptors), and histocompatibility molecules in man, and Lyb antigen receptor in mice. In the resting or unactivated B-cell, 90% of the synthesized Ig molecules are surface-bound as receptors.[197] According to Unanue and Benacerraf,[197] these molecules are linked to the cell's cytoskeleton, and when antigens (ligands) bind with the Ig molecule, they form complexes that move and coalesce at one pole of the cell, causing "capping". Following capping, most of the Ig-ligand complexes are internalized by endocytosis and eventually digested, resulting in a decrease in the Ig molecules on the plasma membrane. However, afer a few hours, the cell regenerates the lost Ig receptors. Note that only when the B-cell stimulation by the Ig-ligand complex is activated by a cooperative interaction by the T-cells and/or macrophages, do the B-cells differentiate into plasma cells and synthesize and secrete antibodies, which have the same specificity as the surface Ig molecules. Note also that the binding of the antigen by the B-cell is necessary to initiate the immune response. Stimulation of the B-cells by the Ig-ligand complex results in extensive development of the endoplasmic reticulum, where the antibodies are synthesized.

T-lymphocytes or T-cells — Although the T-cells do not produce antibodies, they are as antigen-specific as the B-cells.[197] Following stimulation by the antigens, T-cells divide and differentiate into three well-defined (on the basis of the surface antigens they express) subpopulations (effector cells), which have different functions and modulate the cellular immune response in mammals.

1. Helper-Inducer T-Lymphocytes

These cells are essential for B-cell activation, division, and differentiation into plasma cells, and synthesis of antibodies; they also activate cytotoxic and suppressor T-cells, thus causing them to respond to antigens.

It is the helper T-cells, which, following activation by antigens, secrete several factors, called lymphokines or interleukins (between leukocytes), which, unlike the antibodies, do not specifically react with the antigen that induces their secretion.[56] One of these factors, the MIF (macrophage migration inhibition factor), activates the macrophages to engage in more vigorous phagocytosis and by inhibiting their migration, causes their aggregation in the region of the MIF-secreting helper T-cells. Another important lymphokine is interleukin 2 (IL-2) or T-cell growth factor, which binds to the receptors on the surface of activated T-cells and stimulates them to proliferate.

2. Suppressor T-Lymphocytes

These T-cells suppress the immune response by inactivating the B-cells and the cytotoxic T-cells, and thus maintain a balance between the two subsets of T-cells (helper T-cells and cytotoxic T-cells). The helper and suppressor cells are called regulatory T-cells because they act as regulators of the immune system. These regulatory cells recognize the special B- and T-cells that they influence by recognizing the foreign antigen attached to the interacting cell.

3. Cytotoxic T-Lymphocytes

These are also called effector cells and attack nonself tissue when activated by it. They are especially active against virus-infected cells, which they kill, thus preventing multiplication of the virus.

The three T-cell subsets can be categorized by monoclonal antibodies. For example, the helper-inducer T-cells can be labeled with OKT4 and the suppressor and cytotoxic T-cells (and their precursors) with OKT8.[479] More importantly, it is known that the ratio of helper to suppressor cells (H/S ratio), as defined by the surface antigens OKT4 and OKT8 is characteristically altered in some human diseases. For example, the H/S ratio increases in patients with active multiple sclerosis and decreases in autoimmune deficiency syndrome (AIDS).[479] According to him,[479] this ratio is not specific for a particular disease, such as AIDS, and may be altered by other factors.

B. FUNCTIONAL ANALOGIES WITH INSECT IMMUNOCYTES

The following hypotheses in this section are adopted from my earlier review,[48] with several modifications.

1. The Primitive Immunocyte: GR

On the basis of the known functions of the two arthropod immunocytes and their comparisons with those of the B- and T-lymphocytes, it can be hypothesized that the primitive GR, as is found in the most primitive arthropod, *Limulus,* as well as some higher arthropods (e.g., insects) performs some of the functions of both the B- and T-lymphocytes (Figure 29A), as well as of the macrophages and platelets. This hypothesis was recently confirmed by Miyan.[480,481] In a study of the degenerating eclosion muscles in *Glossina morsitans,*[480] he independently confirmed that degeneration was primarily mediated through the "action of a macrophage-like cell" (undoubtedly a GR; see his Figure 10b) "in a manner that resembles the attack of a specific target by T lymphocytes". He has now concluded[481] that the immunocytes are activated by neural input, because severing the nerve to the muscle results in the survival of the muscle, but inactivates the immunocytes. Earlier, Rabinovitch and De Stefano[178] had concluded, on the basis of phagocytosis of sheep erythrocytes, that the cell membranes of the phagocytes of *Galleria mellonella* and the macrophages of mice recognized foreign antigens in a similar fashion. More importantly, Stefano et al.[12] (see Section VII.E) concluded that the cell-mediated immune reactions by insect immunocytes (GRs) are similar to those by vertebrate T-lymphocytes.

Thus, the ubiquitous GR can be regarded to produce and release into the hemolymph at least five (perhaps more) factors that are functionally analogous to those of the B- and T-lymphocytes, macrophages, and platelets (Figure 29A). Factor I represents the hemag-

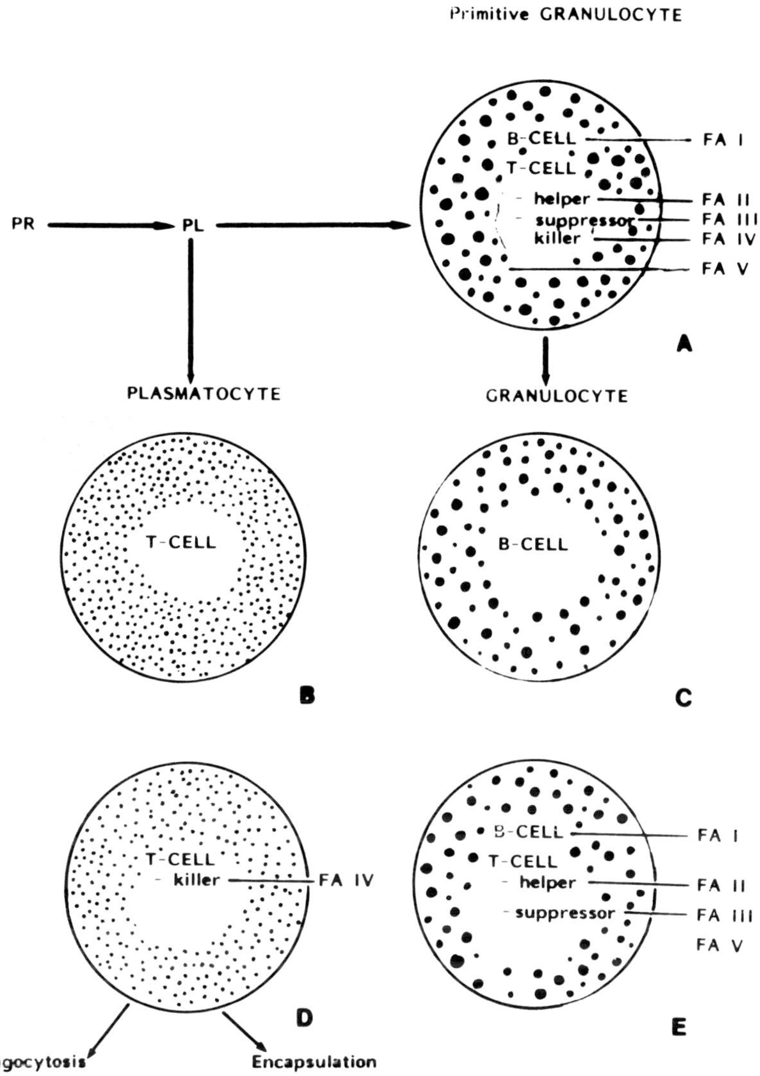

FIGURE 29. Diagram explaining the hypotheses regarding the functional analogies of arthropod immunocytes and those of vertebrate B- and T-lymphocytes, macrophages, and human platelets. (A) The most primitive granulocyte (e.g., in *Limulus*) that is functionally analogous to both B- and T-lymphocytes, as well as to macrophages, and produces at least five (FA I—V) factors. At this evolutionary stage, the plasmatocyte is evanescent and has not differentiated into a distinct hemocyte type. (B and C) The first alternative of the hypothesis, according to which the plasmatocyte (B) evolved into a distinct type, and acquired all the functions of the vertebrate T-lymphocyte, with the granulocyte (C) retaining the functions of the B-lymphocyte. (D and E) The second alternative of the hypothesis, according to which only the killer function of the primitive granulocyte (A) was shifted to the plasmatocyte (D) and all the functions of the B-lymphocyte and the helper and suppressor functions of the T-lymphocyte were retained by the granulocyte (E); thus, functionally, the granulocyte is a hybrid of B- and T-lymphocytes and of the macrophage. FA I represents functions of nonself recognition and hemagglutinin synthesis and release; FA II represents the function of causing proliferation and/or activation of plasmatocytes and/or granulocytes; FA III represents the function of maintaining a balance between FA I and FA II by suppressing their overactivities; FA IV represents the functions of activating plasmatocytes and/or granulocytes to attach to nonself tissue; FA V represents the functions of regulating the titer of hemolymph hemagglutinins, complement-like factors, and opsonins, coagulation, activation of the prophenoloxidase system, melanin production, and proteolytic enzymes, to mention a few. (Adapted from Gupta, A. P., *Hemocytic and Humoral Immunity in Arthropods*, Gupta, A. P., Ed., John Wiley & Sons, New York, 1986, 3. With permission.)

glutinins and/or opsonins that aid in the recognition of nonself tissue (comparable to the two major functions of the B-lymphocytes, namely, production of antibodies and nonself tissue recognition; see Section VIII.B). Factor II (referred to as the so-called recognition factors and lymphokine-like factors in insects) causes proliferation and/or activation of PLs (or GRs), as occurs in phagocytosis, encapsulation, and nodule formation (comparable to the functions of the helper/inducer T-lymphocytes; see Section VIII.A). Factor III (so far unidentified in arthropods) maintains the balance between factors I and II by suppressing their overactivity (comparable to the function of the suppressor T-lymphocyte). Factor IV (so far unidentified in arthropods; PAS may be a likely candidate) activates the PLs and/or GRs to attack nonself tissue (comparable to the function of the cytotoxic T-lymphocyte), and Factor V (so far identified in arthropods) regulates the levels of hemolymph hemagglutinin and/or opsonins and the occurrence of coagulation, melanin synthesis, activation of polyphenoloxidase, complement-like factors, proteolytic enzymes, and other humoral activities. As a matter of fact, according to Söderhäll and Smith,[395] recognition and phagocytosis of bacteria by the hemocytes of the decapod, *Carcinus maenas* is mediated by the prophenoloxidase-activating system, residing exclusively in the GRs.

2. The Differentiated PL and Its Functions

It is generally recognized that other types of hemocytes evolved or differentiated from the primitive GR.[14] Now, what changes occurred in the functions of the primitive GR, when it gave rise to other hemocyte types in higher arthropods? Let us consider two rather simple alternatives.

Let us suppose (first alternative) that when the PL evolved and became a distinct type, it acquired all the functions of the T-lymphocyte (Figure 29B), while the GR retained the functions of the B-lymphocyte (Figure 29C) (see Section VIII.A). If we choose this alternative, however, it would be inconsistent with the known functions of both the GRs and the PLs. From what we know of the functions of PLs in arthropods, it seems that they did not acquire all the three functions (helper, suppressor, and killer) of the T-lymphocytes, as this alternative would suggest. For example, PLs are not known to cause proliferation of GRs, instead the latter are known to cause proliferation of PLs upon lysis. Furthermore, both the PLs and GRs are known to be involved in phagocytosis and encapsulation in higher arthropods.

It seems more reasonable to suppose (second alternative) that only the killer or cytotoxic function was shifted to the PL (Figure 32D), and all the functions of the B-lymphocyte and the helper and suppressor functions of the T-cell were retained by the GR (Figure 29E). Thus, functionally, the GR is a hybrid of B- and T-lymphocytes, as well as of macrophages and platelets. It is generally accepted that, like the platelets, the GR (or CO of other authors) participates in clotting.

According to this alternative, the GRs in most arthropods perform the two important functions of the vertebrate B-lymphocytes, namely, the production of antibodies or immunoglobulins (functionally analogous to arthropod hemagglutinins) and recognition of nonself tissue. In addition, they perform the helper and suppressor functions of the T-lymphocytes. For example, we know[14] that following recognition of a foreign antigen and after contacting it, the GRs lyse and release into the hemolymph the so-called recognition factors that cause the proliferation of PLs/GRs and attract them to the foreign antigen (= comparable to helper T-cell function) to phagocytose and/or encapsulate the foreign antigen (= comparable to the killer T-cell function). The suppressor function of keeping the production of hemagglutinins and proliferation of PLs in balance by suppressing their overactivity would be necessary to maintain the effectiveness of the arthropod immune system.

Thus, the PLs (Figure 29D) perform primarily the killer function of the cytotoxic T-lymphocyte as well as the so-called HIT cells (= helper cell-independent T-cells) by phagocytosis, thus acting as sort of double-duty cells. According to Lackie,[257] PLs are "capable of recognizing and phagocytosing foreign particles without the help of granulocytes". In addition, they also act as macrophages.[399,400] When they become abnormal (as do killer T-cells in some diseases, e.g., rheumatoid arthritis), they turn against their self tissue (as in the case of the mutant malignant PLs in *Drosophila*.[482] According to Gateff,[482] mutant genes cause malignant transformation of larval PLs that fail to recognize self tissue and invade it. Matz[483] reported that tumors produced by severing recurrent nerves in *L. maderae* and *L. migratoria* were perceived as foreign tissue and were encapsulated by the GRs (not PLs).

Thus, in summary: (1) to some degree, the arthropod GRs and PLs are functionally comparable to the vertebrate (mammalian) B- and T-lymphocytes, the former being a functional hybrid of both B- and T-lymphocytes; (2) in those arthropods (e.g., *Limulus*) that possess only GRs, these cells normally perform the functions of both B- and T-lymphocytes; (3) in arthropods that possess both GRs and PLs, it seems the killer function of the T-lymphocytes has been shifted primarily to the PLs, the GRs primarily retaining the functions of the B-lymphocytes and the helper and suppressor functions of the T-lymphocytes; (4) under certain conditions (e.g., in response to a native tumor), the PLs assume the functions of HIT calls (helper cell-independent cytotoxic T-lymphocyte), as reported by Widmer and Bach[484] in mice. The PLs also perform the functions of vertebrate macrophages.

In a preliminary test, we found that the hemocyte sample from the cockroach, *G. portentosa,* treated with the monoclonal antibody OKT11, specific for the peripheral human T-lymphocytes, showed 2 to 10% positivity of the hemocytes. One important reason for the low positivity could be the fact that only short homologues of the amino acid sequences of the human T-lymphocyte antibodies are present in the surface receptors of the arthropod immunocytes. This does not sound so far-fetched if we accept the fact that immunoglobulins of higher vertebrates or mammals evolved as modifications of a common ancestral protein. And if this were true, the amino acid homology would suggest at the molecular level, the evolutionary antiquity of the origin of the defense mechanism in all animals. As a matter of fact, Alberts et al.,[56] pointing out the homology between domains of immunoglobulins, have suggested that in evolution, a primordial gene coded for a single 110-amino acid domain of unknown function, and it is from this chain that the immunoglobulin chains of higher animals evolved by a series of gene duplications. Shepherd et al.[485] showed that the homoeotic genes of the bithorax complex (BX-C) that determine the identity of the thoracic and abdominal segments and those of the antennapedia complex (ANT-C) in *D. melanogaster* share brief sequences of 60 amino acids (of the "Homoeo domain") with those of a frog, *Xenopus laevis;* and, according to them, this domain may also be present in earthworm, beetle, chicken, mouse, and human genomes. Vasta et al.[486] have reported a similar sharing of short stretches of common amino acid sequences between immunoglobulin and nonimmunoglobulin molecules.

IX. SOME STRUCTURAL AND FUNCTIONAL ANALOGIES BETWEEN INSECT IMMUNOCYTES AND HUMAN MACROPHAGES AND PLATELETS

A. MACROPHAGES

These mononuclear cells, like the arthropod GRs/PLs are highly phagocytic cells, and because they phagocytose large particles, Metchnikoff[170] called them *macrophages.* These large (15 to 20 μm in diameter) cells are found in both tissues and blood, originate from promonocytes, and after passing through the monocyte stage mature as macrophages.[197] The two major functions of these cells are phagocytosis of large antigens or particles that are coated with antibody and complement, and presentation of histocompatibility antigen molecules (class I or II) to the T-cells (see discussion on Histocompatibility, Section X).

Macrophages have two distinct Fc receptors and C3 receptors and bind to such carbohydrate residues of glycoproteins as mannose, *N*-acetylglucosamine, and fucose.[197] Note that the lectin-binding receptors that bind these carbohydrates are also present on the plasma membrane of the *L. polyphemus* GR.[11]

After binding the antigen to its surface, the macrophage internalizes it by endocytosis (an energy-requiring process) into an endocytic vesicle formed by the plasma membrane. This eventually fuses with the primary lysosome and forms phagolysosome in which the intracellular digestion of the engulfed material commences, aided by a large number of enzymes (i.e., acid phosphatases, β-glucuronidase, β-glucosaminidase, β-galactosides, acid ribonuclease, acid deoxyribonuclease, and cholesterol esterase, to mention a few) in the lysosome. Note that according to Unanue and Benacerraf,[197] lysosomes are not involved in killing bacteria, and phagocytosis by macrophages increases oxygen consumption in glucose oxidation, during which, glucose is oxidized to CO_2 and 5-carbon sugar.

In addition to their role in phagocytosis, the macrophages secrete: (1) enzymes that affect extracellular proteins, such as plasminogen activators (28 kDa and 48 kDa serine proteases that cause lysis of fibrin), collagenase and elastase (that lyse collagen and elastic tissue, respectively), and lysosomal proteases (e.g., acid phosphatase, β-glucuronidase); (2) factors involved in defense processes, such as the 14-kDa lysozyme, complement protein (C2 through C5 and Factor B), and 45-kDa interferon (a group of activated proteins); (3) factors that modulate functions of lymphocytes, such as interleukin 1 (IL-1) or lymphocyte-activating factor (15-kDa proteins that are secreted especially after interaction with bacterial products or T-cells and induce fever following infection, and immunologic reaction), some growth and differentiating factors, cytotoxins, and low molecular weight products (e.g., cAMP, prostaglandins, and purine and pyrimidine nucleosides).[197]

Because macrophages evolved earlier than lymphocytes, "their nonspecific contribution to defense against exogenous invasion—the release of enzymes, interferon, complement protein, lytic and growth factors—is protean. As lymphocytes emerged in evolution as cells capable of directing specific immune responses, the multifaceted properties of the phagocytes were adapted to the more complex requirements of specific immunity."[197]

B. PLATELETS

These non-nucleated cells are derived from nucleated megakaryocytes, measure 2 to 5 μm in diameter, initiate blood clotting, and agglutinate immediately on contact with tissues, glass, or even air.[157] Their primary role, as that of the GR, is in hemostasis; they first adhere to the injured surface (adherence being aided by the plasma membrane glycoprotein complex) and then begin to spread, which necessitates change in shape and formation of pseudopods, and rapid release of the granules and their contents (ADP, serotonin, vWF [von Willebrand Factor], and fibrinogen);[487] this platelet release reaction has been described by Holmsen.[488] Apparently, following the discharge of the granule contents, the initial layer of platelets is formed, which then interacts with other platelets and causes their aggregation, thus forming the plug.

According to Daimon and Sano-Martins,[212] the platelet aggregation can be brought about by wheat germ agglutinin (WGA)-binding glycoproteins of the plasma membrane; it is known that WGA binds to glycoproteins that are rich in sialic acids (e.g., the Ib glycoprotein complex of the platelet plasma membrane).[489] Daimon and Sano-Martins[212] have demonstrated that WGA preferentially binds to the glycoprotein I complex (GP I) of the platelet surface membrane whose O-linked oligosaccharide chains are rich in sialic acid.[489] More importantly, these authors found that, compared with those on the surface membranes of unstimulated platelets, the GP I molecules were distributed in masses on those of stimulated platelets, thus prompting these authors to conclude that perhaps translocation of the glycoproteins might occur during aggregation reaction. Note that we have demonstrated the presence of WGA-binding lectin receptors (glycoconjugates) on *L. polyphemus* GRs,[11] and also suggested elsewhere[6] that there are some structural and functional similarities between the arthropod GR and human platelet; the platelets, like the GRs, are also flattened, discoid, or slightly elliptical[490] and possess peripheral microtubule bands and intracytoplasmic granules (α and dense granules). In terms of splaying and disorganization of the microtubules in activated platelets,[491] the activated GRs show similar features.

X. HISTOCOMPATIBILITY IN TRANSPLANTATION IMMUNITY

Transplantation involves several kinds of grafts (transfer of live cells, tissues, and organs): *autograft* — from one site to another in the same individual; *isograft or syngeneic graft* — from one member to another of an inbred strain; these are not rejected and take without problems; *allograft* — among genetically disparate individuals of the same species, e.g., man; these are rejected, the rejection depending on differences in histocompatibility antigens; *xenograft* — from one species to another. These various grafts can be conveniently used to study cell-mediated immune reactions among recipients of both phylogenetically unrelated and related taxa.

It is well known that transplantation or graft reactions in animals are controlled by *histocompatibility molecules,* which are encoded by a gene segment, called major histocompatibility complex (MHC), whose genes produce cell surface glycoproteins that regulate recognition of foreign antigens by T-cells.[197] These MHC-produced molecules were discovered in studies of transplantation of allogeneic tissues and cells.

The MHC gene complex can be divided into class I and class II loci, whose products, the cell surface antigens, differ in structure, function, and other properties. The class I and II glycoproteins are also referred to as classical transplantation and Ia (= I-region-associated antigen) antigens, respectively.[197]

The most characteristic feature of these two kinds of molecules is their extraordinary polymorphism. In other words, each locus has many (20 to 40) allelic variants. In the nemertine, *Lineus* sp., however, the MHC gene is monomorphic, i.e., allelic diversity is absent,[492] and might represent a primitive stage. From the functional standpoint, it is important to emphasize that most subsets of T-cells will recognize a foreign antigen *only* when the latter is presented to them along with one of their own histocompatibility molecules on the surface of an antigen-presenting cell (e.g., monocyte or macrophage). In other words, a T-cell will not recognize a foreign antigen alone on the antigen-presenting cell from a different species expressing a different histocompatibility allele; this suggests that histocompatibility glycoproteins *restrict* (called MHC restriction) the T-cell from recognizing a foreign antigen, the restriction being that, in order to recognize the foreign antigen, it must at the same time recognize its own class I and II molecules as self.

Functionally, the class I molecules modulate interaction between cytotoxic T-cells and the target cell, while those of class II regulate the interaction between the helper T-cells and the antigen-presenting cell containing the foreign antigen. The MHC genes have not been reported in arthropods, but must exist because products of such genes (e.g., surface glycoproteins) are present on the plasma membrane of the GR of *L. polyphemus*[11] and some other arthropods.

A. MOLECULAR STRUCTURE AND SURFACE EXPRESSION OF THE MHC-PRODUCED MOLECULES

According to Alberts et al.,[56] the class I glycoproteins are encoded by at least three separate gene loci in both mice and human beings, each locus encoding a single polypeptide molecule (45 kDa with a chain length of about 345 amino acids), which is inserted in the plasma membrane. Each glycoprotein molecule consists of a short cytoplasmic COOH-terminal, a middle hydrophobic segment traversing the lipid bilayer, and a large extracellular NH_2 terminal segment, which represents about 80% of the total mass, and is folded into three separate domains. The class I glycoproteins are noncovalently associated with β_2 microglobulin, have functional preoccupation with cytotoxic T-cells,[197] and constitute about 1% of the plasma membrane proteins.[56] Alberts et al.[56] state that "the MHC glycoproteins help to channel each class of antigen into appropriate immune response pathway", thus functioning "as antigen recognition molecules distinguishing between classes of antigens". The highest concentration of these class I antigens occurs on mature B- and T-cells and macrophages. Since the glycoproteins of class I are found on virtually all nucleated cells (Alberts et al.), it is likely that they exist also on the membranes of the nucleated arthropod immunocytes.

The class II glycoproteins are 24-kDa (α chain) and 28-kDa (β chain) molecules, with a chain length of 130 (α chain) and 220 (β chain) amino acids, and have two extracellular domains and functional occupation with helper T-cells.[197] They are expressed only on lymphoid cells (e.g., B- and T-cells), while those of class I are found not only on lymphoid cells but other cells as well (e.g., RBC). Furthermore, lymphokines (secreted by helper-inducer T-cells; see Section VIII.A) promote expression of class II molecules on the antigen-presenting cells (e.g., macrophages).

B. β_2-MICROGLOBULIN-LIKE MOLECULES IN INSECTS

β_2-Microglobulin was first isolated from human urine,[493] and its precise function is unknown in mammals, although it is associated with MHC heavy chain (HLA or H_2).[494] Because Roch et al.[495] have demonstrated β_2-microglobulin-like molecules on the *Lumbricus terrestris* leukocytes by polyclonal and monoclonal antihuman β_2-microglobulin antibodies, these or similar molecules on the surface of the arthropod immunocytes probably occur.

C. TRANSPLANTATION REACTIONS IN INSECTS

In insects, both integumental and tissue grafts have been attempted with varying degrees of success, depending, among other factors, on the degree of foreignness and phylogenetic closeness of the donor and the recipient.

1. Autograft Reactivity

Reports of autografts (from one site to another in the same individual) in insects are few. It has been reported that cuticular autografts in the cockroach, *B. giganteus* and the stick insect, *Extatosoma* were not rejected.[496,497] However, hemocytic response to autografts, on the basis of hemocytic nuclei (ranging from 51 to 54) has been found on the grafts.[498,499]

2. Syngeneic (Isograft) Reactivity

All instances of implantation of corpora allata, ovaries, or other endocrine or nonendocrine tissues/organs from members of a highly inbred insect colony can be considered syngrafts, because the implanted tissues/organs are not encapsulated. In man, these will be comparable to grafts between identical twins.[500]

3. Allograft (Homograft) Reactivity

Apparently, insects show both nonrejection and rejection of allografts (among genetically disparate individuals of the same species), as evidenced by nonencapsulation and encapsulation of the grafts, respectively. Some examples of nonrejection are corpora allata in *Rhodnius;*[501] ovaries in *D. melanogaster,*[502] mosquitoes,[503] flies,[504] and cockroaches;[505] abdominal ganglia in *P. americana;*[506] nerve cord in *P. americana;*[330,358] and imaginal discs in *D. melanogaster.*[253,502,507]

Examples in which the allografts were encapsulated are those of ovarioles in *Oncopeltus fasciatus,*[504] "lymph gland" (= hemopoietic organ) tumors in *D. melanogaster,* and *Pimpla instigator* eggs by the host.[508] Because one species, *D. melanogaster,* does not encapsulate one tissue (ovaries) but encapsulates another tumor ("lymph glands"), suggests that the latter tissue is perceived as nonself.

4. Xenograft (Heterograft) Reactivity

As with allografts, insects show both nonrejection and rejection of xenografts (from one species to another). There are many instances in which the xenografts are not encapsulated: ovaries from among *Sarcophaga, Phormia, Musca,* and *Calliphora*[504] and between the species *Aedes aegypti* and *Culex molestus;*[503] whereas, corpora allata from *Tenebrio* or *Carausius,*[509] and nerve cords of *Schistocerca* and *Galleria* larvae[393] are encapsulated. This differential hemocyte response is attributed to phylogenetic relatedness, being absent in related species (e.g., Diptera) and present in taxa belonging to different orders (e.g., Coleoptera, Orthoptera, and Lepidoptera). Han and Gupta,[7] however, found that nerve cords from *P. americana* are encapsulated by the GRs of *B. germanica,* both belonging to Dictyoptera.

XI. SUMMARY

Insects and most other arthropods have six types of hemocytes (prohemocyte, plasmatocyte, granulocyte/coagulocyte, spherulocyte, oenocytoid, and adipohemocyte) that can be recognized by light (LM), scanning (SEM), and transmission (TEM) electron microscopy. Of these six, the granulocyte is the plesiomorphic type that is found in all arthropods, and is thought to have evolved into other types in higher arthropods. A key to identify the six types of hemocytes by LM has been provided. It is important to note that in this review, the category of coagulocyte has been merged with the granulocyte, because the latter performs the coagulation function of the former, and both are ultrastructurally similar. The term *immunocyte* has been used for granulocytes and plasmatocytes, because of the roles of these hemocytes in foreign antigen recognition and synthesis and release of several immunologically important humoral factors.

In insects and other arthropods, foreign antigen (nonself tissue) is recognized by the combined actions of the granulocytes (GRs) and plasmatocytes (PLs), several humoral factors (hemagglutinins, lysozymes, attacins, cecropins, Gallysin 1 and 2, defensin, hemolin, complement-like factors, and prophenoloxidase system), and coagulation. Roles of the plasma membranes and its receptors, intramembranous particles, and cytoskeleton in cellular immunity have been discussed.

Phagocytosis and encapsulation are the two most important cellular immune reactions against biotic or abiotic foreign matters. The controversial aspects of these two processes, cells involved, initial stages of the foreign matter recognition, and various stages by which the two processes are accomplished have been discussed. Encapsulation in the German cockroach, *Blattella germanica* has been described in detail. The question of cell-cell communication among the capsule-forming immunocytes has been raised and a possible mechanism via the gap junctions has been suggested. The avoidance strategies to thwart the host's immune system by parasitoids and other microorganisms that permanently reside in their insect hosts have been discussed. Hemagglutinins are the most commonly found surveillance molecules among arthropods; their origin, and functions as surface receptors and opsonin have been presented in detail.

Coagulation in insects and other arthropods essentially involves a proclotting enzyme (zymogen), the clotting enzyme (serine protease), and the clottable protein (coagulogen), which is converted to coagulin (= clot) as a result of polymerization. The cells involved in coagulation, cellular changes in them, and the biochemical aspects have been discussed.

At least in the cockroach, *Leucophaea maderae*, the clot is formed as a result of two coagulogens, one originating in the GRs/PLs and the other in the hemolymph; the hemolymph coagulogen (lipophorin) strengthens the basic clot formed by the hemocytic coagulogen. Ca ions are essential for clotting. They induce both the lysis of the GRs as well as the action of the serine protease.

Functional and some ultrastructural analogies between insect immunocytes (GRs and PLs) and the vertebrate B- and T-lymphocytes, macrophages, and platelets have been suggested in the hope of prompting a thorough search of the insect immunocyte functions comparable in precision and details to those of B- and T-lymphocytes, macrophages, and platelets. On the basis of the known insect immunocyte functions, it has been suggested that (1) to some degree, the GR is a functional hybrid of B- and T-lymphocytes; (2) in arthropods (e.g., *Limulus*) that possess only the GRs, these immunocytes normally perform the functions of both B- and T-lymphocytes, macrophages, and platelets; (3) in arthropods that possess both GRs and PLs, the killer function of T-lymphocytes seems to have shifted primarily to the PLs; (4) under certain conditions (e.g., in response to a native tumor), PLs and/or GRs perform the function of cytotoxic T-lymphocytes, both immunocytes functioning like macrophages.

Finally, histocompatibility in transplantation immunity, including examples of autograft, syngeneic graft (isograft), allograft (homograft), and xenograft (heterograft) have been presented.

ACKNOWLEDGMENT

This is New Jersey Agricultural Experiment Station Publication No. F-08125-01-91, supported by State funds and U.S. Hatch Act funds.

REFERENCES

1. **Gupta, A. P.,** The identity of the so-called crescent cell in the hemolymph of the cockroach, *Gromphadorhina portentosa* (Schaum) (Dictyoptera: Blaberidae), *Cytologia,* 50, 739, 1985.
2. **Chiang, A. S., Gupta, A. P., and Han, S. S.,** Arthropod immune system: I. Comparative light and electron microscopic accounts of immunocytes and other hemocytes of *Blattella germanica* (Dictyoptera: Blattellidae), *J. Morphol.,* 198, 257, 1988.
3. **Gupta, A. P. and Han, S. S.,** Arthropod immune system. III. Septate junctions in the hemocytic capsule of the German cockroach, *Blattella germanica, Tissue Cell,* 20(4), 629, 1988.
4. **Gupta, A. P.,** Insect host immune system and endocytobionts: their avoidance strategies, in *Insect Endocytobiosis: Morphology, Physiology, Genetics, Evolution,* Schwemmler, W., Ed., CRC Press, Boca Raton, FL, 1989, 11.
5. **Han, S. S. and Gupta, A. P.,** Immune reaction of *Blattella germanica* (L.) (Dictyoptera: Blattellidae) immunocyte (granulocyte) to foreign implants, *Proc. 18th Int. Congr. Entomol.,* 1988 (Abstr.), 253, 1988.
6. **Han, S. S. and Gupta, A. P.,** Arthropod immune system. V. Activated immunocytes (granulocytes) of the German cockroach, *Blattella germanica* (L.) (Dictyoptera: Blattellidae) show increased number of microtubules and nuclear pores during immune reaction to foreign tissue, *Cell Struct. Funct.,* 13, 333, 1988.
7. **Han, S. S. and Gupta, A. P.,** Arthropod immune system. II. Encapsulation of implanted nerve cord and "plain gut" surgical suture by granulocytes of *Blattella germanica* (L.) (Dictyoptera: Blattellidae), *Zool. Sci.,* 6, 303, 1989.
8. **Skinner, C. H., DiCicco, D. S., Kim, D., Rosser, R. J., Suckewer, S., Gupta, A. P., and Hirschberg, J. G.,** Contact microscopy with a soft X-ray laser, *J. Microsc.* (*Oxford*), 159, 51, 1990.
9. **Dunn, S. M., Shemlon, S., and Gupta, A. P.,** Interpreting transmission electron micrographs of insect hemocytes, *J. Computer Assist. Microsc.,* 1991, in press.

10. **Gupta, A. P.,** Insect hemocytes: classification and immunologic functions, in *Recent Advances in Insect Physiology and Toxicology*, Gujar, G. T., Ed., Argicole Publishing Academy, New Delhi, 1991, in press.
11. **Gupta, A. P., Orenberg, S. D., Das, Y. T., and Chattopadhyay, S. K.,** Lectin-binding receptors, Na^+, K^+-ATPase, and acetylcholinesterase on immunocyte plasma membrane of *Limulus polyphemus, Exp. Cell Res.*, 194, 83, 1991.
12. **Stefano, G. B., Leung, M. K., Zhao, X., and Scharrer, B.,** Evidence for the involvement of opioid neuropeptides in the adherence and migration of immunocompetent invertebrate hemocytes, *Proc. Natl. Acad. Sci. U.S.A.*, 86, 626, 1989.
13. **Beaulaton, J.,** Ultrastructural visualization of lectin-binding sites and anionic groups on the cell surface of immunocompetent hemocytes in Oak silkworm, *Antheraea pernyi, Dev. Comp. Immunol.*, 9(1) (Abstr.), 164, 1985.
14. **Gupta, A. P.,** Cellular elements in the hemolymph, in *Comprehensive Insect Physiology, Biochemistry and Pharmacology*, Vol. 3, Kerkut, G. A. and Gilbert, L. I., Eds., Pergamon Press, Oxford, 1985, 401.
15. **Gupta, A. P. and Sutherland, D. J.,** *In vitro* transformations of the insect plasmatocyte in some insects, *J. Insect Physiol.*, 12, 1369, 1966.
16. **Howell, W. H.,** Observations upon the chemical composition and coagulation of the blood of *Limulus polyphemus, Callinectes hastatus*, and *Cucumaria* sp., *Johns Hopkins Univ. Cir.*, 5, 4, 1885.
17. **Loeb, L.,** On the spontaneous agglutination of blood cells of arthropods, *Univ. Penn. Med. Bull.*, 16, 441, 1904.
18. **Loeb, L.,** Uber die Blutgerinnung bei Wirbellosen, *Biochem. Z.*, 24, 478, 1910.
19. **Maluf, N. S. R.,** The blood of arthropods, *Q. Rev. Biol.*, 14, 149, 1939.
20. **Dumont, J. N., Anderson, E., and Winner, G.,** Some cytologic characteristics of the hemocytes of *Limulus* during clotting, *J. Morphol.*, 119, 181, 1966.
21. **Murer, E. H., Levin, J., and Holme, R.,** Isolation and studies of the granules of the amoebocytes of *Limulus polyphemus* the horseshoe crab, *J. Cell Physiol.*, 86, 533, 1975.
22. **Gupta, A. P.,** Arthropod hemocytes and phylogeny, in *Arthropod Phylogeny*, Gupta, A. P., Ed., Van Nostrand Reinhold, New York, 1979, 669.
23. **Gupta, A. P.,** Identification key for hemocyte types in hangingdrop preparations, in *Insect Hemocytes*, Gupta, A. P., Ed., Cambridge University Press, Cambridge, 1979, 527.
24. **Grégoire, Ch. and Goffinet, G.,** Controversies about the coagulocyte, in *Insect Hemocytes*, Gupta, A. P., Ed., Cambridge University Press, Cambridge, 1979, 189.
25. **Barracco, M. A. and Loch, C. T.,** Ultrastructural studies of the hemocytes of *Panstrongylus megistus* (Hemiptera: Reduviidae), *Mem. Inst. Oswaldo Cruz*, Rio de Janeiro, 84(2), 171, 1989.
26. **Baerwald, R. J. and Bousch, G. M.,** Fine structure of hemocytes of *Periplaneta americana* (Orthoptera: Blattidae), with particular reference to marginal bundles, *J. Ultrastruct. Res.*, 31, 151, 1970.
27. **Hagopian, M.,** Unique structures in the insect granular hemocytes, *J. Ultrastruct. Res.*, 36, 646, 1971.
28. **Moran, D. T.,** The fine structure of cockroach blood cells, *Tissue Cell*, 3, 413, 1971.
29. **Scharrer, B.,** Cytophysiological features of haemocytes in cockroaches, *Z. Zellforsch.*, 129, 301, 1972.
30. **Raina, A. K.,** Ultrastructure of the larval haemocytes of the pink bollworm *Pectinophora gossypiella* (Lepidoptera: Gelechiidae), *Int. J. Insect Morphol. Embryol.*, 5, 187, 1976.
31. **Devauchelle, G.,** Étude ultrastructure des hémocytes du Coléoptère *Melolontha melolontha* (L.), *J. Ultrastruct. Res.*, 34, 492, 1971.
32. **Landureau, J. C. and Grellet, P.,** New permanent cell lines from cockroach hemocytes: Physiological and ultrastructural characteristics, *J. Insect Physiol.*, 21, 137, 1975.
33. **Hoffmann, J. A., Stoekel, M. E., Porte, A., and Joly, P.,** Ultrastructure des hemocytes de *Locusta migratoria* (Orthoptère), *C.R. Hebd. Séances Acad. Sci.* Paris, 266D, 503, 1968.
34. **Hoffmann, J. A., Porte, A., and Joly, P.,** On the localization of phenoloxidase activity in coagulocytes of *Locusta migratoria* L. (Orthoptera), *C.R. Acad. Sci.* Paris, 270D, 629, 1970.
35. **Stang-Voss, C.,** Zur Ultrastrucktur der Blutzellen wirbelloser Tiere. I. Über die Hämocyten der Larve des Mehlkafers *Tenebrio molitor, Z. Zellforsch.*, 103, 589, 1970.
36. **Ratcliffe, N. A. and Price, C. D.,** Correlation of light and electron microscope hemocyte structure in the Dictyoptera, *J. Morphol.*, 144, 485, 1974.
37. **Akai, H. and Sato, S.,** Ultrastructure of the larval hemocytes of the silkworm, *Bombyx mori* L. (Lepidoptera:Bombycidae), *Int. J. Insect Morphol. Embryol.*, 2, 207, 1973.
38. **Rowley, A. F. and Ratcliffe, N. A.,** Insects, in *Invertebrate Blood Cells*, Vol. 2, Ratcliffe, N. A. and Rowley, A. F., Eds., Academic Press, New York, 1981, 421.
39. **Neuwirth, M.,** The structure of the hemocytes of *Galleria mellonella* (Lepidoptera), *J. Morphol.*, 139(1), 105, 1973.
40. **Costin, N. M.,** Histochemical observation of the haemocytes of *Locusta migratoria, Histochem. J.*, 7, 21, 1975.
41. **François, J.,** Étude ultrastructurale des hémocytes du Thysanoure *Thermobia domestica* (Insecte, Aptérygote), *Pedobiologia*, 14, 157, 1974.

42. **François, J.,** Hémocyte et organe hématopoïètique de *Thermobia domestica* (Packard) (Thysanura: Lepismatidae). *Int. J. Insect Morphol. Embryol.,* 4(6), 477, 1975.
43. **Barracco, M. A., Oliveira, A., and Schlemper, B., Jr.,** The hemocytes of *Panstrongylus megistus* (Hemiptera: Reduviidae), *Mem. Inst. Oswaldo Cruz,* Rio de Janeiro, 82, 431, 1987.
44. **Stang-Voss, C.,** Zur Ultrastruktur der Blutzellen wirbelloser Tiere. V. Über die Hämocyten von *Astacus astacus* (Crustacea), *Z. Zellforsch.,* 122, 68, 1971.
45. **Amirante, G. A.,** Agglutinins and lectins of Crustacea, in *Hemocytic and Humoral Immunity in Arthropods,* Gupta, A. P., Ed., John Wiley & Sons, New York, 1986, 359.
46. **Yeaton, R. W.,** Occurrence of nonlymphatic hemagglutinins in Arthropods and their possible functions, in *Hemocytic and Humoral Immunity in Arthropods,* Gupta, A. P., Ed., John Wiley & Sons, New York, 1986, 505.
47. **Arnold, J. W.,** Controversies about hemocyte types in insects, in *Insect Hemocytes,* Gupta, A. P., Ed., Cambridge University Press, Cambridge, 1979, 231.
48. **Gupta, A. P.,** Arthropod immunocytes: identification, structure, functions, and analogies to the functions of vertebrate B- and T-lymphocytes, in *Hemocytic and Humoral Immunity in Arthropods,* Gupta, A. P., Ed., John Wiley & Sons, New York, 1986, 3.
49. **Brehélin, M. and Zachary, D.,** Insect hemocytes: A new classification to rule out the controversy, in *Immunity in Invertebrates,* Brehélin, M., Ed., Springer-Verlag, Berlin, 1986, 36.
50. **Rowley, A. F.,** The role of the haemocytes of *Clitumnus extradentatus* in haemolymph coagulation, *Cell Tissue Res.,* 182(4), 513, 1977.
51. **Baerwald, R. J.,** Inverted gap and other cell junctions in cockroach hemocyte capsules: a thin section and freeze-fracture study, *Tissue Cell,* 7(3), 575, 1975.
52. **Baerwald, R. J.,** Fine structure of hemocyte membranes and intercellular junctions formed during hemocyte encapsulation, in *Insect Hemocytes,* Gupta, A. P., Ed., Cambridge University Press, Cambridge, 1979, 155.
53. **Shen, B. W., Josephs, R., and Steck, T. L.,** Ultrastructure of unit fragments of the skeleton of the human erythrocyte membrane, *J. Cell Biol.,* 99, 810, 1984.
54. **Darnell, J., Lodish, L., and Baltimore, D.,** *Molecular Cell Biology,* W. H. Freeman, New York, 1986.
55. **Steck, T. L.,** Biological membranes, in *Biochemistry and Biology of Plasma Lipoproteins,* Scanu, A. M. and Spector, A. A., Eds., Marcel Dekker, New York, 1986, 145.
56. **Alberts, B., Bray, D., Lewis, J., Raff, M., Roberts, K., and Watson, J. D.,** *Molecular Biology of the Cell,* 2nd ed., Garland Publishing, New York, 1989.
57. **Wallach, D. F. H.,** *Fundamentals of Receptor Molecular Biology,* Marcel Dekker, New York, 1987.
58. **Bennett, M. V. L. and Spray, D. C., Eds.,** *Gap Junctions,* Cold Spring Harbor Laboratory, New York, 1985.
59. **Hanna, R. B., Keefer, J. S., and Pappas, G. P.,** The fine structure of a rectifying electronic synapse, *J. Cell Biol.,* 79, 764, 1978.
60. **Yahara, I. and Edelman, G. M.,** Electron microscopic analysis of the modulation of lymphocyte receptor mobility, *Exp. Cell Res.,* 91, 125, 1975.
61. **Pereira, M. E. A., Loures, M. A., Villalta, F., and Andrade, A. F. B.,** Lectin receptors as markers for *Trypanosoma cruzi, J. Exp. Med.,* 152, 1375, 1980.
62. **Roch, P. and Valembois, P.,** Evidence for concanavalin A-receptors and their redistribution on lumbricoid leucocytes, *Dev. Comp. Immunol.,* 2, 51, 1978.
63. **Schoenberg, D. A. and Cheng, T. C.,** Lectin-binding specificities of hemocytes from two strains of *Biomphalaria glabrata* as demonstrated by microadsorption assays, *Dev. Comp. Immunol.,* 4, 617, 1980.
64. **Kehoe, J. M. and Seide, R. K.,** Comparative structural studies of limulin, in *Hemocytic and Humoral Immunity in Arthropods,* Gupta, A. P., Ed., John Wiley & Sons, New York, 1986, 345.
65. **Burger, M. M. and Goldberg, A. R.,** Identification of a tumor-specific determinant on neoplastic cell surfaces, *Proc. Natl. Acad. Sci. U.S.A.,* 57, 359, 1967.
66. **Klein, P. J., Vierbuchen, M., Schulz, K. G., Wurz, H., Citoler, P., Uhlenbruch, G., Ortmann, M., and Fisher, R.,** Hormone-dependent lectin binding site. 2. Lectin receptors as an indicator of hormone-sensitive mammary carcinomas, *Tumor Diagnostik,* 2, 240, 1981.
67. **Cohen, E.,** Detection of membrane receptors by arthropod agglutinins, in *Hemocytic and Humoral Immunity in Arthropods,* Gupta, A. P., Ed., John Wiley & Sons, New York, 1986, 493.
68. **Gupta, A. P., Ed.,** *Hemocytic and Humoral Immunity in Arthropods,* John Wiley & Sons, New York, 1986.
69. **Pendland, J. C. and Boucias, D. G.,** Lectin binding characteristics of several entomogenous hyphomycetes: possible relationship to insect hemagglutinins, *Mycologia,* 78(5), 818, 1986.
70. **Schreiner, G. F. and Unanue, E. R.,** Membrane and cytoplasmic changes in B lymphocytes induced by ligand-surface immunoglobulin interaction, *Adv. Immunol.,* 24, 37, 1976.
71. **Dustin, P.,** *Microtubules,* Springer-Verlag, Berlin, 1984.
72. **Stein, W. D. and Bronner, F., Eds.,** *Cell Shape,* Academic Press, New York, 1989.

73. **Bennett, V.**, The molecular basis for membrane-cytoskeleton association in human erythrocytes, *J. Cell Biochem.*, 18, 49, 1982.
74. **Boyles, J., Fox, J. E. B., and Phillips, D. R.**, The filamentous cytoskeleton of the discoid platelet, *J. Cell Biol.*, 97, 274a, 1983.
75. **Bray, D. and White, J. G.**, Cortical flow in animal cells, *Science*, 239, 883, 1988.
76. **Lloyd, C. W., Hyams, J. S., and Warn, R. M., Eds.**, *Cytoskeleton: Cell Function and Organization*, (suppl. 5 of *J. Cell Sci.*), Company of Biologists, Ltd. Cambridge, 1986.
77. **Shay, J. W., Ed.**, *Cell and Molecular Biology of the Cytoskeleton*, Plenum Press, New York, 1986.
78. **Goldman, R. D. and Steinert, P. M., Eds.**, *Cellular and Molecular Biology of the Intermediate Filaments*, Plenum Press, New York, 1990.
79. **Hwang, I-C., Han, S. S., and Gupta, A. P.**, Cytoskeleton of the immunocyte (granulocyte) of the German cockroach, *Blattella germanica* (L.), *Exp. Cell Res.*, 1991, submitted.
80. **Dales, S.**, Concerning the universality of a microtubule antigen in animal cells, *J. Cell Biol.*, 52, 748, 1972.
81. **Durrer, A., Pallas, D., and Solomon, F.**, Molecular analysis of cytoplasmic microtubules *in situ:* identification of both widespread and specific proteins, *Cell*, 24, 203, 1981.
82. **Cohen, W. D. and Nemhauser, I.**, Marginal bands and cytoskeleton in blood cells of marine invertebrates, in *Blood Cells of Marine Invertebrates*, Cohen, W. D., Ed., Alan R. Liss, New York, 1985, 3.
83. **Bernier-Valentin, F. and Rousset, B.**, Interaction of tubulin with liver mitochondria, *J. Biol. Chem.*, 257, 7092, 1982.
84. **Bhattacharya, B. and Wolff, J.**, Polymerization of membrane tubulin, *Nature* (*London*), 264, 756, 1976.
85. **François, J.**, L'encapsulation hémocytaire experimentale chez le lépisme *Thermobia domestica*, *J. Insect Physiol.*, 21, 1535, 1975.
86. **Brehélin, M., Hoffmann, J. A., Matz, G., and Porte, A.**, Encapsulation of implanted foreign bodies by hemocytes in *Locusta migratoria* and *Melolontha melolontha*, *Cell Tissue Res.*, 160, 283, 1975.
87. **Weisenberg, R. C.**, Microtubule formation *in vitro* in solutions containing low calcium concentrations, *Science*, 177, 1104, 1972.
88. **Fulton, C. and Simpson, P. A.**, Tubular pools, synthesis and utilization, in *Microtubules*, Roberts, K. and Hyams, J. J., Eds., Academic Press, New York, 1979, 117.
89. **Olmstead, J. B.**, Tubulin pools in differentiating neuroblastoma cells, *J. Cell Biol.*, 89, 418, 1981.
90. **Ben-Zev'ev, A., Farmer, S. R., and Penman, S.**, Mechanisms of regulating tubulin synthesis in cultured mammalian cells, *Cell*, 17, 319, 1981.
91. **Cleveland, D. W.**, Molecular mechanisms controlling tubulin synthesis, in *Cell and Molecular Biology of the Cytoskeleton*, Shay, S. J. W., Ed., Plenum Press, New York, 1980, 203.
92. **Cleveland, D. W., Lopata, M. A., Sherline, P., and Kirschner, M. W.**, Unpolymerized tubulin modulates the level of tubulin mRNA, *Cell*, 25, 537, 1981.
93. **Cleveland, D. W. and Kirschner, M.**, Autoregulatory control of the expression of α- and β-tubulin: implications for microtubule assembly, *Cold Spring Harbor Symp. Quant. Biol.*, 46, 171, 1982.
94. **Borgers, M., De Brabander, M., Van Reempts, S., Wouters, W. F., and Jacobs, W. A.**, Intranuclear microtubules in lung mast cells of guinea-pigs in anaphylactic shock, *Lab. Invest.*, 37, 1, 1977.
95. **Bergström-Porter, B. and Shelton, E.**, Effect of uncouplers of oxidative phosphorylation on microtubule location and surface structure in murine mast cells, *Anat. Rec.*, 195, 375, 1979.
96. **Wadsworth, P. and Sloboda, R. D.**, Microinjection of fluorescent tubulin into dividing sea urchin cells, *J. Cell Biol.*, 97, 1249, 1983.
97. **Weeks, D. P. and Collis, P. S.**, Induction and synthesis of tubulin during the cell cycle and life cycle of *Chlamydomonas reinhardi*, *Dev. Biol.*, 69, 400, 1979.
98. **Pipeleers, D. G., Pipeleers-Marichal, M. A., and Kipnis, D. M.**, Physiological regulation of total tubulin and polymerized tubulin in tissues, *J. Cell Biol.*, 74, 351, 1977.
99. **Pipeleers, D. G., Pipeleers-Marichal, M. A., Sherline, P., and Kipnis, D. M.**, A sensitive method for measuring polymerized and depolymerized forms of tubulin in tissues, *J. Cell Biol.*, 74, 341, 1977.
100. **Sherline, P. and Mundy, G. R.**, Role of the tubulin-microtubule system in lymphocyte activation, *J. Cell Biol.*, 74, 371, 1977.
101. **Nemhauser, I., Joseph-Silverstein, J., and Cohen, W. D.**, Centrioles as microtubule-organizing centers for marginal bands of molluscan erythrocytes, *J. Cell Biol.*, 96, 979, 1983.
102. **Margolis, R. L. and Wilson, L.**, Microtubule treadmills — possible molecular machinery, *Nature* (*London*), 293, 705, 1981.
103. **Karr, T. L. and Purich, D. L.**, A microtubule assembly disassembly model based on drug effects and depolymerization kinetics after rapid dilution, *J. Biol. Chem.*, 254(21), 10885, 1979.
104. **Allen, C. and Borisy, G. G.**, Structural polarity and directional growth of microtubules of *Chlamydomonas* flagella, *J. Mol. Biol.*, 90, 381, 1974.
105. **Bergen, L. G., Kuriyama, R., and Borisy, G. G.**, Polarity of microtubules nucleated by centrosomes and chromosomes of Chinese hamster ovary cells *in vitro*, *J. Cell Biol.*, 84, 151, 1980.

106. **Borisy, G. G. and Bergen, L. G.**, A direct method for analyzing the polymerization kinetics at the two ends of a microtubule, *Methods Cell Biol.*, 24, 171, 1982.
107. **Jordan, M. A. and Farrell, K. W.**, Differential radiolabeling of opposite microtubule ends: methodology, equilibrium exchange-flux analysis, and drug poisoning, *Anal. Biochem.*, 130(1), 41, 1983.
108. **Wilson, L., Bamburg, J. R., Mizel, S. B., Grisham, L. M., and Creswell, K. M.**, Interaction of drugs with microtubule proteins, *Fed. Proc.*, 33, 158, 1974.
109. **Margolis, R. L., Rauch, C. T., and Wilson, L.**, Mechanisms of colchicine-dimer addition to microtubule ends: implications for the microtubule polymerization mechanisms, *Biochemistry*, 19, 5550, 1980.
110. **Schliwa, M.**, *The Cytoskeleton: An Introductory Survey*, Springer-Verlag, Berlin, 1986.
111. **Wright, M., Moisand, A., and Oustrin, M. L.**, Stabilization of monoasters by taxol in amebae of *Physarum polycephalum* (Myxomycetes), *Protoplasma*, 113, 44, 1982.
112. **Glenney, J. R., Glenney, P., and Weber, K.**, Erythroid spectrin, brain fodrin and intestinal brush border proteins (TW 260/240) are related molecules containing a common calmodulin-binding subunit bound to a variant cell type specific subunit, *Proc. Natl. Acad. Sci. U.S.A.*, 79, 4002, 1982.
113. **Osborn, M., Webster, R. E., and Weber, K.**, Individual microtubules viewed by immunofluorescence and electron microscopy in the same PtK2 cell, *J. Cell Biol.*, 77, R27, 1978.
114. **Albertini, D. F., Berlin, R. D., and Martin, J. M.**, The mechanism of concanavalin A cap formation in leucocytes, *J. Cell Sci.*, 26, 57, 1977.
115. **Gabbiani, G., Chaponnier, C., Zumbe, A., and Vassalli, P.**, Actin and tubulin co-cap with surface immunoglobulins in mouse B lymphocytes, *Nature (London)*, 269, 697, 1977.
116. **Bourguignon, L. Y., Tokuyasu, K. T., and Singer, S. J.**, The capping of lymphocytes and other cells, studied by an improved method for immunofluorescence staining of frozen sections, *J. Cell Physiol.*, 95, 239, 1978.
117. **Bourguignon, L. Y. and Singer, S. J.**, Transmembrane interactions and the mechanism of capping of surface receptors by their specific ligands, *Proc. Natl. Acad. Sci. U.S.A.*, 74, 5031, 1977.
118. **Fischer, T. M.**, A comparison of the flow behavior of disc-shaped versus elliptic red blood cells (RBC), *Blood Cells*, 4, 453, 1978.
119. **Behnke, O.**, Microtubules in disk-shaped blood cells. *Int. Rev. Exp. Pathol.*, 9, 1, 1970.
120. **Ikeda, Y.**, The study of platelet microtubule assembly, *Keio J. Med.*, 25(4), 139, 1976.
121. **Joseph-Silverstein, J. and Cohen, W. D.**, Taxol-induced stabilization and reassembly of functional marginal bands in erythrocytes at 0°C, *J. Cell Biol.*, 97, 212a, 1983.
122. **Arnold, J. W.**, The hemocytes of insects, in *The Physiology of Insecta*, Vol. 5, 2nd ed., Rockstein, M., Ed., Academic Press, New York, 1974, 201.
123. **Yeager, J. F.**, The blood picture of the southern armyworm *(Prodenia eridania), J. Agric. Res.*, 71, 1, 1945.
124. **Jones, J. C.**, A phase contrast study of the blood cells in *Prodenia* larvae (Order Lepidoptera), *Q. J. Microsc. Sci.*, 100(10), 17, 1959.
125. **Gupta, A. P.**, Studies of the blood of Meloidae (Coleoptera). I. The haemocytes of *Epicauta cinerea* (Forster), and a synonymy of haemocyte terminologies, *Cytologia*, 34(2), 300, 1969.
126. **Rizki, M. T. M.**, Experimental analysis of hemocyte morphology in insects, *Am. Zool.*, 2, 247, 1962.
127. **Rowley, A. F. and Ratcliffe, N. A.**, An ultrastructural study of the *in vitro* phagocytosis of *Escherichia coli* by the hemocytes of *Calliphora erythrocephala, J. Ultrastruct. Res.*, 55, 193, 1976.
128. **Zachary, D. and Hoffmann, J. A.**, The haemocytes of *Calliphora erythrocephala* (Meig) (Diptera), *Z. Zellforsch.*, 141, 55, 1973.
129. **Barracco, M. A. and Cestari, A. N.**, The ultrastructure of the larval hemocytes of *Trichosia pubescens* (Diptera: Sciaridae), *Rev. Bras. Genet.*, 10(3), 435, 1987.
130. **Zachary, D., Brehélin, M., and Hoffmann, J. A.**, Role of "thrombocytoids" in capsule formation in the dipteran *Calliphora erythrocephala, Cell Tissue Res.*, 162, 343, 1975.
131. **Gupta, A. P. and Sutherland, D. J.**, Observations on the spherule cells in some Blattaria (Orthoptera), *Bull. Entomol. Soc. Am.*, 11 (Abstr.), 161, 1965.
132. **Lea, M. S. and Gilbert, L. I.**, The hemocytes of *Hyalophora cecropia* (Lepidoptera), *J. Morphol.*, 118(2), 197, 1966.
133. **Gupta, A. P.**, Hemocyte types: their structures, synonymies, interrelationships, and taxonomic significance, in *Insect Hemocytes*, Gupta, A. P., Ed., Cambridge University Press, Cambridge, 1979, 85.
134. **Hazarika, L. K. and Gupta, A. P.**, Variation in hemocyte populations in various developmental stages of *Blattella germanica* (L.) (Dictyoptera: Blattellidae), *Zool. Sci.*, 4, 307, 1987.
135. **Gupta, A. P.**, Origin and affinities of Myriapoda, in *Myriapod Biology*, Camatini, M., Ed., Academic Press, New York, 1979, 373.
136. **Goffinet, G. and Grégoire, Ch.**, Coagulocyte alterations in clotting hemolymph of *Carausius morosus* L., *Arch.. Int. Physiol. Biochem.*, 83 (4), 707, 1975.
137. **Millara, P.**, Contribution a l'étude cytologique et physiologique des leucocytes d'Insectes, *Bull. Biol. Fr. Belg.*, 81, 129, 1947.

138. **Cuénot, L.**, Études physiologiques sur les Orthoptères, *Arch. Biol.*, 14, 293, 1896.
139. **Hollande, A. C.**, Contribution á l'étude du sang des Coléoptères, *Arch. Zool. Exp. Gén.* (Ser 5), 21, 271, 1909.
140. **Hollande, A. C.**, Étude histologiques comparée du sang des insectes á hémorrhèe et des insectes sans hémorrhèe, *Arch. Zool. Exp. Gén.* (Ser 5), 6, 283, 1911.
141. **Wigglesworth, V. B.**, *The Principles of Insect Physiology*, 1st ed., Methuen, London, 1939.
142. **Wigglesworth, V. B.**, Insect blood cells, *Annu. Rev. Entomol.*, 4, 1, 1959.
143. **Jones, J. C.**, Current concepts concerning insect hemocytes, *Am. Zool.*, 2, 209, 1962.
144. **Arvy, L. and Lhoste, J.**, Les variation du leucogramme au cours de la métamorphose chez *Forficula auricularia* L., *Bull. Soc. Zool. Fr.*, 70, 114, 1946.
145. **Andrew, W.**, *Comparative Hematology*, Grune & Stratton, New York, 1965.
146. **Harpaz, F., Kislev, N., and Zelcer, A.**, Electron-microscopic studies on hemocytes of the Egyptian cottonworm, *Spodoptera littoralis* (Boisduval) infected with a nuclear-polyhedrosis virus, as compared to noninfected hemocytes. I. Noninfected hemocytes, *J. Invertebr. Pathol.*, 14, 175, 1969.
147. **Beaulaton, J. and Monpeyssin, M.**, Ultrastructure et cytochimie des hemocytes d'*Antheraea pernyi* Guer. (Lepidoptera: Attacidae). II. Cellules a spherules et oenocytoides, *Biol. Cell.*, 28(1), 13, 1977.
148. **Gupta, A. P. and Sutherland, D. J.**, Phase contrast and histochemical studies of spherule cells in cockroaches (Dictyoptera), *Ann. Entomol. Soc. Am.*, 60(3), 557, 1967.
149. **Dennell, R.**, A study of the insect cuticle: The formation of the puparium of *Sarcophaga falculata* Pand. (Diptera), *Proc. R. Soc.*, B. 134, 79, 1947.
150. **Jones, J. C.**, DDT and the hemocyte picture of the mealworm, *Tenebrio molitor* L., *J. Cell Comp. Physiol.*, 50, 423, 1957.
151. **Srdić, Ž., and Gloor, H.**, Spherule cells in *Drosophila* species, *Experientia*, 35, 1226, 1979.
152. **Ashhurst, D. E.**, Histochemical properties of the spherulocytes of *Galleria mellonella* L. (Lepidoptera: Pyralidae), *Int. J. Insect Morphol. Embryol.*, 11, 285, 1982.
153. **Nittono, Y.**, Studies on the blood cells in the silkworm, *Bombyx mori* L., *Bull. Seric. Exp. Stn. Tokyo*, 16, 171, 1960 (in Japanese, English summary).
154. **Gupta, A. P.**, Midgut lesions in *Epicauta cinerea* (Coleoptera: Meloidae), *Ann. Entomol. Soc. Am.*, 63, 1786, 1970.
155. **Akai, H. and Sato, S.**, Surface and internal ultrastructure of the hemocytes of some insects, in *Insect Hemocytes*, Gupta, A. P., Ed., Cambridge University Press, Cambridge, 1979, 129.
156. **Cook, D., Stolz, D. B., and Pauley, C.**, Purification and preliminary characterization of insect spherulocytes, *Insect Biochem.*, 15(3), 419, 1985.
157. **Bessis, M.**, *Living Blood Cells and Their Ultrastructure*, Springer-Verlag, Berlin, 1973 (translated by R. I. Weed).
158. **Serafini-Fracassini, A., Durward, J. J., and Crawford, J.**, The morphology of the heparin-protein macromolecule and its organization in the mast cell granule, *J. Ultrastruct. Res.*, 28, 131, 1969.
159. **Smith, S. L. and Smith, A. C.**, Sensitization and histamine release by cells of the sand dollar, *Mellita quinquiesperforata*, *Dev. Comp. Immunol.*, 9, 597, 1985.
160. **Chaigneau, J.**, Neurohemal organs in Crustacea, in *Neurohemal Organs of Arthropods*, Gupta, A. P., Ed., Charles C. Thomas, Springfield, IL, 1983, 53.
161. **Scharrer, B.**, Current views on the mechanisms of release of neurosecretory products, in *Neurohemal Organs of Arthropods*, Gupta, A. P., Ed., Charles C. Thomas, Springfield, IL, 1983, 602.
162. **Klemm, N.**, Identification of monoamine-containing neurons in arthropods, in *Arthropod Brain*, Gupta, A. P., Ed., John Wiley & Sons, New York, 1987, 529.
163. **Strolenberg, G. E. C. M. and Van Herp, F.**, Mise en évidence du phénomene d'éxocytose dans la glande du sinus d'*Astacus leptodactylus* (Nordmann), *Cell Tissue Res.*, 180, 203, 1977.
164. **Ritter, H., Jr.**, Blood of a cockroach: unusual cellular behavior, *Science*, 147, 518, 1965.
165. **Scharrer, B.**, The fine structure of an unusual hemocyte in the insect *Gromphadorhina portentosa*, *Life Sci.*, 4, 1741, 1965.
166. **Wigglesworth, V. B.**, Hemocytes and growth in insects, in *Insect Hemocytes*, Gupta, A. P., Ed., Cambridge University Press, Cambridge, 1979, 303.
167. **Wigglesworth, V. B.**, Secretory activities of plasmatocytes and oenocytoids during the moulting cycle in an insect *(Rhodnius)*, *Tissue Cell*, 11(1), 69, 1979.
168. **Ashida, M. and Yoshida, H.**, Activation of two zymogens of proteases by peptidoglycan or β-1, 3-glucan and the conversion of prophenoloxidase into phenoloxidase in insect plasma, *Dev. Comp. Immunol.*, 9(1)(Abstr.), 166, 1985.
169. **Ashhurst, D. E. and Richards, A. G.**, Some histochemical observations on the blood cells of the wax-moth *Galleria mellonella* L., *J. Morphol.*, 114, 247, 1964.
170. **Metchnikoff, E.**, *Leçons sur la Pathologie Comparée de l'Inflammation*, Masson et cie, Paris, 1892. (Reprinted by Dover Publications, New York, 1968, as ''Lectures on the Comparative Pathology of Inflammation''.)

171. **Herbert, D. and Bardossi, F.,** *Secret in the White Cell,* Harper & Row, New York, 1969.
172. **Metchnikoff, E.,** Über eine Sprosspilzkrankheit der Daphnien: Beitrag zur lehre über den Kampf der Phagocyten gegen Krankheitserreger, *Verchow's Arch. Pathol. Anat. Physiol.,* 96, 177, 1884.
173. **Cuénot, L.,** Études physiologiques sur les crustaces decapodes, *Arch. Biol. Liege,* 13, 245, 1895.
174. **Cuénot, L.,** Les globules sanguins et les organes lymphoïdes des Invertebres, (revue critique et nouvelles recherches), *Arch. Anat. Microsc.,* 1, 153, 1897.
175. **Metalnikov, S.,** Recherches expérimentales sur les Chenilles de *Galleria mellonella, Arch. Zool. Exp.* (Ser 4), 8, 489, 1908.
176. **Cameron, G. R.,** Inflammation in the caterpillars of Lepidoptera, *J. Pathol. Bacteriol.,* 38, 441, 1934.
177. **Bettini, S., Sarkaria, D. S., and Patton, R. L.,** Observations on the fate of vertebrate erythrocytes and hemoglobin injected into the blood of the American cockroach *(Periplaneta americana), Science,* 113, 9, 1951.
178. **Rabinovitch, M. and De Stefano, M. J.,** Interactions of red cells with phagocytes of the wax moth *Galleria mellonella* (L.) and mouse, *Exp. Cell. Res.,* 59, 272, 1970.
179. **Chadwick, J. S.,** Hemolymph changes with the infection or induced immunity in insects and ticks, in *Invertebrate Immunity,* Maramorosch, K. and Shope, R. E., Eds., Academic Press, New York, 1975, 241.
180. **Boman, H. G., Faye, I., Pye, A., and Rasmuson, T.,** Inducible immunity system of giant silkmoths, in *Comparative Pathobiology,* Vol. 4, Bulla, L. and Cheng, T. C., Eds., Plenum Press, New York, 1979, 145.
181. **Armstrong, P. B. and Levin, J.,** *In vitro* phagocytosis by *Limulus* blood cells, *J. Invertebr. Pathol.,* 34, 145, 1979.
182. **Wheeler, R. E.,** Studies on the total haemocyte count and haemolymph volume in *Periplaneta americana* (L.) with special reference to the last moulting cycle, *J. Insect Physiol.,* 9, 223, 1962.
183. **Wittig, G.,** Phagocytosis by blood cells in healthy and diseased caterpillers. II. A consideration of the method of making hemocyte counts, *J. Invertebr. Pathol.,* 8, 461, 1966.
184. **Leutenegger, R.,** Early events of *Sericesthis* iridescent virus infection in hemocytes of *Galleria mellonella* (L.), *Virology,* 32, 109, 1967.
185. **Stairs, G. R.,** Changes in the susceptibility of *Galleria mellonella* (Linnaeus) larvae to nuclear-polyhedrosis virus following blockage of phagocytes with India ink, *J. Insect Pathol.,* 6, 373, 1964.
186. **Hoffmann, J. A., Zachary, D., Hoffmann, D., and Brehélin, M.,** Postembryonic development and differentiation: hemopoietic tissues and their functions in some insects, in *Insect Hemocytes,* Gupta, A. P., Ed., Cambridge University Press, Cambridge, 1979, 29.
187. **Sharma, P. R., Tikku, K., and Saxena, B. P.,** An electron microscopic study of normal haemocytes of *Poecilocerus pictus* (Fabr.) and their response to injected yeast cells, *Insect Sci. Appl.,* 7(1), 85, 1986.
188. **Kaaya, G. P., Otieno, L. H., and Dardji, N.,** Cellular and humoral immunity in *Glossina, Dev. Comp. Immunol.,* 9(1)(Abstr.), 177, 1985.
189. **Wago, H. and Ichikawa, Y.,** Changes in the phagocytic rate during the larval development and manner of hemocyte reactions to foreign cells in *Bombyx mori, Appl. Entomol. Zool.,* 14(4), 397, 1979.
190. **Wago, H.,** The role of hemolymph in the initial cellular attachment to foreign cells by the hemocytes of the silkworm, *Bombyx mori, Dev. Comp. Immunol.,* 5, 217, 1981.
191. **Harshbarger, J. C. and Heimpel, A. M.,** Effects on zymosan on phagocytosis in larvae of the greater wax moth, *Galleria mellonella, J. Invertebr. Pathol.,* 10, 176, 1968.
192. **Söderhäll, K. and Smith, V. J.,** Prophenoloxidase-activating cascade as a recognition and defense system in arthropods, in *Hemocytic and Humoral Immunity in Arthropods,* Gupta, A. P., Ed., John Wiley & Sons, New York, 1986, 251.
193. **Ratcliffe, N. A., Rowley, A. F., Fitzgerald, S. W., and Rhodes, C. P.,** Invertebrate immunity: basic concepts and recent advances, *Int. Rev. Cytol.,* 97, 183, 1985.
194. **Pruzanski, W. and Sato, S.,** The influence of natural and synthetic cationic substances on phagocytic activity of human polymorphonuclear cell, *Exp. Cell Res.,* 117, 1, 1978.
195. **Hoffmann, J. A., Porte, A., and Joly, P.,** Présence d'un tissue hématopoïètique au niveau du diaphragme dorsal de *Locusta migratoria* (Orthoptere), *C. R. Séances Acad. Sci. Paris,* 266D, 1882, 1968.
196. **Bang, F. B.,** Phagocytosis in invertebrates, in *Invertebrate Immunity,* Maramorosch, K. and Shope, R. E., Eds., Academic Press, New York, 1975, 135.
197. **Unanue, E. R. and Benacerraf, B.,** *Textbook of Immunology,* 2nd ed., Williams & Wilkins, Baltimore, 1984.
198. **Wago, H.,** Observations of the early processes of phagocytosis by the granular cells of the silkworm, *Bombyx mori, Appl. Entomol. Zool.,* 15, 489, 1980.
199. **Wago, H.,** Humoral factors promoting the adhesive properties of the granular cells and plasmatocytes of the silkworm, *Bombyx mori,* and their possible role in the initial cellular reactions to foreignness, *Cell. Immunol.,* 54, 155, 1980.
200. **Mohrig, W., Schittek, D., and Ehlers, D.,** Mediators in insect immunity, in *Hemocytic and Humoral Immunity In Arthropods,* Gupta, A. P., Ed., John Wiley & Sons, New York, 1986, 431.

201. **Rowley, A. F., Ratcliffe, N. A., Leonard, C. M., Richards, E. H., and Renwrant, L.,** Humoral recognition factors in insects, with particular reference to agglutinins and the phenoloxidase system, in *Hemocytic and Humoral Immunity in Arthropods,* Gupta, A. P., Ed., John Wiley & Sons, New York, 1986, 381.
202. **Zachary, D. and Hoffmann, D.,** Lysozyme is stored on the granules of certain hemocyte types of *Locusta, J. Insect Physiol.,* 30, 405, 1984.
203. **Schwemmler, W. and Müller, H.,** Role of insect lysozymes in endocytobiosis and immunity of leafhoppers, in *Hemocytic and Humoral Immunity in Arthropods,* Gupta, A. P., Ed., John Wiley & Sons, New York, 1986, 449.
204. **Naccache, P. H. and Sha'afi, R. I.,** Arachidonic acid, leukotriene B_4, and neutrophil activation, *Ann. N.Y. Acad. Sci.,* 414, 125, 1982.
205. **Powell, W. S.,** Properties of leukotriene B_4 20-hydroxylase from polymorphonuclear leukocytes, *J. Biol. Chem.,* 259, 3082, 1984.
206. **Rich, A. M., Weissman, G., Anderson, C., Vosshall, L., Haines, K. A., Humphreys, T., and Dunham, P.,** Calcium-dependent aggregation of marine sponge cells is provoked by leukotriene B_4 and inhibited by inhibitors of arachidonic acid oxidation, *Biochem. Biophys. Res. Commun.,* 121, 863, 1984.
207. **Hostetter, R. K. and Cooper, E. L.,** Earthworm coelomocyte immunity, in *Comtemporary Topics in Immunobiology,* Vol. 4, *Invertebrate Immunology,* Cooper, E. L., Ed., Plenum Press, New York, 1974.
208. **Strange, P. G.,** Editor's introduction, in *Cell Surface Receptors,* Strange, P. G., Ed., Ellis Horwood Ltd., Chichester, U.K., 1983, 13.
209. **Kosterlitz, H. W., Paterson, S. J., and Robson, L. E.,** Subtypes of the opiate receptor, in *Cell Surface Receptors,* Strange, P. G., Ed., Ellis Horwood Ltd., Chichester, U.K., 1983, 41.
210. **Langley, J. N.,** On the reaction of cells and of nerve endings to certain poisons, chiefly as regards the reaction of striated muscle to nicotine and curare, *J. Physiol.,* 33, 374, 1905.
211. **Gebbinck, J.,** Some observations on concanavalin A receptors on free cells of marine invertebrates, *Dev. Comp. Immunol.,* 4(1), 33, 1980.
212. **Daimon, T. and Sano-Martins, I. S.,** Alterations of wheat-germ agglutinin binding pattern on cell surface of blood platelets after thrombin stimulation, *Histochemistry,* 91, 461, 1989.
213. **Sharon, N.,** Carbohydrates as recognition determinants in phagocytosis and in lectin-mediated killing target cells, *Biol. Cell.,* 51, 239, 1984.
214. **Bradley, R. S., Stuart, G. S., Stiles, B., and Hapner, K. D.,** Grasshopper haemagglutinin: immunochemical localization in haemocytes and investigation of opsonic property, *J. Insect Physiol.,* 35(5), 353, 1989.
215. **Sharon, N. and Lis, H.,** Lectins as recognition molecules, *Science,* 246, 227, 1989.
216. **Ofek, I. and Sharon, N.,** Lectinophagocytosis: a molecular mechanism of recognition between cell surface sugars and lectins in the phagocytosis of bacteria, *Infect. Immun.,* 56, 539, 1988.
217. **Kan, V. L. and Bennett, J. E.,** Lectin-like attachment sites on murine pulmonary alveolar macrophages bind *Aspergillus fumigatus* conidia, *J. Infect. Dis.,* 158, 407, 1988.
218. **Petri, W. A., Jr., Chapman, M. D., Snodgrass, T., Mann, B., Broman, J., and Ravdin, J. I.,** Subunit structure of the galactose and *N*-acetyl-D-galactose-inhibitable adherence lectin of *Entamoeba histolytica, J. Biol. Chem.,* 264, 3007, 1989.
219. **Ashwell, G. and Morell, A. G.,** The role of surface carbohydrates in the hepatic recognition and transport of circulating glycoproteins, *Adv. Enzymol.,* 41, 99, 1974.
220. **Ashwell, G. and Harford, J.,** Carbohydrate-specific receptors of the liver, *Annu. Rev. Biochem.,* 51, 531, 1982.
221. **Kelm, S. and Schauer, R.,** The galactose-recognizing system of rat peritoneal macrophages: identification and characterization of the receptor molecule, *Biol. Chem. Hoppe-Seyler,* 369, 693, 1988.
222. **Oda, S., Sato, M., Toyoshima, S., and Osawa, T.,** Purification and characterization of a lectin-like molecule specific for galactose *N*-acetyl-galactosamine from tumoricidal macrophages, *J. Biochem.,* 104, 600, 1988.
223. **Araujo-Jorge, T. C. and De Souja, W.,** Interaction of *Trypanosoma cruzi* with macrophages, involvement of surface galactose and *N*-acetyl-D-galactosamine residues in the recognition process, *Acta Trop.,* 45(2), 127, 1988.
224. **Tuppy, H. and Gottschalk, A.,** The structure of sialic acids and their quantitation, in *Glycoproteins: Their Composition, Structure and Functions,* 2nd ed., Gattschalk, A., Ed., Academic Press, New York, 1972, chap. 1.
225. **Rosenberg, A. and Schengrund, C. L., Ed.,** *Biological Roles of Sialic Acid,* Plenum Press, New York, 1976.
226. **McKay, D., Jenkin, C. R., and Rowley, D.,** Immunity in invertebrates. I. Studies on the naturally occurring hemagglutinins in the fluid from invertebrates, *Aust. J. Exp. Med. Sci.,* 47, 125, 1969.

227. **McKay, D. and Jenkin, C. R.,** Immunity in the invertebrates. The role of serum factors in phagocytosis of erythrocytes by haemocytes of the freshwater crayfish *(Parachaeraps bicarinatus), Aust. J. Exp. Biol. Sci.,* 48, 139, 1970.
228. **Paterson, W. D. and Stewart, J. E.,** *In vitro* phagocytosis by hemocytes of the American lobster *(Homarus americanus), J. Fish. Res. Board. Can.,* 31, 1051, 1974.
229. **Schapiro, H. C., Steenburgen, J. F., and Fitzerald, Z. A.,** Hemocytes and phagocytosis in the American lobster, *Homarus americanus,* in *Comparative Pathobiology,* Vol. 3, Bulla, L. A., Jr. and Cheng, T. C. Eds., Plenum Press, New York, 1977.
230. **Seaman, R. and Robert, N. L.,** Immunological response of male cockroaches to injection of *Tetrahymena pyriformis, Science,* 161, 1359, 1968.
231. **Brewer, F. D. and Vinson, S. B.,** Chemicals affecting the encapsulation of foreign material in an insect, *J. Invertebr. Pathol.,* 18, 287, 1971.
232. **Rizki, T. M. and Rizki, R. M.,** The cellular defense reactions of *Drosophila melanogaster,* in *The Ultrastructure and Functioning of Insect Cells,* Akai, H., King, R. C., and Morohishi, S., Eds., Plenum Press, New York, 1982, 173.
233. **Dularay, B. and Lackie, A. M.,** Haemocytic encapsulation and phenoloxidase-activation pathway in the locust, *Schistocerca gregaria,* Forsk., *Insect Biochem.,* 15(6), 827, 1985.
234. **Ratcliffe, N. A., Leonard, C., and Rowley, A. F.,** Prophenoloxidase activation: nonself recognition and cell cooperation in insect immunity, *Science,* 226, 557, 1984.
235. **Durliat, M.,** Clotting processes in Crustacea Decapoda, *Biol. Rev.,* 60, 473, 1985.
236. **Liu, T. Y., Seid, R. C., Tai, J. P., Liang, S. M., Sakmar, T. P., and Robbins, J. B.,** Studies on *Limulus* lysate coagulating system, in *Biomedical Applications of the Horseshoe Crab (Limulidae),* Cohen, E., Ed., Alan R. Liss, New York, 1979, 147.
237. **Harada, T., Morita, T., Iwanaga, S., Nakamura, S., and Niwa, M.,** A new chromogenic substrate method for assay of bacterial endotoxins using *Limulus* hemocyte lysate, in *Biomedical Applications of the Horseshoe Crab (Limulidae),* Cohen, E., Ed., Alan R. Liss, New York, 1979, 209.
238. **Salt, G.,** *The Cellular Defense Reactions of Insects,* Cambridge University Press, Cambridge, 1970.
239. **Ratcliffe, N. A. and Rowley, A. F.,** Role of hemocytes in defense against biological agents, in *Insect Hemocytes,* Gupta, A. P., Ed., Cambridge University Press, Cambridge, 1979, 331.
240. **Werner, R. A. and Jones, J. C.,** Phagocytic haemocytes in unfixed *Galleria mellonella* larvae, *J. Insect Physiol.,* 15, 425, 1969.
241. **Kislev, N., Harpaz, F., and Zelcer, A.,** Electron-microscopic studies on hemocytes of the Egyptian cottonworm, *Spodoptera littoralis* (Boisduval) infected with a nuclear-polyhedrosis virus, as compared to non-infected hemocytes. II. Noninfected hemocytes, *J. Invertebr. Pathol.,* 14, 245, 1969.
242. **Grimstone, A. V., Rotheram, S., and Salt, G.,** An electron microscope study of capsule formation by insect blood cells, *J. Cell Sci.,* 2, 281, 1967.
243. **Pastan, I. H. and Willingham, M. C.,** Journey to the center of the cell: role of receptosome, *Science,* 214, 504, 1981.
244. **Pearse, B. M. F.,** Clathrin: a unique protein associated with intracellular transfer of membrane by coated vesicles, *Proc. Natl. Acad. Sci. U.S.A.,* 73, 1255, 1976.
245. **Crossley, A. C. S.,** The cytophysiology of insect blood, *Adv. Insect Physiol.,* 11, 117, 1975.
246. **Bowers, B.,** Coated vesicles in the pericardial cells of the aphid (*Myzus persicae* Sulz.), *Protoplasma,* 59, 351, 1964.
247. **Crossley, A. C.,** The ultrastructure and function of pericardial cells and other nephrocytes in an insect: *Calliphora erythrocephala, Tissue Cell,* 4, 529, 1972.
248. **Rothman, J. E.,** The Golgi apparatus: two organelles in tandem, *Science,* 213, 1212, 1981.
249. **Götz, P.,** Encapsulation in arthropods, in *Immunity in Invertebrates,* Brehélin, M., Ed., Springer-Verlag, Berlin, 1986, 153.
250. **Götz, P., Roettgen, J., and Lingg, W.,** Encapsulement humoral en tant que réaction de defense chez les Diptères, *Ann. Parasitol. Hum. Comp.,* 52, 1977.
251. **Poinar, G. O., Jr., Hess, R. T., and Peterson, J. J.,** Immune responses of mosquitoes against *Romanomermis culicivora* (Memithida Nematoda), *J. Nematol.,* 11, 110, 1979.
252. **Ratner, R. and Vinson, S. B.,** Phagocytosis and encapsulation: cellular immune response in Arthropoda, *Am. Zool.,* 23, 185, 1983.
253. **Rizki, T. M. and Rizki, R. M.,** Developmental analysis of a temperature-sensitive melanotic tumor mutant in *Drosophila melanogaster, Wilhelm Roux's Arch. Dev. Biol.,* 189, 197, 1980.
254. **Walter, D. R. and Williams, C. M.,** Reaggregation of insect cells as studied by a new method of tissue and organ culture, *Science,* 154, 516, 1966.
255. **Vinson, S. B.,** The role of the foreign surface and female parasitoid secretions on the immune response of an insect, *Parasitology,* 68, 27, 1974.
256. **Dunphy, G. B. and Nolan, R. A.,** Response of Eastern Hemlock Looper haemocytes to selected stages of *Entomophthora egressa* and other foreign particles, *J. Invertebr. Pathol.,* 36, 71, 1980.

257. **Lackie, A. M.,** Immune recognition in insects, *Dev. Comp. Immunol.*, 5, 191, 1981.
258. **Lackie, A. M.,** Immunological recognition of cuticular transplants in insects, *Dev. Comp. Immunol.*, 7, 41, 1983.
259. **Lackie, A. M. and Taklc, G.,** Initiation and cessation of hemocytic encapsulation in the locust *Schistocerca gregaria* and in the cockroach *Periplaneta americana, Dev. Comp. Immunol.*, 9(Abstr.), 178, 1985.
260. **Ratner, S. and Vinson, S. B.,** Encapsulation reactions *in vitro* by haemocytes of *Heliothis virescens, J. Insect Physiol.*, 29, 855, 1983.
261. **Poinar, G. O., Jr., Leutenegger, R., and Götz, P.,** Ultrastructure of the formation of a melanotic capsule in *Diabrotica* (Coleoptera) in response to a parasitic nematode (Mermithidae), *J. Ultrastruct. Res.*, 25, 293, 1968.
262. **Rizki, M. T. M. and Rizki, R. M.,** Cell interactions in hereditary melanotic tumor formation in *Drosophila,* in *Proc. First Int. Colloq. Invertebrate Pathol.*, Angus, T. A., Faulkner, F., and Rosenfield, A., Eds., Queens University, Kingston, Ontario, 1976, 137.
263. **Parish, C. R.,** Simple model for self-nonself discrimination in invertebrates, *Nature (London)*, 267, 711, 1977.
264. **Ferron, P.,** Biological control of insect pests by entomogenous fungi, *Annu. Rev. Entomol.*, 23, 409, 1978.
265. **Nappi, A. J. and Stoffolano, J. G., Jr.,** *Heterotylenchus autumnalis.* Hemocytic reactions and capsule formation in the host, *Musca domestica, Exp. Parasitol.*, 29, 116, 1971.
266. **Nappi, A. J. and Stoffolano, J. G., Jr.,** Haemocytic changes associated wtih the immune reaction of nematode-infected larvae of *Orthellia caesarion, Parasitology*, 65, 295, 1972.
267. **Crossley, A. C.,** Biochemical and ultrastructural aspects of synthesis, storage, and secretion in hemocytes, in *Insect Hemocytes,* Gupta, A. P., Ed., Cambridge University Press, Cambridge, 1979, 423.
268. **Schmit, A. R., Rowley, A. F., and Ratcliffe, N. A.,** The role of *Galleria mellonella* hemocytes in melanin formation, *J. Invertebr. Pathol.*, 29, 232, 1976.
269. **Mazzone, H. M.,** Influence of polyphenol oxidase on hemocyte cultures of the gypsy moth, in *Invertebrate Tissue Culture: Applications in Medicine, Biology and Agriculture,* Kurstak, E. and Maramorosch, K., Eds., Academic Press, New York, 1976, 275.
270. **Caveney, S. and Berdan, R.,** Selectivity in junctional coupling between cells of insect tissue, in *Insect Ultrastructure,* Vol. 1, King, R. C. and Akai, H., Eds., Plenum Press, New York, 1982, 434.
271. **Taylor, R. L.,** A suggested role for the polyphenolphenoloxidase system in invertebrate immunity, *J. Invertebr. Pathol.*, 14, 427, 1969.
272. **Beresky, M. A. and Hall, D. W.,** The influence of phenylthiourea on encapsulation, melanization, and survival in larvae of the mosquito *Aedes aegypti* parasitized by the nematode *Neoplectana carpocapsae, J. Invertebr. Pathol.*, 29(1), 74, 1977.
273. **Kuo, M. J. and Alexander, M.,** Inhibition of the lysis of fungi by melanins, *J. Bacteriol.*, 94, 624, 1967.
274. **Bull, A. T.,** Inhibition of polysaccharides by melanin: enzyme inhibition in relation to mycolysis, *Arch. Biochem. Biophys.*, 137, 345, 1970.
275. **Pawelek, J. M. and Lerner, A. B.,** 5, 6-Dihydroxyindol is a melanin precursor showing potent cytotoxicity, *Nature (London)*, 276, 627, 1978.
276. **Chadwick, J. S. and Dunphy, G. B.,** Antibacterial and antiviral factors in arthropod hemolymph, in *Hemocytic and Humoral Immunity in Arthropods,* Gupta, A. P., Ed., John Wiley & Sons, New York, 1986, 287.
277. **Pearson, G. R., Freeman, B. Z., and Hines, W. D.,** Thin section electron micrographs of monocytes infected with *Brucella suis, J. Bacteriol.*, 86, 1123, 1963.
278. **Gresson, R. A. R. and Threadgold, L. T.,** An electron microscope study of bacteria in the oocytes and follicle cells of *Blatta orientalis, Q. J. Microsc. Sci.*, 101, 295, 1960.
279. **Grinyer, I. and Musgrave, A. J.,** Ultrastructure and peripheral membranes of the mycetomal microorganism of *Sitophilus granarius* (L.), *J. Cell Sci.*, 1, 181, 1966.
280. **Lamb, K. P. and Hinde, R.,** Structure and development of the mycetome in the cabbage aphid *Brevicorynae brassicae, J. Invertebr. Pathol.*, 9, 3, 1967.
281. **Gromov, B. V. and Mamakaeva, K. A.,** Blattabacterium in the fat body of Maritime Territory relic roach, *Cryptocerus relictus, Mikrobiologia,* 49(6), 1005, 1980.
282. **Griffiths, G. W. and Beck, S. D.,** Ultrastructure of the pea aphid mycetocytes: evidence for symbiotic secretion, *Cell Tissue Res.*, 159, 351, 1975.
283. **Brennan, B. M. and T. C. Cheng.,** Resistance of *Moniliformis dubius* to the defense reactions of the American cockroach, *Periplaneta americana, J. Invertebr. Pathol.*, 26, 65, 1975.
284. **Söderhäll, K.,** Fungal cell wall β-1,3-glucans induce clotting and phenoloxidase attachment to foreign surfaces of crayfish hemocyte lysate, *Dev. Comp. Immunol.*, 5, 565, 1981.
285. **Sutherland, D. R., Christensen, B. M., and Forton, K. F.,** Defense reaction of mosquitoes to filarial worm: role of the microbial sheath in the response of mosquitoes to inoculated *Brugia pahangi* microfilariae, *J. Invertebr. Pathol.*, 44, 275, 1984.

286. **Lafond, M. M., Christensen, B. M., and Lasee, B. A.** Defense reactions of mosquitoes to filarial worms: potential mechanism for avoidance of the response by *Brugia pahangi* microfilariae, *J. Invertebr. Pathol.*, 46, 26, 1985.
287. **Miller, L. H., Howard, R. J., Carter, R., Good, M. F., Nussenzweig, U., and Nussenzweig, R. S.,** Research toward malaria vaccines, *Science*, 234, 1349, 1985.
288. **Aikawa, M., Yoshida, N., Nussenzweig, R. S., and Nussenzweig, V.,** The protective antigen of malarial sporozoites *(Plasmodium berghei)* is a differentiation antigen, *J. Immunol.*, 126, 2494, 1981.
289. **Flyg, C. and Xanthopoulos, K. G.,** Insect pathogenic properties of *Serratia marcescens*. Passive and active resistance to insect immunity studied with protease-deficient and phage-resistant mutants, *J. Gen. Microbiol.*, 129, 453, 1983.
290. **Dalhammar, G.,** Studies related to immune inhibitor A. A proteolytic enzyme produced by insect pathogen *Bacillus thuringiensis*, Ph.D. thesis, University of Stockholm, Sweden, 1987.
291. **Siden, I., Dalhammar, G., Telander, B., Boman. H. G., and Sommerville, H.,** Virulence factors in *Bacillus thuringiensis:* purification and properties of a protein inhibitor of immunity in insects, *J. Gen. Microbiol.*, 114, 45, 1979.
292. **Götz, P., Boman, A., and Boman, H. G.,** Interactions between insect immunity and an insect pathogenic nematode with symbiotic bacteria, *Proc. R. Soc. London, Ser. B*, 212, 333, 1981.
293. **Kitano, H.,** Defensive ability of *Apanteles glomeratus* L., (Hymenoptera: Braconidae) to hemocytic reaction of *Pieris rapae crucivora* Boisduval (Lepidoptera: Pieridae), *Appl. Entomol. Zool.*, 4(1), 51, 1969.
294. **Kitano, H.,** Experimental studies on the parasitism of *Apanteles glomeratus* L. with special reference to its encapsulation-inhibiting capacity, *Bull. Tokyo Gekugei Univ.*, 21(4), 95, 1969.
295. **Salt, G.,** Experimental studies in insect parasitism. XIII. The haemocytic reaction of a caterpillar to eggs of its habitual parasite, *Proc. R. Soc. London, Ser. B*, 162, 303, 1965.
296. **Rotheram, S.,** Immune surface of eggs of a parasitic insect, *Nature (London)*, 214, 700, 1967.
297. **Rotheram, S. M.,** The surface of the egg of a parasitic insect. I. The surface of the egg and first instar larvae of *Nemeritis, Proc. R. Soc. London*, 183, 179, 1973.
298. **Rotheram, S. M.,** The surface of the egg of a parasitic insect. II. The ultrastructure of the particulate coat on the egg of *Nemeritis, Proc. R. Soc. London, Ser. B*, 183, 195, 1973.
299. **Bedwin, O.,** An insect glycoprotein: a study of the particles responsible for the resistance of a parasitoid's egg to the defense reaction of its insect host, *Proc. R. Soc. London, Ser. B*, 205, 271, 1979.
300. **Vinson, S. B. and Scott, J. R.,** Particles containing DNA associated with the oocyte of an insect parasitoid, *J. Invertebr. Pathol.*, 25, 375, 1975.
301. **Stoltz, D. B. and Vinson, S. B.,** Baculovirus-like particles in the reproductive tracts of female parasitoid wasps. II. The genus *Apanteles, Can. J. Microbiol.*, 23, 28, 1977.
302. **Stoltz, D. B. and Vinson, S. B.,** Viruses and parasitism in insects, *Virus Res.*, 24, 125, 1979.
303. **Edson, K. M., Vinson, S. B., Stoltz, D. B., and Summers, M. D.,** Virus in a parasitoid wasp: suppression of the cellular immune response in the parasitoid's host, *Science*, 211, 582, 1981.
304. **Stoltz, D. B., Krell, P. J., and Vinson, S. B.,** Polydisperse viral DNAs in ichneumonid ovaries: a survey, *Can. J. Microbiol.*, 27, 123, 1981.
305. **Vinson, S. B. and Scott, J. R.,** Parasitoid egg shell changes in a suitable and unsuitable host, *J. Ultrastruct. Res.*, 47, 1, 1974.
306. **Norton, W. N. and Vinson, S. B.,** Egg membrane synthesis of a hymenopteran parasitoid, *Am. Zool.*, 15, 828, 1976.
307. **Feddersen, I., Sander, K., and Schmidt, O.,** Virus-like antigenic determinants protect an insect parasitoid from encapsulation, *Experientia*, 42, 1278, 1986.
308. **Schmidt, O. and Schuchmann-Feddersen, I.,** Role of virus-like particles in parasitoid-host interaction of insects, *Subcell. Biochem.*, 15, 91, 1989.
309. **Berg, R., Schuchmann-Feddersen, I., and Schmidt, O.,** Bacterial infection induces a moth *(Ephestia kühniella)* protein which has antigenic similarity to virus-like particle proteins of a parasitoid wasp *(Venturia canescens), J. Insect Physiol.*, 34, 437, 1988.
310. **Krell, P. J.,** Replication of long virus-like particles in the reproductive tract of the ichneumonid wasp *Diadegma terebrans, J. Gen. Virol.*, 68, 1477, 1987.
311. **Stoltz, D. B., Guzo, D., Belland, E. R., Lucarotti, C. J., and MacKinnon, E. A.,** Venom promotes uncoating *in vitro* and persistence *in vivo* of DNA from a braconid polydnavirus, *J. Gen. Virol.*, 69, 1477, 1988.
312. **Stoltz, D. B., Krell, P. J., Cook, D., MacKinnon, E. A., and Lucarotti, C. J.,** An unusual virus from the parasitoid wasp *Cotesia malanoscela, Virology*, 162, 311, 1988.
313. **Nappi, A. J. and Streams, F. A.,** Hemocytic reactions of *Drosophila melanogaster* to the parasites *Pseudeucoila mellipes* and *P. bochei, J. Insect Physiol.*, 15, 1551, 1969.
314. **Rizki, R. M. and Rizki, T. M.,** Selective destruction of a host blood cell type by a parasitoid wasp, *Proc. Natl. Acad. Sci. U.S.A.*, 81, 6154, 1984.

315. **Rizki, T. M. and Rizki, R. M.,** Surface changes on hemocytes during encapsulation in *Drosophila melanogaster* Meigen, in *Hemocytic and Humoral Immunity in Arthropods,* Gupta, A. P., Ed., John Wiley & Sons, New York, 1986, 115.
316. **Kurstak, E., Goring, I., and Vago, C.,** Cellular defense in an arthropod in response to infection with a *Salmonella tryphimurium* strain, *J. Microbiol. Serol.,* 35, 45, 1969.
317. **Ratcliffe, N. A. and Walters, J. B.,** Studies on the *in vitro* cellular reactions of insects: clearance of pathogenic and non-pathogenic bacteria in *Galleria mellonella, J. Insect Physiol.,* 29, 407, 1983.
318. **Cheung, P. Y. K., Grula, E. A., and Burton, R. L.,** Hemolymph responses in *Heliothis zea* to inoculation with *Bacillus thuringiensis* or *Micrococcus lysodeikticus, J. Invertebr. Pathol.,* 31, 148, 1978.
319. **Gilliam, M. and Shimanuki, M.,** *In vitro* phagocytosis of *Nosema apis* spores by honey-bee hemocytes, *J. Invertebr. Pathol.,* 9, 387, 1967.
320. **Vey, A., Quiot, J. M., and Vago, C.,** Formation *in vitro* de reaction d'immunite cellulaire chez les insectes, in *Proc. Second Int. Colloq. Invertebrate Tissue Culture,* Barigozzi, C., Ed., Instituto Lombardo di Scienze e Lettere, Milan, 1968, 254.
321. **Gagen, S. J. and Ratcliffe, N. A.,** Studies on the *in vivo* cellular reactions and fate of injected baceria in *Galleria mellonella* and *Pieris brassicae* larvae, *J. Invertebr. Pathol.,* 28 (I), 17, 1976.
322. **Ratcliffe, N. A. and Gagen, S. J.,** Cellular defense reactions of insect hemocytes *in vivo:* nodule formation and development in *Galleria mellonella* and *Pieris brassicae* larvae, *J. Invertebr. Pathol.,* 28(3), 373, 1976.
323. **Ratcliffe, N. A. and Gagen, S. J.,** Studies on the *in vivo* cellular reactions of insects: an ultrastructural analysis of nodule formation in *Galleria mellonella, Tissue Cell,* 9(1), 73, 1977.
324. **Götz, P. and Vey, A.,** Humoral encapsulation in insects, in *Hemocytic and Humoral Immunity in Arthropods,* Gupta, A. P., Ed., John Wiley & Sons, New York, 1986, 407.
325. **Wülker, W.,** Untersuchungen über die Intersexualitat der Chironomiden (Dipt.) nach *Paramermis* infektion, *Arch. Hydrobiol.,* Suppl. 25, 127, 1961.
326. **Bronskill, J. F.,** Encapsulation of rhabitoid nematodes in mosquitoes, *Can. J. Zool.,* 40, 1269, 1962.
327. **Elsinger, J. H. Y.,** Behavior of microfilariae of *Brugia pahangi* in *Anopheles quadrimaculatus, Am. J. Trop. Med. Hyg.,* 1, 749, 1962.
328. **Cooper, E. L.,** Chronic allograft rejection in *Lumbricus terrestris, J. Exp. Zool.* 171, 69, 1969.
329. **Cooper, E. L.,** Allograft rejection in *Eisenia foetida, Transplantation,* 8, 220, 1969.
330. **Scott, M. T.,** A naturally occurring haemagglutinin in the haemolymph of the American cockroach, *Arch. Zool. Exp. Gen.,* 112, 73, 1971.
331. **Bretting, H., Kalthoff, H., and Fehr, S.,** Studies on the relationship between lectins from *Axinella polypoides* agglutinating bacteria and human erythrocytes, *J. Invertebr. Pathol.,* 32, 151, 1978.
332. **Bang, F. B.,** Serological responses among invertebrates other than insects, *Fed. Proc.,* 26, 1680, 1967.
333. **Pereira, M. E. A., Andrade, F. B., and Ribeiro, T. M. C.,** Lectins of distinct specificity in *Rhodnius prolixus* interact selectively with *Trypanosoma cruzi, Science,* 211, 597, 1981.
334. **Ingram, G. A., East, E., and Molyneux, D. H.,** Agglutinins of *Trypanosoma, Leishmania* and *Crithidia* in insect haemolymph, *Dev. Comp. Immunol.,* 7, 649, 1983
335. **Stein, P. C. and Basch, P. F.,** Purification and binding properties of hemagglutinins from *Biomphalaria glabrata, J. Invertebr. Pathol.,* 33, 10, 1979.
336. **Noguchi, H.,** On the multiplicity of the serum haemagglutinins of cold blooded animals, *Zentralb. Bakteriol. I. Abt. Orig.,* 34, 286, 1903.
337. **Tsivion, Y. and Sharon, N.,** Lipid-mediated hemagglutination and its relevance to lectin-mediated agglutination, *Biochem. Biophys. Acta,* 642, 336, 1981.
338. **Ravindranath, M. H. and Cooper, E. L.,** Crab lectins: receptor specificity and biomedical applications, in *Recognition Proteins, Receptors and Probes: Invertebrates,* Cohen, E., Ed., Alan R. Liss, New York, 1984, 83.
339. **Renkonen, K. O.,** Studies on hemagglutinins present in seeds of some representatives of Leguminosae, *Ann. Med. Exp. Fenn.* (Helsinki), 26, 66, 1948.
340. **Boyd, W. C. and Reguera, R. N.,** Hemagglutinating substances in various plants, *J. Immunol.,* 62, 333, 1949.
341. **Goldstein, I. J., Hughes, R. C., Monsigny, M., Osawa, T., and Sharon, N.,** What should be called a lectin?, *Nature (London),* 285, 66, 1980.
342. **Hapner, K. D. and Stebbins, M. R.,.** Biochemistry of arthropod agglutinins, in *Hemocytic and Humoral Immunity in Arthropods,* Gupta, A. P., Ed., John Wiley & Sons, New York, 1986, 227.
343. **Stebbins, M. R. and Hapner, K. D.,** Isolation, characterization and inhibition of arthropod agglutinins, in *Hemocytic and Humoral Immunity in Arthropods,* Gupta, A. P., Ed., John Wiley & Sons, New York, 1986, 463.
344. **Pistole, T. G. and Graf, S. A.,** Antibacterial activity in *Limulus,* in *Hemocytic and Humoral Immunity in Arthropods,* Gupta, A. P., Ed., John Wiley & Sons, New York, 1986, 331.

345. **Cohen, E., Rose, A. W., and Wissler, F. C.,** Heteragglutinins of the horseshoe crab *Limulus polyphemus, Life Sci.,* 4, 2009, 1965.
346. **Yeaton, R. W.,** Invertebrate lectins: I. Occurrence, *Dev. Comp. Immunol.,* 5, 391, 1981
347. **Komano, H., Mazumo, D., and Natori, S.,** A possible mechanism of induction on insect lectin, *J. Biol. Chem.,* 256, 7087, 1981.
348. **Kaplan, R., Li, S. S. L., and Kehoe, J. M.,** Molecular characterization of limulin, a sialic acid binding lectin from the hemolymph of the horseshoe crab, *Biochemistry,* 16, 4297, 1977.
349. **Kehoe, J. M., Kaplan, R., and Li, S. S. L.,** Functional implication of the covalent structure of limulin: an overview, *Prog. Clin. Biol. Res.,* 21, 617, 1979.
350. **Amirante, G. A.,** Production of heteroagglutinins in haemocytes of *Leucophaea maderae* L., *Experientia,* 32(4), 526, 1976.
351. **Valvassori, R. and Amirante, G. A.,** Caracteristiques immunochimiques et ultrastructurales des hemocoytes de *Leucophaea maderae* L. (Insecta Blattoidea), *Monit. Zool. Ital.* (N.S.), 10, 403, 1976.
352. **Amirante, G. A. and Mazzalai, F. G.,** Synthesis and localization of hemagglutinins in haemocytes of the cockroach *Leucophaea maderae* L., *Dev. Comp. Immunol.,* 2(4), 735, 1978.
353. **Marek, M.,** Effect of stress of cooling on the synthesis of immunoglobulin in haemolymph of *Galleria mellonella* L., *Comp. Biochem. Physiol.,* 35, 615, 1970.
354. **Yeaton, R. W.,** Wound response in insects, *Am. Zool.,* 23, 195, 1983.
355. **Cornick, J. W. and Stewart, J. E.,** Partial characterization of a natural agglutinin in the hemolymph of the lobster *Homarus americanus, J. Invertebr. Pathol.,* 21, 255, 1973.
356. **Cornick, J. W. and Stewart, J. E.,** Lobster *(Homarus americanus)* hemocytes: classification, differential counts, and associated agglutinin activity, *J. Invertebr. Pathol.,* 31, 194, 1978.
357. **Furman, R. M. and Pistole, T. G.,** Bactericidal activity of hemolymph from the horseshoe crab, *Limulus polyphemus, J. Invertebr. Pathol.,* 28, 239, 1976.
358. **Lackie, A. M.,** Biochemical characterization and biological role of a serum lectin from the cockroach *Periplaneta americana, Dev. Comp. Immunol.,* 10, 631, 1986.
359. **Stiles, B., Bradley, R. S., Stuart, G. S., and Hapner, K. D.,** Site of synthesis of haemolymph agglutinin of *Melanoplus differentialis* (Acrididae: Orthoptera), *J. Insect Physiol.,* 34, 1077, 1988.
360. **Nelstrop, A. E., Taylor, G., and Collard, P.,** Comparative studies of serum and haemolymph proteins, *Comp. Biochem. Physiol.,* 35, 191, 1970.
361. **Scott, M. T.,** Recognition of foreignness in invertebrates: transplantation studies using the American cockroach, *Periplaneta americana, Transplantation,* 11, 78, 1971.
362. **Anderson, R. S., Holmes, B., and Good, R. A.,** *In vitro* bactericidal capacity of *Blaberus craniifer* hemocytes, *J. Invertebr. Pathol.,* 22(1), 127, 1973.
363. **Kubo, T., Komano, H., Okada, O., and Natori, S.,** Identification of hemagglutinating protein and bactericidal activity in the hemolymph of adult *Sarcophaga peregrina* on injury of the body wall, *Dev. Comp. Immunol.,* 8, 283, 1984.
364. **Hall, J. L. and Rowlands, D. T., Jr.,** Heterogeneity of lobster agglutinins. 1. Purification and physiochemical characterization, *Biochemistry,* 13, 821, 1974.
365. **Hall, J. and Rowlands, D. T. Jr.,** Heterogeneity of lobster agglutinins. 2. Specificity of agglutinin-erythrocyte binding, *Biochemistry,* 13, 828, 1974.
366. **Ibrahim, E. A. R., Ingram, G. A., and Molyneax, D. H.,** Haemagglutinins and parasite agglutinins in haemolymph and gut of *Glossina, Parasitology,* 35, 151, 1984.
367. **Mauchamp, B.,** Purification of an *N*-acetyl-D-glucosamine specific lectin (P.B.A.) from epidermal cell membranes of *Pieris brassicae* L., *Biochemistry,* 64, 1001, 1982.
368. **Denburg, J. L.,** Cockroach muscle hemagglutinins-candidate recognition macromolecules, *Biochem. Biophys. Res. Commun.,* 97, 33, 1980.
369. **Komano, H., Nozawa, R., Mazumo, D., Natori, S.,** Measurement of *Sarcophaga peregrina* lectin under various physiological conditions by radioimmunoassay, *J. Biol. Chem.,* 258, 2143, 1983.
370. **Peters, W., Kolb, H., and Kolb-Bachofen, V.,** Evidence for a sugar receptor (lectin) in the peritrophic membrane of the blowfly larva, *Calliphora erythrocephala* Mg. (Diptera), *J. Insect Physiol.,* 29, 275, 1983.
371. **Marchalonis, J. J. and Edelman, G. M.,** Isolation and characterization of a hemagglutinin from *Limulus polyphemus, J. Mol. Biol.,* 32, 453, 1968.
372. **Pistole, T. G. and Rustam-Abadi, H.,** Lectins from the horseshoe crab, *Limulus polyphemus,* reactive with bacterial lipopolysaccharides, in *Protides of Biological Fluids: Twenty-seventh Colloquium,* Peeters, H., Ed., Pergamon Press, Oxford, 1979, 423.
373. **Rostam-Abadi, H. and Pistole, T. G.,** Sites on the lipopolysaccharide molecule reactive with *Limulus* agglutinins, *Proc. Clin. Biol. Res.* 29, 525, 1979.
374. **Pistole, T. G.,** *Limulus* lectins: analogues of vertebrate immunoglobulins, in *Physiology and Biology of Horseshoe Crabs: Studies on Normal and Environmentally Stressed Animals,* Bonventura, J., Bonventura, C., and Tesh, S., Eds., Alan R. Liss, New York, 1982, 283.

375. **Nicolson, G. L.,** Ultrastructural localization of lectin receptors, in *Advanced Techniques, Biological Electron Microscopy,* Vol. 2, Koehler, J. K., Ed., Springer-Verlag, Berlin, 1978, 1.
376. **Hapner, K. D. and Jermyn, M. A.,** Haemagglutinin activity in the haemolymph of *Teleogryllus commodus* (Walker), *Insect Biochem.,* 11, 287, 1981.
377. **Stynen, D., Vansteenwegen, K., and DeLoof, A.,** Anti-galactose lectins in the haemolymph of *Sarcophaga bullata* and three other calliphorid flies, *Comp. Biochem. Physiol.,* 81B(1), 171, 1985.
378. **Boyd, W. C.,** The lectins: their present status, *Vox Sang,* 8, 1, 1963.
379. **Komano, H., Mazumo, D., and Natori, S.,** Purification of lectin induced in the hemolymph of *Sarcophaga peregrina* larvae on injury, *J. Biol. Chem.,* 255, 2919, 1980.
380. **Suzuki, T. and Natori, S.,** Identification of a protein having hemagglutinating activity in the hemolymph of the silkworm, *Bombyx mori, J. Biochem.,* 93, 583, 1983.
381. **Weir, D. M.,** Surface carbohydrates and lectins in cellular recognition, *Immunol. Today,* 1, 45, 1980.
382. **Tyson, C. J. and Jenkin, C. R.,** The importance of opsonic factors in the removal of bacteria from the circulation of the crayfish (*Parachaeraps bicarinatus*), *Aust. J. Exp. Biol. Med. Sci.,* 51, 609, 1973.
383. **Tyson, C. J. and Jenkin, C. R.,** Phagocytosis of bacteria *in vitro* by *(Parachaeraps bicarinus), Aust. J. Exp. Biol. Med. Sci.,* 52, 341, 1974.
384. **Paterson, W. D., Stewart, J. E., and Zwicker, B. M.,** Phagocytosis as a cellular immune response mechanism in the American lobster, *Homarus americanus, J. Invertebr. Pathol.,* 27, 95, 1976.
385. **Tripp, M. R.,** Hemagglutinin in the blood of oyster *Cassostraca virginica, J. Invertebr. Pathol.,* 8, 478, 1966.
386. **Anderson, R. S. and Good, R. A.,** Opsonic involvement in phagocytosis by mollusk hemocytes, *J. Invertebr. Pathol.,* 27, 57, 1976.
387. **Rowley, A. F. and Ratcliffe, N. A.,** Insect erythrocyte agglutinins: *in vitro* opsonization experiments with *Clitumnus extradentatus* and *Periplaneta americana* haemocytes, *Immunology,* 40, 483, 1980.
388. **Ratcliffe, N. A. and Rowley, A. F.,** Recognition factors in insect hemolymph, *Dev. Comp. Immunol.,* 7, 653, 1983.
389. **Coombe, D. R., Schluter, S. F., Ey, P. L., and Jenkin, C. R.,** Identification of the HA-2 agglutinin in the haemolymph of the ascidian *Botrylloides leachii* as the factor promoting adhesion of sheep erythrocytes to mouse macrophages, *Dev. Comp. Immunol.,* 6(1), 65, 1982.
390. **Schluter, S. F., Ey, P. L., Keough, D. R., and Jenkin, C. R.,** Identification of two carbohydrate-specific erythrocyte agglutinins in the haemolymph of the prochordate *Bottrylloides leachii, Immunology,* 42, 241, 1981.
391. **Chain, B. M. and Anderson, R. S.,** Observations on the cytochemistry of the hemocytes of an insect, *Galleria mellonella, J. Histochem. Cytochem.,* 31, 601, 1983.
392. **Anderson, R. S. and Chain, B. M.,** Macrophage functions in insects, in *Hemocytic and Humoral Immunity in Arthropods,* Gupta, A. P., Ed., John Wiley & Sons, New York, 1986, 77.
393. **Schmit, A. R. and Ratcliffe, N. A.,** The encapsulation of foreign tissue implants in *Galleria mellonella* larvae, *J. Insect Physiol.,* 23, 175, 1977.
394. **Söderhäll, K., Hall, L., Unestam, T., and Nyhlen, L.,** Attachment of phenoloxidase to fungal cell walls in arthropod immunity, *J. Invertebr. Pathol,* 34, 285, 1979.
395. **Söderhäll, K. and Smith, V. J.,** Separation of the haemocyte populations of *Carcinus maenas* and other marine decapods, and phenoloxidase distribution, *Dev. Comp. Immunol.,* 7, 229, 1983.
396. **Majerus, P. W. and Brodie, G. N.,** The binding of phytohemagglutinins to human platelet plasma membranes, *J. Biol. Chem.,* 247, 4253, 1972.
397. **Rizki, T. M. and Rizki, R. M.,** Blood cell surface changes in *Drosophila* mutants with melanotic tumors, *Science,* 220, 73, 1983.
398. **Nappi, A. J. and Silvers, M.,** Cell surface changes associated with cellular immune reactions in *Drosophila, Science,* 225, 1166, 1984.
399. **Anderson, R. S.,** Expression of receptors by insect macrophages, in *Physiology of Thymus and Bone Marrow-Bursa Cells,* Wright, R. W. and Cooper, E. L., Eds., Elsevier-North Holland, Amsterdam, 1976, 27.
400. **Anderson, R. S.,** Macrophage functions in insects, in *Proc. of the First Int. Colloquium of Invertebrate Pathol.,* Angus, T. A., Faulkner, P., and Rosenfield, A., Eds., Queen's University, Kingston, 1976, 215.
401. **Stynen, D., Peterson, M., and DeLoof, A.,** Proteins with hemagglutinin activity in larvae of the Colorado potato beetle *Leptinotarsa decemlineata, J. Insect Physiol.,* 28, 465, 1982.
402. **Berlin, R. D.,** Effect of Con A on phagocytosis, *Nature* (*London*), 235, 44, 1972.
403. **Hankins, C. N. and Shannon, L. M.,** The physical and enzymatic properties of a phytohemagglutinin from mung beans, *J. Biol. Chem.,* 253, 7791, 1978.
404. **Anderson, R. S., Day, N. K. B., and Good, R. A.,** Specific hemagglutinin and a modulator of complement in cockroach hemolymph, *Infect. Immun.,* 5, 55, 1972.
405. **Day, N. K. B., Gewurz, H., Johannsen, R., Finstad, J., and Good, R. A.,** Complement and complement-like activity in lower vertebrates and invertebrates, *J. Exp. Med.,* 132, 941, 1970.

406. **Aston, W. P., Chadwick, J. S., and Henderson, M. J.,** Effect of cobra venom factor on the *in vivo* immune response in *Galleria mellonella* to *Pseudomonas aeruginosa, J. Invertebr. Pathol.,* 27(2), 171, 1976.
407. **Chadwick, J. S., Aston, W. P., and Ricketson, J. R.,** Further studies on the effects and role of cobra venom factor on protective immunity in *Galleria mellonella:* activity in response against *Proteus mirabilis, Dev. Comp. Immunol,* 4, 223, 1980.
408. **D'Cruz, O. J. M.,** Interaction of silkworm hemolymph components with human complement, *Fed. Proc.,* 42, 1237, 1983.
409. **D'Cruz, O. J. M., Burton, N. K., and Day, N. K.,** A new role for the major larval serum protein of *Spodoptera* in modulating mammalian complement, *Fed. Proc.,* 43, 1764, 1984.
410. **Götz, P. and Boman, H. G.,** Cellular and humoral immunity in insects, in *Comprehensive Insect Physiology, Biochemistry and Pharmacology,* Vol. 3, Kerkut, G. A. and Gilbert, L. I., Eds., Pergamon Press, Oxford, 1985, 453.
411. **Boman, H. G., Faye, I., V. Hofsten, P., Kockum, K., Lee, J. Y., Xanthopoulos, K., Bonnich, G. H., Engstrom, A., Merrifield, B. R., and Andreu, D.,** Antibacterial immune proteins in insects — a review of some current perspectives, in *Immunity in Invertebrate,* Brehélin, M., Ed., Springer-Verlag, Berlin, 1986, 63.
412. **Hultmark, D., Steiner, H., Rasmussen, T., and Boman, H. G.,** Insect immunity. Purification and properties of three inducible bactericidal proteins from hemolymph of immunized pupae of *Hyalophora cecropia, Eur. J. Biochem.,* 106, 7, 1980.
413. **Hultmark, D., Engström, A., Bennich, H., Kapur, R., and Boman, H. G.,** Insect Immunity; isolation and structure of cecropin D and four minor antibacterial components from *Cecropia* pupae, *Eur. J. Biochem.,* 127, 207, 1982.
414. **Lee, J., Boman, A., Chaunxin, S., Andersson, M., Jornvall, H., Mutt, V., and Boman, H. G.,** Antibacterial peptides from pig intestine: isolation of a mammalian cecropin, *Proc. Natl. Acad. Sci. U.S.A.,* 86, 9159, 1989.
415. **Lee, S., Mihara, H., Aoyagi, H., Kate, T., Izumiya, N., and Yamasaki, N.,** Relationship between antimicrobial activity and amphiphilic property of basic model peptides, *Biochim. Biophys. Acta,* 862, 211, 1986.
416. **Boman, H. G. and Hultmark, D.,** Cell-free immunity in insects, *Annu. Rev. Microbiol.,* 41, 193-126, 1987.
417. **Okai, Y. and Qu, X.,** Effects of a purified cecropin D from a Chinese silk moth on growth, function and differentiation of murine hemopoietic cells, *Immunol. Lett.,* 20, 127, 1989.
418. **Pye, A. D. and Boman, H. G.,** Insect immunity. II. Purification and partial characterization of immune protein P5 from haemolymph of *Hyalophora cecropia* pupae, *Infect. Immun.,* 17, 408, 1977.
419. **Hultmark, D., Engström, A., Anderson, K., Steiner, H., Bennich, H., and Boman, H. G.,** Insect immunity. Attacins, a family of antibacterial proteins from *Hyalophora cecropia, EMBO J.,* 2, 571, 1983.
420. **Engström, P., Carlsson, A., Engström, A., Tao, Z-L., and Bennich, H.,** The antibacterial effect of attacins from the silk moth *Hyalophora cecropia* is directed against the outer membrane of *Escherichia coli, EMBO J.,* 3, 3347, 1984.
421. **Casteels, P., Ampe, C., Riviere, L., Van Damme, J., Elicone, C., Fleming, M., Jacobs, F., and Tempst, P.,** Isolation and characterization of apadecin, a major antibacterial response peptide in the honeybee *(Apis mellifera), Eur. J. Biochem.,* 187, 381, 1990.
422. **Dimarcq, J-L., Keppi, E., Dunbar, B., Lambert, J., Reichhart, J-M., Hoffmann, D., Rankine, S. M., Fothergill, J. E., and Hoffmann, J. A.,** Insect immunity. Purification and characterization of a family of novel inducible antibacterial proteins from immunized larvae of the dipteran *Phormia terranovae* and complete amino acid sequence of the predominant member, diptericin A, *Eur. J. Biochem.,* 171, 17, 1988.
423. **Houk, E. J. and Griffiths, G. W.,** Intracellular symbiotes of the Homoptera, *Annu. Rev. Entomol.,* 25, 161, 1980.
424. **Mohrig, W. and Messner, B.,** Immunreaktionen bei Insekten. I. Lysozym als grundlegender antibakterieller Faktor im humoralen Absehrmechanismus der Insekten, *Sonderb. Biol. Zentralbl.,* 87, 439, 1968.
425. **Chadwick, J. S.,** Relation of lysozyme concentration to acquired immunity against *Pseudomonas aeruginosa* in *Galleria mellonella, J. Invertebr. Pathol.,* 15, 455, 1970.
426. **Powning, R. F. and Davidson, W. J.,** Studies on insect bacteriolytic enzymes. 1. Lysozyme in hemolymph of *Galleria mellonella* and *Bombyx mori, Comp. Biochem. Physiol.,* 45B, 669, 1973.
427. **Hoffmann, D., Brehélin, M., and Hoffmann, J. A.,** Premiers résultats sur les réactions de défense antibactériennes de larves et d'imagos de *Locusta migratoria, Ann. Parasitol. Hum. Comp.,* 52, 87, 1977.
428. **Dunn, P. E.,** Biochemical aspects of insect immunity, *Annu. Rev. Entomol.,* 31, 321, 1986.
429. **Engström, A., Xanthopoulos, K. G., Boman, H. G., and Bennich, H.,** Aminoacid and cDNA sequences of lysozyme from *Hyalophora cecropia, EMBO J.,* 4, 2119, 1985.

430. **Anderson, R. S. and M. L. Cook.**, Induction of lysozymelike activity in the hemolymph of an insect *Spodoptera eridania, J. Invertebr. Pathol.*, 33, 197, 1979.
431. **Jarosz, J.**, Simultaneous induction of protective immunity and selective synthesis of hemolymph lysozyme protein in larvae of *Galleria mellonella, Biol. Zentralbl.*, 98, 459, 1979.
432. **Phipps, D. J., Chadwick, J. S., Leeder, R. G., and Aston, W. P.**, The hemolytic activity of *Galleria mellonella* hemolymph, *Dev. Comp. Immunol.*, 13, 103, 1989.
433. **Matsuyama, K. and Natori, S.**, Purification of three antibacterial proteins from the culture medium of NIH-Sape-4, an embryonic cell line of *Sarcophaga peregrina, J. Biol. Chem.*, 263, 17112, 1988.
434. **Lambert, J., Keppi, E., Dimascq, J., Wicker, C., Reichhart, J., Dunbar, B., Lepage, P., Dorsselaer, A., Hoffmann, J., Fothergill, J., and Hoffmann, D.**, Insect immunity. Isolation from immune blood of the dipteran *Phormia terranovae* of two insect antibacterial peptides with sequence homology to rabbit lung macrophage bactericidal peptide, *Proc. Natl. Acad. Sci. U.S.A.*, 86, 262, 1989.
435. **Selsted, M. E., Brown, D. M., Delange, R. J., and Lehrer, R. I.**, Primary structures of MCP-1 and MCP-2, natural peptide antibiotics of rabbit lung macrophages, *J. Biol. Chem.*, 258, 14485, 1983.
436. **Selsted, M. E., Brown, D. M., Delange, R. J., Harwig, S. L., and Lehrer, R. I.**, Primary structures of six antimicrobial peptides of rabbit perotoneal neutrophils, *J. Biol. Chem.*, 260, 4579, 1985.
437. **Sun, S-C., Lindström, I., Boman, H. G., Faye, I., and Schmidt, O.**, Hemlin: an insect-immune protein belonging to the immunoglobulin superfamily, *Science*, 250, 1729, 1990.
438. **Rasmuson, T. and Boman, H. G.**, Insect immunity. V. Purification and some properties of immune protein P4 from hemolymph of *Hyalophora cecropia* pupae, *Insect Biochem.*, 9, 259, 1979.
439. **Andersson, K. and Steiner, H.**, Structure and properties of protein P4, the major bacteria-inducible protein in pupae of *Hyalophora cecropia, Insect Biochem.*, 17, 131, 1983.
440. **Unestam, T. and Nylund, J. E.**, Blood reactions *in vitro* in crayfish against a fungal parasite, *Aphanomyces astaci, J. Invertebr. Pathol.*, 19, 94, 1972.
441. **Söderhäll, K.**, Prophenoloxidase activation system and melanization — a recognition mechanism of arthropods? A review, *Dev. Comp. Immunol.*, 6, 601, 1982.
442. **Söderhäll, K. and Ajaxon, R.**, Effect of quinones and melanin on mycelial growth of *Aphanomyces* spp. and extracellular protease of *Aphanomyces astaci,* a parasite on crayfish, *J. Invertebr. Pathol.*, 39, 105, 1982.
443. **Pye, A. E.**, Microbial activation of phenoloxidase from immune insect larvae, *Nature (London)*, 251, 610, 1974.
444. **Unestam, T. and Söderhäll, K.**, Soluble fragment from fungal walls elicit defense reactions in crayfish, *Nature (London)*, 254, 45, 1977.
445. **Ashida, M. and Söderhäll, K.**, The prophenoloxidase activating system in crayfish, *Comp. Biochem. Physiol.*, 77B, 21, 1984.
446. **Ashida, M. and Ohnishi, E.**, Activation of prophenoloxidase in the hemolymph of the silkworm, *Bombyx mori, Arch. Biochem. Biophys.*, 122, 411, 1967.
447. **Ashida, M. and Dohke, K.**, Activation of prophenoloxidase by the activating enzyme of the silkworm, *Bombyx mori, Insect Biochem.*, 10, 37, 1980.
448. **Wybran, J., Appleboom, T., Famaey, J. P., and Govaerts, A.**, Suggestive evidence for receptors for morphine and methionine-enkaphalin on normal human T lymphocytes, *J. Immunol.*, 123(3), 1068, 1979.
449. **Van Epps, D. E. and Saland, L.**, Beta-endorphin and met-enkephalin stimulate human peripheral blood mononuclear cell chemotaxis, *J. Immunol.*, 132(6), 3046, 1984.
450. **Van Epps, D. E. and Kutvirt, S. L.**, Modulation of human neutrophil adherence by beta-endorphin and met-enkephalin, *J. Neuroimmunol.*, 15, 219, 1987.
451. **Jancovic, B. D., Marcovic, B. M., and Spector, N. H., Eds.**, *Neuroimmune Interactions: Proceedings of the Second International Workshop on Neuroimmunomodulation,* Vol. 496, New York Academy of Sciences, New York, 1987.
452. **Perez-Polo, J. R., Bulloch, K., Angeletti, R. H., Hashim, G. A., and de Vellis, J., Eds.**, Neuroimmunomodulation, *J. Neurosci.*, 18(1), 1, 1987.
453. **Cuénot, L.**, Études sur le sang et les glandes lymphatiques dans le série animale, *Arch. Zool. Exp. Gén.*, 9, 365—475, 592—670, 1881.
454. **Heilbrunn, L. V.**, The evolution of the haemostatic mechanism, in *Functions of the Blood,* Macfarlane, R. G. and Robb-Smith, A. H. T., Eds., Academic Press, New York, 1961, 283.
455. **Gilliam, M. and Shimanuki, H.**, Coagulation of hemolymph of the larval honeybee (*Apis mellifera* L.), *Experientia,* 26, 908, 1970.
456. **Bohn, H.**, Hemolymph clotting in insects, in *Immunity in Invertebrates,* Brehélin, M., Ed., Springer-Verlag, Berlin, 1986, 188.
457. **Hoffmann, J. A. and Stoekel, M. E.**, Sur le modification ultrastructurales des coagulocytes au cours de la coagulation de l'hémolymphe chez un insecte Orthopteroïde: *Locusta migratoria, C. R. Séances Sci. Biol. Strasbourg,* 162, 2257, 1968.

458. **Armstrong, P. B. and Rickles, F. F.**, Endotoxin induced degranulation of the *Limulus* amoebocyte, *Exp. Cell Res.*, 140, 15, 1982.
459. **Durliat, M. and Vranckx, R.**, Analyses of the hemocyte proteins from *Astacus leptodactylus, C.R. Acad. Sci. (Paris)*, 282(24) D, 2215, 1976.
460. **Stutman, L. J. and Dolliver, M.**, Mechanism of coagulation in *Gecarcinus lateralis, Am. Zool.*, 8, 481, 1968.
461. **Ghidalia, W., Vendrely, R., Coirault, Y., Montmory, C., and Prouwartelle, O.**, Hemolymph coagulation in *Macropipus puber* (L.) Crustace Decapoda. Respective roles of hemolymph and plasma, *C. R. Acad. Sci. (Paris)*, 284 (I) D, 69, 1977.
462. **Marschall, K. J.**, Bau und Funktionen der Blutzellen des Mehlkafers *Tenebrio molitor* L., *Z. Morphol. Okol. Tiere*, 58, 182, 1966.
463. **Brehélin, M.**, Sur la coagulation de l'hémolymphe chez *Locusta migratoria, Ann. Parasitol. Hum. Comp.*, 52, 98, 1977.
464. **Wigglesworth, V. B.**, The haemocytes and connective tissue formation in an insect, *Rhodnius prolixus* (Hemiptera), *Q. J. Microsc. Sci.*, 97, 89, 1956
465. **Gellissen, G. and Wyatt, G. R.**, Production of lipophorin in the fat body of adult *Locusta migratoria:* comparison with vitellogenin, *Can. J. Biochem.*, 59, 648, 1981.
466. **Bohn, H., Barwig, B., and Bohn, B.**, Immunochemical analysis of hemolymph clotting in the insect *Leucophaea maderae* (Blattaria), *J. Comp. Physiol.*, 143, 169, 1981.
467. **Grégoire, Ch.**, Blood coagulation in Arthropods. III. Reactions of insect haemolymph to coagulation inhibitors of vertebrate blood, *Biol. Bull. (Woods Hole, Mass.)*, 104, 372, 1953
468. **Muttkowski, R. A.**, Studies on the blood of insects. III. The coagulation and clotting of insect blood, *Bull. Brooklyn Entomol. Soc.*, 19, 128, 1924.
469. **Beard, R. L.**, Experimental observations on coagulation of insect hemolymph, *Physiol. Zool.*, 23, 47, 1950.
470. **Shull, W. E., Riley, M. K., and Richardson, C. H.**, Some effects of certain toxic gases on the blood of the cockroach *Periplaneta orientalis* (L.), *J. Econ. Entomol.*, 25, 1070, 1932.
471. **Shull, W. E. and Rice, P. L.**, A method for temporary inhibition of blood coagulation in insects, *J. Econ. Entomol.*, 26, 1083, 1933.
472. **Grégoire, Ch.**, Coagulation de l'hémolymphe chez les insectes irradies par les rayons X, *Arch. Int. Physiol.*, 63(2), 246, 1955.
473. **Bowen, T. J. and Kilby, B. A.**, Electrophoresis of locust haemolymph, *Arch. Int. Physiol.*, 61, 413, 1953.
474. **Whitten, J. M.**, Haemocyte activity in relation to epidermal cell growth, cuticle secretion and cell death in a metamorphosing cyclorrhaphan pupa, *J. Insect Physiol.*, 15, 763, 1968.
475. **Yeager, J. F. and Munson, S. C.**, Changes induced in the blood cells of the southern armyworm *(Prodenia eridania)* by the administration of poisons, *J. Agric. Res.*, 64(6), 307, 1942.
476. **Gupta, A. P. and Sutherland, D. J.**, Effects of sublethal doses of chlordane on the hemocytes and midgut epithelium of *Periplaneta americana, Ann. Entomol. Soc. Am.*, 61(4), 910, 1968.
477. **Yeager, J. F. and Munson, S. C.**, Histochemical detection of glycogen in the blood cells of the southern armyworm *(Prodenia eridania)* and in other tissues, especially midgut epithelium, *J. Agric. Res.*, 63(5), 257, 1941.
478. **Du Pasquier, L.**, Phylogenesis of the vertebrate immune system, in *The Immune System*, Melchers, S. F. and Rajewsky, K., Eds., Springer-Verlag, Berlin, 1976, 101.
479. **Valdiserri, R. O.**, Decreased T-helper/T-suppressor ratios in homosexual men: a non-specific ratio, *Lab. Med.*, 15(7), 472, 1984.
480. **Miyan, J. A.**, 'Killer'-cell-mediated destruction of dipteran eclosion muscles, *Proc. R. Soc. London*, B236, 91, 1989.
481. **Miyan, J. A.**, Neural control in the immunocytotoxic destruction of muscles in Diptera, *Tissue Cell*, 22(5), 673, 1990.
482. **Gateff, E.**, Hematopoietic malignancies of genetic origin in *Drosophila melanogaster, 17th Int. Congr. Entomol.*, 18.1(5)(Abstr.), 760, 1984.
483. **Matz, G.**, Cellular defence mechanisms in *Leucophaea maderae* Babr. and *Locusta migratoria* tumors (Insecta), *Dev. Comp. Immunol.*, 7, 669, 1983.
484. **Widmer, M. B. and Bach, F. H.**, Antigen-driven helper cell-independent cloned cytotoxic T lymphocyte, *Nature (London)* 294, 750, 1981.
485. **Shepherd, J. C. W., McGinnis, W., Carrasco, A. E., De Robertis, E. M., and Gehring, W. J.**, FLy and frog homeo domains show homologues with yeast mating types regulatory protein, *Nature (London)*, 310, 70, 1984.
486. **Vasta, G. R., Marchalonis, J. J., and Kohler, H.**, Invertebrate recognition protein cross-reacts with an immunoglobulin idiotype, *J. Exp. Med.*, 159, 1270, 1984.
487. **Braman, A. M. and Schwartz, K. A.**, Platelet disorders, *Lab. Med.*, 20(12), 831, 1989.

488. **Holmsen, H.,** Biochemistry of the platelet release reaction, in *Ciba Foundation Symposium 35, Biochemistry and Pharmacology of Platelets,* Weatherall, D. J., chairman, Elsevier North-Holland, Amsterdam, 1975, 175.
489. **Clemetson, K. J.,** Glycoproteins of the platelet plasma membrane, in *Platelet Membrane Glycoproteins,* George, J. N., Nurden, A. T., and Philips, D. R., Eds., Plenum Press, New York, 1985, 51.
490. **Behnke, O.,** Further studies on microtubules. A marginal bundle in human and rat thrombocytes, *J. Ultrastruct. Res.,* 13, 469, 1965.
491. **White, J. G.,** Platelet morphology, in *The Circulating Platelet,* Johnson, S. A., Ed., Academic Press, New York, 1971, 447.
492. **Bierne, J.,** Non-self recognition in *Lineus:* allografts succeed because of the monomorphism of the major histocompatibility genes, *Dev. Comp. Immunol.,* 9(1)(Abstr.), 172, 1985.
493. **Berggard, I. and Bearn, A. G.,** Isolation and properties of a low molecular weight β_2-globulin occurring in human biological fluids, *J. Biol. Chem.,* 243, 4095, 1968.
494. **Nakamuro, K., Tanigaki, N., and Pressman, D.,** Multiple common properties of human β_2-microglubulin and the common portion fragment derived from HA-A antigen molecules, *Proc. Natl. Acad. Sci. U.S.A.,* 70, 2863, 1973.
495. **Roch, P. G., Cooper, E. L., and Eskinasi, D. P.,** A β_2-microglubulin-like molecule on earthworm leucocyte membranes, *Eur. J. Immunol.,* 13, 1037, 1983.
496. **Thomas, I. G. and Ratcliffe, N. A.,** Studies on recognition of foreignness in insects utilising integumental transplants, in *Aspects of Developmental and Comparative Immunology,* Solomon, B., Ed., Pergamon Press, Oxford, 1980, 105.
497. **Thomas, I. G. and Ratcliffe, N. A.,** Integumental grafting and immunorecognition in insects, *Dev. Comp. Immunol.,* 6, 643, 1982.
498. **Jones, S. E. and Bell, W. J.,** Cell-mediated immune-type response of the American cockroach, *Dev. Comp. Immunol.,* 6, 35, 1982.
499. **George, J. F., Karp, R. D., and Rheins, L. A.,** Primary integumentary xenograft reactivity in the American cockroach, *Periplaneta americana, Transplantation,* 37, 478, 1984.
500. **Cooper, E. L.,** *Comparative Immunology,* Prentice-Hall, Englewood Cliffs, NJ, 1976.
501. **Wigglesworth, V. B.,** The function of the corpus allatum in the growth and reproduction of *Rhodnius prolixux* (Hemiptera), *Q. J. Microsc. Sci.,* 79, 91, 1937.
502. **Ephrussi, B. and Beadle, G. W.,** A technique for transplantation for *Drosophila, Am. Nat.,* 70, 218, 1936.
503. **Larsen, J. R. and Bodenstein, D.,** Humoral control of egg maturation in mosquitoes, *J. Exp. Zool.,* 140, 343, 1959.
504. **Peterson, L.,** Cellular immune responses of insects to foreign tissue implants, Ph.D. dissertation, University of Illinois, Urbana, IL, 1968.
505. **Bell, W. J.,** Yolk formation by transplanted cockroach oocytes, *J. Exp. Zool.,* 181, 41, 1972.
506. **Guthrie, D. M.,** Physiological competition between host and implanted ganglia in an insect *(Periplaneta americana), Nature (London),* 210, 312, 1966.
507. **Bohn, H.,** The origin of the epidermis in supernumerary regenerates of triple legs in cockroaches (Blattaria), *J. Embryol. Exp. Morphol.,* 28, 185, 1972.
508. **Carton, Y.,** Isogenic, allogenic and xenogenic transplants in an insect species, *Transplantation,* 21, 17, 1976.
509. **Piepho, H.,** Uber das Ausmass der Arlungspezifizitat von Metamorphosphormone bei Insekten, *Biol. Zentralbl.,* 69, 1, 1950.
510. **Wigglesworth, V. B.,** The role of the haemocytes in the growth and moulting of an insect, *Rhodnius prolixus* (Hemiptera), *J. Exp. Biol.,* 32, 649, 1955.
511. **Shrivastava, S. C. and Richards, A. G.,** An autoradiographic study of the relation between hemocytes and connective tissue in the wax moth, *Galleria mellonella* L., *Biol. Bull. (Woods Hole, Mass.),* 128, 337, 1965.
512. **Kollmann, M.,** Recherches sur les leucocytes et le tissue lymphoïde des invertébres, *Ann. Sci. Nat. Zool.,* 9, 1, 1908.
513. **Paillot, A.,** Le karyocinetose, nouvelle réaction d'immunité naturelle observe chez les chenilles de macrolepidoptères, *C.R. Acad. Sci. (Paris),* 169, 396, 1919.
514. **Paillot, A. and Noel, R.,** Recherches histophysiologiques sur les cellules péricardiales et les élements du sang des larves d'insectes (*Bombyx mori* et *Pieris brassicae*), *Bull. Hist. Appl. Physiol. Pathol.,* 5, 105, 1928.
515. **Rizki, M. T. M.,** The larval blood cells of *Drosophila willistoni, J. Exp. Zool,* 123, 397, 1953.
516. **Nappi, A. J.,** Hemocytes of larvae of *Drosophila euronotus* (Diptera: Drosophilidae), *Ann. Entomol. Soc. Am.,* 63, 1217, 1970.
517. **Selman, J.,** The fate of the blood cells during the life history of *Sialis lutaria* L., *J. Insect Physiol.,* 8, 209, 1961.

518. **Müller, K.**, Uber die Korpuskularen Elemente der Blutflussigkeit bei der erwaschsenen Honigbiene *(Apis mellifica), Erlanger Jahr. Blenenkunde,* 3, 5, 1925.
519. **Tuzet, O. and Manier, F. F.**, Recherches sur les hémocytes fusiforme geants (= ''vermiform cells'' de Yeager) de hémolymphe d'insects homolmetaboles et heterometaboles, *Ann. Sci. Nat. Zool. Biol. Anim.*, 12(1), 81, 1959.
520. **Jones, J. C.**, The hemocytes of *Rhodnius prolixus* (Stal), *Biol. Bull.* (Woods Hole), 129, 282, 1965.
521. **Bogojavlensky, K. S.**, The formed elements of the blood of insects, *Arch. Russ. Anat. Hist. Embryol.*, 11, 361, 1932 (in Russian).
522. **Whitten, J. M.**, Haemocytes and the metamorphosing tissues in *Sarcophaga bullata, Drosophila melanogaster,* and other cyclorrhaphous Diptera, *J. Insect Physiol.*, 10, 447, 1963.
523. **Bruntz, L.**, Nouvelles recherches sur l'excretion et la phagocytose chez les Thysanoures, *Arch. Zool. Exp. Gén.*, 38, 471, 1908.
524. **Arvy, L. and Gabe, M.**, Identification des diastrases sanguines chez quelques insectes, *C.R. Soc. Biol.* (Paris), 140, 757, 1946.
525. **Arvy, L.**, Données sur la leucopoïèse chez *Musca domestica* L., *Proc. R. Entomol. Soc. London,* A29, 39, 1954.
526. **Arnold, J. W.**, Larval hemocytes in Noctuidae (Insects: Lepidoptera), *Int. J. Insect Morphol. Embryol.*, 11(3/4), 173, 1982.
527. **Wigglesworth, V. B.**, The physiology of the cuticle and of ecdysis in *Rhodnius prolixus* (Triatomidae, Hemiptera), with special reference to the function of the oenocytes and of the dermal glands, *Q. J. Microsc. Sci.*, 76, 269, 1933.
528. **Metalnikov, S. and Gaschen, H.**, Immunité cellulaire el humoral chez la chenille, *Ann. Inst. Pasteur,* 36, 233, 1922.
529. **Tateiwa, J.**, Le formule leucocytaire du sang des chenilles normales et imunisées de *Galleria mellonella, Ann. Inst. Pasteur,* 42, 791, 1928.
530. **Metalnikov, S. and Chorine. V.**, On the natural and acquired immunity of *Pyrausta nubilalis* Hb., *Int. Corn Borer Invest. (Chicago)*, 2, 22, 1929.
531. **Jones, J. C.**, The normal hemocyte picture of the yellow mealworm *Tenebrio molitor* Linnaeus, *Iowa State Coll. J. Sci.*, 24, 356, 1950.
532. **Jones, J. C. and Tauber, O. E.**, Abnormal hemocytes in mealworms (*Tenebrio molitor* L.), *Ann. Entomol. Soc. Am.*, 47(3), 428, 1954.
533. **Wille, H. and Vecchi, M. A.**, Étude sur l'hémolymphe de l'abeille (*Apis mellifica* L.). I. Les frottis de sang de l'abeille adelle adulte d'éte, *Mitt. Schweiz. Entomol. Ges.*, 34, 69, 1966.
534. **Lutz, K. G.**, Das Bluten der Coccinelliden, *Zool. Anz.*, 18, 244, 1895.
535. **Hrdy, I.**, Comparison of preparation and staining techniques of insect blood cells, *Acta Soc. Entomol. Czech.*, 54, 305, 1957.
536. **Arnold, J. W.**, The haemocytes of the Mediterranean flour moth, *Ephestia kühniella* Zell. (Lepidoptera: Pyralidae), *Can. J. Zool.*, 30, 352, 1952.
537. **Graber, V.**, Ueber die Blutkorperchen der Insekten, *Sitzb. Akad. Math-Nat. Wiss. Wien,* 64, 9, 1871.
538. **Gupta, A. P., Das, Y. T., and Rao, B. R.**, Anatomy of cockroach, in *Cockroaches as Model Systems for Neurobiology: Applications in Biomedical Research,* Huber, I., Masler, E. P., and Rao, B. R., Eds., CRC Press, Boca Raton, FL, 1990, 11.

Chapter 3

ENCAPSULATION IN INSECTS: POLYDNAVIRUSES AND ENCAPSULATION-PROMOTING FACTORS

D. Huw Davies and Mike T. Siva-Jothy

TABLE OF CONTENTS

I. INTRODUCTION

Many organisms that have parasitic associations with insect hosts redirect or suppress their host's hemocytic defense mechanisms in order to evade encapsulation.[1-6] We are only beginning to appreciate the varied mechanisms by which pathogens can subvert their host's cellular and humoral responses. One such mechanism, which centers around a fascinating mutualistic relationship between polydnaviruses and particular superfamilies of parasitic Hymenoptera, may provide invertebrate cell biologists with an important tool with which to understand the molecular and cellular events that lead to encapsulation and other multicellular responses in insects.

In this review, we will first examine a class of novel soluble factors that mediate the adhesive and motile properties of cells involved in encapsulation (so-called cell adhesion factors or encapsulation promoting factors). Interest in these factors has been raised by the possibility that they represent targets for organisms that interfere with encapsulation. We shall then summarize the inhibitory properties of polydnaviruses on the response by insect hemocytes to pathogens.

II. ARTHROPOD RESPONSES TO PATHOGENS

The cellular phenomena that occur when hemocytes encounter and sequester different pathogens is well documented, and despite the diversity of arthropods under study, generalizations about the cellular reactions elicited can be made. The following scheme is based on observations from insects and crustaceans. Although simplified, it serves as a useful framework to appreciate the many stages at which the intervention of a response could occur.

A. A MODEL

We have divided the hemocytes involved in these reactions into two functional types (Figure 1). The initial recognition hemocyte (RH), which is generally granular and phagocytic, is responsible for initiating cellular reactions by adhering to foreign surfaces. Microscopic targets (yeast, bacteria, erythrocytes, India ink pigment particles) are eliminated by phagocytosis, the simplest and most primordial of self/nonself adhesion phenomena. Small numbers of microscopic particles are cleared effectively by phagocytosis alone (possibly by a distinct population of efficient, nongranular phagocytes, such as the hyaline cell of crustaceans). In most cases, however, this is performed by RHs. Recognition of foreign surfaces by these cells promotes the release, by degranulation, of a variety of substances, including components of the phenoloxidase cascade, viscous extracellular matrix, and presumably other factors that recruit additional cells. Large numbers of microscopic pathogens often result in sticky aggregations of RHs and particles, and a similar milieu of RHs and extracellular matrix occurs at the surface of macroscopic targets (parasites, parasitoid larvae, glass beads, xenografts, and implants) which are too large to be internalized by individual cells. This is a common initial phase of multicellular responses (nodulation, encapsulation, and possibly wound repair).

Subsequent sequestration within a multicellular sheath of hemocytes serves to neutralize the pathogen. This is often completed by recruitment of a second functional hemocyte type. Recruitment causes these "sequestration hemocytes" (SHs) to become mutually adhesive and to cooperate to form the multicellular sheath. SHs are usually non- or weakly phagocytic, but capable of extensive flattening down within the multicellular sheath as well as on planar surface *in vitro*. In other cases, sequestration may be performed by more RHs. Thus, the recognition and sequestration events may be performed by distinct hemocyte populations (such as the granulocytes and plasmatocytes of some insects, which results in biphasic multicellular responses), or by the same population (such as the semigranular cells of crus-

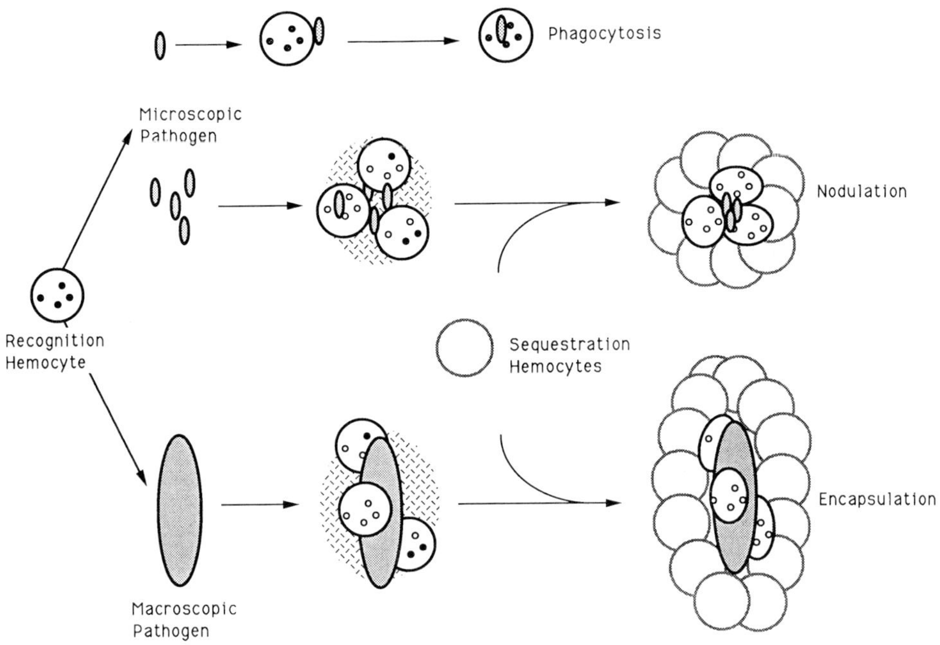

FIGURE 1. A schematic representation of insect cellular responses to pathogens (see text for details). Recognition of foreign surfaces is performed by specialized recognition hemocytes (RHs). Microscopic pathogens may be removed by phagocytosis. Multiple microscopic pathogens and macroscopic pathogens elicit degranulation by multiple RHs, localized coagulation, and recruitment of sequestration hemocytes (SHs). Recruited SHs become mutually adhesive and form a multicellular tissue within which they flatten down to sequester the pathogens. RHs and SHs may be a single multifunctional hemocyte or two separate hemocyte types.

taceans, and the granulocytes of other insects), and which therefore possess the combined abilities required to execute both events.

B. ADHESION-PROMOTING FACTORS

In addition to the conspicuous cellular component of arthropodan responses to pathogens outlined above, there is also an important humoral component. This may include components of the phenoloxidase biosynthetic pathway and/or lectins. In recent years, an additional class of soluble mediators has been described. These factors possess multiple properties associated with the activities of the hemocytes involved in multicellular responses. So far, these factors have been demonstrated only in *in vitro* spreading and encapsulation assays where they promote adhesion to glass and other abiotic surfaces. In insects, the change by SHs from quiescence to a state of enhanced adhesive, and consequently motile, behavior occurs simply by removal from the hemocoel. In general, crustacean hemocytes seem less sensitive to the effects of bleeding. For insects, however, a variety of methods can be used to reverse or inhibit this "activation" process, and in so doing enable the design of assays to detect factors that can cause activation. Precisely what causes this change in adhesive behavior is not clear. One possibility is that epidermal tissue (which is damaged when hemolymph is removed) is a source of factors that initiate the change. The only support for this has come from early studies by Cherbas.[7] By removing hemolymph from diapausing saturniid pupae in such a way as to avoid contact with damaged epidermal tissue, plasmatocytes could be maintained in culture in their quiescent state. Fragments of epidermal tissue or a 50-kDa protein purified therefrom (termed hemokinin) caused the cells to become adhesive and ameboid. These experiments possibly represent the first description of an adhesion-promoting factor, but were unfortunately never pursued.

Another method of obtaining quiescent insect hemocytes is trypsinization.[8,9] *Heliothis virescens* (noctuid) plasmatocytes (PLs) treated in this way lose all adhesive properties and consequently fail to encapsulate *in vitro*. This *in vitro* encapsulation system relies on the natural tendency of freshly withdrawn hemocytes to clump upon agitation. When a suitable target is included, the hemocytes clump around it, and subsequently flatten down to produce a cohesive capsule. Ultrastructurally, *in vitro* capsules are similar to their *in vivo* counterparts, being composed of numerous flattened SHs (PLs).[8] However, the inner RH (granulocyte) layer is absent; evidently the model does not reproduce the complex biphasic response seen *in vivo,* perhaps in part because of SH "activation" during the bleeding process. Nevertheless, encapsulation capacity, which is abolished if the hemolymph is collected into medium supplemented with trypsin, is restored upon adding back cell-free hemolymph. The active component has been termed encapsulation-promoting factor (EPF). Despite being present in the cell-free fraction, EPF is likely to originate in the hemocytes or the damaged integument. Trypsinized PLs are also unable to spread over glass coverslips *in vitro;* similarly, this effect is reversed by EPF. Whether EPF mediates granulocyte-mediated phenomena in *H. virescens* (degranulation, phagocytosis, rosetting of erythrocytes, and participation in the first phase of encapsulation/nodulation) is unresolved and will require the development of suitable assays. EPF has been purified by HPLC, using *in vitro* encapsulation and spreading by PL *in vitro* as bioassays. Although it has not been fully characterized, EPF appears to be a low-molecular-weight polypeptide, and has no hemagglutinating activity.

In another system, trypsinization of monolayers of hemocytes from the crayfish, *Parachaeraps bicarinatus,* abolishes their phagocytic capacities; phagocytosis can be restored by adding cell-free crayfish hemolymph.[10] The trypsin-sensitive proteins that underlie these adhesive phenomena are not known, but are likely to be serum components (such as EPF itself). It is perhaps remarkable that the ligand for such a factor on the PL surface is insensitive to trypsin.

A factor with similar properties has been purified from the lysate of hemocytes from the crayfish *Pacifastacus leniusculus.*[11] It is a 76-kDa protein normally located within the granules of hemocytes, but released by degranulation in response to bacterial LPS or fungal β-1,3-glucans. To date this factor has been reported to promote hemocyte attachment and spreading,[11] degranulation,[12] and encapsulation.[13] The pleotropic nature of this protein may reflect the multifunctional nature of the hemocytes with which it cooperates. Interestingly, the properties of the 76-kDa protein are mimicked by the synthetic pentameric peptide GRGDS but not the analogue GRGES.[14] In higher vertebrates, the motif RGD is crucial for the binding between extracellular matrix (ECM) proteins, such as fibronectin, and their family of ligands (integrins) expressed on many cell types. The implications are that the 76-kDa protein may have the RGD motif and/or crustacean hemocytes bear cell surface integrins.

Owing to a paucity of information regarding the biochemical details of hemocyte adhesion, we have drawn on data from phylogenetically diverse organisms. Despite superficial similarities in the properties of these factors from insects and crustaceans, the homology between them is unknown. Nevertheless, we believe that the existence of soluble factors that mediate the adhesive activities of hemocytes is widespread in invertebrates.

III. INTERFERENCE WITH ENCAPSULATION

Arthropods normally mount an encapsulation response in order to sequester macrobiotic pathogens. However, many species of endoparasitic Hymenoptera (parasitoid wasps) lay their eggs into insect hosts, where the eggs develop and hatch without being subjected to any overt attack from the host's normal responses to pathogens. Parasitoids have therefore evolved mechanisms to avoid this response. Both passive and active strategies have been envisaged, but we still do not fully understand how these mechanisms operate. In most studied examples, it is now apparent that various cofactors associated with oviposition are in part responsible for actively interfering with the host's encapsulation response.[15] These cofactors are proving to be useful tools for understanding the encapsulation responses with which they interfere. Such cofactors include various accessory gland secretions, such as mucins and venoms.[6,16-19] Other sources of active interference may come from factors secreted by the teratocytes[20-22] or the parasitoid egg.[23] These may be supplemented by passive mechanisms.[24,25] However, of all the known ovipositional cofactors, the most intensively studied has been the polydnaviruses and virus-like particles associated with the reproductive tracts of parasitoids.[26,27] The role played by these viruses in a number of different host/parasitoid relationships is currently under study.

A. VIRUS-LIKE PARTICLES

Historically, our understanding of polydnaviruses and evasion of encapsulation by parasitoids has been due to pioneering work by Salt who studied the ability of eggs of the ichneumonid parasitoid *Venturia* (formerly *Nemeritis) canescens* to avoid the hemocytic reactions of the host.[28,29] He established that this property was owing to a protective coat on the surface of the egg acquired in the calyx of the adult parasitoid. Later studies established that the protective coat consisted of a dense suspension of virus-like particles.[30-33] A search for similar protective particles in the reproductive tracts of other wasps[34-36] revealed this to be a widespread phenomenon.[37,38] Ironically, however, the prototype virus-like particle (VLP) from *Venturia* now appears to be somewhat unusual in both its structure and its *modus operandi* when compared with the polydnaviruses as a whole. First, attempts at genomic analysis were thwarted by the apparent absence of significant quantities of nucleic acid associated with these particles.[33] This stands in contrast to the remarkable complexity of polydnavirus genomic organization (see Section III.B.1). As "empty virions", *Venturia* VLPs seemingly interfere with encapsulation by a mechanism that must preclude viral expression within the host. Eggs denuded of surface VLPs by abrasion, are encapsulated, suggesting that VLPs mediate protection only while adherent to the egg surface. This is also in contrast with most polydnaviruses that can operate remotely from the egg. It is believed that *Venturia* VLPs protect either by providing a surface to the parasitoid egg that is not recognized by the host hemocytes as foreign (passive defense), or by a more active method to interfere with encapsulation. Recent studies have addressed these questions, and are now inevitably making inroads into the biochemical details of recognition. The use of mammalian antisera raised to the particles has shown that they share antigenic similarities with a 42-kDa protein found in the tissues of the host *Ephestia*.[39] Immunofluorescence studies indicate it is localized on the surface of tissues lining the hemocoel, and in hemocytes (from where it is possibly deposited onto the tissues). *Venturia* VLPs may therefore confer passive protection by mimicking the host's surfaces. However, this protein may also play a more active role.[40] It is induced upon bacterial infection, an event that is associated with a poor encapsulation response.[41] The evolutionary relationship between *Venturia* VLPs and the polydnaviruses is not known.

B. POLYDNAVIRUSES

Polydnaviruses can be divided into two morphological subgroups.[42] One, found predominantly in braconids, has cylindrical nucleocapsids of varying length and is surrounded by a single membrane. The other, found predominantly in ichneumonids, exhibits various ovoid or bullet-shaped morphologies and is surrounded by two membranes. It is now apparent that polydnaviruses can coexist with other viruses in the reproductive tracts of some ichneumonids.[43,44]

Available evidence (discussed in Section III.B.3). suggests that polydnaviruses modulate the physiology of the host to the benefit of the parasitoid by penetrating host tissues and expressing proteins. Braconid polydnaviruses usually require the synergistic action of parasitoid venom to exert their suppressive effects on encapsulation,[45-51] possibly by promoting the penetration of polydnaviruses into host cells and the uncoating of the nucleocapsids into the cytoplasm.[42] The role of ichneumonid venom in parasitism has not been established; indeed manual injection experiments indicate that polydnavirus in the absence of venom is sufficient to suppress encapsulation.

1. Gene Expression

A characteristic feature of polydnaviruses is complex genomes, consisting of several circular, double-stranded DNA segments. Our understanding of the life cycle of polydnaviruses is incomplete (Figure 2). The most extensively studied polydnavirus is that associated with the ichneumonid *Campoletis sonorensis* and a noctuid host *Heliothis virescens*. The *C. sonorensis* polydnavirus genome consists of at least 28 superhelical, closed circular DNA segments that range in size from 6 to 21 kilobases.[52] The genome is transcriptionally multipartite[52,53] and homologous regions, consisting of a family of repeated elements, exist between a number of the segments.[54] Electron microscopic studies[55,56] show that viral replication occurs within the nuclei of specialized calyx cells in the reproductive tract of the female parasitoid where capsid proteins assemble around the viral DNA and become enveloped. Viral DNA can be detected in most tissues in both male and female wasps, although the vast majority is found in the calyx tissue where viral transcription and replication occurs.[57-59] Mature virus particles are released by budding into the lumen of the oviduct and form the major component of "calyx fluid" (often used by experimenters as a crude source of polydnavirus) which is injected into the host with the egg during oviposition. Upon introduction into the hemocoel of the host during oviposition, virions penetrate a variety of tissues including fat body, hemocytes, muscle, and tracheal epithelium.[60] Viral mRNAs are detected within 2 h of oviposition,[53,61,62] and the pattern of viral mRNA expression is reproduced after the injection of control larvae with purified polydnavirus.[63] Two genes (WHv1 and WHv2), which are vigorously expressed during the early stages of parasitization, have been studied in detail.[62,64] These genes, which exhibit strong homology with each other, map to different superhelical DNA segments of the viral genome, and the transcripts of these two genes are spliced. Expression of viral genes persists throughout development of the parasitoid within the host.[58,61] Viral replication does not appear to occur in infected host tissue, although the concentration of detectable polydnavirus DNA increases rapidly in the developing parasitoid larva immediately prior to emergence from the host.[55,58] The remainder of the viral life cycle is not clear. The presence of viral DNA segments in noncalyx tissue in adult wasps, in both integrated and extrachromosomal forms, indicates the possibility of germ line transmission,[57] although the idea that the virus is acquired by the developing parasitoid larvae by ingestion of infected host tissue has not been refuted.

Similar polydisperse ds superhelical DNA genomes are reported for other polydnaviruses,[27] notably for the ichneumonid *Diadegma terebrans*[65] and the braconid *Hyposoter exiguae*.[66] It has been suggested that *C. sonorensis* polydnavirus organization may have resulted from multiple recombination events and sequence divergence,[54] although our understanding of the evolution of multipartite genomes awaits further comparative data.

2. Gene Products

Studies which report that the effects of polydnaviruses are abolished if the viral DNA is inactivated (such as by UV irradiation/psoralen treatment),[45,67,68] allude to the importance of gene expression in host regulation. Of particular interest in this context are those genes that are transcribed in the parasitized host only. The products of these genes are prime candidates for modulators of host physiology. Comparative SDS-PAGE studies of the parasitized and unparasitized hosts of many different parasitoids reveal the appearance of novel

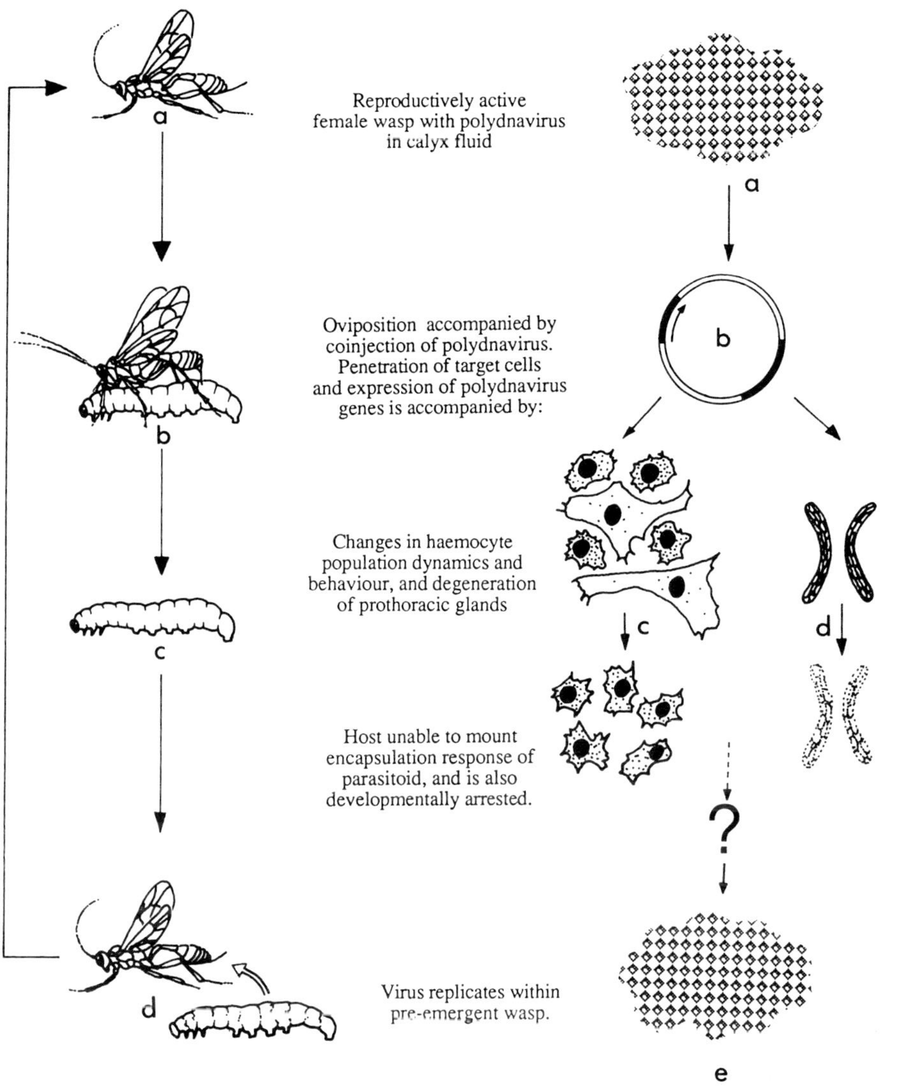

FIGURE 2. A generalized scheme for the key events in the life history of a polydnavirus (see text for details). The left side of the diagram represents host/parasitoid relationships, while the right side represents polydnavirus cell and molecular biology. Egg-laying female wasps (left, a) carry replicating polydnaviruses in the calyces of their reproductive tracts (right, a). Oviposition into a suitable host (left, b) is accompanied by co-injection of polydnavirus which penetrates host tissues wherein viral genes are expressed (right, b). Parasitization is accompanied by physiological changes in the host including perturbations of hemocyte function (right, c) and hormonal control of development (right, d). As a result, host development and encapsulation response is arrested. Replication of the virus occurs in the developing parasitoid shortly before emergence from the host.

proteins in the hemolymph within a few hours of oviposition.[68,70,71] In most cases, this effect of parasitism can be reproduced by injection of viable polydnaviruses in the absence of the parasitoid egg. Moreover, viral inactivation experiments clearly show that some of the novel proteins are products of polydnavirus gene expression.[68,71] In one study, the induction of viral gene products did not occur in nonpermissive hosts of the parasitoid.[71] Together these data strongly suggest that polydnavirus gene expression is essential for it to confer its protective effect toward the parasitoid progeny, and that this protection is mediated by the gene products.

Studies to map the genes encoding the products identified by comparative electrophoretic studies should help clarify their roles. A more direct approach is to generate cDNA from viral mRNA transcripts expressed in parasitized hosts. Only one novel protein is detected in the hemolymph from larvae of *H. virescens* parasitized by *C. sonorensis* using SDS-PAGE.[71] However, at least 12 mRNA species are identified in these larvae, of which some have been cloned and expressed *in vitro*.[63] This technology will allow us to test directly the effects of viral gene products on host physiology. In lieu of this, studies to examine their effects at the morphological level have produced some interesting observations.

3. Effects on Cellular Responses: Hemocyte Behavior *In Vivo* and *In Vitro*

In recent years investigators have noticed that manual injection of polydnaviruses induce a variety of physiological changes in the host that mimic the effects of natural parasitism. One such effect is a suppressive effect on encapsulation. We are still ignorant of precisely how this suppression occurs. However, a common phenomenon has been observed in many host/parasitoid systems, which may underlie the suppression; specific hemocyte phenotypes are often deleted from circulation or have a reduced capacity to spread over planar surfaces *in vitro*. The predominant target hemocyte for this effect is the sequestration hemocyte (SH).

The phenomenon was first described in *Drosophila* larvae parasitized by *Pseudeucoila*.[72,73] These larvae have abnormally low numbers of SHs (lamellocytes). In this case, the parasitoid-derived factor that mediates the effect was later identified as a hemolytic substance ("lamellolysin") found in an accessory gland of the female reproductive tract which specifically eliminates lamellocytes, seemingly without compromising the ability of the host to survive until the parasitoid emerges.[6] In this system, polydnaviruses do not seem to be involved. Lamellocyte destruction is essential for the egg to survive since parasitoids that do not induce lysis fail to survive within the host.[74]

Tent caterpillar *(Malacosoma disstria)* larvae are unable to encapsulate eggs of the ichneumonid *Hyposoter fugitivus*.[5] This effect is mediated by polydnavirus, because eggs also fail to be encapsulated if injected manually in the presence of calyx fluid. Cytological analysis of parasitized or virus-injected larvae revealed a change in hemocyte behavior, associated predominantly with a reduction in the spreading behavior of the PL *in vitro*.

Similarly, the tussock moth, *Orgyia leucostigma* is unable to encapsulate oviposited eggs of *Cotesia melanoscela,* and it exhibits a similar pattern of reduced PL spreading capacity and reduced numbers in circulation.[46] This phenomenon is also reproduced by manual injection of polydnavirus *plus* (typically for braconids) venom.

The same phenomenon occurs in the model of *C. sonorensis* parasitism where wasp calyx fluid is manually injected into larvae of the tobacco budworm, *H. virescens.* Such larvae fail to encapsulate eggs of the parasitoid or abiotic glass implants. There is a rapid depletion of the majority of circulating PLs; those that remain exhibit retarded spreading behavior *in vitro*.[67] Here, the effect is targeted to the SHs (PLs) only. Phagocytosis, erythrocyte rosetting, hemolymph coagulation, and the initial recognition stages of encapsulation and nodulation (phenomena attributed to granulocytes, the RHs) remain unaffected.[75] Interestingly, naturally parasitized larvae show similar hemocytic changes, while retaining the ability to encapsulate glass implants normally.[67]

In another example, larvae of the common armyworm, *Pseudaletia separata,* are successfully parasitized by the braconid *Micropletis mediator*.[76] Such larvae also fail to effect complete encapsulation of co-injected abiotic targets (Sephadex beads). Such capsules appear to have attracted only a layer of RHs (granulocytes); the subsequent sequestration phase fails to occur. The behavior of the SHs (PLs) was not examined, although the granulocytes were affected, displaying a reduced propensity to extend surface processes (filopodia). These effects can be reproduced by injection of calyx fluid plus venom. In contrast, *Pseudaletia separata,* which are also successfully parasitized by the braconid *Apanteles* (= *Cotesia) kariyai*, retain their ability to encapsulate Sephadex beads; the granulocytes also have normal filopodia.[76]

The simplest explanation for the reduced *in vitro* spreading capacity that usually afflicts the SHs from parasitized hosts is that polydnaviruses penetrate these cells with direct pathological consequences. The target tissue for polydnavirus penetration that mediates these effects, however, has not been identified. Attempts to infect *H. virescens* hemocytes with *C. sonorensis* polydnavirus *in vitro* had no effect on the morphology or *in vitro* encapsulation capacity of these cells.[89] There is also no evidence from others working in the field that polydnaviruses mediate their suppressive effect directly. The most plausible alternative explanation is that penetration or infection of some other tissue causes suppression indirectly, mediated by viral gene products.

Whatever the cause, reduced spreading is likely to be the effect of a failure to adhere normally to foreign surfaces. The relevance of adhesion and consequent spreading by PL to planar glass surfaces *in vitro* is not well understood, although, as mentioned in our discussion of encapsulation-promoting factors, it is likely to be mediated by soluble plasma components. We have speculated that these factors are liberated by damaged integument and/or granular RHs. A wider understanding of how adhesion factors operate will allow us to test whether their functions are blocked by the effects of parasitization and in particular by the products of polydnaviruses. Preliminary experiments indicate that the EPF from *H. virescens* is not the target for the suppressive effect on encapsulation caused by the *C. sonorensis* polydnavirus, because cell-free hemolymph from parasitized larvae contain EPF activity in both the *in vitro* spreading and encapsulation bioassays with trypsinized hemocytes.[90]

The mechanical basis of spreading behavior is clearly an *in vitro* manifestation of the natural ability of these cells to interdigitate with other hemocytes to produce a compact multicellular tissue during encapsulation. Treatments to disrupt the filamentous actin organization of the cytoskeleton in PLs interferes with spreading and the compaction of capsules formed *in vitro*,[77] supporting the idea that these are analogous motile phenomena. The similar behavioral changes by SH from parasitized hosts may, therefore, have similar consequences to the function of these cells in response to pathogens *in vivo*. Together these data suggest polydnaviruses may interfere with encapsulation by reducing the adhesive properties of SHs.

A contentious issue concerns those parasitized larvae which retain their abilities to mount encapsulation responses to other targets.[67,76] Precisely how this specificity is achieved is unclear, but it points toward a role for the egg itself.[23,25,39] In other systems, however, more than just the parasitoid progeny escape encapsulation. Here, the ability of parasitized larvae to mount effective encapsulation or other cellular responses, notably nodulation, to bacteria and other pathogens is often severely compromised.[5,46,75,76,78] In these cases, parasitism interferes with multicellular responses in general, and there is no specificity to the suppressive effect. Whether this reduces the survival of the host (with attendant consequences for the developing parasitoid larva) is an interesting paradox.

Although outside the scope of this review, polydnaviruses also have profound effects on the host endocrine system resulting in developmental arrest.[69,79] Many earlier studies have noted that differential and total hemocyte counts fluctuate with the molting cycle, suggesting that hemopoiesis or the activity of hemocytes is under endocrine control. Interestingly, recent evidence has shown that insect hemocyte adhesion and motility is enhanced by neuropeptides.[80] It seems plausible, given the coeffect of polydnavirus infection, that hormones may have important roles in hemocyte development, function, and/or activation.

4. Effects on Humoral Responses

Products of the prophenoloxidase (PO) enzyme cascade have long been implicated as bactericidins, adhesins, and opsonins in the recognition responses to pathogens by arthropods.[13] Released by granular RHs in response to foreign surfaces, the end product of this biosynthetic system is melanin, a pigment often found in insect capsules. The hemolymph of hosts parasitized by certain parasitoids exhibit a reduced ability to melanize in air.[5,81,82] This phenomenon is reproducible by injection of polydnavirus alone[5,81] and may result in the increased stability of the oenocytoids,[5] a cell that provides substrates for melanin biosynthesis. Given the importance of melanin in responses to pathogens it is perhaps not unreasonable to expect parasitoids to evolve ways of interfering with melanization, or indeed any other host recognition phenomenon mediated by soluble factors or molecules.

IV. CONCLUSIONS AND PROSPECTS

The different aspects of invertebrate responses to pathogens are currently studied using a variety of phylogenetically diverse organisms. In some respects, this approach is beneficial, perhaps leading to a greater understanding of common underlying mechanisms of recognition and of the evolution of immune systems in general. Inevitably, however, attempts to correlate such data are problematic and any unifying hypotheses speculative. Nevertheless, we have seen how adhesive phenomena, which precede motile phenomena, such as spreading, encapsulation, and phagocytosis *in vitro,* appear to be mediated by soluble factors. We have also seen how polydnaviruses, which actively interfere with these phenomena, require penetration of target tissue and gene expression within their respective hosts to do this. Parasitization or injection of polydnavirus induces changes in hemocyte behavior, which may contribute to the anomalous encapsulation response by the host. Whether hemocyte behavior is affected through direct infection or via viral gene products produced in other target tissues is not yet known. It is tempting (but very premature) to speculate that these gene products interfere with the soluble factors described so far, such as EPF, or their receptors. Much more needs to be learned about these molecules before these questions can be resolved.

We may learn much from the vertebrates. Preliminary observations[14] suggest that crustacean hemocytes bear integrins. In higher vertebrates, integrins mediate many direct cell-cell adhesion phenomena during embryogenesis, morphogenesis, and other processes, including those of the immune and nervous systems. The interactions mediated by integrin cell adhesion molecules or extracellular matrix (ECM) ligands are typically transient, characteristic of dynamic cellular interactions occurring between migratory cells, and which serve to identify, sort, and traffic particular cells around the body. In many cases, these proteins also transduce intracellular signals and thereby influence the proliferative, migratory, or other outcomes arising from cellular interactions.[83] It is perhaps comforting that insect integrins have been identified (in *Drosophila*) that mediate cell-cell and cell-ECM interactions during embryogenesis.[84] Another family of cell surface molecules that deserves mention at this point is the immunoglobulin (Ig) superfamily. Based on a fundamental "Ig-domain" structure, members of this family have undergone sequence diversification and most contain multiple domains. Members of the Ig superfamily are well characterized in the mammalian immune system where they serve as surface recognition molecules, acting as receptors for other cell surface ligands, antigens, and growth factors. It has been suggested that the Ig superfamily evolved from simpler Ig-like molecules originally involved in primitive cell-cell interactions.[85] At present, members of this family have been described in only a handful of invertebrates.[86-88] Their role in adhesion phenomena that underlie responses to invading pathogens is not known.

The other major question concerns the identity and function of polydnavirus gene products that interfere with host defensive and endocrine systems. Experimentally, this could be achieved by selective blocking of individual viral genes, or to express these genes within the host. The complexity of the polydnavirus genome and the absence of a tissue culture system for viral propagation *in vitro* probably precludes the construction of viruses with mutations in specific genes. Also, the technical problems of selective gene blocking at the whole animal level using antisense technology is presently formidable. A realistic possibility is to express polydnavirus genes from appropriate expression systems from within host larvae, or to generate host larvae expressing polydnavirus transgenes. This would enable us to assess the effect of these genes on host physiology. Clearly, the future application of molecular approaches to understanding how polydnaviruses operate will lead to the development of a new generation of potent, but ecologically sound, pest control strategies, in addition to helping us resolve a fascinating biological problem.

V. SUMMARY

The cellular response by arthropods to penetration of the hemocoel by macroscopic pathogens, or by multiple microscopic pathogens, is a multicellular sequestration reaction mediated by the hemocytes. In some arthropods, initial recognition of the pathogen as foreign and subsequent sequestration within a multicellular sheath, are performed by a single population of multifunctional hemocytes. In others, recognition and sequestration are delegated to two cooperative hemocyte subpopulations, thus resulting in a biphasic cellular reaction. The development of *in vitro* models of encapsulation has led to the discovery (in the insect, *Heliothis virescens*,[8,9] and in the crustacean, *Pacifastacus leniusculus*[11-13]) of soluble factors that promote this and other cellular phenomena *in vitro*. The insect factor, EPF, promotes both the encapsulation and spreading capacities of trypsinized plasmatocytes *in vitro*. The crustacean factor, a 76-kDa protein found in granules within hemocytes, also promotes degranulation *in vitro*, in addition to spreading and encapsulation. Although the homology between these two factors is not known, they may share a common ability to mediate the adhesive properties (and consequent motile behavior) of the hemocytes during multicellular responses. The poor spreading and encapsulation capacity of *H. virescens* PLs in the absence of EPF[9] or after cytoskeletal disruption,[77] is reproduced after parasitization by a parasitoid wasp.[67] These similarities perhaps allude to a common underlying process. Most host-parasitoid relationships are associated with some form of host hemocyte pathology, which is often mediated by polydnaviruses. Carried by the female parasitoid and co-oviposited with her eggs, these complex viruses protect the wasp larva from encapsulation by a mechanism that is only poorly understood. This effect is likely to be directed at the capsule-forming hemocytes (or the soluble factors they require) and mediated by the expression of viral genes within the insect host.

REFERENCES

1. **Nappi, A. J. and Stoffolano, J. G.,** Haemocytic changes associated with the immune reaction of nematode-infected larvae of *Orthellia caesarion*, *Parasitology*, 65, 295, 1972.
2. **Christensen, B. M. and LaFond, M. M.,** Parasite-induced suppression of the immune response in *Aedes aegypti* by *Brugia pahangi*, *J. Parasitol.*, 72, 216, 1986.
3. **Lackie, A. M. and Holt, R. H. F.,** Immunosuppression by larvae of *Moniliformis moniliformis* (Acanthocephala) in their cockroach host *(Periplaneta americana)*, *Parasitology*, 98, 307, 1988.
4. **Christensen, B. M., Huff, B. M., Miranpuri, G. S., Harris, K. L., and Christensen, L. A.,** Hemocyte population changes during the immune response of *Aedes aegypti* to inoculated microfilariae of *Dirofilaria immitis*, *J. Parasitol.*, 75, 119, 1989.
5. **Stoltz, D. B. and Guzo, D.,** Apparent haemocytic transformations associated with parasitoid-induced inhibition of immunity in *Malacosoma disstria* larvae, *J. Insect Physiol.*, 32, 377, 1986.
6. **Rizki, R. M. and Rizki, T. M.,** Selective destruction of a host blood cell type by a parasitoid wasp, *Proc. Natl. Acad. Sci. U.S.A.*, 81, 6154, 1984.

7. **Cherbas, L.**, The induction of an injury reaction in cultured haemocytes from saturniid pupae, *J. Insect Physiol.*, 19, 2011, 1973.
8. **Ratner, S. and Vinson, S. B.**, Encapsulation reactions *in vitro* by haemocytes of *Heliothis virescens, J. Insect Physiol.*, 29, 855, 1983.
9. **Davies, D. H., Hayes, T. K., and Vinson, S. B.**, Preliminary characterization and purification of *in vitro* encapsulation promoting factor: a peptide that mediates insect haemocyte adhesion, *Dev. Comp. Immunol.*, 12, 241, 1988.
10. **Tyson, C. J. and Jenkin, C. R.**, Phagocytosis of bacteria *in vitro* by haemocytes of the crayfish *Parachaeraps bicarinatus, Aust. J. Exp. Biol. Med. Sci.*, 52, 341, 1974.
11. **Johansson, M. W. and Söderhäll, K.**, Isolation and purification of a cell adhesion factor from crayfish blood cells, *J. Cell Biol.*, 106, 1795, 1988.
12. **Johansson, M. W. and Söderhäll, K.**, A cell adhesion factor from crayfish haemocytes has degranulation activity towards crayfish granular cells, *Insect Biochem.*, 19, 183, 1989.
13. **Johansson, M. W. and Söderhäll, K.**, Cellular immunity in crustaceans and the proPO system, *Parasitol. Today*, 5, 171, 1989.
14. **Johansson, M. W. and Söderhäll, K.**, A peptide containing the cell adhesion sequence RGD can mediate degranulation and cell adhesion of crayfish granular haemocytes *in vitro, Insect Biochem.*, 19, 573, 1989.
15. **Stoltz, D. B.**, Interactions between parasitoid-associated products and host insects: a brief overview, *J. Insect Physiol.*, 32, 347, 1986.
16. **Fuhrer, E.**, Mucopolysaccharide im weiblichen Geschlectsapparat parasitischer hymenopteren, *Naturwissenschaften*, 59, 167, 1972.
17. **Fuhrer, E.**, Sekreton von Mucopolysacchariden im weiblichen Geschlectsapparat von *Pimpla turionellae* L. (Hym., Ichneumonidae), *Z. Parasitenkd.*, 41, 207, 1973.
18. **Osman, S. H.**, Die Wirkung der Sekrete der weiblichen Genitalanhangsdrüsen von *Pimpla turionellae* L. (Hym., Ichneumonidae) auf die Hämocyten und die Einkapselungsreaktion von Wirstpuppen, *Z. Parasitenkd.*, 57, 89, 1978.
19. **Osman, S. E. and Führer, E.**, Histochemical analysis of accessory gland secretions in female *Pimpla turionellae* L. (Hymenoptera: Ichneumonidae), *Int. J. Invertebr. Reprod.*, 1, 323, 1979.
20. **Kitano, H.**, Experimental studies on the parasitism of *Apanteles glomeratus* L. with special reference to its encapsulation inhibiting capacity, *Bull. Tokyo Gakuge Univ.*, 21, 95, 1969.
21. **Kitano, H.**, Effects of the parasitization of a braconid, *Apanteles,* on the blood of its host, *Pieris, J. Insect Physiol.*, 20, 315, 1974.
22. **Vinson, S. B.**, Factors involved in successful attack on *Heliothis virescens* by the parasitoid *Cardiochiles nigriceps, J. Invertebr. Pathol.*, 20, 118, 1972.
23. **Kitano, H. and Nakatsuji, N.**, Resistance of *Apanteles* eggs to the haemocytic encapsulation by their habitual host, *Pieris., J. Insect Physiol.*, 24, 261, 1978.
24. **Norton, W. M. and Vinson, S. B.**, Encapsulation of a parasitoid egg within its habitual host: an ultrastructural investigation, *J. Invertebr. Pathol.*, 30, 55, 1977.
25. **Davies, D. H. and Vinson, S. B.**, Passive evasion by eggs of the braconid parasitoid *Cardiochiles nigriceps* from encapsulation *in vitro* by haemocytes of host *Heliothis virescens.* Possible role for fibrous layer in immunity, *J. Insect Physiol.*, 32, 1003, 1986.
26. **Stoltz, D. B. and Krell. P. J.**, Polydnaviridae: a proposed new virus family, Abstr. 81st Annu. Meet. *Am. Soc. Microbiol.*, Dallas, 1981, 232.
27. **Stoltz, D. B., Krell, P., Summers, M. D., and Vinson, S. B.**, Polydnaviridae — a proposed family of insect viruses with segmented, double-stranded, circular DNA genomes, *Intervirology*, 21, 1, 1984.
28. **Salt, G.**, The ichneumonid parasite *Nemeritis canescens* (Gravenhorst) in relation to the wax moth *Galleria mellonella* (L), *Trans. R. Entomol. Soc. London,* 116, 1, 1964.
29. **Salt, G.**, Experimental studies on insect parasitism. XIII. The haemocytic reaction of a caterpillar to eggs of its habitual parasite, *Proc. R. Soc. London, Ser. B,* 162, 308, 1965.
30. **Rotheram, S. M.**, Immune surface of eggs of a parasite insect, *Nature (London),* 214, 700, 1967.
31. **Rotheram, S. M.**, The surface of the egg of a parasitic insect. I. The surface of the egg and first instar larvae of *Nemeritis, Proc. R. Soc. London,* 183, 179, 1973.
32. **Rotheram, S. M.**, The surface of the egg of a parasitic insect II. The ultrastructure of the particulate coat on the egg of *Nemeritis, Proc R. Soc. London, Ser. B,* 183, 195, 1973.
33. **Bedwin O.**, An insect glycoprotein; a study of the particles responsible for the resistance of a parasitoid's egg to the defence reactions of its insect host, *Proc. R. Soc. London, Ser. B,* 205, 271, 1979.
34. **Vinson, S. B. and Scott, J. R.**, Particles containing DNA associated with the oocyte of an insect parasitoid, *J. Invertebr. Pathol.*, 25, 375, 1975.
35. **Stoltz, D. B. and Vinson, S. B.**, Baculovirus-like particles in the reproductive tracts of female parasitoid wasps. II. The genus *Apanteles, Can. J. Microbiol.*, 23, 28, 1977.
36. **Edson, K. M., Vinson, S. B., Stoltz, D. B., and Summers, M. D.**, Virus in a parasitoid wasp: suppression of the cellular immune response in the parasitoid's host, *Science,* 211, 582, 1981.

37. **Stoltz, D. B. and Vinson, S. B.,** Viruses and parasitism in insects, *Virus Res.*, 24, 125, 1979.
38. **Stoltz, D. B., Krell, P. J., and Vinson, S. B.,** Polydisperse viral DNAs in ichneumonid ovaries: a survey, *Can. J. Microbiol.*, 27, 123, 1981.
39. **Feddersen, I., Sander, K., and Schmidt, O.,** Virus-like particles with host protein-like antigenic determinants protect an insect parasitoid from encapsulation, *Experientia*, 42, 1278, 1986.
40. **Schmidt, O. and Schuchmann-Feddersen, I.,** Role of virus-like particles in parasitoid-host interaction of insects, *Subcell. Biochem.*, 15, 91, 1989.
41. **Berg, R., Schuchmann-Feddersen, I., and Schmidt, O.,** Bacterial infection induces a moth *(Ephestia kühniella)* protein which has antigenic similarity to virus-like particle proteins of a parasitoid wasp *(Venturia canescens), J. Insect Physiol.*, 34, 473, 1988.
42. **Stoltz, D. B., Guzo, D., Belland, E. R., Lucarotti, C. J., and MacKinnon, E. A.,** Venom promotes uncoating *in vitro* and persistence *in vivo* of DNA from a braconid polydnavirus, *J. Gen. Virol.*, 69, 903, 1988.
43. **Krell, P. J.,** Replication of long virus-like particles in the reproductive tract of the ichneumonid wasp *Diadegma terebrans, J. Gen. Virol.*, 68, 1477, 1987.
44. **Stoltz, D. B., Krell, P. J., Cook, D., MacKinnon, E. A., and Lucarotti, C. J.,** An unusual virus from the parasitoid wasp *Cotesia melanoscela, Virology*, 162, 311, 1988.
45. **Guzo, D. and Stoltz, D. B.,** Obligatory multiparasitism in the tussock moth, *Orgyia leucostigma, Paratisology*, 90, 1, 1985.
46. **Guzo, D. and Stoltz, D. B.,** Observations on cellular immunity and parasitism in the tussock moth, *J. Insect Physiol.*, 33, 19, 1987.
47. **Kitano, H.,** Effect of the venom of the gregarious parasitoid *Apanteles glomeratus* on its hemocytic encapsulation by the host, *Pieris, J. Invertebr. Pathol.*, 40, 61, 1982.
48. **Kitano, H.,** The role of *Apanteles glomeratus* venom in the defensive response of its host, *Pieris rapae crucivora, J. Insect Physiol.*, 32, 369, 1986.
49. **Tanaka, T.,** Effect of the venom of the endoparasitoid, *Apanteles kariyai* Watanabe, on the cellular defence reaction of the host, *Pseudaletia separata* Walker, *J. Insect Physiol.*, 33, 413, 1987.
50. **Wago, H. and Kitano, H.,** Effects of the venom from *Apanteles glomeratus* on the hemocytes and hemolymph of *Pieris rapae crucivora, Appl. Entomol. Zool.*, 20, 103, 1985.
51. **Wago, H. and Tanaka, T.,** Synergistic effects of calyx fluid and venom of *Apanteles kariyai* Watanabe (Hymenoptera, braconidae) on the granular cells of *Pseudaletia separata* Walker (Lepidoptera, noctuidae), *Zool. Sci.*, 6, 691, 1989.
52. **Krell, P. J., Summers, M. D., and Vinson, S. B.,** . Virus with a multipartite superhelical DNA genome from the ichneumonid parasitoid *Campoletis sonorensis, J. Virol.*, 43, 859, 1982.
53. **Blissard, G. W., Fleming, J. G. W., Vinson, S. B., and Summers, M. D.,** *Campoletis sonorensis* virus: expression in *Heliothis virescens* and identification of expressed sequences, *J. Insect Physiol.*, 32, 351, 1986.
54. **Theilmann, D. A. and Summers, M. D.,** Physical analysis of the *Campoletis sonorensis* virus multipartite genome and identification of a family of tandemly repeated elements, *J. Virol.*, 61, 2589, 1987.
55. **Norton, W. N., Vinson, S. B., and Stoltz, D. B.,** Nuclear secretory particles associated with the calyx cells of the ichneumonid parasitoid *Campoletis sonorensis* (Cameron), *Cell Tissue Res.*, 162, 195, 1975.
56. **Norton W. N. and Vinson S. B.,** Correlating the initiation of virus replication with a specific pupal developmental phase of an ichneumonid parasitoid, *Cell Tissue Res.*, 231, 387, 1983.
57. **Fleming, J. G. W. and Summers, M. D.,** *Campoletis sonorensis* endoparasitic wasps contain forms of *C. sonorensis* virus DNA suggestive of integrated and extrachromasomal polydnavirus DNA, *J. Virol.*, 57, 552, 1986.
58. **Theilmann, D. A. and Summers, M. D.,** Molecular analysis of *Campoletis sonorensis* virus DNA in the Lepidopteran host *Heliothis virescens, J. Gen. Virol.*, 67, 1961, 1986.
59. **Theilmann, D. A. and Summers, M. D.,** Identification and comparison of *Campoletis sonorensis* virus transcripts expressed from four genomic segments in the insect hosts *Campoletis sonorensis* and *Heliothis virescens, Virology*, 167, 329, 1988.
60. **Stoltz, D. B. and Vinson, S. B.,** Penetration into caterpillar cells of virus-like particles injected during oviposition by parasitoid ichneumonid wasps, *Can. J. Microbiol.*, 25, 207, 1979.
61. **Fleming, J. G. W., Blissard, G. W., Summers, M. D., and Vinson, S. B.,** Expression of *Campoletis sonorensis* virus in the parasitized host, *Heliothis virescens, J. Virol.*, 48, 74, 1983.
62. **Blissard, G. W., Smith, O. P., and Summers, M. D.,** Two related viral genes are located on a single superhelical DNA segment of the multipartite *Campoletis sonorensis* virus genome, *Virology*, 160, 44, 1987.
63. **Blissard, G. W., Vinson, S. B., and Summers, M. D.,** Identification, mapping, and *in vitro* translation of *Campoletis sonorensis* virus mRNAs from parasitized *Heliothis virescens* larvae, *J. Virol.*, 57, 318, 1986.

64. **Blissard, G. W., Theilmann, D. A., and Summers, M. D.,** Segment W of *Campoletis sonorensis* virus: expression, gene products, and organization, *Virology,* 169, 78, 1989
65. **Krell, P. J.,** Polydnavirus replication and tissue organization of infected cells in the parasitic wasp *Diadegma terebrans, Can. J. Microbiol.,* 33, 176, 1987.
66. **Krell, P. J. and Stoltz, D. B.,** Virus-like particles in the ovary of an Ichneumonid wasp: purification and preliminary characterization, *Virology,* 101, 408, 1980.
67. **Davies, D. H., Strand, M. R., and Vinson, S. B.,** Changes in differential haemocyte count and *in vitro* behaviour of plasmatocytes from host *Heliothis virescens* caused by *Campoletis sonorensis* polydnavirus, *J. Insect Physiol.,* 33, 143, 1987.
68. **Beckage, N. E., Templeton, T. J., Nielsen, B. D., Cook, D. I., and Stoltz, D. B.,** Parasitism-induced hemolymph polypeptides in *Manduca sexta* (L.) larvae parasitized by the braconid wasp *Cotesia congregata, Insect Biochem.,* 17, 439, 1987.
69. **Dover, B. A., Davies, D. H., Strand, M. R., Gray, R. S., Keeley, L. L., and Vinson, S. B.,** Ecdysteroid-titre reduction and developmental arrest of last-instar *Heliothis virescens* larvae by calyx fluid from the parasitoid *Campoletis sonorensis, J. Insect Physiol.,* 33, 333, 1987.
70. **Ferkovich, S. M., Greany, P. D., and Dillard, C.,** Changes in haemolymph proteins of the fall armyworm, *Spodoptera frugiperda* (J. E. Smith), associated with parasitism by the braconid *Cotesia marginiventris* (Cresson), *J. Insect Physiol.,* 29, 933, 1983.
71. **Cook, D. I., Stoltz, D. B., and Vinson, S. B.,** Induction of a new haemolymph glycoprotein in larvae of permissive hosts parasitized by *Campoletis sonorensis, Insect Biochem.,* 14, 45, 1984.
72. **Streams, F. A. and Greenberg, L.,** Inhibition of the defence reaction of *Drosophila melanogaster* parasitized simultaneously by the wasps *Pseudeucola bochei* and *Pseudeucoila mellipes, J. Invertebr. Pathol.,* 13, 371, 1969.
73. **Nappi, A. J. and Streams, F. A.,** Haemocytic reactions of *Drosophila melanogaster* to the parasites *Pseudeucoila mellipes* and *P. bochei, J. Insect Physiol.,* 15, 1551, 1969.
74. **Rizki, T. M., Rizki, R. M., and Carton, Y.,** *Leptopilina heterotoma* and *L. boulardi:* strategies to avoid cellular defense response of *Drosophila melanogaster, Exp. Parasitol.,* 70, 466, 1990.
75. **Davies, D. H. and Vinson, S. B.,** Interference with function of plasmatocytes of *Heliothis virescens in vivo* by calyx fluid of the parasitoid *Campoletis sonorensis, Cell Tissue Res.,* 251, 467, 1988.
76. **Tanaka, T.,** Morphological changes in haemocytes of the host, *Pseudaletia separata* parasitized by *Microplitis mediator* or *Apanteles kariyai, Dev. Comp. Immunol.,* 11, 57, 1987.
77. **Davies, D. H. and Preston, T. M.,** Effect of disruption of plasmatocyte microfilaments on encapsulation *in vitro, Dev. Comp. Immunol.,* 11, 353, 1987.
78. **Ross D. R. and Dunn, P. E.,** Effect of parasitism by *Cotesia congregata* on the susceptibility of *Manduca sexta* larvae to bacterial infection, *Dev. Comp. Immunol.,* 13, 205, 1989.
79. **Dover, B. A., Davies, D. H., and Vinson, S. B.,** Dose-dependent influence of *Campoletis sonorensis* polydnavirus on the development and ecdysteroid titers of last-instar *Heliothis virescens* larvae, *Arch. Insect Biochem. Physiol.,* 8, 113, 1988.
80. **Steffano, G. B., Leung, M. K., Zhao, X. H., and Scharrer, B.,** Evidence for the involvement of opioid neuropeptides in the adherence and migration of immunocompetent invertebrate hemocytes, *Proc. Natl. Acad. Sci. U.S.A.,* 86, 626, 1989.
81. **Stoltz, D. B. and Cook, D. I.,** Inhibition of host phenoloxidase activity by parasitoid hymenoptera, *Experientia,* 39, 1022, 1983.
82. **Willers, D. and Lehmann-Danzinger, H.,** Überleben endoparasitischer Hymenopteren in Puppen von Lepidopteren durch Hemmung der Phenoloxidase, *Z. Parasitenkd.,* 70, 403, 1984.
83. **Hynes, R. O.,** Integrins: a family of cell surface receptors, *Cell,* 48, 549, 1987.
84. **Sémérevira, M., Naidet, C., Krejci, E., and Gratecos, D.,** Towards the molecular biology of cell adhesion in Drosophila, *Trends Genet.,* 5, 24, 1989.
85. **Williams, A. F.,** A year in the life of the immunoglobulin superfamily, *Immunol. Today,* 8, 298, 1987.
86. **Williams, A. F., Tse, A. G. D., and Gagnon, J.,** Squid glycoproteins with structural similarities to Thy-1 and Ly-6 antigens, *Immunogenetics,* 27, 265, 1988.
87. **Harrelson, A. L. and Goodman, C. S.,** Growth cone guidance in insects: fascilin II is a member of the immunoglobulin superfamily, *Science,* 242, 700, 1988.
88. **Seeger, M. A., Haffley, L., and Kaufman, T. C.,** Characterization of *amalgam:* a member of the immunoglobulin superfamily from *Drosophila, Cell,* 55, 589, 1988.
89. **Davies, D. H. and Strand, M. R.,** unpublished observations.
90. **Davies, D. H.,** unpublished observations.

Chapter 4

GAP CELL JUNCTIONS, CELL ADHESION MOLECULES, AND MOLECULAR BASIS OF ENCAPSULATION*

Ayodhya P. Gupta

TABLE OF CONTENTS

* This is the tenth in a series of papers on arthropod immune systems.

I. INTRODUCTION

Although gap junctions (and other cell junctions) have been reported in a few insect capsules, their functional significance at certain crucial stages of the encapsulation process has not been discussed. We do not know the molecular nature of the gap junction polypeptide of the hemocytic capsule and the roles of the two important and ubiquitous intracellular messengers or mediators (adenosine 3′,5′cyclic monophosphate [cAMP] and Ca^{2+}) in the molecular regulation of junctional communication. Similarly, we have little or no understanding of the roles of the granulocyte plasma membrane glycoproteins in the initial cell-cell adhesion and those of phosphoinositides (PIs) in signal transduction during encapsulation.

The purpose of this chapter is to first briefly review the current state of knowledge about certain aspects of gap junctions, surface adhesion molecules, and PIs in other invertebrate and vertebrate tissues, and then to postulate a molecular basis of encapsulation, in terms of the roles of plasma membrane glycoproteins in cell-cell adhesion, PIs in signal transduction, and $[cAMP]_i$ and $[Ca^{2+}]_i$ ions in gap junctional communication during encapsulation. This hypothesis leans on the extensive information that exists on adhesion molecules, hydrolysis of PIs during signal transduction, and $[cAMP]_i$ and $[Ca^{2+}]_i$ ions in other animals. It also carries the underlying notion that it would stimulate investigations of the nature of the surface adhesion molecules and PIs of arthropod immunocytes and their roles, as well as those of the intracellular messengers, to elucidate the exact molecular basis of encapsulation.

Gap junctions are dynamic organelles, open and close (hence called *gated*) in response to cell activities, and are modulated by their gating mechanism (permeability),[1-3] their turnover rate in the plasma membrane,[4] and turnover of the entire gap junction plaque.[5] In fact, the function of a cell organelle, such as gap junction, can be deduced in part from its turnover rate, physiological factors influencing this turnover rate, and the latter's concomitant relationship to specific cell activity.

Classical studies of gap junctions involve their morphological and biochemical analyses and measurement of the junctional conductance. The latter has been recently accomplished by isolating gap junctions and generating antibodies specific to their protein, which are then injected to inhibit junctional conductance.[6,7] Such blocking provides direct evidence of the role of gap junctions in cell-cell communication. Curiously, there is wide variation in the

M.W. (16 to 54 kDa) of the polypeptide that is thought to compose gap junctions in various tissues (see reviews in References 8 and 9). Finbow et al.[10] have shown, using rabbit antisera raised against isolated gap junctions from lobster hepatopancreas, that in this arthropod, these organelles are composed of 18-kDa proteins.

The biochemical nature of the gap junction polypeptide in the hemocytic capsule of an arthropod is unknown. Are the gap junctions composed of the same glycoproteins that comprise the intramembranous particles (IPs) of the plasma membrane, or of different ones? Most recent work suggests a size difference between gap junction particles and those of the intramembranous particles.[11] Note that capsule formation in the German cockroach, *Blattella germanica* commences 40 to 45 min postimplantation (of pieces of surgical suture), and gap junctions are present as soon as the first few layers are formed,[12] which suggests that they are formed very rapidly. In fact, in reaggregates (loose pellets formed by centrifugation) of Novikoff hepatoma cells, cell junctions are formed within 10 to 15 min.[13]

II. FORMATION AND MOLECULAR STRUCTURE OF GAP JUNCTIONS

Gap junctions are plasma membrane specializations in the form of channels or "connexons" that appear as IPs in freeze-fracture replicas. They are made up of integral membrane proteins and their life history involves transcriptional and translational processes, insertion of the synthesized hemichannels into the plasma membrane, assembling into plaques, coupling with similar plaques of the other cell, and removal, degradation, or reuse.[14]

Formation of gap junctions was first described in Novikoff hepatoma cells in culture.[15] Gap junction particles or connexons in these cells, consist of six identical 27-kDa protein subunits, 9 to 11 nm in diameter, which are rapidly inserted into the plasma membrane when two cells come within 10 nm of each other. These particles soon assemble into primary plaques. The identical plaques of the two apposing cells finally integrate into each other when the two cells come within a 2 nm "gap" of each other, hence the name gap junction. In *B. germanica,* the intercellular space between the junction-forming GRs is 2.5 nm and the intramembranous particles are 20 nm in diameter, there being ca. $640/\mu m^2$.[12]

Gap junctions allow flow of electric current (carried primarily by K^+ ions) and passage of inorganic ions, sugars, amino acids, nucleotides, vitamins, and other small water-soluble molecules (not exceeding 1 to 1.5 kDa, and therefore excluding macromolecules, such as proteins, nucleic acids, and polysaccharides) from one cell to another, thus coupling the two cells for cell-cell communication.[16] The number of open channels and the rates of their formation determine the strength of junctional coupling.[14] Such coupling can be demonstrated by indirect dye and metabolic coupling techniques.

It has been suggested that pools of gap junction precursors (unrecognizable in freeze-fracture replicas) are stored in the plasma membrane and are self-assembled into gap junctions.[17,18] Several hundred channels are assembled to form a gap junction plaque, in which the junctions appear as annular units located in the plane of the membrane. The channels are located at the geometrical center of the annular units.[19] The channel diameter apparently varies, being 1.6 to 2 nm in mammals and 2 to 3 nm in insect cells.[20,21]

Because gap junctions are plasma membrane specializations, the question has been raised as to whether cytoskeletal elements participate in their formation; indirect evidence suggests that none do. In prostatic epithelial cell cultures, colchicine and cytochalasin B (that disrupt microtubules) stimulate gap junction formation. Furthermore, cooling the culture to 0°C for 5 min (this depolymerizes the microtubules) and then returning it to 37°C, also induced gap junction formation, which suggests the dissolution of microtubules and actin microfilaments induces self-assembly of gap junction precursors into gap junctions.[22] Injection of colchicine into mouse hepatocytes also increased gap junction areas.[23] This suggests that gap junctions form rapidly and spontaneously, and cytoskeletal elements (microtubules and actin microfilaments) are apparently not involved in their formation.[13] However, removal of gap junction plaques from membranes is thought to involve actin, clathrin, and lysosomes.[18] Finally, the rate of gap junction protein synthesis and its degradation closely parallel plaque formation and its disappearance, respectively.[18]

III. ULTRAMICROSCOPY OF GAP JUNCTIONS

A. THIN SECTIONS

After conventional fixation and staining, gap junctions appear as areas where the plasma membranes of two cells are separated by a 2- to 3-nm wide gap, which is bridged by regularly arranged particles or connexons (Figures 1A, D and 2). Glutaraldehyde, which is most commonly used as a fixative, causes irreversible uncoupling (blocking) of gap junctions, and is therefore unsuitable for demonstrating open or coupled junctions; it probably also alters the *in situ* arrangement of gap junction plaques.[24]

B. RAPID FREEZE-FRACTURE REPLICAS

Rapid freeze-fracture method is most suitable to study the interior of cell membranes, because this technique overcomes the artifacts that conventional chemical fixation may cause. A freezing rate of 10^4 Ks^{-1} is necessary to preserve an organelle in its native state.[25] However, for gap junctions whose bore diameter is between 1.6 to 3.0 nm, it is doubtful if rapid freezing could discriminate between open and closed junctions.

In freeze-fracture replicas, gap junctions appear as clusters or aggregates on plaques (Figures 1B, C and 3), which may be loosely or tightly packed. Whether the degree of packing on the plaques is correlated with coupled or uncoupled states of gap junctions or the conductivity of the plaques, is uncertain.[24] However, it has been demonstrated in crayfish lateral axons that coupled axons have well-defined junctional plaques of different sizes and very few isolated particles, whereas, uncoupled (induced by 25% acetate) axons contained fewer plaques and numerous dispersed particles.[26] Hülser et al.[24] reported that, although in unfixed cells gap junction packing ranged from loosely to tightly packed plaques, fixed cells lacked loosely packed plaques, the particle diameter remaining the same in both fixed and unfixed cells. Furthermore, they considered tightly packed plaques in both fixed and unfixed cells as the final stage in the life span of gap junctions, after which they are internalized and recycled, in the same manner as receptors are collected in coated pits. Thus, the loosely packed plaques are to be regarded as active junctions that regulate intracellular signals and metabolites. This may also explain why unfixed cells show both loosely and tightly packed plaques. In a 3-day-old capsule in *B. germanica,* we found both loosely and tightly packed junctions (see Figure 10 in Han and Gupta[12]).

In the very few arthropods in which the gap junction particles had been studied by the early 1970s, they were generally reported to be located on the extracellular E-fracture face. Therefore Flower[27] designated these gap junctions as "inverted type", with reference to the mammalian gap junctions, in which the location of the particles is reversed (i.e., on the cytoplasmic P-fracture face). Thus, it was suggested[28,29] that the extracellular E-type gap junction may be characteristic of arthropods, although such junctions have also been found in the coelenterate, *Hydra,*[30] and the nematode, *Trichinella spiralis,*[31] and some mollusks.[32,33] In view of this, I suggested that the E-type probably represents a primitive stage, while the P-type is a more advanced type that originated before the evolution of arthropods, in which both P- and E-types are present.[34] Indeed, P-type junctions are now known to be present in several arthropods.[35-37]

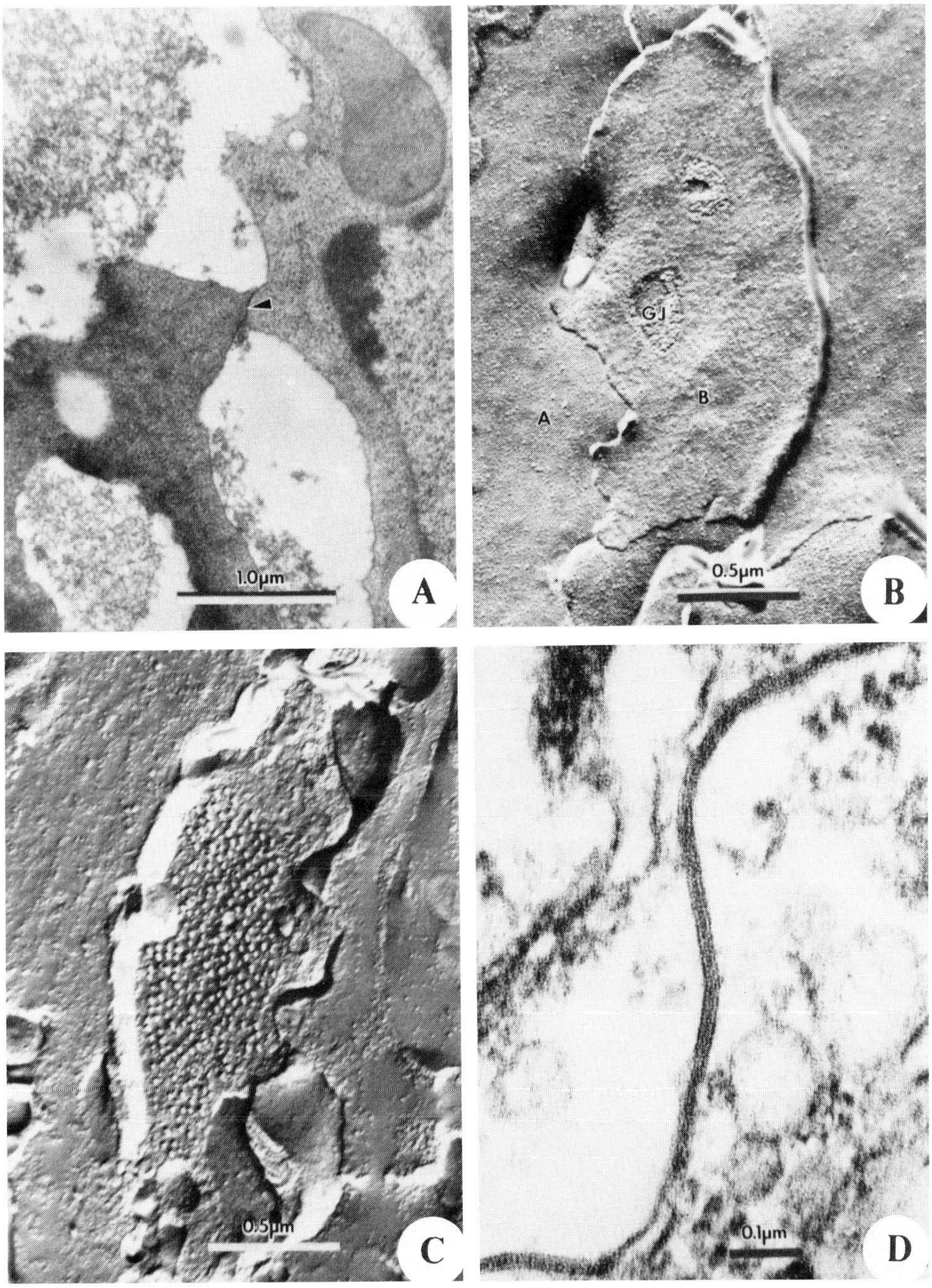

FIGURE 1. (A) Thin section of a 3-day-old capsule of *Blattella germanica,* showing gap junction (arrowhead) in the inner region. (B) Freeze-fracture replica of the inner region of a 3-day-old capsule, showing gap junction (GJ). Note that of the two E-type junctions, one is less tightly packed (i.e., IPs show greater dispersion). A = P-fracture face; B = E-fracture face. (C) Freeze-fracture replica of the outer region of a 3-day-old capsule, showing gap junction. The IPs on the E-fracture face are 20 nm in diameter, and their number is ca $640/\mu m^2$. (D) Thin section of a 3-day-old capsule, showing gap junction in the outer region. The intercellular space between the two junction-forming granulocytes is 2.5 nm. (From Han, S. S. and Gupta, A. P., *Zool. Sci.*, 6, 303, 1989. With permission.)

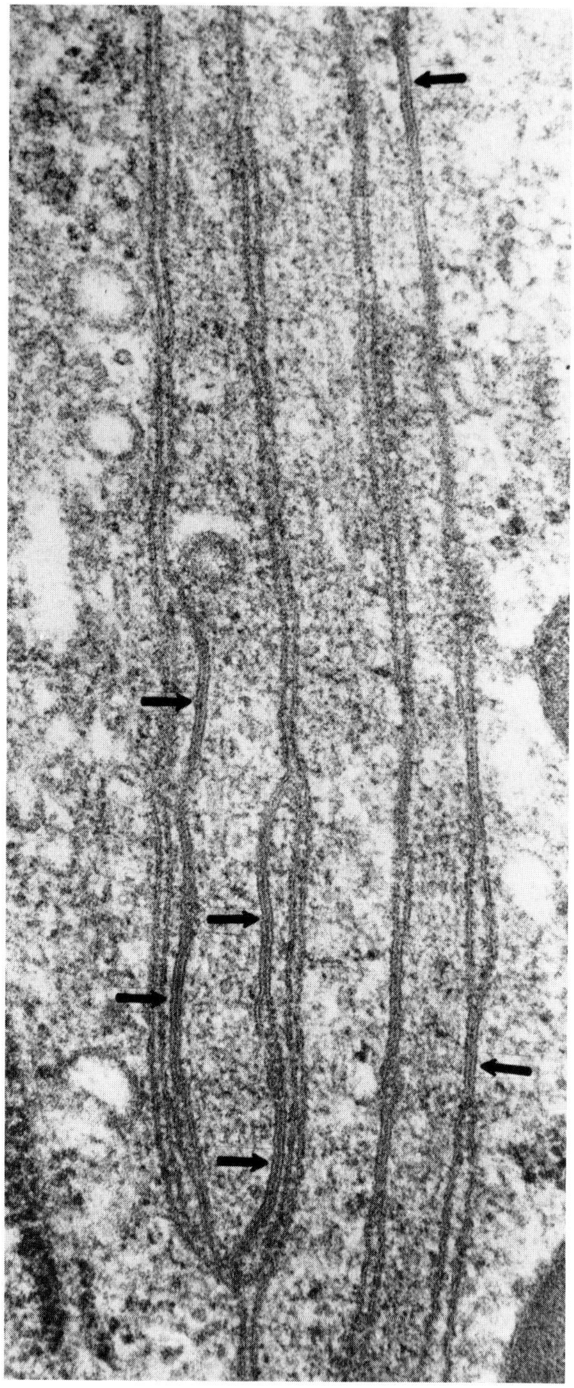

FIGURE 2. Thin section of a capsule in *P. americana,* showing numerous gap junctions seen in cross-section (arrows). (× 74,000) (From Baerwald, R. J., *Tissue Cell,* 7(3), 575, 1975. With permission.)

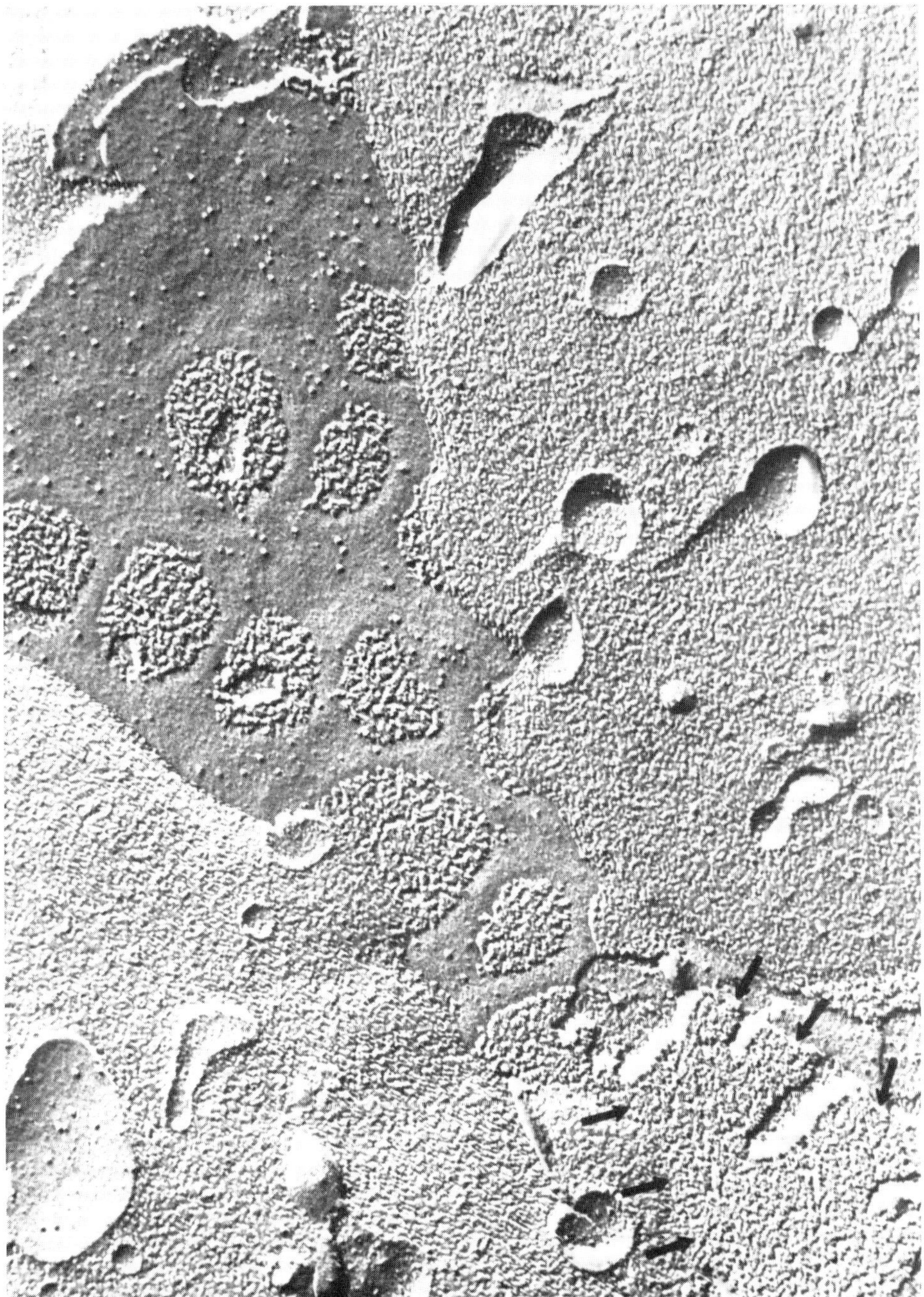

FIGURE 3. Freeze-fracture replica of a 48-h-old capsule of *P. americana*. At least 13 discrete E-type gap junctions are evident. In addition, a much larger junctional region can be seen near bottom of picture outlined by arrow. (× 50,000) (From Baerwald, R. J., *Insect Hemocytes*, Gupta, A. P., Ed., Cambridge University Press, Cambridge, 1979, 155. With permission.)

C. ISOLATED OR EXTRACTED GAP JUNCTIONS

Gap junctions are isolated or extracted from the plasma membranes by using alkali (detergent-independent)[38] and detergent (Triton X-100) (detergent-dependent) methods,[10] which supposedly provide greater yield and versatility.

Isolated gap junctions appear as double-membrane plaques of about 15 nm in overall thickness,[19] which contain ordered domains of connexons (annular units) that are linked to their opposite numbers across the extracellular space between the two membranes.

Isolated gap junctions have been successfully used to determine the organization of the annular units (connexons) and the location of the plaques (which appear as hexagonal arrays in some tissues, e.g., rat liver hepatocytes) across the lipid bilayer, and have been used to produce three-dimensional reconstructions of gap junctions. Isolated gap junctions have been studied in negatively stained preparations or frozen-hydrated specimens.

1. Negatively Stained Preparations

In this method, aliquots of membrane suspensions are deposited on carbon-coated grids as thin films and negatively stained with uranyl acetate,[39] or phosphotungstate (pH 7.4).[10] Methods for negative staining for various tissues, including photographic techniques applied to negative staining procedures, have been discussed by Horne.[40]

2. Frozen-Hydrated Specimens

In this method, unstained membrane suspensions, deposited on carbon-coated grids, are plunged into liquid ethane and then transferred under liquid nitrogen into an electron microscope that maintains the specimen at $-120°C$ during imaging.[19] According to these authors, negative staining yields information about the subunit arrangement at the water-membrane interface (because the uranyl ions are predominantly located at the membrane/water interface), while the frozen-hydrated method provides the same information in the hydrophobic region (because of the differences in the density of protein and lipid inside the membrane) of the bilayer. Thus, the two methods provide a consistent picture of gap channel in its two conformational states.

Using negatively stained[41] and frozen-hydrated sections,[42] Zampighi and Simon,[19] by tilting the specimens (0 to 79° to the incident electron beam in negatively stained junctions, and 0 to 35° in frozen-hydrated sections), produced a series of sections that were parallel to the plane of the membrane. They processed the resulting images by Fourier method, combined the information from different tilt angles to represent the variation caused by electron density across the entire thickness of the junction,[43] and produced three-dimensional maps of junctions from both negatively and frozen-hydrated specimens.

Their results showed that the annulus (connexon) is composed of six *weakly* modulated subunits (hexamers; ca. 5 to 7 nm long in negatively stained preparations), which surround a central cavity that widens to a diameter of ca. 2 nm in the extracellular region. By contrast, in the frozen-hydrated sections, the hexamers were *strongly* modulated. Zampighi and Simon[19] attributed these differences in the appearance of the hexamers to the differences in contrast, because contrast in the two methods is obtained from different portions of the hexamer (central portion of the bilayer in frozen-hydrated method and surface in the negatively stained sections). Thus, the two methods showed that the hexamer can exist in two conformationally distinct forms, corresponding to the *open* state (in which the hexamer subunits are tilted across the bilayer and the hydrophilic cavity at the center spans the entire thickness of the membrane) and *closed* state (subunits are less inclined across the bilayer and the channel is interrupted at its cytoplasmic portion). In this way, the channel permeability is modulated by a coordinated movement among the subunits of the hexamer by closing the channel at its cytoplasmic end.

IV. MODULATION OF GAP JUNCTIONS

Gating of gap junctions and their formation and removal (turnover rate), is rigorously modulated by such factors as junctional conductance, secretory activity of the cells involved, intracellular ions, hormones, and proliferative and growth activities of the cells.

A. GAP JUNCTION GATING MECHANISM OR CONDUCTANCE

Gap junctions are dynamic structures that are not continuously open, but can reversibly open and close in response to external electrical or chemical stimuli or cellular activities over a short time (seconds to minutes). Control beyond a few minutes to the lifetime of the organism, influences turnover rate of the junctions.[14] Membrane perturbations also cause gating.[16] Several kinds of membrane perturbations are known, such as change in voltage across membrane (*voltage-gated* channels), mechanical stimulation (*mechanically gated* channels), and binding of a signal molecule (*ligand-gated* channels). The latter could be an extracellular ligand (e.g., *neurotransmitter-gated* channel), or intracellular, such as an ion (*ion-gated* channel), a nucleotide (*nucleotide-gated* channel), or guanosine triphosphate (GTP)-binding regulatory protein (*G-protein-gated* channel). Approximately 50 ion channels are known.

Because cAMP and Ca^{2+} ions affect gap junction modulation and cell-cell communication, I will now discuss the effects of the intracellular concentrations of these mediators on cell-cell communication, and later, on exocytosis and cell proliferation.

1. Intracellular Messengers: Cyclic AMP $[cAMP]_i$ and $[Ca^{2+}]_i$

$[cAMP]_i$ and $[Ca^{2+}]_i$ are two of the most important and ubiquitous intracellular messengers or mediators, which have negative feedback effects on each other. They act as allosteric effectors (i.e., they bind to specific proteins and change the mediators' conformations) and are generated intracellularly when a plasma membrane receptor is activated by an extracellular signaling molecule (e.g., growth hormone) or a foreign antigen; this activation alters the cytoplasmic domain of the receptor, so that it activates a membrane-bound lipoprotein enzyme, *adenylate cyclase* (AC) (via the G-protein), which catalyzes the synthesis of 3′,5′-AMP from ATP on the cytoplasmic side of the membrane. The AC also produces a soluble mediator that mobilizes Ca^{2+} from microsomes, mitochondria, and a calcium sequestering compartment (CSC) and/or calciosome; apparently, mobilization from the latter is mediated via yet another intracellular mesenger, *inositol triphosphate* (IP_3).[44,45] Volpe et al.[45] showed that calciosomes are discrete organelles, store Ca^{2+}, and are the target of inositol-triphosphate. Note that even mechanical stimulus (stretching the cell) releases Ca^{2+} from its intracellular store[46] (see also discussion in Section VI.E.2).

Note that receptor-mediated hydrolysis of PIs (inositol-containing phospholipids) also leads to $[Ca^{2+}]_i$ mobilization and activation of plasma membrane protein kinase C, both of which play important roles in exocytosis and cell proliferation. The end result of the PI hydrolysis is the production of another intracellular messenger, cyclic GMP (see Section VI.B).

It is important to mention that the appropriate intracellular concentrations of both $[cAMP]_i$ and $[Ca^{2+}]_i$ are rigorously regulated, because, if unchecked, the amplifying and regulating effects of the extracellular signaling molecule on the catalytic cascade of these two messengers would push their respective concentrations to catastrophic proportions. For example, a nanomolar (10^{-9} *M*) concentration of an extracellular signal induces micromolar (10^{-6} *M*) concentrations of $[cAMP]_i$ and $[Ca^{2+}]_i$ ions.[47] That is why the $[cAMP]_i$ is rapidly and continuously destroyed by one or more cAMP phosphodiesterases, which hydrolyze cAMP to 5′-AMP,[16,48] and Ca^{2+} is sequestered in the CSC, bound to a Ca^{2+}-binding protein, *calsequestrin*. Note that CSC is not a part of the rough endoplasmic reticulum of the cell, but a separate entity.[16]

What is important to emphasize here from the standpoint of discussing the molecular basis of encapsulation (see Section VI.E), is the negative feedback effects of intracellular $[Ca^{2+}]_i$ and $[cAMP]_i$ ions on junctional communication.[47] An increase in $[Ca^{2+}]_i$ ions due to Ca^{2+}-mobilizing effects of $[cAMP]_i$,[49,50] reduces the cells to a diminished state of cell-cell communication, and a decrease in $[Ca^{2+}]_i$ ions again activates adenylate cyclase (AC), producing more $[cAMP]_i$ ions, which enhance cell-cell communication.[48] At some point (e.g., when encapsulation is to be terminated), the Ca^{2+}-modulating effect of the $[cAMP]_i$ is switched on (perhaps by some temporal switch mechanism that would reduce $[cAMP]_i$ by activating any of the mechanisms listed in the following section), which results in an increase in $[Ca^{2+}]_i$ ions and concomitant reduction and blocking of cell-cell communication.

a. $[cAMP]_i$ Concentration and Its Modulation

AC activation, in fact, requires the activation of two plasma membrane proteins (receptor protein and the G-protein). The receptor-ligand complex first binds and activates the G-protein (GTP-binding protein, a GTPase) at the cytoplasmic face of the membrane; this changes the conformation of the G-protein, and activates the AC to synthesize $[cAMP]_i$. The GTPase self-inactivates by hydrolyzing its bound GTP to GDP and inorganic phosphate; this process returns the AC to its original inactive state.[16] Note that AC is also activated by β-ecdysone[48] (see Section IV.B.2). Inhibition of the cAMP phosphodiesterase (e.g., experimentally by theophylline) results in increased $[cAMP]_i$ level in salivary gland cells of *Drosophila hydei*.[48] Intracellular concentration of $[cAMP]_i$ can be decreased by (1) inhibiting the AC (by Ca^{2+}-CaM complexes); (2) activating the cAMP phosphodiesterase (that breaks down cAMP to 5′-AMP); and (3) increasing $[Ca^{2+}]_i$ ion concentration (by phosphorylating some Ca-kinases [Ca^{2+}-CaM-dependent, membrane transport proteins] by A-kinase [PK-A]) (see also the following discussion).

b. $[Ca^{2+}]_i$ Concentration and Its Modulation

In a resting cell, the concentration of free Ca^{2+} is extremely low (ca 10^{-7} *M*) and may rise up to 5×10^{-6} *M*, when the cell is activated by an extracellular signal. The concentration of Ca^{2+} in the extracellular fluid may be higher than 10^{-3} *M*, and consequently a strong gradient may exist, which would tend to drive Ca^{2+} ions into the cell across the plasma membrane. The low cytosolic concentration of Ca^{2+} is maintained by pumping out Ca^{2+} by a calcium ion pump (Ca^{2+}-ATPase) and/or by a Na^+-Ca^{2+}-exchanger that couples efflux of Ca^{2+} to influx of Na^+, and is activated only when Ca^{2+} level rises to above ten times that of normal level.[16] The effectiveness of Na^+-Ca^{2+}-exchanger depends on the transmembrane Na^+-gradient that is maintained by the Na^+,K^+-ATPase, the currently accepted stoichiometry being 3 Na^+:1 Ca^{2+}.[51] In fact, Na^+-gradient-dependent transport of Ca^{2+} ions occurs in rabbit T-cell plasma membrane.[52] The process of exchange is reversible and is inhibited by tetracaine and verapamil (inhibitors of Na^+/Ca^{2+} exchange), but not by orthovanadate. We recently demonstrated the existence of the Na^+,K^+-ATPase on the plasma membrane of the granulocyte (GR) of *Limulus polyphemus*,[53] and I have suggested earlier[34] and elsewhere (see Chapter 4) that the arthropod GR is a functional hybrid of vertebrate B- and T-lymphocytes, and also functionally resembles macrophages and human platelets.[54]

Finally, in addition to the Ca^{2+}-ATPase, the microsomes, mitochondria, and CSC and/or calciosome also keep the $[Ca^{2+}]_i$ level low by taking up large amounts of it against steep cytosolic gradients.

In muscle cells, Ca^{2+}-ATPase transports $[Ca^{2+}]_i$ from the cytosol to the sarcoplasmic reticulum (SC) (a network of tubular sacs), which stores $[Ca^{2+}]_i$, as the CSC and/or calciosome do in other cells. Release of $[Ca^{2+}]_i$ ions (as a result of depolarization) from the SC to the cytosol, results in muscle contraction, and transport of $[Ca^{2+}]_i$ ions back into the SC causes relaxation.[55]

The $[Ca^{2+}]_i$-ATPase or the calcium ion pump, like the Na^+,K^+-ATPase, is phosphorylated and dephosphorylated during its pumping action; and for every ATP molecule hydrolyzed, the Ca^{2+}-ATPase pumps two Ca^{2+} ions into the SC of the muscle cell. Because large amino acid homology between Ca^{2+}-ATPase and Na^+,K^+-ATPase exists, these two ion pumps are regarded as related.[16]

Any structure that is subject to Ca^{2+}-dependent regulation must have an affinity constant (K_a) for Ca^{2+} of about 10^6 liters/mole.[16] The Ca^{2+}-binding protein, calmodulin (CaM), fulfills this criterion, is found in all animal and plant cells, and is highly conserved; however, its presence in the insect GR needs to be confirmed. It serves as a multi-purpose intracellular receptor for Ca^{2+}, and mediates most Ca^{2+}-dependent processes. It is a single polypeptide chain of about 150 amino acid residues, with four high-affinity Ca^{2+}-binding sites.[56-58] Its conformation changes when it binds to Ca^{2+}. This allosteric activation of the CaM is analogous to that of A-kinase (cAMP-dependent protein kinase, also known as PK-A) by cAMP, except that Ca^{2+}-CaM complexes, unlike the cAMP-A-kinase complex, have no enzymic activity themselves. They activate many enzymes and membrane transport proteins (Ca^{2+}-CaM-dependent protein kinase), called Ca-kinases, which, like the A-kinases, phosphorylate serine and threonine residues on proteins.[16] $[Ca^{2+}]_i$ acts as an intracellular messenger in cell secretion and proliferation (see discussion in Section VI.C).

2. Effects of $[cAMP]_i$ on Gap Junctional Communication

Apparently, arthropod hemocytes synthesize and secrete cAMP into the hemolymph, and consequently the increased level of hemolymph cAMP might be responsible for increased gap junctional conductance and even GR/PL (plasmatocyte) proliferation during encapsulation, because cAMP is known to modulate cell proliferation[59] (see discussion in Section VI.C). Furthermore, increased levels of cAMP in the hemolymph probably also prevent any degranulation of GRs, while the latter are engaged in forming capsule layers (see discussion in Section VI.E). Battelle and Kravitz[60] and Gole et al.[61] have reported the presence of AC receptors on the hemocytes of the American lobster, *Homarus americanus* and the forest tent caterpillar, *Malacosoma disstria,* respectively.

Injection of octopamine into third-instar larvae and pupae of *M. disstria* produced rapid elevation of hemolymph cAMP levels; this action of octopamine was attributed to the hemocytes, on the basis of its effect on a hemocyte cell line (Md 66).[61] The effect was simulated by the adrenergic agonist, synephrine, and was inhibited by α-adrenergic blocking agent, phentolamine. Gole et al.[61] concluded that the hemocyte AC receptor system resembles the equivalent vertebrate system, and that it is activated by sodium fluoride and requires GTP for the activation of the AC. Note that starvation and low cell density also elevate $[cAMP]_i$. Finally, cAMP-increasing agents increase junctional coupling and gap junction formation in several cell lines.[62] How does $[cAMP]_i$ work?

a. Mode of $[cAMP]_i$ *Action*

It is thought that $[cAMP]_i$ activates a cAMP receptor, which is a protein kinase (PK), also called A-kinase (PK-A), which in its inactive state, consists of a complex of a pair of regulatory subunits (that bind cAMP and occur in two isoenzyme forms, I and II) and a pair of catabolic subunits, which are invariant.[16,63,64] The cAMP-binding alters the conformation of the regulatory subunits, which then dissociate from the complex, activate and release the two catabolic subunits, which then phosphorylate the serine and threonine residues of the junctional protein, which acts as a substrate for the PK-A. This is supported by Saez et al.,[65] who suggested that it is the longer cytoplasmic carboxyl domain of the 27-kDa gap junctional protein (the amino terminal is also cytoplasmic but shorter) that is the site of serine residue phosphorylation through the action of cAMP-dependent PK-A. Spray et al.[14] suggested that "cAMP and Ca-dependent kinase systems converge to phosphorylate the prominent gap junction channel protein (27 kDa) at common sites . . . cAMP increases junctional conductance within minutes (by direct phosphorylation) and prolongs the lifetime of channels in the membrane by several hours through effects on turnover of the protein.''

b. [cAMP]$_i$ Modulates Gap Junctional Communication

cAMP-dependent PK-A is considered a physiological modulator of gap junctions, because its activation by hormones or pharmacological agents, causes an increase in junctional permeability and an increase in the number of gap junctions.[66-68]

It is known that intracellular injection of cAMP increases electrical coupling of mammalian cardiac cells within 30 sec.[69] Earlier, Hax et al.[48] had reported that cAMP enhanced junctional conductance in the cells of the *Drosophila* salivary gland. Azarnia et al.[70] have discussed the role of another PK-A that phosphorylates the same(?) or another junctional protein on the tyrosine residue, and down-regulates junctional communication. This PK-A is the product of cellular *src* gene (c-src), which is ''highly conserved in evolution and present in all animal species''. cAMP stimulates gap junctional permeability in a murine cancer cell line (C1-ID), whose cells are normally deficient in gap junctions.[71,72]

In addition to surface receptors, the hormonal system is integrated with the intercellular communication via gap junctions, and this results in a direct exchange of chemical and electrical messages between cells. In *D. hydei* salivary gland cells[48] and the epithelial cells of *Tenebrio*,[73-75] β-ecdysone causes increased junctional communication; however, the involvement of cAMP in this effect is unconfirmed.

Increased levels of cAMP retard LPS-induced degranulation (exocytosis) of the GRs of the American Horseshoe crab, *L. polyphemus* (see Chapter 1).

Finally, demonstration of gating requires evidence that the gap junctions remain in the membrane in the nonconducting (blocked) state, and in this respect, gating is different from formation and removal of gap junctions (turnover rate).[76] Although persistence of gap junctions in the membrane can be inferred from speed and reversibility of conductance change (often in single channel recordings of macroscopic currents), morphological evidence of conductance (i.e., between open and closed configurations of gap junctions) has not been generally possible because of fixation artifacts, even with fast freeze-fracture methods[77] (see also Section III.B).

3. Effects of [Ca^{2+}]$_i$ and [H^+]$_i$ Ions

a. [Ca^{2+}]$_i$ Modulates Gap Junctional Communication

Compared with [cAMP]$_i$, very little is known about the precise site of action of [Ca^{2+}]$_i$; and unlike [cAMP]$_i$, [Ca^{2+}]$_i$ ions block gap junctional communication and are known to cause exocytosis or degranulation in human platelets and the GRs of *L. polyphemus*. Furthermore, unlike the [cAMP]$_i$, [Ca^{2+}]$_i$ ions are necessary for cell-cell adhesion, coagulation, and prophenoloxidase activation.

Loewenstein[78] was the first to suggest that Ca^{2+} ions regulate junctional communication. About a decade later, the question was raised whether some of the effects attributed to [Ca^{2+}]$_i$ were those of H^+.[79] Indeed, it has been shown that intracellular [pH]$_i$ falls when [Ca^{2+}]$_i$ level is raised in snail neurons,[80] and that intracellular acidification causes reversible uncoupling of gap junctions in the embryo of the African clawed frog, *Xenopus laevis*.[81] Thus, we now know that increased free [Ca^{2+}]$_i$ concentration,[82-84] cytoplasmic acidification,[85] and establishment of transjunctional voltage,[86] all regulate channel closing. Note, however, that it has been shown, by simultaneously monitoring the intracellular [pH]$_i$ and [Ca^{2+}]$_i$ in the salivary glands of *Chironomus thumni,* that [Ca^{2+}]$_i$ caused junctional uncoupling at constant or even alkaline [pH]$_i$.[87] This suggests that [Ca^{2+}]$_i$ alone was sufficient to cause channel closure.[88] Of the two ions, H^+ binds nonspecifically to many proteins, but Ca^{2+} does so with high specificity.[88]

Various accounts in the literature suggest that different concentrations (10^{-6} to 10^{-3} M) of Ca^{2+} are involved in channel closure in different tissues,[89] and similarly different levels of cytoplasmic acidification are required to reduce junctional conductance.[89,90] It is unknown, however, whether these variations in $[Ca^{2+}]_i$ and $[H^+]_i$ concentrations are due to differences in junctional proteins in various tissues or if they simply reflect different responses of a single ''archaic gating'' according to the differing cytoplasmic milieu of various tissues.[89] It was found that the endogenous channels in the oocytes of *Xenopus* are highly sensitive to cytoplasmic acidification, because junctional conductance was reduced by injections of H^+ buffer solution at pH 7.0, and acidification via CO_2 (which reduced intracellular pH) abolished conductance with CO_2 levels as low as 10%.[89] Furthermore, the junctions were totally refractory to injection of Ca^{2+}-buffers with Ca^{2+} concentration of 10^{-4} M.

B. GAP JUNCTION TURNOVER (FORMATION AND REMOVAL)

The formation and degradation of gap junctions are significant for several biological phenomena, and their removal and loss is associated with terminal differentiation in certain cell types (e.g., mammalian oocytes).[18] Gap junction turnover is modulated by the cell's secretory and mitotic activities, hormones, and other compounds, such as heparin and proteoglycans.

1. Effect of Secretory and Mitotic Activities

Secretory activity (e.g., β-cells of pancreatic islets when stimulated to release insulin) increases frequency of gap junctions by 54 to 56%, compared with 27 to 48% in controls.[91] Exposure to a secretagogue, such as glucose, for a short duration increased the frequency of small-sized junctions, while a prolonged stimulation by glibenclamide (another secretagogue) caused both numerical increase and enlargement of gap junctions.[92]

SW-13 cultured tumor cells (from human adrenalcortical carcinoma) form monolayers (typical of normal epithelial cells), but some of the cells detach from the monolayer and ''pile up'' (typical of tumor cell growth). It has been shown that the number of gap junctions per cell in the monolayer increases ca 24-fold between 2 to 24 h;[18,93] the cells that detach from the monolayer and ''pile up'' are devoid of junctional membrane (showing gap junction down-regulation). This loss of gap junctions in the detached cells is associated with increased mitotic activity, because 88% of these cells divide within 5.6 h, average time between rounding and division being 2.3 h. In fact, gap junction loss is associated with cell proliferation in several tissues, such as regenerative liver and epidermal proliferative skin diseases.[18,93] In other words, cells remain in contact during periods of low mobility and non-proliferation, and separate when migration and mitosis are needed (see also discussion on cell proliferation in Section VI.C).

2. Effect of Hormones and Other Factors

Estrogen and prostaglandins stimulate gap junction formation in the myometrium (smooth muscle layer of the uterus in humans and other vertebrates), and progesterone inhibits it.[94] Glucagon (hormone that stimulates liver cells to breakdown glycogen and release glucose into the blood) increases gap junction mRNA, and insulin decreases it.[14] Furthermore, treatment with heparin or proteoglycans (composed of long glycosaminoglycan chains covalently linked to a core protein, and found in the extracellular matrix), especially heparin sulfate proteoglycan in the extracellular matrix of the liver, enhances re-expression of gap junctions between liver cells in culture. In the absence of glucagon, heparins have no effect on the level of gap junction mRNA.

Ashhurst,[95] on the basis of histochemical tests on spherulocytes (which also contain mucopolysaccharides) of the greater wax moth, *Galleria mellonella,* suggested that the presence of sulfated glycosaminoglycan polymer (confirmed by very long periodate oxidation) indicated the presence of heparin-like molecules in these hemocytes. Although the roles of these molecules in gap junctional conductance remains to be investigated, it can be suggested that they perhaps modulate the rapid formation of gap junctions among encapsulating GRs.

In a preliminary work on encapsulation of pieces of surgical sutures in *H. americanus*, we have observed spherulocytes (SPs) on the 3-day-old implants. A re-evaluation of the hemocyte types in this animal will be presented elsewhere. Note, however, that SPs are present in *G. germanica*[96] and in most other insects (see Chapter 2).

In insects, β-ecdysone (molting hormone) is reported to increase junctional coupling in *D. hydei* salivary gland cells,[48] epithelial cells in *Tenebrio*,[73-75,97] dispersion of junctional particles in the perineurial sheath (the outer sheath surrounding the glial-neuronal mass) cells in the central nervous system of *Manduca sexta*,[98] and an increase in the gap junctions in the midgut cells in the Horseshoe crab, *L. polyphemus*.[15] In this animal, the gap junctions were irregular and enlarged, with loosely packed particles. Whether the gap junctional communication is normally endocrine-system-driven or not in insects and other arthropods is unknown, although the morphogenetic hormones of arthropods regulate the histogenesis, organogenesis, and morphogenesis of many tissues and organs in these animals (see reviews in Gupta[99-101]).

C. MODULATION OF GAP JUNCTION SIZE AND FUNCTIONAL AREA

Gap junction populations are known to vary within a single tissue, depending on its physiological and developmental state.[91,102,103] The amount of gap junction protein also may vary within the same system over time.[4]

Growth in the size of gap junction plaque is attributed to aggregations of smaller plaques into larger ones;[18,93] and because in mammalian follicle cells, larger junctions showed evidence of endocytosis, and were the first to disappear during down-regulation of gap junctions, these authors considered large plaques as terminal stages. Endocytosis commences by the formation of a large junctional invagination into one of the two cells, which later pinches off in the form of a bimembranous vesicle within that cell, enclosing some cytoplasm from the other cell. The pinched-off membrane consists of clathrin and actin.[104] As in phagocytic endocytosis, the bimembranous vesicle later fuses with the lysosomes and is eventually degraded and recycled.

Note that junctional degradation is accomplished not only by an internalization mechanism, but also by dispersion and lateral diffusion of the junctional particles in the plasma membrane, thus becoming unrecognizable[98] in freeze-fractured replicas. These authors report that when the larva of *M. sexta* enters the pupal stage, within 24 to 48 h, the gap junctions in the glial cells (that ensheath axons) in the central nervous system "are pulled out into linear strands" at several points, and the strands become associated with "particle-free aisles" that are straddled by membranous particles. The membrane sinks into "grooves" (depressions, not "furrows") in the E face along which the particles stream out away from the plaque periphery. The appearance of the grooves was thought to be a cellular response to stimulation by β-ecdysone. Later, the particle-free aisles widen and the particles are progressively dispersed on the E face, instead of being internalized by endocytosis.

V. CELL SURFACE ADHESION MOLECULES

Plasma membrane glycoproteins are known to participate in cell-cell adhesion. Cell adhesion molecules, called *cadherins* (a family of glycoproteins) have been reported by Takeichi,[105] who found three subtypes (N-, P-, and E-cadherins) of these molecules that cause Ca^{2+}-dependent cell-cell adhesion in vertebrate tissues. Antibodies of these molecules induced dissociation of cell layers, which indicates their importance in a multicellular structure. These cadherins are integral membrane proteins (deduced from cDNA encoding) of 723 to 748 amino acids, and resemble neural cell adhesion molecules (N-CAM) as transmembrane proteins, except that in the absence of Ca^{2+}, they (the cadherins) undergo conformational change and are degraded by proteolytic enzymes.

What is most interesting from the standpoint of cessation of GR-GR adhesion during the latter part of encapsulation, is Takeichi's[105] observation that "the expression of each cadherin subclass is spatiotemporally regulated and associated with a variety of morphogenetic events; e.g., the termination or initiation of expression of a cadherin subclass in a given cell collective is correlated with its segregation from or connection with other cell collectives".

Although, it remains to be demonstrated whether during the initial stage of encapsulation, GR-GR adhesion is Ca^{2+}-dependent, Ca^{2+} ions are known to mediate cell-cell adhesion in keratinocytes (located in the basal layer of human epidermis).[106] These cells normally proliferate continuously and release their progeny into the upper layer of the epidermis. However, among dissociated keratinocytes in a culture medium, in which Ca^{2+} is kept abnormally low (0.02 to 0.1 m*M*; normal level in culture ranges from 1.2 to 1.8 m*M*), cell adhesion fails and the cells form a monolayer in which proliferating and differentiating cells are found mixed together. If the Ca^{2+} level is raised (1.2 m*M*, the monolayer becomes a multilayerd epithelium, in which the basal layer is composed of proliferating cells and the upper one by differentiating cells, just as is the case in normal skin.[106]

It should be mentioned, however, that Ca^{2+}-independent cell-cell adhesion also occurs. For example, in neurons and glial cells, one of the plasma membrane glycoproteins, called N-CAM, binds them together. This binding is independent of Ca^{2+}, and is due to homophilic interaction in which two N-CAM molecules are joined together. Cunningham et al.[107] have reported that N-CAM is a long peptide chain that is folded into five immunoglobulin-like, transmembrane domains. It was suggested on this basis that N-CAM belonged to an ancient superfamily of recognition proteins to which antibodies also belong. Furthermore, it has been suggested that the immunoglobulin superfamily evolved from simpler immunoglobulin-like molecules, which originated in primitive cell-cell interactions.[108]

It is known that dissociated cells with tissue- or cell-specific recognition systems (e.g., in liver and embryonic cells and tissues) tend to adhere more readily with their own aggregates than to those of others. It has been shown in Leghorn chicken embryos that liver and neural retina cell aggregates of isotypic combinations (aggregates and suspensions of the same tissue type) collected many more labeled (with tritiated thymidine) cells than those in heterotypic (aggregates and suspensions of different types) combinations.[109] They concluded that (1) specificity exists between cells of the same kind; (2) this specificity exists between cells before they get into aggregates; and (3) isotypic aggregates are more stable than heterotypic ones. This may be the reason why activated GRs, following invasion by an antigen, tend to adhere to aggregates of other GRs and not to fat body cells that are found in the hemolymph or around organs and tissues.

VI. MOLECULAR BASIS OF ENCAPSULATION

A. SIGNAL TRANSDUCTION BY GR SURFACE RECEPTORS

According to Kikkawa et al.,[110] most tissues seem to have two classes of surface receptors that transduce information across the plasma membrane: (1) those related to cAMP production via adenylate cyclase (AC) and kinase A (e.g., platelets and lymphocytes), and (2) those related to cGMP production by the activation of inositol-containing phospholipids or phosphoinositides (PIs) of the plasma membrane via guanylate cyclase (GC) and kinase C (e.g., mast cells). It should be mentioned that signals that produce cAMP, antagonize cGMP production and vice versa; in other words, cAMP and cGMP have negative feedback effects on each other. I will now discuss in some detail the signal transduction mechanism via the PI pathway, since it is not generally mentioned in accounts of encapsulation in insects.

Receptor-mediated (e.g., Con A receptor) hydrolysis of PIs (a Ca^{2+}-mediated process) triggers an integrated cascade of intracellular processes that sequentially releases 1,2-diacylglycerol (produced transiently in the membrane and degraded within a few seconds) to activate protein kinase C (a Ca^{2+}-dependent process),[111,112] mobilizes Ca^{2+} (once mobilized, it quickly returns to its basal or below basal levels),[113] releases arachidonic acid (AA),[114] activates GC,[115] and ultimately produces cGMP.[116] Note that phosphorylation of diacylglycerol (DG) produces phosphatidic acid, which acts as an ionophore and facilitates entry of extracellular Ca^{2+}. Furthermore, AA may also be derived from PIs as well as phosphatidylcholine by the activation of a nonspecific phospholipase A_2, for which Ca^{2+} is necessary. This pathway for AA production is well developed in neutrophils and platelets.[117] Finally, like the cAMP, kinase C is known to be activated by the G-protein via the IP hydrolysis pathway.[16]

Receptor-specific hydrolysis of the PI is ubiquitous; a small portion of the PI is phosphorylated at 4-position to form PI-4-phosphate or PIP, which is phosphorylated at 5-position to produce PI-4,5-phosphate or PIP_2, which seems to be the primary substrate for the receptor mechanism;[117] as it is degraded, the PIP_2 is constantly replenished by phosphorylation of PI. High intracellular concentration of inositol is necessary for resynthesis of PI, and it is maintained by (1) recycling of inositol head groups, (2) *de novo* synthesis from glucose-6-phosphate, and (3) uptake of plasma inositol derived from diet.[117] Con A receptors use PIs as a transducing mechanism in mast cells and lymphocytes.[117] The GRs of *L. polyphemus* have Con A-binding receptors,[53] which suggests that these immunocytes probably use membrane PI-transducing mechanism to mobilize $[Ca^{2+}]_i$ and to generate cGMP.

Rapid hydrolysis of PIP_2 by agonists (e.g., Con A) is known to occur in liver cells,[118-120] platelets,[121,122] and insect salivary gland cells.[123] When the salivary gland was stimulated with 5-HT (serotonin), within 5 sec the level of IP_3 (1,4,5-triphosphate), derived from PIP_2, increased fourfold, there being no change in the levels of inositol 1-phosphate or free inositol. Stimulation of platelets by thrombin also showed a marked increase in IP_3 level.[124] Two molecules of ATP are required to phosphorylate PIP_2, which suggests that the breakdown of membrane PI is energy dependent, requiring a constant supply of ATP.

Exogenous DG is known to stimulate platelet secretion without causing any change in $[Ca^{2+}]_i$ levels.[117] In fact, a synthetic DG, 1,oleoyl-2-acetylglycerol, has been used to elucidate the link between inositolphospholipid turnover and protein kinase C activation; it is readily incorporated into the cell membrane and dispersed into the phospholipid bilayer to activate the kinase C directly.[125] Furthermore, using monoclonal antibodies raised against protein kinase C, the presence of this enzyme in the granules of bone marrow granulocytes,[110] and also in the periphery of the brain cell nuclei (see discussion on page 210 in Kikkawa et al.[110]) has been demonstrated; its presence in the nuclear membrane, implicates kinase C in cell proliferation.

Protein kinase C has been shown to be involved in $[Ca^{2+}]_i$ extrusion (ATPase is necessary for this process) after the latter's mobilization in the cytosol, because $[Ca^{2+}]_i$ is decreased by adding TPA (12-*O*-tetradecanoylphorbol-13-acetate). TPA is a tumor-promoting phorbol ester that is very similar to DG in its molecular structure, and activates kinase C both *in vivo* and *in vitro*.[126-131]

Protein kinase C, like kinase A (which is cAMP-dependent), phosphorylates seryl and threonyl residues, but not the tyrosyl residues of endogenous proteins[110] (see discussion in Section IV.A.2.a). All seryl and threonyl residues that are phosphorylated by these two kinases have basic amino acyl residues on both amino and carboxyl terminal sides.[132] Thus, although the two kinases transduce different information into the cell, sometimes their signal pathways may produce similar responses, depending on the structure of the substrate protein molecule.

Finally, cGMP is present in animal cells in lower quantity than cAMP, probably because at a physiological concentration of substrates, cGMP is hydrolyzed at a greater rate than cAMP,[133] and both are regulated by $[Ca^{2+}]_i$.[134] Unlike thc AC, GC is not always membrane bound.[16] Note that cytoplasmic phosphodiesterase also catalyzes the hydrolysis of cGMP.

B. EXOCYTOSIS OF GRs

Exocytosis/degranulation of GRs has been reported in most all cases of encapsulation in insects, and has been associated with the release of the so-called recognition, growth, opsonin-like, and coagulation factors, and prophenoloxidase.

Ca^{2+} ions (intra- or extracellular) provide the stimulus for exocytosis.[134-137] The granules in the GRs of *L. polyphemus* are thought to contain Ca^{2+}[138] (see also Chapter 1). Furthermore, $[Ca^{2+}]_i$ ions in the GRs of this animal rise when stimulated by the endotoxin of Gram-negative bacteria.[137] It should be mentioned here that Cannon et al.[139] found Mg^{2+} to be more effective than Ca^{2+} in stimulating isolated GRs of *Limulus* to degranulate.

What is rarely mentioned in connection with exocytosis is the role of receptor-mediated hydrolysis (a Ca^{2+}-mediated process) of membrane PIs, which almost simultaneously generates DG (that activates the protein kinase C) and mobilizes $[Ca^{2+}]_i$. Both the activation of protein kinase C by DG and Ca^{2+} mobilization are necessary for exocytosis in neutrophils, basophils, and mast cells, and for the release of serotonin from platelets.[110,116,140]

C. PROLIFERATION OF GRs/PLs

It is often mentioned that when the GRs degranulate on contact with the foreign antigen, they release growth factors that cause proliferation of other GRs/PLs that are needed in large numbers to form layers. However, how this proliferation is accomplished is rarely discussed. $[cAMP]_i$, $[Ca^{2+}]_i$, and protein kinase C are known to modulate cell proliferation directly or indirectly in various tissues and cells.

In addition to increasing intercellular communication (see discussion in Section IV.A.2), $[cAMP]_i$ is known to modulate cell proliferation and neoplastic development.[59,67] Note that thrombin (involved in clotting and wound healing) is thought to act as a growth factor,[141] and specific high-affinity thrombin receptors have been found on mouse embryo cells. Thrombin causes a transient increase in the level of $[cAMP]_i$ in NIL hamster cells and other types of cells,[141] and is therefore thought to act as a possible modulator of mitogenic signaling via the transmembrane thrombin receptors.

$[Ca^{2+}]_i$ and calmodulin (CaM) (a membrane-associated protein) also modulate cell proliferation. High concentration of $[Ca^{2+}]_i$ and a CaM-like protein[142,143] are known to block gap junction permeability (see discussion in Section IV.A.3), which in turn regulates proliferative and migratory activities of the cells involved. It has been shown in neonatal rats that gap junction protein in hepatocytes disappears with the loss of gap junction plaques, which in turn immediately initiates rapid cell proliferation.[4] When the liver is reconstituted as a result of cell proliferation, gap junctions rapidly reappear.[144]

What is significant to mention here are the effects of protein kinase C activation and Ca^{2+} mobilization in gene expression and cell proliferation.[110] In fact, it has been shown that both Ca^{2+} mobilization and kinase C pathways are essential, and both act synergistically for mitogenic DNA synthesis in peripheral human T-lymphocytes,[110] and in B-lymphocyte proliferation.[146] However, for rapid proliferation, such as during encapsulation, some growth factors (e.g., mammalian interleukin 2), even in low concentrations, are necessary. Presumably, kinase C plays some role in the gene expression of some proteins and enzymes related to cell growth and differentiation.[110]

D. CELL JUNCTIONS IN HEMOCYTIC CAPSULES

Gap junctions in hemocytic capsules have been reported in only a few insects: *E.*

kühniella,[147] *P. americana,*[29,148] and *B. germanica.*[12] However, their functional significance in capsule formation has been rarely discussed.

Desmosomes have been reported in the capsules of *C. morosus,*[149] *L. migratoria* and *M. melolontha,*[150] *P. americana,*[29] and *G. mellonella,*[151] and septate junctions in *B. germanica*[152] (Figures 4A,B and 5A,B).

It seems the desmosomes and septate junctions help maintain the structural integrity of the capsule. Desmosomes are known to anchor cells together as localized "weld spots", and are most abundant in tissues that are subject to severe mechanical stress (e.g., cardiac muscle, skin, and neck of uterus).[16] It has been suggested that septate junctions in invertebrates are functionally analogous to vertebrate tight junctions,[30] or to a combination of tight and intermediate junctions.[153] We suggested that septate junctions in the hemocytic capsules of *B. germanica*[152] were involved in anchoring the layer-forming GRs together to insure a tightly formed capsule that would withstand the mechanical stress and the shearing forces that might be exerted by a foreign antigen that is being encapsulated, especially a live one. We concluded that the compactness of the capsule in this insect was confirmed by the fact that both lanthanum and ruthenium red could not impregnate it.

E. MOLECULAR BASIS OF ENCAPSULATION

It is well established that $[Ca^{2+}]_i$ and $[cAMP]_i$ play significant roles (albeit opposite) in junctional conductance among cells, including those of growing tissues. On the basis of these proven roles, coupled with the facts that plasma membrane glycoproteins are involved in cell-cell adhesion and phosphoinositides in signal transduction, I postulate the following roles of these ions/molecules in encapsulation (Figure 6). In this discussion, the capsule will be regarded as a growing tissue, and therefore, expected to behave as such during its morphogenesis. The roles of only the major immunocyte, the granulocyte (GR), and occasionally the plasmatocyte (PL) will be discussed; both GRs and PLs are known to participate in encapsulation in many insects. For convenience of discussion, the encapsulation process has been divided into three phases and the capsule in *B. germanica* used as a reference.

1. Foreign Antigen Recognition Phase: Exocytosis/Degranulation of the GR and GR-GR Adhesion

During this phase, the $[Ca^{2+}]_i$ level is high enough to cause exocytosis of GRs. By contrast, the $[cAMP]_i$ level is low, because the high $[Ca^{2+}]_i$ level inhibits the plasma membrane adenylate cyclase (AC), and thus keeps $[cAMP]_i$ low. Furthermore, because Ca^{2+} stimulates phosphodiesterase activity, it degrades cAMP. Thus, $[Ca^{2+}]_i$ is an important regulator of $[cAMP]_i$ homeostasis, participating in not only its degradation, but also its formation (see discussion in the next section). The surface glycoproteins facilitate GR-GR adhesion, which is a Ca^{2+}-dependent process.

When a foreign biotic or abiotic antigen (ligand) enters the immune system, it randomly comes in contact or collides with the GR, which is activated (Figure 6) and whose surface receptor (RE) form a receptor-foreign antigen (FA) (ligand)-complex. The latter triggers the mobilization of $[Ca^{2+}]_i$ from the cytosolic Ca^{2+}-buffering systems (microsomes,[134,154] mitochondria,[155] calcium-sequestering-compartment,[16] and/or calciosome[45]) by any of the following two pathways: (1) by activating the AC (via the G-protein, or (2) by activating protein kinase C (via the hydrolysis of the phosphoinositides) (see discussion in Section VI.A). Which pathway (AC or kinase C, or both) actually operates in *B. germanica* or other insects is unknown. At any rate, encounter with the foreign antigen, formation of the receptor-ligand complex, and the resulting reactions, cause a rapid transient increase of $[Ca^{2+}]_i$ ions, which are necessary for exocytosis or degranulation of the GRs (see discussion in Section VI.B). I have mentioned earlier that the appropriate concentration of $[Ca^{2+}]_i$ is rigorously maintained by the cell, and any excess amount may be extruded, as happens in human platelets.

Most recently, it was demonstrated that in stretch stimulated single epithelial cells in rat lungs, there was a 3.5-fold increase in $[Ca^{2+}]_i$ and that calcium was mobilized from intracellular rather than extracellular stores (via the plasma membrane Ca^{2+} channels).[46] Furthermore, the transient increase (less than 60 sec) in $[Ca^{2+}]_i$ ions triggered sustained (15 to 30 min) stimulation of surfactant secretion and release by exocytosis. It is interesting to recall here that in *B. germanica*, as encapsulation commences, the GRs flatten and stretch (see Figure 25 in Chapter 2), with a concomitant increase in peripheral microtubules[12,54] (see also Figure 12 in Chapter 2). It may be that the flattening-stretching stimulus also causes the proposed transient increase in the $[Ca^{2+}]_i$ ions in the GRs, and probably a concomitant rapid synthesis and release of coagulation, recognition, opsonin-like, and growth factors. The hypothesis assumes that the transient increase in $[Ca^{2+}]_i$ at this point in time inhibits any increase in the $[cAMP]_i$ level by inhibiting the AC by Ca^{2+}-CaM complexes and/or activating the cAMP phosphodiesterase (see Section 4.A.1.a). In addition, kinase C has a negative feedback effect on cAMP. The cGMP level is also kept low. Note that exocytosis is also caused by bacterial endotoxin or LPS.

By using the ionophore A23187 (a Ca^{2+}-selective [not specific] ionophore that mimics the effects of microinjected Ca^{2+} into cells), we can experimentally circumvent the receptor-activated effects and can demonstrate whether Ca^{2+} is the main mediator during this phase. This property of A23187 allows one to study the effects of Ca^{2+} on a variety of cellular functions, because in each case where Ca^{2+} is the primary mediator, the final response of A23187 mimics its role,[134] and is known to culminate in exocytosis, as reported in platelet aggregation.[156] It must be noted, however, that because A23187 circumvents the receptor activation, it fails to increase the $[cAMP]_i$ level in the cell. In fact, it may decrease its level,[134] because $[Ca^{2+}]_i$ depresses AC. Another ionophore, ionomycin, is known to have a greater affinity for $[Ca^{2+}]_i$.[157]

The exocytosis, in turn: (1) causes localized coagulation (a Ca^{2+}-dependent process) trapping more GRs and facilitating their aggregation; (2) releases the so-called recognition factors (or host factors; see discussion on phagocytosis in Chapter 2), which attract other GRs/PLs to the site of action; (3) releases growth factors, which facilitate GR/PL proliferation (protein kinase C also facilitates DNA synthesis and cell proliferation); and (4) releases prophenoloxidase, which is then activated (a Ca^{2+}-dependent process) and produces melanization that is always found around a foreign antigen which is being encapsulated.[12] Opsonin-like sticky factors are supposedly also released at this time.

At the initial stage of GR aggregation at the site of action, cell junctions are not formed; instead, isotypic GR-GR contact is facilitated by cadherin-like plasma membrane glycoproteins (agglutinins?) on the GRs, which function as cell-cell recognition or adhesion molecules (a Ca^{2+}-dependent process). It is very likely that some of the lectin-binding receptors[53] on the GR plasma membrane act as adhesion molecules and facilitate adhesion of activated GR-GR. The initial nonjunctional GR-GR contact (Figure 7A, B) is ideal for limited cell locomotion, but not tight enough to immobilize the GRs to form cell junctions.

2. Layer Formation Phase: Formation of Cell Junctions and Maintenance of GR-GR Communication via Gap Junctions

During this phase, gap and other cell junctions are formed and the $[cAMP]_i$ level rises to maintain the gap junctional communication among the encapsulating GRs; inversely, the $[Ca^{2+}]_i$ level recedes, which, together with high $[cAMP]_i$ level prevents any exocytosis in the GRs. High $[cAMP]_i$ level is also known to retard LPS-induced degranulation of the GR in *L. polyphemus* (see discussion in Section IV.A.2.b). Finally, the capsule formation nears completion.

Ca^{2+} plays a direct role in enhancing the $[cAMP]_i$ level by binding with CaM, which acts as its receptor and activates AC; it is found in all animal and plant cells and highly

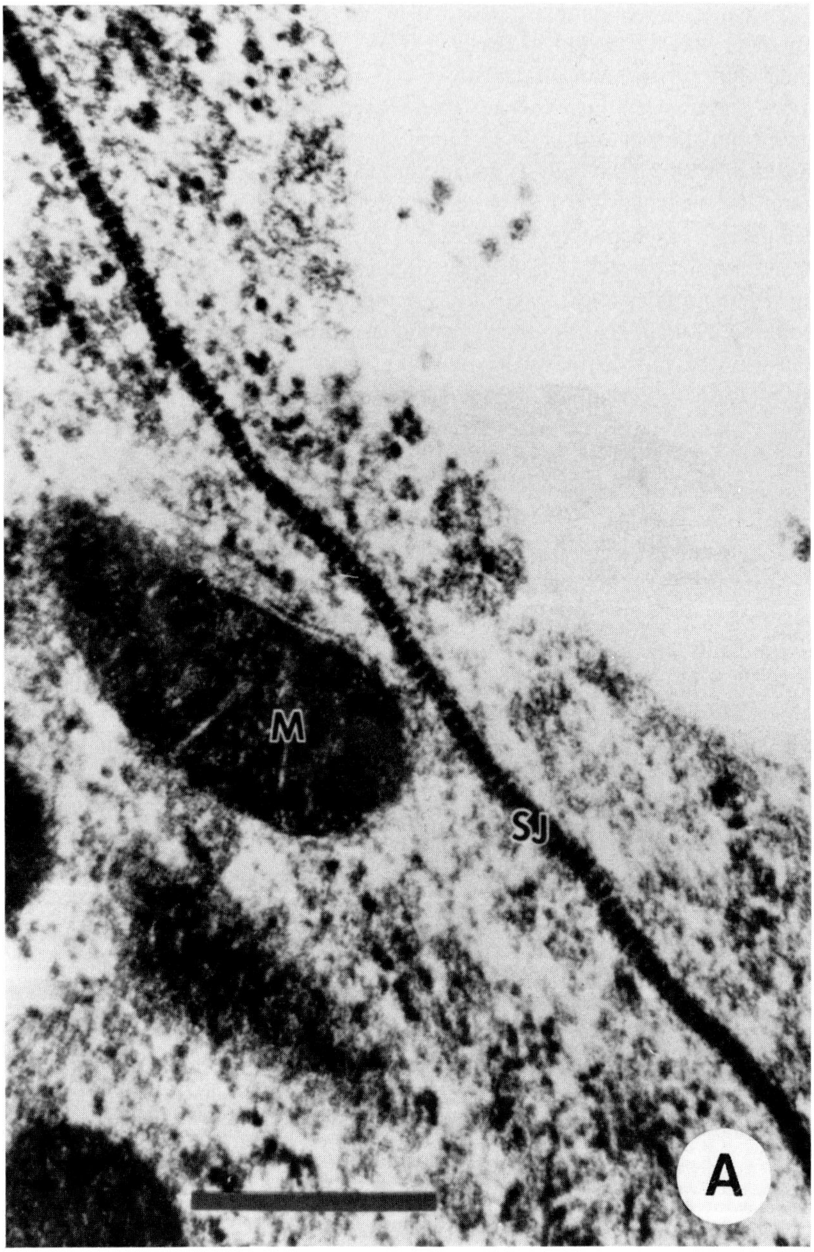

FIGURE 4. (A) Thin section of a 7-day-old capsule of *B. germanica,* showing septate junctions (SJ). Note mitochondria (M) in close proximity to the junctions (× 48,000). (B) High magnification of septate junction, showing 27-nm wide intercellular space and the faintly "scalloped" nature of septa (× 110,000). (From Han, S. S. and Gupta, A. P., *Cell Struct. Funct.*, 13, 333, 1988. With permission.)

conserved.[134] In addition, CaM also inhibits cAMP phosphodiesterase and protein kinase C, both inhibitions contributing to enhance the $[cAMP]_i$ level.

The low $[Ca^{2+}]_i$ during this phase is probably regulated by pumping Ca^{2+}-ions out by the calcium ion pump (Ca^{2+}-ATPase) (as happens in human platelets),[134] perhaps in conjunction with the Na^+-Ca^{2+}-exchanger, and/or by sequestering Ca^{2+} ions into microsomes,

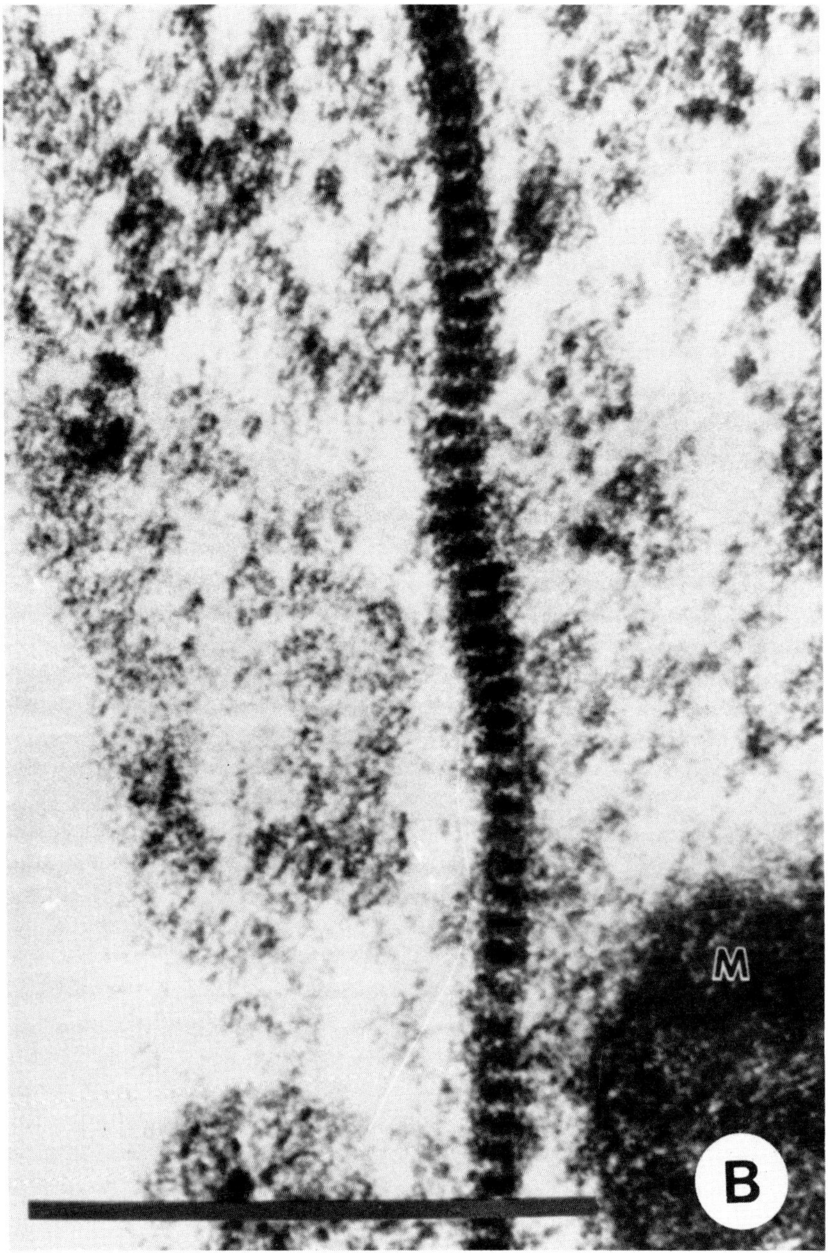

FIGURE 4B.

mitochondria, and CSC- and calciosome-like cytosolic storage organelles (see discussion in Section IV.1.B). How precisely low $[Ca^{2+}]_i$ level is maintained in *B. germanica* and other insects is unknown.

In addition to CSC- and calciosome-like organelles, mitochondria and microsomes are the two major Ca^{2+}-buffering systems in cells.[134] Microsomes are derived from both smooth and rough endoplasmic reticulum; those derived from the latter are still capable of protein and lipid synthesis, and protein glycosylation.[16] In addition, microvesicles that contain ATP, can accumulate Ca^{2+} by an ATP-dependent process,[158] and also rid the cell of excess Ca^{2+},

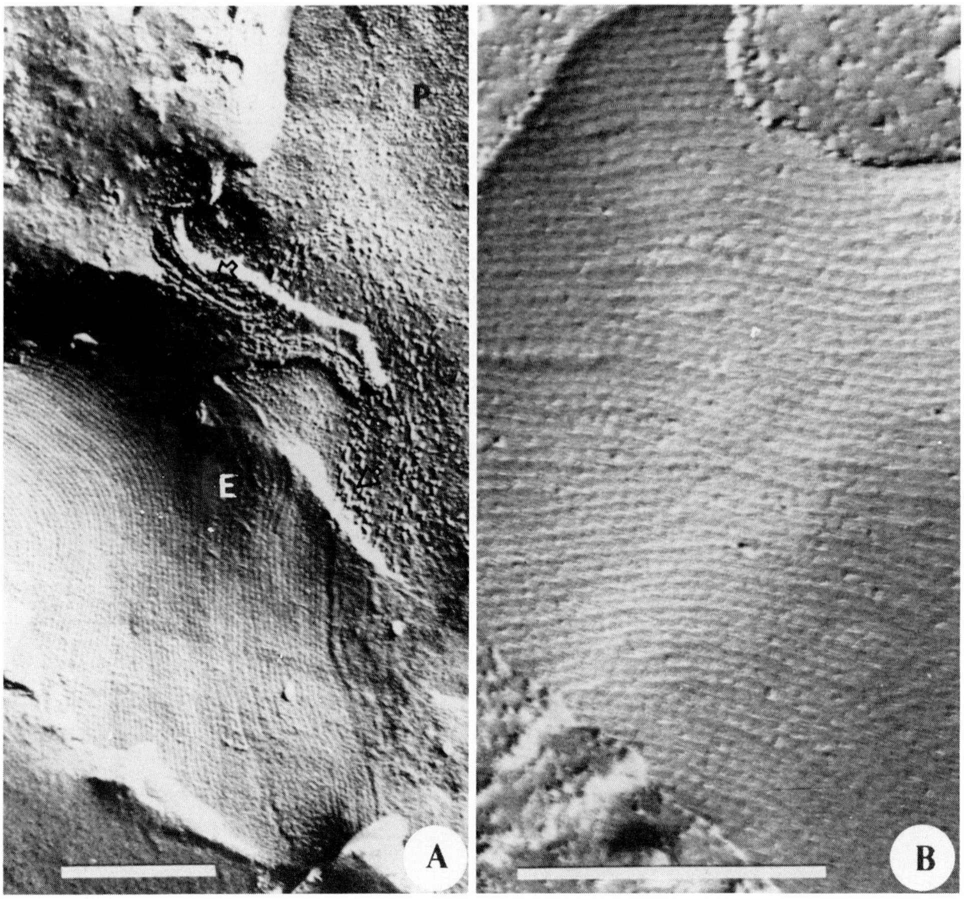

FIGURE 5. (A) Freeze-fracture replica of a 7-day-old capsule of *B. germanica,* showing septate junction. The junctions show furrows on the extracellular B (= E) fracture face and a part of the regular rows of IPs (arrows) and nonjunctional region of the A (= P) fracture face (× 29,670). (B) Freeze-fracture replica, showing furrows (13 nm wide, 23 nm center to center) on E fracture face (× 60,000). (From Han, S. S. and Gupta, A. P., *Cell Struct. Funct.,* 13, 333, 1988. With permission.)

as secretory granules do, by discharging it into the extracellular space.[134] We have demonstrated the presence of active and enlarged mitochondria[152] (Figure 4A), rough endoplasmic reticulum, and microvesicles (see Figure 27 in Chapter 2) in the activated GRs of *B. germanica* (see Figures 6 and 8 in Han and Gupta[12]). It should be mentioned that chromaffin granules of platelets are also known to uptake Ca^{2+}.[159] Thus, it is likely that the granules of the GRs provide another Ca^{2+}-buffering system to lower Ca^{2+} in *B. germanica* by releasing these cations to the outside.

The layer formation, of necessity, involves cell movement to contact neighboring GRs, followed by their immobilization to form a compact capsule, layer by layer. Because capsule formation is essentially a morphogenetic process, the GRs that are engaged in this process, probably detect even small adhering differences in other GRs and move toward them. In fact, it has been shown that motile neutrophils can detect differences in adherence as small as 1% on either side, and steadily move to contact neighboring cells to which they adhere intimately.[160] In this way, the GRs form the first layer around the foreign antigen.

As is the case in any growing tissue, any further layering would require immobilization of the GRs, especially of those forming the inner layers (mobilization always being limited to GRs on the outermost layers). This is accomplished by the formation of cell junctions at this stage. It is known that the plasma membranes of embryonic cells, while moving over each other come within 10 to 20 nm of each other, which is close enough to trigger the synthesis of gap junctional hemichannels (see Section II.), yet not narrow enough (2 to 3 nm) to allow linking with their counterparts of other cells. In *B. germanica* capsules, we found the intercellular space between junction-forming GRs to be 2.5 nm.[12] It is likely that some of the projecting plasma membrane glycoproteins of the GRs are spontaneously activated to initiate synthesis of channel proteins as soon as the GRs begin to form layers. In fact, it is known that the hemagglutinin glycoprotein of the influenza virus (first plasma membrane glycoprotein whose three-dimensional structure was determined) projects from the plasma membrane for about 13 nm.[16] It has been hypothesized that in growing embryos, "positional information" is set up by movement of ions from a region of a tissue where its intracellular concentration is high to a region of lesser concentration, "thereby setting up a smooth concentration gradient", which could serve as a reference system to determine the locations of cells within the growing embryo.[16] Such a gradient, if indeed it operates in the capsule, would instruct each GR to attach in a certain position. The hypothesis assumes that $[cAMP]_i$ generates this gradient via the gap junctions.

While gap junctions provide junctional communication among the layer-forming GRs, desmosomes, desmosome-like-junctions, and septate junctions maintain the structural integrity of the capsule (see Section V.A).

Three important questions need to be addressed here: (1) Does indeed junctional communication occur among encapsulating GRs? (2) Once established, how is this junctional communication maintained? and (3) Precisely when and how is it blocked?

Although junctional conductance remains to be experimentally demonstrated in *B. germanica* and other insects, carboxyfluorescein, when microinjected into the 72-h-old hemocyte capsule of *P. americana,* "moved readily from cell to cell" via the gap junction.[148] Note that hemocytes are generally monodispersed and free in the hemolymph, and form junctions only when activated during capsule formation. Note also that activated GRs show other ultrastructural changes and are engaged in the synthesis of tubulin.[54] Finally, cell-cell communication via gap junctions has been demonstrated in many tissues (see reviews in Bennett and Spray,[8] Hertzberg and Johnson,[9] and Sperelakis and Cole[161]), and must occur among the encapsulating GRs and/or PLs.

Once commenced, the junctional conductance is maintained by a high level of intracellular (probably also extracellular) concentration of $[cAMP]_i$, as has been reported in the cells of the *Drosophila* salivary gland. In this way, $[cAMP]_i$ controls "metabolic cooperation between adjacent cells in multicellular masses and organisms".[48]

3. Cessation of Layer Formation: GR-GR Uncoupling and Formation of the Outer Sheath

This phase would require reversing the respective concentrations of $[Ca^{2+}]_i$ and $[cAMP]_i$ ions in contrast to their concentrations in the previous phase, i.e., lower levels of $[cAMP]_i$ and higher levels of $[Ca^{2+}]_i$. Kinase C perhaps plays no role in $[Ca^{2+}]_i$ mobilization at this stage. The outer sheath is formed and encapsulation ends.

The higher $[Ca^{2+}]_i$ level reduces and finally blocks junctional conductance; consequently, the GRs revert back to a noncommunicative state. Furthermore, it causes some exocytosis, which releases coagulation factors and prophenoloxidase, the former facilitating the formation of the outer sheath and the latter producing the random melanin deposits observed in the *B. germanica* capsule[12] (Figure 8). How is the high $[Ca^{2+}]_i$ achieved?

It is known that $[cAMP]_i$ in fact mobilizes Ca^{2+} ions and thus increases the latter's concentration,[49] which in turn would inhibit AC activation,[162-166] and consequently reduce $[cAMP]_i$ ions (see Section IV.A.1). It is probably in this way that $[cAMP]_i$ and $[Ca^{2+}]_i$ ions

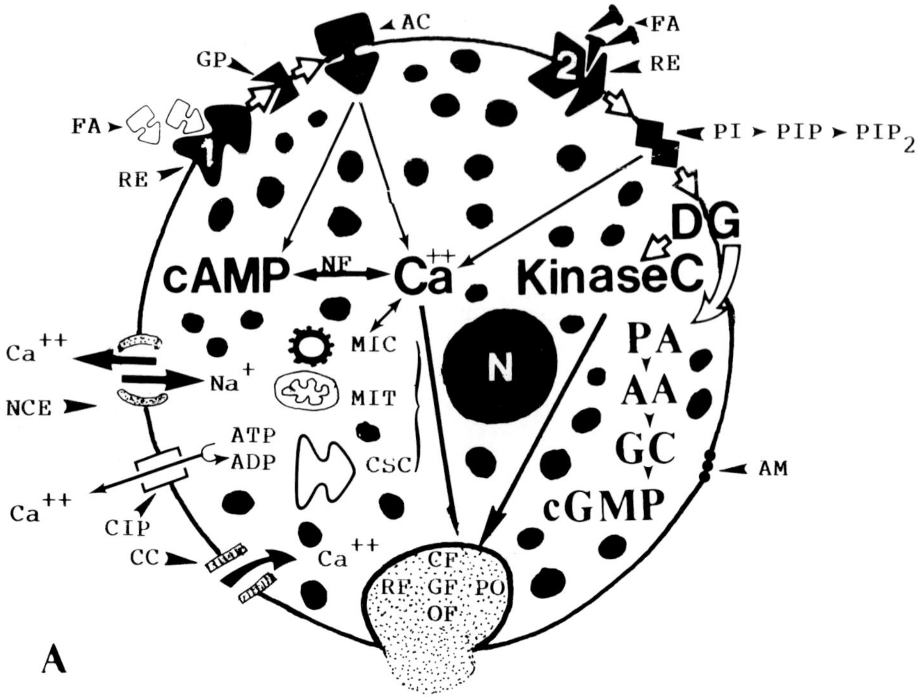
AC
FA
RE
GP
FA
RE
1
2
PI ➤ PIP ➤ PIP2
DG
cAMP
NF
Ca++
KinaseC
Ca++
Na+
NCE
MIC
MIT
N
PA
AA
GC
cGMP
ATP
ADP
CSC
AM
Ca++
CIP
CC
Ca++
CF
RF GF PO
OF
A

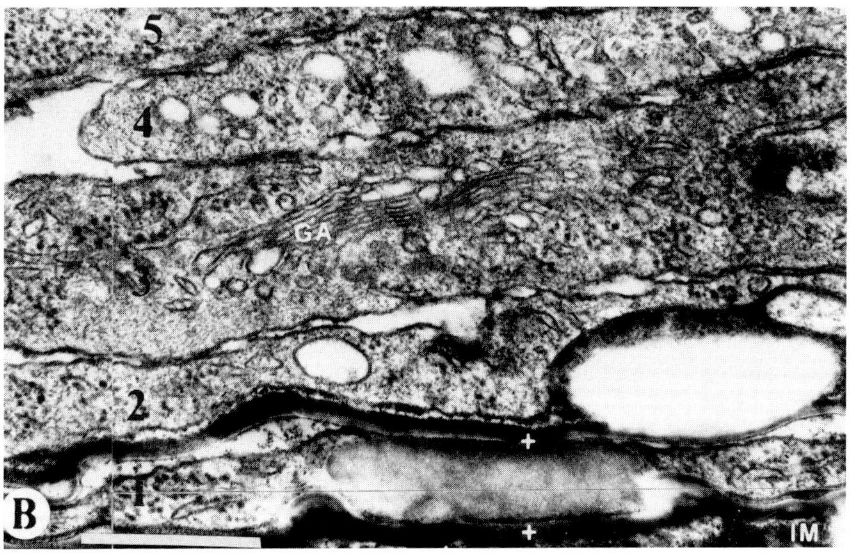
5
4
GA
3
2
1
B
IM

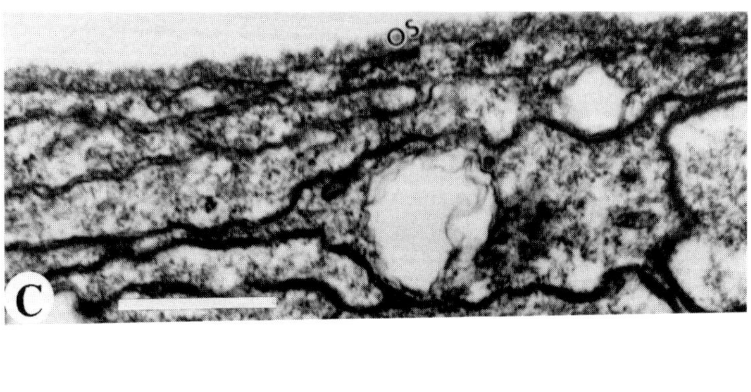
OS
C

FIGURE 6. (A) *Phase 1,* showing the postulated molecular events during the initial stage of encapsulation. The circle represents an activated granulocyte (GR), in which signal transduction by the receptor (RE)-foreign antigen (FA) (ligand) complex may occur by any (or both) of two modes: (1) by activating adenylate cyclase (AC) via the G-protein (GP) and/or (2) by activating the protein kinase C via the hydrolysis of the membrane phosphoinositides (PI). Hydrolysis of the PI (a Ca^{2+}-mediated process) (sequence being PI $\rightarrow$ PIP $\rightarrow$ PIP_2) triggers an integrated cascade of intracellular processes that sequentially releases 1,2-diacylglycerol (DG) (produced transiently in the membrane and degraded within a few seconds) to activate protein kinase C (a Ca^{2+}-dependent process), mobilizes Ca^{2+}, releases arachidonic acid (AA) via the production of phosphatidic acid (PA), activates guanylate cyclase (GC), and ultimately produces cyclic guanosine monophosphate (cGMP). Note that the PA, produced by the phosphorylation of DG, acts as an ionophore that may facilitate entry of extracellular Ca^{2+} through the Ca^{2+} channel (CC). Both modes mobilize Ca^{2+} from the intracellular Ca^{2+}-buffering systems, namely, microsomes (MIC), mitochondria (MIT), calcium-sequestering-compartment (CSC), and/or calciosome (not shown). The first mode ends in the production of adenosine 3′5, cyclic monophosphate (cAMP), which is kept at a low level during this phase by the Ca^{2+} ions by activating cAMP phosphodiesterase activity and inactivating the AC via Ca^{2+}-calmodulin complex (not shown); kinase C also has a negative feedback (NF) effect on the cAMP, and latter on cGMP, whose level is also kept low. cAMP and Ca^{2+} have NF effects on each other. Note that PI hydrolysis simultaneously generates DG and mobilizes Ca^{2+}, which is a major effector of Phase 1. The transient increase in Ca^{2+}, triggers a rapid synthesis of coagulation factors (CF), the so-called recognition factors (RF), growth factors (GF), prophenoloxidase (PO), and perhaps opsonin-like factors (OF). A high $[Ca^{2+}]_i$ level is necessary to facilitate exocytosis/degranulation of the GR, which releases CF, which causes localized coagulation (a Ca^{2+}-dependent process), trapping more GRs and facilitating their aggregation around the FA; RF (which attracts other GRs/PLs to the site of action); GF (which facilitates proliferation of GRs/PLs; note that Ca^{2+} and kinase C are needed for DNA synthesis and cell proliferation); PO which is then activated (a Ca^{2+}-dependent process) and produces melanization that is always present around the FA (see B); and perhaps OF. Note that excess Ca^{2+} may be expelled to the exterior of the GR by the calcium ion pump (CIP) (Ca^{2+}-ATPase) and/or the Na^+-Ca^{2+} exchanger (NCE), which couples efflux of Ca^{2+} to the influx of Na^+; this exchange is reversible and can be experimentally inhibited by tetracaine and verapamil. The effectiveness of the NCE depends on transmembrane Na^+-gradient that is maintained by Na^+,K^+-ATPase, which is present on the plasma membrane of the GR of *L. polyphemus,* and most probably also exists on those of insect GRs that participate in encapsulation. During Phase 1, no cell junctions are present and GR-GR isotypic adhesion (a Ca^{2+}-dependent process) is facilitated by the surface adhesion molecules (AM) (which may be surface agglutinin molecules). (B) *Phase 2.* During this phase gap and other cell junctions are formed and the intracellular molecular milieu is reversed, i.e., the $[cAMP]_i$ level rises to maintain gap junctional communication among the encapsulating GRs; and the $[Ca^{2+}]_i$ level recedes owing to its sequestration into the MIC, MIT, CSC, and/or calciosome, and perhaps also by efflux by the CIP and NCE. The micrograph shows a section through a 3-day-old capsule in *B. germanica.* Note the first five (1 through 5) layers around the implant (IM). Melanization (+ +) around the IM occurs concurrently with the formation of the first layer, and continues throughout encapsulation, as indicated by random deposits of melanin in the completed (10-day-old) capsule. Encapsulation is nearly completed during this phase. (Bar = 0.5 μm.) (C) *Phase 3.* During this phase, an outer sheath (OS) is formed around the completed capsule and the intracellular molecular milieu reverses to low $[cAMP]_i$ and high $[Ca^{2+}]_i$ levels. It is postulated that at a certain critical time, the $[cAMP]_i$ mobilizes $[Ca^{2+}]_i$ from the Ca-buffering systems (an an internal temporal switch mechanism). The high $[Ca^{2+}]_i$ level, by its NF effect, keeps $[cAMP]_i$ level low, reduces and blocks gap junctional communication among the encapsulating GRs, and causes some exocytosis, thus facilitating the formation of the OS. It is suggested that the OS is formed because of the blocking of gap junctional communication during this phase, although one may argue that the latter occurs as a result of the formation of the OS. (Bar = 0.5 μm.) 1 = AC mode; 2 = kinase C mode; AA = arachidonic acid; AC = adenylate cyclase; AM = adhesion molecules (agglutinin?); cAMP = adenosine 3′5′cyclic monophosphate; CC = Ca^{2+} channel; CF = coagulation factors; cGMP = cyclic guanosine monophosphate; CIP = calcium ion pump (Ca^{2+}-ATPase); CSC = calcium-sequestering-compartment (and/or calsiosome, not shown); DG = 1,2,diacylglycerol; FA = foreign antigen or ligand; GC = guanylate cyclase; GF = growth factor; GP = G-protein; GR = granulocyte; IM = implant of surgical suture; MIC = microsome; MIT = mitochondria; N = nucleus; NCE = Na^+-Ca^{2+}- exchanger; NF = negative feedback effect; OF = opsonin-like factors; OS = outer sheath; PA = phosphatidic acid; PI = phosphoinositide; PIP = PI-4-phosphate; PIP_2 = PI-4,5-phosphate; PO = prophenoloxidase; RE = receptor; RF = recognition factors. (Parts B and C from Han, S. S. and Gupta, A. P., *Zool. Sci.,* 6, 303, 1989. With permission.)

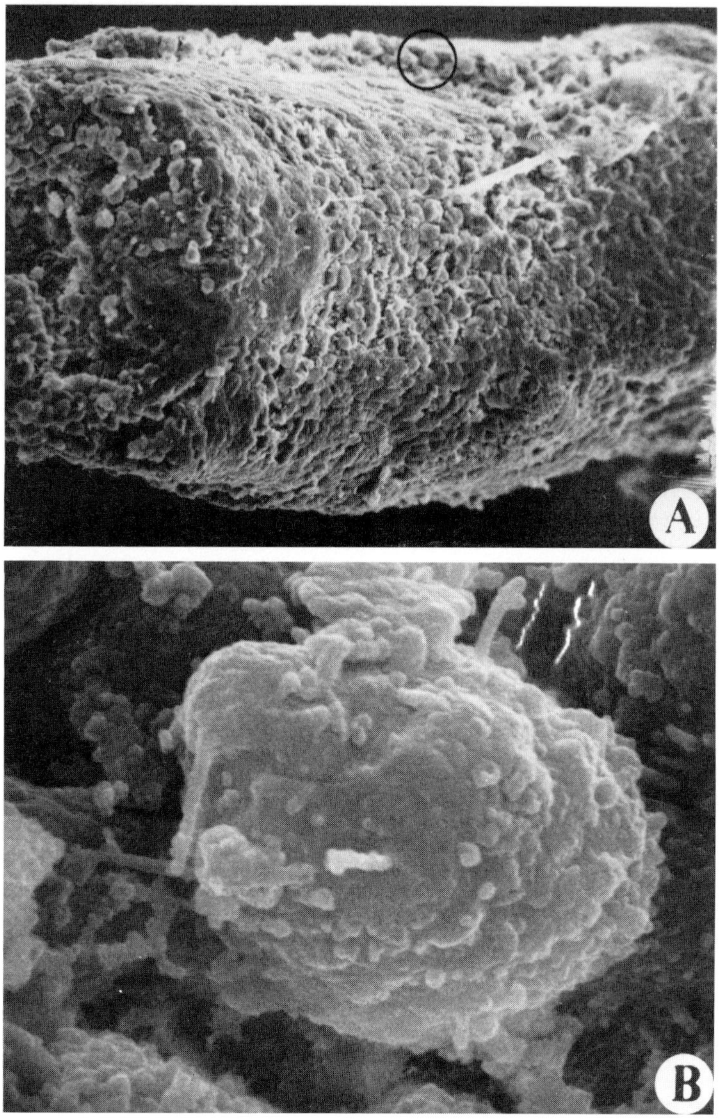

FIGURE 7. (A) A 2-h-old implant from *B. germanica*, showing layering by the GRs (× 1,000). (B) Magnified view of the GR in circle in (A), showing its activated state. Note many pseudopods (× 25,000).

are mutually involved in negative feedbacks, and thereby maintain junctional communication among the GRs until a certain time, when $[cAMP]_i$ concentration decreases and that of $[Ca^{2+}]_i$ increases, thus allowing the GRs to cease cell-cell communication. The timing of the cessation of cellular communication appears to be an inherent temporal feature (an internal switch mechanism) of a growing multicellular tissue.

Note that temporal and spatial regulation of cell-cell communication has been described in mice embryo[167] and in the epidermis during larval development in some insects.[97] In addition, temporal blockage of gap junctions has been reported in cultured multicellular spheroids of mammary tumor cells of the marshall rat; 2 days after the formation of the spheroids, the pores in the gap junctions are blocked (indicated by the fact that injected Lucifer Yellow dye [diam. 1.2 nm] was retained in the injected cell and did not spread to

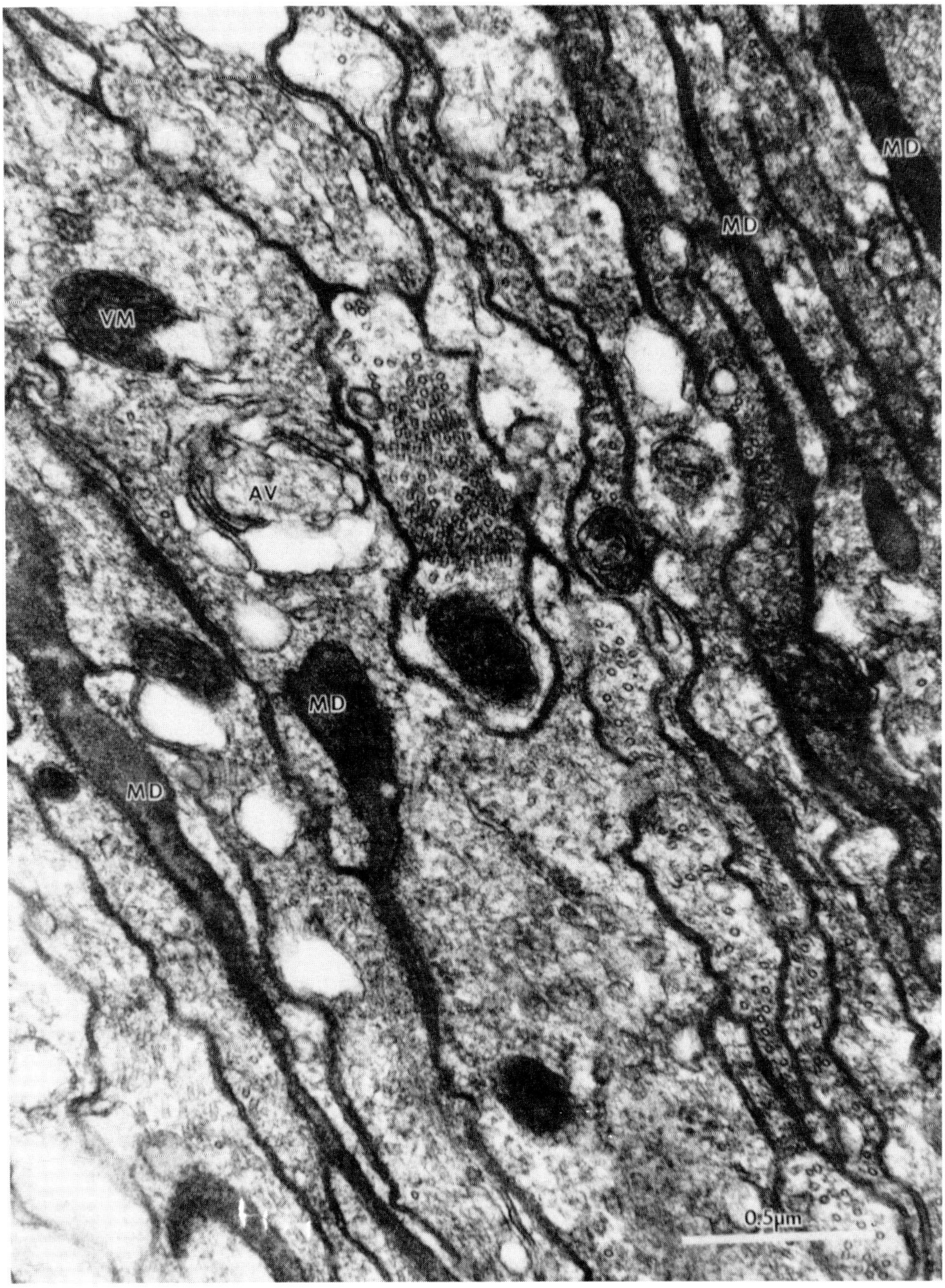

FIGURE 8. Section of a 10-day-old capsule (TEM) of *B. germanica* showing random deposits of melanin (MD), vacuolated mitochondria (VM), and autophagic vacuoles (AV). (From Han, S. S. and Gupta, A. P., *Zool. Sci.*, 6, 303, 1989. With permission.)

other cells of the spheroid).[168] I have suggested elsewhere[34] that if, indeed, similar cessation of cell-cell communication occurred during capsule formation, it should explain the progressive retardation of the attachment of immunocytes to the developing capsule as its thickness increases.

Given the fact that in *B. germanica,* after the capsule reaches a certain thickness, an outer sheath is gradually formed around it (see Figure 28 in Chapter 2), which completely blocks further recruitment and attachment of GRs to the capsule, it would be interesting to determine whether cell-cell communication ceases prior to or after the formation of the outer sheath. Furthermore, because there are two defined regions (compartments?) in the capsule of *B. germanica,* which are completed at different times, it should be interesting to find out if there are any temporal and spatial patterns in junctional communication, as has been shown in segments (compartments) of developing tissues.[97,167,169-171] Also, it would be interesting to find out if cessation or inactivation of cell adhesion molecules occurs, as suggested by Takeichi.[105] Whichever mechanism(s) regulates encapsulation in *B. germanica,* it ceases to operate when the capsule reaches a certain thickness or age, because no further recruitment and attachment of immunocytes occur. In fact, this cessation of attachment is not abrupt, but seems to occur gradually. How this mechanism is turned off in this and other insects is unknown.

Alternatively, one may argue that in *B. germanica,* the outer sheath develops in the completed capsule (10-day-old) because by this time, it is composed of the host material, which renders the capsule as a self tissue, and therefore does not trigger any further immunoreaction among the GRs. If this were the case, the sheath would act as a barrier and interrupt any further cell-cell communication through gap junctions. Because we found this sheath only when we allowed encapsulation to proceed at least until 10 days, it is likely that the occurrence of a sheath was missed in the capsules of other insects, which were not allowed to fully mature.

VII. SUMMARY

Although gap junctions (and other cell junctions) have been reported in a few insect capsules, how they regulate cell-cell communication during encapsulation via the two intracellular messengers (adenosine 3′,5′-cyclic monophosphate [cAMP] and Ca^{2+}) is unknown. Also unknown is the role of some of the immunocyte plasma membrane glycoproteins in cell-cell adhesion and that of phosphoinositides in signal transduction. On the basis of the proven roles of these ions/molecules in other invertebrate and vertebrate tissues, a molecular basis of encapsulation has been postulated.

Gap junctions are plasma membrane specializations in the form of channels or "connexons" (appearing as intramembranous particles in freeze-fracture replicas) that are formed when two adjacent cells come within a 2-nm space of each other (hence the name "gap".) The channels reversibly open and close (hence called "gated") in response to external stimuli and/or cell activities, and allow passage of ions and metabolites between the coupled cells. They are made up of six identical 18- to 27- kDa protein subunits, 9 to 11 nm in diameter. Gap junctions have been studied in thin sections, freeze-fracture replicas, and as extracted or isolated units, and are present in the hemocytic capsules of *B. germanica* and those of a few other insects.

Intracellular $[cAMP]_i$ and $[Ca^{2+}]_i$ ions, which have negative feedback effects on each other, determine the conductance of gap junctions, and hence, along with the surface adhesion molecules, play significant roles in encapsulation. It is no wonder, therefore, that intracellular concentrations of $[cAMP]_i$ and $[Ca^{2+}]_i$ are rigorously controlled concomitantly with their roles during morphogenesis of multicellular tissues. Cells engaged in cell-cell communication seem to have evolved mechanisms to determine how much of each of these molecules/ions will be present within the cell at any given time.

For the convenience of discussion, three phases of encapsulation have been recognized: (1) foreign antigen recognition phase, (2) layer formation phase, and (3) cessation of layer formation phase. During the first phase, the $[Ca^{2+}]_i$ level is kept high and that of $[cAMP]_i$ low within the GRs. High $[Ca^{2+}]_i$ concentration facilitates exocytosis of the GRs, which causes localized coagulation (a Ca^{2+}-dependent process), release of recognition and growth factors, the prophenoloxidase (its activation is a Ca^{2+}-dependent process), and perhaps also opsonin-like molecules. Cell junctions are not formed during this phase and isotypic GR-GR adhesion is due to surface glycoproteins.

During the second phase, gap and other cell junctions are formed and the $[cAMP]_i$ level, of necessity, rises to maintain junctional communication among the encapsulating GRs. Inversely, the $[Ca^{2+}]_i$ level is reduced, which prevents exocytosis. The capsule attains a certain thickness.

During a third phase, the $[Ca^{2+}]_i$ level rises again and reduces and blocks junctional conductance, because cell-cell communication is no longer necessary, the capsule having completed its growth. High $[Ca^{2+}]_i$ also induces some exocytosis and coagulation, which aid in the formation of the outer sheath that terminates encapsulation. It still needs to be investigated as to whether the blocking of the gap junctional communication results in the formation of the outer sheath in *B. germanica,* and is also responsible for eventual retardation of cell recruitment and attachment to the capsule in other insect capsules, in which the presence of the outer sheath has not been reported.

ACKNOWLEDGMENT

This is New Jersey Agricultural Experiment Station Publication No. F-08125-03-90, supported by State funds and U.S. Hatch Act funds. I am grateful to Roy J. Baerwald for providing the prints for Figures 2 and 3.

REFERENCES

1. **Sheridan, J. D. and Atkinson, M. M.,** Physiological roles of permeable junctions: some possibilities, *Annu. Rev. Physiol.,* 47, 337, 1985.
2. **Spray, D. C. and Bennett, M. V. L.,** Physiology and pharmacology of gap junctions, *Annu. Rev. Physiol.,* 47, 281, 1985.
3. **Neyton, J. and Trautmann, A.,** Physiological modulation of gap junction permeability, *J. Exp. Biol.,* 124, 93, 1986.
4. **Traub, O., Druge, P. M., and Willecke, K.,** Degradation and resynthesis of gap junction protein in plasma membranes of regenerating liver after partial hepatectomy or cholestasis, *Proc. Natl. Acad. Sci. U.S.A.,* 80, 755, 1983.
5. **Yancey, S. B., Edens, J. E., Trosko, J. E., Chang, C. C., and Revel, J. P.,** Decreased incidence of gap junctions between Chinese hamster V-79 cells upon exposure to the tumor promoter 12-*O*-tetradecanoylphorbol 13-acetate, *Exp. Cell Res.,* 139, 329, 1982.
6. **Hertzberg, E. L. and Spray, D. C.,** Studies of gap junctions: biochemical analysis and use of antibody probes, in *Gap Junctions,* Bennett, M. V. L. and Spray, D. C., Eds., Cold Spring Harbor Laboratory, New York, 1985, 57.
7. **Hertzberg, E. L., Spray, D. C., and Bennett, M. V. L.,** Reduction of gap junctional conductance by microinjection of antibodies against the 27-kDa liver gap junction polypeptide, *Proc. Natl. Acad. Sci. U.S.A.,* 82, 2412, 1985.
8. **Bennett, M. V. L. and Spray, D. C., Eds.,** *Gap Junctions,* Cold Spring Harbor Laboratory, New York, 1985.
9. **Hertzberg, E. L. and Johnson, R. G., Eds.,** *Gap Junctions,* Vol. 7, in *Modern Cell Biology,* Satir, B. H., Series Ed., Alan R. Liss, New York, 1988.
10. **Finbow, M. E., Buultjens, T. E. J., Lane, N. J., Shuttleworth, J., and Pitts, J. D.,** Isolation and characterization of arthropod gap junctions, *EMBO J.,* 3, 2271, 1984.
11. **Ramon, F., Rivera, A., Simon, A. S., and Zampighi, G. A.,** Is the number of gap junction channels related to the cell's input resistance? in *Cell Interactions and Gap Junction,* Vol. 1, Sperelakis, N. and Cole, W. C., Eds., CRC Press, Boca Raton, FL, 1989, 225.

12. **Han, S. S. and Gupta, A. P.,** Arthropod immune system. II. Encapsulation of implanted nerve cord and "plain gut" surgical suture by granulocytes of *Blattella germanica* (L) (Dictyoptera: Blattellidae), *Zool. Sci.*, 6, 303, 1989.
13. **Johnson, R. G., Meyer, R. A., and Lampe, P. A. D.,** Gap junction formation: a "self-assembly" model involving membrane domains of lipid and protein, in *Cell Interactions and Gap Junctions*, Vol. 1, Sperelakis, N. and Cole, W. C., Eds., CRC Press, Boca Raton, FL, 1989, 159.
14. **Spray, D. C., Saez, J. C., Burt, J. M., Watanbe, T., Reed, L. M., Hertzberg, E. L., and Bennett, M. V. L.,** Gap junctional conductance: multiple sites of regulation, in *Gap Junctions*, Vol. 7, Hertzberg, E. L. and Johnson, R. G., Eds., in *Modern Cell Biology*, Satir, B. H., Series Ed., Alan R. Liss, New York, 1988, 227.
15. **Johnson, R. G., Hammer, M., Sheridan, J. D., and Revel, J. P.,** Gap junction formation between reaggregated Novikoff hepatoma cells, *Proc. Natl. Acad. Sci. U.S.A.*, 71, 4536, 1974.
16. **Alberts, B., Bray, D., Lewis, J., Raff, M., Roberts, K., and Watson, J. D.,** *Molecular Biology of the Cell*, 2nd ed., Garland Publishing, New York, 1989.
17. **Preus, D., Kam, E., Sheridan, J. D., and Johnson, R. G.,** Formation of functional gap junctions at 4 C, *J. Cell Biol.*, 97, 83a, 1983.
18. **Larsen, W. J.,** Mechanism of gap junction modulation, in *Cell Interactions and Gap Junctions*, Sperelakis, N. and Cole, W. C., Eds., CRC Press, Boca Raton, FL, 1989, 3.
19. **Zampighi, G. A. and Simon, S. A.,** The structure of gap junctions as revealed by electron microscopy, in *Gap Junctions*, Bennett, M. V. L. and Spray, D. C., Eds., Cold Spring Harbor Laboratory, New York, 1985, 13.
20. **Flagg-Newton, J. L., Simpson, I., and Loewenstein, W. R.,** Permeability of the cell-to-cell membrane channels in mammalian cell junction, *Science*, 205, 404, 1979.
21. **Schwarzmann, G. O. H., Wiegandt, H., Rose, B., Zimmerman, A., Ben-Haim, D., and Loewenstein, W. R.,** Diameter of the cell-to-cell junctional membrane channels, as probed with neutral molecules, *Science*, 213, 551, 1981.
22. **Tadvalkar, G. and Pinto de Silva, P.,** *In vitro* rapid assembly of gap junctions is induced by cytoskeletal disruptors, *J. Cell Biol.*, 96, 1279, 1983.
23. **Russat, J., Robenek, H., and Themann, H.,** Alterations of tight and gap junctions in mouse hepatocytes following administration of colchicine, *Cell Tissue Res.*, 223, 187, 1982.
24. **Hulser, D. F., Paschke, D., and Greule, J.,** Gap junctions: correlated electrophysiological recordings and ultrastructural analysis by fast freezing and freeze-fracturing, in *Electron Microscopy of Subcellular Dynamics*, Plattner, H., Ed., CRC Press, Boca Raton, FL, 1989, 33.
25. **Moore, H.,** Theory and practice of high pressure freezing, in *Cryotechniques in Biological Electron Microscopy*, Steinbrecht, R. A. and Zierold, K., Eds., Springer-Verlag, Berlin, 1986, 175.
26. **Zampighi, G., Kreman, M., Ramon, F., Moreno, A. L., and Simon, S. A.,** Structural characteristics of gap junctions. I. Channel number in coupled and uncoupled conditions, *J. Cell Biol.*, 106, 1667, 1988.
27. **Flower, N. E.,** A new junctional structure in the epithelia of insects of the order Dictyoptera, *J. Cell Sci.*, 10, 683, 1972.
28. **Staehelin, L. A.,** Structure and function of intercellular junctions, *Int. Rev. Cytol.*, 39, 191, 1974.
29. **Baerwald, R. J.,** Inverted gap and other cell junctions in cockroach hemocyte capsule: a thin section and freeze-fracture study, *Tissue Cell*, 7(3), 575, 1975.
30. **Wood, R. L.,** The cell junctions of *Hydra* as viewed by freeze-fracture replication, *J. Ultrastruct. Res.*, 25, 461, 1977.
31. **Wright, K. A. and Lee, D. L.,** Membrane specializations in a nematode revealed by freeze-fracture technique, *Zeiss Inf.*, 1984 (3), 35, 1984.
32. **Flower, N. E.,** Septate and gap junctions between the epithelial cells of an invertebrate, the mollusk *Cominella maculosa*, *J. Ultrastruct. Res.*, 37, 259, 1971.
33. **Gilula, N. P. and Satir, P.,** Septate and gap junctions in molluscan gill epithelium, *J. Cell Biol.*, 51, 869, 1971.
34. **Gupta, A. P.,** Arthropod immunocytes: identification, structure, functions, and analogies to the functions of vertebrate B- and T-lymphocytes, in *Hemocytic and Humoral Immunity in Arthropods*, Gupta, A. P., Ed., John Wiley & Sons, New York, 1986, 3.
35. **Peracchia, C.,** Excitable membrane ultrastructure. 1. Freeze-fracture of crayfish axons, *J. Cell Biol.*, 61, 107, 1974.
36. **Shaw, S. R. and Stowe, S.,** Freeze-fracture evidence for gap junctions connecting the axon terminals of dipteran photoreceptors, *J. Cell Sci.*, 53, 115, 1982.
37. **St. Marie, R. L. and Carlson, S. D.,** Synaptic vesicle activity in stimulated and unstimulated photoreceptor axons in the housefly: a freeze-fracture study, *J. Neurocytol.*, 11, 747, 1982.
38. **Hertzberg, E. L.,** A detergent-independent procedure for the isolation of gap junctions from rat liver, *Biol. Chem.*, 259(15), 9936, 1984.

39. **Zampighi, G. and Unwin, P. N. T.,** Two forms of isolated gap junctions, *J. Mol. Biol.,* 135, 451, 1979.
40. **Horne, R. W.,** Negative staining methods, in *Techniques for Electron Microscopy,* Kay, D., Ed., Academic Press, New York, 1965, 328, chap. 11.
41. **Unwin, P. N. T. and Zampighi, G.,** Structure of the junctions between communicating cells, *Nature (London),* 283, 545, 1980.
42. **Unwin, P. N. T. and Ennis, P. D.,** Two configurations of a channel forming membrane protein, *Nature (London),* 307, 609, 1984.
43. **Henderson, R. and Unwin, P. N. T.,** Three-dimensional model of purple membrane obtained by electron microscopy, *Nature (London),* 257, 28, 1975.
44. **Carafoli, E.,** Intracellular calcium homeostasis, *Annu. Rev. Biochem.,* 56, 395, 1987.
45. **Volpe, P., Krause, K. H., Hashimoto, S., Zorzato, F., Pozzan, T., Meldolesi, J., and Lew, D. P.,** "Calciosome", a cytoplasmic organelle: the inositol 1,4,5-triphosphate-senstive Ca^{2+} store of non-muscle cells?, *Proc. Natl. Acad. Sci. U.S.A.,* 85, 1091, 1988.
46. **Wirtz, H. R. W. and Dobbs, L. G.,** Calcium mobilization and exocytosis after one mechanical stretch of lung epithelial cells, *Science,* 250, 1266, 1990.
47. **Cohen, P.,** Protein phosphorylation and hormone action, *Proc. R. Soc. London,* (Biol.), 234, 115, 1988.
48. **Hax, W. M. A., van Venrooij, G. E. P. M., and Vossenberg, J. B.,** Cell communication: a cyclic-AMP mediated phenomenon, *J. Membr. Biol.,* 19, 253, 1974.
49. **Friedmann, N. and Park, C. R.,** Early effects of 3′,5′ adenosine monophosphate on fluxes of calcium and potassium in the perfused liver of normal and adrenalectomized rats, *Proc. Natl. Acad. Sci. U.S.A.,* 61, 504, 1968.
50. **Borle, A. B.,** Cyclic AMP stimulation of calcium efflux from kidney, liver, and heart mitochondria, *J. Membr. Biol.,* 16, 221, 1974.
51. **Hesketh, R., Smith, G. A., and Metcalfe, J. C.,** Intracellular calcium and normal eukaryotic cell growth, in *Control of Animal Cell Proliferation,* Boyanton, A. L. and Leffert, H. J., Eds., Academic Press, New York, 1987, 395.
52. **Ueda, T.,** Na^+-Ca^{2+} exchange activity in rabbit lymphocyte plasma membrane, *Biochim. Biophys. Acta,* 734, 342, 1983.
53. **Gupta, A. P., Orenberg, S. D., Das, Y. T., and Chattopadhyaya, S. K.,** Lectin-binding receptors, Na^+, K^+-ATPase, and acetylcholinesterase on immunocyte plasma membrane of *Limulus polyphemus, Exp. Cell Res.,* 194, 83, 1991.
54. **Han, S. S. and Gupta, A. P.,** Arthropod immune system. V. Activated immunocytes (granulocytes) of the German cockroach, *Blattella germanica* (L.) (Dictyoptera: Blattellidae) show increased number of microtubules and nuclear pores during immune reaction to foreign tissue, *Cell Struct. Funct.,* 13, 333, 1988.
55. **Darnell, J., Lodish, H., and Baltimore, D.,** *Molecular Cell Biology,* Scientific American Books, New York, 1986.
56. **Klee, C. B., Crouch, T. H., and Richman, P. G.,** Calmodulin, *Annu. Rev. Biochem.,* 49, 489, 1980.
57. **Babu, Y. S., Sack, J. S., Greenhough, T. J., Bugg, C. F., Means, A. R., and Cook, W. J.,** Three-dimensional structure of calmodulin, *Nature (London),* 315, 37, 1985.
58. **Gerday, C., Gilles, R., and Bolis, L., Eds.,** *Calcium and Calcium-Binding Protein,* Springer-Verlag, Berlin, 1988.
59. **Whitfield, J. F., Durkin, J. P., Franks, D. J., Kleine, L. P., Raptes, L., Rixon, R. H., Sikorska, A., and Walker, P. R.,** Calcium, cyclic AMP and protein kinase C — partners in mitogenesis, *Cancer Meta. Rev.,* 5, 205, 1987.
60. **Battelle, B. A. and Kravitz, E. A.,** Targets of octopamine action in the lobster: cyclic nucleotide changes and physiological effects in hemolymph, heart and exoskeletal muscle, *J. Pharmacol. Exp. Ther.,* 205, 438, 1978.
61. **Gole, J. W. D., Downer, R. G. H., and Sohi, S. S.,** Octopamine-sensitive adenylate cyclase in haemocytes of the forest tent caterpillar, *Malacosoma disstria* Hubner (Lepidoptera: Lasiocampidae), *Can. J. Zool.,* 60, 825, 1982.
62. **Flagg-Newton, J. L., Dahl, G., and Loewenstein, W. R.,** Cell junction and cyclic AMP. I. Upregulation of junctional membrane permeability and junctional membrane particles by administration of cyclic nucleotide or phosphodiesterase inhibitor, *J. Membr. Biol.,* 63, 105, 1981.
63. **De Mello, W. C.,** The role of cAMP on the modulation of junctional conductance. An integrated hypothesis, *Cell Biol. Int. Rep.,* 7, 1033, 1983.
64. **Wiener, E. C. and Loewenstein, W. R.,** Correction of cell-cell communication defect by introduction of a protein kinase into mutant cells, *Nature (London),* 305, 433, 1983.
65. **Saez, J. C., Spray, D. C., Nairn, A. C., Hertzberg, E. L., and Greengard, P.,** cAMP increases junctional conductance and stimulates phosphorylation of the 27-kDa principal gap junction polypeptide, *Proc. Natl. Acad. Sci. U.S.A.,* 83, 2473, 1986.

66. **Azarnia, R., Dahl, G., and Loewenstein, W. R.,** Cell junction and cyclic AMP. III. Promotion of junctional membrane permeability and junctional membrane particles in a junction-deficient cell type, *J. Membr. Biol.*, 63, 133, 1981.
67. **Flagg-Newton, J. L. and Loewenstein, W. R.,** Cell junctions and cyclic AMP. II. Modulation of junctional membrane permeability, dependent on serum and cell density, *J. Membr. Biol.*, 63, 123, 1981.
68. **Murray, A. W. and Gainer, H. St. C.,** Regulation of gap junctional communication by protein kinases, in *Cell Interactions and Gap Junctions,* Sperelakis, N. and Cole, W. C., Eds., CRC Press, Boca Raton, FL, 1989, 97.
69. **De Mello, W. C.,** Effect of intracellular injection of cAMP on the electrical coupling of mammalian cardiac cells, *Biochem. Biophys. Res. Commun.*, 119, 1001, 1984.
70. **Azarnia, R., Reddy, S., Kmiecik, T. E., Shalloway, D., and Loewenstein, W. R.,** The cellular *src* gene: regulation of communication and growth, in *Gap Junctions,* Vol. 7, Hetzber, E. L. and Johnson, R. G., Eds., in *Modern Cell Biology,* Satir, B. H., Series Ed., Alan R. Liss, New York, 1988, 423.
71. **Azarnia, R., Larsen, W. J., and Loewenstein, W. R.,** The membrane junctions in communicating and non-communicating cells, their hybrids and segregants, *Proc. Natl. Acad. Sci. U.S.A.*, 71, 880, 1974.
72. **Larsen, W. J., Azarnia, R., and Loewenstein, W. R.,** Intercellular communication and tissue growth. IX. Junctional membrane structure of hybrids between communication-competent and communication-incompetent cell, *J. Membr. Biol.*, 34, 39, 1977.
73. **Caveney, S.,** Intercellular communication in insect development is hormonally controlled, *Science,* 199, 192, 1978.
74. **Caveney, S. and Blennerhassett, M. G.,** Elevation of ionic conductance between insect epidermal cells by β-ecdysone *in vitro, J. Insect Physiol.*, 26, 13, 1979.
75. **Caveney, S., Berdan, R. C., and McLean, S.,** Cell-to-cell ionic communication stimulated by 20-hydroxyecdysone occurs in the absence of protein synthesis and gap junction growth, *J. Insect Physiol.*, 26, 557, 1980.
76. **Bennett, M. V. L., Verselis, V., White, R. L., and Spray, D. C.,** Gap junctional conductance: gating, in *Gap Junctions,* Vol. 7, Hertzberg, E. L. and Johnson, R. G., Eds., in *Modern Cell Biology,* Satir, B. H., Series Ed., Alan R. Liss, New York, 1988, 287.
77. **Hanna, R. B., Ornberg, R. L., and Reese, T. S.,** Structural details of rapidly frozen gap junctions, in *Gap Junction,* Bennett, M. V. L. and Spray, D. C., Eds., Cold Spring Harbor Laboratory, New York, 1985, 23.
78. **Loewenstein, W. R.,** Permeability of membrane junctions, *Ann. N.Y. Acad. Sci.*, 137, 441, 1966.
79. **De Weer, P.,** Aspects of the recovery processes in nerve, in *International Review of Science,* Vol. 3, Hunt, C. C., Ed., Butterworths, London, 1975, 231.
80. **Meech, R. W. and Thomas, R. C.,** The effect of calcium injection on the intracellular sodium and pH of snail neurons, *J. Physiol.*, 265, 867, 1977.
81. **Turin, L. and Warner, A. E.,** Intracellular pH in early *Xenopus* embryos: its effects on current flow between blastomeres, *J. Physiol.*, 300, 489, 1980.
82. **Rose, B. and Loewenstein, W. R.,** Permeability of cell junctions depends on local cytoplasmic calcium activity, *Nature (London)*, 254, 250, 1975.
83. **Peracchia, C.,** Permeability and regulation of gap junction channels in cells and artificial lipid bilayers, in *Intercellular Communication — New Perspectives,* De Mellow, W. C., Ed., Plenum Publishing Co., New York, 1987, 65.
84. **Loewenstein, W. R.,** The cell-to-cell channel of gap junctions, *Cell,* 48, 725, 1987.
85. **Turin, L. and Warner, A. E.,** Carbon dioxide reversibly abolishes ionic communication between cells of early amphibian embryo, *Nature (London)*, 270, 56, 1978.
86. **Spray, D. C., Harris, A. L., and Bennett, M. V. L.,** Physiology and pharmacology of gap junctions, *Annu. Rev. Physiol.*, 47, 281, 1985.
87. **Rose, B. and Rick, R.,** Intracellular pH, intracellular free Ca and junctional cell-cell coupling, *J. Membr. Biol.*, 44, 377, 1978.
88. **Loewenstein, W. R.,** Introduction to the regulation of junctional channel, in *Gap Junctions,* Vol. 7, Hertzberg, E. L. and Johnson, R. G., Eds., in *Modern Cell Biology,* Satir, B. H., Series Ed., Alan R. Liss, New York, 1988, 221.
89. **Dahl, G., Werner, R., and Levine, E.,** Paired oocytes: an expression system for cell-cell channel, in *Gap Junctions,* Vol. 7, Hertzberg, E. L. and Spray, D. C., Eds., in *Modern Cell Biology,* Satir, B. H., Series Ed., Alan R. Liss, New York, 1988, 183.
90. **Noma, A. and Tsuboi, N.,** Dependence of junctional conductance on proton, calcium and magnesium ions in cardiac paired cells of guinea-pig, *J. Physiol.*, 382, 193, 1987.
91. **Meda, P., Perrelet, A., and Orci, L.,** Gap junctions and cell to cell coupling, in *Modern Cell Biology,* Vol. 3, Satir, B. H., Ed., Alan R. Liss, New York, 1984, 131.
92. **Meda, P., Perrelet, A., and Orci, L.,** Increase of gap junctions between pancreatic beta-cells during stimulation of insulin secretion, *J. Cell Biol.*, 82, 441, 1979.

93. **Larsen, W. J., Skowron Lomneth, C., and Carron, C. P.,** Gap junction modulation: possible role in tumor cell behavior, in *Biochemical Mechanisms and Regulation of Intercellular Communication,* Milman, H. A. and Elmore, E., Eds., Princeton Scientific Publ. Co., Princeton, 1988, 151.
94. **Cole, W. C. and Garfield, R. C.,** Evidence for physiological regulation of myometrial gap junction permeability, *Am. J. Physiol.,* 251 C411, 1988.
95. **Ashhurst, D. E.,** Histochemical properties of the spherulocytes of *Galleria mellonella* L. (Lepidoptera: Pyralidae), *Int. J. Insect Morphol. Embryol.,* 11, 285, 1982.
96. **Hazarika, L. K. and Gupta, A. P.,** Variations in hemocyte populations during various developmental stages of *Blattella germanica* (L.) Dictyoptera: Blattellidae), *Zool. Sci.,* 4, 307, 1987.
97. **Caveney, S. and Sanfranyos, R.,** Developmental physiology of insect epidermal gap junctions, in *Cell Interactions and Gap Junctions,* Vol. 1, Sperelakis, N. and Cole, W. C., Eds., CRC Press, Boca Raton, FL, 1989, 107, chap. 7.
98. **Lane, N. J. and Swales, L. S.,** Dispersal of junctional particles, not internalization, during *in vitro* disappearance of gap junctions, *Cell,* 19, 579, 1980.
99. **Gupta, A. P., Ed.,** *Morphogenetic Hormones of Arthropods,* Part 1. *Discoveries, Syntheses, Metabolism, Evolution, Modes of Action, and Techniques,* Rutgers University Press, New Brunswick, NJ, 1990a.
100. **Gupta, A. P.,** *Morphogenetic Hormones of Arthropods,* Part 2. *Embryonic and Postembryonic Sources,* Rutgers University Press, New Brunswick, NJ, 1990b.
101. **Gupta, A. P.,** *Morphogenetic Hormones of Arthropod,* Part 3. *Roles in Histogenesis, Organogenesis, and Morphogenesis,* Rutgers University Press, New Brunswick, NJ, 1991.
102. **Larsen, W. J.,** Relating the population dynamics of gap junctions to cellular function, in *Gap Junctions,* Bennett, M. V. L. and Spray, D. C., Eds., Cold Spring Harbor Laboratory, New York, 1985, 289.
103. **Larsen, W. J. and Risinger, M. A.,** The dynamic life histories of intercellular membrane junction, in *Modern Cell Biology,* Vol. 4, Satir, B. H., Ed., Alan R. Liss, New York, 1985, 151.
104. **Larsen, W. J., Tung, H. N., Murray, S. A., and Swenson, C. A.,** Evidence for the participation of actin microfilaments and bristle coats in the internalization of gap junction membrane, *J. Cell Biol.,* 83, 567, 1979.
105. **Takeichi, M.,** The cadherins: cell-cell adhesion molecules controlling animal morphogenesis, *Development,* 102, 639, 1988.
106. **Hennings, H. and Holbrook, K. A.,** Calcium regulation of cell-cell contact and differentiation of epidermal cells in culture, *Exp. Cell Res.,* 143, 127, 1983.
107. **Cunningham, B. A., Hemperly, J. J., Murray, B. A., Prediger, E. A., Brackenbury, R., and Edelman, G. M.,** Neural cell adhesion molecule: structure, immunoglobulin-like domains, cell surface modulation and alternative RNA splicing, *Science,* 236, 799, 1987.
108. **Williams, A. F.,** A year in the life of the immunoglobulin superfamily, *Immunol. Today,* 8, 298, 1987.
109. **Roth, S. A. and Weston, J. A.,** The measurement of intracellular adhesion, *Proc. Natl. Acad. Sci. U.S.A.,* 58, 974, 1967.
110. **Kikkawa, U., Kitano, T., Saito, N., Kishimoto, A. M., Taniyama, K., Tanaka, C., and Nishizuka, Y.,** Role of protein kinase C in calcium-mediated signal transduction, in *Calcium and the Cell,* Evered, D. and Whelan, J., Eds., Ciba Foundation Symposium 122, John Wiley & Sons, New York, 1986, 197.
111. **Kaibuchi, K., Sano, K., Hoshijima, M., Takai, Y., and Nishizuka, Y.,** Phosphatidylinositol turnover in platelet activation, calcium mobilization and protein phosphorylation, *Cell Calcium,* 3, 323, 1982.
112. **Nishizuka, Y.,** Phospholipid degradation and signal translation for protein phosphorylation, *Trends Biochem. Sci.,* 8, 13, 1983.
113. **Michell, R. H.,** Inositol phospholipids and cell surface receptor function, *Biochim. Biophys. Acta,* 415, 81, 1975.
114. **Fain, J. N.,** Involvement of phosphatidylinositol breakdown in elevation of cytosol Ca^{2+} by hormones and relationship to prostaglandin formation, in *Hormone Receptors,* Kohn, L. D., Ed., John Wiley & Sons, New York, 1982, 237.
115. **Gerzer, R., Hamet, P., Ross, A. H., Lawson, J. A., and Hardman, J. G.,** Calcium-induced release from platelet membranes of fatty acids that modulate soluble guanylate cyclase, *J. Pharmacol. Exp. Ther.,* 226, 180, 1983.
116. **Nishizuka, Y.,** Turnover of inositol phospholipids and signal transduction, *Science,* 225, 1365, 1984b.
117. **Berridge, M.,** Calcium-mobilizing receptors: membrane phosphoinositides and signal transduction, in *Calcium in Biological Systems,* Rubin, R. P., Weiss, G. B., and Putney, Jr., Eds., Plenum Press, New York, 1985, 37.
118. **Kirk, C. J., Creba, J. A., Downes, C. P., and Michell, R. H.,** Hormone-stimulated metabolism of inositol lipids and its relationship to hepatic receptor function, *Biochem. Soc. Trans.,* 9, 377, 1981.
119. **Rhodes, D., Prpic, V., Exton, J. H., and Blackmore, P. F.,** Stimulation of phosphatidylinositol 4,5-biphosphate hydrolysis in hepatocytes by vasopressin, *J. Biol. Chem.,* 258, 2770, 1983.

120. **Thomas, A. P., Marks, J. S., Coll, K. E., and Williamson, J. R.,** Quantitative and early kinetics of inositol lipid changes induced by vasopression in isolated and cultured hepatocytes, *J. Biol. Chem.*, 258, 5716, 1983.
121. **Billah, M. M., Leptina, E. G., and Cuatrecasas, P.,** Phosphatidylinositol-specific phospholipase-C of platelets. Association with 1,2 diacylglyceride-kinase and inhibition of cyclic AMP, *Biochem. Biophys. Res. Commun.*, 90, 92, 1979.
122. **Holmsen, H., Kaplan, K. L., and Dangekmaier, C. A.,** Differential energy requirements for platelet responses, *Biochem. J.*, 208, 9, 1982.
123. **Berridge, M. J.,** Rapid accumulation of inositol triphosphate reveals that agonists hydrolyse polyphosphoinositides instead of phosphatidylinositol, *Biochem. J.*, 212, 848, 1983.
124. **Agranoff, B. W., Murthy, P., and Seguin, E. B.,** Thrombin-induced phosphodiesterase cleavage of phosphatidyl biophosphate in human platelets, *J. Biol. Chem.*, 258, 2076, 1983.
125. **Nishizuka, Y.,** The role of protein kinase C in cell surface signal transduction and tumour promotion, *Nature (London)*, 308, 693, 1984a.
126. **Tsien, R. Y., Pozzan, T., and Rink, T. J.,** T-cell mitogenes cause early changes in cytoplasmic free Ca^{2+} and membrane potential in lymphocytes, *Nature (London)*, 295, 68, 1982.
127. **Moolenaar, W. H., Tertoolen, L. G. J., and de Laat, S. W.,** Phorbol ester and diacylglycerol mimic growth factors in raising cytoplasmic pH, *Nature (London)*, 312, 371, 1984.
128. **Lagast, H., Pozzan, T., Waldvogel, F. A., and Pew, P. D.,** Phorbol myristate acetate stimulates ATP-dependent calcium transport by the plasma membrane of neutrophils, *L. Clin. Invest.*, 73, 878, 1984.
129. **Albert, P. R. and Tashjian, A. H.,** Dual actions of phorbol esters on cytosolic free Ca^{2+} concentrations and reconstitution with isomucin of acute thyrotropin-releasing hormone responses, *J. Biol. Chem.*, 260, 8746, 1985.
130. **Drummond, A. H.,** Bidirectional control of cytosolic free calcium by thyrotropin-releasing hormone in pituitary cells, *Nature (London)*, 315, 752, 185.
131. **Sagi-Eisenberg, R., Lieman, H., and Pecht, I.,** Protein kinase C regulation of the receptor-coupled calcium signal in histamine-secreting rat basophilic leukemia cells, *Nature (London)*, 313, 59, 1985.
132. **Kishimoto, A., Nishiyama, K., Nakanishi, H., Uratsuji, Y., Nomura, H., Takeyama, Y., and Nishizuka, Y.,** Studies on the phosphorylation of myelin basic protein by protein kinase C and adenosine-3′:5′-monophosphate-dependent protein kinase, *J. Biol. Chem.*, 260, 12492, 1985.
133. **Cheung, W. Y.,** Calmodulin plays a pivotal role in cellular regulation, *Science*, 207, 19, 1979.
134. **Rubin, R. P.,** *Calcium and Cellular Secretion*, Plenum Press, New York, 1982.
135. **Douglas, W. W.,** Stimulus-secretion coupling: variations on the theme of calcium-activated exocytosis involving cellular and extracellular sources of calcium, in *Respiratory Tract Mucus*, Ciba Foundation Symposium 54, Elsevier/North-Holland, Amsterdam, 1978, 61.
136. **Baker, P. F., Knight, D. E., and Whittaker, M. J.,** Calcium and the control of exocytosis, in *Calcium Binding Proteins*, Siegel, F., Ed., Elsevier/North-Holland, Amsterdam, 1980, 47.
137. **Ornberg, R. L.,** Exocytosis in *Limulus* amebocytes, in *Blood Cells of Marine Invertebrates*, Cohen, W. D., Ed., Alan R. Liss, New York, 1985, 127.
138. **Ornberg, R. L. and Reese, T. S.,** Quick freezing and freeze-free-substitution for X-ray microanalysis of calcium, in *Microprobe Analysis of Biological Systems*, Hutchinson, T. E. and Somlyo, A. P., Eds., Academic Press, New York, 1981, 213.
139. **Cannon, G. W., Tsuchiya, M., Rittschof, D., and Bonaventura, J.,** Magnesium dependence of endotoxin-induced degranulation of *Limulus* amebocytes, *Biol. Bull., (Woods Hole, Mass.)*, 171, 330, 1986.
140. **Takai, Y., Kaibuchi, K., Tsuda, T., and Hoshijima, M.,** Role of protein kinase C in transmembrane signaling, in *Membrane Skeleton and Cytoskeletal Membrane Associations*, Bennett, V., Cohen, C. M., Lux, S. E., and Palek, J., Eds., Alan R. Liss, New York, 1985, 47.
141. **Carney, D. H.,** Characterization of the thrombin receptor and its involvement in initiation of cell proliferation, in *Control of Animal Cell Proliferation*, Vol. 2, Boynton, A. L. and Leffert, H. L., Eds., Academic Press, New York, 1987, 265.
142. **Peracchia, C.,** The calmodulin hypothesis for gap junction regulation six years later, in *Gap Junctions*, Vol. 7, Hertzberg, E. and Johnson, R. G., Eds., in *Modern Cell Biology*, Alan R. Liss, New York, 1988, 267.
143. **Perrachia, C.,** Control of gap junction permeability and calmodulin-like proteins, in *Cell Interactions and Gap Junctions*, Sperelakis, N. and Cole, W. C., Eds., CRC Press, Boca Raton, FL, 1989, 125.
144. **Meyer, D. J., Yancey, S. B., and Revel, J. P.,** Intracellular communication in normal and regenerating rat liver: a quantitative analysis, *J. Cell Biol.*, 91, 505, 1981.
145. **Kaibuchi, K., Takai, Y., and Nishizuka, Y.,** Protein kinase C and calcium ion in mitogenic response of macrophage-depleted human peripheral lymphocytes, *J. Biol. Chem.*, 260, 1366, 1985.
146. **Guy, G. R., Gordon, J., Michell, R. H., and Brown, G.,** Synergism between diacylglycerols and calcium ionophore in the induction of human B cell proliferation mimics the inositiol lipid polyphosphate breakdown signals induced by crosslinking surface immunoglobulin, *Biochem. Biophys. Res. Commun.*, 131, 484, 1985.

147. **Grimstone, A. V., Rotheram, S., and Salt, G.,** An electronmicroscope study of capsule formation by insect blood cells, *J. Cell Sci.*, 2, 281, 1967.
148. **Caveney, S. and Berdan, R.,** Selectivity in junctional coupling between cells of insect tissue, in *Insect Ultrastructure*, Vol. 1, King, R. C. and Akai, H., Eds., Plenum Press, New York, 1982, 434.
149. **Smith, D. S.,** *Insect Cells: Their Structure and Function*, Oliver & Boyd, Edinburgh, 1968.
150. **Brehélin, M., Hoffmann, J. A., Matz, G., and Porte, A.,** Encapsulation of implanted foreign bodies by hemocytes in *Locusta migratoria* and *Melolontha melolontha*, *Cell Tissue Res.*, 160, 283, 1975.
151. **Ratcliffe, N. A. and Gagen, S. J.,** Studies on the *in vitro* cellular reactions of insects: an ultrastructural analysis of nodule formation in *Galleria mellonella*, *Tissue Cell*, 9(1), 73, 1977.
152. **Gupta, A. P. and Han, S. S.,** Arthropod immune system. III. Septate junctions in the hemocytic capsule of the German cockroach, *Blattella germanica*, *Tissue Cell*, 20(4), 629, 1988.
153. **Oschman, J. L. and Berridge, M. J.,** Structural and functional aspect of salivary fluid secretion in *Calliphora*, *Tissue Cell*, 2, 281, 1970.
154. **Feinstein, M. B.,** Release of intracellular membrane-bound calcium precedes the onset of stimulation-induced exocytosis in platelets, *Biochem. Biophys. Res. Commun.*, 93, 593, 1980.
155. **Lehninger, A. L., Reynafarje, B., Versesi, A., and Tew, W. P.,** Transport and accumulation of calcium in mitochondria, *Ann. N.Y. Acad. Sci.*, 307, 160, 1978.
156. **Kagayama, M. and Douglas, W. W.,** Electron microscopic evidence of calcium-induced exocytosis in mast cells treated with 48/80 or ionophores A-23187 and X-537A, *J. Cell Biol.*, 62, 519, 1974.
157. **Kauffman, R. F., Taylor, R. W., and Pfeiffer, R. D.,** Cation transport and specificity of ionomycin, *J. Biol. Chem.*, 255, 2735, 1980.
158. **Torp-Pedersen, C., Saermark, T., Bundgaard, M., and Thorn, N. A.,** ATP-dependent Ca^{2+} accumulation by microvesicles isolated from bovine neurohypophyses, *J. Neurochem.*, 35, 552, 1980.
159. **Häusler, R., Burger, A., and Niedermaier, W.,** Evidence for an inherent ATP-stimulated uptake of calcium into chromaffin granules, *Naunyn-Schmiedelberg's Arch. Pharmacol.*, 315, 255, 1981.
160. **Devreotes, P. and Zigmond, S. H.,** Chemotaxis in eukaryotic cells: a focus on leucocytes and *Dictyostelium*, *Annu. Rev. Cell Biol.*, 4, 649, 1988.
161. **Sperelakis, N. and Cole, W. C., Eds.,** *Cell Interactions and Gap Junctions*, Vol. 1, CRC Press, Boca Raton, FL, 1989.
162. **Bar, H. P. and Hechter, O.,** Adenyl cyclase and hormone action. III. Calcium requirement for ACTH stimulation of adenyl cyclase, *Biochem. Biophys. Res. Commun.*, 35, 681, 1969.
163. **Rasmussen H.,** Cell communication, calcium ion and cyclic adenosine monophosphate, *Science*, 170, 404, 1970.
164. **Bradham, L. S., Holt, D. A., and Sims, M.,** The effect of Ca^{2+} on the adenyl cyclase of calf brain, *Biochim. Biophys. Acta*, 201, 250, 1970.
165. **Kelly, L. A. and Koritz, S. B.,** Bovine adrenal cortical adenyl cyclase and its stimulation by adrenocorticotropic hormone and NaF, *Biochim. Biophys. Acta*, 237, 141, 1971.
166. **Rubin, R. P., Carchman, R. A., and Jaanus, S. B.,** Role of calcium and adenosine cyclic 3′,5′ phosphate in action of adrenocorticotropin, *Nature (London)*, 240, 150, 1972.
167. **Lo, C. W.,** Communication compartments: a conserved role in pattern formation, in *Cell Interactions and Gap Junctions*, Sperelakis, N. and Cole, W. C., Eds., CRC Press, Boca Raton, FL, 1989, 85.
168. **Brümmer, F. and Hülser, D. F.,** Gap junctions in multicell spheroids, *Zeiss Inf.*, 1982, 34, 1982.
169. **Caveney, S. and Sanfranyos, R.,** Control of molecular movement within a developmental compartment, in *Gap Junctions*, Bennett, M. V. L. and Spray, D. C., Eds., Cold Spring Harbor Laboratory, New York, 1985, 265.
170. **Lo, C. W.,** Communication compartmentation and pattern formation in development, in *Gap Junctions*, Bennett, M. V. L. and Spray, D. C., Eds., Cold Spring Harbor Laboratory, New York, 1986, 251.
171. **Warner, A. E.,** Antibodies to gap junction protein: probes for studying cell interactions during development, in *Gap Junctions*, Cold Spring Harbor Laboratory, New York, 1985, 275.
172. **Baerwald, R. J.,** Fine structure of hemocyte membranes and intercellular junctions formed during hemocyte encapsulation, in *Insect Hemocytes*, Gupta, A. P., Ed., Cambridge University Press, Cambridge, 1979, 155.

Chapter 5

AN INTRODUCTION TO IMAGE ANALYSIS OF HEMOCYTE SERIAL SECTIONS

Stanley Dunn

TABLE OF CONTENTS

I. INTRODUCTION

The difficulty with human image interpretation of microscope section images is the wide variability in the criteria and rules used to identify the structures present. Nowhere is this more evident than in arthropod hematology where inconsistent terminology has led to disagreement among researchers in the field over nomenclature and classification of the forms and functions of the hemocytes and their organelles.

There is a need for a computer image interpretation system that will provide a definitive analysis and classification. Such a system will have the advantage of uniformly applying decision criteria, leading to a reproducible classification with quantifiable error. This will allow the users to concentrate on developing accurate models for cells and their organelles, leading to a classification that is uniformly agreed upon.

This chapter outlines the framework of an automatic image interpretation system and its application in the interpretation of transmission electron micrographs (TEMs) of arthropod hemocytes. Image interpretation constitutes the extraction of useful, accurate, and reliable information about the objects that are present in a scene captured in a given image, and the use of this information in *understanding* the scene. The first major task in image understanding systems is to *segment* the image into meaningful entities, which are given a characterization that is suitable for machine recognition. For this task to be accomplished, the intermediate level tasks that must be performed should be well planned. A good segmentation plan requires knowledge of the image formation process, knowledge of the domain of analysis, and some qualitative description of the objects that are present in the image. The translation of domain and image formation knowledge and the qualitative object descriptions into a quantitative object model to guide the segmentation process, constitutes what we call *segmentation with models of expected structure*. The interpretation paradigm is based on using object models that describe the *expected presentation* of those objects likely to be present in the images. The user inputs the image, information about the image formation process, and a description of the domain, and the image is segmented to yield regions that most closely resemble the models of the expected structure.

These experiments with TEM sections of arthropod immunocytes (granulocytes) yielded promising results, supporting our belief that this image interpretation paradigm and knowledge representation scheme can be applied to other image processing tasks in other natural and/or biological domains. We present results in cell discrimination by a human expert, based on the computer results. Our long term goal is to automate this process and produce a computer-based cell identification and classification system.

A. IMAGE ANALYSIS

Image processing and interpretation[1-7] is an engineering discipline of gathering pictures from one or more sources, performing a series of operations, and interpreting the results to answer a question. The sources of these images can be from television cameras in normal room lighting, infrared cameras, magnetic resonance imaging or ultrasound, as well as electron microscopy. The applications are just as numerous, ranging from medicine to industrial inspection to autonomous navigation.

What all these environments and imaging systems have in common is the processing between image acquisition and answer extraction. The field of computer (or machine) vision has evolved to become the study of general processing algorithms and recognition strategies without regard for the particular domain or image acquisition method. This has met with moderate success; the difficulty is the experimental verification which comes from the use of these general methods in military, industrial and medical applications. At this point in time one of the greatest stumbling blocks is the lack of computational power available to complete the desired task in a reasonable amount of time. Inspection and navigation tasks

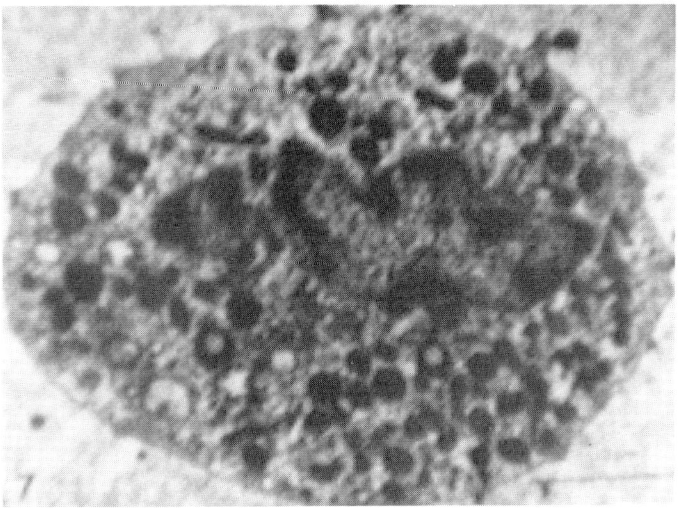

FIGURE 1. A digital image of a transmission electron micrograph.

require response in a second or less from algorithms that typically take minutes or hours to complete on a single processor machine. Thus, there is a strong motivation to study how to do image processing on advanced computers. The emphasis in this chapter is on the development of working systems; details on computer architectures for image analysis can be found in Reference 8.

B. IMAGE PROCESSING

Image processing operations are those performed on digitized pictures designed to detect the presence of features of interest. These features in the image should correspond to physical characteristics of interest, such as intensity highlights, repeated patterns, and edges. Regions of little to no intensity variation are identified by performing image processing to remove intensity variations due to noise.

Operations on pictures that fall into this category are characterized as 2D signal processing operations. These may be expressed by a convolution operation, functions of the entire intensity distribution, or as nonlinear filtering operations in 2D. Figure 1 is a sample transmission electron micrograph with a number of organelles, and Figures 2 to 4 show some sample image processing operations on Figure 1. Figure 2 is the result of a convolution operation to extract edges; Figure 3 is a thresholding operation to extract the organelles using the global intensity distribution; and Figure 4 is a noise removal process based on a nonlinear filtering operation.

In all of these cases, however, the operations map an image to an image. The result of an image processing operation is an image of the same size (the border elements may be handled differently for reasons that will be discussed later). The computational complexity of these operations is proportional to the number of picture elements (pixels) multiplied by the size of the local neighborhood used in the operation (which may range from 1 to 225, typically). Images are typically rectangular and range in size from 64 × 64 (65,536 elements) to 2048 × 2048 (4,194,304 elements). As the images increase in size there is a greater demand for more computational resources. On a typical computer the image is processed in a raster fashion: row by row from top to bottom. As the size increases the execution time increases in proportion to the square of one of the dimensions. To curb this growth in execution time, either more processors or special purpose hardware must be added.

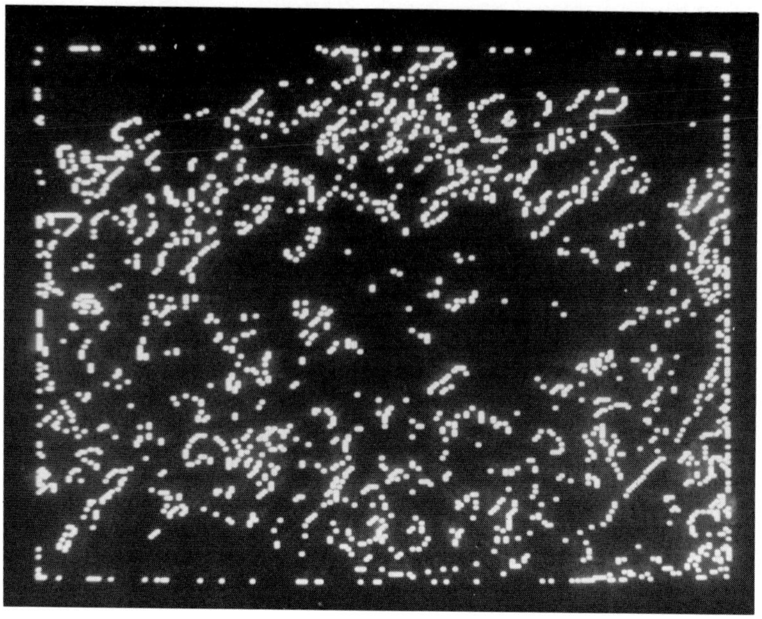

FIGURE 2. The edges located automatically in Figure 1.

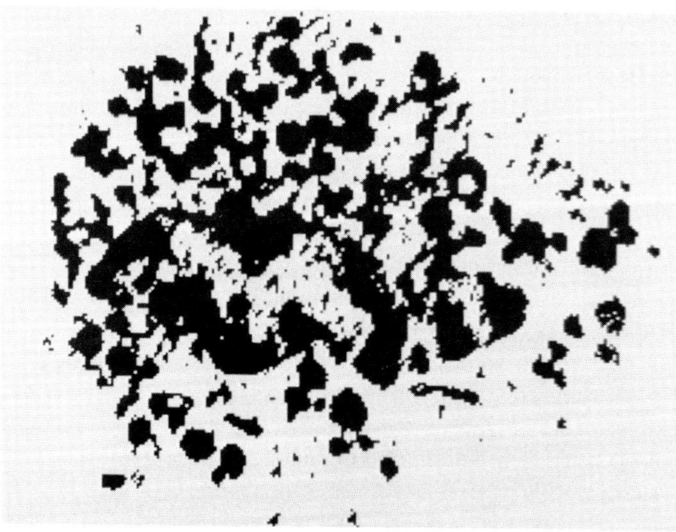

FIGURE 3. The image in Figure 1 thresholded.

C. IMAGE UNDERSTANDING

Operations on image data that fall into this category differ from image processing in that these are operations on the features to synthesize information that is not immediately available. These include the problems of motion in a sequence of images, computing surface shape, segmentation of the image into meaningful regions, and assigning labels to these regions. While the first two items in the list are mathematical, they involve typical optimization procedures and are computational expensive. The latter two items in the list of

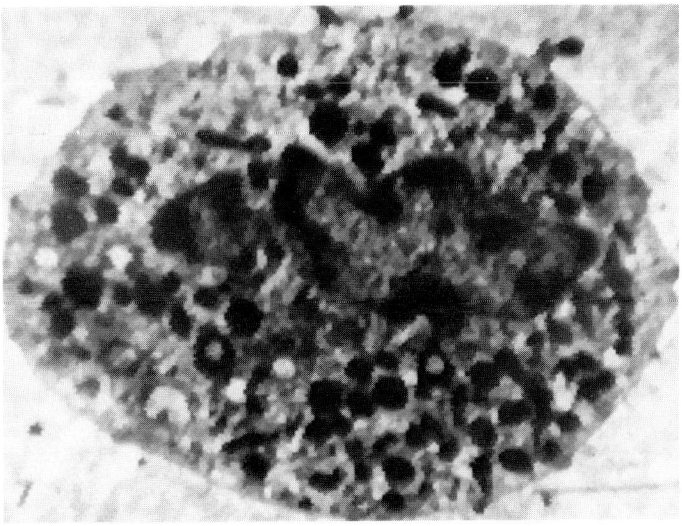

FIGURE 4. The result of smoothing Figure 1.

examples require symbolic computation. The optimization methods can be as elementary as finding the least squares solution for a set of parameters or involve surface fitting using a regularization algorithm or one of its alternatives.

Image understanding tasks differ from image processing in that much of the intermediate data and final results are symbolic rather than numeric. As is true of many symbolic processes, the data structures are not necessarily a fixed regular size. To contrast image understanding with image processing where the operations are image to image, image understanding operations are image to "list" transformations. These "lists" can be arbitrarily long and there can be arbitrarily complex sublists that represent image properties and regions, knowledge about the world, or knowledge about how to process an image. As expected, the complexity of the image understanding operations is dependent on the complexity and contents of the image.

There are three major computational problems limiting advances in research and applications of image processing and understanding (together they shall be referred to as computer vision). The first problem is the sheer volume of the data required to complete a task. It is not uncommon for images to be 480 rows by 512 columns or 512 rows by 512 columns. Each pixel is typically 8 to 16 bit integer data, some applications such as NMR spectroscopy using 32 bit floating point numbers. Although each individual image by itself may not be a great deal of raw data, the application may require that they be processed very rapidly. Inspection tasks, where an image processing system may be placed on an assembly line or an autonomous vehicle, may require that an image be processed in a very small fraction of a second. In medical imaging, the volume of data comes from attempts to build 3D reconstructions of internal organs of the body. To obtain relatively accurate reconstructions, the indivual images correspond to thin sections and consequently there may be several hundred (typically 256) in a reconstruction.

A second problem is just a consequence of dealing with spatial data. Many computer vision algorithms including local processing and interpretation require some knowledge of local properties (either numeric or symbolic) to complete the operation. Finally, computer vision problems are unique in that they require both numeric and symbolic computation. If an inspection task is metrology, then the computation may be entirely numeric; interpretation tasks require some level of symbols.

The purpose of this chapter is threefold: first, to introduce the reader to, and provide a general overview of, image processing and image understanding; second, to provide a summary of how computer image processing can be used in insect hematology and other biological applications; and third as a reference to literature in this area. The aim is to provide material from which one can assess the state of the art, identify research opportunities, and learn about the classical image processing operations on which people are working and have working systems.

II. BACKGROUND ON COMPUTER IMAGE ANALYSIS IN CYTOLOGY

Although the problem of automatically classifying human blood cell images has been studied for over 20 years, there has been little work in developing comprehensive research tools which would allow the researcher to study the complete internal and external structure of the cell as well as automatically classifying the cell structure. Our goal is to combine these three necessary functions into one computer-based imaging system.

One of the earliest efforts in automatic cell classification was the work of Prewitt and Mendelsohn.[9] A group of 22 blood cells were prepared with a special staining procedure, digitized, and processed. The cell samples were classified into one of four different categories. By 1971, Young[10] and Bacus[11] extended this classification to all six major categories of normal cells. They used an image enhancement technique (the whitening transformation)[12] that was quite suitable for this type of analysis when the advantage of color is added. They assumed a bivariate distribution of image pixels, and using two filtered monochromatic images (of different colors) from the same cell, they produced a density image (black and white) and a color image of the cell. They then proceeded to use these results to segment the images. They obtained a better segmentation compared with simple gray-level histogram thresholding.

By the end of the 1970s, Bacus[12] and others extended the area to classify over 17 types of blood cells. Red blood cells could be analyzed. Bacus et al.[13] had developed methods to identify subcategories of normal and abnormal blood cells. They proposed an objective measure for assessing the quality of their segmentations.[14,15] They used an error measure that quantified the disagreement between the "true" and test segmentations, and the errors in pixel classification. Their measure was a function of the proportion of misclassified pixels and the spatial distribution of pixels within a given class. The measure was image-domain independent, image-size independent, and easy to compute. With this measure, they could compare various segmentation algorithms and correct or modify some of their heuristic techniques, leading to better classifications. Mui et al.[16] later proposed a feature selection and statistical pattern recognition approach to the automatic classification problem.

The work of Bacus et al. led to the development of the leucocyte automatic recognition computer (LARC)[17,18] to classify leucocytes in human blood specimens. Bacus[17] gives a comprehensive historical analysis of the development of automated differential systems. Furthermore, Trobaugh and Bacus[18] examined the design and performance of the LARC. Clinical evaluations of this system and other similar systems have confirmed the theoretical results, and have also exposed some weaknesses both in the physics of the designs[19] and in the image analysis and classification algorithms.

Little progress has been made on the problem of automatic classification since then. The early heuristic research efforts of the middle 1960s evolved into some interesting theoretical results in the 1970s. In the same period, there has been remarkable progress in the fields of computer vision, pattern recognition, computer graphics, and artificial intelligence. In addition, more sophisticated microscopes have been developed to produce images with such high resolution that we can now make a comprehensive analysis of the cell's

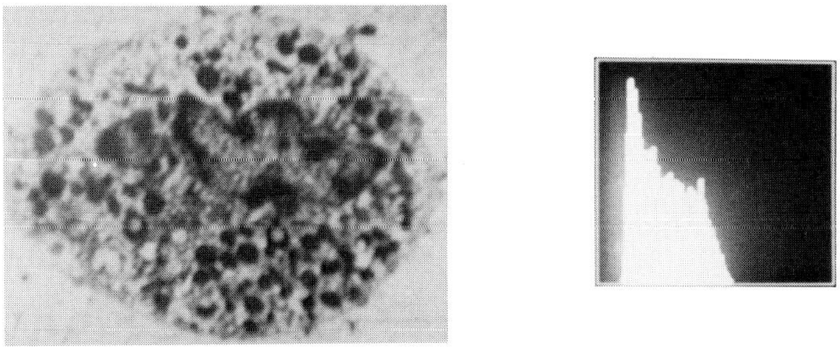

FIGURE 5. An image of a transmission electron micrograph and its histogram.

internal and external structure, thus increasing the chances for a more accurate identification and classification.

III. BASIC IMAGE ANALYSIS

A. PIXEL LEVEL OPERATIONS

Typically, the first operations one will perform on images will be those that are uniformly applied to each point of the image. The *point operations* are applied to extract either global information about the entire image or local information about properties at each point. Sometimes the second group of operations are called local operations, since the computation involves intensities of points in a given neighborhood. For purposes of this review they have been grouped together with operations on individual points since the operation applied to each pixel is the same. That is, the instruction stream is the same and applied to the multiple data in the image.

There are two subclasses of (individual) point operations. The first group are those that produce a global result and the second group are those that are binary operations of pairs of images producing a single image as a result. The operations in the first group are based on computing the histogram and will be considered first; the second group shall be considered separately since pairs of images are used.

1. Histograms

A common example of a global point operation is computing the gray level intensity histogram. The histogram is the frequency distribution of image intensities in the image, and is formed by incrementing an array element that corresponds to image intensities. The image is scanned, and for each observed intensity, the corresponding histogram array location is incremented. Figure 5 is an image of a transmission electron micrograph and its intensity histogram. This is considered a point operation, since the same operation of fetching the intensity and incrementing the histogram is applied to each point.

The histogram is an intermediate result that is often used for global thresholding purposes to segment the image so as to rapidly identify objects of interest. Notice in Figure 5 that the two peaks are well separated. If a gray level between the two peaks is chosen as a threshold, then the objects are well separated from the background. This global histogram can also be used for contrast enhancement or correction and interactive smoothing.

2. Local Operations

The second type of point operations is the larger class of *local operations* where the value of a specific property is estimated at each point in the image based on the intensities at the location and a surrounding neighborhood. The most common of these are edge operators whose output is some measure of edge strength at each point in the image. Other samples include either low or high pass filtering to remove noise.

Local operations are typically implemented as convolutional operators in two dimensions using the general equation

$$h_{ij} = \sum_{k=-m}^{m} \sum_{l=-n}^{n} f_{ij} g_{i-k,j-l} \qquad (1)$$

where f_{ij} are the points in the input image and the elements g_{kl} are the coefficients of the convolutional kernel. The limits m and n are the size of the neighborhood used and are typically equal.

3. Edge Operators

The classical model of an edge is a discontinuity in the image intensity. The intensity discontinuity is detected by locating pixels in the image where the gray level gradient is not zero. An edge operator is an operator with a small spatial extent designed to detect the presence of local edges in the image function. Different operators will perform well in different domains; ideally the local edges should correspond to relevant boundaries in the scene being imaged.

There are three classes of edge operators: first, there are edge operators based on intensity gradients; second, those based on template matching at different orientations; and third, edge operators where the intensities are fit with parametric edge models to model the local surface structures. Examples of all these will be given in turn.

Operators in the first class are the gradient and the laplacian.[4] Both are digital approximations of the continuous derivatives. Let the image be represented by the function f(x,y). Then

$$\Delta_1 = f(x + n, y) - f(x, y)$$

$$\Delta_2 = f(x, y + n) - f(x, y)$$

and the gradient magnitude and direction are given by

$$s(x, y) = \sqrt{\Delta_1^2 + \Delta_2^2}$$

$$\phi(x, y) = \tan^{-1}(\Delta_2/\Delta_1)$$

respectively. These differences, Δ_1 and Δ_2, have the digital convolution kernels

$$\begin{bmatrix} -1 & 0 & 1 \\ -1 & 0 & 1 \\ -1 & 0 & 1 \end{bmatrix} \quad \begin{bmatrix} 1 & 1 & 1 \\ 0 & 0 & 0 \\ -1 & -1 & -1 \end{bmatrix}$$

for 3 × 3 neighborhoods. These will detect edges that go through the center pixel whereas

$$\begin{bmatrix} 0 & 1 \\ -1 & 0 \end{bmatrix} \quad \begin{bmatrix} 1 & 0 \\ 0 & -1 \end{bmatrix}$$

will detect edges that pass through the diagonal.

The laplacian is used to locate edges by finding those points that satisfy the equation

$$\frac{\partial^2 f}{\partial x^2} + \frac{\partial^2 f}{\partial y^2} = 0$$

that is, where the laplacian is zero. The digital version is

$$L(x, y) = f(x, y) - \frac{1}{4}[f(x, y + 1) + f(x, y - 1) + f(x + 1, y) + f(x - 1, y)] \quad (2)$$

with the convolution kernel

$$\begin{bmatrix} 0 & -1/4 & 0 \\ -1/4 & 1 & -1/4 \\ 0 & -1/4 & 0 \end{bmatrix}$$

A variation of this operator, the Marr-Hildreth[3] operator, is used since the second derivative in Equation 2 above is very noisy. In the Marr-Hildreth operator the image is smoothed (with a Gaussian) and then sharpened to locate the edges as in

$$\nabla^2 G(x, y) = -\frac{1}{2\pi\sigma^4}\left[2 - \frac{x^2 + y^2}{\sigma^2}\right] e^{-x2 + y2/\sigma 2}$$

In all cases the edges will be wide since the gradients will not be exactly zero near the actual discontinuities. Edges can be thinned either by thresholding the output of these operators or by thinning the edges in the direction of the gradient.

This problem of imprecise edge localization has led people to consider using templates to locate edges, with a template matching strategy. One example is the Kirsch operator, whose output is given by

$$S(x) = \max\left[1, \max \sum_{k-1}^{k+1} f(x, y)_k\right]$$

where $(x,y)_k$ are the eight neighboring pixels of (x,y). The summation and the outer maximum can be implemented by template matching in a series of directions given by the inner maximum. This is why the Kirsch operator is considered to be in the class of template matching operators.[1]

Not all methods for detecting and localizing edges are based on the conventional filtering approach. The advantage of the parametric approach is that additional knowledge is used in the process. The operator described by Hueckel (see Reference 4 for details) is a scheme where a surface is fit to the intensity function in the least squares sense. Other notable edge detection methods in this category are Nevatia and Haralick.[5] In all cases, they provide more detailed edge structure but at a cost of increased computation.

This is the classical categorization of edge operators and detection schemes. There are many variations on the methods described here. See any of the texts by Ballard and Brown,[1] Rosenfeld and Kak,[4] or Horn[20] for more detail. In the next section, region-based image processing operations are studied.

B. IMAGE REGION OPERATIONS

The image processing operations described in this section are the operations that group pixels into intermediate level tokens suitable for matching, recognition, or other higher level cognitive tasks including path planning or obstacle avoidance. These grouping operators use the results of the low level operations and rarely use the gray level intensity information directly (see gray level connected components below). This is a departure from the low level operations where the results were in the form of an array; rather, the results are tokens representing regions or curves in the images.

The operations perform *grouping* of individual pixels into collections. These collections can be either regions or curves in the image plane. These grouping operations segment the image into a set of features that represent features in the image that appear in the scene.

The first type of operations are region-based grouping or *segmentation* operations. The basic idea is to form connected components of pixels by analyzing their spatial connectivity; spatially adjacent points are likely to be part of the same region if the difference in intensity is small. If the intensity difference is less than some threshold of the tolerable noise, then the two pixels are part of a region that represents a surface in the scene.

The second type of operation is edge grouping into lines and curves. The basic idea is to fit parametric curves to observed edge points, but since the edge detectors do have noise output it is best to fit the most likely curves in the image. One common technique is called the Hough transform, a method of detecting the most likely lines (or other space curves) in an image.

These two algorithms were selected as representative of intermediate level grouping operations since they are commonly used image processing operations in practice.

In the previous section, one of the example point operations was to form the gray level histogram. The purpose was to compute the frequency distributions of intensities in the image. There are two main uses of this information.

The first application is to enhance the contrasts by expanding the histogram to use the full dynamic range of gray levels. The histogram, which approximates the probability distribution, is used to first form the cumulative mass function and then map this cumulative mass function to the full dynamic range of available gray levels. This has the net effect of increasing the contrast where there is little and decreasing the contrast where it is great.[1,4]

The second use of this gray level is in a global thresholding technique. First the gray level histogram is formed, and then a decision criterion is used to classify points as belonging to one object or another using the gray level as the deciding criteria. Often, a simplifying assumption that the histogram is bimodal has to be made first, then the histogram separating the two peaks is found by some means and a pixel is classified as belonging to the object if its intensity is greater than this threshold.[4]

This description of thresholding is included as an illustration of segmentation or separation. The purpose of the global thresholding technique is to separate the pixels into regions using only global information in the form of a gray level histogram. This works well if the intensity distribution of the objects are well separated and the objects themselves are compact and convex. The technique may not work if the peaks are not well separated and/or the objects are concave. The results of segmentation via global thresholding may not yield connected regions.

1. Connected Components

Blob coloring, or connected component algorithms, are region-based segmentation algorithms that group pixels into regions based on their local connectivity. Groups of pixels that are connected are assigned a unique region number, and then each region can be analyzed separately.

There are two types of connectivity used in the blob coloring algorithms: 4 or 8 con-

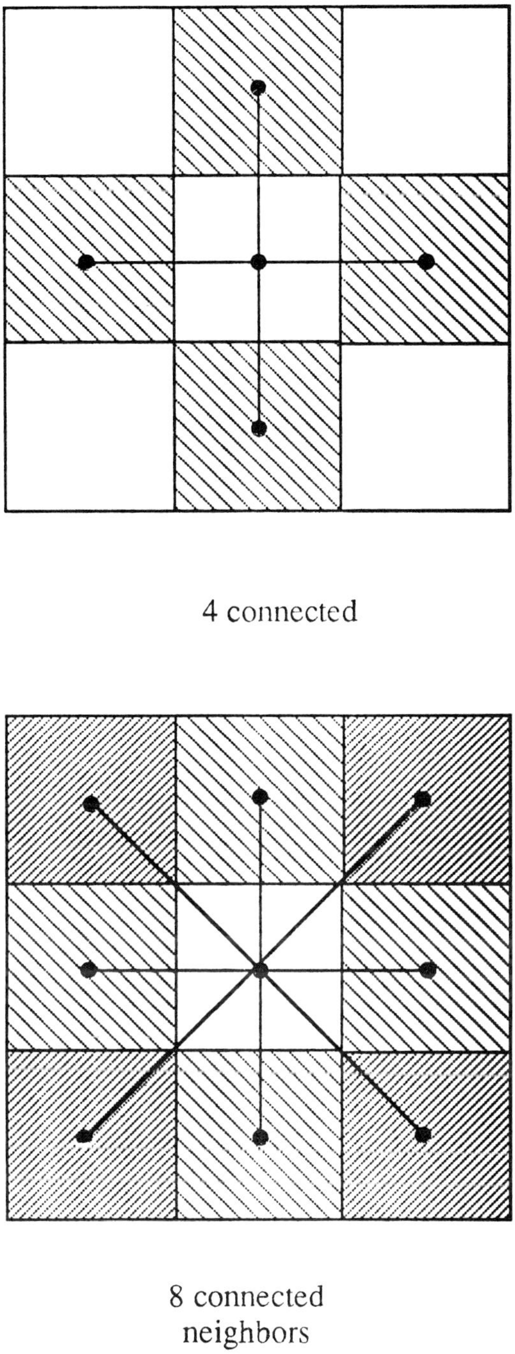

FIGURE 6. 4 and 8 connectivity.

nectivity. The four pixels adjacent to the center pixel of a neighborhood that are above, below, to the left, and to the right of the center are its 4 connected neighbors. Together with the four pixels at the corners of the 3×3 neighborhood, the center pixel, and one of these eight are said to be 8 connected neighbors. Figure 6 shows the 4 connected and 8 connected neighbors of a pixel.

The simplest connected component is for binary images and assumes that the regions

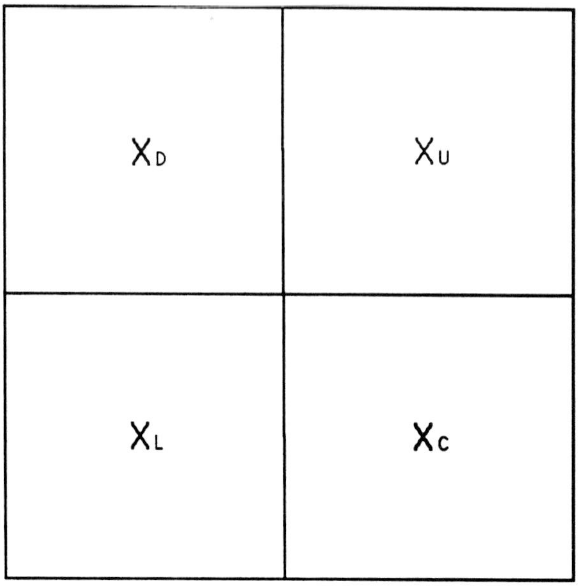

FIGURE 7. Computing connected components.

in the image are 4 connected. The blob coloring algorithm scans the image in a raster fashion and attempts to assign a label (i.e., a region number or "color") to a pixel, if it is 4 connected to a pixel that is already labeled. Since the image is processed in a raster fashion, it is not necessary to look at all four neighbors, but only at the two directly to the left and above. The 4 connected neighbors to the right and below will be labeled in subsequent iterations of the algorithm. There are two passes to the algorithm: in the first pass, each pixel that is part of some region is assigned a label; in the second pass, regions that contain pixels with different labels have those differences resolved.

Figure 7 shows the configuration used. Each iteration proceeds as follows:

If ($x_c = 1$) then:

if $x_u = 1$ and $x_l = 0$ then $x_c = x_u$

if $x_l = 1$ and $x_u = 0$ then $x_c = x_l$

if $x_l = 1$ and $x_u = 1$ then $x_c = x_l$ and $x_u = x_l$

if $x_l = 0$ and $x_u = 0$ then $x_c = x_u$; $k = k + 1$

The first two conditions check to see if the pixel at the center x_c is part of a region that already contains one (and only one) of its neighbors to the left (x_l) and above (x_u). In either case the label already assigned to the respective point is propagated to x_c. If both x_l and x_u are each part of a region already, then the different labels have to be noted and are resolved in second pass. If neither x_l or x_u are object points, the x_c starts a new region and is assigned a new label. This procedure is repeated for all points in the image processed in a raster fashion. The second pass of the algorithm also processes the image in a raster fashion and checks for regions that have been assigned multiple labels.

This algorithm is easily extended to 8 connected regions by including in the test a diagonal point x_d. Instead of just checking the four combinations of the pairs, addition checks on triples of points will have to be made as follows:

If (x_c = 1) then

if x_d = n then x_c = n

if x_u = n then x_c = n

if x_l = n then x_c = n

if x_l = n and x_u = n then x_c = n

if x_l = m and x_u = n then x_c = n and m = n

The equivalence that m = n is resolved in the second pass of the algorithm. Other statistics on region size and shape can be accumulated at the same time that the equivalences are resolved, providing a set of features that can be used for object recognition.

Both of these algorithms for connected components are a good basis for a local connectivity segmentation procedure, but as you see, suffer from the restriction that the input image must be binary. Either one must begin with a binary image or an image that is to be thresholded. To make a procedure like this work on gray level images is still today a very difficult problem to solve.

The most natural extension to gray level imagery is to introduce a tolerance band for gray level differences. In the above two algorithms the gray level tolerance is zero; if one assumes that pixels that are part of the same region have intensities that are similar, then the difference in gray level should be very small.

This approach is commonly used, but one caution is that there will be misleading results in cases where there is a monotonic shift in intensity in the image. If the monotonic shift is slow enough so that the difference in intensity between adjacent pixels is less than the tolerance threshold, large regions of the image will be mistaken for large connected components. This procedure can be modified to include other point or local features such as edges,[21] but an optimal segmentation is still far away.

The connected component algorithms can be used to compute many features for object or pattern recognition. For each connected component it is easy to compute the area, location, and moments after each pixel has been labeled in the image. For many applications involving object or particle counting or distribution this region analysis may be sufficient, but very often boundary information is needed for 3D recognition and interpretation problems.

2. Hough Transforms

Boundary direction can be considered as another grouping operation, only this time edge elements detected by local operators are grouped into one dimensional edge features. There are five basic approaches for boundary grouping: searching near an approximate location; graph searching; dynamic programming; contour following, and the Hough transform.[1] Only the last of the five will be explained here.

Hough transforms are used to detect boundaries whose shapes are given in an analytical or tabular form. The traditional Hough transform uses analytical representations, whereas the generalized Hough transform uses tables and is capable of recognizing arbitrary boundaries in the image plane.

In the traditional Hough transform the boundary shape is determined as a parametric curve. The localization of the curve is identified in a robust manner; the Hough transform is unaffected by gaps or noise (where many of the search or optimization methods would fail). The traditional case can be illustrated with an example of detecting a straight line.

Figure 8 is an example of a straight line in the x-y plane. The traditional representation of the line is the equation

$$y = mx + b$$

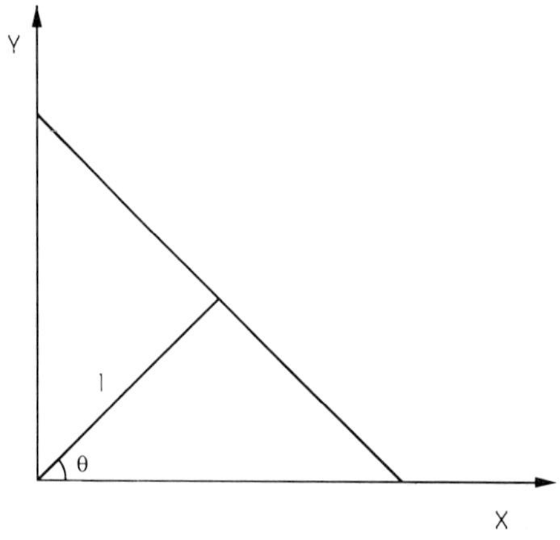

FIGURE 8. Parameterization of a straight line.

However, for lines parallel to the y axis, this formula has a similarity that will pose a problem later in the Hough algorithm. An alternate is to parameterize the line by the length and orientation of the perpendicular foot of the line to the origin. The alternative parametric equation is

$$\rho = x \cos\phi + y \sin\phi \tag{3}$$

where ρ and ϕ are shown in Figure 8, and x and y are the coordinates of points along the straight line.

Lines will be represented by the parameters (ρ,ϕ), and the purpose of the Hough transform is to find the parameters (ρ,ϕ) of the most likely lines in the plane given the locations (x,y) of edge points that were detected.

For each point in the image, there is a set of lines that pass through the point, one at each angle (ϕ). As the angle (ϕ) is varied, Equation 3 can be solved for the length of the perpendicular (ρ) of the line. An accumulator array is used to keep track of the (ρ,ϕ) combinations that satisfy Equation 3; if there is a line in the image, then all points that are on that line will give evidence that the line exists. The accumulator array is used to count the number of points that satisfy the equation of a line whose parameters are the index of a cell in the accumulator array.

The algorithm proceeds as follows:

```
for each x
   for each y
      if each image (x,y) is an edge point
         for each φ = 0 to π (or 2π)
            ρ = x cosφ + y sinφ
            increment through (ρ,φ)
```

After computing the Hough array in this manner, the lines that appear in the image are identified by locating the most likely lines or "peaks" in the Hough space. Local maxima in the Hough accumulator array are colinear points in the array.

There are many variations of the basic algorithm that locate lines by using slightly different features. Instead of incrementing by 1, i.e., just counting the number of points along the line, often the gradient or other measure of edge strength is used as evidence of an edge point in the image. Those edge points with stronger edge magnitude are more likely to lie on a line and not be noise on the image.

Other parametric curves can be located in the image plane by setting up the Hough accumulator array for the parameters in the curve. For example, to locate circles in the image plane

$$(x - a)^2 + (y - b)^2 = r^2$$

requires a Hough space for the parameters a, b, and r. As two are fixed for each point, a third is solved for and the cell H(a,b,r) is incremented. The drawback is that the dimensionality of the Hough space increases and will become prohibitive to implement at some point.

An alternative to using analytic representations and high dimensional Hough spaces (accumulator arrays) is the generalized Hough transform (GHT).[1] In the GHT, the goal is to find the location of a reference point for the curve instead of finding its analytic representation. No analytic representations are used; instead, for each curve a table is precomputed that gives the location of the reference point as a function of the gradient direction at the boundary points. This way, arbitrary curves can be detected even if there is no analytic representation.

The look-up table for each space curve is computed as follows. First a reference point for the curve is chosen. Then, for each boundary point, the vector from the boundary point to the reference point is recorded in the table. After this is repeated for all points on the boundary one has a table, indexed by gradient direction, that contains vectors to possible locations of the reference point of the curve. Thus the accumulator array for the GHT is only two dimensions and will yield the location of the reference point for the space curve.

The GHT algorithm is as follows:

```
for each x
  for each y
    if the image (x,y) is an edge
      compute the gradient direction φ(x,y)
        for each entry in the R table for φ(x,y)
          compute x_c = x + r(φ)cos[α(φ)]
          compute y_c = x + r(φ)cos[α(φ)]
          increment H(x_c,y_c) = H(x_c,y_c) + 1
```

The GHT algorithm, as presented, will locate translated versions of the curve. The dimensionality of the Hough space can be increased to account for rotation and/or scaling parameters that can be included in the equations to compute the location of the reference point. Figure 9 is an example of the geometry of the GHT.

To summarize, in this section some of the basic image processing and analysis techniques have been described. In all quantitification problems, many of the pixel level operations are used; more often than not, both connected components and some sort of model matching (like the Hough transforms) are used. In Section VI, there is a description of a typical image analysis system that uses these techniques to reproducibly quantify characteristics of arthropod hemocytes.

IV. CLASSIFYING ARTHROPOD HEMOCYTES

Arthropods (lobster, shrimp, insects, spiders, and others) have six kinds of hemocytes or blood cells: prohemocytes, plasmatocytes, granulocytes/coagulocytes, spherulocytes, adipohemocytes, and oenocytoids. Two of these, the granulocyte (GR) and the plasmatocyte (PL), provide cellular immunity (phagocytosis and encapsulation) to these animals. In rec-

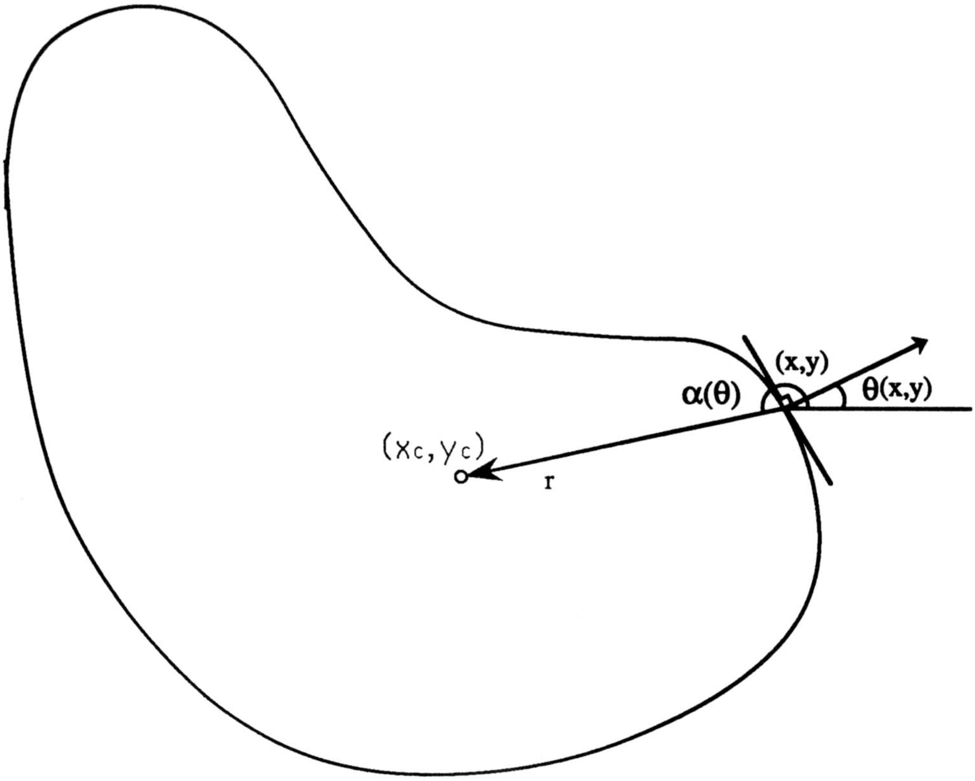

FIGURE 9. Computing a generalized Hough transform.

ognition of the ability of the GRs and PLs to discriminate between self and nonself tissues, Gupta[22] coined the term *immunocyte* for these hemocytes. One of the nagging problems in arthropod hematology is the lack of reliable, error-free criteria to correctly identify and classify these various types of hemocytes. Often, the hemocytes are classified on the basis of their morphological characteristics, staining reactions, and functions, and it is not uncommon to discover that the same hemocyte has been differently or incorrectly identified by two or more authors. This has inevitably resulted in confusing terminology,[22] and it has not been possible to consistently attribute the same immunologic function to the same hemocyte types. In addition, there are long-standing controversies regarding such hemocytes as coagulocytes, the so-called crystal cells, adipohemocytes, and others that are found in arthropods.

Computer-based identification of blood cells is routinely used in clinical laboratories, and we have used classical image-processing methods[1,4] to determine a robust set of image features to recognize some of these cells. The features describe mainly the size, shape, textural characteristics, and external boundary continuity of cell organelles like mitochondria, granules, and vacuoles. We selected the GR for this study because it is found in all major arthropod taxa[23] and is their main immunocyte[22] (see also Chapters 2 and 4). It often has a flat discoid shape and characteristic intracytoplasmic granules.

Our goal is to develop a computer vision system to visualize complete 3D images of both internal and external features of arthropod blood cells by light, transmission, and scanning electron microscopy. By manipulating such 3D images, we would study the role of such organelles as marginal microtubule bands, which are crucial in maintaining the flat and discoid nature of these cells (in the German cockroach) and increase sevenfold in number in an activated state.[24] Some preliminary results of computer-assisted identification of arthropod blood cells were presented in Dunn and Gupta.[25]

In the examples in this chapter we shall be concentrating on 2D segmentation of serial sections because this is the basis for successful 3D analysis. With a proper segmentation of a consecutive series of microscope slices and a good identification of the objects present in these slices, we can proceed to a 3D reconstruction of the internal and external structures of both the cells and their organelles.

V. MODEL BASED IMAGE ANALYSIS

The objective of an image interpretation system is to construct an explicit description of the scene captured in a given image. To do this, a computer image interpretation system must solve two problems. First, the system must extract an explicit description of the scene from the image; and second, it must decide what interpretative techniques should be used to complete the vision task (arrive at an understanding of the scene). Thus the first major task in any image understanding system is to *segment* the scene into meaningful entities, which are given a characterization suitable for machine recognition. The task of understanding becomes much easier if the first task is done well.

A good segmentation scheme is the key to a good image interpretation system. Constructing meaningful entities from *iconic* data (pixel raster representation with gray levels) poses problems that are generally linked to the type of objects to be identified. Restraining segmentation algorithms to a particular class of images may simplify the process, but the loss of retargetability is a very serious handicap of such a solution. The complex human visual system first extracts reliable intrinsic information from the input image and successfully processes the extracted information with some domain knowledge input as the interpretation progresses from the retinal (acquisition) to the cognitive (interpretative) levels.[26,27] A near optimal segmentation scheme should adopt this approach. Many solutions have been proposed for the segmentation problem,[1,3,4] but few utilize the formal mechnism of reasoning that is apparent in the biological visual system. A few of these techniques have been quite successful in a variety of specific instances, but do relatively poorly in the general case. Computer vision systems built to analyze scenes of man-made objects are successful to a large extent because most objects in the scene can be easily modeled to a high degree of accuracy. When it comes to natural objects with large variations in structure, edge-based or region-based segmentation algorithms[28] present serious inadequacies.

The most general image segmentation scheme that would be suitable for images of both man-made and natural objects would be data driven. That is, it would use only the raw image data and make no reference to the domain of analysis. Such an ideal system has been found to be unachievable even in the most complex systems. It has been shown experimentally[29] that the best such a computer system can achieve is preliminary grouping of pixels and initial partitioning of regions. Obtaining a satisfactory final segmentation requires domain-specific knowledge. The efficient organization and optimal insertion of this knowledge in the control processes and tasks to be performed is the key to building a good image understanding system.

However, data-driven algorithms can be used to build an initial low-level segmentation module that will serve as a basis for a general image interpretation scheme.[30] Such a model would act as an input processor that accepts iconic data from any domain and produces the initial partitioning into primitive elements and their abstract description for use by later processes in the interpretation hierarchy. At the intermediate and higher levels, the processes could be more specialized and domain specific. This has been the guiding principle in the work we are reporting here. The complex problems of scene analysis to be solved here require a combination of regular image segmentation techniques and artificial intelligence (AI) techniques for reasoning, planning, and learning to be applied with a proper stratification of domain knowledge use.[31,32]

A. HEMOCYTE MODELING

Several examples of the use of models in image segmentation and interpretation can be cited in the literature.[33-35] In most of these examples, model-driven image interpretation tasks have mostly been limited to high level matching and recognition. The use of models in their abstract symbolic form needs to be included in the segmentation process to permit greater control of the lower level processes, allow greater flexibility, and most importantly as an aid or filter in reducing the amount of noise in the output of the low-level operators. Perhaps the greatest challenge in building image interpretation systems for images of biological structure is discriminating between biological variability and noise in the output of low-level image operators. In most cases where the use of models has been attempted, only elementary cues like the curvature of edges and gray level similarity are used.[36] Higher and more sophisticated characterizations of models of scene objects need to be exploited. Reynolds et al.[37] propose a flexible intermediate level representation for multiresolution segmentation using multiresolution models. They propose foveation (focus of attention) as a means of speeding up the process. The system is, however, limited by the fact that it uses model descriptions that are suitable only for man-made objects. In the same line, Suetens et al.[35] propose a system with hierarchical instantiation of models. They implement matching and interpretation as integrated processes. The system has the inconvenience that knowledge representation is still primitive (at the iconic level). In their work, Kapouleas and Kulikowski[38] describe a system with solid models for 3D vision. They propose a very good example of the use of expected biological structure in 3D. With this system, they are able to segment and interpret MRI images of the human brain. They start with an initial low-level segmentation, which they successively refine with increased use of domain knowledge.

In our system, we use models in both the segmentation and interpretation processes. Our segmentation is successively refined as we go from coarser to more complete model descriptions. In this process we successively drop candidates that do not fit the descriptions of expected objects or from which we are least likely to *build* the expected structures.[29,30] The process consists of determining the goodness of fit with one object model parameter at a time and dropping the worst fits. In this way, biological variability does not have to be explicitly modeled. Rather, candidates that are least likely to build the expected structures are assumed to be outside the limits of possible biological variability for the given expected structure and are dropped.

VI. AN EXAMPLE HEMOCYTE IMAGE ANALYSIS SYSTEM

A. GLOBAL SYSTEM DESCRIPTION

The organization of the image models, domain models, and image processing tasks is shown in Figure 10.

The digitized image is processed in three parallel paths. The first is *quantized thresholding,* to extract features to be matched against the models of expected structure, extract principal boundaries *(forbidden boundaries)* for use in assessing the quality of the initial segmentation, and to define control cues for region growing. The second is *quantized smoothed histogram equalization,* to extract cues for computing texture parameters to be used in comparing image object texture features. A histogram of the original image is filtered to obtain a smooth curve and histogram equalization is done around the maxima and minima of the filtered histogram (see Dunn and Shemlon[30] for details). The last is *data-driven segmentation* where edge and image intensity information is used to generate an initial fine segmentation without any image type or domain-knowledge use. The image is first filtered. After edge detection, pixels within this filtered image are grouped using a modified version of the connected components algorithm (see Dunn and Shemlon[30] for details). Information about regions resulting from this initial segmentation is generated in the form of a global data base.[21]

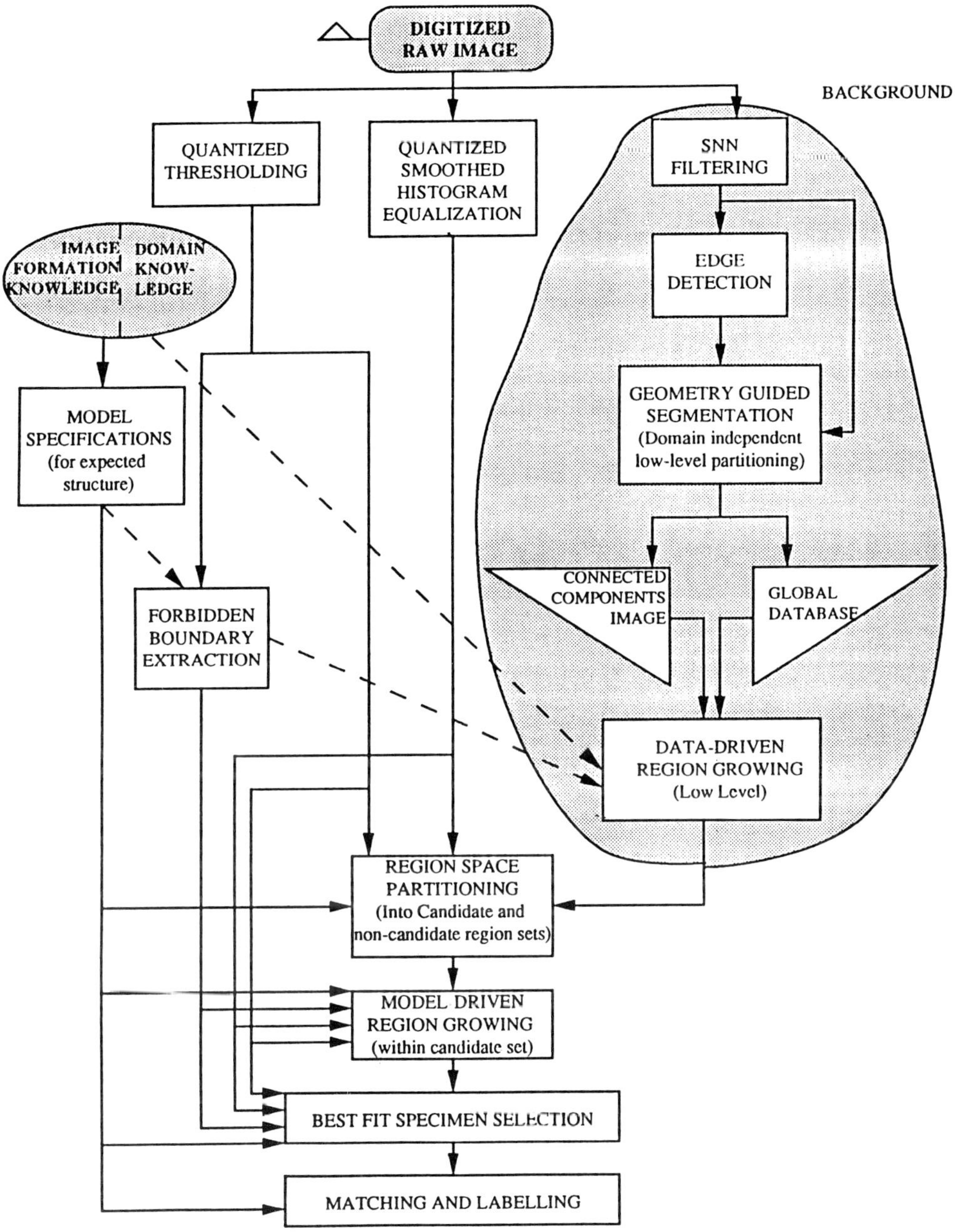

FIGURE 10. Flow chart of the example TEM processing system.

After this initial segmentation, domain-independent data-driven region-growing algorithms are used to obtain a finer segmentation. The algorithms use two segmentations of the given scene (a coarser and finer segmentation) to obtain an intermediate segmentation that is more refined that the coarser segmentation and devoid of the unnecessary detail in the finer segmentation (see Dunn et al.[29,30] for details).

The results of these three parallel paths are then combined in *model-driven region growing,* where domain knowledge and information about the image formation process are used to guide the merging of small regions to form candidates that most closely match one (or possibly more) of the models of expected structures. Knowledge of the domain of analysis (for example, insect hematology) and of the image formation process (for example, transmission electron microscopy [TEM]) is exploited in the construction of quantitative object descriptions that characterize structures expected to be present in such images, from qualitative object model specifications.

After region growing, the best fits in the final segmentation (final regions whose characteristics best match model specifications) are selected as specimens. The other regions in the final set are compared to these best fits, and the closest matches are selected and labeled along with the specimens as the targeted objects.

The whole segmentation process, from raw data to object extraction, is automatic. No human intervention is needed for any threshold selection. All thresholds are selected automatically on the basis of knowledge about the data being analyzed. All that the system needs is knowledge of the domain (insect hemocytes), the objects we are looking for (granules, mitochondria, or other organelle), and information on the image formation process (TEM).

The segmentation process is speeded up by using prior knowledge of the final characteristics of the regions we are building. The initial segmentation region space can be split into two classes: one of likely candidates (from which the desired structures can be grown) and the second constituting the background. This reduction in the search space increases the probability of obtaining a near-optimal final segmentation of desired objects. Merging rules, grouping criteria, model specifications, and search space constraints are successively enhanced as the region growing progresses (more complete model descriptions and stronger constraints are used) until a final set of most likely candidate regions is obtained. As the segmentation progresses, the size of the candidate region space shrinks (since many regions are merged into one). This will insure that the time complexity of the algorithm is not exponential in the number of regions.

B. MODEL SPECIFICATIONS

Regions are grown to fit structures that we expect to find in the image. If the final segmentation step produces regions that match the expected description, they are extracted and the remaining regions are considered either to belong to another object's candidate region space or to be part of background. Quantitative object descriptions result from combining qualitative object models in a given domain (arthropod hematology in this case) and image formation models. These quantitative parameters describe object appearance in the image. For the moment, we shall concentrate on the form of the object descriptions and how they are used. Object descriptions take the form of mathematical expressions or statistical parameters that characterize image regions. Regions are characterized by their shape and intensity parameters. The brightness of a region is measured by comparing its average gray level to a computed threshold. The region's position in the image is given by its centroid. The chain code encoding the region's bounding contour is used to extract information on the region's concavity. It can be used to characterize the fact that a region is lobed, protruded, or has a smooth outline. The contour profile obtained from the chain code can be used as a measure of shape irregularity. A more explicit characterization of shape, using moments and other descriptors to compute geometric invariants, follows. Parameters for characterizing texture are also examined.

The visible regions in the image are characterized by their shape and intensity parameters. We characterize shape using geometric invariants for various 2D shapes. Texture is characterized using statistical parameters computed from the gray scale distribution of pixels within a region.[29,30]

We have used moments, areas, and perimeters as shape descriptors to compute some geometric invariants for simple 2D shapes that model some of the objects we expect to find in the images being analyzed. As an example, the following 2D shapes can be quantitatively modeled as follows:

Qualitative shape description	Quantitative shape models
Round	$\frac{P^2 M_{xx} M_{yy}}{A^2} = 1$
Elliptical	$\frac{A^2(M_{xx}^2 + M_{yy}^2)}{2\ PM_{xx}^2 M_{yy}^2} \approx 1$
Square	$\frac{3\ P^2 M_{xx} M_{yy}}{4\ A^2} = 1$
Rectangular	$\frac{P^2}{8(6(M_{xx}^2 + M_{yy}^2) + A)} = 1$

where

A = area,
P = perimeter,
M_{xx} = square root of the normalized second moment about the x axis, and
M_{yy} = square root of the normalized second moment about the y axis.

We characterize texture using very simple statistical parameters. The average gray level of a region (AGL), its standard deviation (SD), the percentage of pixels with gray level value greater than 2 standard deviations above the mean (TP2), between 1 and 2 standard deviations above the mean (TP1), and less than 1 standard deviation above the mean (TP0) are all used to characterize a region's texture. These texture parameters are computed from a texture-enhanced version of the original image that is obtained using the quantized smoothed histogram equalization and quantized thresholding techniques (see Figure 10).

C. CREATING QUANTITATIVE MODELS OF ORGANELLE

The algorithms described above were used to process and interpret images of transmission electron micrographs of insect hemocytes. The following are some of the features we use to describe regions in our segmentation paradigm:

- Shape: triangular, rectangular, round, elliptical, trapezoidal, or irregular
- Position: specification of location or bounding window
- Relative size: very small, small, average, large, or very large (relative to the sizes of similar objects in the image)
- Relative texture: highly, medium, or nontextured (relative to the textures of similar objects in the image)
- Intensity specification: very dark, dark, medium, bright, or very bright

As an example, the following objects (organelles) often present in TEM images of insect hemocytes[22,23] could be qualitatively modeled as follows:

Granules	Mitochondria
Dark or very dark	Dark
Round or elliptical in shape	Round or elliptical in shape
Average to large in size	Average to large in size
Relatively nontextured	Textured
Located everywhere except the nucleus	Located everywhere except the nucleus

These qualitative descriptions were converted into quantitative measures that are usable by the segmentation system. The descriptions of "dark" or "light" are translated into criteria that compare the average gray level of image regions to gray level thresholds obtained by quantized thresholding. Qualitative descriptions of shape are translated into quantitative measures using the shape parameters and 2D geometric invariants. The classification "textured" or "nontextured" is quantitatively expressed using the texture parameters that are computed from the quantized smoothed histogram equalized image.

With these quantitative object descriptions, a TEM serial section is segmented using the following model-based, region-growing procedure. All dark regions in the primary segmentation will constitute the initial class of candidates that can be grown to nuclear chromatin, mitochondria, granules, or nucleoli. We can use the shape constraint to extract a subset of most likely granules, mitochondria, or nucleoli. Since granules are generally nontextured, a subset of most likely granules can easily be extracted leaving us with another subset of most likely nucleoli and mitochondria. From the subset of most likely granules (if nonempty), the best fits will be chosen to act as granule specimens. If the subset of most likely nucleoli and mitochondria is nonempty, it can be separated into two subsets of most likely nucleoli and most likely mitochondria using texture discrimination with the parameters defined previously. The best fits in these two subsets also become specimens. Other objects present in the image are matched to the selected specimens to generate the final class of objects that best fit the models of the structure (organelle) we are looking for.

D. EXPERIMENTS AND RESULTS

These algorithms were used to process and interpret TEM serial sections of insect hemocytes. Figure 11 shows some of the original TEM images used in these experiments. The images were first processed to eliminate noise and enhance edge features. The filtering algorithm also eliminated some of the fine texture in the granules, mitochondria, and chromatin. To avoid this inconvenience, the filtered images were used only for edge detection and data-driven segmentation. The parameters used in model-driven, region-growing specimen selection and matching were computed from images obtained by quantized smoothed histogram equalization and quantized thresholding.

As we can see in Figure 12 (upper right), all dark regions are extracted as candidates to be merged to form granules, nucleoli, chromatin, and mitochondria. Figure 12 (lower left) shows the targets around which primary region growing will be done. Here the criterion used to select these regions is size, since we know our expected structures are larger than the average region produced by the early segmentation process. In Figure 12 (lower right), we show the final segmentation.

Using the computed shape parameters, we evaluated the various invariant expressions for each shape for a given region, and expressed the degree of approximation of the region's shape to the given shape as a percentile. This percentile is the ratio of the computed value to the expected value. Figure 13 shows some sample regions extracted from an insect granulocyte. Table 1 shows the computed shape descriptors (area, perimeter, and moments) and the geometric invariants, which express how close the given regions approximate circles, ellipses, squares, and rectangles. The most likely candidates for granules and mitochondria are those closest in shape to ellipses or circles, for example, regions a, e, g, i, j, and k. We remark here that due to digitization errors a cartesian plane approximates figures with straight edges better than it does for curves. It is therefore no surprise that the computed geometric invariants should be better for squares and rectangles than for circles and ellipses.

In Figure 14 (upper right and lower left), candidates that best fit the description of either granules or mitochondria are highlighted. The best fits are separated into the granules and mitochondria specimens. Figure 14 shows the final results after matching and labeling other objects in the scene. The granules are highlighted brighter than the mitochondria.

In Figures 15, 16, and 17 (see originals in Figure 11), we show preliminary experimental

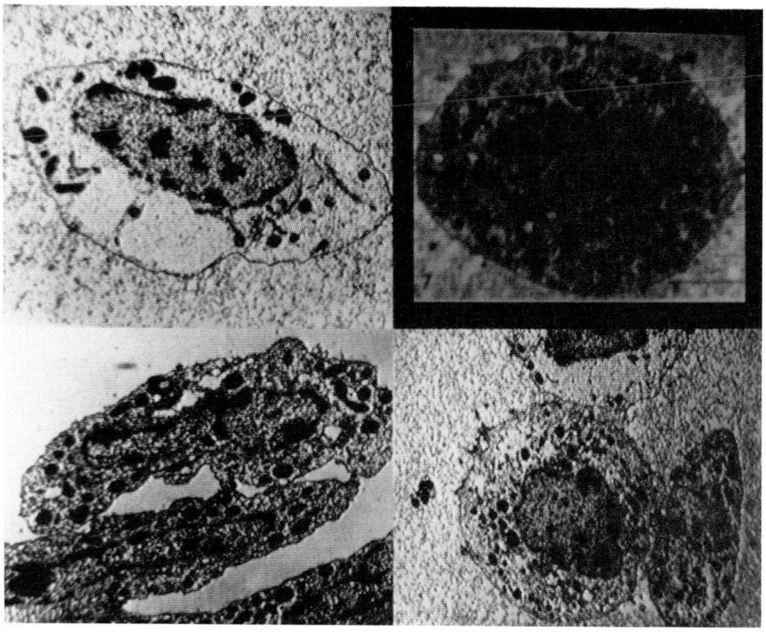

FIGURE 11. Sample TEM images processed.

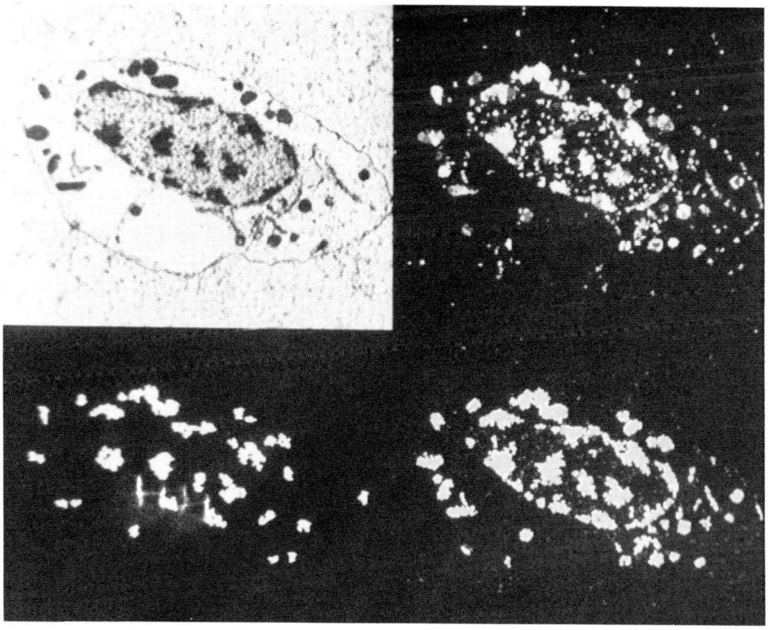

FIGURE 12. Results of the sequence of processing steps.

results of using organelle labeling to differentiate between different types of insect hemocytes. Based on the granules and mitochondria extracted compared to the number of organelle present, we can label granulocytes in Figure 15, a plasmatocyte in Figure 16 (top), a partial cross section of a spherulocyte in Figure 16 (bottom), and coagulocytes (considered to be granulocytes by Gupta[39]) in Figure 17 (top). Figure 17 (bottom) contains more than one cell

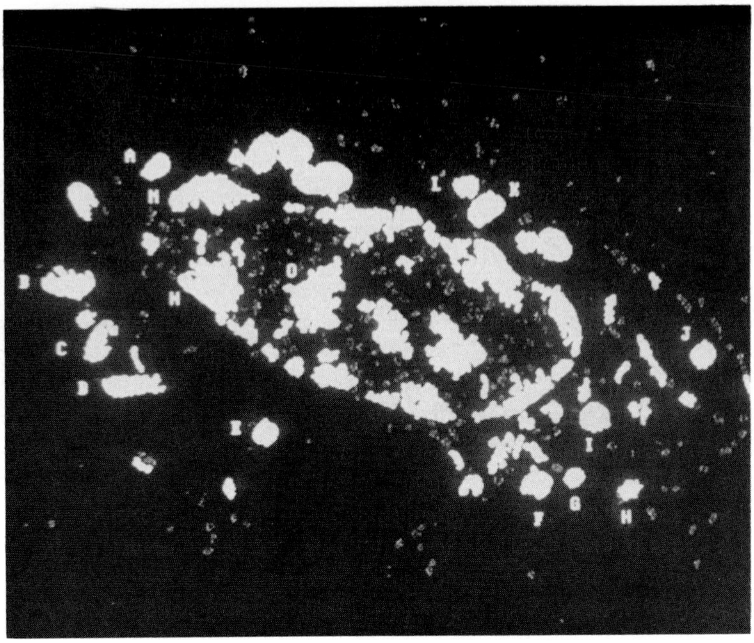

FIGURE 13. Sample regions detected in a granulocyte.

TABLE 1
Shape Descriptors and Values of Geometric Invariants for Regions in Figure 13

Region	A	P	M_{xx}	M_{yy}	% Circle fit	% Ellipse fit	% Square fit	% Rectangle fit
a	185.50	58.18	4.44	4.33	53	53	71	98
b	432.00	131.54	4.98	8.78	25	29	33	48
c	265.00	112.57	7.85	4.52	16	18	21	48
d	385.00	117.40	3.76	9.97	29	43	38	62
e	192.00	55.94	4.51	3.89	67	68	89	97
f	240.00	73.60	4.61	5.00	46	46	62	76
g	117.00	41.80	3.48	3.18	71	71	94	87
h	150.50	58.83	4.16	4.18	38	38	51	84
i	275.50	74.53	5.23	4.84	54	54	72	84
j	216.00	58.77	4.72	4.15	69	70	92	95
k	352.50	78.33	5.75	5.87	60	60	80	99
l	168.00	62.77	4.27	3.90	43	43	57	75
m	713.00	218.45	6.29	13.6	12	16	17	35
n	989.00	209.28	10.64	9.35	22	23	30	40
o	1014.00	213.28	12.22	8.66	21	23	28	42

(two normal granulocytes and a degranulated granulocyte), as can be concluded from the numerous mitochondria and granules identified.

Cell discrimination at this stage is not yet automatic. The decisions are still made by a human expert, but they are based on the computer results. There is work in progress to automate this process.

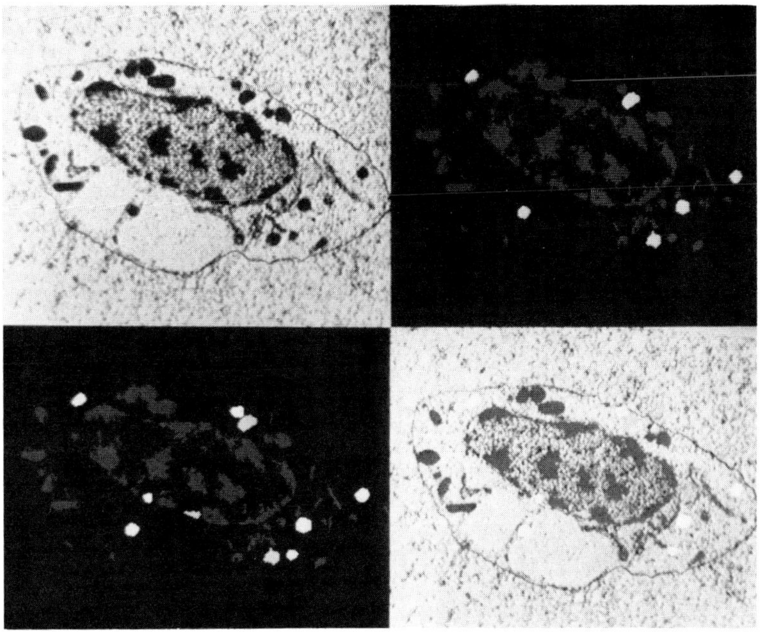

FIGURE 14. Granules and mitochondria detected.

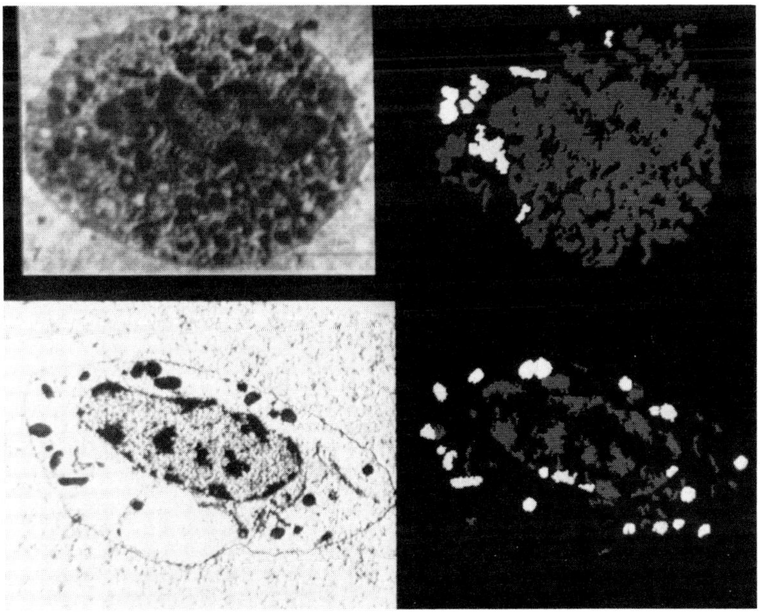

FIGURE 15. Images classified as granulocytes.

E. ANALYSIS AND DISCUSSION

To judge the accuracy of our segmentation, we present in Table 2 a summarized analysis of the experimental results on nine images. In these experiments we were interested in extracting only granules and mitochondria. The segmentation is in error if: objects present in the scene are not identified either because they have been merged with close-by nonobjects

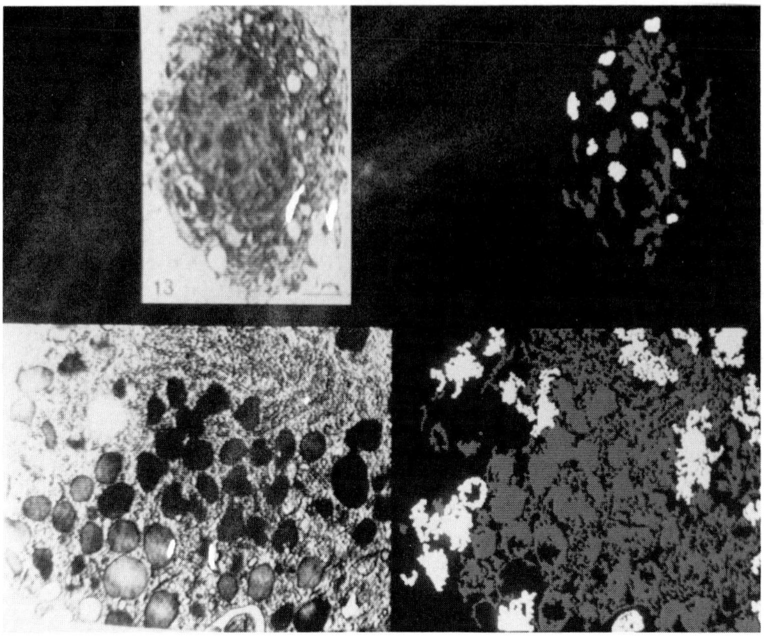

FIGURE 16. Images classified as plasmatocytes and spherulocytes.

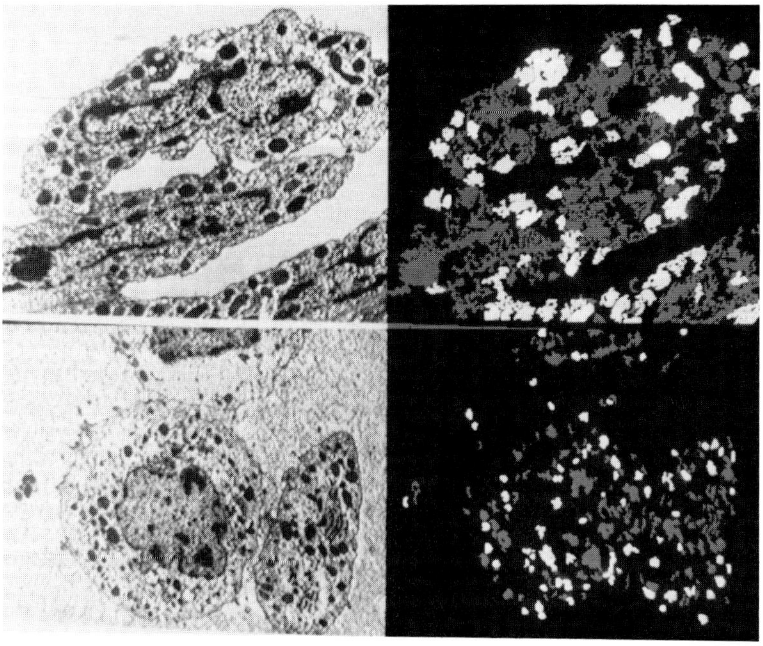

FIGURE 17. Images classified as coagulocytes (= granulocytes).

or because they fell outside the range of close matches (for example a granule that is too close to the nucleus could be merged to nuclear chromatin, or a lightly stained granule would not be dark enough to be considered a close match). Objects are identified but incorrectly labeled (for example identifying a granule and labeling it as mitochondrion) either because two close-by objects (of different types) were incorrectly merged together, or because sim-

TABLE 2
Three-Way Scatter Matrix for Error Analysis

Real	Classification: Granules	Mitochondria	Other
Granules	12.17	0.17	9.67
	(55%)	(1%)	(44%)
Mitochondria	0.00	10.17	6.17
	(0%)	(62%)	(38%)
Other	1.17	1.33	19.83
	(5%)	(6%)	(89%)

ilarities among different objects introduced errors; nonobjects are incorrectly identified and incorrectly labeled (for example, labeling nuclear chromatin, which is not one of the objects we are searching for, as either granule or mitochondrion).

Globally, we have an object extraction and identification rate of about 58% for organelles in insect hemocytes (55% for granules and 62% for mitochondria). In the specimen extraction phase, it is near 100%, as all the specimen choices are almost always perfect. In all our experiments, only one case of wrong specimen choice was reported. The image in this case was very unevenly stained. About 41% of objects present in insect hemocytes were not extracted (44% for granules and 38% for mitochondria). In Table 2, *other* represents nucleoli or dark nuclear chromatin, two structures which can easily be confused with either granules or mitochondria. Of the identified granules 5% were not granules, and about 6% of the mitochondria identified were false positives.

The mislabeling rate was quite low. Only about 1% of the granules were mislabeled as mitochondria. This demonstrates the reliability of our texture measures in texture discrimination. Less than 1% of regions with very little or no texture were misclassified as texture regions. There is no misclassification of textured regions (mitochondria) as nontextured regions (granules). However, such results cannot be guaranteed if the images are noise contaminated or if the serial sections are unevenly stained.

In the matching of other objects in the scene to the selected specimens, various errors can be introduced, for the reasons cited previously. Errors can also be introduced by other discrepancies. In cases of bad iconic data (uneven staining, for example, in Figure 16 [bottom]), the system can fail to produce specimens for a scene containing expected structure. Such bad data can also introduce errors such as identifying parts of a single object as two different objects or identifying only parts of objects. In cases of very noisy images (Figures 15 [top] and 16 [top]), many objects present in the scene can be missed and the probability for mislabeling and identifying nonobjects increases.

VII. SUMMARY AND CONCLUSIONS

One major problem often encountered in building computer systems to analyze and interpret images of natural scenes common in biology and medicine is the difficulty in characterizing objects present in these scenes. The objects often appear in such varied forms that it seems impossible to design a single complete model to represent a given object and all its variations, but it is possible to create less complete models that produce satisfactory results. The main problems to be solved in the model design phase are how to represent deformation and parameterization of shape and textural structure. The system must be able to recognize similar objects in the scene regardless of variations in size, position, or orientation. For example, granules in a TEM serial section come in various sizes, shapes, and orientations, and can be anywhere in the image except the nucleus. The system must also allow for variations in the textural characteristics of the same object. For example, the mitochondria in a TEM serial section do not necessarily have exactly the same texture, gray level, or shape.

The system described in this chapter on image analysis uses *models of expected structure*. These are computer models that represent what the user expects to see in the microscope sections. The objects are characterized by quantitative parameters computed from the image based on qualitative descriptions of these objects that are found in the biological literature. The processes that determine permissible observed variations in these parameters are described in Dunn et al.[29] In Dunn and Shemlon,[30] there is an example of how to build these models for objects other than arthropod hemocyte organelle.

The process of model-driven image understanding encompasses three phases. First, it is necessary to pre-process the images to bring out salient features like edges, fine texture, and other cues that are necessary for the later processes. The second phase includes building quantitative object descriptions from qualitative object models and models of the image formation process, and the use of these object descriptions in model-guided segmentation. The final phase is object extraction, which constitutes object specimen selection and matching.

The expected presentation of objects in an image can be learned for a given class of images that are captured by any image formation process in any given domain. Such learning is useful for further optimizing the segmentation process and perfecting the object recognition processes.

Microscopy presents unique opportunities for model-driven image analysis. First, the human interpretation is done using two-dimensional information. Second, images of a sample can be gathered from more than one imaging system (such as light and electron microscopy or electron and scanning confocal microscopy), illustrating the need for explicitly separating the domain and image formation models. Third, the image analysis must ultimately produce accurate and reproducible quantitative information.

We have presented in this study a practical implementation of a proposed segmentation scheme to be part of an image interpretation system. This system is general enough for a wide variety of microscope section image analysis problems and applications in other areas of image analysis. The system uses a judicious combination of data- and model-driven algorithms to segment a serial section image into regions that most closely approximate objects that are likely to be present in these images. Data-driven segmentation algorithms are applied at the earlier stages of the process where only the intrinsic information (such as intensity, edges, or boundaries) present in the images is used. It should, however, be remarked here that the merging criteria at the earlier stages could also be model driven with some domain-dependent information like shape, linearity, position, neighborhood characteristics, and texture being used, but this will defeat the purpose of domain-independent region growing and limit the use of the algorithm to only a specified class of images.

We proposed a systematic procedure for going from iconic information to a symbolic description. We introduced the symbolic representation that can be used for this description and then proceeded to use this in model-driven region growing to obtain a good segmentation of the images being analyzed. After segmentation, we proceeded to model matching and interpretation. In Section VI, we present an example that explains the interface between the image features and the corresponding symbolic object descriptions. Initially, the symbolic description models are used to generate the model-driven segmentation, but later, they are used to interpret the contents of the images.

In conclusion, our error analysis demonstrates the reliability of this approach and we therefore propose this segmentation paradigm as a better model for consideration. Future work will consist of building structural, functional, geometric, and morphological model descriptions for all organelles. Our model descriptions have not been optimized, so the identification is not perfect and the success in classification is quite low, as can be seen from the results. In future work, we hope to optimize both the model descriptions and the implementation of the extraction and labeling algorithms, leading to better classifications.

Some of the crucial problems that still remain to be solved include the modeling of complex objects that can have completely different morphologies, depending on how the data are registered. A good example is the golgi complex. In TEM images of insect hemocyte cross sections, this structure presents itself like microtubules or microfilalae, depending on whether the cell is sliced transversely or longitudinally. A complex structure like the nucleus, which is an amalgamation of many scene objects, is also not easy to model, because its presentation is also data-registration dependent. Depending on where the cell is sliced, the cross section obtained can either be without a nucleus or contain a slice of the nucleus with or without a nucleolus or other nuclear structures. Some of the errors cited in the analysis of the results can be corrected by optimizing the model descriptions. This is, however, a difficult problem because the ideal model is unachievable for such natural structures which are in perpetual evolution. It is very difficult to model all the forms of granules in a cell undergoing degranulation or to completely model the nucleus of a cell that is undergoing cell division. Other organelles in such a cell would also have structures that are distortions of the normal when the cell is undergoing cell division. However, near optimal model descriptions must be developed by experts in the field. With experimentation, we can also optimize the matching algorithms by learning what parameters should be relaxed and which should be prioritized for particular data. This can help reduce the number of objects that are missed, wrongly labeled, or wrongly extracted.

The problem of bad data registration has also not been completely addressed. Uneven staining of microscope sections for a TEM cross section can produce images in which known structure can very easily be mislabeled. Regular methods for normalizing intensity have been proved not to be very satisfactory solutions to this problem. Images obtained using equipment that is not of the resolution of a TEM need to be processed with a different image formation model. The resolution and complexity of the object description and image data must correspond for such a system to be guaranteed to give good results. As an example, the descriptions of granules and mitochondria would necessarily change if one uses light microscopy in place of a TEM. The object model remains the same, but as one changes the image formation process the object description must change. This, however, does not require a change in the organization of the segmentation system, since the paradigm is not changed. Our experiments with micrographs, radiographs, and aerial images support this assertion.

In the future, we shall consider the problems of reconstruction and interpretation of 3D structures from a series of well-segmented, consecutive 2D serial sections. From the 3D volume and surface data we can synthesize an arbitrary section containing the objects of interest. The long term goal of this research is to produce a computer-based cell identification and classification system.

ACKNOWLEDGMENTS

The support of NIH Research Resource Grant RR-2483 is gratefully acknowledged. The micrographs used in this paper were produced by Dr. S. S. Han, when he spent his sabbatical leave in A. P. Gupta's laboratory, and we are very grateful for his permission to use them. The help of Mr. Stephen Shemlon in implementing the programs is gratefully acknowledged.

REFERENCES

1. **Ballard, D. H. and Brown, C. M.,** *Computer Vision,* Prentice-Hall, Englewood Cliffs, NJ, 1982.
2. **Russ, J. C.,** *Computer-Assisted Microscopy: The Measurement and Analysis of Images,* Plenum Press, New York, 1990.
3. **Marr, D.,** *Vision,* W. H. Freeman, New York, 1982.
4. **Rosenfeld, A. and Kak, A. C.,** *Digital Picture Processing,* Vols. 1 and 2, Academic Press, New York, 1982.
5. **Nevatia, R.,** *Machine Perception,* Prentice-Hall, Englewood Cliffs, NJ, 1982.
6. **Serra, J.,** *Image Analysis and Mathematical Morphology,* Academic Press, New York, 1982.
7. **Pavlidis, T.,** *Algorithms for Graphics and Image Processing,* Computer Science Press, Rockville, 1982.
8. **Reeves, A. P.,** Parallel computer architectures for image processing, *Comp. Vision, Graph. Image Process.,* 25, 68, 1984.
9. **Prewitt, J. M. S. and Mendelsohn, M. L.,** The analysis of cell images, *Ann. N.Y. Acad. Sci.,* 128, 1035, 1966.
10. **Young, I. T.,** Automated Leukocyte Recognition, Ph.D. thesis, Massachusettes Institute of Technology, Cambridge, MA, 1969.
11. **Bacus, J. W.,** An Automated Classification of the Peripheral Blood Leukocytes by Means of Digital Image Processing, Ph.D. thesis, University of Illinois Medical Center, Chicago, 1971.
12. **Bacus, J. W.,** A whitening transformation for two-color blood cell images, *Pattern Recog.,* 8, 53, 1976.
13. **Bacus, J. W., Belanger, M. G., Aggarwal, R. K., and Trobaugh, F. E.,** Image processing for automated erythrocyte classification, *J. Histochem. Cytochem.,* 24, 195, 1976.
14. **Yasnoff, W. A., Mui, J. K., and Bacus, J. W.,** Error measures for scene segmentation, *Pattern Recog.,* 9, 217, 1977.
15. **Yasnoff, W. A., Galbraith, W., and Bacus, J. W.,** Error measures for objective assessment of scene segmentation algorithms, *Analy. Quant. Cytol. J.,* 1, 107, 1979.
16. **Mui, J. K., Fu, K. S., and Bacus, J. W.,** Feature selection in automated classification of blood cell neutrophils, in *Proc. IEEE Pattern Recog. Image Process.,* IEEE, 1978, 486.
17. **Bacus, J. W.,** The development of automated differential systems, in *Differential Leucocyte Counting,* 1, College of American Pathologists, 1979, 95.
18. **Trobaugh, F. E. and Bacus, J. W.,** Design and performance of the *LARC* automated leucocyte classifier, in *Differential Leucocyte Counting,* 2, College of American Pathologists, 1979, 119.
19. **Bacus, J. W., Watt, S., and Trobaugh, F. E.,** Clinical evaluation of a new electrical impedance instrument for counting platelets in whole blood, *J. Clin. Pathol.,* 73, 655, 1980.
20. **Horn, B. K. P.,** *Robot Vision,* MIT Press, Cambridge, 1986.
21. **Liang, T.,** Geometry Guided Image Segmentation, Master's thesis, Rutgers University, New Brunswick, NJ, 1989.
22. **Gupta, A. P.,** Arthropod immunocytes: identification, structure, functions, and analogies to the functions of vertebrate B- and T-lymphocytes, in *Hemocytic and Humoral Immunity in Arthropods,* Gupta, A. P., Ed., John Wiley & Sons, New York, 1986, 3.
23. **Gupta, A. P.,** Arthropod hemocytes and phylogeny, in *Arthropod Phylogeny,* Gupta, A. P., Eds., Van Nostrand Reinhold, New York, 1979, 669.
24. **Han, S. S. and Gupta, A. P.,** Arthropod immune system. V. Activated immunocytes (granulocytes) of the German cockroach, *Blattella germanica* (L.) *(Dictyoptera: Blattellidae)* show increased number of microtubules and nuclear pores during immune reaction to foreign tissue, *Cell Struct. Funct.,* 13, 333, 1988.
25. **Dunn, S. M. and Gupta, A. P.,** Computer imaging of arthropod immunocytes, presented at 18th International Congress of Entomology, Vancouver, July 1 to 4, 1988.
26. **Cohen, D. H. and Sherman, S. M.,** The nervous system, in *Physiology,* Berne, R. M. and Levy, M. N., Eds., C. V. Mosby, St. Louis, 1983, 5—21 and 69—356.
27. **Hubel, D. H. and Wiesel, T. N.,** Functional architecture of macaque monkey visual cortex, *Proc. Philos. Trans. R. Soc.,* 198, 1, 1977.
28. **Haralick, R. M. and Shapiro, L. G.,** Survey: image segmentation techniques, *Comp. Vision Graph. Image Process.,* 29, 100, 1985.
29. **Dunn, S. M., Shemlon, S., and Gupta, A. P.,** Interpreting transmission electron micrographs of insect hemocytes, *J. Comp. Assist. Micros.,* 2, 1991, in press.
30. **Dunn, S. M. and Shemlon, S.,** Expectation driven image segmentation, *Visual Commun. Image Represent.,* 2, 1991 (in press).
31. **Barry, M., Cyrluk, D., Kapur, D., Mundy, J., and Nguyen, V.,** A multi-level geometric reasoning system for vision, *Artif. Intell.,* 37, 291, 1988.
32. **Suetens, P. and Oosterlinck, A.,** Using expert systems for image understanding, *Int. J. Pattern Recog. Artif. Intell.,* 1, 237, 1987.

33. **Ejiri, M., Yoda, H., Sakou, H., and Sakamoto, Y.,** Knowledge-directed inspection for complex multilayered patterns, *Machine Vision Appl.*, 2, 155, 1989.
34. **Mitchell, B. T. and Gillies, A. M.,** A model-based computer vision system for recognizing handwritten zip codes, *Machine Vision Appl.*, 2, 231, 1989.
35. **Suetens, P., Smets, C., Van de Werf, F., and Oosterlinck, A.,** Recognition of the coronary blood vessels on angiograms using hierarchical model-based iconic search, *Proc. IEEE Comp. Soc. Confer. Comp. Vision Pattern Recog. (CVPR),* IEEE, 1989, 576.
36. **Pavlidis, T.,** A vectorizer and feature extractor for document recognition, *Comp. Vision Graph. Image Process.*, 35, 111, 1986.
37. **Reynolds, G., Irwin, N., Hanson, A., and Riseman, E.,** Heirarchical knowledge-directed object extraction using a combined region and line representation, *Proc. IEEE Comp. Soc. Confer. Comp. Vision Pattern Recog. (CVPR),* IEEE, 1984, 238.
38. **Kapouleas, I. and Kulikowski, C. A.,** A model based system for the interpretation of MR human brain scans, *Proc. SPIE Med. Imaging II Confer.*, Newport Beach, 1988.
39. **Gupta, A. P.,** Insect hemocytes: classification and immunologic functions, in *Recent Advances in Insect Physiology and Toxicology,* Gujar, G. T., Ed., Agricole Publishing Academy, New Delhi, 1991, in press.

Chapter 6

PHAGOCYTOSIS IN CRUSTACEA: THE ROLE OF OPSONINS

Peter L. Ey

TABLE OF CONTENTS

I. INTRODUCTION

The invertebrates comprise a highly diverse group of animals that are able to eliminate pathogens and maintain self-integrity through the operation of an efficient phagocytic system.[1] These animals cannot produce the highly specific antibodies made by vertebrates; and the various means by which phagocytic cells from different invertebrates recognize foreign antigen have yet to be properly elucidated.[2-5]

Samples of hemolymph from different invertebrate species have been shown in many cases, but not all, to contain soluble factors (opsonins), which bind to foreign particles (e.g., bacteria, yeast, erythrocytes) and promote the ingestion of the particles by phagocytic cells.[6-12] The nature and identity of the opsonic factors remain unknown, except in the case of the vineyard snail, *Helix pomatia,* in which the clearance of particles from the circulation has been shown to be promoted by an *N*-acetylgalactosamine-specific (anti-A) agglutinin that is present at low concentration in the hemolymph,[13-15] and which probably originates from the albumin gland.[16] Because opsonins may have evolved independently within the different taxa of invertebrates, one should not be surprised if significant structural and binding-site differences are found to occur between the molecules isolated from, e.g., mollusks and arthropods, or even from animals grouped within these divisions.

Agglutinins also have been detected in the body fluids of many invertebrates,[17-21] and it has been widely speculated that some of these molecules, many of them sugar specific, may function as opsonins, as seems to be the case in *H. pomatia.*[22] This view is strengthened by the detection of lectin-like molecules, antigenically-related to hemolymph-derived agglutinins, on the surface of hemocytes from the cockroach *Leucophaea maderae,*[23-25] the snail *Lymnaea stagnalis,*[26] and the oyster *Crassostrea virginica.*[27] However, any functional or structural links that may exist between serum agglutinins and opsonic factors will only be convincingly demonstrated by purifying the molecules responsible for each activity and rigorously comparing their biological and chemical properties. So far, only a few invertebrate agglutinins have been purified and structurally characterized in any detail.[3,17,18,20] Moreover, with the exception of the *H. pomatia* anti-A lectin, it remains to be shown whether these purified agglutinins do indeed function as opsonins *within* the animals from which they are derived.

Crustaceans, now grouped within a separate phylum,[28] include some of the more familiar arthropods, such as crabs, crayfish, lobsters, and shrimps. These animals have been utilized extensively for research on invertebrate immunity and self/nonself discrimination,[4] their commercial importance perhaps underpinning the intellectual interest in elucidating the means by which they resist infection. Although only a few crustacean species have been examined for phagocytic recognition or the presence of agglutinins in the hemolymph, these studies have contributed significantly to our understanding of invertebrate immunity generally.

II. EVIDENCE FOR OPSONINS IN CRUSTACEANS

As has been observed in other invertebrates, the distinction of self from nonself and the recognition of damaged or modified self components in crustaceans appears to be mediated, at least in part, by a system of humoral and cell surface-associated factors, which are functionally analogous to the immunoglobulin and complement molecules of vertebrates (see also Chapter 14). The existence of these recognition factors has been examined *in vivo* as well as *in vitro* by measuring the uptake and removal of foreign particles by hemocytes or other cells and evaluating the role of depletable humoral factors in this process.

A. *IN VIVO* CLEARANCE STUDIES

One of the most direct means of assessing the capacity of an animal to distinguish self from nonself is to study the clearance of labeled probes introduced into the circulation.

TABLE 1
Studies on *In Vivo* Clearance of Foreign Material in Crustaceans and Phagocytosis by Isolated Hemocytes *In Vitro*

Animal	Species	Type of study	Parameter studied	Probe used	Opsonins detected	Ref.
Crayfish	*Cambarus virilis*	*In vivo*	Clearance	Phage, albumin	nt[a]	29
	Cherax destructor	*In vivo*	Clearance	Carbon	nt	30
	Cherax destructor	*In vitro*	Phagocytosis	Rbc[a]	+	37,38
	Cherax destructor	*In vivo*	Clearance	Bacteria	+	31
	Cherax destructor	*In vivo*	Clearance	Bacteria	+[b]	32
	Cherax destructor	*In vitro*	Phagocytosis	Bacteria	+[b]	9
	Procambarus clarkii	*In vivo*	Clearance	Proteins	nt	33
	Orconectes virilis	*In vivo*	Clearance	Latex	nt	34
Lobster	*Homarus americanus*	*In vitro*	Phagocytosis	Rbc	+[b]	41,42
	Homarus americanus	*In vitro*	Phagocytosis, activation	Rbc	+	43,44
Crab	*Callinectes sapidus*	*In vivo*	Clearance	Viruses, proteins	nt	35
	Carcinus maenas	*In vivo*	Clearance	Bacteria	nt	4

[a] Rbc, red blood cells; nt, not tested.
[b] Opsonins detected in hemocyte lysates.

Differences in the rates of clearance between one probe and another, can provide information on the extent to which a given probe is recognized as "foreign"; accumulation in particular organs or cells may reveal how the foreign material is removed and processed; and injection of a large amount of unlabeled probe prior to injection of the labeled test dose, allows one to investigate the role of humoral opsonic factors in the removal process. Where cells or macroscopic particles are used as probes, however, aggregation (nonspecific or, e.g., agglutinin-mediated) may lead to rapid elimination without necessarily involving attachment *per se* to fixed cells. The active involvement of humoral agents and their ability to mediate particle attachment to isolated hemocytes or epithelial cells, is therefore a requisite for concluding that opsonins play any role in the clearance of injected material.

Several parameters may be identified as being important in determining how rapidly injected foreign matter is cleared from the circulation, and in which organs the material accumulates: (1) the site of entry, (2) the chemical nature, particle size, and dose of the introduced material, (3) the presence/absence of opsonins that can bind this material, (4) the activity of circulating phagocytes, and (5) the location and number of fixed phagocytic cells.

In vivo clearance studies have been carried out on crustaceans, as listed in Table 1. Similar work has been done on mollusks, in considerable detail in *Helix* spp.[3,13-15] The crustaceans studied include the crayfish *Cambarus virilis,*[29] *Cherax destructor,*[30-32] *Procambarus clarkii,*[33] and *Orconectes virilis,*[34] and the crabs *Callinectes sapidus*[35] and *Carcinus maenas.*[4,36]

1. *Cambarus virilis*

Teague and Friou[29] monitored the fate of ^{131}I-labeled human serum albumin and bacteriophage ϕX174 in *C. virilis*, following intracardiac injection. They observed rapid disappearance of albumin from the crayfish circulation, but only slow removal of phage particles. There was no evidence of accelerated removal of albumin in crayfish that had been previously "immunized" with this protein, indicating that these animals lacked the capacity to mount an anamnestic response. Although the results suggested that the albumin was rapidly degraded, the authors found no localization of radioactivity within any particular organ, and commented that " . . . the rapid disappearance of HSA-I^{131} from crayfish serum, even after the first injection, makes it difficult to exclude the existence of 'natural antibody' before injection.''

2. *Cherax destructor*

The recognition of foreignness by phagocytic cells from the crayfish *Cherax destructor* (previously *Parachaeraps bicarinatus*) was initially examined by Reade.[30] He showed that carbon particles, cleared from the circulation, accumulated in fixed phagocytes predominantly within the hepatopancreas. The process of phagocytosis appeared to be similar to that seen in mammalian macrophages, although the cells had more of the characteristics of mesothelium than of mammalian-type endothelium. Subsequent studies by McKay and Jenkin[31] and Tyson and Jenkin[32] demonstrated that *in vivo* the elimination of bacteria from the circulation of the crayfish was dependent on recognition factors circulating in the hemolymph. The data indicating the importance of circulating opsonic factors in promoting the removal of bacteria, were derived from ''blockading''-type experiments, which had been used successfully in vertebrates. For example, if a vertebrate that has just eliminated a primary dose of bacteria from the circulation is given a second similar dose, the latter is removed more slowly than the primary dose. However, if the bacteria in the second dose are first pretreated (opsonized) with specific antibody, their rate of clearance may be returned to normal or even enhanced, which indicates that the ''blockade'' effected by the primary dose is due to a depletion of circulating antibody. Experiments similar to these were carried out in the crayfish, and it was found that blockade in these animals could be reversed by pretreating the second dose of bacteria with crayfish serum *in vitro*. Prior absorption of the serum used for opsonization of the second dose with the same strain of bacteria prevented the reversal of the blockade. Moreover, if the primary (blockading) dose of bacteria was pretreated with serum prior to injection, thus minimizing the removal of circulating recognition factors, the second dose was cleared at normal rates. These results indicated quite unambiguously that a group of soluble opsonins was present in crayfish hemolymph, and that these molecules bound in a specific manner to injected bacteria, thereby promoting the clearance of the cells from the circulation.

3. *Procambarus clarkii*

The ability of this crayfish to recognize various proteins as foreign and to clear them from its circulation, was investigated by Sloan and colleagues.[33] Their experiments were based on the premise that if the animal possessed recognition molecules, whether free or cellbound, they must be of finite number (and thus saturable) and have some discernible degree of specificity. By following the kinetics of clearance of 5-μg amounts of ^{125}I-labeled proteins injected intracardially, this crayfish was found to be capable of discriminating between two of its own proteins (hemocyanin and another uncharacterized hemolymph protein, neither of which was eliminated to any significant extent from the circulation in the 24-h period after injection), and several foreign proteins (bovine serum albumin [BSA] and human gamma globulin [HGG]), which were cleared with considerable rapidity. BSA was removed faster than HGG, clearance of 75% of the radioactivity requiring 4 and 24 h, respectively. The radioactivity became concentrated mainly in the gills during this period.

The specificity of this discriminatory system was investigated by injecting crayfish with 5 μg of ^{125}I-labeled protein mixed with 5 mg of an unlabeled protein. It was found that the clearance of labeled BSA was retarded by excess BSA and also by other serum albumins (human, rabbit, and horse), but not by HGG, bovine gamma globulin (BGG), or keyhole limpet hemocyanin (KLH). Both HGG and BGG slowed down the initial elimination of HGG, whereas BSA and KLH had no effect. The clearance of labeled KLH was similarly specifically inhibited by KLH, but not by BSA or BGG. These results clearly supported the conclusion that this crayfish possesses at least three groups of naturally occurring recognition molecules, which are specific for albumins, gamma globulins, and hemocyanins, respectively. However, the data provide no clue as to whether any of these recognition molecules are cell associated or occur as opsonins in the hemolymph.

4. *Orconectes virilis*

Merrill, Mongeon, and Fisher[34] investigated the clearance of fluorescent (rhodamine-labeled) latex particles (1.48 $\pm$ 0.15 μm diameter, 3 $\times$ 10^7 particles per dose) injected into either the midventral or pericardial sinus of the freshwater crayfish *O. virilis*. The particles were cleared rapidly from the hemolymph, >97% being removed within 1 h of injection at either site. The fate of the latex particles was determined by removing various organs from crayfish killed at various times after injection, digesting the tissues by hot acid hydrolysis, and then measuring the amount of fluorescent material in each sample. Accumulation occurred predominantly in the heart, antennal glands, gills, and to a lesser extent, the hepatopancreas. The possible involvement of opsonins in the clearance of latex particles was not addressed.

5. *Callinectes sapidus*

The discriminatory capability of the blue crab has been examined by McCumber and Clem.[35] They studied the clearance and subsequent organ distribution of viruses and xenogeneic proteins following injection into the infrabranchial sinus. Blue crab hemocyanin, injected as a control (self) protein, showed no significant clearance, except for an initial period of equilibrium; and hemocyanins from closely related crustaceans exhibited only slight to negligible clearance. In contrast, hemocyanins from more distantly related species (*Limulus* and the keyhole limpet) and other foreign proteins (BSA, BGG) were, with some differences in rate, rapidly removed from the circulation. Interestingly, poliovirus and bacteriophages T2 and T4 were rapidly cleared, whereas phages T3, T7, and ϕII remained at or near initial levels for much longer time periods.

Analysis of various organs 1 h after injection revealed that cleared proteins became concentrated in the gills (as was reported for the crayfish *P. clarkii*[33]) where they were apparently degraded, whereas T2 and T4 phages were concentrated in the hepatopancreas. The basis for this difference in localization is not known. Attempts to enhance the clearance rates of proteins and viruses by injecting crabs with 0.1 to 20 mg amounts of BSA and BGG, or with 10^8 plaque-forming units of T2 phage two weeks prior to challenge, were unsuccessful. Unfortunately, the possible role of opsonins in the clearance of these substances in *C. callinectes* was not investigated, nor was any "blockade"-type experiment (*cf.* Sloan, et al.[33]) attempted to determine whether the clearance involved receptors with a restricted range of specificity.

6. *Carcinus maenas*

The elimination of various bacteria from the circulation of this crab was studied by Smith and Ratcliffe.[36] Clearance was very rapid, up to 70% of an injected dose of 1 $\times$ 10^8 cells being removed within 5 min of injection. Associated with a marked fall in circulating hemocyte numbers, live radiolabeled bacteria were found to concentrate within the first hour in the gills, hepatopancreas, and other organs, principally within clumps of hemocytes.[4] Much of the label remained in the clumps and nodules for at least 18 d, indicating that breakdown was slow. Addressing the question of whether bacterial sequestration within aggregates was a result of phagocytosis, the authors concluded from *in vitro* experiments with hemocytes that ingestion is unlikely to account for the observed rates of clearance, but that " . . . association of bacteria with the hemocyte surface, followed by sequestration within cell clumps, may represent the major route of removal from the circulation in *C. maenas* . . . " (Ratcliffe et al.[4]). The role of humoral factors in clearance was not examined.

B. PHAGOCYTOSIS BY HEMOCYTES *IN VITRO*

1. *Cherax destructor*

McKay and Jenkin[37] found that the uptake and ingestion of vertebrate erythrocytes *in vitro* by *C. destructor* hemocytes was dependent on humoral factors in the hemolymph. These molecules appeared to be phylum specific, insofar as hemolymph from mollusks such as the snail *Helix aspersa* and the mussel *Velesunio ambiguus,* although possessing agglutinins for sheep erythrocytes, would not promote the phagocytosis of these cells by phagocytes from the crayfish, whereas hemolymph from the reef crab *Ozius truncatus* did promote phagocytosis. A rabbit anti-(sheep erythrocyte) serum was also inactive, a finding similar to that reported by Stuart[7] using phagocytic cells from the octopus, *Eledone cirrosa.* There appeared to be some degree of specificity in this reaction, since crayfish hemolymph absorbed with human erythrocytes remained opsonic for sheep erythrocytes.[37] In a subsequent study,[38] it was shown that the phagocytic activity of hemocytes was enhanced if the cells were obtained from crayfish that had been injected with endotoxin or killed cells of a pathogenic *Pseudomonas* species. Phagocytosis still depended upon opsonization of the erythrocytes, however, and injected crayfish showed no increase in the level of agglutinins or opsonins.

Further investigations concerned with the phagocytosis of various strains of bacteria indicated that these cells, unlike erythrocytes, could be recognized and ingested by crayfish hemocytes in the absence of hemolymph.[9] These *in vitro* findings were somewhat contradictory to previous *in vivo* data[31,32] (see Section II.A.2), from which the elimination of bacteria from the circulation of the crayfish had been seen to depend on recognition factors circulating in the hemolymph. The apparent ability of hemocytes from *C. destructor* to recognize bacteria but not erythrocytes *in vitro* in the absence of hemolymph, may be determined by the density of the specific factor(s) associated with the surface of the phagocytes. There is a great difference between the surface area of an erythrocyte and that of a bacterium, and it is possible that the number of recognition molecules present on a hemocyte is insufficient to bind erythrocytes. Experiments in the vertebrates would support this suggestion; Miescher et al.[39] found that the amount of antibody required to facilitate the uptake of erythrocytes by phagocytes is greater than that required for uptake of bacteria.

Additional *in vitro* studies[9] showed that treating crayfish hemocytes with trypsin destroyed their ability to recognize bacteria unless the bacteria had been previously opsonized with hemolymph. These results suggested that recognition factors for bacteria were both free in the hemolymph and associated with the plasma membrane of the phagocytic cells, raising the possibility that a single factor may be involved. Preliminary experiments indicated that the hemolymph and hemocyte factors were similar. Monolayers of hemocytes were treated with trypsin and divided into two groups. One group was incubated with a tissue culture medium supplemented with crayfish serum, while the other group was incubated in medium alone. The cells were then washed *in situ,* following which, unopsonized bacteria were added to the monolayers. Only those cells preincubated with serum were able to ingest the bacteria. It could be argued that incubating the hemocytes in the presence of serum provided a nutritionally richer environment, which enabled the cells to resynthesize their membrane receptors. However, this seems unlikely since hemocytes incubated in serum, which had previously been absorbed with bacteria, did not regain their ability to recognize unopsonized bacteria.

Tyson and Jenkin (unpublished data) partially characterized the hemagglutinins and recognition factors for bacteria in the hemolymph from *C. destructor.* Rabbits were hyperimmunized with sheep erythrocytes that had been pretreated with crayfish serum and subsequently washed. The immunoglobulin fraction of the rabbit antiserum was first passed down a Bio-Gel A50M column containing covalently linked hemocyanin to remove anti-hemocyanin antibodies and finally coupled to agarose for use as an immunoadsorbent. The latter removed not only all the hemagglutinating activity from crayfish serum, but also the opsonic activity for both bacteria and erythrocytes. Earlier work had indicated that there were opsonins displaying some specificity for various strains of bacteria as well as for erythrocytes from different species of vertebrates.[9,40] The adsorption of all these factors to the immunoadsorbent suggested that they were antigenically related. Both the hemagglutinating and the opsonic activity required Ca^{2+} ions. Fractionation of the material eluted in 3 *M* NaSCN from the affinity column on a calibrated Bio-Gel P-150 column, indicated that both the opsonins and the hemagglutinins were molecules with a molecular weight of $\simeq$ 81 kDa.

2. *Homarus americanus*

The capability of hemocytes from the lobster *Homarus americanus* to ingest sheep erythrocytes *in vitro* was examined by Paterson and Stewart.[41] They found that erythrocytes became attached to lobster hemocytes and were phagocytosed. They also showed that lobster hemolymph contained opsonins, which promoted uptake and ingestion. Hemocyte lysates were found to exhibit opsonic activity, pointing to the possibility that these cells may synthesize and secrete opsonins into the hemolymph and/or bear membrane-bound recognition molecules adsorbed from the hemolymph. These findings are consistent with the demonstration of opsonic factors in lysates of hemocytes from *C. destructor*[9,32] and the mollusk *Helix pomatia*.[26]

In subsequent work, Paterson et al.[42] showed that *H. americanus* injected with killed bacteria or endotoxin exhibited increased phagocytic activity of hemocytes towards erythrocytes and bacteria. The activation of *H. americanus* hemocytes has been studied in detail by Goldenberg et al.[43] These investigators demonstrated that hemocytes could be activated for *in vitro* phagocytosis of sheep erythrocytes by incubating the lobster cells with bacterial lipopolysaccharide, or by prolonged adherence to glass coverslips (3 to 6 h). Activation enhanced by 15- to 20-fold the recognition and phagocytosis of erythrocytes by hemocytes in the presence of hemolymph opsonins. The increased activity apparently resulted from both the recruitment of a quiescent precursor population during activation, and an increase in the availability of opsonin binding sites on hemocyte membranes. Following activation, up to 32% of total hemocytes were actively phagocytic. Morphological changes which preceded activation included spreading, formation of filopodia and pseudopodia, granular darkening and dispersion, and vacuolation.[43,44]

These observations with *C. destructor* and *H. americanus* are consistent, in that hemocyte phagocytosis is largely dependent upon opsonization of the target cells, and it can be enhanced significantly by "activation" of the hemocyte population. There are, nonetheless, examples of phagocytosis occurring upon hemocyte-foreign body contact without the mediation of hemolymph or serum-derived humoral factors.[9,31,32,37,40,41,43,45] The phagocytosis observed in the absence of hemolymph is frequently minimal, however, and as has been pointed out by Ratcliffe,[4] some failures to detect opsonins in crustacean sera may be due to losses incurred by the clotting process, and the use of plasma may yield different results (e.g., Schapiro et al.[46]). The role of cell-bound recognition molecules ("receptors") on phagocytes, and the relationship (functional and structural) of these receptors to opsonins, where these can be shown to be present, remains to be properly defined.

III. AGGLUTININS

As has been mentioned in the preceding sections, agglutinins for bacteria, erythrocytes, or other types of cells have been detected in samples of hemolymph from a variety of crustaceans. Many of these molecules are lectins, i.e., they have binding sites specific for oligosaccharide moieties. Although the natural functions of agglutinins within invertebrate

TABLE 2
Studies on Agglutinins in the Hemolymph from Crustaceans

Animal	Species	Particles agglutinated	Agglutinins Charact.[a]	Agglutinins Purif.[a]	Agglutinins Specificity	Ref.
Crab	*Birgus latro*	Rbc	Part.[a]	—	Sugars	60
	Callinectes sapidus	Rbc, bacteria	Part.	Part.	—	56,61
	Callinectes sapidus	Rbc[b]	Yes	—	Sugars	62
	Cancer antennarius	Rbc	Yes	Yes	Sialic acids	63
Crayfish	*Cherax destructor*	Rbc, bacteria	No	—	—	40
	Procambarus clarkii	Rbc, bacteria	Part.	—	—	55,56
Lobster	*Panulirus argus*	Rbc	Yes	Yes	—	57
	Panulirus interruptus	Rbc	Part.	Part.	—	47,48
	Homarus americanus	Rbc[b]	Yes	—	Sugars	50
	Homarus americanus	Rbc[b]	Yes	Yes	Sugars	51,52
	Homarus americanus	Rbc	Yes	Yes	Sugars	53
	Homarus americanus	Rbc	Yes	Yes	Sialic acids	54
Prawn	*Macrobranchium rosenbergii*	Rbc	Part.	—	Sugars	58

[a] Charact. = characterization; purif. = purification; part. = partial.
[b] Agglutinins detected in hemocyte lysates.

organisms have yet to be elucidated and are likely to differ substantially between groups of animals, there is clearly some basis for considering their role as opsonins or "recognition molecules". For this reason, it seems pertinent to discuss briefly the information available on agglutinins from crustaceans. This topic has been covered by a number of reviews[3,17,18,20,21,25] to which the reader is referred for a comprehensive discussion.

A. GENERAL

The studies of Tyler and associates[47,48] on agglutinins from *Birgus latro* and *Panulirus interruptus* extended the much earlier work of Cantacuzène.[49] Their results indicated that *P. interruptus* hemolymph may contain as many as 8 to 10 different agglutinins. The number of published studies on crustacean agglutinins is not large; most of these are summarized in Table 2. Only a few species have been examined, with attention being concentrated largely on the lobster *H. americanus*.

The agglutinins from *H. americanus* were partially characterized by Cornick and Stewart.[50] They tested the capacity of various sugars to inhibit the agglutinating activity of lobster serum for vertebrate erythrocytes, and concluded that the serum contained several agglutinins of differing specificity. They also found that the hemocytes were a rich source of hemagglutinins, a point relevant to the detection of opsonins in *H. americanus* hemocytes by Paterson and Stewart.[41]

The physicochemical characteristics of the *H. americanus* agglutinins have been investigated in detail. Hall and Rowlands[51] found that the yield of agglutinating activity for human and mouse erythrocytes was highest from both hemolymph and hemocyte lysates, the latter being the richer source on a total protein basis. The hemagglutinins, partially purified by precipation with ammonium sulfate in the presence of $CaCl_2$, were redissolved and subjected to chromatography on Sepharose 4B. This yielded two peaks of activity, the first (designated L Ag-1) agglutinating human erythrocytes and the second (L Ag-2) agglutinating mouse erythrocytes. L Ag-1 reacted with human, mouse, horse, hamster, chicken, and sheep erythrocytes, while L Ag-2 reacted with only mouse, horse, and hamster erythrocytes. L Ag-1 was much larger than L Ag-2 ($s_{20,w}$ ~ 11 S) and the two agglutinins were shown to be antigenically distinct. Human erythrocytes, sensitized with L Ag-1, bound to lobster hemocytes, and there was some indication of phagocytosis though this did not appear to be striking.[52]

Hall and Rowlands[52] examined the binding specificities of the hemagglutinins present in the L Ag-1 and L Ag-2 fractions. A wide variety of cell (erythrocytes, lymphocytes, and a species of *Achromobacter*) were agglutinated by these preparations. In preliminary experiments using hemolymph, it was found that the agglutination of human erythrocytes could be inhibited by several simple sugars, the most potent being *N*-acetylneuraminic acid (AcNeu). The agglutination of mouse erythrocytes by purified L Ag-2 was inhibited by *N*-acetylgalactosamine and, to a lesser extent, by *N*-acetylmannosamine or *N*-acetylglucosamine. In contrast, the agglutination of mouse erythrocytes by purified L Ag-1 was not inhibited by a variety of simple sugars, nor by a combination of the aforementioned acetylated sugars.

Further work on the agglutinins from *H. americanus* has been carried out by Hartman et al.[53] and Abel et al.[54] These studies confirmed the occurrence of at least two distinct agglutinins and demonstrated that L Ag-1 is a sialic acid-binding lectin.

Other studies (Table 2) have demonstrated the occurrence of agglutinins in the crayfish *Procambarus clarkii*[55,56] and *Cherax destructor,*[37] the spiny lobster *Panulirus argus,*[57] the prawn *Macrobranchium rosenbergii,*[58] and the crabs *Birgus latro,*[56,60] *Callinectes sapidus,*[56,61,62] and *Cancer antennarius.*[63] Most of these studies were directed at determining whether one or several agglutinins were present in hemolymph, binding specificities being examined in some cases by sugar-inhibition tests. Few crustacean agglutinins have been properly purified, those published being restricted to the *P. argus* agglutinin, the L Ag-1 agglutinin of *H. americanus*, and a sialic acid-binding agglutinin from *C. antennarius,* which was purified to homogeneity by affinity chromatography using bovine submaxillary mucin.[63]

B. SPECIFICITY

In several of the studies mentioned above, there is evidence that multiple agglutinins occur within serum samples from individual crustaceans.[51,52,60,62] These conclusions are based on experiments demonstrating differential agglutination of erythrocytes, cross-absorption studies, and sugar-inhibition profiles. Agglutinins inhibitable by *N*-amino-sugars appear to be common. Sialic acid-binding agglutinins have been demonstrated in *H. americanus,*[54] *B. latro,*[60] *M. rosenbergii,*[58] *C. sapidus,*[64] and *C. antennarius.*[63] Agglutinins specific for sialic acids are common in other arthropods, those in horseshoe crabs (Chelicerata) being well-characterized.[17,18] Because arthropods lack sialic acids,[65] the apparent common occurrence of sialic acid-binding agglutinins in both Chelicerata and Crustacea species has further stimulated conjecture that the function of these proteins is to bind bacteria and other foreign organisms that bear exposed sialic acid-related residues. There is clearly much experimental work to be done before the validity of this hypothesis can be properly assessed.

C. HEMOCYTE-ASSOCIATED LECTINS AND AGGLUTININS

Agglutinins have been detected in lysates of hemocytes from several crustacea: *H. americanus,*[50,51] *C. destructor,*[9] the shrimp *Squilla mantis,*[66] and *C. sapidus.*[62] The agglutinating activity of the hemocyte lysate is associated with the membrane (microsomal) fraction of the cells in *C. sapidus,* and both serum- and cell-derived agglutinins were inhibited by low concentrations of *N*-amino sugars, suggesting that they are related.[62]

It is apparent that more work needs to be done on these cell-associated lectins, not only with regard to their relationship with serum agglutinins and opsonins, but especially in terms of their biosynthesis, membrane localization (as extracellular receptors?), heterogeneity, and turnover (release/secretion). The data on hemocyte-derived lectins in mollusks[26,27] and insects[24] suggest that hemocytes from quite diverse invertebrate species may synthesize lectins/agglutinins; the outstanding question relating to these molecules is whether or not they function as recognition molecules, either in a cell-bound state or in a soluble form, to mediate self/nonself recognition.

IV. BACTERICIDINS AND NEUTRALIZING FACTORS

There are various reports of bactericidal factors in crustacean sera.[67,68] This subject has been reviewed by Dales.[69] Since there is no evidence to indicate that these factors are opsonic, there seems no reason to give more than a cursory mention of their existence here. It is worth noting, however, that the sera of both *Cherax destructor,* which contains opsonins, and *C. maenas,* which appears to lack opsonins, have been reported to lack bactericidal or bacteriostatic activity.[31,45]

Also of interest is the finding that *Callinectes sapidus* possesses a humoral factor, which neutralizes T2 bacteriophage in relatively specific fashion.[70] The molecule responsible for this neutralization was isolated from crab plasma and shown to be distinct from hemocyanin and the agglutinin for mouse erythrocytes. It remains to be shown whether it is this factor that mediates the clearance of T2 phage *in vivo*[35] (Section II.A.5).

V. SUMMARY

The phagocytic cells of animals display a high degree of selectivity in what they will ingest. In the vertebrates, immunoglobulin molecules (antibodies) and components of the complement system provide this selectivity by opsonizing target particles, thereby mediating attachment to phagocytes via cell-surface Fc receptors or C3 receptors. The data accumulated on invertebrates indicate that these animals cannot make immunoglobulins or respond anamnestically to foreign substances. Nonetheless, it has been found that the phagocytic cells of invertebrates react efficiently to a variety of foreign particles and the recognition frequently involves the prior interaction of the foreign material with factors in the hemolymph.[1,3,69] Self/nonself discrimination by invertebrate phagocytes therefore seems to involve a system functionally analogous to that found in the vertebrates.[71] The possible role of complement in invertebrates warrants further attention.[72,73]

The evidence for opsonins in crustaceans, like that for mollusks, has been derived from a relatively small number of papers dealing with only a few species. Nonetheless, these studies clearly indicate that the ability of phagocytes from different crustaceans to discriminate between self and nonself is determined by factors that, in some species at least, are free in the hemolymph. Two experimental approaches have been used: (1) binding and ingestion by hemocytes *in vitro* of foreign cells before or after sensitization with crustacean hemolymph or serum; and (2) the clearance *in vivo* of injected proteins or particles. The importance of humoral opsonic factors in the recognition/binding process has been examined in both systems by absorption of serum *in vitro* or by competition with unlabeled (or different) probes *in vivo.*

There is strong evidence that "recognition" factors, some of which may be cell-adsorbed opsonins, are exposed on the surface membrane of crustacean hemocytes. There is speculation but as yet no definitive data concerning the site(s) of synthesis of these molecules, whether they exist on hemocytes as integral, membrane-bound receptors or, perhaps after being secreted by hemocytes or other cells, they bind (as opsonins or opsonin-foreign body complexes) to various types of cells in a manner functionally analogous to the binding of vertebrate immunoglobulins to macrophage Fc or C3 receptors. Such binding to phagocytes may occur via the sugar(?)-specific ''recognition'' site(s) of the opsonin or through secondary (''Fc''-type) interactions.

If these issues are to be properly resolved, further rigorous, detailed studies need to be undertaken in a number of different crustacean species. This may yield a perspective, presently unobtainable, of features that may be unique to particular crustaceans or common to arthropods and other invertebrates. The recent discovery[22] that mollusk *(H. pomatia)* hemocytes possess mannose-6-phosphate-specific receptors which recognize cryptic sites on the homologous serum and albumin gland agglutinins (the receptors bind opsonized erythrocytes only if the attached agglutinins are specifically ligand-bound via *N*-acetylgalactosamine residues) should whet our appetite for more clearly elucidating the role of opsonins in crustaceans.

ACKNOWLEDGMENT

I wish to express my gratitude to Dr. Charles Jenkin for stimulating my interest in self/nonself discrimination in invertebrates and for innumerable, enjoyable discussions on this subject over many years. I am a recipient of a Career Research Fellowship from the National Health and Medical Research Council of Australia.

REFERENCES

1. **Bang, F. B.,** Phagocytosis in invertebrates, in *Invertebrate Immunity,* Maramorosch, K. and Shope, R. E., Eds., Academic Press, New York, 1975, 137.
2. **Jenkin, C. R.,** Factors involved in the recognition of foreign material by phagocytic cells from invertebrates, in *Comparative Immunology,* Marchalonis, J. J., Ed., Blackwell, Oxford, 1976, 80.
3. **Ey, P. L. and Jenkin, C. R.,** Molecular basis of self/non-self discrimination in the Invertebrata, in *The Reticuloendothelial System: A Comprehensive Treatise. Phylogeny and Ontogeny,* Vol. 3, Cohen, N. and Sigel, M. M., Eds., Plenum Press, New York, 1982, 321.
4. **Ratcliffe, N. A., White, K. N., Rowley, A. F., and Walters, J. B.,** Cellular defense systems of the Arthropoda, in *The Reticuloendothelial System: A Comprehensive Treatise. Phylogeny and Ontogeny,* Vol. 3, Cohen, N. and Sigel, M. M., Plenum Press, New York, 1982, 167.
5. **Coombe, D. R., Ey, P. L., and Jenkin, C. R.,** Self/non-self recognition in invertebrates, *Q. Rev. Biol.,* 59, 231, 1984.
6. **Tripp, M. R. and Kent, V. E.,** Studies on the oyster cellular immunity, *In Vitro,* 3, 129, 1967.
7. **Stuart, A. E.,** The reticuloendothelial apparatus of the lesser octopus, *Eledone cirrosa, J. Pathol. Bacteriol.,* 96, 401, 1968.
8. **Prowse, R. H. and Tait, N. N.,** In vitro phagocytosis by amoebocytes from the hemolymph of *Helix aspersa* (Müller). I. Evidence for opsonic factor(s) in serum, *Immunology,* 17, 437, 1969.
9. **Tyson, C. J. and Jenkin, C. R.,** Phagocytosis of bacteria *in vitro* by hemocytes from the crayfish *(Parachaeraps bicarinatus,) Aust. J. Exp. Biol. Med. Sci.,* 52, 341, 1974.
10. **Anderson, R. S. and Good, R. A.,** Opsonic involvement in phagocytosis by mollusc hemocytes, *J. Invertebr. Pathol.,* 27, 57, 1976.
11. **Sminia, T., van der Knaap, W. P. W., and Edelenbosch, P.,** The role of serum factors in phagocytosis of foreign particles by blood cells of the freshwater snail *Lymnaea stagnalis, Dev. Comp. Immunol.,* 3, 37, 1979.
12. **Pendland, J. C., Health, M. A., and Boucias, D. G.,** Function of a galactose-binding lectin from *Spodoptera exigua* larval hemolymph: opsonisation of blastospores from entomogeneous hyphomycetes, *J. Insect Physiol.,* 34, 533, 1988.
13. **Renwrantz, L. and Mohr, W.,** Opsonizing effect of serum and albumin gland extracts on the elimination of human erythrocytes from the circulation of *Helix pomatia, J. Invertebr. Pathol.,* 31, 164, 1978.
14. **Harm, H. and Renwrantz, L.,** The inhibition of serum opsonins by a carbohydrate and the opsonizing effect of purified agglutinin on the clearance of nonself particles from the circulation of *Helix pomatia, J. Invertebr. Pathol.,* 36, 64, 1980.

15. **Renwrantz, L., Schäncke, W., Harm, H., Erl, H., Liebsch, H., and Gercken, J.,** Discriminative ability and function of the immunological recognition system of the snail *Helix pomatia, J. Comp. Physiol. B,* 141, 477, 1981.
16. **Mullainadhan, P. and Renwrantz, L.,** Comparative analysis of agglutinins from hemolymph and albumin gland of *Helix pomatia, J. Comp. Physiol.,* 159, 443, 1989.
17. **Vasta, G. R. and Marchalonis, J. J.,** Humoral recognition factors in the Arthropoda. The specificity of Chelicerata serum lectins, *Am. Zool.,* 23, 157, 1983.
18. **Vasta, G. R. and Marchalonis, J. J.,** Invertebrate agglutinins and the evolution of humoral and cellular recognition factors, in *Invertebrate Models. Cell Receptors and Cell Communication,* Greenberg, A. H., Ed., Karger, Basel, 1987, 104.
19. **Coombe, D. R., Ey, P. L., and Jenkin, C. R.,** Ascidian hemagglutinins: incidence in various species, binding specificities and preliminary characterisation of selected agglutinins, *Comp. Biochem. Physiol.,* 77B, 811, 1984.
20. **Stebbins, M. R. and Hapner, K. D.,** Isolation, characterization, and inhibition of arthropod agglutinins, in *Hemocytic and Humoral Immunity in Arthropods,* Gupta, A. P., Ed., John Wiley & Sons, New York, 1986, chap. 18.
21. **Hapner, K. D. and Stebbins, M. R.,** Biochemistry of arthropod agglutinins, in *Hemocytic and Humoral Immunity in Arthropods,* Gupta, A. P., Ed., John Wiley & Sons, New York, 1986, chap. 8.
22. **Richards, E. H. and Renwrantz, L. R.,** Two lectins on the surface of *Helix pomatia* hemocytes: a Ca^{2+}-dependent GalN Ac-specific lectin and a Ca^{2+}-independent, mannose 6-phosphate-specific lectin which recognises activated homologous opsonins, *J. Comp. Physiol.,* 161B, 43, 1991.
23. **Amirante, G. A.,** Production of heteroagglutinins in hemocytes of *Leucophaea maderae L., Experientia,* 32, 526, 1976.
24. **Amirante, G. A. and Mazzalai, F. G.,** Synthesis and localization of hemagglutinins in hemocytes of the cockroach *Leucophaea maderae* L., *Dev. Comp. Immmunol.,* 2, 735, 1978.
25. **Amirante, G. A.,** Agglutinins and lectins of crustacea: their composition, synthesis, and functions, in *Hemocytic and Humoral Immunity in Arthropods,* Gupta, A. P., Ed., John Wiley & Sons, New York, 1986, chap. 13.
26. **Renwrantz, L. and Stahmer, A.,** Opsonizing properties of an isolated hemolymph agglutinin and demonstration of lectin-like recognition molecules at the surface of hemocytes from *Mytilus edulis, J. Comp. Physiol. B,* 149, 535, 1983.
27. **Vasta, G. R., Cheng, T. C., and Marchalonis, J. J.,** A lectin on the hemocyte membrane of the oyster, *(Crassostrea virginica), Cell. Immunol.,* 88, 475, 1984.
28. **Manton, S. M.,** *The Arthropoda: Habits, Functional Morphology, and Evolution,* Oxford University Press, London, 1977.
29. **Teague, P. O. and Friou, G. J.,** Lack of immunological responses by an invertebrate, *Comp. Biochem. Physiol.,* 12, 471, 1964.
30. **Reade, P. C.,** Phagocytosis in invertebrates, *Aust. J. Exp. Med. Biol. Sci.,* 46, 219, 1968.
31. **McKay, D. and Jenkin, C. R.,** Immunity in the invertebrates: the fate and distribution of bacteria in normal and immunised crayfish *(Parachaeraps bicarinatus), Aust. J. Exp. Biol. Med. Sci.,* 48, 599, 1970.
32. **Tyson, C. J. and Jenkin, C. R.,** The importance of opsonic factors in the removal of bacteria from the circulation of the crayfish *(Parachaeraps bicarinatus), Aust. J. Exp. Biol. Med. Sci.,* 51, 609, 1973.
33. **Sloan, B., Yocum, C., and Clem, L. W.,** Recognition of self from non-self in crustaceans, *Nature (London),* 258, 521, 1975.
34. **Merrill, D. P., Mongeon, S. A., and Fisher, S.,** Distribution of fluorescent latex particles following clearance from the hemolymph of the freshwater crayfish, *Orconectes virilis* (Hagen), *J. Comp. Physiol.,* 132, 363, 1979.
35. **McCumber, L. J. and Clem, L. W.,** Recognition of viruses and xenogeneic proteins by the blue crab, *Callinectes sapidus:* I. Clearance and organ concentration, *Dev. Comp. Immunol.,* 1, 5, 1977.
36. **Smith, V. J. and Ratcliffe, N. A.,** Host defence reactions of the shore crab, *Carcinus maenas* (L.): Clearance and distribution of injected test particles, *J. Mar. Biol. Assoc. U.K.,* 60, 89, 1980.
37. **McKay, D. and Jenkin, C. R.,** Immunity in the invertebrates. The role of serum factors in the phagocytosis of erythrocytes by haemocytes of the freshwater crayfish *(Parachaeraps bicarinatus), Aust. J. Exp. Biol. Med. Sci.,* 48, 139, 1970.
38. **McKay, D. and Jenkin, C. R.,** Immunity in the invertebrates. Correlation of the phagocytic activity of haemocytes with resistance to infection in the crayfish *(Parachaeraps bicarinatus), Aust. J. Exp. Biol. Med. Sci.,* 48, 609, 1970.
39. **Miescher, P. A., Spiegelberg, H., and Benacerraf, B.,** Studies on the mechanism of immune phagocytosis of sensitised bacteria and red cells by the reticulo-endothelial system in mice, in *Role du Système Réticulo-Endothélial dans l'Immunité Antibactérienne et Antitumorale,* Halpern, B. N., Ed., CNRS, Paris, 1963, 463.

40. **McKay, D. and Jenkin, C. R.,** Immunity in the invertebrates. II. Adaptive immunity in the crayfish *(Parachaeraps bicarinatus), Immunology,* 17, 127, 1969.
41. **Paterson, W. D. and Stewart, J. E.,** *In vitro* phagocytosis by hemocytes of the American lobster, *Homarus americanus, J. Fish. Res. Board. Can.,* 31, 1051, 1974.
42. **Paterson, W. D., Stewart, J. E., and Zwicker, B. M.,** Phagocytosis as a cellular immune response mechanism in the American lobster, *Homarus americanus, J. Invertebr. Pathol.,* 27, 95, 1976.
43. **Goldenberg, P. Z., Huebner, E., and Greenberg, A. H.,** Activation of lobster hemocytes for phagocytosis, *J. Invertebr. Pathol.,* 43, 77, 1984.
44. **Goldenberg, P. Z., Greenberg, A. H., and Gerrard, J. M.,** Activation of lobster hemocytes: cytoarchitectural aspects, *J. Invertebr. Pathol.,* 47, 143, 1986.
45. **Smith, V. J. and Ratcliffe, N. A.,** Host defence reactions of the shore crab *Carcinus maenas* (L.) *in vitro, J. Mar. Biol. Assoc. U.K.,* 58, 367, 1978.
46. **Schapiro, H. C., Steenbergen, J. F., and Fitzgerald, Z. A.,** Hemocytes and phagocytosis in the American Lobster, *Homarus americanus,* in *Comparative Immunology,* Vol. 3 *(Invertebrate Immune Responses,)* Bulla, L. A., Jr. and Cheng, T. C., Eds., Plenum Press, New York, 1977, 126.
47. **Tyler, A. and Metz, C. B.,** Natural heteroagglutinins in the serum of the spiny lobster, *Panulirus interruptus.* I. Taxonomic range of activities, electrophoretic and immunizing properties, *J. Exp. Zool.,* 100, 387, 1945.
48. **Tyler, A. and Scheer, B. T.,** Natural heteroagglutinins in the serum of the spiny lobster *(Panulirus interruptus).* II. Chemical and antigenic relation to blood proteins, *Biol. Bull.,* 89(3), 193, 1945.
49. **Cantacuzène, J.,** Anticorps normaux et expèrimentaux chez qualques invertebrès marins, *C.R. Soc. Biol.,* 82, 1087, 1919.
50. **Cornick, J. W. and Stewart, J. E.,** Partial characterisation of a natural agglutinin in the hemolymph of the lobster, *Homarus americanus, J. Invertebr. Pathol.,* 21, 255, 1973.
51. **Hall, J. L. and Rowlands, D. T., Jr.,** Heterogeneity of lobster agglutinins. I. Purification and physiochemical characterisation, *Biochemistry,* 13, 821, 1974.
52. **Hall, J. L. and Rowlands, D. T., Jr.,** Heterogeneity of lobster agglutinins. II. Specificity of agglutinin-erythrocyte binding, *Biochemistry,* 13, 828, 1974.
53. **Hartman, A. L., Campbell, P. A., and Abel, C. A.,** An improved method for the isolation of lobster lectins, *Dev. Comp. Immunol.,* 2, 617, 1978.
54. **Abel, C. A., Campbell, P. A., Vanderwall, J., and Hartman, A. L.,** Studies on the structure and carbohydrate binding properties of lobster agglutinin 1 (LAg1), a sialic acid-binding lectin, *Prog. Clin. Biol. Res.,* 157, 103, 1984.
55. **Miller, V. H., Ballback, R. S., Pauley, G. B., and Krassner, S. M.,** A preliminary physicochemical characterization of an agglutinin found in the hemolymph of the crayfish *Procambarus clarkii, J. Invertebr. Pathol.,* 19, 83, 1972.
56. **Pauley, G. B.,** Physicochemical properties of the natural agglutinins of some mollusks and crustaceans, *Ann. N.Y. Acad. Sci.,* 234, 145, 1974.
57. **Acton, R. T., Weinheimer, P. F., and Niedermeier, W.,** The carbohydrate composition of invertebrate hemagglutinin subunits isolated from the lobster *Panulirus argus* and the oyster *Crassostrea virginica, Comp. Biochem. Physiol. B,* 44, 185, 1973.
58. **Vasta, G. R., Warr, G. W., and Marchalonis, J. J.,** Serological characterization of humoral lectins from the freshwater prawn *Macrobrachium rosenbergii, Dev. Comp. Immunol.,* 7, 13, 1983.
59. **Cohen, E.,** Immunologic observations of the agglutinins of the hemolymph of *Limulus polyphemus* and *Birgus latro, Trans. N.Y. Acad. Sci.,* 30, 427, 1968.
60. **Vasta, G. R. and Cohen, E.,** Carbohydrate specificities of *Birgus latro* (coconut crab) serum lectins, *Dev. Comp. Immunol.,* 8, 197, 1984.
61. **Pauley, G. B.,** Comparison of a natural agglutinin in the hemolymph of the blue crab, *Callinectes sapidus,* with agglutinins of other invertebrates, in *Contemporary Topics in Immunology,* Vol. 4, Cooper, E. L., Ed., Plenum Press, New York, 1974, 241.
62. **Cassels, F. J., Marchalonis, J. J., and Vasta, G. R.,** Heterogeneous humoral and hemocyte-associated lectins with *N*-acylaminosugar specificities from the blue crab, *Callinectes sapidus* Rathburn, *Comp. Biochem. Physiol. B,* 85, 23, 1986.
63. **Ravindranath, M. H., Higa, H. H., Cooper, E. L., and Paulson, J. C.,** Purification and characterization of an *O*-acetylsialic acid-specific lectin from a marine crab *Cancer antennarius, J. Biol. Chem.,* 260, 8850, 1985.
64. **Vasta, G. R. and Cassels, F. J.,** *N*-acylaminosugar-specific lectins in the serum and hemocyte microsomal fraction from the blue crab *Callinectes sapidus, Am. Zool.,* 23, 901, 1983.
65. **Warren, L.,** The distribution of sialic acids in nature, *Comp. Biochem. Physiol.,* 10, 153, 1963.
66. **Amirante, G. A. and Basso, V.,** Analytical study of lectins in *Squilla mantis* L. (Crustacea, Stomatopoda) using monoclonal antibodies, *Dev. Comp. Immunol.,* 6, 721, 1984.

67. **Evans, E. E., Painter, B., Evans, M. L., Weinheimer, P. F., and Acton, R. T.,** An induced bactericidin in the spiny lobster, *Panulirus argus, Proc. Soc. Exp. Biol. Med.,* 128, 394, 1968.
68. **Acton, R. T., Weinheimer, P. F., and Evans, E. E.,** A bactericidal system in the lobster *Homarus americanus, J. Invertebr. Pathol.,* 13, 463, 1969.
69. **Dales, R. P.,** Defence of invertebrates against bacterial infection, *J. R. Soc. Med.,* 72, 688, 1979.
70. **McCumber, L. J., Hoffmann, E. M., and Clem, L. W.,** Recognition of viruses and xenogeneic proteins by the blue crab, *Callinectes sapidus:* a humoral receptor for T_2 bacteriophage, *J. Invertebr. Pathol.,* 33, 1, 1979.
71. **Gupta, A. P.,** Arthropod immunocytes: identification, structure, functions, and analogies to the functions of vertebrate B- and T-lymphocytes, in *Hemocytic and Humoral Immunity in Arthropods,* Gupta, A. P., Ed., John Wiley & Sons, New York, 1986.
72. **Day, N. K. B., Gewurz, H., Johannsen, R., Finstad, J., and Good, R. A.,** Complement and complement-like activity in lower vertebrates and invertebrates, *J. Exp. Med.,* 132, 941, 1970.
73. **Bertheussen, K.,** Receptors for complement on echinoid phagocytes. II. Purified human complement mediates echinoid phagocytosis, *Dev. Comp. Immunol.,* 6, 635, 1982.

Chapter 7

PHAGOCYTIC RECOGNITION IN *BOMBYX MORI*

Haruhisa Wago

TABLE OF CONTENTS

I. INTRODUCTION

It is well known that the host defense mechanism in vertebrates consists of nonspecific humoral and cellular factors, and specific cellular and humoral ones. For example, hosts are able to respond against invasion of foreign objects first by nonspecific elements, such as lysozyme, complements, interferon, lysin, transferrin, and then by typical immune reactions of a clonal nature. We also know that macrophages are involved in part or all of these sequential defense processes both as phagocytes and antigen-presenting cells in some ways, whereas T- and B-lymphocytes are engaged in only the later stages of specific host defense reactions (Figure 1). Insects, on the other hand, are capable of surviving largely owing to nonspecific defense elements. They are known to possess several defense factors in the hemolymph,[1] such as antibacterial and antiviral activities, lectin (agglutinin), prophenoloxidase-activating system, and complement-like factors. Phagocytosis, nodule formation, and encapsulation by the hemocytes are also important cellular reactions in insects.

The granulocyte (GR) in the silkworm *Bombyx mori* functions as the main phagocyte, and is responsible for the cellular reactions, such as phagocytosis, nodule formation, and encapsulation.[2-6] Additionally, amebocyte-like plasmatocytes (PLs), which have great adhesive properties, are also involved in the later stage of encapsulation.[4] Hence, the GRs and PLs, which may be regarded as immunocompetent cells, will be assumed to play separate roles in the cellular host defense reactions. Here, five major questions need to be addressed for analyzing the aspects of cellular defense reactions by the GRs and PLs, which have been appropriately referred to as immunocytes.[1,7]

1. How do phagocytosis and encapsulation reactions commence in the body cavity?
2. How does the GR recognize foreign materials?
3. What is the real role of cytoplasmic processes, such as filopodia or lamellipodia in recognition and elimination of foreign substances?
4. Are there humoral factors, which promote the cellular reactions, including attachment and ingestion phases?
5. Do aspects of cellular reactions or of humoral factors change in relation to ontogeny?

This chapter considers these questions largely on the basis of my own experiments in *B. mori* or other lepidopteran species. It will be described that (1) phagocytic recognition by the GRs is closely related to the recognition of self components; (2) only the GRs among hemocytes express lectin receptor on the surface, and also secrete lectin into the hemolymph, particularly at the time of pupation; and (3) the attachment phase of the GRs is completely dependent on the presence of phenoloxidase. Furthermore, it will be discussed that PL adhesion is rapidly provoked even against foreign materials which have no components in their structure that trigger prophenoloxidase-activating system, and there are insect species that supply prophenoloxidase into the hemolymph by rapid oenocytoid lysis, which occurs after invasion of the foreign materials.

II. HEMOCYTE PROFILES IN *BOMBYX MORI*

A. GRANULOCYTES (GRS)

As previously reported,[3,8-11] the GRs are one of the most common hemocytes in *Bombyx mori*. They are oval cells, 4 to 13 μm in diameter, with a central nucleus accounting for 20 to 40% of the cell volume, and contain numerous cytoplasmic inclusions. In the scanning electron microscopic (SEM) study, the cells in the larval silkworms show many filopodia on the cell surface[6] (Figure 2A), although they undergo some morphological changes after pupation: they increase the cell size and lose their filopodia. However, after emergence the

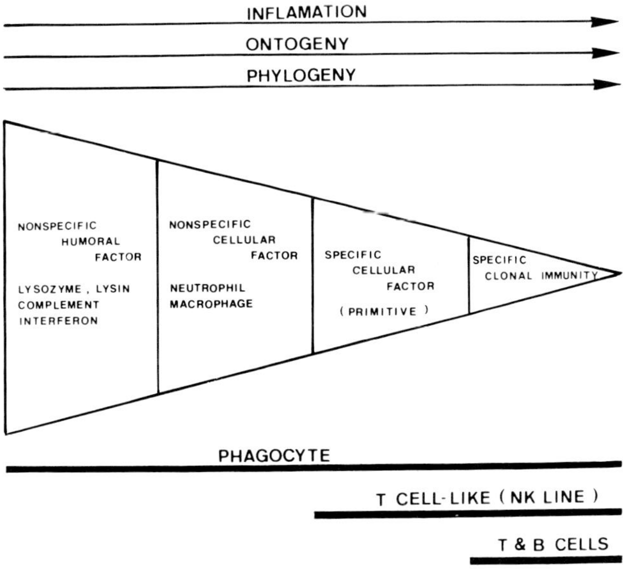

FIGURE 1. A scheme for the host defense mechanisms induced in animals. The sequential steps from nonspecific elements to the specific cellular and humoral immunities are provoked in inflammation and ontogeny, and are likewise reflected phylogenetically. Note that each immunocompetent cell is responsible for the different step in this defense system.

cells possess the same morphology as the larval ones. The filopodia greatly vary in shape, length, or number, and are observed to be folded soon after transferring to a low temperature. Furthermore, the cells are seen to have numerous electron-dense granules in the cytoplasm when examined by transmission electron microscope (TEM) (Figure 2B). A number of cytoplasmic inclusions are supposed to be mucopolysaccharide.[2]

B. PLASMATOCYTES (PLS)

PLs are usually spindle- or leaf-shaped, or oval cells, especially characterized by a central well-stained nucleus occupying about 50 to 70% of the cell volume. PL is also one of the most common hemocytes and rapidly undergoes the extensive morphological transformation. Though mostly spindle-shaped in circulation, they gradually retract their spindle end and round up within 10 sec after withdrawal. Then, after attachment to the substratum, they start to produce filamentous or lamelliform processes, and show random locomotion. Thus, PLs are considered to be the only cells, which are capable of movement,[12-14] just like amoeba or mammalian macrophages. Since they have a combination of filopodia and lamellipodia, they are very adhesive (Figure 3A). Like the GRs, PLs are found throughout the larval to adult stages. However, they drastically change in morphology, cell size at the time of pupation; they increase in size and become rugged, probably due to the increase in the granule size and number.

Though PLs resemble mammalian macrophages in morphology (Figure 3B) and movement, they scarcely show the phagocytic function *in vivo* and *in vitro*. Recent work revealed that like the mammalian macrophages, they use microfilaments and microtubules in spreading behavior.[14] The initial attachment is followed by a rounding-up with a filopod production, which is inhibited by cytochalasin B (CB) (Figure 4A), and the spreading proceeds with the formation of lamellae, which expand within the interfilopodial space. This spreading is also greatly inhibited by microtubule inhibitors, such as colchicine or vinblastine (Figure 4B). This cellular aspect is observed in other lepidopteran species,[12] and CB is reported to produce

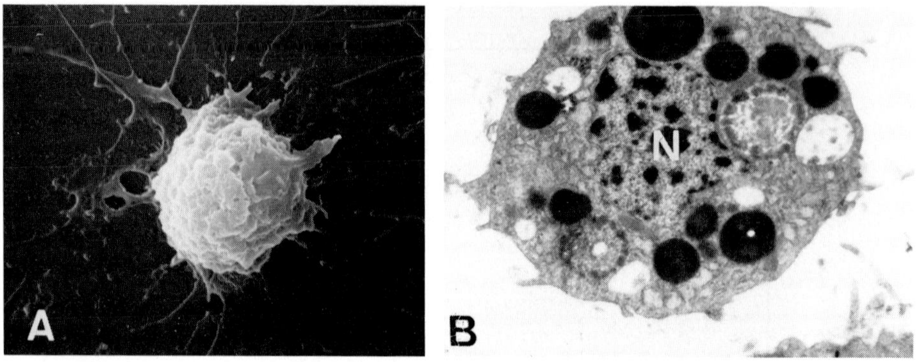

FIGURE 2. (A) SEM micrograph of the larval granulocyte of *Bombyx mori* with many elongated filopodia at the cell periphery (× 4500). (B) TEM micrograph of the larval granulocyte of *B. mori* with numerous electron-dense granules in the cytoplasm and a central nucleus (N) (× 9000).

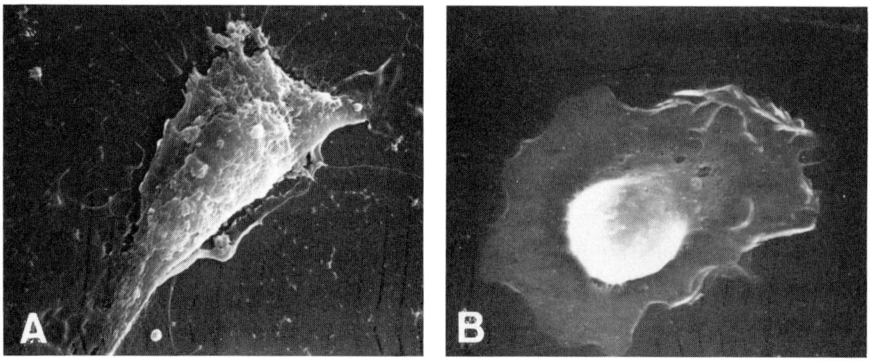

FIGURE 3. (A) SEM micrograph of the larval plasmatocyte of *B. mori* with the lamellipodia and fine filopodia (× 4000). (B) SEM micrograph of rat peritoneal macrophage with the lamellipodia extensively spread (× 3600).

a reversible and dose-dependent effect on the rate of spreading. According to these authors, microfilament-dependent motility is important for the encapsulation process by PLs.[15] Additionally, it is found that *B. mori* PLs, although almost similar to mammalian macrophages in terms of sensitivity to temperature and osmotic shock and in the requirement of divalent cation for attachment, they possess only nonspecific esterase as a lysosomal enzyme (Wago, unpublished). Thus, the presence of a low lysosomal enzyme activity seems to be related to very scarce or absent phagocytic activity.

C. OTHER HEMOCYTES

Although Price and Ratcliffe[16] classified insect hemocytes into six main categories (prohemocytes, plasmatocytes, granulocytes, spherulocytes, coagulocytes, and oenocytoids), *B. mori* has only five types, coagulocytes being absent.[2] These are produced in the hemopoietic organs, located near the imaginal wing discs in the thorax.[8]

Since *in vitro* incubated hemocytes on a glass surface are shown to have their typical surface features,[17] it is easy to identify other hemocytes than the GRs and PLs. Prohemocytes are oval or round in shape and the smallest among hemocytes, with 3 to 7 μm in diameter (Figure 5A). They have a smooth surface, showing little distinct morphology. In addition, they are characterized by a low endoplasmic reticulum and cytoplasmic inclusions.[3] The

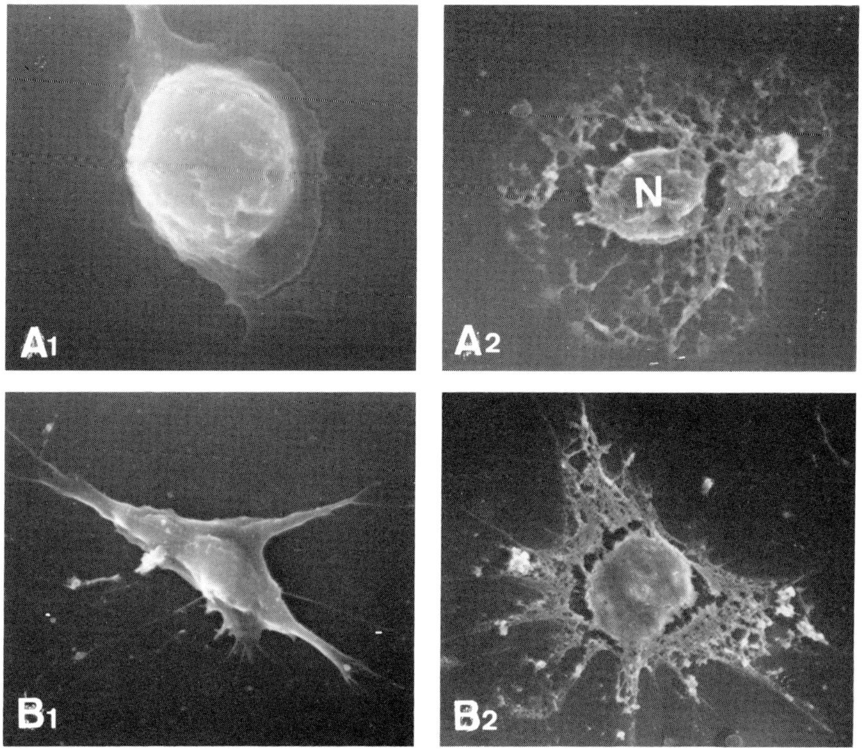

FIGURE 4. (A1) SEM micrograph of the larval plasmatocyte of *B. mori*, which was treated with cytochalasin B. Note that the cell fails to spread or elongate its cytoplasmic processes (× 6000). (A2) SEM micrograph of the larval plasmatocyte of *B. mori*, which was treated with cytochalasin B and then extracted with Triton-X 100. Note that the cell has a central nucleus (N) and fails to spread the cytoskeletal components, which are adhered to the substratum (× 6000). (B1) SEM micrograph of the larval plasmatocyte of *B. mori*, which was treated with colchicine. Note that the cell fails to spread the interfilopodial space and shows the sharp spike-like processes (× 4500). (B2) SEM micrograph of the larval plasmatocyte of *B. mori*, which was treated with colchicine and then extracted with Triton-X 100. Note that microfilament bundles are not inhibited (× 6000).

spherulocytes (SPs) are characterized by several spherules 2 to 4 μm in diameter (Figure 5B). These are known to possess neutral mucopolysaccharides and mucoprotein,[2] and the fine granular materials are released into the hemolymph by the matured SPs.[3] In the adult, SPs become elliptical or oval, 8 to 15 μm in diameter. Their cytoplasm is filled with a number of spherules, which are smaller and more numerous than those in larvae. Finally, oenocytoids (OEs) are the largest and spherical hemocytes with a size varying from 8 to 20 μm in diameter (Figure 5C), and characterized by large cytoplasmic inclusions. As will be seen, OE has an important role in the host defense system, particularly in secreting prophenoloxidase into the plasma.

III. DEVELOPMENTAL CHANGES IN A HOST DEFENSE STRATEGY

Our recent work suggests that in *B. mori* the immune system changes as the larva progresses toward the pupal stage.

A. SWITCHING OF HOST DEFENSE STRATEGY FROM LARVA TO PUPA

It is of great importance to learn how the host defense strategy of insects changes in relation to ontogeny, especially in holometaborous insects. Although the complement-like

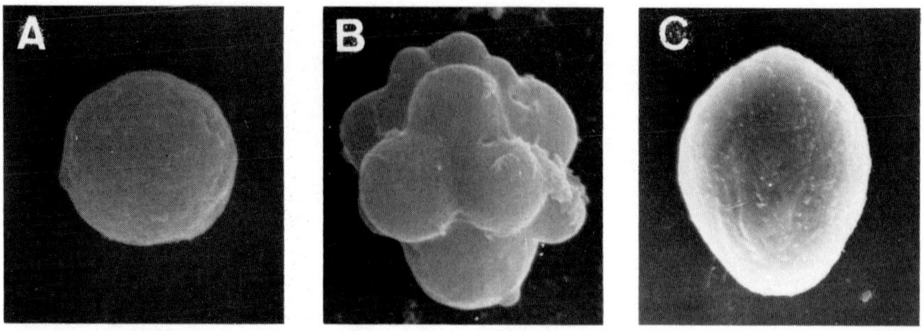

FIGURE 5. (A) SEM micrograph of the larval prohemocyte of *B. mori* with smooth surface (× 6500). (B) SEM micrograph of larval spherulocyte of *B. mori* with nodulose surface (× 8200). (C) SEM micrograph of the larval oenocytoid of *B. mori* (× 4400).

factors,[18] lectin,[19] antibacterial activity,[20] and prophenoloxidase-activating system[21] are present in the larval hemolymph, their quantitative changes during development are unknown. Similarly, though the GRs and PLs, called immunocytes,[7] are important cellular components, particularly involved in phagocytosis and encapsulation,[11] it is important to elucidate the developmental changes in the size and population of these immunocytes as well as the appearance of their surface receptors for lectin, complement-like factors, or other humoral elements. In the German cockroach *Blattella germanica,* there is a gradual increase not only in the total hemocyte count, but also in the differential hemocyte count of the two immunocytes as the insect progresses toward the adult stage.[22] Cell population is also shown to increase during larval development of *Euxoa declaratsa.*[23] These examples suggest that the cellular defense system becomes stronger as the insect develops, because the effectiveness of the cellular reactions depends largely on the number of immunocytes, as indicated by Gupta.[1] This is supported by the situation in *B. mori,* because there is a gradual increase in the total hemocyte count and the available hemolymph volume as larval development progresses.[5] Furthermore, the ability to produce a primary cytoadherence response by the GRs is shown to increase with larval age *in vitro,*[24] and the uptake rate of foreign cells by the GRs increases similarly.[25] As the number of GRs increases with development, it seems likely that the enhancement of hemocytic reactions to foreign cells, such as avian erythrocytes, is due to the increase in the number of these immunocytes. Immunocytes of *B. mori* bind to avian or mammalian erythrocytes *in vitro* even in the absence of hemolymph. Furthermore, the number of these cells binding to erythrocytes and the frequency of erythrocytes bound per immunocyte increase during the larval development,[24] which suggests that there is a progressive increase in the surface receptors of the immunocytes, and the late-instar larvae are more immunocompetent than they were in the earlier stages.

Additionally, my recent work on *B. mori, B. mandarina,* and *Pieris rapae crucivora*[26] has shown that the pupal GRs increase in size (approximately twice as large as the larval ones) and lose their filopodia (Figure 6D), while the number of the total hemocytes after pupation greatly decreases to almost one-third of that in the last larval stage. In this manner, the number and morphology of GRs drastically change in relation to pupation. Interestingly, at this stage, a strong lectin activity appears in the plasma, and lectin receptors are simultaneously produced on the surface of only the GRs. Although age-related occurrence of natural agglutinins was reported in the erisilkworm *Philosamia ricini,*[27] little is known about the accurate developmental characteristics of lectins. The relationship between the lectin activity and metamorphosis has been examined in *Sarcophaga peregrina*[28] and in *B. mori.*[19] In the latter case, the lectin is mostly produced in relation to larval molting and pupation; in general, it is normally present in the larval hemolymph, but declines in the last instar. However, this situation generally differs among *B. mori* strains. In *P. rapae crucivora,* both

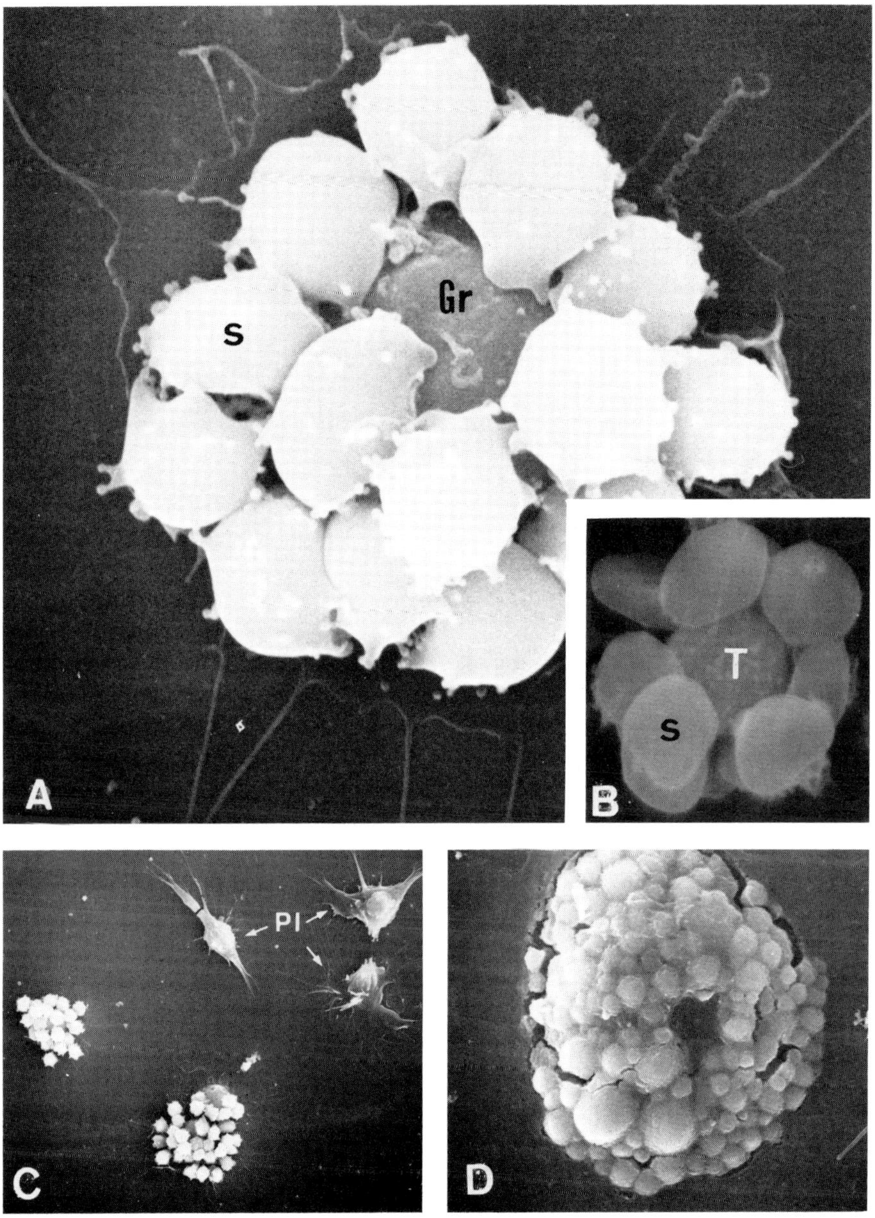

FIGURE 6. (A) SEM micrograph of a granulocyte (Gr) obtained from prepupal stage of *Pieris rapae crucivora*, which adsorbed numerous sheep erythrocytes (S) on its surface. This phenomenon is observed when the lectin activity is present in the hemolymph and the granulocyte expresses the lectin receptors. (B) SEM micrograph of human T-lymphocyte (T) which formed the spontaneous rosette with sheep erythrocytes (S) (× 7000). (C) SEM micrograph of the granulocytes and plasmatocytes (Pl). Note that Pl fails to adsorb the sheep erythrocytes on its surface. This phenomenon is observed when the lectin activity is produced in the hemolymph (× 1100). (D) SEM micrograph of a pupal granulocyte of *B. mori*. Note that the cell and cytoplasmic granule increase the size and lose the filopodia (× 4800).

the lectin activity and the expression of lectin receptors on GRs are directly related to the gut-purge and puparium formation, and found during the pupal stage. Though the change of lectin activity differs from species to species, the lectin production is thought to be closely associated with pupation in the holometaborous insects.[26,28] The presence of lectin in the hemolymph greatly enhanced the binding frequency of sheep erythrocytes to only the GRs

(Figure 6A,C), which resembled the spontaneous rosette formation between T-lymphocytes and sheep erythrocytes (Figure 6B). Interestingly, this opsonic effect is found in a related species, such as *P. melete,* but not in an unrelated one, such as *Papilio xuthus* or *P. machaon* (Wago, unpublished). This implies that the lectin or its receptors are similar in structure; this in turn suggests that only the GRs produce lectin receptors on the cell membrane and that lectin acts virtually as opsonin at a certain stage of development in lepidopteran species, particularly at a stage where the larval tissues disintegrate and adult tissues or organs are being formed. More recently, we found that only the GRs, which were obtained from the late last-instar larvae before pupation in *B. mori* and *P. rapae crucivora,* secreted lectin into the culture medium. This indicated that the plasma lectin is also synthesized by the GRs (Wago, unpublished), which is also the case in the cockroach *Leucophaea maderae.*[29]

In *S. peregrina,* the lectin is induced under two different physiological conditions (on injury of the body wall and on pupation[30]) and the lectin gene is also activated in the embryos.[31] From these studies, it is suggested that lectin participates in removal of foreign substances as well as the larval tissue fragment produced during metamorphosis.

From ligation experiments, we have shown that only the hemolymph obtained from pupated halves contained lectin activity and the GRs found in that part increased their size, lost filopodia, and had lectin receptors on the surface.[32] This implies that these changes are under endocrine control, and may be one of the host defense strategies in Lepidoptera. Moreover, treatment of GRs with trypsin decreases both the spontaneous and lectin-mediated binding frequencies of mammalian erythrocytes. Furthermore, some monosaccharides, such as D-glucose, D-fucose, and glucuronic acid inhibit not only agglutination, but also lectin-mediated binding. In this sense, it will be assumed that the plasma membrane of the phagocytic GR has a trypsin-sensitive structure, which binds to lectin playing a key role as a specific recognition molecule for the GRs to bind to the carbohydrates. In fact, from lectin-binding studies, it is reported that *Blaberus* and *Extatosoma* PLs and spreading GRs in *Extatosoma* possess surface *N*-acetyl-D-glucosamine, D-galactose, and mannose moieties; *Extatosoma* cystocytes (= coagulocytes) have D-galactose and mannose; and *Blaberus* and *Extatosoma* GRs have mannose determinants.[33]

It is possible that lepidopteran species compensate for the decreased number of GRs by greater cell size, and for the disappearance of filopodia on the surface of the possession of lectin receptors. Since the time taken by GRs to phagocytose foreign cells and to eliminate them from circulation is almost the same during larval and pupal stages, the change of the host defense system from larval to pupal stage does not affect the degree of elimination of foreign materials.

IV. ASPECTS OF CELLULAR HOST DEFENSE REACTIONS

A. ESSENTIAL ROLES OF THE GRS AND CELLULAR RECOGNITION

As described in Section II, both the GRs and PLs are considered to be responsible for phagocytosis and encapsulation in many insect species.[34] However, in *B. mori*, only the GRs function as phagocytes and engage in the early phase of encapsulation.[2-6,9] PLs show little or no phagocytic activity and participate only in the later phase of encapsulation. It remains unclear why GRs and/or PLs differ in function from species to species. However, it is certain that the GRs of insects serve as scavenger cells, having the capability to express lectin receptors, synthesize lectin, and store lysozyme, which is an antibacterial factor;[35] they probably also secrete complement-like factors.[1,36] Thus, it is concluded that the GRs are the major immunocytes in insects.

Prior to the study of phagocytic recognition of foreignness, it is necessary to clarify the phagocytic and encapsulation processes shown by the GRs, and to determine the limit of recognition by injecting various biotic or abiotic objects into the body cavity. In order to

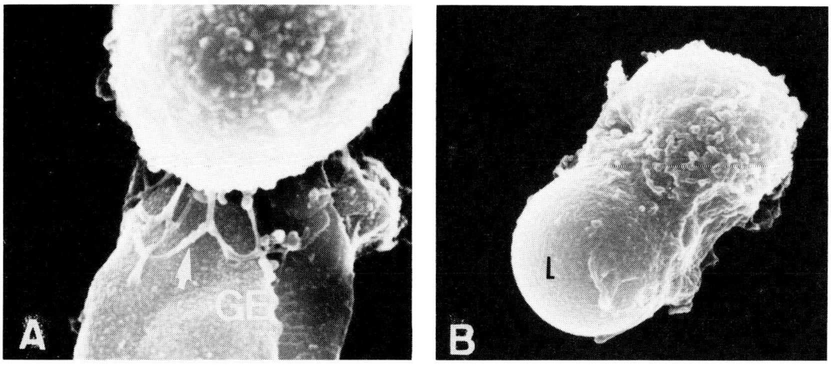

FIGURE 7. (A) SEM micrograph of a larval granulocyte of *B. mori,* which attached to a goose erythrocyte (GE) and elongate filopodia (arrow) over the surface of erythrocyte (× 9600). (B) SEM micrograph of a larval granulocyte of *B. mori,* which extended the veil-like membrane processes to internalize a 5.7 μm latex particle (L) (× 6200). (From Wago, H., *Appl. Ent. Zool.*, 15, 489, 1980; Wago, H., *Cell Immunol.*, 54, 155, 1980. With permission.)

determine the size of foreign materials, which *B. mori* GRs are capable of internalizing, and to obtain much information on the early phases of phagocytosis, various sizes of latex particles, glass beads, and avian and mammalian erythrocytes were introduced into the larval hemocoel. We found that the attached GRs could internalize almost the same size of foreign material as a phagocytic GR, and that the phagocytic process is composed of three phases: attachment, filopodial elongation, and actual internalization by veil-like membrane processes[6,17] (Figure 7). In contrast, phagocytosis by PLs in *Galleria mellonella* occur by pseudopod formation, and the uptake of larger foreign substances is accompanied by invagination of cell membrane.[37] Phagocytosis is usually accomplished in three stages, including immunorecognition, ingestion, and clearance from the body.[34]

The larval GRs in *B. mori* are 4 to 13 μm in diameter. Interestingly, phagocytosis and early encapsulation are concurrently observed when the size of the injected foreign materials is almost the same as that of the phagocyte; for example, 5.7 μm latex particles or goose erythrocytes. In this case, a smaller GR could not take up a latex or goose erythrocyte, and the first phase of encapsulation was seen.[25] According to Kawakami,[38] reproducing fungi are phagocytosed by a single larger GR, whereas the capsule is formed by the smaller ones. Therefore, it is concluded that the size of foreign materials determines the initiation of phagocytosis or the early phase of encapsulation.

Hemocytic encapsulation is an important mechanism to isolate invading large foreign materials. In most cases of encapsulation, the GRs first make contact with the surface of the large foreign materials, and then capsule formation and melanization by the GRs and/ or PLs gradually occurs.[34,39,40] At the time of the initial phase of encapsulation, the discharge of some cytoplasmic materials from the attached GRs (probably glycoprotein and mucopolysaccharide contained in the granules) occurs.[2,41-43] In any case, encapsulation reaction in insects is supposed to have three important immunologic events: (1) recognition of both nonself and its large size by the GRs, (2) cell-cell interactions mediated by some chemotactic factors released from the adhered GRs as a result of discrimination of large size, and (3) the mechanism whereby further hemocyte accumulation ceases after the completion of encapsulation.

In *B. mori*, when abiotic foreign materials, such as large latex particles (25.7 μm in diameter), are injected, hemocyte types other than GRs participate in the first phase of encapsulation, depending on their adhesiveness (Figure 8A), and the hemocytic reaction is irregular and slow. This suggests that the adhesive PLs or other hemocyte types can bind

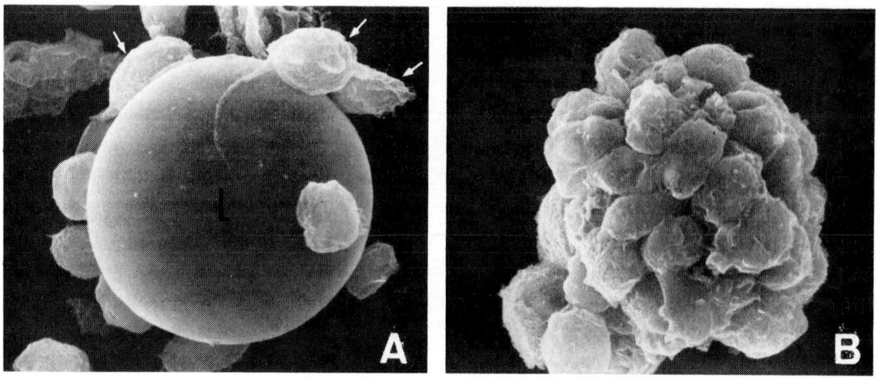

FIGURE 8. (A) SEM micrograph of the early phase of encapsulation by the larval hemocytes of *B. mori* against a 25.7 μm latex particle (L). Note that the attachment of plasmatocytes (arrows) are seen (× 2400). (B) SEM micrograph of the completed phase of encapsulation by the larval hemocytes of *B. mori* against a large pollen (× 1800).

to a large foreign surface by random contact during the first phase, and not in response to some chemotactic stimuli released from the foreign materials, suggesting the important role of the adhesiveness of hemocytes in attachment. Additionally, it was shown that though hemocyte type involved in spontaneous cytoadherence was mainly the GR, both PLs and prohemocytes were also observed to adhere to foreign cells *in vitro*.[44] In contrast, the introduction of a biotic foreign material, such as a large pollen particle about 33 μm in diameter into the body cavity, showed more rapid and regular encapsulation reaction mainly by the GRs (Figure 8B). As will be mentioned, since the foreign materials that trigger the prophenoloxidase-activating system enhance the attachment of only the GRs to them, the production of phenoloxidase by the foreign materials in the plasma or on the foreign surface may determine the hemocytic reaction. As a result, though various types of hemocytes have a chance to adhere to the foreign substances via random contact, depending largely on their adhesiveness and their proportions in the total hemocyte population, the active attachment phase of the GRs to large foreign substances seems to be dependent on the presence of phenoloxidase (Wago, unpublished). This strongly suggests the essential role of phenoloxidase also in the phagocytic recognition by the GRs.

What is the range of discrimination capacity between self and nonself by the GRs in *B. mori*? Generally speaking, the recognition processes in insects have been examined by investigating the fate of implants or cuticular grafts obtained from intraspecific (allogeneic) and interspecific (xenogeneic) sources. That is to say, if phagocytosis or encapsulation does not occur, it is assumed that the surface of transplants has not been recognized as foreign. As demonstrated by many investigators, allografts are not rejected in insects, although the hemocytes respond only to the damaged tissue in the same manner, as the wounded self material.[39,45-48] The same is true in *B. mori*.[49] The allografts, such as brain or subesophageal ganglion, remained unencapsulated, except for the injured sites or the cut end. Hence, it is generally accepted that allograft recognition is absent in insects.

On the other hand, xenografts are known to be rejected and usually become melanized.[50] However, transplants between closely related species are more likely to be accepted than those between distantly separated genera.[51] It is also thought that the phylogenetic relatedness of the recipient and the donor determines the extent of recognition, and that whether intact xenograft tissue is recognized depends on the discriminatory ability of the recipient species itself.[50]

B. ROLE OF CYTOPLASMIC PROCESSES IN FOREIGNNESS RECOGNITION AND IN EXPRESSION OF PHAGOCYTIC FUNCTION

As already mentioned, the initial cellular reactions by *B. mori* larval GRs to foreign objects consist of attachment and filopodial elongation. These two steps are common in both phagocytosis and encapsulation. However, the degree of filopodial elongation is different, depending on the size of foreign materials, which suggests that the total phagocytic signals obtained during filopodial elongation may be important for phagocytic function. In vertebrate phagocytes, the continued interactions between plasma membrane receptors of the phagocytes and foreign particle-bound ligands are required for the repeated generation of intracellular phagocytic signal during the entire ingestion process.[52] From a series of experiments with regard to filopodial function of larval GRs in *B. mori*, it is shown that (1) filopodial elongation is a temperature-sensitive process,[53] and (2) microfilaments are involved in filopodial function.[54] A low temperature highly inhibits the elongation process; however, if the incubation hemolymph contains phenoloxidase, the attachment is not inhibited by a low temperature. Cytochalasin B (CB), which disrupts microfilament function, causes a marked inhibition of filopodial elongation even at an adequate temperature. CB is also known to inhibit both phagocytosis by mammalian macrophages[55-57] and macrophage-lymphocyte interactions.[57,58] Finally, rosette formation with avian and mammalian erythrocytes is inhibited by CB in *Spodoptera eridania*.[59] Thus, microfilament structures have significant roles in a wide range of immunologic phenomena.

Because of the role of filopodial elongation in triggering or regulating cellular reactions, the significance of elongation both for recognizing foreign objects and in the progress of phagocytosis was investigated by introducting elongation-inhibiting systems, and by using both self and nonself materials. As a result, the incubation of GRs on self or intraspecific materials was shown to decrease the filopodial elongation.[60] Moreover, elongation-inhibited cells by CB were able to ingest the foreign cells even when the appropriate temperature was provided. Conversely, cells having more or less the ability to extend their filopodia against the attached erythrocytes could phagocytose them[61] (Figure 9). These results suggest that filopodial elongation is stimulated by nonself recognition, and a certain level of intracellular phagocytic signal obtained during elongation process is required to complete phagocytosis. Though the GRs lose their filopodia at the time of pupation, they can ingest or encapsulate the invading foreign materials. In this sense, a membrane-associated recognition unit on the filopodial or cell surface may be important for obtaining the phagocytic signal during the spreading of the membrane. In any case, if the attached materials should be apparently nonself components, the GRs may start elongating their filopodia in the larval stage or spreading the membrane processes in the pupal stage, the quantity of the phagocytic signal might determine the triggering of cellular reactions, phagocytosis, or the release of some chemotactic factors after their breakdown (early encapsulation response). (For a discussion on the significance of cell junctions and the molecular basis of encapsulation, see Chapter 4.)

The recognition of foreign materials by the GRs of *B. mori* seems to take place directly via cell-cell contact, irrespective of the serum components. However, it is certain that hemolymph factors, such as lectin and prophenoloxidase-activating system, or various receptors for their humoral factors on the immunocytes, are responsible for the cellular reactions to eliminate foreign materials by the GRs.

Recognition of nonself in the simplest way might depend on the physiological differences of nonself surface, such as surface charge and its pattern, or on the degree of hydrophobicity. In fact, it is shown that the charge and wettability of a foreign surface, and also of the immunocyte surface, may play some role in recognition and hemocyte adhesion.[62] In *B. mori*, experiments performed by removing the effect of prophenoloxidase-activating system, using serine protease inhibitor, showed some indication of the role of surface charge of

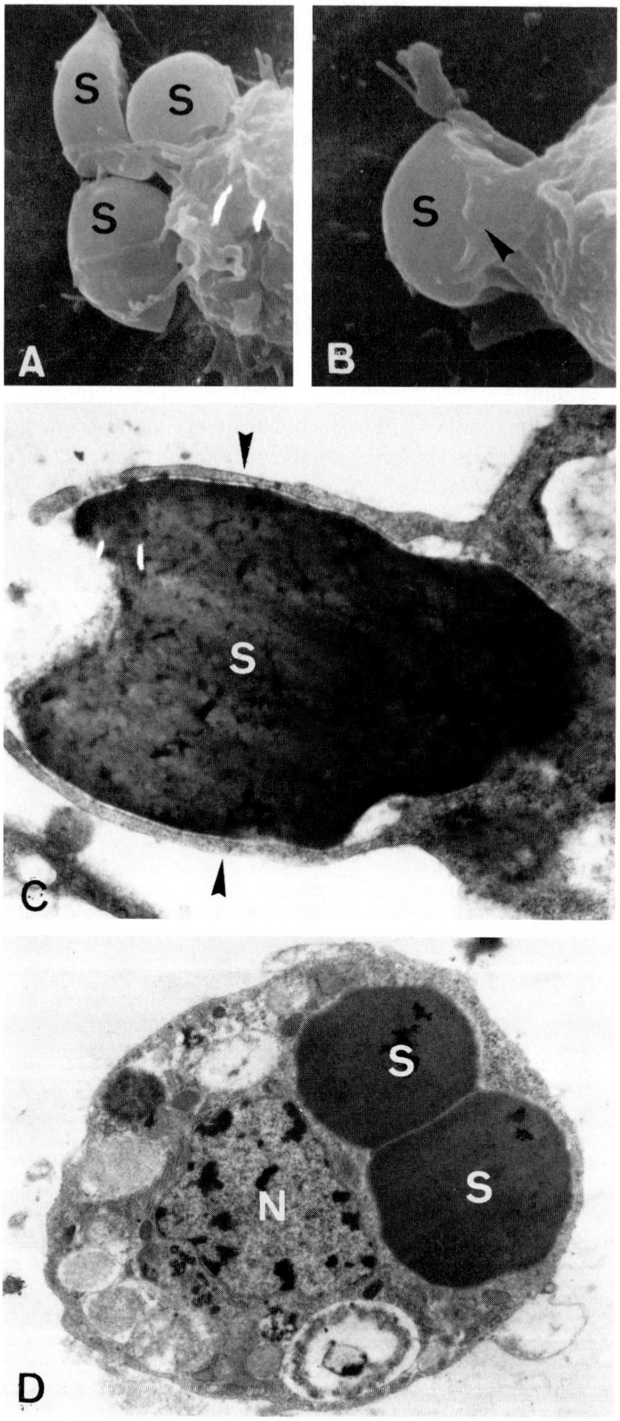

FIGURE 9. (A) SEM micrograph of a granulocyte of *B. mori,* which extended several filopodia against the three attached sheep erythrocytes (S) (× 7,700). (B) SEM micrograph of a granulocyte of *B. mori,* which extended the veil-like membrane processes (arrow) to internalize a sheep erythrocyte (S) (× 8,500). (C) TEM micrograph of a granulocyte of *B. mori,* which extended the fine membrane processes (arrow) to engulf a sheep erythrocyte (× 18,000). (D) TEM micrograph of a granulocyte of *B. mori,* with two internalized erythrocytes (S) in the cytoplasm (× 9,500). (From Wago, H., *The Ultrastructure and Functioning of Insect Cells,* Akai, H., King, R. C., and Morohoshi, S., Eds., The Society for Insect Cells Japan, 1982, 177—180. With permission.)

GRs. In addition, recognition of nonself may be attributable to the absence of certain components on the nonself material, which are commonly present on the self surface (and self-marker), as suggested by Salt.[63] If the phagocytes should possess any unit, which recognizes self-marker (self-recognition unit) on the surface of the immunocyte filopodia or membrane, the binding of its unit to the self-marker would inhibit the phagocytic reactions, including filopodial elongation and the spreading of the membrane processes. In this way, the recognition of forcignness seems to rely on the surface phenomena produced by phagocyte-foreign surface contacts. Since the damaged, abraded, or heat-treated surface of implants from the intraspecific individuals are reported to receive stronger hemocytic reactions,[46,63] these physical treatments must have the destructive effects on self-marker.

According to Salt[63] and Wigglesworth,[64] it is the connective tissue sheath or the basement membrane common to self surface that the phagocytes associate with self. The mucopolysaccharide component on the self is supposed to be the most important in regulating the phagocytic reactions. In fact, in the melanotic tumor mutants of *Drosophila melanogaster,* the encapsulation response of hemocytes is directed against tissue surface, which shows the structural changes in the basement membrane.[65-67] Since a monoclonal antibody specific for *D. melanogaster* basement membrane does not bind to *D. virilis* basement membrane, there are indeed antigenic differences between the basement membranes of these two species, and the hemocytes are considered to be able to recognize the difference between the molecular architectures of *D. melanogaster* and *D. virilis*. For this reason, *D. virilis* is supposed to be rejected or encapsulated by the melanotic tumor larvae of *D. melanogaster*.[68] The results of our experiment to examine the effects of collagenase and lecithinase C on encapsulation of allogeneic brain transplants of *B. mori* larvae support the results of Rizki and associates.[49,68] However, all tissue transplants are shown to be covered in a negatively charged layer of connective tissue with proteoglycans. So, recognition of a transplant by the phagocytes would be considered to be due to some recognition system of self molecular components.[50]

According to Anderson,[69] serum-independent attachment by insect phagocytes to foreign objects is mediated by the interactions with nonspecific surface receptors on the phagocytes. Therefore, it may be assumed that phagocytes more or less possess both self-recognition unit and nonspecific receptors on some part of their membranes. If this interpretation would be accepted, the mechanism of the recognition of foreign materials would be intelligibly accounted for in the following way: if the attachment should occur only via nonspecific receptors by contacting with the nonself substances, this signal might elicit phagocytosis or early encapsulation by the phagocytes. Conversely, if phagocytes should obtain two kinds of signals from nonspecific receptors and self-recognition unit, then neither cellular reaction would take place. Scott[70] reported that the uptake of sheep and chicken erythrocytes onto the surface of cockroach phagocytcs arc abolished by pretreatment of phagocytes with trypsin. Thus, if nonspecific receptors should bc present, they might have a trypsin-sensitive protein structure.

V. PROPHENOLOXIDASE-ACTIVATING SYSTEM (PROPO-AS) AND THE CELLULAR REACTIONS BY GRANULOCYTES

A. SIGNIFICANCE OF PROPO-AS IN HOST DEFENSE

The proPO-AS, which is well known as the enzyme cascade responsible for converting prophenoloxidase(proPO) into active phenoloxidase(PO) during melanization, plays an essential role in the host defense mechanism, including nonself recognition in insects.[71-73] PO promotes melanin formation, and melanin deposition is related to nodule formation and encapsulation by the GRs in *B. mori*. An activating protease is present in the cuticle,[21,74] and brings about the cleavage of hemolymph proPO, resulting in the release of a peptide of molecular weight 5 kDa.[21,75]

In vertebrates, macrophages and polymorphonuclear leukocytes play an important role in the primary host defense reactions against invaders. Mammalian macrophages are also known to possess the specific binding receptors for the Fc fragment to the immunoglobulin (IgG) and for the activated complement component (C3b), whose presence promotes the specific binding to foreign materials coated with IgG or C3b. Therefore, these serum factors (IgG and C3b) act as ligands to mediate the specific attachment. However, Ig has not been found in invertebrates. Instead, lectins are known to serve as opsonins (see Chapter 6) and to mediate the cellular reactions.[76-78] However, in insects, lectin is not known to have an opsonic activity.[70,79] My recent work on lepidopteran species has clearly shown that the GRs obtained from pupae agglutinate mammalian erythrocytes in the presence of lectin.[32] Accordingly, at a certain developmental stage, lectin appears to play an important role as a recognition molecule in the elimination of foreign particles, which have some carbohydrate moiety that lectin can recognize and thus serve as an opsonic factor for the phagocytic GRs.

In addition, even the hemolymph without lectin activity but with phenoloxidase activity, is shown to promote filopodial elongation of *B. mori* GRs, and consequently to increase their adhesive properites;[17] the binding frequency of GRs to foreign cells is also augmented by the hemolymph factor.[44] It is unclear what component in the plasma has these influences, although it is possible that the components of the proPO-AS produce such effects. Because proPO-AS plays a role in nonself recognition in crustaceans and the proPO cascade is reported to contain some factors serving as opsonins for the hemocytes[80,81] (see also Chapter 6), the relationship between proPO-AS and the initial cellular attachment by *B. mori* GRs was studied, using a serine protease inhibitor (nitrophenyl-guanidobenzoate: NPGB) and an oxidase inhibitor (phenylthiourea:PTU). The components of the proPO-AS are known to be stimulated by a serine protease enzyme,[21,82] and the plasma of *B. mori* contains the proPO-AS whose activation is initiated by elicitors, such as glucans with β-1,3-glycosidic linkage (zymosan).[21] Moreover, β-1,3-glucan recognition protein occurs in the plasma, and the interaction of its recognition protein with elicitor is known to trigger the proPO-AS in the silkworm plasma.[83]

Injection of zymosan clearly showed that the attachment of only the GRs was dependent on the formation of PO, and that erythrophagocytosis by the GRs was also enhanced. However, PLs were able to attach and spread extensively even in the presence of NPGB.[14,84] Though NPGB-preinjection suppressed *in vivo* erythrophagocytosis by the GRs, it did not affect the adhesion phase of PLs. In this manner, the attachment of the GRs is dependent at least on the presence of PO in the plasma or on the foreign surface. Conversely, PTU had no inhibitory effect on the attachment of the two immunocytes. However, PLs, incubated with NPGB for over 120 min, rounded up and began to detach with time. GRs could not adhere, at least during a 60-min incubation with NPGB, although the bleeding method greatly affected the adherence of GRs. So, it was possible to separate PLs from GRs by introducing such an incubation system with NPGB. Then, GRs were obtained from other unadhesive hemocytes by incubating the detached hemocytes, following the replacement of the first medium with NPGB-free insect saline. Such a method enabled us to obtain monolayers of PLs and GRs. As a result, PO formation by the two immunocytes and the effects of various kinds of drugs on it were examined. In any case, an opsonic role for the proPO-AS in the hemolymph of *B. mori* is suggested; PO formation after invasion of foreign materials is an essential step for the rapid attachment of the GRs.

B. FORMATION OF PO FROM THE GRS AND POSSIBLE ROLE OF PLS IN HOST DEFENSE

PO activity is located in the GRs of *Locusta migratoria* and *Galleria mellonella*.[85,86] Confirmatory biochemical data for the cellular origin of PO activity was also obtained in *Blaberus craniifer*.[72] Furthermore, the GRs of insects are known to react by extensively discharging their cytoplasmic contents over nonself materials in contact with the foreign surface;[34] some components of proPO-AS are thought to be released from the GRs by exocytosis, leading to coating of the nonself materials.[73] We examined whether the separated GRs or PLs of *B. mori* would produce proPO or PO after adherence to foreign surface on the exterior of the cell, and whether zymosan and its major subcomponents such as β-1,3-glucan and mannan would augment the immediate formation of melanin pigment from the cell. Since the activation of proPO by β-1,3-glucan is reported to increase the hemocytic encapsulation in crustaceans,[87] the effects of β-1,3-glucan on *in vitro* erythrophagocytosis by the GRs was also investigated.

We found that the separated GRs could form melanin after addition of dihydroxyphenylalanin (L-dopa) (Figure 10A). Interestingly, even the presence of NPGB did not completely inhibit the PO activity in the incubation medium of the GR's monolayer; this suggests that PO was not newly produced at the outside of the GRs by splitting proPO by a serine protease. It has been demonstrated that proPO is synthesized by the oenocytoids and secreted into the hemolymph[88] and that the GRs are so labile that the spontaneous release of the proPO occurs during handling or bleeding.[89] In view of these facts and my result that NPGB does not always suppress the PO production, the GRs may be able to sequester the PO, which is converted from proPO during bleeding, at the cell surface or membrane. As recent study on degranulation of GRs revealed that they degranulated much stronger in the presence of PO than in the absence of PO,[90] GRs may sequester the PO during degranulation.

The aggregated GRs showed much stronger melanization response than a single cell, and the neuron-like processes were observed between the masses of aggregated cells (Figure 10B). Moreoever, it was shown by TEM that the melanized GR was surrounded by melanized coagulum, and the discharge of granular materials was related to its formation to some extent (Figure 10C). It was interesting that these phenomena were similar to the *in vivo* early encapsulation reaction. Addition of EGTA or EDTA to the incubation medium completely inhibited melanin formation by the attached GRs, indicating the calcium requirement for melanin formation by the GRs. Moreover, it was noted that there was an initial rapid generation of PO, following exposure to β-1,3-glucan, which was then followed by a prolonged stimulation of PO formation. On the other hand, β-1,3-glucan augmented erythrophagocytosis by the GRs of *B. mori*. The addition of laminarin is also known to enhance the phagocytic activity of GRs towards the bacteria, and the GRs incubated with laminarin were found to have numerous bacteria on the surface.[73] In my studies, β-1,3-glucan induced PO production from the GRs and enhanced phagocytosis, whereas mannan, the second major subcomponent of zymosan, had no effect on either PO production or phagocytosis. Therefore, the active component in zymosan in inducing PO generation from the attached GRs, and in stimulating the phagocytosis, is considered to be β-1,3-glucan. This aspect was similar to proPO-AS in the plasma of *B. mori*.

On the other hand, the PO activity was scarcely detected in separated PLs; however, a cell population of larger PLs (2%) showed melanization 20 h after the onset of incubation. β-1,3-glucan had no effect on PO generation in PLs. The attachment of the GRs is dependent on the presence of PO, while that of PLs is independent of the presence of PO, and calcium ion is required for PL's attachment. So, if the invading foreign objects have no proPO-AS in their structures, such as latex particles, the cellular reaction by the PLs will be evoked more rapidly than that by the GRs. This behavior of PLs will serve in isolation of foreign materials. In contrast, if proPO-AS is initiated by the foreign materials with an elicitor such as β-1,3-glucan in the surface structure, they will be efficiently reacted by mainly the GRs and consequently melanized. As the attached GRs form the PO more and more by degranulation, more melanin will be produced at the reacted site of foreign materials. In this manner, The GRs and PLs have a separate role in the cellular reactions, especially in encapsulation. Because *B. mori* PLs show little phagocytic activity and lysosomal enzyme

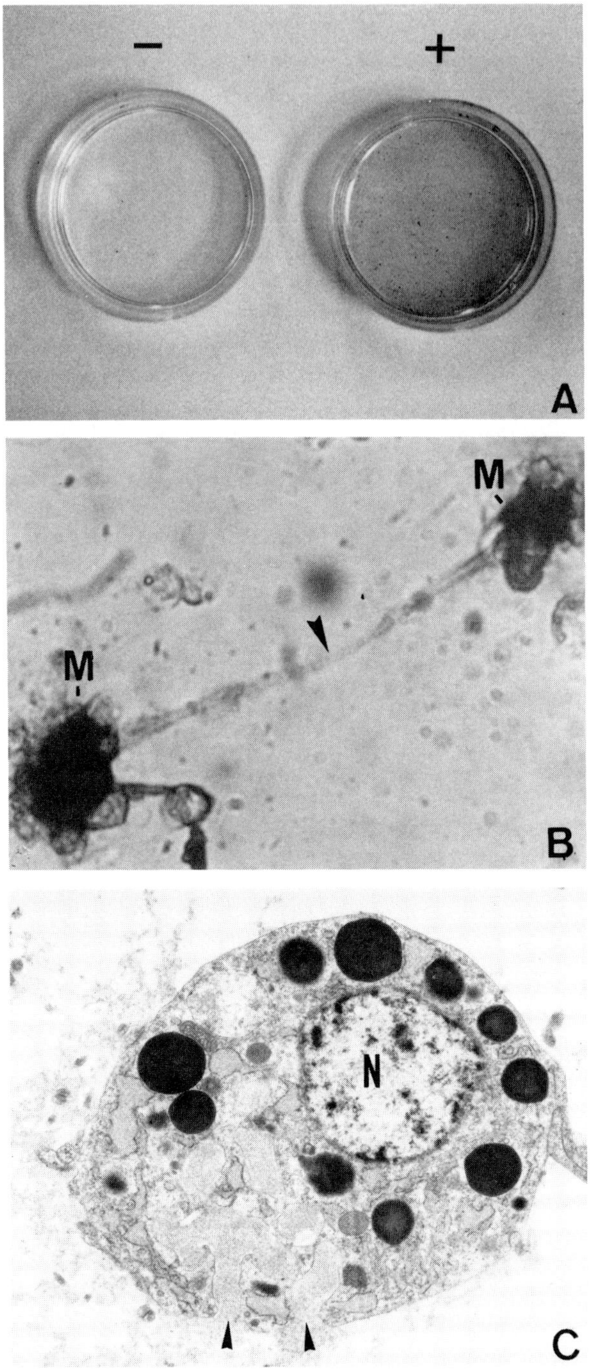

FIGURE 10. (A) The separated and incubated granulocytes form melanin *in vitro* in the presence of L-dopa (+), whereas they show no melanization without L-dopa (−). (B) The degree of melanization (M) by the aggregated granulocytes is more notable than a single cell, and the neuron-like processes (arrow) are formed between the cell masses. (C) TEM micrograph of a larval granulocyte of *B. mori* incubated in the presence of L-dopa. Note that the cell is surrounded by melanized coagulum and the discharge of granular materials is related to its formation (× 9000). (From Wago, H. and Ichikawa, Y., *Invertebrate and Fish Tissue Culture,* Kuroda, Y., Kurstal, E., and Maramorosch, K., Eds., Japan Scientific Societies Press, Tokyo/Springer-Verlag, Berlin, 1988, 185—188. With permission.)

activity, it is considered that the primary adhesion of PLs is directed rather for reducing the movement of invaders. In any case, since the GRs may possess some receptors for proPO-AS, the actual nature of its receptor is now under investigation.

C. OENOCYTOIDS AS A SUPPLIER OF PROPO AND THEIR INVOLVEMENT IN HOST DEFENSE

In *B. mori*, proPO occurs in the plasma,[82,91] as in other insects.[92-95] However, cells or tissues producing the proPO enzyme are not very well known. This problem is important, because PO involved in protection of insects against invaders is usually present in the hemolymph as a proenzyme in the form of proPO. As the proPO discharge by crustacean hemocytes is caused by a variety of nonself molecules,[81] it is probable that the presence of proPO in the plasma could be due to its discharge from or secretion by hemocytes in insects. In fact, a cytochemical study showed that PO originated in the oenocytoids (OEs) and GRs of *Galleria mellonella*.[86] In *B. mori*, proPO was demonstrated by immunohistochemical technique to be synthesized by OEs.[88,96]

More recently, we investigated the role of OEs in cellular reactions by the GRs in *Spodoptera litura*. The OEs of this species are very labile and easily break down within one minute after bleeding or by the injection of foreign materials into the body cavity;[97] this phenomenon is hardly observed in *B. mori*. However, in *B. mori* also, a small population of OEs is seen to become fragile with time after bleeding. This OE-lysis is inhibited by CO_2 and an acidic pH region (pH 4), so it is possible to keep OEs intact even after bleeding by introducing these physiological conditions. Intact OEs obtained in this way break down *in vitro* by the addition of foreign materials. When zymosan was added, much PO activity was detected in the supernatant. The supernatant with OE-lysis had much higher PO activity than that without lysis.[98] This phenomenon is similar to the large-scale lysis of the GRs *in vivo* by LPS, thereby releasing the cytoplasmic contents into the hemolymph of *Limulus*.[99,100] These observations strongly suggest that the proPO or some humoral components involved in proPO-AS may be directly supplied by OE-lysis, and is associated with nonself recognition. Hence, this reaction is considered to be an initial cellular response in insects. If the reactions by GR against the invading microorganisms or parasites will be more enhanced by the discharge of proPO through OE-lysis, it is possible to call OEs an important immunocompetent cell. Recent work on this aspect indicates that some chemical mediator originating from GRs, which reacted to the foreign materials, participates in OE-lysis (Kurihara and Wago, unpublished).

VI. SUMMARY

In *B. mori*, the GRs play a key role as a phagocyte to eliminate the foreign materials invading the hemocoel; and the elongation of filopodia on the surface is very important in phagocytic function in the larval stage. A series of experiments have shown that (1) filopodial elongation is closely related to nonself recognition, (2) microfilaments are involved in filopodial function, and (3) a certain level of an intracellular phagocytic signal obtained during the elongation process is required for evoking the phagocyte behavior. Moreover, recent work suggests that the rapid attachment of GRs greatly depends on the presence of PO in the plasma or on the foreign surface, and once adhered, GRs form melanin from the substrate by sequestering proPO-AS in the coagulum generated on the cell surface after degranulation; thus, PO serves as opsonin in the attachment of GRs. The presence or absence of PO has nothing to do with the PL attachment. Thus, PLs and GRs play separate roles in host defense in *B. mori*. Additionally, lectin receptors are induced on only the GRs simultaneously with the appearance of lectin activity in the plasma at the time of pupation. Therefore, this recognition system is also essential in the elimination of lectin-binding foreign objects as well as disintegrated self components produced at pupation. On the other hand, since both the foreign materials, except for the allografts and the injured sites of self or allografts, are treated as nonself by GRs in *B. mori*, irrespective of the absence or presence of lectin, the GRs seem to be able to recognize allografts as a self component by direct contact. In addition, OEs are an important hemocyte type engaged in supplying hemolymph proPO, and are therefore regarded as immunocompetent cells.

ACKNOWLEDGMENTS

I would like to thank Professor A. P. Gupta of Rutgers University for critical reading of this chapter. This work was supported in part by a Grant-in-Aid (Bio Media Program) from the Ministry of Agriculture, Forestry, and Fisheries (BMP-90-IV-2-7).

REFERENCES

1. **Gupta, A. P.,** Arthropod immunocytes: identification, structure, functions, and analogies to the functions of vertebrate B- and T-lymphocytes, in *Hemocyte and Humoral Immunity in Arthropods,* Gupta, A. P., Ed., Wiley-Interscience, John Wiley & Sons, 1986, 3.
2. **Nittono, Y.,** Studies on the blood cells in the silkworm, *Bombyx mori* L., *Bull. Sericult. Exp. Sta. Tokyo,* 16, 171, 1960.
3. **Akai, H. and Sato, S.,** Ultrastructure of the larval hemocytes of the silkworm, *Bombyx mori* L. (Lepidoptera:Bombycidae), *Int. J. Insect Morphol. Embryol.,* 2, 207, 1973.
4. **Sato, S., Akai, H., and Sawada, H.,** An ultrastructural study of capsule formation by *Bombyx mori, Annot. Zool. Jpn.,* 49, 177, 1976.
5. **Wago, H. and Ichikawa, Y.,** Hemocytic reactions to foreign cells in the silkworm, *Bombyx mori,* during post-embryonic development, *Appl. Ent. Zool.,* 14, 36, 1979.
6. **Wago, H.,** Observation of early processes of phagocytosis by the granular cells of the silkworm, *Bombyx mori, Appl. Ent. Zool.,* 15, 489, 1980.
7. **Gupta, A. P.,** Cellular elements in the hemolymph, in *Comprehensive Insect Physiology, Biochemistry and Pharmacology,* Vol. 3, Kerkut, G. A. and Gilbert, L. I., Eds., Pergamon Press, Oxford, 1985, 401—451.
8. **Akai, H. and Sato, S.,** An ultrastructural study of the hemopoietic organs of the silkworm, *Bombyx mori, J. Insect Physiol.,* 17, 1665, 1971.
9. **Akai, H. and Sato, S.,** Surface ultrastructure of the larval hemocytes of the silkworm, *Bombyx mori* L. (Lepidoptera:Bombycidae), *Int. J. Insect Morphol. Embryol.,* 5, 17, 1976.
10. **Sato, S. and Akai, H.,** Development of the hemopoietic organs of the silkworm, *Bombyx mori* L., *J. Seric. Sci. Jpn.,* 46, 397, 1977.
11. **Wago, H.,** Cellular recognition of foreign materials by *Bombyx mori* phagocytes, I. Immunocompetent cells, *Dev. Comp. Immunol.,* 6, 591, 1982.
12. **Davies, D. H. and Preston, T. M.,** The behaviour of insect plasmatocytes *in vitro:* changes in morphology during settling, spreading, and locomotion, *J. Exp. Zool.,* 236, 71, 1985.
13. **Wago, H. and Kitano, H.,** Morphological and functional characterization of the larval hemocytes of the cabbage white butterfly *Pieris rapae crucivora, Appl. Ent. Zool.,* 20, 1, 1985.
14. **Wago, H.,** The role of microfilaments and microtubules in the spreading and motile behaviour of plasmatocytes of *Bombyx mori, Bull. Saitama Medical School Junior College,* 1, 3, 1990.
15. **Davies, D. H. and Preston, T. M.,** Effect of disruption of plasmatocyte microfilaments on encapsulation *in vitro, Dev. Comp. Immunol.,* 11, 353, 1987.
16. **Price, C. D. and Ratcliffe, N. A.,** A reappraisal of insect haemocytic classification by the examination of blood from fifteen insect orders, *Z. Zellforsch.,* 147, 537, 1974.
17. **Wago, H.,** Humoral factors promoting the adhesive properties of the granular cells and plasmatocytes of the silkworm *Bombyx mori,* and their possible role in the initial cellular reactions to foreignness, *Cell. Immunol.,* 54, 155, 1980.
18. **Noguchi, A., Baba, A., and Tamura, N.,** Silkworm haemolymph factor(s) interacting with vertebrate complement system, *Dev. Comp. Immunol.,* 5, 111, 1981.

19. **Suzuki, T. and Natori, S.,** Identification of a protein having hemagglutinating activity in the hemolymph of the silkworm, *Bombyx mori, J. Biochem.,* 93, 583, 1983.
20. **Kinoshita, T. and Inoue, K.,** Bactericidal activity of the normal, cell-free hemolymph of silkworm *(Bombyx mori), Infect. Immun.,* 16, 32, 1977.
21. **Ashida, M. and Dohke, K.,** Activation of prophenoloxidase by the activating enzyme of the silkworm, *Bombyx mori, Insect Biochem.,* 10, 34, 1980.
22. **Hazarika, L. K. and Gupta, A. P.,** Variation in hemocyte populations during various developmental stages of *Blattera germanica* (L.) (Dictyoptera:Blattelidae), *Zool. Sci.,* 4, 307, 1987.
23. **Arnold, J. W. and Hinks, C. F.,** Haemopoiesis in Lepidoptera, I. The multiplication of circulating hemocytes, *Can. J. Zool.,* 54, 1003, 1976.
24. **Wago, H. and Ichikawa, Y.,** *In vitro* analyses of cyto-adherence phenomenon between *Bombyx mori* hemocytes and goose erythrocytes in relation to the larval development, *Appl. Ent. Zool.,* 14, 256, 1979.
25. **Wago, H. and Ichikawa, Y.,** Changes in the phagocytic rate during the larval development and manner of hemocytic reactions to foreign cells in *Bombyx mori, Appl. Ent. Zool.,* 14, 397, 1979.
26. **Wago, H.,** Switching of host defense mechanisms in relation to metamorphosis of insects, in *Proc. Jpn. Assoc. Dev. Comp. Immunol.,* Wago, H., Ed., Tokyo, Japan, 1990, 12 (in Japanese).
27. **Bellah, M. E. M. and Ridi, R. E.,** Age-related occurrence of natural agglutinins in the erisilkworm, *Philosamia ricini, Dev. Comp. Immunol.,* 12, 707, 1988.
28. **Komano, H., Nozawa, R., Mizuno, D., and Natori, S.,** Measurement of *Sarcophaga peregrina* lectin under various physiological conditions by radioimmunoassay, *J. Biol. Chem.,* 258, 2143, 1983.
29. **Amirante, G. A. and Mazzalai, F. G.,** Synthesis and localization of hemagglutinins in hemocytes of the cockroach *Leucophaea maderae* L., *Dev. Comp. Immunol.,* 2, 735, 1978.
30. **Komano, H. and Natori, S.,** Participation of *Sarcophaga peregrina* lectin in the lysis of sheep red blood cells injected into the body cavity of larvae, *Dev. Comp. Immunol.,* 9, 31, 1985.
31. **Takahashi, H., Komano, H., and Natori, S.,** Expression of the lectin gene in *Sarcophaga peregrina* during normal development and under conditions where the defence mechanism is activated, *J. Insect Physiol.,* 32, 771, 1986.
32. **Wago, H.,** Characteristics of some elements involved in host defense of *Pieris rapae crucivora,* and effects of ligation on their production, *Zool. Sci.,* 2, 903, 1985.
33. **Richards, E. H., Ratcliffe, N. A., and Renwrantz, L.,** The binding of lectins to carbohydrate moieties on haemocytes of the insects, *Blaberus craniifer* (Dictyoptera) and *Extatosoma tiaratum* (Phasmida), *Cell Tissue Res.,* 257, 445, 1989.
34. **Ratcliffe, N. A. and Rowley, A. F.,** Role of hemocytes in defense against biological agents, in *Insect Hemocytes,* Gupta, A. P., Ed., Cambridge University Press, Cambridge, 1979, 331.
35. **Zachary, D. and Hoffmann, D.,** Lysozyme is stored on the granules of certain haemocyte types of *Locusta, J. Insect Physiol.,* 30, 405, 1984.
36. **Gupta, A. P.,** Arthropod hemocytes and phylogeny, in *Arthropod Phylogeny,* Gupta, A. P., Ed., Van Nostrand Reinhold, New York, 1979, 669.
37. **Ratcliffe, N. A. and Rowley, A. F.,** *In vitro* phagocytosis of bacteria by insect blood cells, *Nature (London),* 252, 391, 1974.
38. **Kawakami, K.,** Phagocytosis in muscadin-diseased larvae of the silkworm, *Bombyx mori* (Linnaeus), *J. Invertebr. Pathol.,* 7, 203, 1965.
39. **Lackie, A. M.,** Cellular recognition of foreignness in two insect species, the American cockroach and the desert locust, *Immunology,* 36, 909, 1979.
40. **Han, S. S. and Gupta, A. P.,** Arthropod immune system. II. Encapsulation of implanted nerve cord and "plain gut" surgical suture by granulocytes of *Blattella germanica* (L.) (Dictyoptera:Blattellidae), *Zool. Sci.,* 6, 303, 1989.
41. **Neuwirth, M.,** The structure of the hemocytes of *Galleria mellonella* (Lepidoptera), *J. Morphol.,* 139, 105, 1973.
42. **Rizki, T. M. and Rizki, R. M.,** Cell interactions in hereditary melanotic tumors of *Drosophila melanogaster,* in *Proc. First Int. Colloquium Invertebr. Pathol.,* Queen's University, Kingston, Canada, 1976, 137.
43. **Schmit, A. R. and Ratcliffe, N. A.,** The encapsulation of foreign tissue implants in *Galleria mellonella* larvae, *J. Insect Physiol.,* 23, 175, 1977.
44. **Wago, H.,** The role of hemolymph in the initial cellular attachment to foreign cells by the hemocytes of the silkworm, *Bombyx mori, Dev. Comp. Immunol.,* 5, 217, 1981.
45. **Salt, G.,** The haemocytic reaction of insects to foreign bodies, in *The Cell and the Organism,* Ramsay, J. A. and Wigglesworth, V. B., Eds., Cambridge University Press, Cambridge, 1961, 175.
46. **Scott, M. T.,** Recognition of foreignness in invertebrates: transplantation studies using the American cockroach *(Periplaneta americana), Transplantation,* 11, 78, 1971.
47. **Bell, W. J.,** Yolk formation by transplanted cockroach oocytes, *J. Exp. Zool.,* 181, 41, 1972.

48. **Thomas, I. G. and Ratcliffe, N. A.,** Integumental grafting and immunorecognition in insects, *Dev. Comp. Immunol.,* 6, 643, 1982.
49. **Wago, H.,** Immune systems in insects, *Nikkei Science,* 15, 22, 1986 (in Japanese).
50. **Lackie, A. M.,** Transplantation: the limits of recognition, in *Hemocytic and Humoral Immunity in Arthropods,* Gupta, A. P., Ed., Wiley-Interscience, John Wiley & Sons, 1986, 191.
51. **Jones, S. E. and Bell, W. J.,** Cell-mediated immune-type response of the American cockroach, *Dev. Comp. Immunol.,* 6, 35, 1982.
52. **Shaw, D. R. and Griffin, D. R.,** Phagocytosis requires repeated triggering of macrophage phagocytic receptors during particle ingestion, *Nature (London),* 289, 409, 1981.
53. **Wago, H.,** Regulatory effects of temperature and hemolymph on inhibition and enhancement of attachment and filopodia function of phagocytic granular cells of the silkworm *Bombyx mori, Dev. Comp. Immunol.,* 6, 231, 1982.
54. **Wago, H.,** Involvement of microfilaments in filopodial function of phagocytic granular cells of the silkworm, *Bombyx mori. Dev. Comp. Immunol.,* 6, 655, 1982.
55. **Allison, A. C., Davies, P. D., and dePetris, S.,** Role of contractile microfilaments in macrophage movement and endocytosis, *Nature (London),* 232, 153, 1970.
56. **Malawista, S. E., Gee, J. B., and Bensch, K.,** Cytochalasin B reversibly inhibits phagocytosis: functional, metabolic, and ultrastructural effects in human blood leukocytes and rabbit alveolar macrophages, *Yale J. Biol. Med.,* 44, 286, 1971.
57. **Henney, C. S. and Bubbers, J. E.,** Antigen-T lymphocyte interactions: inhibition by cytochalasin B, *J. Immunol.,* 111, 85, 1973.
58. **Lipsky, P. E. and Rosenthal, A. S.,** Macrophage-lymphocyte interactions: antigen-independent binding of guinea pig lymph node lymphocytes by macrophages, *J. Immunol.,* 115, 440, 1975.
59. **Anderson, R. S.,** Rosette formation by insect macrophages, inhibition by cytochalasin B, *Cell. Immunol.,* 29, 331, 1977.
60. **Wago, H.,** The important significance of filopodial elongation of phagocytic granular cells of the silkworm, *Bombyx mori,* in recognition of foreignness, *Dev. Comp. Immunol.,* 7, 445, 1983.
61. **Wago, H.,** *In vitro* evidence for the requirement of filopodial elongation for the progress of phagocytosis by phagocytic granular cells of the silkworm, *Bombyx mori, Dev. Comp. Immunol.,* 8, 7, 1984.
62. **Lackie, A. M.,** Effect of substratum wettability and charge on adhesion *in vitro* and encapsulation *in vivo, J. Cell Sci.,* 63, 181, 1983.
63. **Salt, G.,** *The Cellular Defense Reactions of Insects,* Cambridge University Press, Cambridge, 1970, 31.
64. **Wigglesworth, V. B.,** *The Principles of Insect Physiology,* 7th ed., Chapman and Hall, London, 1972, 411—475.
65. **Rizki, T. M. and Rizki, R. M.,** Cell interactions in the differentiation of a melanotic tumor in *Drosphila, Differentiation,* 12, 167, 1979.
66. **Rizki, T. M. and Rizki, R. M.,** Developmental analysis of a temperature-sensitive melanotic tumor mutant in *Drosophila melanogaster, Wilhelm Roux's Arch. Dev. Biol.,* 189, 197, 1980.
67. **Rizki, T. M. and Rizki, R. M.,** The cellular defense reactions of *Drosophila melanogaster,* in *The Ultrastructure and Functioning of Insect Cells,* Akai, H. and Morohoshi, S., Eds., Society for Insect Cells, 1982, 173.
68. **Rizki, R. M., Rizki, T. M., Bebbington, C. R., and Roberts, D. B.,** *Drosophila* larval fat body surfaces changes in transplant compatibility during development, *Roux's Arch. Dev. Biol.,* 192, 1, 1983.
69. **Anderson, R. S.,** Expression of receptors by insect macrophages, in *Phylogeny of Thymus and Bone Marrow-Bursa Cells,* Wright, R. K. and Cooper, E. L., Eds., Elsevier North Holland, Amsterdam, 1976, 27.
70. **Scott, M. T.,** Recognition of foreignness in invertebrates. II. *In vitro* studies of cockroach phagocytic haemocytes, *Immunology,* 21, 817, 1971.
71. **Ratcliffe, N. A., Leonard, C., and Rowley, A. F.,** Prophenoloxidase activation: nonself recognition and cell cooperation in insect immunity, *Science,* 226, 557, 1984.
72. **Leonard, C. M., Söderhäll, K., and Ratcliffe, N. A.,** Studies on prophenoloxidase and protease activity of *Blaberus craniifer* haemocytes, *Insect Biochem.,* 15, 803, 1985.
73. **Leonard, C. M., Ratcliffe, N. A., and Rowley, A. F.,** The role of prophenoloxidase activation in non-self recognition and phagocytosis by insect blood cells, *J. Insect Physiol.,* 31, 789, 1985.
74. **Dohke, K.,** Studies on the prophenoloxidase activating enzyme from cuticle of the silkworm, *Bombyx mori.* I. Activation reaction by the enzyme, *Arch. Biochem. Biophys.,* 157, 203, 1973.
75. **Ashida, M., Dohke, K., and Ohnishi, E.,** Activation of prophenoloxidase. 3. Release of peptide from prophenoloxidase by the activating enzyme, *Biochem. Biophys. Res. Commun.,* 57, 1089, 1974.
76. **Renwrantz, L. and Cheng, T. C.,** Agglutinin-mediated attachment of erythrocytes to hemocytes of *Helix pomatia, J. Invertebr. Pathol.,* 29, 97, 1977.
77. **Renwrantz, L. and Mohr, W.,** Opsonizing effect of serum and albumin gland extracts on the elimination of human erythrocytes from the circulation of *Helix pomatia, J. Invertebr. Pathol.,* 31, 164, 1978.

78. **Sminia, T., van der Knaap, W. P. W., and Edelenbosch, P.,** The role of serum factors in the phagocytosis of foreign particles by blood cells of the freshwater snail *Lymnaea stagnalis, Dev. Comp. Immunol.,* 3, 37, 1979.
79. **Rowley, A. F. and Ratcliffe, N. A.,** Insect erythrocyte agglutinins: *in vitro* opsonization experiments with *Clitumunus extradentatus* and *Periplaneta americana* haemocytes, *Immunology,* 40, 483, 1980.
80. **Smith, V. and Söderhäll, K.,** β-1,3-glucan activation of crustacean hemocytes *in vitro* and *in vivo, Biol. Bull.,* 164, 299, 1983.
81. **Smith, V. and Söderhäll, K.,** Induction of degranulation and lysis of haemocytes in the freshwater crayfish *Astacus astacus* by components of the prophenolocxidase activating system *in vitro, Cell Tissue Res.,* 233, 295, 1983.
82. **Ashida, M., Ishizaki, Y., and Iwahana, M.,** Activation of prophenoloxidase by bacterial cell walls or by β-1,3-glucans in plasma of the silkworm, *Bombyx mori, Biochem. Biophys. Res. Commun.,* 113, 562, 1983.
83. **Yoshida, H., Ochiai, M., and Ashida, M.,** β-1,3-glucan receptor and peptideglycan receptor are present as separate entities within insect prophenoloxidase activating system, *Biochem. Biophys. Res. Commun.,* 114, 1177, 1986.
84. **Wago, H. and Ichikawa, Y.,** The role of phenoloxidase in the cellular defense reactions to foreign materials by the granular cells of the silkworm, *Bombyx mori,* in *Invertebrate and Fish Tissue Culture,* Kuroda, Y., Kurstal, E., and Maramorosch, K., Eds., Japan Scientific Societies Press, Tokyo/Springer-Verlag, Berlin, 1988, 185.
85. **Hoffmann, J. A., Porte, A., and Joly, P.,** On the localization of phenoloxidase activity in coagulation of *Locusta migratoria* (L.) (Orthoptera), *C. R. Séances Acad. Sci.,* (III), 270A, 629, 1970.
86. **Schmit, A. R., Rowley, A. F., and Ratcliffe, N. A.,** The role of *Galleria mellonella* hemocytes in melanin formation, *J. Invertebr. Pathol.,* 29, 232, 1977.
87. **Söderhäll, K., Vey, A., and Ramstedt, M.,** Hemocyte lysate enhancement of fungal spore encapsulation by crayfish hemocytes, *Dev. Comp. Immunol.,* 8, 23, 1984.
88. **Iwama, R. and Ashida, M.,** Biosynthesis of prophenoloxidase in hemocytes of larval hemolymph of the silkworm, *Bombyx mori, Insect Biochem.,* 16, 547, 1986.
89. **Rowley, A. F. and Ratcliffe, N. A.,** Insects, in *Invertebrate Blood Cells,* Vol. 2., Ratcliffe, N. A. and Rowley, A. F., Eds., Academic Press, New York, 1981, 422.
90. **Wago, H.,** Effects of prophenoloxidase-activating system on attachment and degranulation of the granular cells in the silkworm *Bombyx mori, Dev. Comp. Immunol.,* 14, 262, 1990.
91. **Ashida, M. A.,** Cane sugar factor suppressing activation of prophenoloxidase in hemolymph of the silkworm, *Bombyx mori, Insect Biochem.,* 11, 57, 1981.
92. **Götz, P. and Vey, A.,** Humoral encapsulation in Diptera(Insecta): defense reactions of *Chironomus* larvae against fungi, *Parasitology,* 68, 193, 1974.
93. **Pye, A. E.,** Microbial activation of prophenoloxidase from immune insect larvae, *Nature* (*London*), 251, 610, 1974.
94. **Vey, A. and Götz, P.,** Humoral encapsulation in Diptera (Insecta): comparative studies *in vitro, Parasitology,* 70, 77, 1975.
95. **Saul, S. J., Bin, L., and Sugumaran, M.,** The majority of prophenoloxidase in the hemolymph of *Manduca sexta* is present in the plasma and not in the hemocytes, *Dev. Comp. Immunol.,* 11, 479, 1987.
96. **Ashida, M., Ochiai, M., and Niki, T.,** Immunolocalization of prophenoloxidase among hemocytes of the silkworm, *Bombyx mori, Tissue Cell,* 20, 599, 1988.
97. **Kurihara, H., Shimazu, T., Tsuchidate, T., and Wago, H.,** What does oenocytoid lysis of *Spodoptera litura* mean? in *Proc. Jpn. Soc. Appl. Ent. Zool.,* Vol. 31, Jpn. Soc. Appl. Ent. Zool., 1987, 111 (in Japanese).
98. **Kurihara, H., Shimazu, T., Tsuchidate, T., and Wago, H.,** Physiological conditions which inhibit oenocytoid-lysis of *Spodoptera litura,* in *Proc. Jpn. Soc. Appl. Ent. Zool.,* Vol. 32, Jpn. Soc. Appl. Ent. Zool., 1988, 81, (in Japanese).
99. **Armstrong, P. B. and Rickles, F. R.,** Endotoxin induced degranulation of the *Limulus* amoebocytes, *Exp. Cell Res.,* 140, 15, 1982.
100. **Armstrong, P. B.,** Adhesion and motility of the blood cells of *Limulus,* in *Blood Cells of Marine Invertebrates,* Cohen, W. D., Ed., Alan R. Liss, New York, 1985, 77.
101. **Wago, H.,** Possible role of filopodial elongation in the cellular recognition of foreign materials by phagocytic granular cells in *Bombyx mori,* in *The Ultrastructure and Functioning of Insect Cells,* Akai, H., King, R. C., and Morohoshi, S., Eds., The Society for Insect Cells Japan, 1983, 177.

Coagulation

Chapter 8

COAGULATION IN CRUSTACEA*

Michèle Durliat

TABLE OF CONTENTS

* Reproduced with permission from Durliat, M., "Clotting processes in Crustacea Decapoda," *Biol. Rev.*, 60, 473, 1985.

I. INTRODUCTION

Hemolymph coagulation in Crustacea has been reviewed by several authors.[1-6] However, the majority of these reviews are mainly concerned with the cytological aspects of clotting. Crustacean hemocytes, described as early as 1824 by Carus,[7] have been studied along two lines: (1) morphological characterization of different types of cells, and (2) determination of their role in blood clotting without precise chemical identification of the factors involved in the process. Recently, numerous papers have included biochemical approach, and considerable progress has now been made in the analysis of the biochemical events occurring during coagulation. In this paper, the latter aspect and the respective roles of cellular and plasma factors involved in the clotting process will be emphasized. The more important reports deals with the analyses performed in Decapoda; these have been compared with the multienzyme cascade systems analyzed in detail in Limulidae.

II. HISTORICAL OUTLINE

First observations on coagulation were made by the microscope. Lowit (1889)[8] established that the lysis of blood cells was related to the coagulation process. Hardy,[9] Tait[10,11] and Tait and Gunn[12] discovered a category of very fragile blood cells (explosive cells) in Crustacea, which first initiated coagulation of the plasma, forming islands around these cells, and then spreading out with varying degrees of rapidity throughout the hemolymph. These authors suggested that lysis of these unstable hemocytes and release of their granules provided thromboplastic substances (''fibrin'' ferment) that affected a clottable ''fibrinogen''-like factor in the plasma. Three different patterns of coagulation were defined,[10] depending on simultaneous cell agglutination and gelation of the plasma: type A — agglutination of hemocytes without any subsequent gelation of the plasma; type B — cell agglutination followed by a general gelation of the plasma; type C — clotting process limited at first to the periphery of a few insignificant cell clots, subsequently affecting the remaining plasma.

Thus, *in vitro* reactions of crustacean blood are characterized by a series of patterns showing a gradation from a single cellular agglutination without gelation of plasma to firm plasma gelation. Coagulation has been reported to occur either in two successive distinct and independent steps, or as a single continuous process.[3]

It has been difficult to establish a classification of blood cells, because hemocytes are very reactive cells that immediately undergo alterations when removed from the hemocoel. However, during the last 15 years, the use of suitable anticoagulants has allowed to define three main morphological types by both light and electron microscopes:[5] *hyaline cells* with few small granules, *granulocytes* whose cytoplasm is filled with large membrane-bounded granules, and *semigranular cells,* which are an intermediate cell type between the other two (Figure 1). These three types apparently form a continuous differentiation series with many intermediate forms, although in the hemolymph of certain Crustacea, only hemocytes with a definite number of granules have been observed.[13-15] In *Carcinus maenas,* all hemocytes derive from a single cell line, which proliferates in the lymphocytogenic nodules of the gizzard walls. The stem cells are called hemoblasts: they differentiate into circulating hemocytes and into cyanoblasts.[16] These cells keep some affinity for mesenchymal structures, such as the connective tissue.

All three types of hemocytes seem to be implicated in the clotting process. Gelation of plasma occurs following the release of material contained in the gradually dissolving granules.[17] In the crayfish *Orconectes virilis,* the factor(s) released from hemocytes was identified as a glyco- or mucoprotein.[18,19] It seems to be related to Hardy's[9] ''fibrin-ferment'' and Nolf's[20] ''coagulin''. In fact, an unstable and thermolabile protein was first identified in

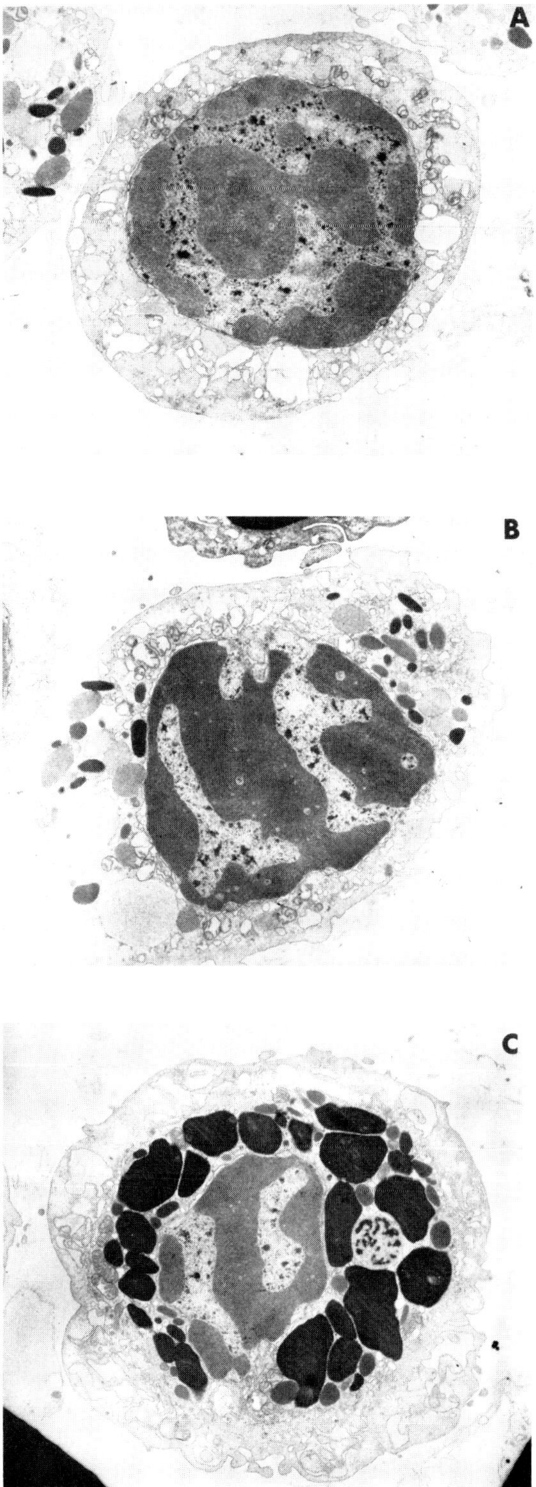

FIGURE 1. Electron micrographs of hemocytes from *Astacus leptodactylus*. (A) Hyaline cell. (B) Semigranular cell with granules of different sizes. (C) Granulocyte with large granules of different shapes (× 3960). (From Durliat and Pochon, unpublished results.)

lobster as a Ca^{2+}-dependent cellular transglutaminase, which acts on a plasma "fibrinogen"-like factor.[21-24] Plasma gelation was extensively studied in macrurans.[25-33] The clottable protein (i.e., plasma coagulogen, as defined by Needham[34]), permitting this gelation, is a high-M.W. plasma protein, which can be converted to a covalently crosslinked gel by the action of Ca^{2+}-dependent cellular enzymes. This plasma coagulogen appears in the form of heterogeneous polymers on polyacrylamide gradient gels[35] (Figure 2), but shows a single precipitate arc on tandem-crossed immunoelectrophoreses carried out with blood from *Astacus leptodactylus*[36] (Figure 3), *Homarus vulgaris,* and several spiny lobster species.[37] Although no evident plasma gelation was observed in the hemolymph of *Cancer pagurus* or *Maia squinado,* a small amount of plasma coagulogen was detected.[38,39] This is not surprising because large variations in the concentration of plasma coagulogen between species have been reported. Moreover, variations in the number of circulating hemocytes, correlated with the intensity of hemolymph coagulation, were also observed, depending on the species and the stage in the molting cycle.[40-42] This would provide a convenient explanation for the different coagulation types proposed by Tait.[11] In Type A, the plasma coagulogen titer would be very low and strong cell agglutination would appear, while in Types B and C, the amount of plasma coagulogen would be high enough to allow the formation of an extensive clot, whereas the cellular reaction would be more moderate. Since plasma-independent cellular clumping always occurs, it suggests that another clottable protein might be present in the hemocytes of all Crustacea. Such a cellular coagulogen different from the well-defined plasma coagulogen was found in the hemocyte granules of *H. vulgaris.*[43,44] Similarly, in the sand crab *Ovalipes bipustulatus,* a cellular protein, clottable by serine protease enzymes, has been isolated and characterized.[45,46] Moreover, serine protease clotting enzyme(s) is present in the hemocyte lysate of *Astacus astacus,*[47-49] and in the blood of certain Decapoda.[37] Cell lysates of Decapoda seem to contain a coagulogen and an enzymatic system able to convert this clottable factor into a gel, which includes the aggregated cell mass. These data suggest the existence of two distinct coagulogens, one of cellular origin and the other present only in cell-free hemolymph. The cellular coagulogen might also be the precursor of plasma coagulogen or they are two distinct proteins, which may or may not interact during the final stage of coagulation.

In the following sections, various aspects of this system, the characteristics and the origin of these two coagulogens, conditions for their activation as well as their interactions with other blood proteins during the defense mechanisms, will be described and discussed in more detail.

III. HUMORAL BLOOD PROTEINS

A. MAIN PLASMA PROTEINS

Hemolymph proteins of Crustacea have been extensively investigated. Quantitative variations in total protein content were studied, depending on osmoregulation, temperature, starvation, and molting cycle. The main components of the hemolymph are contained in the following subsections.

1. Hemocyanins

Hemocyanins constitute the major proteins and act as carriers of oxygen and storage proteins. They contain 1 to 4% of carbohydrates and form large oligomeric fractions with a heterogeneous structure.[50-53] They are constituted of different subunits with an average molecular weight between 70 and 85 kDa. Hemocyanins exhibit different aggregation states, depending on the species studied and experimental conditions, with various sedimentation coefficients. The sedimentation coefficient of the main native hemocyanin is about 25 S in crayfish, lobster, several crabs, and 16 S in spiny lobster (see review by Van Holde and Miller[54]).

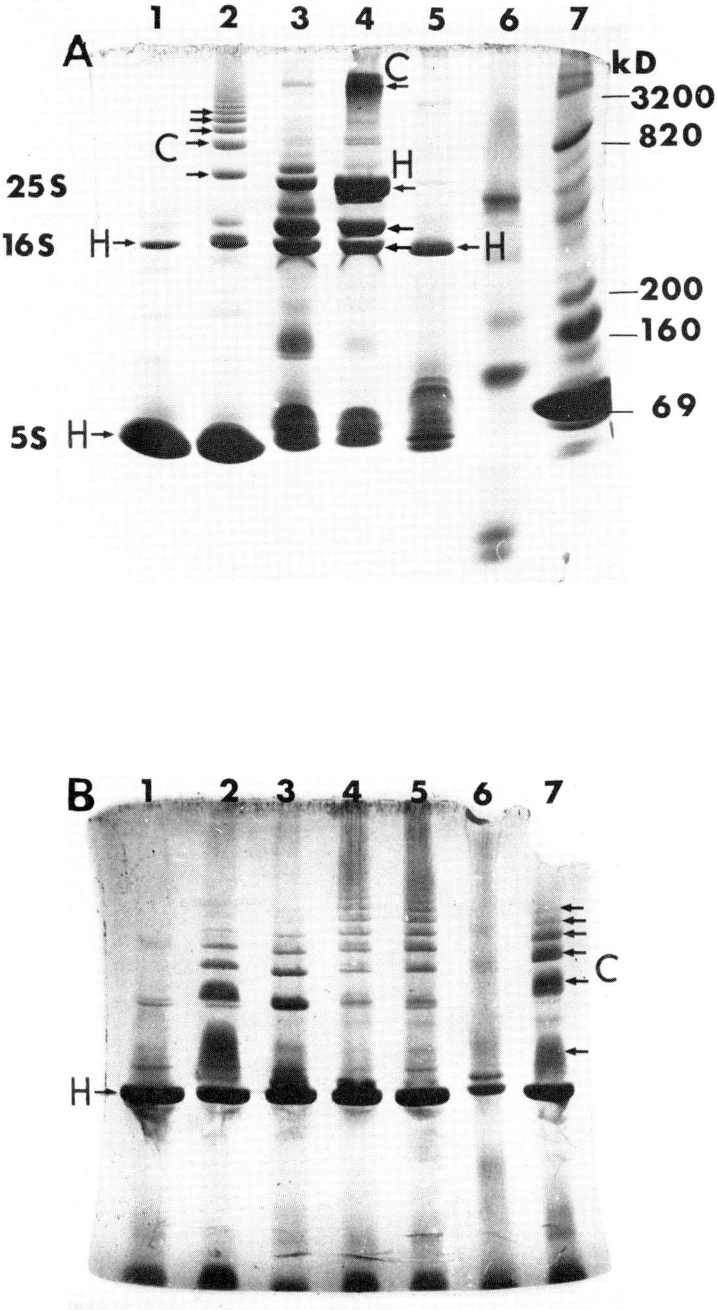

FIGURE 2. Electrophoretic patterns of the hemolymph from Arthropoda, on polyacrylamide gradient gels. (A) Experiments were carried out on 4 to 30% gels. From left to right: serum (lane 1), plasma (lane 2) of *Panulirus vulgaris,* serum (lane 3), plasma (lane 4) of *Astacus leptodactylus*, hemolymph of *Carcinus maenas* (lane 5), and hemolymph of *Carausius morosus* (lane 6); extreme right (lane 7): human serum proteins as standards. (B) Hemolymph samples from *Homarus vulgaris*. Experiments were performed on 2 to 16% gel. From left to right: one serum (lane 1) and six different plasmas obtained respectively with sodium citrate (lane 2), NEM-sodium citrate (lane 3), GME (lane 4), propranolol (lane 5), NEM-PMSF (lane 6), and DC (lane 7). H shows the different aggregation states (5S, 16S, 25S) of hemocyanin. (C) the different fractions of plasma coagulogen.

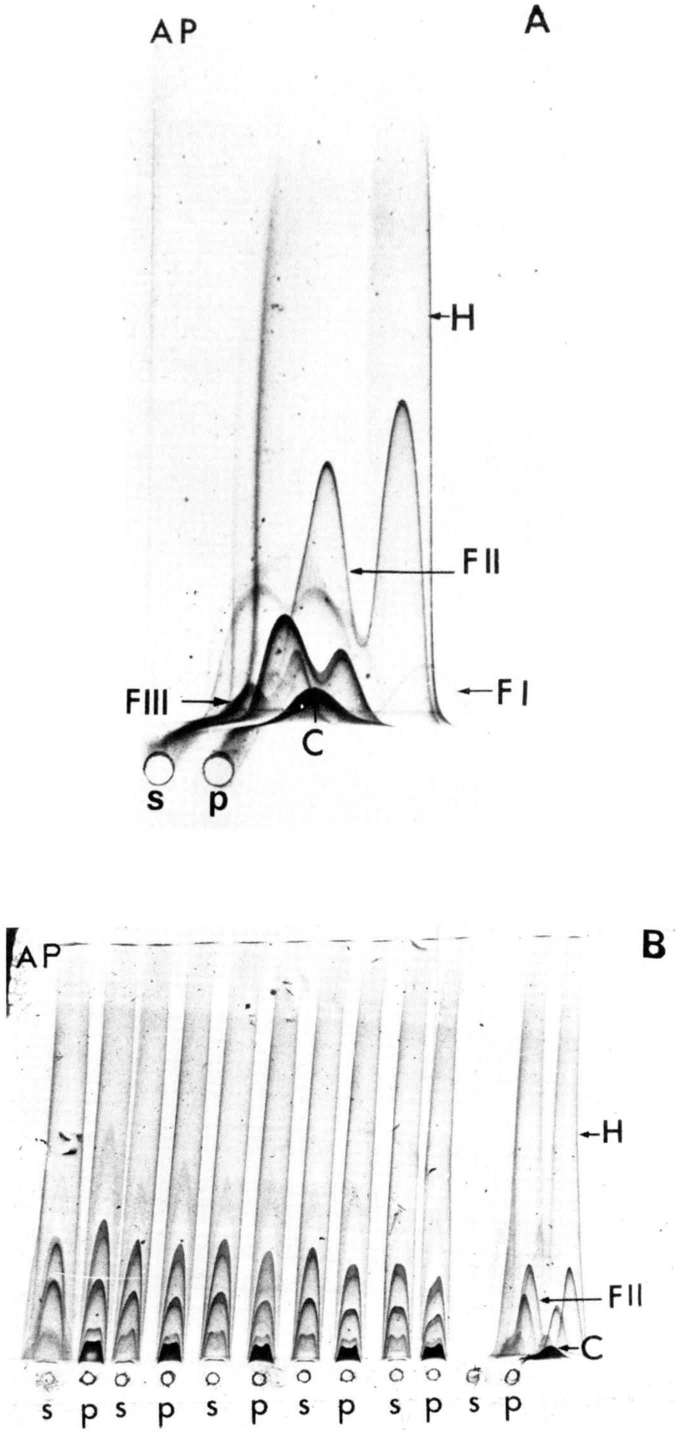

FIGURE 3. Tandem-crossed immunoelectrophoresis (A) and rocket immunoelectrophoresis (B) of samples of serum (s) and plasma (p) from *Astacus leptodactylus*. Experiments were run with an antiplasma (AP) poured into the gel. FI, FII, and FIII indicate molt-related proteins. Other symbols are as in Figure 2.

2. Plasma Coagulogen

The plasma coagulogen or "fibrinogen" is a glycoprotein of about 400 kDa.[30,31] It appears in the hemolymph as several soluble polymers constituted by a variable number of uncovalently linked monomers.[36,37]

3. Heteroagglutinins

Heteroagglutinins agglutinate many foreign particles, and also act as opsonins.[55-59] Their sedimentation coefficient varies between 10 and 19 S. They are constituted of uncovalently linked subunits of a molecular weight of 55 to 68 kDa. Their reactive sites specifically bind sugars.[60] Thus, one agglutinin reacting with sialic acid and another with *N*-acetylgalactosamine were identified in *Homarus americanus*.[56,61] Many lectins were found in serum of *Birgus latro*[62] and *Macrobrachium rosenbergii*,[63] including sialoconjugate binding lectins. For details of agglutinins in Crustacea see Amirante[64] and those in arthropods in general see Hapner and Stebbins.[65]

4. Vitellogenins

Vitellogenins related to oogenesis processes are lipoglycocaroteno-proteins with a M.W. between 300 and 400 kDa (see review in Meusy, 1980).[66]

5. Proteins Related to Molting Cycle

Proteins related to the molting cycle[67-72] were also detected within the epidermis[73,74] and in the calcified cuticle.[75,76]

6. Site of Synthesis

The site of synthesis of the main plasma proteins has not been established. It is known that vitellogenins are synthesized in the subepidermal adipose tissues.[66] However, it seems that other cells from connective tissue, dispersed in different restricted body regions of Decapoda, especially between the hepatopancreas tubules, are implicated in the production of phospholipids and lipoproteins.[77] Numerous works prove that cyanoblasts and cyanocytes are involved in the synthesis of hemocyanins.[14,16,78] The synthesis observed by certain authors within the hepatopancreas[79,80] might be explained by the fact that cyanocytes are closely connected with the connective tissue surrounding this digestive gland. Cyanocytes would derive from the same cell lines (hemoblasts) as circulating hemocytes.[16]

Agglutinins are synthesized by the lobster granulocytes, then released into the blood.[60,81,82] Likewise, it seems that some molt-related proteins might be discharged from the hemocyte population.[83]

Plasma coagulogen of Decapoda does not seem to originate from hepatopancreas cells,[84,85] as fibrinogen originates from liver in vertebrates. Some experiments carried out with different clotting inhibitors do not show the constant presence of this factor within hemocyte lysates. The coagulogen seems to be linked in various degrees to the cell membrane, depending on the composition of the medium (Durliat and Vranckx, unpublished results). It might be produced by certain "differentiated" hemocytes or by lipoprotein cells described by many authors in both the hemocoel and connective tissues of Decapoda.[77,86-93] It is possible that lipoprotein cells originate from the same stem-cells as blood corpuscles and/or that some kinds of circulating cells differentiate from the subepidermal connective tissue.[77] In any case, all these cells certainly have a mesenchymal origin. This cyclic connective tissue seems to exert many important functions in crustacean metabolism (production of proteins, phospholipids, hemocytes, storage of Ca^{2+}); in fact, some "hepatic" functions, analogous to those of the vertebrate liver. These considerations could well validate the concept of hemocyte acting as "free circulating hepatocytes" put forward by Johnston et al.[94] Yet hemocytes could be only a part of the general connective tissues.

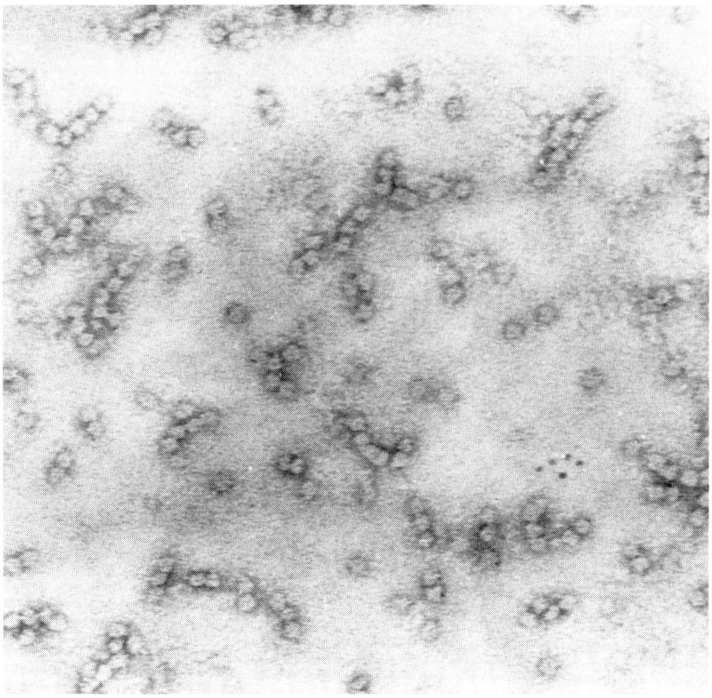

FIGURE 4. Electron micrograph showing chains of plasma coagulogen molecules of crayfish (× 140,000).

B. PLASMA COAGULOGEN

1. Characteristics of Coagulogen

When coagulation involves a complete plasma gelation, the firm transparent clot is unretracted. The plasma coagulogen is a mixture of soluble uncovalently crosslinked polymers. Both 21 and 17 to 18 S components were found in the blood of *Homarus*,[27,28] 14.5 and 19.4 S components in *Panulirus interruptus*[30] and several polymers from about 400 to 1200 kDa in the blood of *H. vulgaris* and some spiny lobsters[37] (Figure 2). In electrophoresis, the lobster monomer moved in the 200 kDa area,[37] and when viewed by electron microscope, was spherical with an average diameter of about 10 nm.[95] Preparations of coagulogen, obtained by salting out procedures from citrated plasma of *A. leptodactylus*, showed elongated chains of globular elements, which resembled simple chains of lobster fibrinogen molecules (Figure 4) (unpublished results). The basic component of the plasma coagulogen is made up of two 200-kDa subunits.[30] These subunits seem to be identical, because they fuse on a single precipitate line in immunoelectrophoresis[36,37] (Figures 3 and 5). In fact, the polymers of this circulating coagulogen do not seem to form a coagulogen but rather a soluble coagulin. Two types of observations support this hypothesis. Firstly, the clotting time of lobster plasma is considerably extended by amines that inhibit the crosslinking of vertebrate fibrin.[96-98] When low doses of amines are used, the higher number of bands detected in the polymer scale indicates that coagulation has been initiated. Secondly, after treatment with soybean trypsin inhibitor, the bands of polymers are present but appear finer and slightly shifted towards higher M.W., perhaps as a result of slight proteolysis. These latter polymers might represent aggregates of coagulogen even though the other polymers could be uncrosslinked fibrin polymers.[37]

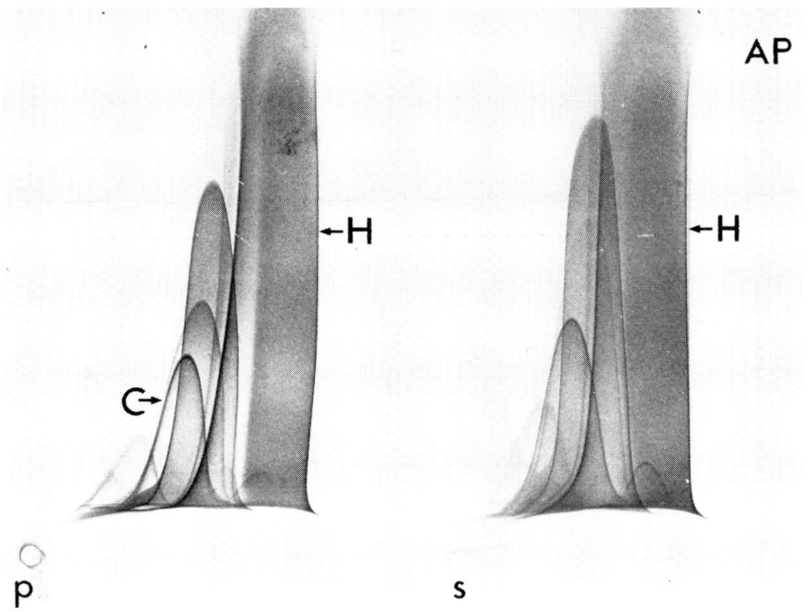

FIGURE 5. Crossed immunoelectrophoresis of hemolymph from *Homarus vulgaris*. Symbols are as in Figures 2 and 3.

2. Amount of Coagulogen

In healthy, freshly caught animals, the clotting processes occur rapidly. In many cases, the existence of the clottable plasma protein has been demonstrated, while in other cases, it could not be found by electrophoresis. These conflicting data could be explained if plasma coagulogen titers would sometimes be too low to be easily detected. In fact, the clottable factor seems to be present in all animals, but large variations in concentrations between species have been reported, from traces in certain crabs to 20% of total blood proteins in lobsters. Moreover, significant fluctuations have been noted during the molting cycle. In *Gecarcinus lateralis,* hemolymph coagulation was reported to be strongest during the premolt stages;[68] identical observations have been made in different Decapoda. However, in *A. leptodactylus,* the concentration of the clottable factor is not related to the molting stage.[71] On the other hand, it should be noted that the amounts of Ca^{2+} and Mg^{2+} ions increased during the premolt stages (review in Jeuniaux[29]). Also, the number of circulating hemocytes increased significantly in *Eriocheir sinensis* under definite physiological conditions, such as molting.[100]

The plasma gelation can be disturbed or abolished, depending on nutritional status, irradiations undergone by the animal, or internal parasites. It is known that starvation in Crustacea induces a decrease in total protein content of blood, including plasma coagulogen,[101-103] that explains the alterations observed in the clotting process. After irradiation, the total protein content of the hemolymph appears unchanged in *A. leptodactylus.* Although plasma clotting is abolished, the amount of coagulogen remains identical to that of the control animals[104] (Figure 6). On the other hand, the number of hemocytes, especially granulocytes decreases. During the active infestation by certain parasites, the hemolymph does not clot, probably because the host exhibits a defense reaction, consisting of widely internal coagulation. In these animals, the amount of remaining plasma coagulogen is not high enough to permit clotting when blood is withdrawn[105-109] or in wound repair.

In *Homarus,* the clotting defects, which are always observed through the trial period of infection by the parasite *Aerococcus viridans,* sometimes depend on the concentration of

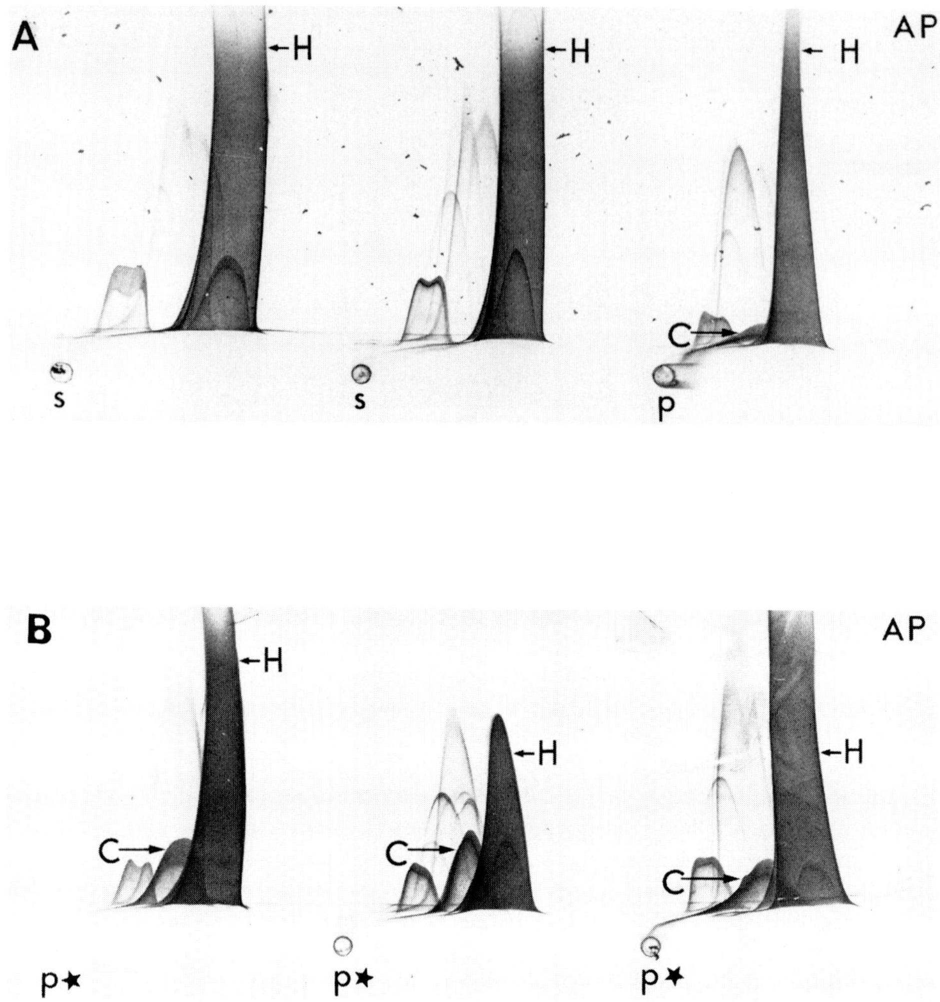

FIGURE 6. Crossed immunoelectrophoresis of hemolymphs from irradiated *Astacus leptodactylus*. (A) Sera (s) and plasma (p) from control animals. (B) From left to right, plasmas (p*) of animals irradiated with 2000, 4000, or 6000 R γ rays.

clottable factor[41,42] (Figure 7). In fact, in lobsters, a close relationship was found between plasma clotting time and number of hemocytes.[40,110] Moreover, the intensity of gelation was directly correlated with the number of circulating hemocytes,[40] particularly the percentage of granulocytes.[41] It has also been noted that the number of hemocytes, especially granulocytes, increases during postmolt stages C2-C3, and mainly in premolt stage D2.

These variations indicate that granulocytes play a key role in the synthesis or release of factors involved in these processes.

3. Other Functions of Coagulogen

In the crayfish *A. leptodactylus,* plasma coagulogen as well as the main molt-related proteins have been found at a relatively high level in compartments other than hemolymph, such as soft integument and calcified cuticle[74,76] (Figures 8 and 9). Sometimes, the coagulogen and the molt-related fraction, named F2, can be melanized in the soft integument extracts.[111] Dialysis of *Astacus* plasma against distilled water yields a precipitate, comprising carotenoproteins and glycoproteins, including the coagulogen.[112] These authors noted that this

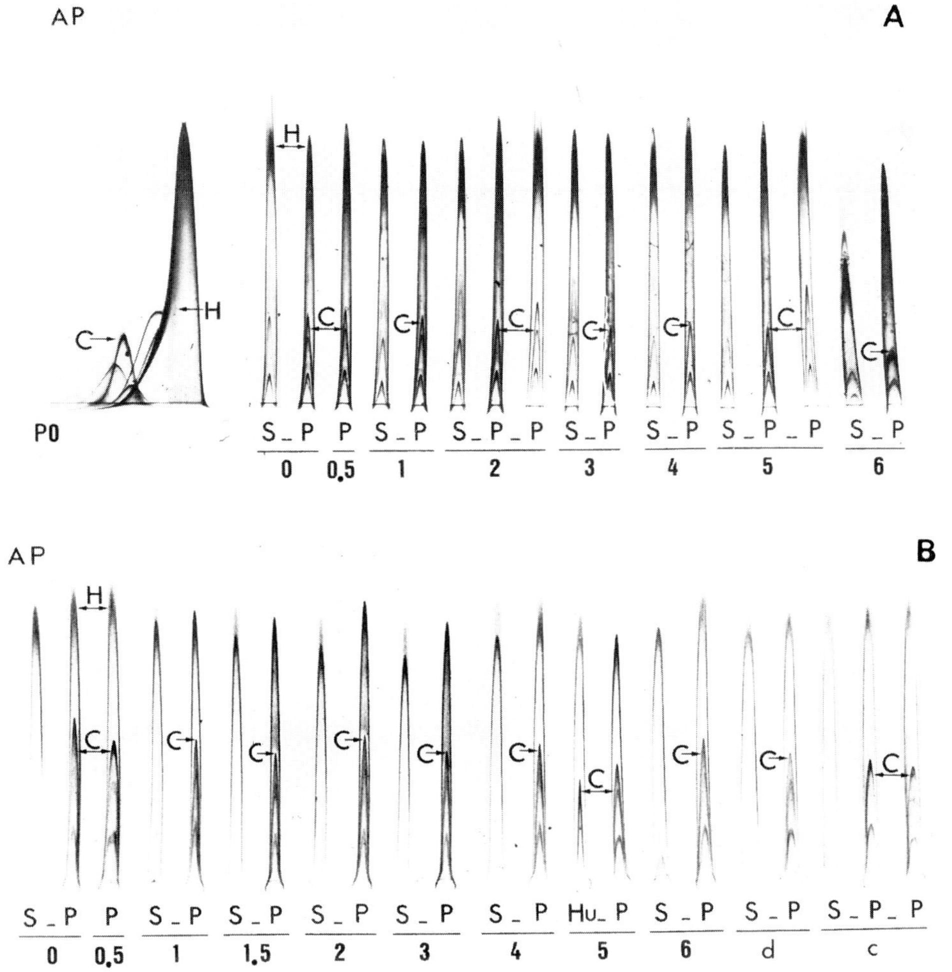

FIGURE 7. Effects of the pathogenic *Aerococcus viridans* on blood proteins of *Homarus vulgaris*. (A) Crossed and rocket immunoelectrophoresis of hemolymph from control lobsters kept at 10°C. From left to right: plasma at time 0, and serum and plasma on day 0; plasma at time 0.5 day, serum and plasma on day 1, serum, plasma with sodium citrate and plasma with NEM on day 2, serum and plasma on day 3, serum and plasma on day 4, serum, plasma with sodium citrate, and plasma with NEM on day 5, serum and plasma on day 6. (B) Rocket immunoelectrophoresis of hemolymph from *Homarus vulgaris* inoculated with *Aerococcus viridans* and kept at 10°C. From left to right: serum and plasma on day 0, plasma 0.5 d after infection, serum and plasma 1 d postinfection, serum and plasma 1.5 d after infection, serum and plasma 2, 3, and 4 d postinfection, uncoagulated hemolymph (Hu) and plasma 5 d postinfection, serum and plasma 6 d postinfection, serum and plasma from just after death lobsters (d), serum and two plasmas from control lobsters (c). Other symbols are as in the preceding figures.

precipitate may be brown, depending on the season. Thus, these proteins that sometimes turn brown during the experiments, might correspond to the 400 and 150 kDa proteins observed in the blood of *Uca pugilator,* which act as carriers of tanning agents into the cuticle.[113,114] These authors also reported that in the same species, granulocytes contain protein-end groups in their cytoplasm as well as granules that might constitute portions of the protein-carrier complex.[115] It is noteworthy that in *P. interruptus* and *A. leptodactylus,* the plasma coagulogen is a glycoprotein with glucose-mannose residues.[30,111] The main molt-related protein also exhibits numerous glucose-mannose residues. These two proteins strongly react with the lectin concanavalin A (Con A)[111] (Figure 10).

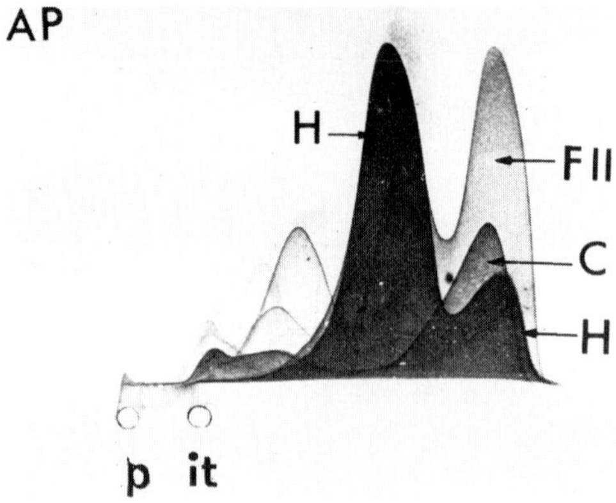

FIGURE 8. Tandem-crossed immunoelectrophoresis of proteins from plasma (p) and premolt soft integument extracts (it) of *Astacus leptodactylus*. Other labels are as in the preceding figures.

In *H. vulgaris*, plasma coagulogen appeared sometimes together with the agglutinin fraction (Figure 11) and the two were probably linked by the interaction of their carbohydrate moieties. Moreover, physicochemical similarities between agglutinin, "fibrinogen", and opsonin were found in lobsters.[25,30,31,116] It is possible that many of these humoral proteins are structurally and/or functionally related and interact during the defense mechanisms in Crustacea.

Plasma coagulogen stains with various degrees of intensity after nitroblue tetrazolium or Sudan black treatment. Its content in both lipid and sugar moieties varies according to the species investigated and the molting stage (unpublished results). This component might act as a lipid carrier.

IV. CELLULAR FACTORS

Recent data indicate that the basic mechanisms of cellular blood clotting in Crustacea and Limulidae are so similar that some of the *Limulus* data may be directly relevant to Crustacea. Four levels of coagulation have been investigated in Crustacea, and appear to be reminiscent of the mechanisms exhibited by Limulidae.

A. HEMOCYTES

All three types of hemocytes are implicated in the clotting processes. Hyaline and semigranular cells are the first to react, and are the most active; although the reaction appears to be somewhat slower, but is more sustained in granulocytes. These three types of cells may be separated from whole hemolymph by density gradient centrifugation.[117,118] The analysis of the different cell fractions shows that *in vitro* these cells differ not only in morphology, but also in biochemical composition and behavior.[119]

Hyaline cells lack phenoloxidase activity, but phagocytosis is their basic property.[120] They appear to be unaffected by foreign molecules, such as microbial products, β-1,3-glucans or lipopolysaccharides (LPS), but their phagocytic activity is stimulated by the prophenoloxidase (proPO)-activating components *in vitro*.

The semigranular cells contain only small amounts of proPO. Phagocytic activity is weak and abolished upon degranulation. Foreign molecules trigger degranulation of these cells, and lysis commonly occurs with proPO-activating components. The semigranular cells of the freshwater crayfish *A. leptodactylus* are also principally responsible for the encapsulation process *in vitro*.[121]

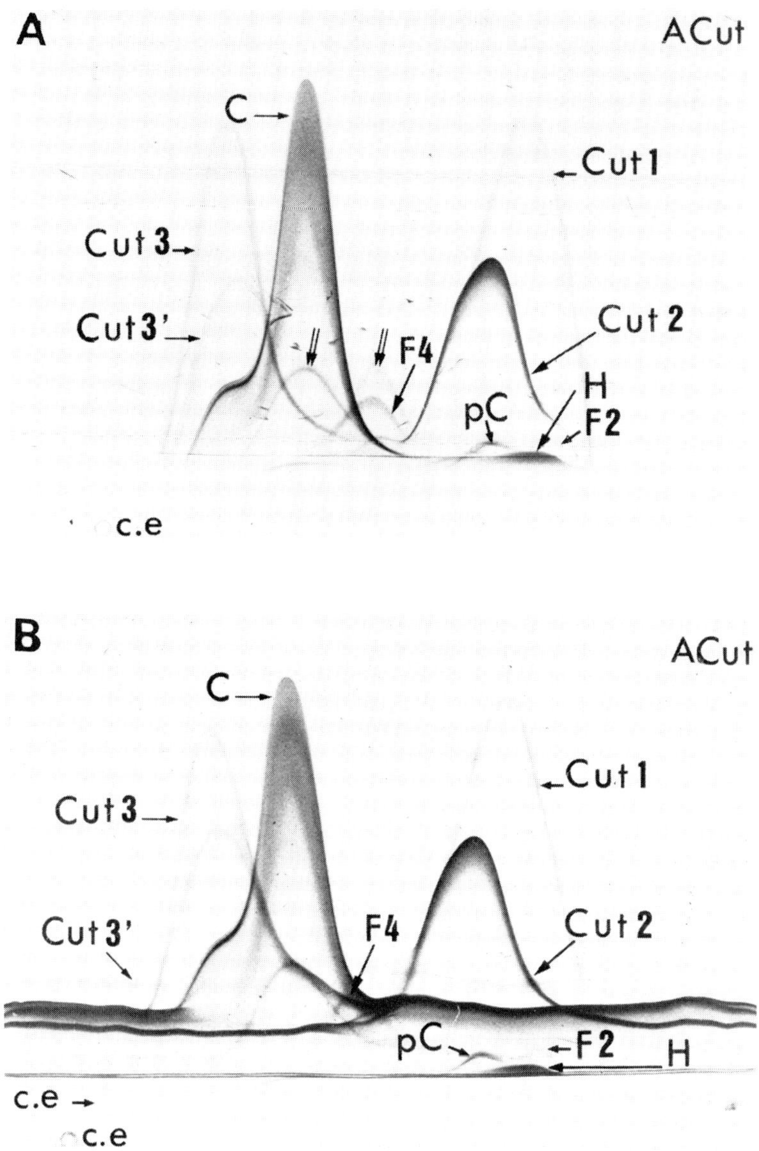

FIGURE 9. Crossed immunoelectrophoresis (A) and crossed-line immunoelectrophoresis (B) of cuticular proteins extracted from the calcified shell of *Astacus leptodactylus*. Symbols: A Cut = anticuticular extract serum poured into the gel; c.e = sample of cuticular extracts deposited in the circular well and in the line; Cut 1, Cut 2, Cut 3, and Cut 3′, show the main specific cuticular proteins; C = crustacyanins; pC = plasma coagulogen; H = hemocyanin; F2 = molt-related protein; and F4 = an undefined factor.

Lastly, the granulocytes contain large amounts of the proPO-activating components (PPA) most likely within secretory vesicles.[118-123] These cells are nonphagocytic and degranulation is usually not observed after treatment with glucans or LPS, but in the presence of active proPO, they undergo exocytosis.

Besides Hardy's[9] ''fibrin-ferment'', which seems to be transglutaminase that acts on the ''fibrinogen''-like factor, all three types of hemocytes contain several other factors involved in different steps of coagulation. Thus, a clottable protein was found within the hemocyte

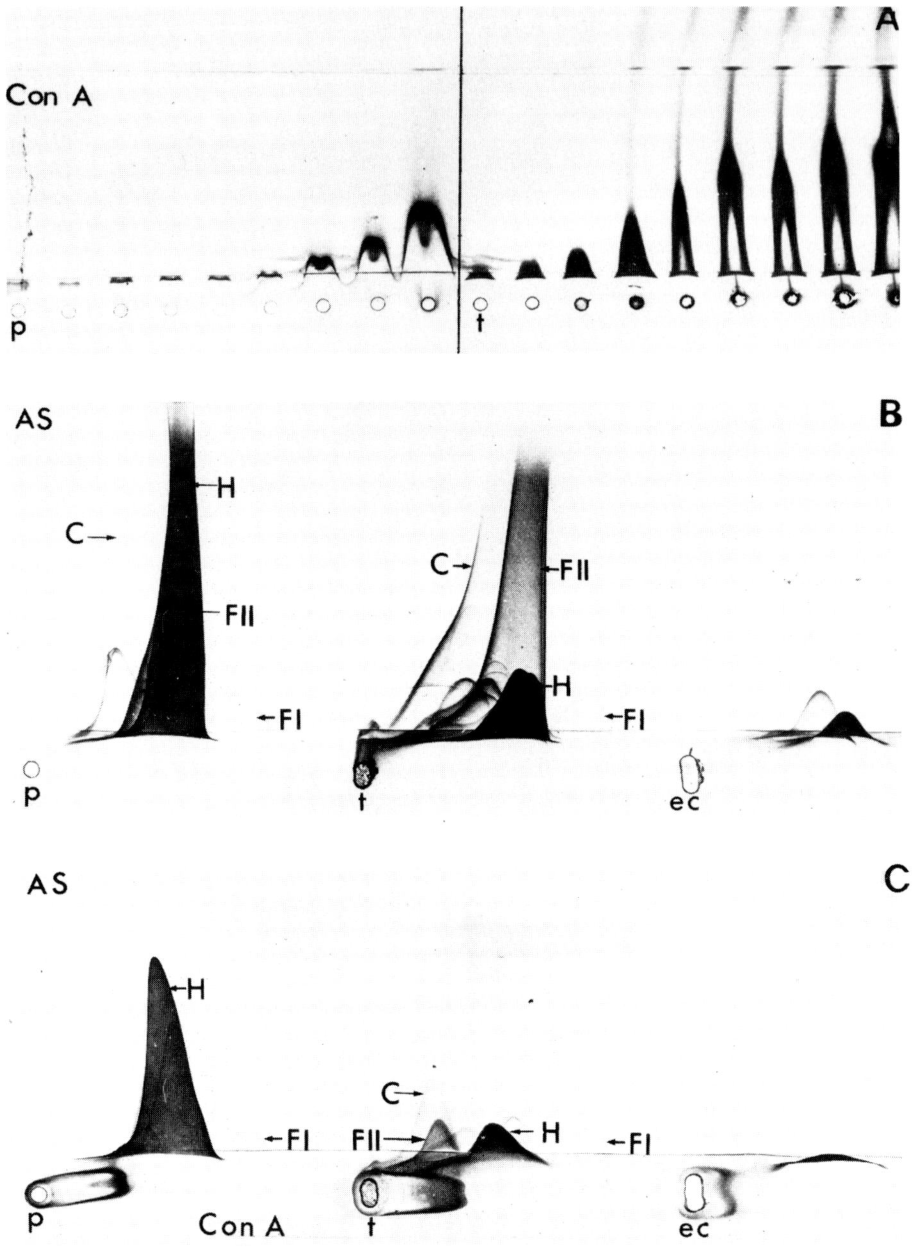

FIGURE 10. Affinoelectrophoreses of samples of hemolymph and integument extracts from crayfish into a gel containing the Con A lectin. (A) From left to right, nine increasing amounts of plasma (p), then nine increasing amounts of soft integument extracts (t). The gel contained 250 μg/cm² of lectin. (B) and (C) Crossed immuno-affinoelectrophoresis of plasma (p) and different cuticular extracts (t, ec) tested against an antiserum (AS) after migration with Con A in the first dimension. In (B), the gel contained 10 μg/cm² concanavalin; in (C), the gel contained 100 μg/cm² lectin). C = coagulogen, H = hemocyanin, and FI and FII = the molt-related fractions.

granules obtained from the whole hemolymph of *H. vulgaris*[43,44] and within the hemocytes of *O. bipustulatus*.[45,46] Moreover, serine protease enzymes are present in hemocytes of *A. astacus*[48,49] and *Carcinus maenas*.[122] Serine proteases have been observed as restricted to the semigranular and granular cells only in several marine Decapoda.[124]

Finally, in crustaceans, as already noted, the complex cascade of proteins constituting

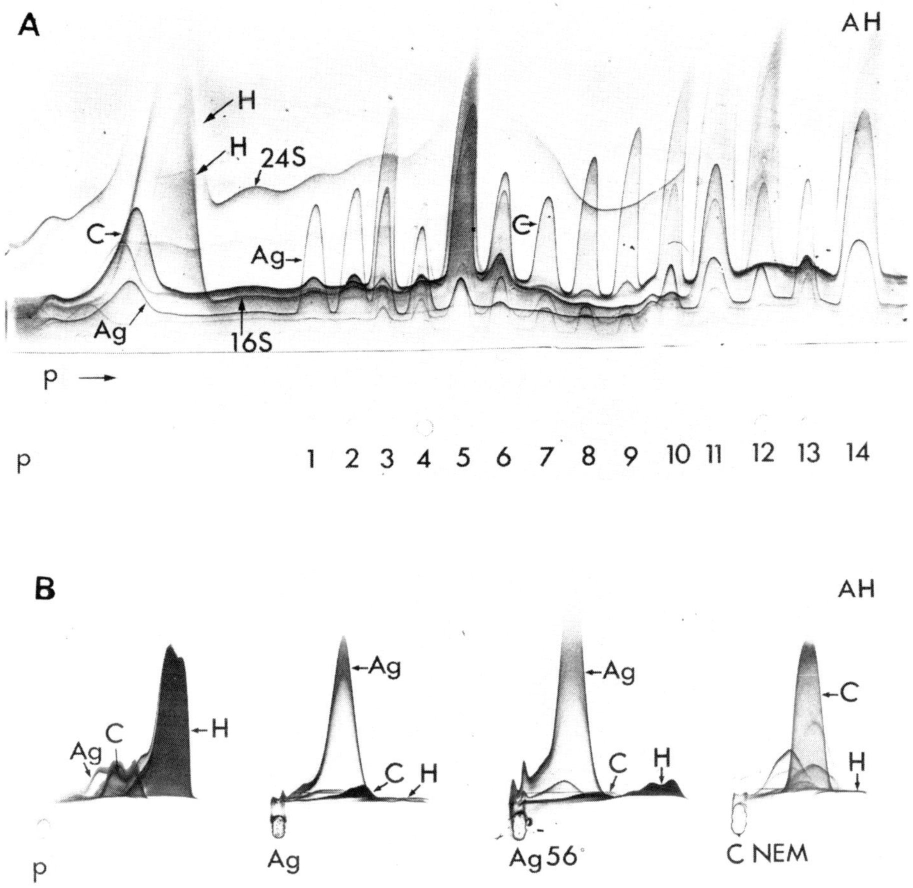

FIGURE 11. Crossed immunoelectrophoreses of samples prepared from the hemolymph of *Homarus vulgaris*. (A) Combined crossed-line and rocket-line immunoelectrophoresis. Experiment was carried out with an anti-hemolymph serum (AH) in the gel and with a plasma (p) poured into the line. From left to right: agglutinins (1, 2), plasma coagulogen with cysteine hydrochloride (3), two coagulogen preparations without inhibitors (4, 5), plasma coagulogen with cysteine hydrochloride (6), NEM (7), sodium citrate (8, 9), and potassium oxalate (10); plasma (11), serum (12), agglutinin (13), and plasma (14). (B) Crossed immunoelectrophoresis performed with an antihemolymph serum-containing gel (AH). From left to right: plasma (p); agglutinin (Ag); agglutinin heated to 56°C (Ag 56°); and plasma coagulogen with NEM (CNEM).

the proPO-activating system, is contained mainly in the granulocytes. However, the proPO cascade can be released from the semigranular cells by exocytosis, if challenged with LPS or β-1,3-glucans. Opsonin-like substances appear to be generated upon proPO system activation[126] as well as a cell-adhesive glycoprotein,[123] and the degranulating factor, which may constitute the "encapsulating promoting factor".[125] The cell-adhesion factor, associated with the proPO system was purified from the hemocyte lysates of *Pacifastacus leniusculus*. It is a 76 kDa glycoprotein which amplifies the release of the proPO system, by degranulating both semigranulocytes and granulocytes in which it is confined.[127] Upon activation of the proPO system, a phagocytosis (opsonic) factor is also produced that stimulates phagocytosis by the hyaline cells.[128] Note also that the degranulating and cell adhesion factors are indeed the same protein.[129] Precisely which of the adhesive proteins in crustacean hemocyte lysates have opsonic properties has not been established. ProPO activation seems to participate also in clotting since hemocyte lysate supernatants reacting with β-1,3-glucans can undergo a gelation reaction *in vitro*.[47]

B. CELLULAR COAGULOGEN

The presence of clottable proteins different from the "fibrinogen" fraction was first reported in the eosinophilic granules of blood cells (supposedly the granulocytes) of *Homarus*.[44] Under certain conditions, the protein content of the granules forms a gel. The suspension of granules can be dissolved by 8 *M* urea in very acidic or basic medium.

A cellular coagulogen has been identified in the hemocyte granules of the sand crab *O. bipustulatus*.[45,46] This crab coagulogen appears to be a single-chain protein with disulfide bonds. This 70-kDa protein is hydrolyzed in the presence of trypsin, and clots in the presence of human thrombin, human factor XIII, and Ca^{2+} ions. As in vertebrates, the sand crab fibrin superstructure is stabilized by covalent bond formation between the fibrin monomers, and this is accomplished by a transglutaminase enzyme, mainly present in the cell-free hemolymph.

Madaras et al.[45,46] seem to be the only ones to obtain such precise biochemical characterization of a crustacean cellular coagulogen. In many cases, only indirect observations suggested its presence. Thus, we observed that the hemocytes of various Decapoda form a cell clump involving clot retraction. Also, propanolol, which in *Limulus polyphemus* preserves the granule membrane but does not prevent amebocyte disruption,[130] delayed plasma gelation, which remained incomplete in *H. vulgaris*.[37] Moreover, in our experiments, urea and acid-treated hemocyte lysates of *A. leptodactylus* or *H. vulgaris* exhibited one or two faint factors, which precipitated around the well or moved towards the cathode on immunoelectrophoresis plates and on 6% acrylamide plates. The cathode components (basic proteins) disappeared by gelling in the tube after a thawing-freezing cycle, and in basic medium. However, we were unable to isolate these components because precipitates and aggregates appeared during the experiments, when the medium pH was no longer acidic. Direct coagulation of these hemocyte lysates with serine protease enzymes, in the presence of factor XIII and Ca^{2+} ions, did not occur, as was achieved with *Limulus* amebocyte lysates.[131] At most, conspicuous flocculation appeared in the samples we tested. The firm retracted clot obtained after freezing-thawing cycles of hemocyte lysates stored in nonsterile tubes, could be separated from the supernatant, and after washings, dissolved in SDS and mercaptoethanol buffer. This cellular "coagulin" exhibited many bands in electrophoresis, including one especially clear fraction located at about 23 kDa. It was not observed in the supernatants of clotted hemocyte lysates. In supernatants of hemocyte lysates prepared in pyrogen-free conditions and treated by the acidification procedure of Solum,[131] several fractions were also observed between 14 and 30 kDa. One of them, of about 29 kDa, was visible in all these supernatants[132] (Figure 12).

These fractions of 29 and 23 kDa recall by their molecular weights those found for the coagulogen and coagulin of Limulidae.[131,133-136] However, it was not possible to ascertain the identity of these 29- and 23-kDa proteins, since we were unable to show that after clotting, the 29-kDa band always disappeared from the supernatants, whereas the 23-kDa fraction appeared in the precipitate.

In Crustacea, the presence of a cellular coagulogen undergoing limited proteolysis through the activation of several enzymes still requires confirmation.

C. SERINE PROTEASE ENZYMES

The cellular coagulogen of *O. bipustulatus* appears to be directly clottable by an enzyme belonging to the serine protease type, such as trypsin and thrombin.[45,46] Identical observations have been reported for the flocculation by trypsin of *A. astacus* hemocyte lysates.[47] Indirect evidence of the involvement of these enzymes is that, in lobsters, certain serine protease inhibitors, such as PMSF (phenylmethane sulfonyl fluoride), DFP (diisopropyl fluorophosphate), or STI (soybean trypsin inhibitor) impair clotting and interact with cellular clumping.[37] The soft gel, when formed, is easily dissociated, probably owing to clotting defects.

In two crayfish species (*A. astacus* and *P. leniusculus*), the coagulation process in hemocyte lysates is also inhibited by STI or DFP and is both Ca^{2+} and temperature dependent.[47] It should be noted that the acidification procedures used to inhibit coagulation in *Limulus* amebocyte (= granulocytes) lysates[131] considerably delayed the gelation of hemocyte lysates from *A. leptodactylus* and *H. vulgaris*.[132] The serine protease identified in the hemocytes of *A. astacus*[48] is highly specific, and hydrolyzes chromogenic substrates, such as the anilide substrate for Factor Xa, and other glycine-arginine-containing anilide substrates with four amino acids in their peptide chain. However, synthetic peptides with three amino acids and containing glycine-arginine are not hydrolyzed.[49] Thus, crayfish hemocyte serine protease has about the same specificity as Limulidae protease. An analogous serine protease enzyme is present in the hemocytes of *C. maenas*[122] and certain other marine Decapoda.[124] It is noteworthy that an analogous activating enzyme, also a serine protease, has been purified from both *Bombyx mori*[137] and *A. astacus* cuticles.[49] Another serine protease S implicated in the activation of prophenoloxidase was also found in hemocyte lysates from *A. astacus*[125] that elicited a clotting reaction.

D. ACTIVATION OF SERINE PROTEASE ENZYMES

The serine protease enzymes involved in the coagulation can be activated via two independent pathways (Table 1).

1. LPS-Mediated Coagulation

Numerous data indicate that endotoxins from various pathogens activate the clotting processes (Table 2). Thus, the endotoxin of an unidentified Gram-negative marine vibrio, known to be pathogenic to *L. polyphemus*, the lipopolysaccharide B from *Escherichia coli*, and the heat-killed Gram-positive bacterium *Gaffkya homari* (= *Aerococcus viridans*), can all activate the coagulation system in the parasitic crustacean *Sacculina carcini*.[106] However, in the plasma of *Sacculina*, neither cells nor platelet-like bodies are observed, and the clotting process does not appear to be Ca^{2+} dependent. Further, not clot retraction or lysis is demonstrable *in vitro* or *in vivo*. An explanation of this unusual system might be that *S. carcini* is a highly specialized parasite with an exceptional coagulation system.[106] In lobsters, the administration of endotoxin from Gram-negative *Vibrio*, reduces the concentration of circulating hemocytes, and *Shigella* toxin produces amebocytopenia, intravascular coagulation, incoagulable plasma, and death of animals.[107] In *C. maenas* and *M. squinado*, infection with bacteria also causes amebocytopenia and abnormalities of coagulation.[107] In several crabs (*M. squinado*, *C. maenas*, and *Portunus depurator*), the blood loses its coagulability, following accidental or experimental infections by the ciliate *Anophrys*, a virus, or a bacterium. The blood of these crabs does not coagulate during active infection, apparently because intravascular clotting has taken place in the live animal.[108,109,138] Analogous intravascular coagulation has been observed in *Callinectes sapidus* infected with *Paramoeba perniciosa*.[139] In *Cancer irroratus*, a *Vibrio* sp. is pathogenic over a wide range of temperatures. It seems that high mortalities in experimentally infected crabs result from hemocyte clumping and extensive intravascular clotting triggered by an endotoxin.[140] The case of the Gram-positive bacterium *Aerococcus viridans* (formerly *Gaffkya homari*) seems appropriate here. This bacterium induces clotting defects in lobsters.[40,141] Hemocyte counts always decline, but depending on environmental temperature, the amount of plasma coagulogen detected is either normal or less than normal (Figure 7), perhaps because of the intravascular destruction induced by widespread bacteremia.[41,42] Early workers in this field suggested that the endotoxin of *A. viridans* is not involved in the defects observed during infection. Yet the alterations we observed in hemocyte patterns recall those seen in vertebrate platelets, following different protocols of infusion of endotoxin from Gram-negative bacteria.[142] Pathogenic strains of *A. viridans* possess a capsule with an alcianophilic acidic polysaccharide,

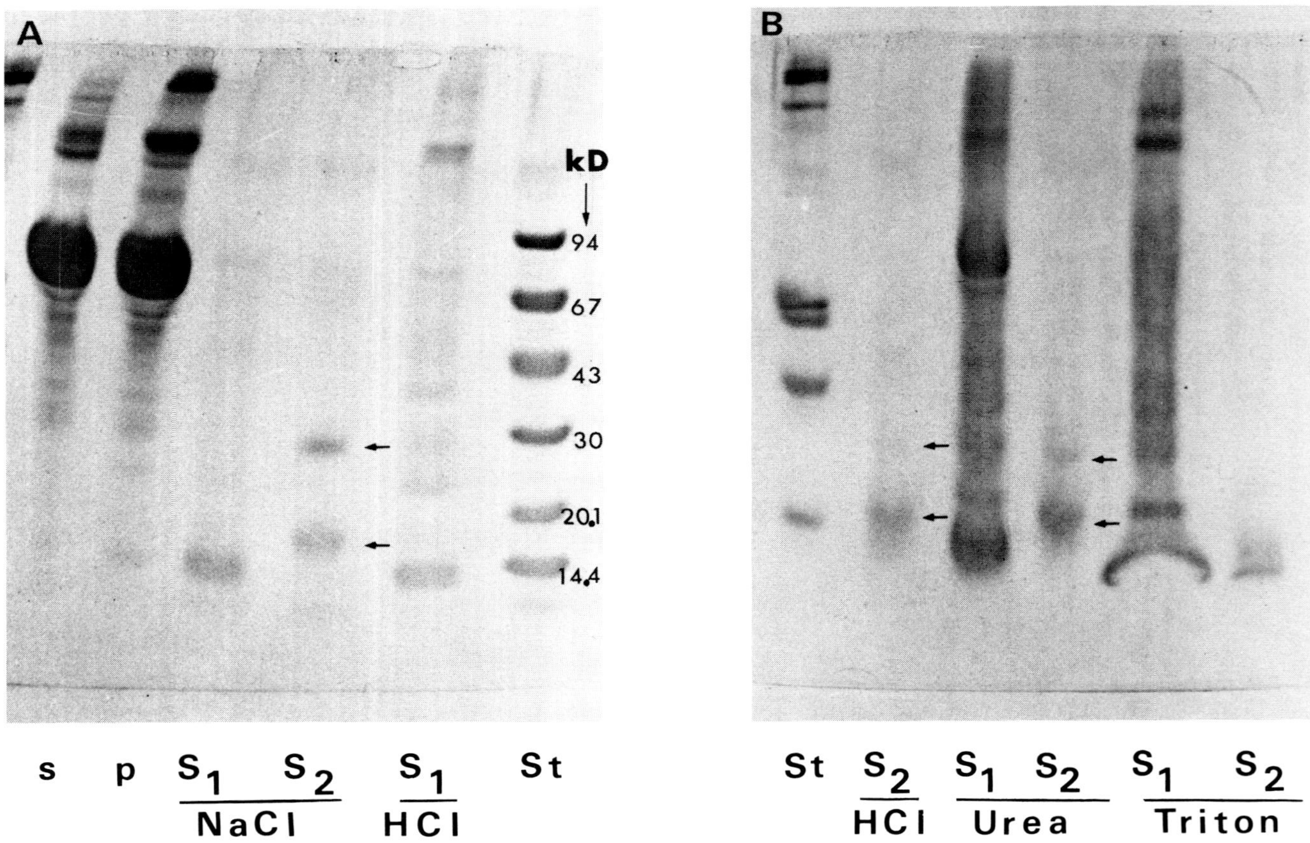

FIGURE 12. SDS-PAGE on 4/30 gradient gels of samples obtained from the hemolymph of *Homarus vulgaris*. (A) From left to right: serum (s), plasma (p), three different supernatants (S1, S2, S1) of hemocyte lysates with EDTA submitted to acidification, and LMW standards (St). (B) From left to right: HMW standards (St), five supernatants (S2, S1, S2, S1, S2) of hemocyte lysates with EDTA submitted to acidification.

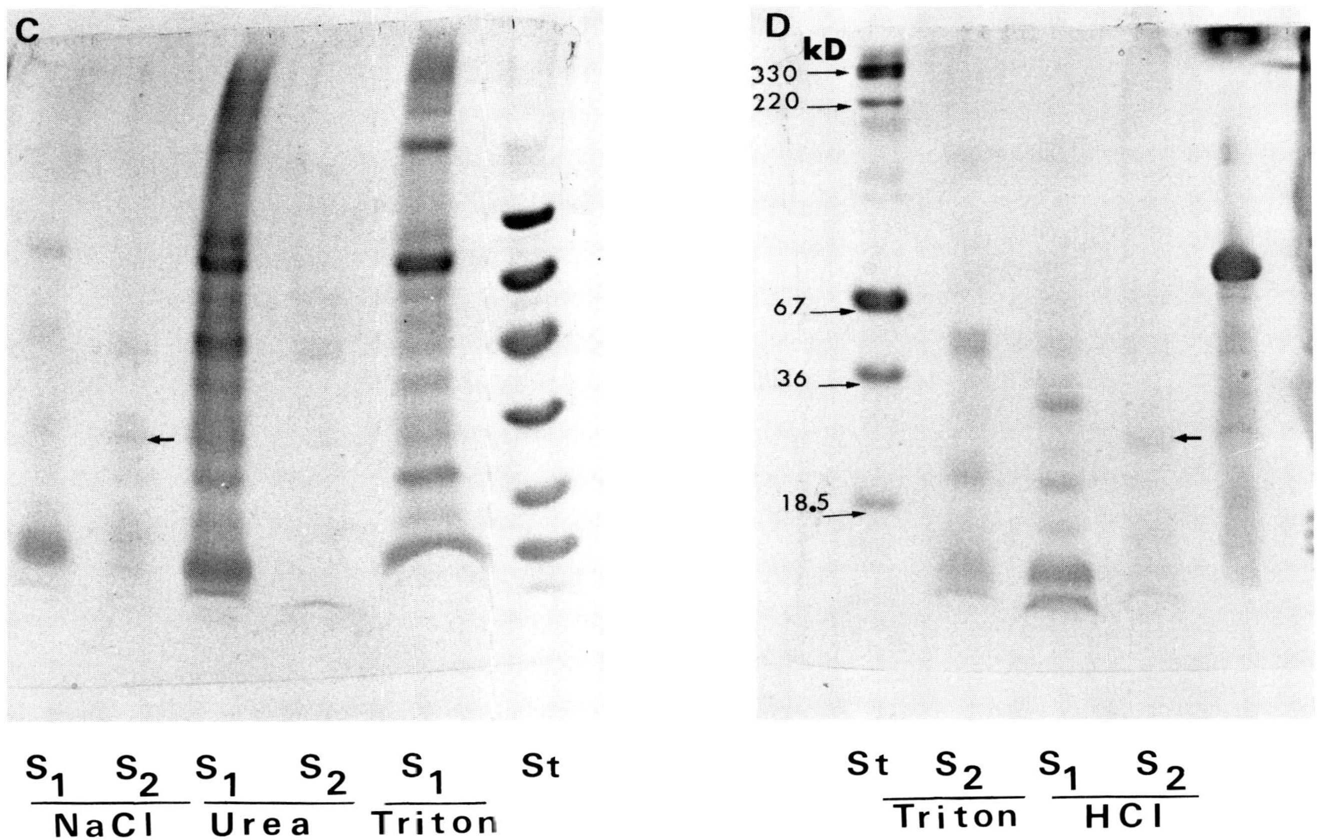

FIGURE 12. (C) Five different samples (S1, S2, S1, S2, S1) with NEM submitted to acidification and LMW standards (St). (D) HMW standards (St), and three supernatants (S2, S1, S2) in NEM submitted to acidification. The arrows show the fractions of 23 and 29 kDa.

TABLE 1
Schematic Representation of Possible Clotting Processes in Decapoda Hemolymph

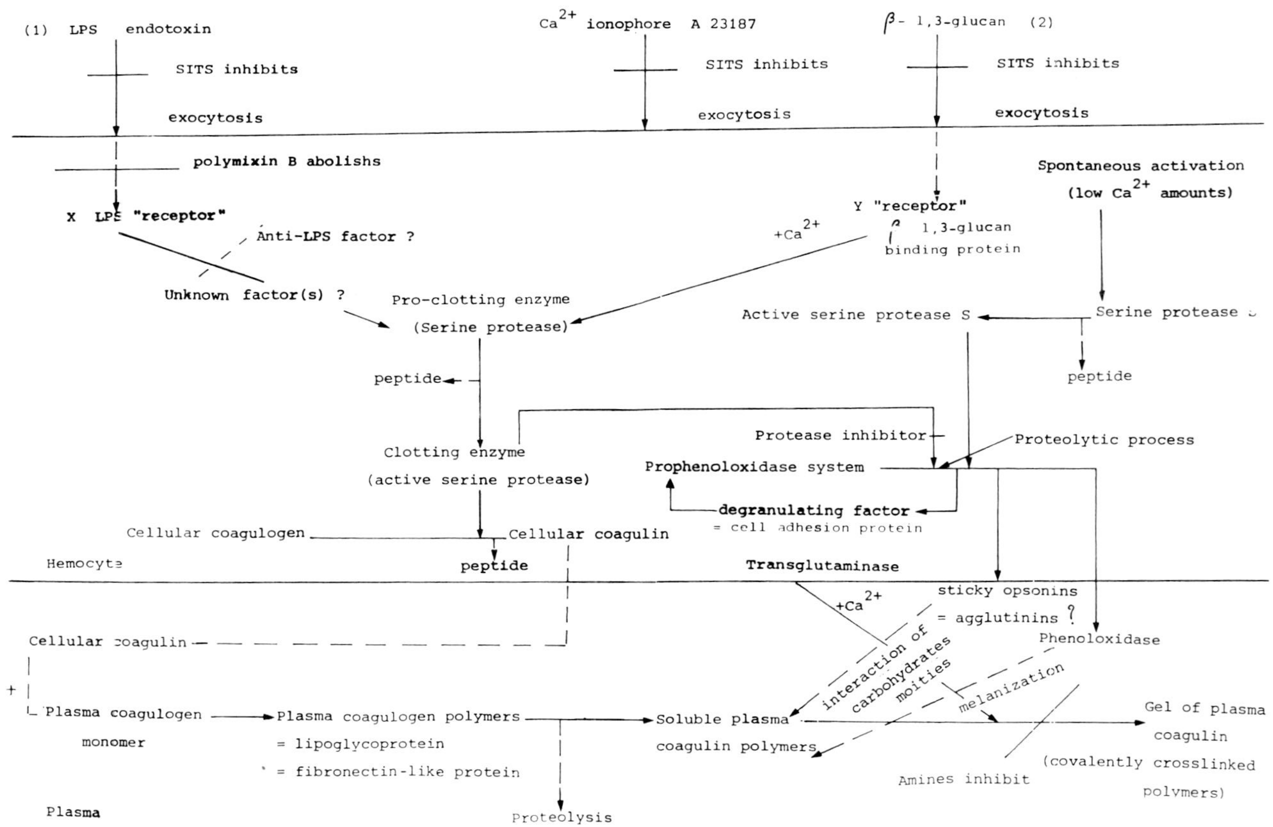

Note: Dashed lines indicate known processes and dotted lines indicate hypothetical processes.

TABLE 2
Examples of LPS-Induced Intravascular Coagulation in Crustacea

	Pathogens						
Hosts	**Gram-negative *Vibrio* or its endotoxin**	**LPS from *Escherichia coli***	**Gram-positive *Gaffkya homari* = *Aerococcus viridans***	***Shigella* toxin**	***Paramoeba perniciosa***	**Unidentified bacteria**	**LPS from Gram-negative bacteria: *Salmonella abortus* *Serratia marcescens***
Sacculina carcini	+ + Barker & Bang (1966)	+ + Barker & Bang (1966)	+ + Barker & Bang (1966)				
Homarus vulgaris *Homarus americanus*	± Levin (1967)		± Durliat & Vranckx (1983)	+ + Levin (1967)			
Carcinus maenas						+ + Levin (1967); Bang (1967, 1970a,b)	
Maia squinado						+ + Levin (1967); Bang (1967, 1970a,b)	
Portunus depurator						+ + Bang (1967, 1970a,b)	
Callinectes sapidus					+ + Johnson (1976)		
Cancer irroratus	+ Newman & Feng (1982)						
Astacus astacus		+ + Söderhäll & Häll (1984)					+ + Söderhäll & Häll (1984)

Note: + + indicates a strong intravascular coagulation; + represents a moderate reaction; ± indicates a doubtful reaction.

but the nonvirulent strains do not.[143] Perhaps the alterations observed in *Homarus* blood after infection by *A. viridans* might be due to the polysaccharide moiety of this capsule, probably a glucan, and not to the endotoxin.

It, therefore, seems more than likely that endotoxins from diverse bacteria can activate clotting processes in Decapoda. Other experiments have, in fact, demonstrated that LPS from Gram-negative bacteria activated both protease S and prophenoloxidase, and elicited a clotting reaction in hemocyte lysates from *A. astacus*. The remaining inactive proteases might then have been activated by the addition of LPS or β-1,3-glucan. Further, when these lipopolysaccharides are treated by polymixin B, they do not produce any protease activity in crayfish hemocyte lysates.[144] In *Limulus,* it is known that polymixin B binds to LPS, and thereby destroys its pyrogenic activity, thus preventing the induction of clotting enzyme activity. Curiously, concentrations of LPS above 10^{-4}g ml^{-1} prevent enzyme activation *in vitro,*[144] but the significance of this result remains obscure.

LPS also acts as the proPO-activating system of crayfish hemocyte lysates by inducing degranulation (exocytosis) and subsequent lysis *in vitro* of monolayers of semigranular hemocytes from *P. leniusculus,* whereas the granulocytes remain unaffected.[123] Limited degranulation of granulocytes of the crab *C. maenas* is also induced by LPS[126] and morphological changes were observed in *H. americanus* hemocytes incubated with LPS *in vitro*.[145] This recalls the degranulation induced *in vitro* by LPS in *Limulus* amebocytes.[146] In arthropods, LPS, serves as indicator for Gram-negative bacteria. It is noteworthy that exocytosis of semigranular or granular cells *in vitro* can also be triggered by the Ca^{2+}-ionophore A 23187. In this case, the proPO system is released in its inactive form, which could be activated if LPS or β-1,3-glucans are added.[123] In crayfish, separate receptors exist for LPS and glucans, and LPS enters the cascade of the proPO-activating system through a different pathway than β-1,3-glucans, since a LPS-inhibited system could be activated by addition of β-1,3-glucans.[144,147]

At the present time, the whole repertoire of proteins, which participate in the conversion of soluble coagulogen into coagulin and then into the final clot has not been established, but all these recent works prove that in Crustacea, LPS from Gram-negative bacteria activates at least a serine protein enzyme involved in coagulation processes. In particular, it appears that the serine proteases are activated by limited proteolysis to produce active enzymes and a number of small, but undefined peptides.[119]

Some ten years ago, it was observed that the hemolymph from two crustaceans (*Astacus nigrescens, Homarus* sp.) did not inactivate endotoxin from *E. coli,* as do the sera from animals higher in the phylogenetic tree.[148] However, in the blood of crayfish, bactericidal activity seems to exist against endotoxins. Two facts support this hypothesis. First, clotting tests performed with Ca^{2+} solutions on *Astacus* hemolymph, withdrawn in the presence of Ca^{2+} ion chelators, are always negative when the anticoagulant is prepared in distilled water, whereas these tests are positive when the same anticoagulant is prepared in the presence of NaCl and KCl.[37] Cellular clotting appears to be mediated by some endotoxins, which might activate the proclotting enzyme in the presence of only Na^+ or K^+ ions, as in *Limulus*. Secondly, doses of endotoxin from *Salmonella typhimurium* (Gram-negative bacterium) increase the resistance of *Parachaeraps bicarinatus* to infection by the pathogenic *Pseudomonas* CP.[149] These findings were confirmed in the lobster *H. americanus* by Paterson et al.[58] who reported increased phagocytosis of bacteria, if lobsters were injected with endotoxin or killed bacteria. The inactivation of endotoxins is indispensable to prevent the induction of intravascular coagulation.

It is interesting that in Limulidae, it was demonstrated that an anticoagulant (anti-LPS factor) inhibiting LPS-induced coagulation is present in the amebocytes.[150]

2. β-1,3-Glucan-Induced Coagulation

For the majority of Crustacea, the proPO system resides in the hemocytes.[151] In *A. astacus,* proPO and serine protease enzymes are only present in the semigranular and granular cells.[117,118,124] All parts of the proPO system seem to be present in the secretory vesicles and the system as a whole is released by exocytosis.[123] In crayfish hemocyte lysate, this proPO is specifically activated by β-1,3-glucan, common to most fungal groups.[151,152] Thus, β-1,3-glucans and the aggregating proteins of the activated proPO system induce rapid degranulation of these two kinds of cells *in vitro.*[118,122] *In vivo,* a marked decline in the number of circulating hemocytes, which is concomitant with the formation of discrete hemocyte clumps in the blood sinuses and gills, is induced in *C. maenas* by injection of β-1,3-glucans but not other soluble glycans.[153] In crayfish, these β-1,3-glucans activate a complex activating pathway, including an enzyme cascade with at least one serine protease, which splits synthetic peptides with the structure R-gly-arg-paranitro-anilide, resulting in a limited proteolytic process.[47-49] This activation causes clotting and PO as well as three or four other proteins to become "sticky".[48,49,125] The degranulation factor produced upon activation of the proPO system, amplifies further release and activation of this system. Furthermore, certain "sticky" proteins may also act as opsonins for crayfish.[122,125,126] An opsonic function for the attaching proteins of a β-1,3-glucan-activated hemocyte lysate was also demonstrated using the phagocytic hyaline cells from the crab *C. maenas,* but PO appeared to lack opsonic properties.[126] These aggregating proteins have not been identified except the cell adhesion factor[127] but, as already noted, it is likely that some of these factors might account for some agglutination activity in crustacean hemolymph.[119] Because the activity of the glucan-treated protease is short, it is an important factor in regulating the coagulation process and proPO activation.[49] As this activating cascade produces fungistatic compounds,[154] it can play an important part in host defense reactions.

Moreover, melanization is visible on at least one blood clotting factor, plasma coagulogen,[111] which sometimes appears linked to a cellular agglutinin fraction. Perhaps these aggregating proteins present in both the hemocytes and the plasma, are closely linked when hemocytes are damaged after hemolymph withdrawal. The same process of melanization occurs in crayfish soft integument, which contains analogous "attaching" factors, i.e., plasma coagulogen,[74] an activating serine protease and a proPO.[49] Consequently, the proPO-activating system and PO might be part of a recognition system.

In addition to β-1,3-glucans, proPO in *A. astacus* is elicited by SDS, trypsin or heat, while the serine protease inhibitor *p*-nitrophenyl *p'*-guanidobenzoate prevents activation of proPO.[155] Note also that a β-1,3-glucan-binding glycoprotein has recently been purified from the plasma and the hemocyte lysates of *P. leniusculus.*[156] This 100-kDa glycoprotein does not show any peptidase or PO activity, but enhances the activation of hemocyte-derived peptidase and proPO only in the presence of β-1,3-glucans. Thus, the action of this β-1,3-glucan-binding protein seems to be at an early stage of activation of the proPO activating system. The high amount of this protein in the plasma may indicate that it may also be involved in other processes in crayfish such as to bind to foreign invaders carrying β-1,3-glucans in their cell walls, and function as an opsonic protein in the removal of these microorganisms from the circulation.[156]

E. INACTIVATION OF SERINE PROTEASE ENZYMES

It is important that Crustacea possess a way of controlling or regulating activations of pro-serine protease enzymes to avoid inappropriate aggregation and clotting reactions. Such regulations have now been shown to occur in crayfish, at both the plasma and cellular levels, which may assist in causing the defense reactions to be restricted to the site of injury.

In the plasma of *Cancer borealis* and *Libinia emarginata,*[157] as well as in crayfish,[158] an α_2-macroglobulin (α_2M)-like activity is present (see also Chapter 11). This purified α_2-macroglobulin is, however, unable to inhibit the proPO-activating enzyme. An analogous situation was reported for *Limulus* α_2-macroglobulin, which does not inhibit the clotting enzyme of ameobocytes (granulocytes)[159] (see also Chapter 10). This α_2-macroglobulin-like protein probably functions in a manner similar to α_2-macroglobulin in vertebrates, i.e., to remove the excess of released proteases, such as those of complement, from the circulation.[158] In the plasma of *Homarus americanus,* a thiolester-containing α_2-macroglobulin homologue was also isolated. It presents properties similar to human α_2M, but consists of only two disulfide-linked subunits of 180 kDa.[160] In *P. leniusculus,* this α_2M-like protein is a disulfide-linked dimer of 190 kDa subunits[161] which contains also thiolester sequences very similar to those of human α_2M and related proteins.[162] This α_2M blocks the ppA (proPO activating enzyme) stored in an inactive form in the secretory granules of the semigranulocytes and granulocytes of crayfish.[163]

On the other hand, another 155 kDa protease inhibitor has been isolated from crayfish plasma.[164] This purified protein inhibits the serine proteases trypsin and chymotrypsin, but not thrombin or subtilisin. It is also able to decrease the activity of the proPO-activating serine protease enzyme. This trypsin inhibitor may represent an important regulator of the proPO-activating system. Thus, it seems that the proPO system is under strict regulation *in vivo* by the plasma protease inhibitors and by the rapid self-inactivation of the serine proteases. However, little is known about the direct regulation of activation of the pro-clotting enzyme, which can be induced by LPS or by β-1,3-glucans.

It is also notable that melittin of bee venom induces degranulation and lysis of isolated granulocytes from crayfish. Yet melittin is an efficient inhibitor of proPO activation.[165] Then, Söderhäll speculated that this type of compound (amphiphilic peptides, which interact with phospholipids) may enable parasites to escape recognition within the hemocoel of their hosts.

A second source of proteinase inhibitors are the blood cells. In crayfish hemocytes, protease inhibitors have been identified. These inhibitors, which are of the subtilisin type, cannot inhibit the proPO-activating enzyme,[166] and may therefore be aimed at proteases from parasitic microorganisms.

Besides the inhibitors found in hemocytes, an analogous protease inhibitor is also present in the soft cuticle of *A. astacus.*[166] It is an active inhibitor of subtilisin, pronase and protease from the fungus *Aphanomyces astaci.*

No inhibitor of the clotting enzyme has been identified in the hemocytes so far. Yet in *L. polyphemus,* during exocytosis, the granulocytes release a series of inhibitors including an acid-labile inhibitor, which is able to inhibit the clotting enzyme[159] (see also Chapter 1). Another acid-stable factor is also liberated from the granulocytes which is active against both serine (trypsin, chymotrypsin) and metal (thermolysin) proteinases.[167]

V. FEATURES OF CLOTTING CELLS

Both granulocytes and semigranulocytes of decapods degranulate *in vitro* when β-1,3-glucans or LPS are present, as do the granulocytes of Limulidae *in vitro* when the cascade coagulation system is activated. Further, marginal microtubules bands (MBs) are present in the clotting cells of several species of marine crabs[168] and those of *Homarus,*[169] as well as in the granulocytes of *Limulus.*[170] Under microtubule-stabilizing conditions, lysis produces granule-free clots in which almost every nucleus has an associated MB. It is striking that both arthropod granulocytes and mammalian platelets contain MBs, since they are both functionally involved in blood clotting. Both contain clotting factors inside cytoplasmic granules, which are released after activation of the coagulation system, and the morphology of both platelets and granulocytes changes drastically when they come into contact with surfaces and during clotting.[170,171]

In *L. polyphemus,* degranulation-inhibition experiments were performed using one of the following three agents: prostaglandin I2, which stimulates adenylate cyclase; db-cAMP (3′,5′-dibutyryl cyclic adenosine monophosphate), which crosses the membrane and raises cAMP levels; or theophylline, which inhibits phosphodiesterase. The results strongly suggest that cAMP has a role in mediating exocytosis in *Limulus* amebocytes (granulocytes) in response to bacterial endotoxin.[146] These three drugs prevent endotoxin-induced exocytosis, and, when administered before cell spreading, inhibit all the changes in shape connected with motility and spreading on the substrate.[146] Moreover, an endotoxin-binding protein that possesses calmodulin-like activity was isolated from the amebocyte plasma membrane of *Limulus*. This endotoxin-binding protein stimulates adenylate cyclase of amebocyte membrane, and increasing levels of intracellular cyclic AMP promote secretion by the amebocytes.[172]

In vertebrates, the cytoplasmic concentration of free calcium in intact platelets is controlled by stimulators of cAMP production, such as prostaglandins or forskolin.[173] c-AMP prevents nearly all platelet activations. In contrast, stimulation of thrombin increases free Ca^{2+} in platelets.[174] Cell coagulation in Decapoda also greatly depends on the concentration of Ca^{2+} ions, which might be regulated by cyclic nucleotide levels. Few data deal withthis point but curiously, in *H. americanus,* a highly specific octopamine-stimulated increase in cAMP in hemolymph raises the rate, and changes the nature of the clotting reaction.[175] In crayfish, inhibition of exocytosis by the calmodulin (calcium-binding protein) antagonist calmidazolium was demonstrated only for the granulocytes. Calmidazolium does not influence the biochemical activation of the proPO cascade by LPS or β-1,3-glucan.[123] In fact, calmodulin mediates exocytosis, which might be a Ca^{2+}-dependent process, but does not seem to be involved in the proPO cascade activation. According to these authors, "the biochemical activation of the proPO enzymes, thus, does not involve any component of the plasma membrane. On the other hand, to evoke degranulation of semigranular cells, putative receptors may be present on the plasmalemma".

Probably, there also exists in crustacean hemocytes a synexin-like protein. In some mammalian tissues, such as human platelets, Ca^{2+} action in exocytosis may be dependent on the function of synexin, a protein, which mediates vesicle plasma membrane fusion. Synexin might be the intracellular receptor for Ca^{2+} in the process of exocytosis.[176] Further investigations are necessary to determine how the rise in hemocyte cytoplasmic Ca^{2+} is regulated in these arthropods and the possible role of cAMP.

Moreover, a variety of agents able to block the transport of monovalent anions across cell membranes, also delay induced exocytosis. Thus, in crayfish hemocytes, the anion channel blocker SITS (4 acetamido-4′-isothiocyanato-stilbene-2,2′-disulfonic acid, disodium salt) inhibits degranulation of granulocytes as well as degranulation and lysis of semigranular cells, irrespective of the exocytotic stimulus (LPS, glucan, or ionophore).[123,126] In contrast, mellitin, which induces degranulation, may render the pores in membrane more permeable to anions.[165] It is known that in the erythrocyte membrane, anion exchange is mediated by a specific granule membrane protein of about 100 kDa, and is competitively inhibited by several specific drugs, such as SITS.[176]

The possible general importance of synexin, anion exchange sites, and cyclic nucleotide amounts in the exocytosis processes needs to be determined in future studies and remains open for investigation.

VI. CONCLUSIONS

In Crustacea, hemolymph coagulation represents a complex enzyme cascade involving cellular and plasma factors. In certain Decapoda, a plasma clotting is observed, which in some respects resembles the final step of vertebrate blood coagulation, i.e., the transformation of soluble fibrin polymers by Factor XIII in the presence of Ca^{2+} ions, into covalently crosslinked fibrin polymers. This plasma clotting process probably occurs in all Decapoda, but as the amount of plasma coagulogen greatly varies according to the species investigated, such clotting may not always be evident. Furthermore, in all Decapoda showing obvious plasma gelation, clot firmness and clotting time greatly vary, depending on the molting stages, physiological and nutritional states, and sex of the animals.[37] This suggests that the mechanism of clotting, which is uniform throughout the vertebrate phylum, is likely to have evolved from a simpler mechanism, possibly inherited from a common ancestor of vertebrates and invertebrates. In Crustacea, the coagulation processes range from a single cellular mechanism to complete plasma gelation. Therefore these animals provide good material for studying the evolution of the clotting processes.

Plasma coagulogen has different functions: it is involved in the final steps of hemolymph coagulation, and seems to act as a protein carrier for tanning agents, and as a reserve protein used during starvation periods. The proportions of both lipids and carbohydrates that it contains, can change depending on environmental conditions. It probably plays other still unknown roles, such as lipid carrier. Plasma coagulogen is also found in compartments other than hemolymph, such as soft integument and calcified cuticle in *A. leptodactylus*. In fact, this crustacean clottable factor does not seem to resemble vertebrate fibrinogen. Its properties (molecular weight, carbohydrate moieties, and presence in several compartments) are similar to those of other substrates, such as fibronectins.[42,177] It is notable that its soluble circulating polymers can be covalently crosslinked by a Ca^{2+}-dependent transglutaminase, just as the fibrin polymers are crosslinked by Factor XIII. However, in the presence of Ca^{2+} ions, plasma transglutaminase can covalently crosslink various proteins not necessarily involved in coagulation processes, such as collagens, fibronectins, and several membrane proteins (review: Folk[180]). Some of these proteins can also be crosslinked by factors other than transglutaminase. Thus, the gelation of any protein in the presence of Ca^{2+}-dependent transglutaminase does not prove the presence of a fibrinogen-like factor.

Moreover, the plasma coagulogen of Decapoda does not seem to originate from hepatopancreas cells,[84,85] unlike fibrinogen originating from liver in vertebrates. The site of its synthesis is not known, but it might be produced by certain differentiated hemocytes or by lipoprotein cells.

Although not definitely confirmed, a cellular coagulogen seems to be present in hemocyte lysates of Crustacea. Its characteristics resemble those of the coagulogen of Limulidae. It is likely that both cellular and plasma coagulogens must interconnect into a common network or complex, as demonstrated in insects.[179-181] These works showed, by using immunochemical tests, that hemocyte and plasma coagulogens have no immunological identity and that no plasma coagulogen is detectable in any type of hemocytes in *Leucophaea maderae*.[180] In the Macrura we studied, plasma coagulogen is not found in hemocyte lysates, but is often linked to a cellular agglutinin, probably by interaction of its carbohydrate moieties, when the blood cells are damaged. Moreover, this plasma coagulogen, which is always present in cell-free hemolymph, cannot gel without prior cellular clumping. However, hemocyte lysates can gel without plasma factors when prepared under apyrogen conditions.

Clotting processes in the hemocytes of Crustacea resemble those observed during activation of both vertebrate platelets and granulocytes of Limulidae, and comprise an enzyme cascade with at least two serine proteases, the cellular proclotting enzyme and the serine protease S, which can be activated by LPS from bacterial endotoxins or by β-1,3-glucans from fungi. These activated serine proteases further induce activation of the complex proPO system, and also elicit a clotting reaction in hemocyte lysates. This latter pathway seems to correspond to an alternate one, found in the mammalian complement system. Activation of the proPO system, elicits the release of several "sticky" proteins, which can act as opsonins, and might correspond to the agglutinins, which attach to plasma coagulogen (see also Chapter 6). This activation also produces the melanization of different components (plasma coagulogen and molt-related fraction F2) in both hemolymph and integument. Melanization constitutes a defense reaction, since quinones, which are intermediates in melanin synthesis from phenolic substances, are cytotoxic and may act as bactericidal or fungistatic agents. In addition, the reaction of melanin with proteins surrounding foreign materials might constitute a barrier (pseudo-cuticle) which serves to isolate the animals from parasites.

Curiously, the proPO-activating system of the silkworm hemolymph is triggered by elicitors, such as Gram-negative or Gram-positive bacterial cell walls, β-1,3-glucans, and denatured lipophorin (a silkworm plasma protein, which might be analogous to a coagulogen) but not by LPS.[182] Thus, the activation of the proPO system by β-1,3-glucans seems similar in Crustacea and Insecta, but further studies are necessary to elucidate the cascade reactions triggering the activation of the proclotting enzyme by LPS.

It should be noted that all these results have been obtained in Decapoda. In other groups of Crustacea, few data concern the biochemical events occurring during clotting processes, probably because of the small size of animals and therefore the reduced volume of available hemolymph. It is more than likely that the clotting reactions are similar to those studied in Decapoda.

VII. SUMMARY

In Decapoda Crustacea, clotting factors are found in both the cell-free plasma and the hemocytes. Plasma coagulogen is a 400-kDa protein with both lipid and carbohydrate moieties. The basic component of the circulating clottable factor is composed of two 200-kDa subunits. Its soluble polymers are converted into covalently crosslinked polymers of coagulin by Ca^{2+}-dependent transglutaminase. In hemolymph, when hemocytes are damaged, plasma coagulogen often appears to be linked to aggregating proteins, such as agglutinins.

In crayfish, plasma coagulogen is also found in other tissues, such as premolt soft integument and premolt calcified cuticle. Its concentration varies greatly with the species investigated and also with the physiological and nutritional states of animals. It seems to possess many diversified functions, such as plasma coagulation, protein transport of tanning agents, lipid and sugar transport, and protein storage, and resembles fibronectin. The site of its synthesis is not known: it might be produced by certain differentiated hemocytes or by lipoprotein cells.

All three types of blood cells: hyaline cells, semigranular and granular hemocytes are implicated in the clotting processes. A type of cellular coagulogen is present in the hemocyte lysates of Decapoda. It is soluble in acid medium, and exhibits a cathodic migration in electrophoresis. It can be converted to a gel by a serine protease proclotting enzyme. This proenzyme can be activated by either LPS or by β-1,3-glucans, which at first elicit exocytosis of semigranular and granular cells. The proPO contained in these two cell types is further activated after a proteolytic process into a degranulating factor, phenoloxidase and several opsonin-like "sticky" proteins, including agglutinins. This activation produces the melanization of different components in both the hemolymph and the integument. It seems that both cellular and plasma coagulogens must interact to form a common complex. However, this is not entirely clear yet in Crustacea.

ACKNOWLEDGMENTS

I thank Jacqueline Defoort for typing this manuscript. The credit for preparing the illustrations of this article goes to "Service d'Accueil de Microscopie Electronique" and to Miss Bigler from the Centre National de la Recherche Scientifique.

REFERENCES

1. **Grégoire, C.,** Blood coagulation in Arthropods. VI. A study by phase contrast microscopy of blood reactions *in vitro* in *Onychophora* and in various groups of arthropods, *Arch. Biol.,* 66, 489, 1955.
2. **Grégoire, C.,** Hemolymph coagulation in Arthropods, *Symp. Zool. Soc. London,* 27, 45, 1970.
3. **Grégoire, C.,** Hemolymph coagulation in Arthropods, in *Chemical Zoology,* Vol. 6, part B, Florkin, M. and Scheer, B. T., Eds., Academic Press, New York, 1971, 145.
4. **Ravindranath, M. H.,** Haemocytes in haemolymph coagulation of Arthropods, *Biol. Rev.,* 55, 139, 1980.
5. **Bauchau, A. G.,** Crustaceans, in *Invertebrate Blood Cells,* Vol. 2, Ratcliffe, N. A. and Rowley, A. F., Eds., Academic Press, New York, 1981, 385.
6. **Durliat, M.,** Clotting processes in Crustacea Decapoda, *Biol. Rev.,* 60, 473, 1985.
7. **Carus, D. C. G.,** Von den auszeren lebensbedingungen der heisz- und kaltblutigen Tiere, Leipzig, 1824, 80.
8. **Lowit, M.,** Uber die Beziehungen der weissen Blutkorperchen zur Blutgerinnung, *Ziegler's Beitr. Z. Pathol. Anat.,* 5, 469, 1889.
9. **Hardy, W. B.,** The blood corpuscles of the Crustacea, together with a suggestion as to the origin of the crustacea fibrin-ferment, *J. Physiol.,* 13, 165, 1892.
10. **Tait, J.,** Crustacean blood coagulation as studied in the Arthrostraca, *Q. J. Exp. Physiol.,* 3, 1, 1910.
11. **Tait, J.,** Types of crustacean blood coagulation, *J. Mar. Biol., Assoc. U.K.,* 9, 191, 1911.
12. **Tait, J. and Gunn, D. J.,** The blood of *Astacus fluviatilis:* a study in crustacean blood, with special reference to coagulation and phagocytosis, *Q. J. Exp. Physiol.,* 12, 35, 1918.
13. **Hearing, V. and Vernick, S. H.,** Fine structure of the blood cells of the lobster *Homarus americanus, Chesapeake Sci.,* 8, 170, 1967.
14. **Stang-Voss, C.,** Studies on the ultrastructure of invertebrate hemocytes. V. On the hemocytes of *Astacus astacus* (Crustacea), *Z. Zellforsch. Mikrosk. Anat.,* 122, 68, 1971.
15. **Vranckx, R. and Durliat, M.,** Etude des hémocytes circulants de quelques Crustacés Décapodes. Corrélations entre polymorphisme et mode de prélèvement, *C. R. Acad. Sci. Paris,* 285D, 1045, 1977.
16. **Ghiretti-Magaldi, A., Milanesi, C., and Tognon, G.,** Hemopoiesis in Crustacea Decapoda: origin and evolution, of hemocytes and cyanocytes of *Carcinus maenas, Cell Differ.,* 6, 167, 1977.
17. **Bauchau, A. G. and de Brouwer, M. B.,** Etude ultrastructurale de la coagulation de l'hémolymphe chez les Crustacés, *J. Microsc. (Paris),* 19, 37, 1974.
18. **Wood, P. J. and Visentin, L. P.,** Histological and histochemical observations of the hemolymph cells in the crayfish *Orconectes virilis, J. Morphol.,* 123, 559, 1967.
19. **Wood, P. J., Podlewski, J., and Shenk, T.,** Cytochemical observations of hemolymph cells during coagulation in the crayfish *Orconectes virilis, J. Morphol.,* 134, 479, 1971.
20. **Nolf, P.,** Contribution à l'étude de la coagulation du sang. La coagulation chez les Crustacés, *Arch. Int. Physiol.,* 7, 411, 1909.
21. **Doolittle, R. F. and Lorand, L.,** Inhibitors of lobster blood clotting, *Biol. Bull. (Woods Hole, Mass.),* 123, 481, 1962.
22. **Bruner-Lorand, J., Urayama, T., and Lorand, L.,** Transglutaminase as a blood clotting enzyme, *Biochem. Biophys. Res. Commun.,* 23, 828, 1966.
23. **Myhrman, R. and Bruner-Lorand, J.,** Lobster muscle transpeptidase, *Methods Enzymol.,* 19, 765, 1970.
24. **Lorand, L.,** Fibrinoligase: the fibrin-stabilizing factor system of blood plasma, *Ann. N.Y. Acad. Sci.,* 202, 6, 1972.
25. **Tyler, A. and Scheer, B. T.,** Natural heteroagglutinins in the serum of the spiny lobster, *Panulirus interruptus.* II. Chemical and antigenic relation to blood proteins, *Biol. Bull. (Woods Hole, Mass.),* 89, 193, 1945.
26. **Glavind, J.,** in *Studies on the Coagulation of Crustacean Blood,* Busch, A., Ed., Nyt Nordisk, Forlag, Copenhagen, 1948, 1.
27. **Duchateau, G. and Florkin, M.,** La coagulation du sang des Arthropodes. IV. Sur le firbinogène et sur la "coaguline" musculaire du homard, *Bull. Soc. Chim. Biol.,* 36, 295, 1954.
28. **Stewart, J. E., Dingle, J. R., and Odense, P. H.,** Constituents of the hemolymph of the lobster *Homarus americanus* Milne Edwards, *Can. J. Biochem.,* 44, 1447, 1966.
29. **Fuller, G. M.,** The study of an invertebrate fibrinogen and its mode of gelation, *Contribution from the Scripps Inst. Oceano.,* 38, 1851, 1968.
30. **Fuller, G. M. and Doolittle, R. F.,** Studies of invertebrate fibrinogen. I. Purification and characterization of fibrinogen from spiny lobster, *Biochemistry,* 10, 1305, 1971a.
31. **Fuller, G. M. and Doolittle, R. F.,** Studies of invertebrate fibrinogen. II. Transformation of lobster fibrinogen into fibrin, *Biochemistry,* 10, 1311, 1971b.
32. **Doolittle, R. F. and Fuller, G. M.,** Sodium dodecyl sulfate polyacrylamide gel electrophoresis studies on lobster fibrinogen and fibrin, *Biochim. Biophys. Acta,* 263, 805, 1972.

33. **Durliat, M.,** Etude de la coagulation de l'hémolymphe chez les Crustacés Décapodes. Analyse électrophorétique des protéines du sérum et du plasma d'écrevisse *(Astacus leptodactylus), C. R. Acad. Sci. Paris,* 278D, 473, 1974.
34. **Needham, A. E.,** Haemostatic mechanisms in the Invertebra, *Symp. Zool. Soc. London,* 27, 19, 1970.
35. **Durliat, M., Vranckx, R., Herberts, C., and Lachaise, F.,** Effets de la congélation sur la séparation électrophorétique des protéines de l'hémolymphe de quelques Arthropodes, *C. R. Soc. Biol.,* 169, 862, 1975.
36. **Durliat, M. and Vranckx, R.,** Coagulation in the crayfish *Astacus leptodactylus:* attempts to identify a fibrinogen-like factor in the hemolymph, *Biol. Bull. (Woods Hole, Mass.),* 151, 467, 1976.
37. **Durliat, M. and Vranckx, R.,** Action of various anticoagulants on hemolymphs of lobsters and spiny lobsters, *Biol. Bull. (Woods Hole, Mass.),* 160, 55, 1981.
38. **Vendrely, R., Ghidalia, W., Coirault, Y., and Montmory, C.,** Coagulation de l'hémolymphe chez les Crustacés. Mise en évidence par électrophorèse d'un composant plasmatique intervenant dans l'hémostase chez un Crustacé du groupe A *(Cancer pagurus), C.R. Acad. Sci. Paris,* 289D, 849, 1979.
39. **Ghidalia, W., Vendrely, R., Montmory, C., Coirault, Y., and Brouard, M. O.,** Coagulation in Decapod Crustacea. Comparative studies of the clotting process in species from groups A, B, and C, *J. Comp. Physiol.,* 142, 473, 1981.
40. **Stewart, J. E., Arie, B., Zwicker, B. M., and Dingle, J. R.,** Gaffkemia, a bacterial disease of the lobster *Homarus americanus:* effects of the pathogen, *Gaffkya homari,* on the physiology of the host, *Can. J. Microbiol.,* 15, 925, 1969.
41. **Durliat, M. and Vranckx, R.,** Analysis of clotting defects in diseased lobsters. 1. Alterations in blood parameters, *Comp. Biochem. Physiol.,* 76A, 95, 1983a.
42. **Durliat, M. and Vranckx, R.,** Analysis of clotting defects in diseased lobsters. 2. Proteins of diseased lobster hemolymph, *Comp. Biochem. Physiol.,* 76A, 103, 1983b.
43. **Mengeot, J. C., Bauchau, A. G., De Brouwer, M. B., and Passelecq-Gérin, E.,** Séparation des granules présents dans les hémocytes des Crustacés par exocytose provoquée, *Comp. Biochem. Physiol.,* 54A, 145, 1976.
44. **Mengeot, J. C., Bauchau, A. G., De Brouwer, M. B., and Passelecq-Gérin, E.,** Isolement des granules des hémocytes de *Homarus vulgaris.* Examens électrophorétiques du contenu protéique des granules, *Comp. Biochem. Physiol.,* 58A, 393, 1977.
45. **Madaras, F., Parkin, J. D., and Castaldi, P. A.,** Coagulation in the sand crab *(Ovalipes bipustulatus), Thromb. Haemostasis,* 42, 734, 1979.
46. **Madaras, F., Chew, M. Y., and Parkin, J. D.,** Purification and characterization of the sand crab *(Ovalipes bipustulatus)* coagulogen (fibrinogen), *Thromb. Haemostasis,* 45, 77, 1981.
47. **Söderhäll, K.,** Fungal cell wall β-1,3-glucans induce clotting and phenoloxidase attachment to foreign surfaces of crayfish hemocyte lysate, *Dev. Comp. Immunol.,* 5, 565, 1981.
48. **Söderhäll, K.,** Prophenoloxidase activating system and melanization. A recognition mechanism of Arthropods? A review, *Dev. Comp. Immunol.,* 6, 601, 1982.
49. **Söderhäll, K.,** β-1,3-glucan enhancement of protease activity in crayfish hemocyte lysate, *Comp. Biochem. Physiol.,* 74B, 221, 1983.
50. **Jeffrey, P. D., Shaw, D. C., and Treacy, G. B.,** Hemocyanin from the Australian freshwater crayfish *Cherax destructor.* Studies of two different monomers and their participation in the formation of multiple hexamers, *Biochemistry,* 15, 5527, 1976.
51. **Jeffrey, P. D., Shaw, D. C., and Treacy, G. B.,** Hemocyanin from the Australian freshwater crayfish *Cherax destructor.* Characterization of a dimeric subunit and its involvement in the formation of the 25 S component, *Biochemistry,* 17, 3078, 1978.
52. **Markl, J., Hofer, A., Bauer, G., Markl, A., Kempter, B., Brenzinger, M., and Linzen, B.,** Subunit heterogeneity in arthropod hemocyanins. II. Crustacea, *J. Comp. Physiol.,* 133, 167, 1979.
53. **Markl, J. and Kempter, B.,** Subunit heterogeneity in crustacean hemocyanins as deduced by two-dimensional immunoelectrophoresis, *J. Comp. Physiol.,* 141B, 495, 1981.
54. **Van Holde, K. E. and Miller, K. I.,** Haemocyanins, *Q. Rev. Biophys.,* 15, 1, 1982.
55. **McKay, D. and Jenkin, C. R.,** Immunity in the Invertebrates. The role of serum factors in the phagocytosis of erythrocytes by haemocytes of the freshwater crayfish *(Parachaeraps bicarinatus), Aust. J. Exp. Biol. Med. Sci.,* 48, 139, 1970.
56. **Hall, J. L. and Rowlands, Jr., D. T.,** Heterogeneity of lobster agglutinins. II. Specificity of agglutinin-erythrocyte binding, *Biochemistry,* 13, 828, 1974a.
57. **Paterson, W. D. and Stewart, J. E.,** *In vitro* phagocytosis by hemocytes of the American lobster *(Homarus americanus), J. Fish. Res. Board Can.,* 31, 1051, 1974.
58. **Paterson, W. D., Stewart J. E., and Zwicker, B. M.,** Phagocytosis as a cellular immune response mechanism in the American lobster *Homarus americanus, J. Invertebr. Pathol.,* 27, 95, 1976.
59. **Paterson, W. D. and Stewart, J. E.,** Rate and duration of phagocytic increase in lobsters induced by *Pseudomonas perolens* endotoxin, *Dev. Comp. Immunol.,* 3, 353, 1979.

60. **Cornick, J. W. and Stewart, J. E.**, Partial characterization of a natural agglutinin in the hemolymph of the lobster, *Homarus americanus, J. Invertebr. Pathol.*, 21, 255, 1973.
61. **Hartman, A. L., Campbell, P. A., and Abel, C. A.**, An improved method for the isolation of lobster lectins, *Dev. Comp. Immunol.*, 2, 617, 1978.
62. **Vasta, G. R. and Cohen, E.**, Carbohydrate specificities of *Birgus latro* (coconut crab) serum lectins, *Dev. Comp. Immunol.*, 8, 197, 1984.
63. **Vasta, G. R., Warr, G. W., and Marchalonis, J. J.**, Serological characterization of humoral lectins from the freshwater prawn *Macrobrachium rosenbergii, Dev. Comp. Immunol.*, 7, 13, 1983.
64. **Amirante, G. A.**, Agglutinins and lectins of Crustacea: Their composition, synthesis and functions, in *Hemocytic and Humoral Immunity in Arthropods*, Gupta, A. P., Ed., John Wiley & Sons, New York, 1986, chap. 13.
65. **Hapner, K. D. and Stebbins, M. R.**, Biochemistry of arthropod agglutinins, in *Hemocytic and Humoral Immunity in Arthropods*, Gupta, A. P., Ed., John Wiley & Sons, New York, 1986, chap. 8.
66. **Meusy, J. J.**, Vitellogenin, the extraovarian precursor of the protein yolk in Crustacea: a review, *Reprod. Nutr. Dev.*, 20, 1, 1980.
67. **Crowley, G. J.**, Studies in arthropod serology. I. Changes in the hemolymph composition as related to the ecdysal cycle, *Wasmann J. Biol.*, 21, 177, 1963.
68. **Stutman, L. J. and Dolliver, M.**, Mechanism of coagulation in *Gecarcinus lateralis, Am. Zool.*, 8, 481, 1968.
69. **Barlow, J. and Ridgway, G. J.**, Changes in serum proteins during the molt and reproductive cycles of the american lobster, *(Homarus americanus), J. Fish. Res. Board Can.*, 26, 2101, 1969.
70. **Ghidalia, W., Vendrely, R., and De Monty De Rezé, M.**, Analyse électrophorétique de la variation des fractions protéiques du sérum de *Macropipus puber* (L.) mâle au cours du cycle d'intermue, *Arch. Zool. Exp. Gén.*, 111, 77, 1970.
71. **Vranckx, R. and Durliat, M.**, Analyse immunologique des variations des protéines de l'hémolymphe d' *Astacus leptodactylus* pendant la prémue, *C. R. Acad. Sci. Paris*, 282D, 1305, 1976.
72. **Chaix, J. C., Marvaldi, J., and Secchi, J.**, Variations of ecdysone titer and hemolymph major proteins during the molt cycle of the spider crab *Acanthonyx lunulatus, Comp. Biochem. Physiol.*, 69B, 709, 1981.
73. **Ghidalia, W., Montmory, C., Vicomte, M., Lambin, P., and Fine, J. M.**, Etude électrophorétique et immunochimique de certains constituants tégumentaires et de leur relation avec les protéines sériques chez *Macropipus puber* (Décapode Brachyoure), *Cah. Biol. Mar.*, 17, 157, 1976.
74. **Durliat, M. and Vranckx, R.**, Changes in water soluble proteins from integument of *Astacus leptodactylus* during the molt cycle, *Comp. Biochem. Physiol.*, 59B, 123, 1978.
75. **Durliat, M., Vranckx, R., and Denoyelle, M.**, Cuticular protein analysis from *Astacus leptodactylus*. 2. Immunological identities of proteins from shell and hemolymph, *Comp. Biochem. Physiol.*, 65B, 65, 1980.
76. **Durliat, M. and Vranckx, R.**, Cuticular proteins in the crayfish *Astacus leptodactylus*. II. Relationships between proteins in plasma, soft integument, and calcified cuticle, *Comp. Biochem. Physiol.*, 85B, 267, 1986.
77. **Vranckx, R. and Durliat, M.**, The "ventral organ" of *Jasus paulensis* (Crustacea, Decapoda), *J. Exp. Zool.*, 222, 37, 1982.
78. **Ghiretti-Magaldi, A., Milanesi, C., and Salvato, B.**, Identification of hemocyanin in the cyanocytes of *Carcinus maenas, Experientia*, 29, 1265, 1973.
79. **Denucé, J. M.**, Phosphatases and esterases in the digestive gland of the crayfish *Orconectes virilis, Arch. Int. Physiol. Biochem.*, 75, 159, 1967.
80. **Senkbeil, E. G. and Wriston, J. C., Jr.**, Hemocyanin synthesis in the American lobster, *Homarus americanus, Comp. Biochem. Physiol.*, 68B, 163, 1981.
81. **Cornick, J. W. and Stewart, J. E.**, Lobster *(Homarus americanus)* hemocytes: classification, differential counts, and associated agglutinin activity, *J. Invertebr. Pathol.*, 31, 194, 1978.
82. **Hall, J. L. and Rowlands, D. T., Jr.**, Heterogeneity of lobster agglutinins. I. Purification and physiochemical characterization, *Biochemistry*, 13, 821, 1974b.
83. **Ferrero, E. A., Graziosi, G., Marzari, R., and Mosco, A.**, Protein pattern variability of the hemolymph of mantis shrimp *Squilla mantis* L. (Crustacea, Stomatopoda), *J. Exp. Zool.*, 225, 341, 1983.
84. **Durliat, M. and Vranckx, R.**, Proteins of aqueous extracts from the hepatopancreas of *Astacus leptodactylus*. 1. Changes in proteins during the molt cycle, *Comp. Biochem. Physiol.*, 71B, 155, 1982a.
85. **Durliat, M. and Vranckx, R.**, Proteins of aqueous extracts from the hepatopancreas of *Astacus leptodactylus*. 2. Immunological identities of proteins from hepatopancreas and blood, *Comp. Biochem. Physiol.*, 71B, 165, 1982b.
86. **Sewell, J. M.**, Lipoprotein cells in the blood of *Carcinus maenas* and their cycle of activity correlated with moult, *Q. J. Microsc. Sci.*, 96, 73, 1955.
87. **Travis, D. F.**, The molting cycle of the spiny lobster, *Panulirus argus* Latreille. IV. Post-ecdysial histological and histochemical changes in the hepatopancreas and integumental tissues, *Biol. Bull. (Woods Hole, Mass.)*, 113, 451, 1957.

88. **Skinner, D. M.,** The structure and metabolism of a crustacean integumentary tissue during a molt cycle, *Biol. Bull. (Woods Hole, Mass.)*, 123, 635, 1962.
89. **Bliss, D. E.,** The pericardial sacs of terrestrial Brachyura. Phylogeny and Evolution of Crustacea, *Museum of Comp. Zool. (Harvard)*, Special publication, 59, 1963.
90. **Keller, R. and Adelung, D.,** Vergleichende morphologische und Untersuchungen des Integumentgewebes und des Häutungshormongehaltes beim Flußkrebs *Orconectes limosus* während eines Häutungszyklus, *Wilhem Roux'Arch.*, 164, 209, 1970.
91. **Bauchau, A. G. and De Brouwer, M. B.,** Ultrastructure des hémocytes d'*Eriocheir sinensis,* Crustacé Décapode Brachyoure, *J. Microsc. (Paris),* 15, 171, 1972.
92. **Vranckx, R. and Durliat, M.,** Comparison of the gradient of setal development of uropods and of scaphognathites in *Astacus leptodactylus, Biol. Bull. (Woods Hole, Mass.*), 155, 627, 1978.
93. **Scheidegger, G.,** Die Ultrastruktur der Reservezellen bei Postembryonalstadien von *Pacifastacus leniusculus* (Crustacea, Decapoda), *Zool. Jahrb. Abt. Anat. Ontog. Tiere,* 107, 321, 1982.
94. **Johnston, M. A., Davies, P. S., and Elder, H. Y.,** Possible hepatic function for crustacean blood cells, *Nature (London*), 230, 471, 1971.
95. **Fuller, G. M., McLean, J. D., and Doolittle, R. F.,** Electron microscopy of lobster fibrinogen and fibrin, *Biochemistry,* 10 Appendix, 1915, 1971.
96. **Lorand, L., Doolittle, R. F., Konishi, K., and Riggs, S. K.,** A new class of blood coagulation inhibitors, *Arch. Biochem. Biophys.*, 102, 71, 1963.
97. **Lorand, L., Jacobsen, A., and Schuel, A.,** Inhibitors of fibrin polymerization, *Biol. Bull. (Woods Hole, Mass.),* 127, 379, 1964.
98. **Lorand, L., Pilkington, T. R. E., and Bruner-Lorand, J.,** A search for new inhibitors of fibrin cross-linkings, *Biol. Bull. (Woods Hole, Mass.*), 129, 413, 1965.
99. **Jeuniaux, C.,** Hemolymph. Arthropoda, in *Chemical Zoology,* Vol. 6, Part B, Florkin, M. and Scheer, B. T., Eds., Academic Press, New York, 1971, 64.
100. **Bauchau, A. G. and Plaquet, J. C.,** Variation du nombre des hémocytes chez les Crustacés Brachyoures, *Crustaceana,* 24, 215, 1973.
101. **Stewart, J. E., Cornick, J. W., and Dingle, J. R.,** An electronic method for counting lobster (*Homarus americanus* Milne Edwards) hemocytes and the influence of diet on hemocyte numbers and hemolymph proteins, *Can. J. Zool.*, 45, 291, 1967.
102. **Busselen, P.,** Effects of moutling cycle and nutritional conditions on haemolymph proteins in *Carcinus maenas, Comp. Biochem. Physiol.*, 37, 73, 1970.
103. **Djangmah, J. S.,** The effects of feeding and starvation on copper in the blood and hepatopancreas, and on blood proteins of *Grangon vulgaris* (Fabricus), *Comp. Biochem. Physiol.*, 32, 709, 1970.
104. **Durliat, M. and Vranckx, R.,** Effects des irradiations sur les facteurs hémolymphatiques d'*Astacus leptodactylus, C. R. Acad. Sci. Paris, Ser. III,* 294, 119, 1982c.
105. **Bang, F. B.,** A bacterial disease of *Limulus polyphemus, Bull. Johns Hopkins Hosp.*, 98, 325, 1956.
106. **Barker, W. H., Jr. and Bang, F. B.,** The effects of infection by Gram-negative bacteria, and their endotoxins, on the blood clotting mechanisms of the crustacean *Sacculina carcini,* a parasite of the crab *Carcinus maenas, J. Invertebr. Pathol.*, 8, 88, 1966.
107. **Levin, J.,** Blood coagulation and endotoxin in invertebrates, *Fed. Proc.*, 26, 1707, 1967.
108. **Bang, F. B.,** A virus disease of the shore crab *Carcinus maenas, Biol. Bull. (Woods Hole, Mass.*), 139, 414, 1970a.
109. **Bang, F. B.,** Disease mechanisms in crustacean and marine arthropods. Symposium diseases of fishes and schellfishes, Special Publication, American Fisheries Society, Washington, 5, 383, 1970b.
110. **Rabin, H. and Hughes, J. T.,** Studies on host-parasite relationships in gaffkemia, *J. Invertebr. Pathol.*, 10, 335, 1968.
111. **Vranckx, R. and Durliat, M.,** Soluble glycoprotein complexes in the blood and integumental extracts of *Astacus leptodactylus, Mol. Physiol.*, 1, 339, 1981.
112. **Pilz, I., Goral, K., Hoylaerts, M., Witters, R., and Lontie, R.,** Studies by small-angle X-ray scattering of the quaternary structure of the 24-S component of the haemocyanin of *Astacus leptodactylus* in solution, *Eur. J. Biochem.*, 105, 539, 1980.
113. **Vacca, L. L. and Fingerman, M.,** The mechanism of tanning in the fiddler crab, *Uca pugilator.* I. Tanning agents and protein carriers in the blood during ecdysis, *Comp. Biochem. Physiol.*, 51B, 475, 1975a.
114. **Vacca, L. L. and Fingerman, M.,** The mechanism of tanning in the fiddler crab, *Uca pugilator.* II. The cyclic appearance of tanning agents and attached carrier proteins in the blood during the molting cycle, *Comp. Biochem. Physiol.*, 51B, 483, 1975b.
115. **Vacca, L. L. and Fingerman, M.,** The roles of hemocytes in tanning during the molting cycle: a histochemical study of the fiddler crab, *Uca pugilator, Biol. Bull. (Woods Hole, Mass.),* 165, 758, 1983.
116. **Goldenberg, P. Z. and Greenberg, A. H.,** Functional heterogeneity of carbohydrate-binding hemolymph proteins: evidence of a nonagglutinating opsonin in *Homarus americanus, J. Invertebr. Pathol.*, 42, 33, 1983.

117. **Söderhäll, K. and Smith, V. J.,** The prophenoloxidase activating system — a complement-like pathway in Arthropods?, in *Infection Process of Fungi,* Aist, J. and Roberts, D. W., Eds., Rockefeller Foundation Study, Bellagio, Italy, 1983a, 1.
118. **Smith, V. J. and Söderhäll, K.,** Induction of degranulation and lysis of haemocytes in the freshwater crayfish *Astacus astacus* by components of the prophenoloxidase activating system *in vitro, Cell Tissue Res.,* 233, 295, 1983b.
119. **Smith, V. J. and Söderhäll, K.,** Cellular immune mechanisms in the Crustacea, in *Immune Mechanisms in Invertebrate Vectors,* Vol. 56, Lackie, A. M., Ed., Oxford University Press, Oxford, 1986, 59.
120. **Smith, V. J. and Ratcliffe, N. A.,** Host defence reactions of the shore crab, *Carcinus maenas* (L.) *in vitro, J. Mar. Biol. Assoc. U.K.,* 58, 367, 1978.
121. **Persson, M., Vey, A., and Söderhäll, K.,** Encapsulation of foreign particles *in vitro* by separated blood cells from crayfish, *Astacus leptodactylus, Cell Tissue Res.,* 247, 409, 1987.
122. **Smith, V. J. and Söderhäll, K.,** β-1,3-glucan activation of crustacean hemocytes *in vitro* and *in vivo, Biol. Bull. (Woods Hole, Mass.),* 164, 299, 1983a.
123. **Johansson, M. W. and Söderhäll, K.,** Exocytosis of the prophenoloxidase activating system from crayfish hemocytes, *J. Comp. Physiol.,* 156B, 175, 1985.
124. **Söderhäll, K. and Smith, V. J.,** Separation of the haemocyte populations of *Carcinus maenas* and other marine Decapods, and prophenoloxidase distribution, *Dev. Comp. Immunol.,* 7, 229, 1983b.
125. **Söderhäll, K., Vey, A., and Ramstedt, M.,** Haemocyte lysate enhancement of fungal spore encapsulation by crayfish haemocytes, *Dev. Comp. Immunol.,* 8, 23, 1984.
126. **Söderhäll, K., Smith, V. J., and Johansson, M. W.,** Exocytosis and uptake of bacteria by isolated haemocyte populations of two crustaceans: evidence for cellular cooperation in the defence reactions of Arthropods, *Cell Tissue Res.,* 245, 43, 1986.
127. **Johansson, M. W. and Söderhäll, K.,** Isolation and purification of a cell adhesion factor from crayfish blood cells, *J. Cell Biol.,* 106, 1795, 1988.
128. **Johansson, M. W. and Söderhäll, K.,** Cellular immunity in Crustaceans and the proPO system, *Parasitol. Today,* 5, 171, 1989a.
129. **Johansson, M. W. and Söderhäll, K.,** A cell adhesion factor from crayfish haemocytes has degranulating activity towards crayfish granular cells, *Insect Biochem.,* 19, 183, 1989b.
130. **Mürer, E. H., Levin, J., and Holme, R.,** Isolation and studies of the amoebocytes of *Limulus polyphemus,* the horseshoe crab, *J. Cell Physiol.,* 86, 533, 1975.
131. **Solum, N. O.,** The coagulation of *Limulus polyphemus* hemocytes. A comparison of the clotted and non-clotted forms of the molecule, *Thromb. Res.,* 2, 55, 1973.
132. **Durliat, M. and Vranckx, R.,** Relationships between plasma and hemocyte proteins in Decapoda, *Comp. Biochem. Physiol.,* 92B, 595, 1989.
133. **Gaffin, S. L.,** The clotting of the lysed white cells of *Limulus* induced by endotoxin. I. Preparation and characterization of clot forming proteins, *Biorheology,* 13, 273, 1976.
134. **Nakamura, S., Iwanaga, S., Harada, T., and Niwa, M.,** A clottable protein (coagulogen) from amoebocyte lysate of Japanese horseshoe crab *(Tachypleus tridentatus).* Its isolation and biochemical properties, *J. Biochem.,* 80, 1011, 1976.
135. **Tai, J. Y. and Liu, T. Y.,** Studies on *Limulus* amoebocyte lysate. I. isolation of proclotting enzyme, *J. Biol. Chem.,* 252, 2178, 1977.
136. **Liu, T. Y., Seid, R. C., Jr., Tai, J. Y., Liang, S. M., Sakmar, T. P., and Robbins, J. B.,** Studies on *Limulus* lysate coagulating system, in *Biomedical Applications of the Horseshoe Crab* (Limulidae), Cohen, E., Ed., Alan R. Liss, New York, 1979, 147.
137. **Dohke, K.,** Studies on prophenoloxidase-activating enzyme from cuticle of the silkworm *Bombyx mori.* I. Activation reaction by the enzyme, *Arch. Biochem. Biophys.,* 157, 203, 1973.
138. **Bang, F. B.,** Serological responses among invertebrates other than insects, *Fed. Proc.,* 26, 1680, 1967.
139. **Johnson, P. T.,** Bacterial infection in the blue crab, *Callinectes sapidus:* course of infection and histopathology, *J. Invertebr. Pathol.,* 28, 25, 1976.
140. **Newman, M. C. and Feng, S. Y.,** Susceptibility and resistance of the rock crab, *Cancer irroratus,* to natural and experimental bacterial infection, *J. Invertebr. Pathol.,* 40, 77, 1982.
141. **Cornick, J. W. and Stewart, J. E.,** Interaction of the pathogen *Gaffkya homari* with natural defence mechanisms of *Homarus americanus, J. Fish. Res. Board. Can.,* 25, 695, 1968.
142. **Beller, F. K.,** The role of endotoxin in disseminated intravascular coagulation, *Thromb. Diath. Haemorr.,* Suppl. 36, 125, 1969.
143. **Johnson, P. T., Stewart, J. E., and Arie, B.,** Histopathology of *Aerococcus viridans* var. *homari* infection (Gaffkemia) in the lobster *Homarus americanus* and a comparison with histological reactions to a Gram-negative species, *Pseudomonas perolens, J. Invertebr. Pathol.,* 38, 127, 1981.
144. **Söderhäll, K. and Hall, L.,** Lipopolysaccharide induced activation of prophenoloxidase activating system in crayfish haemocyte lysate, *Biochim. Biophys. Acta,* 797, 99, 1984.

145. **Goldenberg, P. Z., Huebner, E., and Greenberg, A. H.**, Activation of lobster hemocytes for phagocytosis, *J. Invertebr. Pathol.*, 43, 77, 1984.
146. **Armstrong, P. B. and Rickles, F. F.**, Endotoxin induced degranulation of the *Limulus* amoebocyte, *Cell Res.*, 140, 15, 1982.
147. **Söderhäll, K. and Smith, V. J.**, The prophenoloxidase activating system: the biochemistry of its activation and role in Arthropod cellular immunity, with special reference to crustaceans, in *Immunity in Invertebrates*, Brehélin, M., Ed., Springer-Verlag, Berlin, 1986, chap. 15.
148. **Von Eschen, K. B. and Rudbach, J. A.**, Inactivation of endotoxin by serum: a phylogenetic study, *J. Infect. Dis.*, 129, 21, 1974.
149. **McKay, D. and Jenkin, C. R.**, Immunity in the Invertebrates. II. Adaptive immunity in the crayfish *(Parachaeraps bicarinatus), Immunology,* 17, 127, 1969.
150. **Tanaka, S., Nakamura, T., Morita, T., and Iwanaga, S.**, *Limulus* anti-LPS factor: an anticoagulant which inhibits the endotoxin mediated activation of the *Limulus* coagulation system, *Biochem. Biophys. Res. Commun.*, 105, 717, 1982.
151. **Söderhäll, K. and Unestam, T.**, Activation of crayfish serum prophenoloxidase in arthropod immunity. The specificity of cell wall glucan activation and activation by purified fungal glycoproteins, *Can. J. Microbiol.*, 25, 406, 1979.
152. **Unestam, T. and Söderhäll, K.**, Soluble fragments from fungal cells walls elicit defence reactions in crayfish, *Nature (London),* 267, 45, 1977.
153. **Smith, V. J., Söderhäll, K., and Hamilton, M.**, β-1,3-glucan induced cellular defence reactions in the shore crab, *Carcinus maenas, Comp. Biochem. Physiol.*, 77A, 635, 1984.
154. **Söderhäll, K. and Ajaxon, R.**, Effect of quinones and melanin on mycelial growth of *Aphanomyces* sp. and extracellular protease of *Aphanomyces astaci,* a parasite of crayfish, *J. Invertebr. Pathol.*, 39, 105, 1982.
155. **Ashida, M. and Söderhäll, K.**, The prophenoloxidase activating system in crayfish, *Comp. Biochem. Physiol.*, 77B, 21, 1984.
156. **Duvic, B. and Söderhäll, K.**, Purification and characterization of a β-1,3-glucan binding protein from plasma of the crayfish *Pacifastacus leniusculus, J. Biol. Chem.*, 265, 9327, 1990.
157. **Armstrong, P. B., Rossner, M. T., and Quigley, J. P.**, An α_2-macroglobulin-like activity in the blood of Chelicerate and mandibulate Arthropods, *J. Exp. Zool.*, 236, 1, 1985.
158. **Hergenhahn, H.-G. and Söderhäll, K.**, α_2-Macroglobulin-like activity in plasma of the crayfish *Pacifastacus leniusculus, Comp. Biochem. Physiol.*, 81B, 833, 1985.
159. **Armstrong, P. B., Levin, J., and Quigley, J. P.**, Role of endogenous proteinase inhibitors in the regulation of the blood clotting system of the horseshoe crab, *Limulus polyphemus, Thromb. Haemostasis,* 52, 117, 1984.
160. **Spycher, S. E., Arya, S., Isenman, D. E., and Painter, R. H.**, A functional, thiolester-containing α_2-macroglobulin homologue isolated from the hemolymph of the American lobster *(Homarus americanus), J. Biol. Chem.*, 262, 14606, 1987.
161. **Hergenhahn, H.-G., Hall, M., and Söderhäll, K.**, Purification and characterization of an α_2-macroglobulin-like proteinase inhibitor from plasma of the crayfish *Pacifastacus leniusculus, Biochem. J.*, 255, 801, 1988.
162. **Hall, M., Söderhäll, K., and Sottrup-Jensen, L.**, Amino acid sequence around the thiolester of α_2-macroglobulin from plasma of the crayfish, *Pacifastacus leniusculus, FEBS Lett.*, 254, 111, 1989.
163. **Johansson, M. W. and Söderhäll, K.**, Cellular immunity in crustaceans and the proPO system, *Parasitology Today,* 5, 171, 1989.
164. **Hergenhahn, H.-G., Aspan, A., and Söderhäll, K.**, Purification and characterization of a high-M_r proteinase inhibitor of pro-phenol-oxidase activation from crayfish plasma, *Biochem. J.*, 248, 223, 1987.
165. **Söderhäll, K.**, The bee venom melittin induces lysis of arthropod granular cells and inhibits activation of the prophenoloxidase-activating system, *FEBS Lett.*, 192, 109, 1985.
166. **Hall, L. and Söderhäll, K.**, Purification and properties of a protein inhibitor from crayfish hemolymph, *J. Invertebr. Pathol.*, 39, 29, 1982.
167. **Armstrong, P. B. and Quigley, J. P.**, Proteinase inhibitory activity released from the horseshoe crab blood cells during exocytosis, *Biochim. Biophys. Acta,* 827, 453, 1985.
168. **Cohen, W. D., Nemhauser, I., and Jaeger, R.**, Rapid visualization of the marginal band system in blood cells of marine species, *Biol. Bull. (Woods Hole, Mass.),* 153, 420, 1977.
169. **Cohen, W. D., Nemhauser, I., and Cohen, M. F.**, Marginal bands of lobster blood cells: disappearance associated with changes in cell morphology, *Biol. Bull. (Woods Hole, Mass.),* 164, 50, 1983.
170. **Nemhauser, I., Ornberg, R., and Cohen, W. D.**, Marginal bands in blood cells of Invertebrates, *J. Ultrastruct. Res.*, 70, 308, 1980.
171. **Ilan, S. S. and Gupta, A. P.**, Arthropod immune System. V. Activated immunocytes (granulocytes) of the German cockroach, *Blattella germanica* (L.) (Dictyoptera:Blattellidae) show increased number of microtubules and nuclear pores during immune reaction to foreign tissue, *Cell Struct. Funct.*, 13, 333, 1988.

172. **Liang, S.-M., Hascall, G., Liu, T., and Liu, T.-Y.,** *Limulus* amebocyte clotting cascade: roles of endotoxin and adenylate cyclase, *J. Protein Chem.*, 4, 151, 1985.
173. **Feinstein, M. B., Egan, J. J., Sha'Afi, R. I., and White, J.,** The cytoplasmic concentration of the calcium in platelets is controlled by stimulators of cyclic AMP production (PGD2, PGE1, forskolin), *Biochem. Biophys. Res. Commun.*, 113, 598, 1983.
174. **Rink, T. J., Smith, S. W., and Tsien, R. Y.,** Cytoplasmic free Ca^{2+} in human platelets: Ca^{2+} thresholds and Ca-independent activation for shape-change and secretion, *FEBS Lett.*, 148, 21, 1982.
175. **Battelle, B. A. and Kravitz, E. A.,** Targets of octopamine action in the lobster: cyclic nucleotide changes and physiological effects in hemolymph, heart and exoskeletal muscle, *J. Pharmacol. Exp. Ther.*, 205, 438, 1978.
176. **Pollard, H. B., Pazoles, C. J., Creutz, C. E., and Zinder, O.,** The chromaffin granule and possible mechanisms of exocytosis, *Int. Rev. Cytol.*, 58, 159, 1979.
177. **Fuller, G. M.,** Are lobster fibrinogen and cold-insoluble globulin related molecules?, *Ann. N.Y. Acad. Sci.*, 312, 31, 1978.
178. **Folk, J. E.,** Transglutaminases, *Annu. Rev. Biochem.*, 49, 517, 1980.
179. **Barwig, B. and Bohn, H.,** Evidence for presence of two clotting proteins in Insects, *Naturwissenschaften*, 67, 47, 1980.
180. **Bohn, H., Barwig, B., and Bohn, B.,** Immunochemical analysis of hemolymph clotting in the insect *Leucophaea maderae* (Blattaria), *J. Comp. Physiol.*, 143B, 169, 1981.
181. **Barwig, B.,** Isolation and characterization of plasma coagulogen (PC) of the cockroach *Leucophaea maderae* (Blattaria), *J. Comp. Physiol.*, 155, 135, 1985.
182. **Ashida, M., Ishizaki, Y., and Iwahana, H.,** Activation of prophenoloxidase by bacterial cell walls or β-1,3-glucans in plasma of the silkworm, *Bombyx mori*, *Biochem. Biophys. Res. Commun.*, 113, 562, 1983.

Chapter 9

BACTERIAL CHALLENGE AND SIGNIFICANCE OF AN INDUCIBLE PROTEIN (M13) IN COAGULATION AND CELL AGGLUTINATION IN *MANDUCA SEXTA* (LEPIDOPTERA)

Kemet D. Spence and Michael F. Minnick

TABLE OF CONTENTS

I. INTRODUCTION

Insects and their bacterial pathogens have co-evolved over the millennia, and each has adapted to the other's defensive and invasive innovations, respectively, to a degree which permits survival of the interrelationship. It is clear that if one or the other dominates to too high a degree, one or the other of the protagonists, or at least their interrelationship, must perish. For example, a perfect insect defense would terminate the interaction with the pathogen, while conversely, a highly virulent bacterium might eliminate an insect species. Fortunately for all life forms, such abrupt shifts rarely occur. The changes in a host-parasite interrelationship are commonly gradual, and are certainly responsible for the slow alteration in the fundamental nature of both the host and its bacterial antagonist. In this chapter, we will discuss those interrelationships which have been observed between bacteria and *M. sexta* and other selected insects.

II. MECHANICAL DEFENSE BARRIERS

A. INTEGUMENT, FOREGUT, AND HINDGUT

The integument is composed of the cuticle and the epidermis. The cuticular layers consist of an outer lipoidal epicuticle which retards water loss, and the inner exocuticle and endocuticle are composed in part of relatively tough, interlinked chitin and protein. The foregut and hindgut contain a similar cuticular lining. The integument, foregut, and hindgut are apparently formidable barriers; no reports of primary bacterial penetration of these structures have been made. On the other hand, it is well known that a secondary bacterial invasion is common following the rupture of the integument, either mechanically or by fungi. The fungi are the only known microbial agents which can penetrate intact integument, which they apparently accomplish through a combination of mechanical pressure and the dissolution of the various integumental layers with a range of enzymes which must include lipase, chitinase, and protease activities.[1]

B. MIDGUT EPITHELIUM AND THE INNER TRACHEAL TERMINI

The possibility that bacteria invade the insect hemocoel via the spiracles is largely inferential. The spiracular openings[2] into most lepidopteran tracheae would easily admit particles the size of bacteria. The tracheae[2] of most insects are composed of essentially the same cuticular materials, and of structures analogous to those of the body surface. There are notable exceptions, including the Diptera and *Apis mellifera* (Hymenoptera) in which even the large tracheae contain no chitin. The much more delicate-appearing smallest branches, or tracheoles, of insects thus far reported appear to contain no chitin. The tracheoles range in diameter from 2 to 5 μm down to 0.2 μm at their ends, large enough to admit both viruses and bacteria. The tracheoles, often fluid filled, may be a more important route of infection than one suspects. Unfortunately, this potentially important route of entry has apparently not received significant attention.

The lumen surface of the midgut of larval lepidopterans is composed primarily of a single layer of columnar and goblet cells anchored to a basal membrane that partitions the midgut lumen from the hemolymph of the insect. The digestive function of the midgut cells is obviously not compatible with a relatively impermeable cuticular covering like that of the integument, foregut, and hindgut. In most insects the exposed midgut cells are instead separated from the luminal environment (containing ingested food, microbes) by a separable, selectively permeable, noncellular peritrophic membrane. The extent to which the peritrophic membrane protects the insect from microbial attack is discussed below. Among bacteria known to be capable of penetrating the peritrophic membrane, several are known to also penetrate the midgut epithelium and enter the hemocoel.

TABLE 1
Mechanical Barriers to Bacterial Invasion

Structure	Penetration by bacteria[a]
Integument	No
Trachea	No[b]
Tracheoles	Inferred
Foregut	No
Midgut	Yes
Peritrophic membrane	Yes
Hindgut	No

[a] It is presumed that the relatively tough cuticle or cuticular-type covering or lining of the integument, tracheae, foregut, and hindgut prevent the penetration of bacteria.

[b] The trachea of some insects (see text) lack a cuticular material in their structure. In most insects, the cuticular covering is lacking only in the tracheoles or their distensions.

C. PERITROPHIC MEMBRANE

The peritrophic membrane of lepidopterans and most other insects is composed of multilayers of chitin (and/or other polysaccharides and/or mucopolysaccharides) in a mesh-like base structure together with a protein superstructure. In lepidopterans it is secreted in a precursor form by columnar cells along the entire midgut, and maturates into the mature cross-linked form near and above the tips of the columnar cell microvilli. The peritrophic membrane lines of the midgut, which unlike most other body surfaces, is not otherwise protected by a relatively tough, cuticular covering. The peritrophic membrane plays a number of different roles. Although numerous other functions are known (e.g., the formation of cocoon silk and digestive compartmentalization of the midgut), it is now apparent that it also plays a role in protecting the underlying midgut from microbial attack. A summary of the significance of various structures as mechanical barriers to bacterial invasion is shown in Table 1.

III. BACTERIA-INDUCED CHANGES IN THE MIDGUT AND PERITROPHIC MEMBRANE

A. ALTERATIONS IN MIDGUT STRUCTURE AND PROTEINS

Certainly the most rapid and dramatic changes of the midgut resulting from the action of a microbial agent are those observed when a lepidopteran ingests the crystal endotoxin produced by *Bacillus thuringiensis*. A few studies have described the cellular response of the midgut epithelial cells to sublethal doses of the crystal endotoxin.[3-5] However, most of the data with this toxin have been obtained from insects receiving a lethal dose, which rapidly causes lysis of midgut columnar and goblet cells, loss of electrolyte balance between the hemolymph and gut, larval paralysis, and finally death of the insect.[6-12] Under these rapidly lethal conditions one would be unable to evaluate the significance of an immune response, if indeed there was one.

To determine whether the toxin-treated insect could, if given time, marshal some type of inducible defense to the *B. thuringiensis* endotoxin, it was necessary to administer the

TABLE 2
Alterations in *M. sexta* Midgut and Peritrophic Membrane (PM) after *per os* Administration of a Sublethal Dose of *B. thuringiensis* Crystal Endotoxin

Component/ structure	State of larvae		
	Untreated	Toxin-treated	Recovered
Gut juice proteins	9 major proteins	9 original + ~23 new proteins[a]	9 original
PM proteins	PM1-9, PM11,12	PM5-12[b]	PM1-12
Midgut	Normal	Swelling, microvilli lost, columnar and goblet cells burst	Indistinguishable from untreated larvae

[a] The source of the new proteins is tentatively assumed to be ruptured midgut epithelial cells.
[b] A new protein (PM10) appears in newly synthesized PM following toxin treatment. PM10 remains in the PM throughout the remainder of the larval stage.

toxin in sublethal doses. The effect of sublethal toxin on the midgut cells of fifth instar *M. sexta* larvae, including the subsequent full recovery of the tissue, has been recorded,[5] and is summarized in Table 2. Following oral administration of the toxin, the tissue initially swells. Subsequently, the microvilli disappear from the columnar cells followed (in the anterior area of the midgut) by the rupture of many columnar cells and a few goblet cells. The damage is progressively less severe in the more posterior areas of the midgut. Within 24 h proteins, presumably released by the damaged cells, disappear from the gut juice at approximately the time when the midgut tissue itself has begun recovery. After a delay in developing during the initial intoxification period, new cells rise to replace damaged cells, which are in turn being extruded from the epithelium (into the gut lumen). Eventually the toxin-treated *M. sexta* larvae fully recover. The midguts of toxin-treated larvae at the time of the larval-pupal ecdysis were indistiguishable from those of the untreated insects, and subsequent development of the toxin-treated insects proceeded normally.[5]

The gut juice collected from *M. sexta* larvae following sublethal toxin treatment initially contained a large number of new proteins, most of which were tentatively assumed to be released by toxin-damaged midgut cells.[13] As the midgut healed, the soluble toxin-generated proteins fell in concentration and finally disappeared. Even though no permanent changes appeared to occur in the protein composition of the gut juice, samples of the juice from both control and toxin-treated larvae were removed from a number of larvae and introduced into microcultures with a number of bacteria to determine whether any antibacterial factors might have been released into the soluble fraction of the lumen by the insect. To determine if the toxin treatment induced the bactericidal proteins of the hemolymph, and whether antibacterial activity might be associated with the peritrophic membrane, both hemolymph and homogenized peritrophic membranes from the same insects were also analyzed at the same time. Table 3 shows a summary of the results of these tests. There was a low level of antibacterial activity for two of the three Gram-positive bacteria in the gut juice of pretreated larvae. Following toxin ingestion there was no increase in the antibacterial effect of the gut juice for the Gram-positive bacteria. None of the Gram-negative bacteria tested were inhibited by gut juice from either untreated or toxin-treated larvae. The nature of the antibacterial action of the gut juice has not been further investigated. No antibacterial activity was detected in cultures containing the washed, homogenized peritrophic membranes, nor did the toxin induce the synthesis of the bactericidal proteins (although the nonbactericidal protein M13 was induced) in the hemolymph.

B. ALTERATIONS IN THE PERITROPHIC MEMBRANE

The pore sizes commonly reported for a large number of insect peritrophic membranes, which range in diameter from approximately 2 to 10 nm, are too small for bacteria to pass

TABLE 3
Antibacterial Activity of the Midgut Juice, Minced Peritrophic Membrane (PM) Preparations, and the Hemolymph from Untreated and Toxin-Treated Larvae

	Growth inhibition[a] (naive/toxin-treated)		
	Gut juice	PM[b]	Hemolymph
Escherichia coli	–/–	–/–	–/–
Pseudomonas maltophilia	–/–	NT/NT	NT/NT
P. aeruginosa	–/–	NT/NT	NT/NT
Acinetobacter calcoaceticus	–/–	NT/NT	NT/NT
Bacillus megaterium	1:8/1:8	–/–	–/–
B. thuringiensis	–/–	–/–	–/–
Micrococcus luteus	1:8/1:8	–/–	–/–

[a] All materials were added to and diluted in T-soy broth medium. 10^6 bacterial cells were added to each well at zero time and growth determined after 24 h. The lowest dilution was 1:1 (gut juice or hemolymph:growth medium). Growth was recorded as maximum dilution giving inhibition, no inhibition (–), or not tested (NT).

[b] Peritrophic membranes were removed and rinsed in three changes of iced 0.1 *M* phosphate buffer, pH 7.2. They appeared to be free of food particles. The PM was then lyophilized. A suspension of 10 mg lyophilized PM per ml medium was prepared, minced, then further diluted in series, and bacteria added as with the other materials tested.

through.[14-17] Even the larger pores, which appeared to exist in *Orgyia pseudotsugata,* would not permit the penetration of most common bacterial pathogens.[18] However, since both bacteria and viruses are known to penetrate the midgut epithelium, one of three possibilities must exist: (1) there are at least a few pores or discontinuities in the peritrophic membrane of some insects, which are of sufficient size to allow the passage of microbial agents; (2) the microbe possesses agents (e.g., enzymes) that modify the structure of the peritrophic membrane to permit microbial penetration; and/or, (3) the microbe modifies the synthesis of the peritrophic membrane, which results in the production of a weakened, more porous structure. Depending upon the insect and the bacterial pathogen, it now appears that one or more of these possibilities may exist.

A number of observations of peritrophic membrane permeability to pathogens have been reported. Ingested bacteria, such as *Bacillus popilliae,* do make their way through the peritrophic membrane and midgut epithelium and into the hemocoel during the infection of the European chafer, *Amphimallon majalis.*[19,20] The peritrophic membrane of the honeybee, *A. mellifera.*[21] is penetrated during infection by *Bacillus larvae,* the causative agent of American foulbrood disease in the honey bee.[22] On the other hand, the bacterium, *Bacillus sphaericus* was unable to penetrate the peritrophic membrane of *Culex pipiens quinquefasciatus* larvae until after the toxin of the microbe had killed the host.[23]

The inability of microbial agents to quickly cross the peritrophic membrane has also been observed with insect viruses. For example, most of the ingested mosquito iridescent virus was destroyed in the midgut of *Aedes taeniorhynchus* larvae, apparently due to its inability to penetrate the peritrophic membrane; no virus could be detected in the ectoperitrophic space.[24] However, at too low a level for detection, enough of the iridescent virus did penetrate the *Aedes* membrane in numbers sufficient to infect the insects.[25] These workers observed no discontinuities large enough to admit a virus through the peritrophic membrane of this mosquito. On the other hand, random discontinuities were observed in the peritrophic membranes of two lepidopteran hosts, *Trichoplusia ni* and *Spodoptera frugiperda;* both granulosis and nuclear polyhedrosis viruses were seen to readily penetrate the membranes of these lepidopterans.[25] In another study, the larger pore sizes of the peritrophic membrane of *O. pseudotsugata* did permit the passage of a 35 nm bacteriophage, which is in the size range of some insect viruses, but was impermeable to the spores of *Bacillus thuringiensis*.[18] These workers detected discontinuities and overlapping layers associated with pore-like openings in the membrane of this insect, which may explain the larger pore sizes measured.[18]

The presence of possible membrane-disrupting enzymes associated with microorganisms has suggested that these molecules may facilitate the peritrophic membrane penetration by bacteria[26] and viruses.[25] The peritrophic membrane of *O. pseudotsugata* was susceptible to the action of both protease and chitinase; treatment of the membrane with a protease both hydrolyzed the proteins[26] and greatly increased the permeability of the peritrophic membrane.

The synthesis of the *M. sexta* peritrophic membrane is altered by the ingestion of the *Bacillus thuringiensis* crystal endotoxin.[13,28] At first the peritrophic membrane becomes detached from the midgut epithelium and is structurally much weaker than the normal peritrophic membrane. The three major glycoproteins of the peritrophic membrane are sharply reduced or temporarily disappear. For a time, peritrophic membrane synthesis stops. Subsequently, peritrophic membrane synthesis is reinitiated and the strength of the peritrophic membrane is regained as the midgut recovers from the effects of the toxin. During recovery the three large glycoproteins initially lost return to the structure. In addition, a new protein is incorporated into the peritrophic membrane as the insect recovers. The SDS-dissociated peptide of this protein, termed PM10, has an approximate $M_r = 32$ kDa. PM10 synthesis and incorporation into the peritrophic membrane continues until peritrophic membrane synthesis normally ceases at the onset of wandering. The significance of this protein is unknown. It has been found that the permeability of newly synthesized peritrophic membrane, following recovery from the crystal toxin, is reduced.[28] It is possible that PM10 may in some way influence the permeability of the peritrophic membrane, perhaps by reducing pore size. The possibility that the *B. thuringiensis* crystal toxin stimulated elements of antibacterial defense in the peritrophic membrane or hemolymph was also tested. Although it has been noted that some type of soluble antibacterial activity appeared in the gut juice of larvae treated with a sublethal level of crystal toxin, a medium supplemented with buffer-rinsed and homogenized peritrophic membrane from toxin-treated larvae did not affect the growth of the test bacteria (Table 3).

While these studies showed that the sublethal toxin treatment did not induce the bactericidal proteins in the *M. sexta* hemolymph, induction of the nonbactericidal M13 protein, which appears to be somehow involved in coagulation and cellular adhesion response, did occur.[13] As expected, no inhibitory effect on bacterial survival was detected when hemolymph from toxin-treated larvae was supplemented to the medium (Table 3). Infrequently, the other inducible proteins of the hemolymph were synthesized after toxin treatment. It is presently assumed that in these few cases, the midgut damage was sufficient to allow normal bacterial flora of the midgut to enter the hemolymph and induce these proteins.

IV. DEFENSE SYSTEMS OF THE HEMOLYMPH

A. CONSTITUTIVE DEFENSE PROTEINS

1. Bactericidal Proteins

As a rule, the bacteria-inducible proteins of the *M. sexta* larval hemolymph are either absent or below the level of detection on Coomassie blue or silver-stained gels.[29,30] This is consistent with results showing that the bactericidal action of naive insects is either very low or absent from the hemolymph. However, naive *M. sexta* pupae sometimes possess low, but easily detectable levels of some inducible proteins. For example, although most other inducible proteins were undetectable in naive pupae, low levels of lysozyme were commonly detected in the hemolymph.[30] Subclinical contact with normal flora may be responsible for the low constitutive levels observed. Nevertheless, it is fair to state that the sera of naive insects, whether larvae or pupae, do not contain sufficient antibacterial activity to defend against large numbers of pathogenic bacteria. The inducible nature of the antibacterial hemolymph proteins will be discussed in the following section.

2. Cell-Related Defenses

Recent biochemical and immunologic studies make it clear that the cellular and humoral immune responses of insects are often interdependent (i.e., soluble proteins are often involved with cellular systems). This interdependence has recently been reviewed,[31] so is only briefly treated here.

An efficient melanization system is present in the hemolymph of naive insects.[32-37] Since in many insects the nodules containing bacteria or encapsulated nematodes eventually become melanized, it is assumed that melanization is an important participant in the process of eliminating invaders from the hemocoel. The extensive cross-linkage which takes place during melanization is assumed to aid in the walling-off process that occurs. The constitutive melanization system in lepidopterans is known to involve inactive prophenoloxidases that are converted under appropriate conditions to the active phenoloxidase by a protease or certain polysaccharides. The protease activity is in turn controlled by the presence or absence of protease inhibitors. The actual inducibility of the proteins involved has not been analyzed in detail.

The fact that the nodulation of bacteria in *M. sexta* occurs within a very few minutes after injection[38] shows that a functional system is in place which does not require the induction of proteins. However, this does not exclude the possibility that the response might be enhanced or optimized by the appearance of new inducible defense proteins. Nodulation[38-41] and encapsulation[40,42-44] appear to be closely related processes. Although encapsulation of large invaders (e.g., nematodes) may very well commence within the same time frame as nodulation, the size of the foreign body makes this possibility much less obvious. Many bacteria-induced hemolymph proteins of *M. sexta* larvae are detectable on polyacrylamide gels with standard staining techniques within an hour following injection. If inducible proteins are involved in the nodulation/encapsulation response, this rate of induction would probably be rapid enough to facilitate the nodulation, and if an appropriate inducer were present, encapsulation. The involvement, although not the regulation, of a protein which mediates hemocyte adhesion in the encapsulation process has been reported[45] (see also Chapter 3).

The constitutive phagocytic response to injected bacteria is rapid and efficient.[38,41,46] We have found that following the injection of 10^6 bacteria into fifth instar *M. sexta* larvae, the number of circulating bacteria will often drop by two logs in the first 15 to 30 min following injection. Although it might seem unclear how such an efficient, constitutive phagocytic system could be improved or influenced by the appearance of new, induced proteins, it has been observed that prior exposure of an insect to polymer beads did increase the phagocytic ability.[47-48] It seems possible that inducible molecules may be involved in these systems, especially since this ability could be passively transferred.[48] It is quite easy to imagine how the appearance of even the inducible bactericidal proteins might indirectly help (i.e., dead bacteria would be expected to be more easily disposed of than live ones).

3. Hemolymph Coagulation and Wound Response

Another constitutive defense mechanism in nonimmunized insects is the coagulation system.[49-52] This mechanism is perhaps most obviously seen in the healing of wounds where, for example, the hemolymph around a needle puncture wound penetrating the integument coagulates or thickens, sealing the wound in a very few seconds. In addition to the soluble serum components, the process involves attraction and adhesion of hemocytes, including granulocytes and plasmatocytes, some of which show differentiation (e.g., formation of filopodial extensions). The coagulum formed in this process appears to be primarily a filamentous proteinaceous network involving serum proteins and perhaps also hemocyte proteins as well as nonproteinaceous materials. The wound sealing process begins virtually immediately, while the cellular changes associated with the initial sealing of the wound can be observed in less than a minute.

An inducible protein termed M13 has been detected in *M. sexta* larvae, which appears to be involved in the coagulative process.[53] Although this protein cannot be detected in the hemolymph of naive larvae with standard staining techniques,[30] a small amount is present, and can be detected by the highly sensitive immunoblot analysis.[54] Whether the small amount of M13 present in control hemolymph is employed and is necessary for the coagulative response, or whether more than one stimulus can trigger the response, is not known at this time. A more thorough discussion of the M13 protein is given in the following sections.

B. INDUCIBLE DEFENSE SYSTEMS

1. General Comments

The injection of bacteria into the hemocoel of *H. cecropia* or *M. sexta* induces the synthesis of a number of proteins in the hemolymph.[29,30,55-59] In general, naive insects either lack these proteins or maintain one or more of them at very low levels. In a detailed study with *M. sexta* in which the various bacterial components were employed, it was found that the peptidoglycan of the cell wall was the bacterial component which induced the response.[60] Most detailed investigations of the inducible proteins have been performed with lepidopterans, although several dipterans[61-65] and a coleopteran[66] have been shown to induce antibacterial proteins. Data also suggest that the cockroach can be induced to synthesize proteins with specificity for the antigen employed[67] (see also Chapter 15). The immune response includes the induction of both bactericidal and nonbactericidal proteins. Although the action of the bactericidal proteins has been investigated in some detail, the function of most of the nonbactericidal proteins remains unknown.

It should be further noted that the hemolymph immune response for all stages of a given insect are not necessarily identical. In *M. sexta*, a comparison of the protein profiles of larvae and pupae has shown that there is a differential (stage-specific) synthesis of inducible proteins.[29,30] Although many inducible proteins can be elicited in both larvae and pupae, some induced proteins are unique to each developmental stage.[30] This suggests that there are peculiarities or defensive needs within each stage which require some specialization in the immune response.

2. Inducible Bactericidal Proteins

The first antibacterial insect protein reported was lysozyme.[68,69] It was subsequently found that insects produced a wide array of other inducible proteins in response to the injection of bacteria. At least three families of bactericidal proteins have now been described. These include the lysozyme, cecropin, and attacin groups[30,56-58,66] (see also Chapters 12 and 13). The cecropins and attacins were first reported in *Hyalophora cecropia*. The same types of proteins have subsequently been isolated from a number of different insects and are usually designated by the aforementioned names or as cecropin-like or attacin-like proteins. For example, *M. sexta* synthesizes cecropins,[30,56,57,60] attacins,[66] and lysozymes,[30,56] which are analogous to those of *H. cecropia*. In one case, a cecropin-like protein has been termed sarcotoxin I (from *Sarcophaga*).[64] For our discussion here, the bactericidal proteins will be noted by the simple terms cecropin, attacin, and lysozyme.

A range of both Gram-positive and Gram-negative bacteria are killed by cecropins.[70] A few Gram-negative bacteria were found to be sensitive to the attacins.[58,59] As would be expected, Gram-positive bacteria appear to be the major target of lysozyme.[69]

TABLE 4
Character and Function of Nonbactericidal M13 Protein of *M. sexta* Larvae

Hemagglutinin	Agglutinates sheep RBCs; inhibition of agglutination by glucose indicates that M13 is a glucose-binding lectin
Coagulation	Addition to naive hemolymph causes coagulation of hemolymph and aggregation of hemocytes on coagulum
Protein character	Native M_r = 72 kDa; composed of two equally sized subunits of approximately M_r = 36 kDa; a glycoprotein
Inducibility	Induced in the hemolymph of feeding larvae but not wanderers, prepupae, or pupae; detected in the midgut and epidermis but not the fat body

3. Inducible Nonbactericidal Proteins

A number of workers have identified bacteria-induced proteins for which no activity can be ascribed. Since these proteins have no direct bactericidal action, it is necessary to analyze these proteins for possible participation in known defense systems as well as for possible involvement in novel insect defenses. Among the known systems discussed earlier are nodulation, encapsulation, wound sealing, and wound healing. The coagulation response is of immediate significance in wound sealing, and may certainly be involved in the other defense responses noted. Another series of presumed defensive reactions is melanization. None of the inducible proteins commonly seen by polyacrylamide gel electrophoresis have been reported to be members of the melanization series. However, we found that the hemolymph from immunized insects usually melanizes more rapidly than that from naive or saline-injected insects. Whether this reflects the induction of one or more of the protein components necessary for melanization or whether, for example, the protease inhibitor concentration has been reduced is not known. Since the extent to which these unknown proteins have been screened for other activities is not always clear from reports, one or more of the immune proteins may yet be identified as a component of the melanization system.

The inducible P4 protein of *H. cecropia*[71] has been studied in some detail.[72] Although the function of this protein has not been specifically determined, immunostaining has shown that it is associated with fat body basal membranes, certain hemocytes, and nodules in bacteria-injected pupae.[72] These data may suggest that P4 is involved in some aspect of the cell-based response to bacterial infection. The P4 protein may correspond to the M11 protein[30] of *M. sexta*.

Another nonbactericidal protein, the M13 protein of *M. sexta*, appears to trigger hemolymph coagulation.[53] M13 is one of the differentially synthesized proteins, which is inducible in all feeding larvae until late in the final, fifth instar.[30,73] Approximately coinciding with a low level pulse of 20-hydroxyecdysone in the late fifth instar, and during the subsequent wandering larval period of the fifth instar and pupal stage, an injection of bacteria does not induce M13.[30,73] It seems apparent that in addition to being inducibly regulated by bacterial components, M13 is also under hormonal control. Most other inducible proteins, including all the three bactericidal groups, can be induced during these later periods. Immunologic analysis of various insect tissues including fat body, midgut, and integument suggest that M13 is synthesized in the epidermis and midgut.[54] A number of other nonimmunologically associated proteins, including the blue insecticyanin, are synthesized by the epidermis of *M. sexta*. Synthesis of the latter protein ceases apparently in response to the pulse of 20-hydroxyecdysone mentioned,[74] and at roughly the same time as M13.

The introduction of purified M13 into naive hemolymph triggers a coagulation response, including an acceleration of filipod formation on plasmatocytes, a filamentous network in the hemolymph and the adhesion of other hemocytes to one another within the agglomerate.[53] The specific function of M13 is unclear, although a number of characteristics of this protein suggest several possibilities (Table 4).

M13 agglutinates sheep red blood cells (sRBCs).[53] Among a number of sugars tested, glucose is most effective in neutralizing the agglutinating activity of this protein. It is therefore presently assumed that the protein is a glucose-binding lectin. Partial proteolysis of M13 with trypsin reduces the weight of each subunit by approximately 2 kDa (M13 is a dimer of two approximately 36-kDa subunits).[30,59] The proteolytically modified M13 still retained its sRBC agglutinating activity, but would no longer stimulate coagulation of the hemolymph. This suggests that the protein has two activities, a coagulative as well as a lectin activity. Since the addition of glucose to the hemolymph can inhibit the coagulation stimulated by M13, it appears that the coagulative activity is controlled by the degree of glucose binding to the lectin portion of the protein.

Since there are very low levels of M13 in the hemolymph of naive larvae,[54] one may question whether M13 is required for the coagulation response under all circumstances. In view of the uncertainty, it is possible that there is more than one stimulus that can trigger the coagulation response. In vertebrates, such multiple mechanisms are known to initiate the coagulation response, often via alternate but interlinked pathways. In vertebrates, the cellular component principally involves platelets. It is perhaps safest to conclude that M13 can under certain conditions trigger coagulation but may very well share this ability with other chemical or even physiochemical stimuli.

Several of the major tissues of *M. sexta* have been extracted and analyzed for the presence of M13.[54] In addition to being found in the hemolymph, M13 is always detectable in the epidermis; it is usually detectable in relatively small amounts in midgut tissues. Both the latter tissues represent the exterior of the insect with respect to environmental exposure. Data obtained from in vitro translation of isolated mRNA from fat body, epidermis and midgut tissues, have shown that only the latter two tissues synthesize M13. The analysis of fat body organ cultures for the synthesis of other inducible proteins has shown that of other proteins analyzed, all were synthesized by the fat body.[60,75]

In addition to M13 induction by bacteria, induction and modification of M13 apparently also result from tissue damage. The oral administration of sublethal levels of the *B. thuringiensis* crystal endotoxin, which results in extensive damage to the midgut, commonly (among the other inducible proteins) induces only the formation of M13 in the hemolymph.[13] Further, mechanical damage to the epidermis results in the conversion of M13 to a smaller form of the protein; in this conversion approximately 3 kDa is proteolytically cleaved from each of the approximately 36-kDa subunits of M13 (the modified form has been termed 33K-CRP).[54] Nondenaturing gels suggest that 33K-CRP has a native M_r of 66 kDa. This observation suggests that M13 may be a precursor, while the smaller form of the protein is, in fact, the active form, although the latter could simply be a degradative product. The 33K-CRP has not been purified in large enough quantities to allow answering this question.

In summary, M13 protein of *M. sexta* is found in tissues which interface with the environment. It is induced by bacterial injection into the hemocoel or by toxin damage to the midgut. It appears to influence the coagulation process in the insect. Its presence in the hemolymph of only feeding-larvae, suggests that it has a function peculiar to the defense of this larval state. Two obvious differences between larvae and pupae involve the integument and the midgut. The pupa is protected by a tough sclerotized cuticle, and is also no longer ingesting food, potentially containing microbial pathogens; its midgut has metamorphosed into a much less dynamic organ. The feeding larva, on the other hand, is more exposed to external injury by both small and large predators, and from pathogens contained in the diet. The data suggest that the M13 protein, by virtue of its ability to stimulate the coagulative processes, may be associated with wound-healing, such as aiding in limiting or correcting bacterial breaches in the midgut or sealing predator punctures to the integument through which bacteria might enter the hemocoel.

V. BACTERIAL COUNTERACTION OF INSECT IMMUNE DEFENSES

In a number of cases, bacteria have been found to synthesize molecules that destroy or somehow neutralize the action of the immune bactericidal proteins. *B. thuringiensis*, which produces perhaps the widest array of pesticidal molecules, also produces two anti-immune factors, termed InA and InB.[76]

InA has been more extensively studied,[76,77] and was initially found to be highly toxic for *Drosophila*. It was a single peptide protease with an apparent M_r of 78 kDa. The proteolytic activity appears specific for the cecropins and several other small peptides.

The InB is heat resistant and trichloroacetic acid-soluble, which suggests that it is a nonprotein molecule.[76] The normally strong *in vitro* immune, bactericidal action of the hemolymph for *B. cereus* can be completely neutralized by the addition of InB to the growth medium.

Two proteases have been detected in *Serratia* which may also protect the bacterium to some degree from the bactericidal action of immune proteins.[78]

One of the most interesting accounts of an anti-immune process involves the symbiotic relationship between the nematode, *Neoaplectana carpocapsae* and the bacterium, *Xenorhabdus nematophilus*.[79] The bacterium is a normal inhabitant of the *N. carpocapsae* digestive system. After the nematode is ingested by the larva, it penetrates into the hemolymph where the *X. nematophilus* is released. The nematode produces an anti-immune factor, which the data suggest is a rather specific protease. The bacterium is protected from the insect bactericidal proteins by the presumed protease, which hydrolyzes cecropins A and B and the attacins, the immune proteins which quickly destroy *X. nematophilus* in the absence of the nematode. The bacterium increases the susceptibility of the insect to the nematode. Whereas the LD_{50} was 10 nematodes when *X. nematophilus* was present, the LD_{50} increased to 500 nematodes in axenic nematodes.

VI. SUMMARY

Of the insect surfaces which come into contact with the environment, only the midgut and trachea appear to be especially vulnerable to microbial penetration. There is no cuticular lining in the midgut or the tracheoles. Although there is only inferential information with respect to microbial invasion via the tracheal route, a number of studies have investigated the midgut with respect to microbial damage and as a portal of entry for microbial pathogens into the insect hemocoel.

The peritrophic membrane, a selectively permeable, noncellular membrane, protects the relatively sensitive epithelial midgut cells. The pores of most peritrophic membranes are too small to allow penetration of common bacterial pathogens. However, in several instances, discontinuities much larger than the smaller pores offer easy passage to microbes. The insect might limit the growth of certain bacteria by producing antibacterial activity in the gut juice, and prevent or delay invasion by reducing peritrophic membrane permeability. At least one new protein may be involved in these responses; it is induced in the peritrophic membrane by ingested bacterial toxin.

The hemolymph of the insect also contains both constitutive and inducible defense systems. The constitutive systems, which are functional in nonimmune insects, include phagocytosis, nodulation, and encapsulation; melanization is usually involved with the latter two, presumably to strengthen the cellular shell surrounding the invader.

Three families of bactericidal proteins, designated cecropins, attacins, and lysozymes, are induced in the hemolymph of many insects. Among the inducible proteins which have no bactericidal activity is an inducible glucose-binding lectin of *M. sexta*, M13, which appears to be involved in coagulation. The data suggest that M13 function is intimately related to glucose metabolism and glucose levels in the tissues. M13 is also present in the epidermis and midgut tissues; its presence in tissues facing the external environment suggests that its activity at these sites is important.

Some bacteria-induced proteins of *M. sexta* are differentially synthesized during development. Although most are induced in all stages, some are only induced in certain stages. Protein M13 is induced only in the feeding-larva stage. This suggests that the insect response is modified to deal with vulnerabilities unique to the various stages.

Some pathogens produce molecules which destroy bactericidal insect proteins. Two such agents, including a protease and a nonproteinaceous molecule, have been isolated from *B. thuringiensis*. Two proteases have been isolated from *Serratia,* which neutralize the bactericidal proteins. The bacterium *Xenorhabdus* is protected from cecropins and attacins by the protease of its nematode symbiont.

ACKNOWLEDGMENTS

The authors wish to thank Dr. Ronald E. Hurlbert of Washington State University and Dr. Ayo Gupta of Rutgers University for their review of this manuscript.

REFERENCES

1. **Roberts, D. W.,** Toxins of entomopathogenic fungi, in *Microbial Control of Pests and Plant Diseases 1970—1980,* Burges, H. D., Ed., Academic Press, New York, 1981, 441.
2. **Wigglesworth, V. B.,** *The Principles of Insect Physiology,* Chapman and Hall, London, 1972, 357.
3. **de Lello, E., Hanton, W. K., Bishoff, S. T., and Misch, D. W.,** Histopathological effects of *Bacillus thuringiensis* on the midgut of tobacco hornworm larvae *(Manduca sexta):* low doses compared with fasting, *J. Invertebr. Pathol.,* 43, 169, 1984.
4. **Luthy, P. and Ebersold, H. R.,** Bacillus thuringiensis delta-endotoxin: histopathology and molecular mode of action, in *Pathogenesis of Invertebrate Microbial Diseases,* Davidson, E. W., Ed., Allanheld, Osmun Publ., Totowa, NJ, 1981, 235.
5. **Spies, A. G. and Spence, K. D.,** Effect of sublethal *Bacillus thuringiensis* crystal endotoxin treatment on the larval midgut of a moth, *Manduca:* SEM study, *Tissue Cell,* 17, 379, 1985.
6. **Endo, Y. and Nishiitsutsuji-Uwo, J.,** Mode of action of *Bacillus thuringiensis* delta-toxin: histopathological changes in the silkworm midgut, *J. Invertebr. Pathol.,* 36, 90, 1980.
7. **Fast, P. G. and Morrison, I. K.,** The delta-endotoxin of *Bacillus thuringiensis.* IV. The effect of delta-endotoxin on ion regulation by midgut tissue of *Bombyx mori* larvae, *J. Invertebr. Pathol.,* 30, 208, 1977.
8. **Griego, V. M., Fancher, L. J., and Spence, K. D.,** Scanning electron microscopy of the disruption of tobacco hornworm, *Manduca sexta,* midgut by *Bacillus thuringiensis* endotoxin, *J. Invertebr. Pathol.,* 35, 186, 1980.
9. **Griego, V. M., Moffett, D., and Spence, K. D.,** Inhibition of active K^+ transport in the tobacco hornworm (*Manduca sexta*) midgut after the ingestion of *Bacillus thuringiensis* endotoxin, *J. Insect Physiol.,* 25, 283, 1979.
10. **Heimpel, A. M. and Angus, T. A.,** The site of action of crystalliferous bacteria in lepidopteran larvae, *J. Insect Pathol.,* 1, 152, 1959.
11. **Nishiitsutsuji-Uwo, J. and Endo, Y.,** Mode of action of *Bacillus thuringiensis* delta-endotoxin: general characteristics of intoxicated *Bombyx* larvae, *J. Invertebr. Pathol.,* 35, 219, 1980.
12. **Sutter, G. R. and Raun, E. S.,** Histopathology of European corn borer larvae treated with *Bacillus thuringiensis, J. Invertebr. Pathol.,* 9, 90, 1967.
13. **Rupp, R. A. and Spence, K. D.,** Protein alterations in *Manduca sexta* midgut and haemolymph following treatment with a sublethal dose of *Bacillus thuringiensis* crystal endotoxin, *Insect Biochem.,* 15, 147, 1985.
14. **Peters, W. and Wiese, B.,** Permeability of the peritrophic membranes of some Diptera to labelled dextrans, *J. Insect Physiol.,* 32, 43, 1986.
15. **Santos, C. D. and Terra, W. R.,** Distribution and characterization of oligomeric digestive enzymes from *Erinnyis ello* larvae and inferences concerning secretory mechanisms and the permeability of the peritrophic membrane, *Insect Biochem.,* 16, 691, 1986.

16. **Zhuzhikov, D. P.**, Permeability of the peritrophic membrane in the larvae of *Aedes aegypt, J. Insect Physiol.*, 16, 1193, 1970.
17. **Zimmerman, U. and Mehlan, D.**, Water transport across peritrophic membranes of *Calliphora erythrocephala, Comp. Biochem. Physiol.*, 55A, 119, 1976.
18. **Adang, M. J. and Spence, K. D.**, Permeability of the peritrophic membrane of the Douglas fir tussock moth *(Orgyia pseudotsugata), Comp. Biochem. Physiol.*, 75A, 233, 1983.
19. **Kawanishi, C. Y., Splittstoesser, C. M., and Tashiro, H.**, Infection of the European chafer, *Amphimallon majalis*, by *Bacillus popilliae:* ultrastructure, *J. Invertebr. Pathol.*, 31, 91, 1978.
20. **Splittstoesser, C. M., Kawanishi, C. Y., and Tashiro, H.**, Infection of the European chafer, *Amphimallon majalis* by *Bacillus popilliae:* light and electron microscope observations, *J. Invertebr. Pathol.*, 31, 84, 1978.
21. **Davidson, E. W.**, Ultrastructure of peritrophic membrane development in larvae of the worker honey bee *(Apis mellifera), J. Invertebr. Pathol.*, 15, 451, 1970.
22. **Davidson, E. W.**, Ultrastructure of American foulbrood disease pathogenesis in larvae of the worker honey bee, *Apis mellifera, J. Invertebr. Pathol.*, 21, 53, 1973.
23. **Davidson, E. W.**, Ultrastructure of midgut events in the pathogenesis of *Bacillus sphaericus* strain SSH-1 infections of *Culex pipiens quinquefasciatus* larvae, *Can. J. Microbiol.*, 25, 178, 1979.
24. **Paschke, J. D. and Summers, M. D.**, Early events in the infection of the arthropod gut by pathogenic insect viruses, in *Invertebrate Immunity*, Maramorosch, K. E. and Shope, R. E., Eds., Academic Press, New York, 1975, 75.
25. **Stoltz, D. B. and Summers, M. D.**, Pathway of infection of mosquito iridescent virus. I. Preliminary observations on the fate of ingested virus, *J. Virol.*, 8, 900, 1971.
26. **Brandt, C. R., Adang, M. J., and Spence, K. D.**, The peritrophic membrane: ultrastructure analysis and function as a mechanical barrier to microbial infection in *Orgyia pseudotsugata, J. Invertebr. Pathol.*, 32, 12, 1978.
27. **Adang, M. J. and Spence, K. D.**, Biochemical comparisons of the peritrophic membranes of the lepidopteran, *Orgyia pseudotsugata* and *Manduca sexta, Comp. Biochem. Physiol.*, 73B, 645, 1982.
28. **Rupp, R. A.**, The role of the peritrophic membrane of *Manduca sexta* in insect defense, Ph.D. thesis, Washington State University, Pullman, WA, 1986.
29. **Hughes, J. A., Hurlbert, R. E., Rupp, R. A., and Spence, K. D.**, Bacteria-induced haemolymph proteins of *Manduca sexta* pupae and larvae, *J. Insect Physiol.*, 29, 625, 1983.
30. **Hurlbert, R. E., Karlinsey, J. E., and Spence, K. D.**, Differential synthesis of bacteria-induced proteins of *Manduca sexta* pupae and larvae, *J. Insect Physiol.*, 31, 205, 1985.
31. **Lackie, A. M.**, Haemocyte behavior, *Adv. Insect Physiol.*, 21, 86, 1988.
32. **Ahmed, R. F., Hopkins, T. L., and Kramer, K. J.**, Tyrosine and tyrosine glucoside titres in whole animals and tissues during development of the tobacco hornworm *Manduca sexta, Insect Biochem.*, 13, 369, 1983.
33. **Ashida, M. and Dohke, K.**, Activation of prophenoloxidase by the activating enzyme of the silkworm *Bombyx mori, Insect Biochem.*, 10, 37, 1980.
34. **Aso, Y., Kramer, K. J., Hopkins, T. L., and Whetzel, S. F.**, Properties of tyrosine and dopa quinone imine conversion factor from pharate pupal cuticle of *Manduca sexta, Insect Biochem.*, 14, 463, 1984.
35. **Komano, H., Mizuno, D., and Natori, S.**, Purification of a lectin induced in the haemolymph of *Sarcophaga peregrina* larvae on injury, *J. Biol. Chem.*, 255, 2919, 1980.
36. **Ohnishi, E., Dohke, K., and Ashida, M.**, Activation of prophenoloxidase. II. Activation by alpha-chymotrypsin, *Archs. Biochem. Biophys.*, 139, 143, 1970.
37. **Sugumaran, M., Saul, S. J., and Ramesh, N.**, Endogenous protease inhibitors prevent undesired activation of prophenolase in insect hemolymph, *Biochem. Biophys. Res. Commun.*, 132, 1124, 1985.
38. **Horohov, D. W. and Dunn, P. E.**, Phagocytosis and nodule formation by hemocytes of *Manduca sexta* larvae following injection of *Pseudomonas aeruginosa, J. Invertebr. Pathol.*, 41, 203, 1983.
39. **Dunphy, G. B. and Nolan, R. A.**, Cellular immune responses of spruce budworm larvae to *Entomophthora egressa* protoplasts and other test particles, *J. Invertebr. Pathol.*, 39, 81, 1982.
40. **Ratcliffe, N. A. and Rowley, A. F.**, Cellular defense reactions of insect hemocytes *in vitro:* phagocytosis in a new suspension culture system, *J. Invertebr. Pathol.*, 26, 255, 1975.
41. **Ratcliffe, N. A. and Walters, J. B.**, Studies on the *in vivo* cellular reaction of insects: clearance of pathogenic and non-pathogenic bacteria in *Galleria mellonella* larvae, *J. Insect Physiol.*, 29, 407, 1983.
42. **Lackie, A. M.**, Effect of substratrum wettability and charge on adhesion *in vitro* and encapsulation *in vivo* by insect haemocytes, *J. Cell Sci.*, 63, 181, 1983.
43. **Ratner, S. and Vinson, S. B.**, Encapsulation reactions *in vitro* by haemocytes of *Heliothis virescens, J. Insect Physiol.*, 29, 855, 1983.
44. **Wago, H.**, Cellular recognition of foreign materials by *Bombyx mori* phagocytes: I. Immunocompetent cell, *Dev. Comp. Immunol.*, 6, 591, 1982.

45. **Davies, D. H., Hayes, T. K., and Vinson, S. B.,** Preliminary characterization and purification of *in vitro* encapsulation promoting factor. A peptide that mediates insect hemocyte adhesion, *Dev. Comp. Immunol.*, 12, 241, 1988.
46. **Dunn, P. E. and Drake, D. R.,** Fate of bacteria injected into naive and immunized larvae of the tobacco hornworm *Manduca sexta, J. Invertebr. Pathol.*, 41, 77, 1983.
47. **Dularay, B. and Lackie, A. M.,** The effect of biotic and abiotic implants on the recognition of *Blatta orientalis* cuticular transplants by the cockroach *Periplaneta americana, Dev. Comp. Immunol.*, 11, 69, 1987.
48. **Mohrig, W., Schittek, D., and Hanschke, R.,** Immunological activation of phagocytic cells in *Galleria mellonella, J. Invertebr. Pathol.*, 34, 84, 1979.
49. **Barwig, B.,** Isolation and characterisation of plasma coagulogen (PC) of the cockroach *Leucophaea maderae* (Blatteria), *J. Comp. Physiol.*, 155B, 135, 1985.
50. **Brehélin, M.,** Hemolymph coagulation in *Locusta migratoria:* evidence for a functional equivalent of fibrinogen, *Comp. Biochem. Physiol.*, 62B, 329, 1979.
51. **Geng, C. and Dunn, P. E.,** Hemostasis in larvae of *Manduca sexta:* formation of a fibrous coagulum by hemolymph proteins, *Biochem. Biophys. Res. Commun.*, 155, 1060, 1988.
52. **Grégoire, C. and Goffinet, G.,** Controversies about the coagulocyte, in *Insect Hemocytes,* Gupta, A. P., Ed., Cambridge University Press, Cambridge, 1970, 189.
53. **Minnick, M. F., Rupp, R. A., and Spence, K. D.,** A bacterial-induced lectin which triggers hemocyte coagulation in *Manduca sexta, Biochem. Biophys. Res. Commun.*, 137, 729, 1986.
54. **Minnick, M. F. and Spence, K. D.,** Tissue site and modification of a bacteria-induced coagulation protein from *Manduca sexta, Insect Biochem.*, 18, 637, 1988.
55. **Boman, H. G.,** A molecular approach to immunity and pathogenicity in an insect-bacterial system, in *Chemical Recognition in Biology,* Chapeville, F. and Haenni, A. L., Eds., Springer-Verlag, Heidelberg, 1980, 217.
56. **Dunn, P. E.,** Biochemical aspects of insect immunology, *Annu. Rev. Entomol.*, 31, 321, 1986.
57. **Hoffmann, D., Hultmark, D., and Boman, H. G.,** Insect immunity: *Galleria mellonella* and other lepidoptera have cecropia-P9-like factors active against Gram-negative bacteria, *Insect Biochem.*, 11, 537, 1981.
58. **Hultmark, D., Engström, A., Anderson, K., Steiner, H., Bennich, H., and Bowman, H. G.,** Insect immunity: attacins, a family of antibacterial proteins form *Hyalophora cecropia, EMBO J.*, 2, 571, 1983.
59. **Spies, A. G., Karlinsey, J. E., and Spence, K. D.,** Antibacterial hemolymph proteins of *Manduca sexta, Comp. Biochem. Physiol.*, 83B, 125, 1986.
60. **Dunn, P. E., Dai, W., Kanost, M. R., and Geng, C.,** Soluble peptidoglycan fragments stimulate antibacterial protein synthesis by fat body from larvae of *Manduca sexta, Dev. Comp. Immunol.*, 9, 559, 1985.
61. **Boman, H. G., Nilsson, I., and Rasmuson, B.,** Inducible antibacterial defence system in *Drosophila, Nature (London)*, 237, 232, 1972.
62. **Kaaya, G. P., Flyg, C., and Bowman, H. G.,** Induction of cecropin and attacin-like antibacterial factors in the haemolymph of *Glossina morsitans morsitans, Insect Biochem.*, 17, 309, 1987.
63. **Keppi, E., Zachary, D., Robertson, M., Hoffmann, D., and Hoffmann, J. A.,** Induced antibacterial proteins in the haemolymph of *Phormia terranovae* (Diptera), *Insect Biochem.*, 16, 395, 1986.
64. **Okada, M. and Natori, S.,** Primary structure of sarcotoxin I, an antibacterial protein induced in the hemolymph of *Sarcophaga pregrina* (flesh fly) larvae, *J. Biol. Chem.*, 260, 7174, 1985.
65. **Postlethwait, J. H., Saul, S. H., and Postlethwait, J. A.,** The antibacterial immune response of the medfly, *Ceratitis capitata, J. Insect Physiol.*, 34, 91, 1988.
66. **Spies, A. G., Karlinsey, J. E., and Spence, K. D.,** The immune proteins of the darkling beetle, *Eleodes* (Coleoptera:Tenebrionidae), *J. Invertebr. Pathol.*, 47, 234, 1986.
67. **Karp, R. D. and Rheins, L. A.,** Induction of specific humoral immunity to soluble proteins in the American cockroach *(Periplaneta americana), Dev. Comp. Immunol.*, 4, 620, 1980.
68. **Chadwick, J. S.,** Relation of lysozyme concentration to acquired immunity against *Pseudomonas aeruginosa* in *Galleria mellonella, J. Invertebr. Pathol.*, 15, 455, 1970.
69. **Powning, R. F. and Davidson, W. J.,** Studies on insect bacteriolytic enzymes. I. Lysozyme in hemolymph of *Galleria mellonella* and *Bombyx mori, Comp. Biochem. Physiol.*, 45B, 669, 1973.
70. **Steiner, H., Hultmark, D., Engström, A., Bennich, H., and Boman, H. G.,** Sequence and specificity of two antibacterial proteins involved in insect immunity, *Nature (London)*, 292, 246, 1981.
71. **Rasmuson, T. and Boman, H. G.,** Insect immunity. V. Purification and some properties of immune protein P4 from haemolymph of *Hyalophora cecropia, Insect Biochem.*, 9, 259, 1979.
72. **Andersson, K. and Steiner, H.,** Structure and properties of protein P4, the major bacteria-inducible protein in pupae of *Hyalophora cecropia, Insect Biochem.*, 17, 133, 1987.
73. **Karlinsey, J. E.,** Differential synthesis of bacteria-induced hemolymph proteins in *Manduca sexta,* M.S. thesis, Washington State University, Pullman, WA, 1984.

74. **Riddiford, L. M.,** Changes in translatable mRNAs during larval-pupal transformation of the epidermis of the tobacco hornworm, *Dev. Biol.,* 92, 330, 1982.
75. **Faye, I. and Wyatt, G. R.,** The synthesis of antibacterial proteins in isolated fat body from *Cecropia* silk moth pupae, *Experientia,* 36, 1325, 1980.
76. **Edlund, T., Siden, I., and Boman, H. G.,** Evidence for two immune inhibitors from *Bacillus thuringiensis* interfering with the humoral defense system of Saturniid pupae, *Infect. Immun.,* 14, 934, 1976.
77. **Siden, I., Dalhammar, G., Telander, B., Boman, H. G., and Somerville, H.,** Virulence factors in *Bacillus thuringiensis:* purification and properties of a protein inhibitor of immunity in insects, *J. Gen. Microbiol.,* 114, 45, 1979.
78. **Flyg, C., Kenne, K., and Boman, H. G.,** Insect pathogenic properties of *Serratia marcescens:* phage resistant mutants with a decreased resistance to *Cecropia* immunity and a decreased virulence to Drosophila, *J. Gen. Microbiol.,* 120, 173, 1980.
79. **Gotz, P., Boman, A., and Boman, H. G.,** Interactions between insect immunity and an insect-pathogenic nematode with symbiotic bacteria, *Proc. R. Soc. London,* B212, 333, 1981.

Humoral Factors

Chapter 10

α_2-MACROGLOBULIN: A RECENTLY DISCOVERED DEFENSE SYSTEM IN ARTHROPODS

Peter B. Armstrong and James P. Quigley

TABLE OF CONTENTS

I. INTRODUCTION

Proteases play important physiological roles in a number of processes, including digestion, protein turnover, tissue remodeling, and the conversion of a variety of proteins that are secreted from cells in inactive forms to the active forms.[1,2] In several of the humoral defense systems, including the kinin,[3] complement,[4] and clotting[5,6] systems of mammals, the pathways of conversion involve sequential activation of proteases by other proteases, with the newly activated species themselves acting on target proteins further along in the activation pathways. A variety of proteases are released by neutrophil leukocytes and monocytes during their activities in inflammatory responses.[7,8] Proteases are also involved in a variety of pathological processes, including various connective tissue disorders,[9] neoplastic invasion,[10-12] and the invasion and pathogenesis of bacteria, molds, and multicellular parasites. There are, thus, a variety of sources for proteases in the internal milieu. It is no wonder that animals have evolved regulatory processes for inactivating proteases of endogenous and exogenous origin to limit potentially destructive side effects of the endogenous proteases and the overtly destructive effects of the proteases associated with infective organisms and other pathological processes. Most important of the regulatory molecules of animals are peptide protease inhibitors. These can be classified into two groups, the active-site inhibitors and the α_2-macroglobulin-like inhibitors.[13,14] The former suppress the enzymatic activity of proteases by binding to the active sites of the enzymes,[13,14] whereas the latter do not affect the active site, but instead reduce or eliminate the ability of the target enzyme to interact with macromolecular substrates (e.g., large proteins).[15-18] A diagnostic distinction between the two classes of inhibitors is their effect on the ability of the target protease to hydrolyze the amide or ester bonds of low-molecular-mass substrates: active site inhibitors block this activity in parallel with their ability to block proteolytic activity, whereas the α_2-macroglobulin-like inhibitors block proteolysis with little or no effect on the amidolytic or esterolytic activity directed against low-molecular-mass substrates (Figure 1A, B).[19-24] It is our purpose in this article to review briefly the salient features of the α_2-macroglobulin class of inhibitors with special reference to recent discoveries of the presence of representatives of this class of molecules in the hemolymph of arthropods. Of particular interest will be the possible involvement of α_2-macroglobulin in the humoral immunity of arthropods.

II. CHARACTERISTICS OF THE α_2-MACROGLOBULIN FAMILY OF PROTEINS

As indicated above, the α_2-macroglobulin class of protease inhibitors is unique in reducing or abolishing the proteolytic activity of proteases without affecting the active site of the affected enzymes. The molecular trap hypothesis, proposed initially by Barrett and Starkey,[26] proposes that the α_2-macroglobulin-like inhibitors suppress proteolysis by forming a molecular "cage" around the target protease without altering the active site of the enzyme. Substrate molecules small enough to diffuse into the cage are actively hydrolyzed, whereas large proteins are excluded from access to the active site, and are thus protected from degradation. The suggested sequence of steps in interaction of protease with α_2 macroglobulin begins with the proteolytic cleavage of sensitive peptide bonds located in a defined region of α_2-macroglobulin polypeptide (the "bait" region)[26-33] and continues with a rapid change in conformation of the α_2-macroglobulin[15,29,30,34,35] in a manner that encloses the protease within the coils of the α_2-macroglobulin molecule (Figure 2). Although a number of interesting additional features of the interaction of protease with α_2-macroglobulin have emerged since the hypothesis was initially proposed, the major tenants have withstood experimental investigation. The physical entrapment of a target enzyme by a peptide inhibitor is, to our knowledge, unique to the α_2-macroglobulin-protease interaction. The interaction of inhibitor

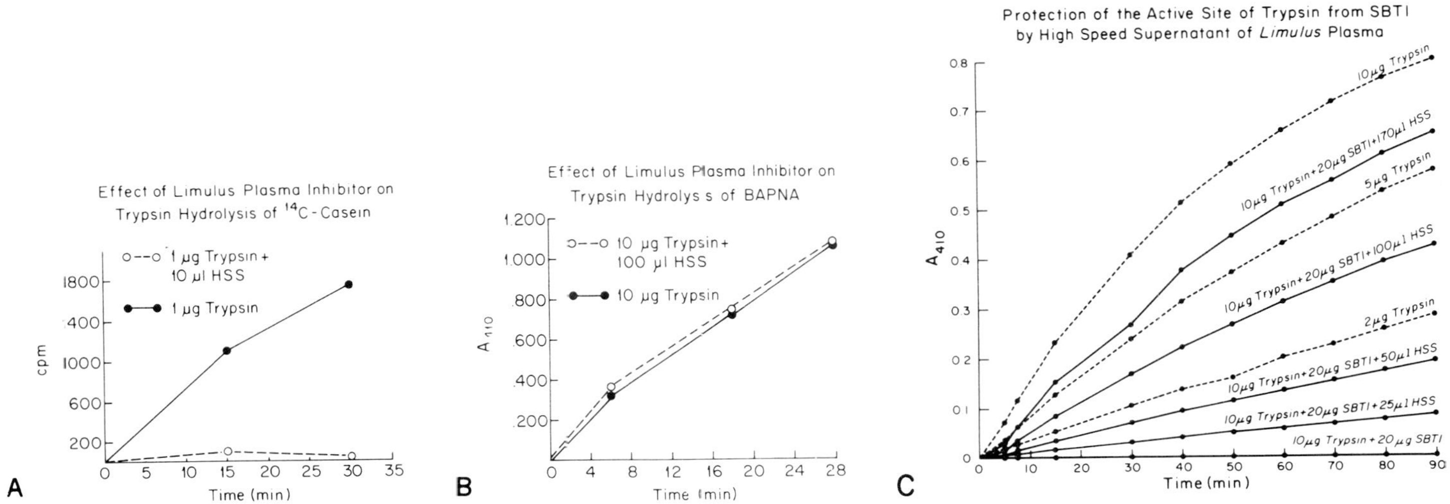

FIGURE 1. Presence of a functional homologue to α_2-macroglobulin in the plasma of *Limulus polyphemus*. Preparations of plasma freed of hemocyanin by ultracentrifugation (high speed supernatant, HSS) inhibit hydrolysis of macromolecular substrata by trypsin ([^{14}C]casein, Figure 1A), but fail to inhibit the amidolytic activity of trypsin directed against the low-molecular-mass substrate, BAPNA (Figure 1B), and protect the amidolytic activity of trypsin for BAPNA from inhibition by saturating quantities of the macromolecular active site inhibitor, soybean trypsin inhibitor (SBTI), (solid lines, Figure 1C). Hydrolysis of [^{14}C]casein (Figure 1A) is followed by the release of acid-soluble [^{14}C]. Hydrolysis of BAPNA is followed by the increase in absorbance at 410 nm. Assay conditions are maintained in which both assays are approximately linear for the amount of enzyme and duration of incubation. The SBTI-protection assay (Figure 1C) employs saturating amounts of enzyme and twice-saturating amounts of SBTI. Trypsin is preincubated for 15 min with plasma, and then for 15 min with SBTI. The assay is initiated by adding substrate (BAPNA). The presence of an α_2-macroglobulin-like inhibitory activity is revealed by the protection of increasing amounts of trypsin by increasing amounts of plasma. The amount of α_2-macroglobulin-like activity, in units of µg of trypsin bound/ml of plasma, is estimated by comparison of the hydrolytic rates in the presence of plasma with the rates of hydrolysis of BAPNA by different concentrations of trypsin in the absence of plasma and SBTI (dashed lines), and from knowledge of the fractional activity of the preparation of trypsin, as determined by titration with *p*-nitrophenyl *p*′-guanidinobenzoate hydrochloride.[25] If endogenous low-molecular-mass active site trypsin inhibitors are present in the plasma, they will reduce the amount of active trypsin and will cause artifactually low levels of α_2-macroglobulin activity in a SBTI-protection assay. Their effects can be estimated by the assay illustrated in Figure 1B, which shows that *Limulus* plasma does not contain active-site trypsin inhibitors. (Parts A and B from Quigley, J. P. and Armstrong, P. B., *J. Biol. Chem.*, 258, 7903, 1983; Part C from Armstrong, P. B., Rossner, M. T., and Quigley, J. P., *J. Exp. Zool.*, 236, 1, 1985. With permission.)

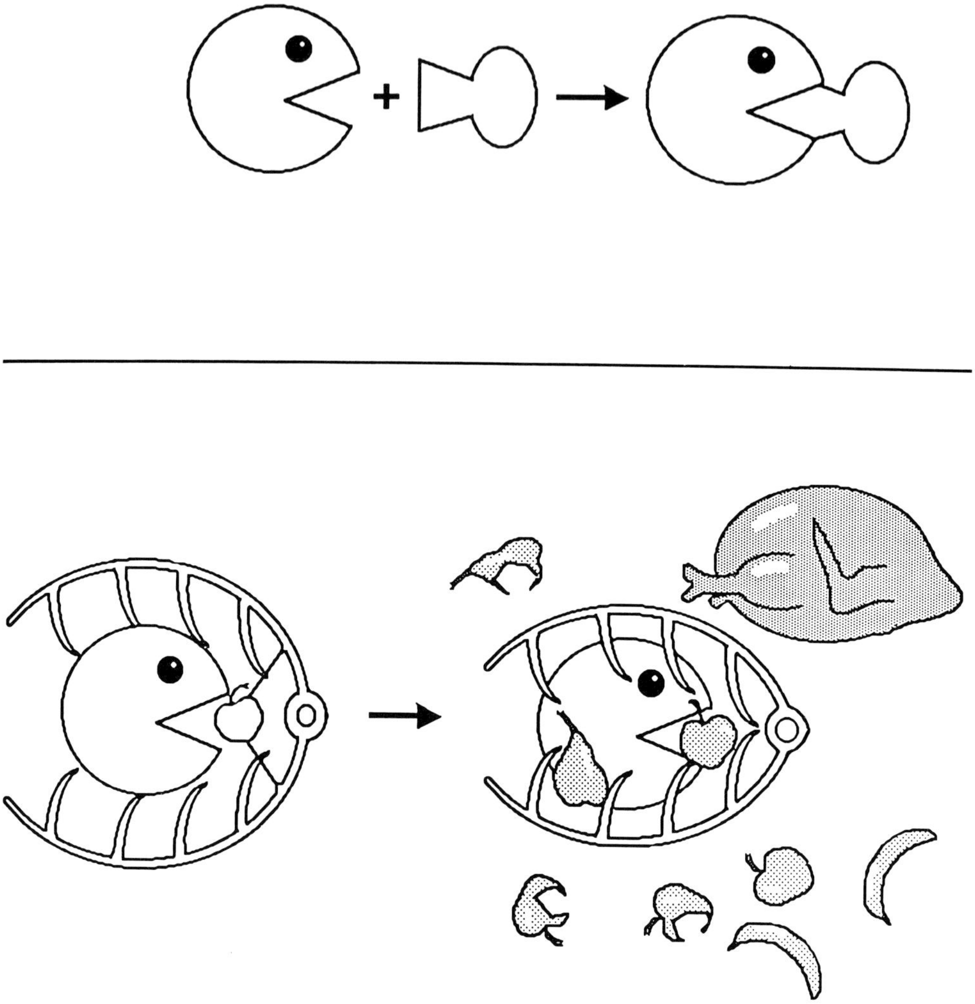

FIGURE 2. Cartoon depicting the differences between the action of active-site protease inhibitors (top panel) and α_2-macroglobulin-type inhibitors (bottom panel). Active site inhibitors (champagne cork, top panel) inactivate target proteases (the computer-animated figure) by binding to and inactivating the active site of the enzyme (the mouth of the figure). In contrast, the α_2-macroglobulin-type inhibitors operate by a very different mechanism. As envisioned by the molecular trap hypothesis, interaction involves cleavage of a peptide bond in the bait region of the α_2-macroglobulin peptide (first figure, bottom panel), which triggers a rapid conformational change that entraps the protease molecule (right figure, bottom panel). The active site of the enzyme is unaffected and can continue to cleave low-molecular-mass substrates (the apples, pears, and bananas of the cartoon: note that several of the substrate molecules have already been hydrolyzed). Entrapped proteases are, however, unable to access large-molecular-mass substrates (the roast turkey of the cartoon), which is the apparent mechanism by which α_2-macroglobulin inhibits proteolysis.

with the active site of the target enzyme, as exemplified by the active site protease inhibitors, is the usual mode of action of naturally occurring enzyme inhibitors.

A second unique feature of the ancestral members of the α_2-macroglobulin family of proteins is the possession of an internal thiol ester bond that links the γ-carbonyl of a glutamine residue and the sulfhydryl of a nearby cysteine.[36]

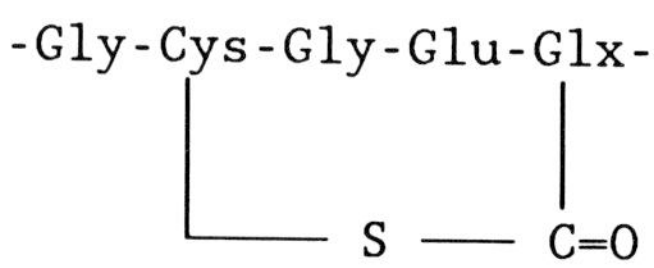

In human α_2-macroglobulin, the cysteine is residue number 949 of a peptide chain that is 1451 amino acids long.[37] Based on similarities of amino acid sequence, the α_2-macroglobulin family of evolutionarily related proteins includes a group of proteins functionally homologous to mammalian plasma α_2-macroglobulin and C3, C4, and C5, all of which are components of the complement cascade.[18] The complement system in vertebrates forms an important element in humoral immunity.[4,38,39] Not all proteins of this larger family preserve the internal thiol ester bond. For example, ovostatin, an α_2-macroglobulin from chicken egg white,[40] and C5 of the complement pathway[18,41] lack the thiol ester bond. Arguments to be presented indicate that the evolutionary ancestor of this family of related proteins possessed this unique functional group, suggesting that its absence in certain members of the α_2-macroglobulin family is a derived character. During reaction of α_2-macroglobulin with protease, the thiol ester bond is activated and establishes covalent bonds with nucleophilic groups, which may result in the formation of covalent ester or amide bonds with the protease molecule.[31,42-52] A similar activation of the thiol ester and resultant isopeptide bond formation is seen when C3 and C4 are activated.[53-58]

Characteristically, active-site peptide protease inhibitors are reactive with only a limited variety of the whole spectrum of proteases, whereas the α_2-macroglobulins are notably promiscuous, reacting with members of all classes of endopeptidases.[26,59] At the molecular level, this is a consequence of the diversity of protease-sensitive peptide bonds in the bait region that presents at least one reactive target peptide bond for almost every endopeptidase.[32,33] For mammalian collagenase the reactivity for the bait region of human α_2-macroglobulin is greater than for type I collagen, the normal substrate.[60] Functionally, the diversity of target peptide bonds in the bait region permits α_2-macroglobulin to serve as an all-purpose inhibitor of proteases of almost any identity and from almost any source.

Although α_2-macroglobulin is usually thought of as a protease inhibitor, it might better be considered as a protease-binding molecule that functions in a receptor-mediated endocytotic clearance pathway for proteases.[17] In mammals, where the endocytotic clearance pathway has been extensively studied, native, unreacted α_2-macroglobulin is stable in the plasma and fails to react with cell surface receptors. After reaction with protease, the α_2-macroglobulin-protease complex is susceptible to rapid binding to surface membrane receptors present on a variety of cell types and to subsequent internalization into secondary lysosomes.[61-77] The present interpretation of these observations is that the cell-surface α_2-macroglobulin receptor recognizes a specific site on reacted α_2-macroglobulin that is masked in the native, unreacted molecule.[78-80] Based on these observations, it can be suggested that the primary function of α_2-macroglobulin in mammals is to serve as an omnifunctional protease carrier molecule that renders endopeptidases sensitive to removal from the body fluids by the receptor-mediated endocytotic pathway. Based on information to be presented in the following sections, it can further be suggested that the α_2-macroglobulin class of proteins is of very ancient origin and, thus, is an example of a primitive system of immunity.

III. BASIC METHODOLOGY FOR THE DISCOVERY OF α_2-MACROGLOBULIN-LIKE PROTEASE INHIBITORS

As indicated, the molecular trap hypothesis proposes that α_2-macroglobulin limits the ability of bound proteases to hydrolyze large-molecular-mass proteins by the formation of a cage around the protease, preventing contact of enzyme with protein substrate. The cage is penetrable by small molecules. This concept also accounts for a surprising feature of the α_2-macroglobulin-protease interaction, viz., the ability of α_2-macroglobulin to protect the active site of bound proteases from inactivation by large-molecular-mass active-site protease inhibitors.[19-24] Apparently, just as the bound protease is shielded from potential protein substrates, so is it shielded from inactivation by large peptide inhibitors (Figure 3). This

FIGURE 3. Specific assay for α_2-macroglobulin activity in a sample containing a mixture of proteins. The sample thought to contain α_2-macroglobulin is first reacted with an excess of trypsin. All of the α_2-macroglobulin reacts with trypsin (top panel). The mixture is then incubated with the high-molecular-mass active site inhibitor, soybean trypsin inhibitor (SBTI), (champagne corks, middle panel). This inactivates all of the unreacted trypsin, but does not react with the trypsin that had previously reacted with α_2-macroglobulin. That latter fraction of trypsin retains the ability to hydrolyze low-molecular-mass substrates, such as BAPNA (apples, lower panel). The rate of hydrolysis of BAPNA provides a measure of the amount of α_2-macroglobulin-bound trypsin, which is a measure of the relative amounts of α_2-macroglobulin in different samples (lower panel).

feature of α_2-macroglobulin has been exploited in the development of a specific assay for its presence in complex mixtures of proteins.[19] The sample is incubated with a protease, usually trypsin, and then reacted with saturating quantities of a large-molecular-mass active site inhibitor of that protease. When trypsin is the protease, soybean trypsin inhibitor (M_r = 25 kDa) is used. Soybean trypsin inhibitor inactivates all free trypsin but does not interact with trypsin sequestered within α_2-macroglobulin. The presence of α_2-macroglobulin in the initial sample is demonstrated by the retention of the ability of a fraction of the added trypsin (e.g., that fraction bound to the α_2-macroglobulin) to hydrolyze a low-molecular-mass substrate, such as N^a-benzoyl-DL-arginine *p*-nitroanilide (BAPNA). In the absence of the protecting effects of α_2-macroglobulin in the sample, all activity to BAPNA is destroyed by the soybean trypsin inhibitor. The amount of α_2-macroglobulin in the sample can be estimated by comparing the amidolytic activity in the reaction mixture to that of control preparations (e.g., preparation of trypsin incubated with varying amounts of the unknown sample in the absence of soybean trypsin inhibitor; this will correct for any diminution of tryptic activity consequent to binding to α_2-macroglobulin and for the presence of low-molecular-mass inhibitors in the sample, but is not a satisfactory control for high-molecular-mass inhibitors that may be present). An example of this analysis is shown in Figure 1B, which demonstrates a lack of effect of *Limulus* plasma on the amidolytic activity of trypsin, and in Figure 1C, which documents the presence of α_2-macroglobulin in the plasma. The plasma sample assayed in Figure 1C contained approximately 20 units per milliliter of α_2-macroglobulin activity, where 1 unit of activity is that capable of protecting 1 μg of trypsin from inactivation by soybean trypsin inhibitor. The fractional activity of the particular trypsin preparation (determined by method of Chase and Shaw[25]) was included in the calculation. In this manner, an α_2-macroglobulin-like activity has been demonstrated in the hemolymph of arthropods (*Limulus*, crustaceans), mollusks, and vertebrates. This methodology is applicable to the search for α_2-macroglobulin-like proteins in other, as yet unstudied, samples. It suffers from two important defects: (1) if bleeding occasions the release or activation of previously latent proteases, these can consume most or all of the α_2-macroglobulin before it has a chance to react with the exogenously added protease; and (2) if low-molecular-mass protease inhibitors are present in the sample, these can enter the α_2-macroglobulin cage surrounding the entrapped protease and inactivate it, resulting in an underestimation of the content of α_2-macroglobulin. However, a positive result is a good indicator of the presence of an α_2-macroglobulin-like species since the assay is apparently quite specific for α_2-macroglobulin.

IV. CHARACTERIZATION OF α_2-MACROGLOBULIN FROM ARTHROPODS

As reviewed above, α_2-macroglobulin is thought to play an important role in the removal of potentially destructive proteases from the body fluids, and, has such, as received extensive study in mammals and other vertebrates. In the summer of 1982, we discovered the existence of a protease inhibitor in the plasma of the horseshoe crab, *Limulus,* that we proposed was a functional and molecular homologue to mammalian α_2-macroglobulin.[81,82] Subsequent studies have substantiated the proposal that α_2-macroglobulin is present in the blood of selected invertebrates[83-88] and is a potentially important element in the humoral defense of arthropods. It is our purpose in this review to outline the current understanding of the properties of the α_2-macroglobulins from arthropods and to discuss some of the directions for future research that appear particularly promising.

Characterization of the protease inhibitor of the plasma of *Limulus* was facilitated by our ability to prepare large quantities of plasma uncontaminated by products released from the blood cells (for techniques, see Armstrong[89,90]). This latter feature was important because

TABLE 1
Inhibition of the Proteolytic Activity of Different Proteases against [^{14}C]Casein and [^{125}I]Fibrin by *Limulus* Plasma

Enzyme	Supernatant addition	Proteolytic activity (cpm) [^{14}C]casein hydrolysis	Proteolytic activity (cpm) [^{125}I]fibrin hydrolysis
Trypsin	−	1,957	12,559
Trypsin	+	194	823
Chymotrypsin	−	1,384	
Chymotrypsin	+	132	
Plasmin	−		15,250
Plasmin	+		620
Elastase	−		22,359
Elastase	+		603
Subtilisin	−	3,187	
Subtilisin	+	270	
Thermolysin	−	2,541	
Thermolysis	+	173	
None	None	146	

the only protease inhibitor detectable in the plasma is the α_2-macroglobulin homologue,[82] whereas the blood cells release both proteases[91-98] and several active site protease inhibitors[83,99] that complicate the analysis of serum from this animal. Our initial characterization of the plasma α_2-macroglobulin included the demonstration that *Limulus* plasma contained an activity that abolished the ability of a diverse array of proteases to digest [^{14}C]-casein, a protein substrate (Table 1), but failed to affect the amidolytic activity of trypsin against the low-molecular-mass substrate, BAPNA (Figure 1A). The protease inhibitory activity was abolished if the plasma was treated with the primary amine, methylamine, consistent with the possibility that inhibition is dependent on an internal thiol ester bond (see the following discussion). Finally, we demonstrated that plasma protected the amidolytic activity of trypsin against the macromolecular active site inhibitor, soybean trypsin inhibitor (Figure 1C; Armstrong et al.[84]). The only protease inhibitor that had been described with these characteristics was α_2-macroglobulin, which was known principally from mammals, but that had been shown also to be present in representatives from all classes of vertebrates.[100-102] Subsequent studies have shown that a protease inhibitor with most of these same characteristics is present in the blood of crustaceans,[84,85,88] (see also Chapter 11) in bivalve, gastropod, and cephalopod mollusks,[103] and in the eggs of certain echinoderms.[103]

The α_2-macroglobulins from *Limulus*[86] and the crustacean, *Homarus* (the American lobster),[88] have been purified by ultracentrifugation to remove hemocyanin, differential precipitation with polyethylene glycol, gel filtration, and ion exchange chromatography. We have not been successful with application to the *Limulus* system of Zn^{2+}-chelate chromatography, the present method of choice for purification of mammalian α_2-macroglobulin:[37,52,104,105] under conditions that were effective for the isolation of human α_2-macroglobulin, *Limulus* α_2-macroglobulin was not bound by the Zn^{2+}-charged resin. Purification of *Limulus* α_2-macroglobulin was facilitated by the relative simplicity of the pattern of proteins in the plasma: there are only three abundant proteins, hemocyanin (about 95% of the total protein), C-reactive protein, and α_2-macroglobulin. The α_2-macroglobulins isolated from *Limulus* and *Homarus* have subunit molecular weights, estimated by SDS-polyacrylamide gel electrophoresis (SDS-PAGE) conducted under reducing conditions, that are very similar to that of human α_2-macroglobulin (viz., 180 to 185 kDa).[86,88] The native molecular weight of human α_2-macroglobulin is 720 kDa,[29,106] consistent with a homotetrameric structure.

Human α_2-macroglobulin shows an apparent molecular weight of 360 kDa by nonreducing SDS-PAGE[27,30] consistent with the proposal that the tetramer is a noncovalently bonded dimer of homodimers of 180 kDa subunits that are linked in the homodimer by disulfide bonds.[30,106,107] Certain of the α_2-macroglobulins of lower vertebrates are apparently homodimers of subunits with a molecular mass close to 180 kDa.[77] Ultracentrifugation, scanning transmission electron microscope, and peptide sequencing studies have shown that *Homarus* and *Limulus* α_2-macroglobulin are homodimers.[88,108] Interestingly, the apparent native molecular masses of both α_2-macroglobulins, as estimated by analytical size exclusion chromatography, were close to that expected for a homotrimer,[86,88] both at low ionic strength and at high (0.5 *M*) salt (viz., 500 to 550 kDa).

One of the most unique features of human α_2-macroglobulin is the presence of a reactive thiol ester bond.[109,110] This bond is broken when the α_2-macroglobulin reacts with proteases, resulting in the exposure of a new thiol group on the cysteine and a highly reactive γ-carbonyl group on the glutamine.[52,110-114] This latter group reacts rapidly with nucleophiles, including amino groups of the entrapped protease molecule.[31,42-51,115] Significant fractions of entrapped trypsin become covalently linked to α_2-macroglobulin by this reaction. The thiol ester bond in native α_2-macroglobulin is reactive to low-molecular-mass primary amines, such as methylamine, resulting in loss of the ability to react subsequently with proteases.[30,42-45,74,102,110,116-118] The thiol ester is also responsible for the susceptibility of α_2-macroglobulin to a specific fragmentation of the peptide chain in response to mild denaturing conditions.[30,44,105,118-120] This apparently results from a nucleophilic attack by the amino group of the glutamic acid residue on the γ-carbonyl group, resulting in rupture of the peptide chain at this site. The presence of thiol ester bonds in candidate proteins can be established by the reactivity with methylamine, the exposure of new thiols during reaction, and the specific fragmentation pattern during mild denaturation. Interestingly, only certain of the members of the α_2-macroglobulin family of proteins contain the thiol ester bond. For example, the bond is present in C3 and C4 but is absent in C5[41] and ovostatin, an α_2-macroglobulin homologue from the egg white of chickens.[40]

Application of the standard tests to the α_2-macroglobulin homologues from crustaceans and *Limulus* indicate that these proteins do contain the internal thiol ester bond and that this bond is necessary for function.[84,85,87,88,121] The protease binding activity of these forms of α_2-macroglobulin is inactivated by exposure to low-molecular-mass primary amines, such as methylamine.[82,84,88] *Limulus* and *Homarus* α_2-macroglobulin also experience autolytic fragmentation during mild thermal denaturation, yielding fragments of approximate molecular masses of 125 and 55 kDa, whereas the methylamine-treated protein is stable under these conditions. Reaction with protease results in the rapid generation of new thiols.[87,88] In addition, the amino acid sequence of the thiol ester of *Limulus* and *Homarus* α_2-macroglobulin is identical to that of human α_2-macroglobulin.[88,121] Since it is unlikely that so unique a functional group as the intra-chain thiol ester could have evolved independently in the α_2-macroglobulins of arthropods and mammals, it is concluded that it is a primitive feature of the molecule; one that must have been present in the α_2-macroglobulin of the common ancestor of mammals and arthropods. Since the lineages leading to the vertebrates and the arthropods diverged approximately 0.55 billion years ago,[122-124] the α_2-macroglobulin molecule and the intra-chain thiol ester bond are truly ancient.

V. STRUCTURE-FUNCTION RELATIONS FOR α_2-MACROGLOBULIN

The mode of interaction of α_2-macroglobulin with target enzymes, as expressed in the molecular trap hypothesis, is unique among enzyme inhibitors. It is assumed that this unusual mode of protein-protein interaction must depend on an unusual architecture of the α_2-

```
Limulus α2M:     L P T G C G E Q N M I K
                 • *   * * * * * * * •
Human α2M :      M P Y G C G E Q N M V L

Lobster α2M:     M P Y G C G E Q N M V N

Human C3:        T P S G C G E Q N M I G

Human C4:        L P R G C G E Q T M I Y
```

FIGURE 4. Peptide sequence surrounding the thiol ester domain of α_2-macroglobulin from *Limulus*, *Homarus*, and humans and the complement proteins, C3 and C4, from humans. The thiol ester in human α_2M is formed from the β-SH group of Cys-949 (C in the single letter code) and the γ-carbonyl group of Gln-952 (Q in the single letter code).[43,126] Amino acid identity is indicated by asterisks, conservative replacements by dots.

macroglobulin molecule itself. α_2-Macroglobulin requires a suitably engineered bait region that is capable of proteolytic cleavage by all classes of proteases and must be capable of a rapid change in folded structure once a peptide bond in the bait region is proteolyzed, so as to enclose the protease responsible for that cleavage. The functional significance of the thiol ester domain of α_2-macroglobulin is still unclear since variants of α_2-macroglobulin that lack this bond are still capable of binding proteases.[40] In addition, the reacted form of α_2-macroglobulin must present a novel domain that is recognized by cell-surface receptors involved in the receptor-mediated endocytotic clearance of the α_2-macroglobulin-protease complex. The realization that α_2-macroglobulin variants that also operate by the molecular trap mechanism are present in organisms as distant in evolution as mammals and arthropods presents us with an opportunity for a comparative approach to a detailed analysis of the structural features of the α_2-macroglobulin molecule necessary for the performance of its unusual functions. An initial step in such an analysis will be to determine the peptide sequence of the protein, for comparison with the already-known sequence of human α_2-macroglobulin.[37,125] Presumably, domains of the molecule irrelevant to shared functions will have experienced extensive divergence in peptide sequence, while domains essential for one or another function will have retained sequence similarity. A striking confirmation of this notion is the similarity of the peptide sequence of the section surrounding the thiol ester domain of human, *Limulus*, and *Homarus* α_2-macroglobulin[88,109,121,126] (Figure 4). In contrast, the amino-terminus of the molecule shows limited sequence similarity between arthropods and humans[121] (Figure 5). Extension of this analysis to the remainder of the molecule should allow us to identify the functionally significant domains of the polypeptide chain and will serve as one avenue to a satisfactory understanding of the manner in which the molecular architecture of the molecule produces its unique function.

VI. INTERACTION OF α_2-MACROGLOBULIN WITH PROTEASES OF THE CLOTTING SYSTEM

Similar to the situation in vertebrates, the blood clotting system of *Limulus* involves activation of a soluble apo form of the principal structural protein of the clot by a specific proteolytic cleavage administered by a particular protease.[91,92,95,96,127,128] The apo form of the structural protein of the clot is known as coagulogen; the proteolyzed form as coagulin.[129-135] Following proteolysis, coagulin polymerizes to form the fibrillar elements of the clot.[91,129,131-133,135-137] The protease involved in cleavage of coagulin is known as clotting enzyme[91,92,127,128,138,139] and is the terminal member of a proteolytic cascade that includes at

```
lobster α2M¹:                    S Y I I T T P R M W V A G
                                   • * • *   * •         *
Limulus α2M:                 K S G F I L T A P K S L T P G K
                             *     • • •     *     *
human α2M²:        S V S G M K P Q Y M V L V P S L L H T E T
```

FIGURE 5. Amino-terminal sequence of α_2-macroglobulin from *Limulus, Homarus,* and humans. Amino acid identity is indicated by asterisks, conservative replacements by dots. ([1]From Spycher, S. E. et al., *J. Biol. Chem.*, 262, 14606, 1987; [2]from Sottrup-Jensen, L. et al., *Ann. N.Y. Acad. Sci.*, 421, 41, 1983. With permission.)

least two enzymes more proximate in the reaction sequence.[93,94,96-98] Clotting occurs, of course, in the presence of plasma, which means in the presence of relatively high concentrations of α_2-macroglobulin, a protease-binding species that reacts nearly instantaneously with other proteases. *Limulus* α_2-macroglobulin fails to inhibit the proteolytic activity of clotting enzyme against its natural substrate even when clotting enzyme and α_2-macroglobulin are incubated prior to exposure to coagulogen.[140] This failure is surprising for two reasons. First, although the *Limulus* and human forms of α_2-macroglobulin show extensive functional similarity, they differ in their interaction with *Limulus* clotting enzyme: human α_2-macroglobulin inhibits the activity of the enzyme against coagulogen while, as indicated above, *Limulus* α_2-macroglobulin fails to inhibit.[140] Second, *Limulus* α_2-macroglobulin is active against a broad spectrum of proteases (e.g., representatives of all categories of serine proteases, metal proteases, and thiol proteases; proteases from vertebrates, plants, and bacteria).[82,86] In fact, the only protease that we have examined whose proteolytic activity is not abolished by *Limulus* α_2-macroglobulin is *Limulus* clotting enzyme. A similar situation may occur in the blood of the crayfish, *Pacifastacus,* in which the proteases involved in activation of the prophenoloxidase system, a component of the systems involved in humoral immunity in crustaceans, are unaffected by the α_2-macroglobulin found in the plasma of that organism.[141]

VII. ROLE OF α-MACROGLOBULIN IN THE CLEARANCE OF PROTEASES FROM THE CIRCULATION

The *in vitro* reactivity of α_2-macroglobulin with endopeptidases suggests that protease binding is part of the normal functional importance of the molecule *in vivo*. Indeed, in mammals, a major function of α_2-macroglobulin appears to be as a carrier of endopeptidases[17] and peptide growth factors.[142-144] Clearance and inactivation of circulating endopeptidases is thought to occur in secondary lysosomes, following upon the receptor-mediated binding to cells of the α_2-macroglobulin-protease complex and its internalization via coated pits.[61-77,145] The half-life of proteases in the circulation of mammals is typically only a few minutes,[70,72,74,146-152] with the α_2-macroglobulin-mediated clearance pathway presumably contributing to the removal of proteases.

Of present interest is the extent of similarity in physiologic function of α_2-macroglobulin between mammals and arthropods. Does arthropod α_2-macroglobulin participate in the removal of humoral proteases? If so, does this removal involve the specific recognition of reacted forms of the α_2-macroglobulin molecule by cell surface receptors, and is receptor-mediated endocytosis involved? The experimental attack on these questions is conceptually straightforward. The clearance of proteases can be investigated by determining the longevity of artifically introduced protease molecules in the circulation. An important negative control is the determination of clearance rates of enzymatically inactive forms of the same protease: since α_2-macroglobulin fails to bind inactive proteases,[153,154] any clearance of inactive molecules must be by an α_2-macroglobulin-independent pathway. If α_2-macroglobulin is the agent responsible for clearance, then α_2-macroglobulin-protease complexes formed *in vitro* and then injected into the animal should be cleared at rates comparable to the clearance rates of injected protease. *Limulus* has an important advantage over vertebrates for the analysis of the physiologic functions of plasma α_2-macroglobulin because the composition of the plasma is much more simple. While mammals possess a plethora of different plasma protease inhibitors,[13,14] *Limulus* has only α_2-macroglobulin. There are just fewer potential players in *Limulus* plasma than are present in mammals. The presence of cell surface receptors for an α_2-macroglobulin-protease complex can be studied in several ways. The standard method of analysis determines the extent of specific binding of a labeled α_2-macroglobulin-protease complex to a suitable population of cells. Methods of labeling include radioiodination and labeling with a fluorescent dye (in which case, detection of binding involves fluorescence microscopy). Specific binding is that that can be displaced by an excess of unlabeled α_2-macroglobulin-protease complex. A negative control is determination of binding of labeled, unreacted α_2-macroglobulin. At least in the mammalian system, only protease- or methylamine-reacted α_2-macroglobulin is capable of binding to the α_2-macroglobulin-protease receptor.[65,73,145] One can look for receptor-mediated binding of *Limulus* α_2-macroglobulin-protease to *Limulus* cells *in vitro,* and one can ask whether *Limulus* α_2-macroglobulin is capable of binding to the receptors of mammalian cells, either by presenting labeled *Limulus* α_2-macroglobulin-protease complex to the mammalian cells or by determining if unlabeled, protease-reacted *Limulus* α_2-macroglobulin can compete for binding with labeled human α_2-macroglobulin-protease complexes. It will be most interesting if it turns out that α_2-macroglobulin participates in a receptor-mediated endocytotic protease clearance pathway in arthropods similar to that demonstrated in mammals.

VIII. α_2-MACROGLOBULIN IN THE BLOOD CELLS

Limulus has only a single cell type in the general circulation, the granular amebocyte.[89] All of the components of the blood clotting system (the clottable protein, coagulogen, and the cascade of proteases that are responsible for its proteolytic modification) are contained in the granules and are released by exocytosis at the initiation of the clotting response. Recently, we have demonstrated that α_2-macroglobulin is also released during degranulation of the amebocyte.[155] Functionally active α_2-macroglobulin can be detected by the SBTI-protection assay in the environment of degranulated cells (Figure 6), and α_2-macroglobulin antigen is detectable by immune staining of Western blots of the materials released by exocytosis.[155] Release requires exocytosis: α_2-macroglobulin is undetectable in the penultimate wash buffer and is not released if the cells are not stimulated to degranulate (Figure 6). Exocytosis does not involve cell lysis: although the cells contain an abundance of the cytoplasmic marker enzyme, lactate dehydrogenase, none is released when the cells are stimulated to degranulate.[84] Data from immunoblotting show that the form of α_2-macroglobulin contained within the blood cells has a subunit molecular mass identical to that of the plasma form. This situation is similar to that of mammals, where the blood platelet, the thrombocytic cell of mammalian blood, releases α_2-macroglobulin and α_1-protease inhibitor during exocytotic degranulation.[156,157] Our present hypothesis is that the α_2-macroglobulin released from the amebocyte may function in the cellular plug formed by aggregated amebocytes at sites of damage to the vascular wall to inactivate the proteases that are also released from the blood cells and from bacteria entrapped in the clot. The gradual degranulation of the blood cells would ensure a gradual release of α_2-macroglobulin in the depths of the cellular clot.

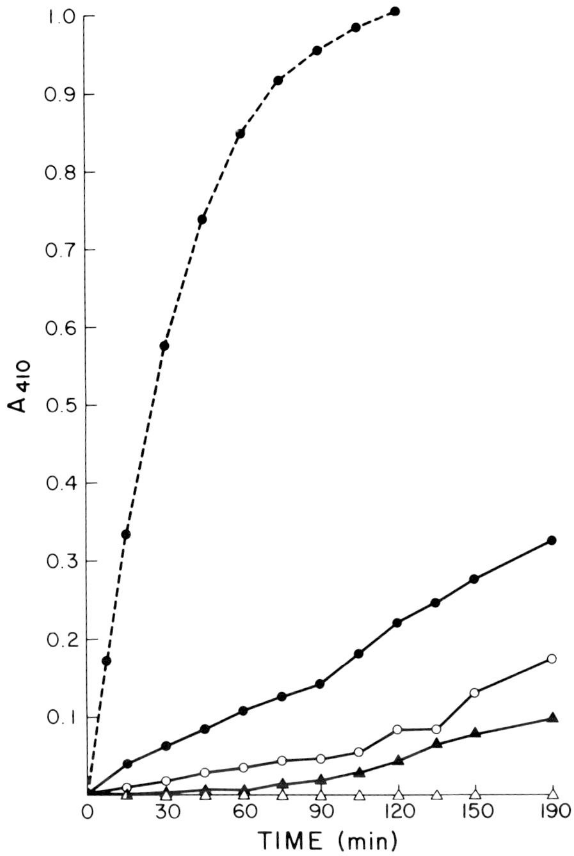

FIGURE 6. Demonstration of α_2-macroglobulin in materials released by saline-washed *Limulus* amebocytes stimulated to degranulate by exposure to the ionophore, A23187. The released materials protect added trypsin from inactivation by the high-molecular-mass active site inhibitor, SBTI, in a dose-dependent fashion. 10-μg samples of trypsin were preincubated with volumes of the released materials, and then an excess of SBTI was added and the ability of the preparation to hydrolyze the low-molecular-mass trypsin substrate, BAPNA, was determined by the increase with time in the optical absorbance at 410 nm. In the absence of the materials released from the blood cells, the activity of trypsin was completely inhibited by the SBTI (not shown). As increasing volumes of the released materials were added, the amidase activity was progressively higher. To estimate the degree of inadvertent contamination with plasma α_2-macroglobulin, the trypsin-protecting activity was determined in saline incubated for 1 h with cells that had not been stimulated to undergo exocytosis (△——△). This preparation lacks trypsin-protecting activity, indicating that all trypsin-protecting, α_2-macroglobulin-like activity in preparations of material released during exocytosis is dependent on exocytotic release and is not due to contamination by plasma α_2-macroglobulin. (●---●): 10 μg trypsin only. (●——●): 10 μg trypsin, 20 μg SBTI, secretory product of 5.4 μl of packed cells. (○——○): 10 μg trypsin, 20 μg SBTI, secretory produce of 2.7 μl of packed cells (▲——▲): 10 μg trypsin, 20 μl SBTI, secretory 1.8 μl of packed cells (△——△): 10 μg trypsin, 20 μg SBTI, materials in the bathing medium of 9.6 μl of packed cells that were not stimulated to degranulate. (From Armstrong, P. B. and Quigley, J. P., *Biochim. Biophys. Acta,* 827, 453, 1985. With permission.)

IX. SUMMARY

Many large arthropods, including *Limulus* and *Homarus*, are relatively long-lived (e.g., in excess of 10 years).[158-160] One requirement for such longevity is the ability to defend against potential pathogens that gain access to the internal milieu. Characterization of the internal defense processes in arthropods has received its impetus from a basic interest in comparative immunity and from efforts to improve veterinary care of aquacultured arthropod of culinary importance and to improve the efficacy of biological control methods for arthropods that are agricultural pests or disease vectors. This article reviews recent information that arthropods share with vertebrates a system of potential importance in humoral immunity, the α_2-macroglobulin system of protease-binding proteins. Although the investigation of α_2-macroglobulin in arthropods is still very much in its infancy, several avenues of future study appear very promising. An understanding of the role of α_2-macroglobulin should, on the one hand, further our knowledge of its place in the complex of defense reactions in arthropods and, on the other, clarify fundamental questions of how α_2-macroglobulin functions wherever it is found.

ACKNOWLEDGMENT

Original research described in this review was supported by Grant No. GM35185 from the National Institutes of Health.

REFERENCES

1. **Barrett, A. J., Ed.,** *Proteinases in Mammalian Cells and Tissues,* North-Holland Publishing Co., Amsterdam, New York, 1977.
2. **Neurath, H.,** The versatility of proteolytic enzymes, *J. Cell Biochem.,* 32, 35, 1986.
3. **Kaplan, A. P., Ghebrehiwet, B., Silverberg, M., and Sealey, J. E.,** The intrinsic coagulation-kinin pathway, complement cascades, plasma renin-angiotensin system, and their interrelationships, *Crit. Rev. Immunol.,* 3, 75, 1981.
4. **Reid, K. B. M. and Porter, R. R.,** The proteolytic activation systems of complement, *Annu. Rev. Biochem.,* 50, 433, 1981.
5. **Davie, E. W., Fujikawa, K., Kurachi, K., and Kisiel, W.,** The role of serine proteases in the blood coagulation cascade, *Adv. Enzymol.,* 48, 277, 1979.
6. **Furie, B. and Furie, B. C.,** The molecular basis of blood coagulation, *Cell,* 53, 505, 1988.
7. **Cohn, Z.,** The role of proteases in macrophage physiology, in *Proteases and Biological Control, Cold Spring Harbor Conferences on Cell Proliferation,* Reich, E., Rifkin, D. B., and Shaw, E., Eds., Cold Spring Harbor Laboratory, NY, 1975, 2, 483.
8. **Havemann, K. and Janoff, A., Eds.,** *Neutral Proteases of Human Polymorphonuclear Leukocytes,* Urbran and Schwarzenberg, Baltimore, Munich, 1978.
9. **Janoff, A.,** Elastase in tissue injury, *Annu. Rev. Med.,* 36, 207, 1985.
10. **Liotta, L. A., Thorgeirsson, U. P., and Garbisa, S.,** Role of collagenases in tumor cell invasion, *Cancer Metastasis Rev.,* 1, 277, 1982.
11. **Mullins, D. E. and Rohrlich, S. T.,** The role of proteinasis in cellular invasiveness, *Biochim. Biophys. Acta,* 695, 177, 1983.
12. **Sloane, B. R. and Honn, K. V.,** Cysteine proteinasis and metastasis, *Cancer Metastasis Rev.,* 3, 249, 1984.
13. **Laskowski, M. and Kato, I.,** Protein inhibitors of proteinases, *Annu. Rev. Biochem.,* 49, 593, 1980.
14. **Travis, J. and Salvesen, G. S.,** Human plasma proteinase inhibitors, *Annu. Rev. Biochem.,* 52, 655, 1983.
15. **Starkey, P. M. and Barrett, A. J.,** α_2-Macroglobulin, a physiological regulator of proteinase activity, in *Proteinases in Mammalian Cells and Tissues,* Barrett, A. J., Ed., Elsevier/North-Holland Biomedical Press, Amsterdam, 1977, 663.
16. **Barrett, A. J.,** α_2-Macroglobulin, *Methods Enzymol.,* 80, 737, 1981.
17. **Van Leuven, F.,** Human α_2-macroglobulin. Primary amines and the mechanisms of endoprotease inhibition and receptor-mediated endocytosis, *Mol. Cell. Biochem.,* 58, 121, 1984.
18. **Sottrup-Jensen, L.,** α_2-Macroglobulin and related thiol ester plasma proteins, in *The Plasma Proteins, Structure, Function, and Genetic Control,* 2nd ed., Putnam, F. W., Ed., Academic Press, New York, 1987, 5, 191.
19. **Ganrot, P. O.,** Determination of α_2-macroglobulin as trypsin-protein esterase, *Clin. Chim. Acta,* 14, 493, 1966.

20. **Berthillier, G., Got, R., and Bertagnolio, G.,** Biochimie de l'α_2-macroglobulin de lapin IV. Effet sur l'activite esterasique de la trypsine et de la chymotrypsine, *Biochim. Biophys. Acta,* 170, 140, 1968.
21. **Bieth, J., Pinchoir, M., and Metais, P.,** The influence of α_2-macroglobulin on the elastolytic and esterolytic activity of elastase, *FEBS Lett.,* 8, 319, 1970.
22. **Harpel, P. C.,** Human plasma alpha 2-macroglobulin. An inhibitor of plasma kallikrein, *J. Exp. Med.,* 132, 329, 1970.
23. **Harpel, P. C. and Mosesson, M. W.,** Degradation of human fibrinogen by plasma α_2-macroglobulin complexes, *J. Clin. Invest.,* 52, 2175, 1973.
24. **Beatty, K., Travis, J., and Bieth, J.,** The effect of α_2-macroglobulin on the interaction of a_1-proteinase inhibitor with porcine trypsin, *Biochim. Biophys. Acta,* 704, 221, 1982.
25. **Chase, T. and Shaw, E.,** Titration of trypsin, plasmin, and thrombin with p-nitrophenyl p'-guanidinobenzoate HCl, *Methods Enzymol.,* 19, 20, 1970.
26. **Barrett, A. J. and Starkey, P. M.,** The interaction of α_2-macroglobulin with proteinases. Characteristics and specificity of the reaction and a hypothesis concerning its molecular mechanism, *Biochem. J.,* 133, 709, 1973.
27. **Harpel, P. C.,** Studies on human plasma α_2-macroglobulin-enzyme interactions. Evidence for proteolytic modification of the subunit chain structure, *J. Exp. Med.,* 138, 508, 1973.
28. **Tsuru, D., Tomimatsu, M., Fujiwara, K., and Kawahara, K.,** A neutral subtilopeptidase inhibitor from porcine serum, Some evidence for α_2-macroglobulin, *J. Biochem.,* 77, 1305, 1975.
29. **Hall, P. K. and Roberts, R. C.,** Physical and chemical properties of human plasma α_2-macroglobulin, *Biochem. J.,* 173, 27, 1978.
30. **Barrett, A. J., Brown, M. A., and Sayers, C. A.,** The electrophoretically "slow" and "fast" forms of the α_2-macroglobulin molecule, *Biochem. J.,* 181, 401, 1979.
31. **Salvesen, G. S. and Barrett, A. J.,** Covalent binding of proteinases in their reaction with α_2-macroglobulin, *Biochem. J.,* 187, 695, 1980.
32. **Sottrup-Jensen, L., Lonblad, P. B., Stephanik, T. M., Petersen, T. E., Magnusson, S., and Jornvall, H.,** Primary structure of the "bait" region for proteinases in α_2-macroglobulin. Nature of the complex, *FEBS Lett.,* 127, 167, 1981.
33. **Roberts, R. C. and Hall, P. K.,** Specificity of proteinases for the "bait" region of α_2-macroglobulin, *Ann. N.Y. Acad. Sci.,* 421, 61, 1983.
34. **Nelles, L. P., Hall, P. K., and Robers, R. C.,** Human alpha-2-macroglobulin. Studies on the electrophoretic heterogeneity, *Biochim. Biophys. Acta,* 623, 46, 1980.
35. **Cummings, H. S., Pizzo, S. V., Strickland, D. K., and Castellino, F. J.,** Effect of methylamine and plasmin on the conformation of human α_2-macroglobulin as revealed by differential scanning calorimetric analysis, *Biophys. J.,* 45, 721, 1984.
36. **Howard, J. B.,** Reactive centers in α_2-macroglobulin, *Ann. N.Y. Acad. Sci.,* 421, 69, 1983.
37. **Sottrup-Jensen, L., Stephanik, T. M., Wierzbicki, D. M., Jones, C. M., Lonblad, P. B., Kristensen, T., Mortensen, S. B., Petersin, T. E., and Magnusson, S.,** The primary structure of α_2-macroglobulin and localization of a factor XIIIa cross-linking site, *Ann. N.Y. Acad. Sci.,* 421, 41, 1983.
38. **Porter, R. R. and Reid, K. B. M.,** Activation of the complement system by antibody-antigen complexes: the classical pathway, *Adv. Protein Chem.,* 33, 1, 1979.
39. **Müller-Eberhardt, H. J. and Schreiber, R. D.,** Molecular biology and chemistry of the alternative pathway of complement, *Adv. Immunol.,* 29, 1, 1980.
40. **Nagase, H., Harris, E. D., Woessner, J. F., and Brew, K.,** Ovostatin: a novel proteinase inhibitor from chicken egg white. I. Purification, physicochemical properties and tissue distribution of ovostatin, *J. Biol. Chem.,* 258, 7481, 1983.
41. **Tack, B. F.,** The β-Cys-γ-Glu thiolester bond in human C3, C4, and α_2-macroglobulin, *Springer Semin. Immunopathol.,* 6, 259, 1983.
42. **Swenson, R. P. and Howard, J. B.,** Characterization of alkylamine-sensitive site in α_2-macroglobulin, *Proc. Natl. Acad. Sci. U.S.A.,* 76, 4313, 1979.
43. **Sottrup-Jensen, L., Peterson, T. E., and Magnusson, L.,** A thiol-ester in α_2-macroglobulin cleaved during proteinase complex formation, *FEBS Lett.,* 121, 275, 1980.
44. **Howard, J. B.,** Reactive site in human α_2-macroglobulin: circumstantial evidence for a thiolester, *Proc. Natl. Acad. Sci. U.S.A.,* 78, 2235, 1981.
45. **Salvesen, G. S., Sayers, C. A., and Barrett, A. J.,** Further characterization of the covalent linking reaction of α_2-macroglobulin, *Biochem. J.,* 195, 453, 1981.
46. **Sottrup-Jensen, L., Petersen, T. E., and Magnusson, L.,** Trypsin-induced activation of the thiol esters in α_2-macroglobulin generates a short-lived intermediate ("nascent" α_2M) that can react rapidly to incorporate not only methylamine or putrescine but also proteins lacking proteinase activity, *FEBS Lett.,* 128, 123, 1981.
47. **Sottrup-Jensen, L., Petersen, T. E., and Magnusson, L.,** Mechanism of proteinase complex formation with α_2-macroglobulin. Three modes of trypsin binding, *FEBS Lett.,* 128, 127, 1981.

48. **Van Leuven, F., Cassiman, J. J., and Van den Berghe, H.,** Functional modifications of α_2-macroglobulin by primary amines. II. Inhibition of covalent binding of trypsin to α_2M by methylamine and other primary amines, *J. Biol. Chem.*, 256, 9023, 1981.
49. **Wang, D., Wu, K., and Feinman, R. D.,** The reaction of α_2-macroglobulin-bound trypsin with soybean trypsin inhibitor, *J. Biol. Chem.*, 256, 10934, 1981.
50. **Wang, D., Wu, K., and Feinman, R. D.,** α_2-macroglobulin protease reactions: relationship of covalent bond formation, methylamine reactivity, and specific proteolysis, *Arch. Biochem. Biophys.*, 211, 500, 1981.
51. **Wang, D., Yuan, A. I., and Feinman, R. D.,** Covalent thrombin-α_2-macroglobulin complexes. Evidence for bivalent cross-linking of inhibitor chains by a single enzyme molecule, *Biochemistry,* 23, 2807, 1984.
52. **Feinman, R. D., Yuan, A. I., Windwer, S. R., and Wang, D.,** Kinetics of the reaction of thrombin and α_2-macroglobulin, *Biochem. J.*, 231, 417, 1985.
53. **Law, S. K., Lichtenberg, N. A., and Levine, R. P.,** Evidence for an ester linkage between the labile binding site of C3b and receptive surfaces, *J. Immunol.*, 123, 1388, 1979.
54. **Campbell, R. D., Dodds, A. W., and Porter, R. R.,** The binding of human complement C4 to antibody-antigen aggregates, *Biochem. J.*, 189, 67, 1980.
55. **Law, S. K., Lichtenberg, N. A., Holcombe, F. H., and Levine, R. P.,** Interaction between the labile binding sites of the fourth (C4) and fifth (C5) human complement proteins and erythrocyte cell membranes, *J. Immunol.*, 125, 634, 1980.
56. **Pangburn, M. K. and Müller-Eberhard, H. J.,** Relation of a putative thioester bond in C3 to activation of the alternative pathway and the binding of C3b to biological targets of complement, *J. Exp. Med.*, 152, 1102, 1980.
57. **Gadd, K. J. and Reid, K. B. M.,** The binding of complement component C3 to antibody-antigen aggregates after activation of the alternative pathway in human serum, *Biochem. J.*, 195, 471, 1981.
58. **Sim, R. B., Twose, T. M., Paterson, D. S., and Sim, E.,** The covalent-binding reaction of complement component C3, *Biochem. J.*, 193, 115, 1981.
59. **Howell, J. B., Beck, T., Bates, B., and Hunter, M. J.,** Interaction of α_2-macroglobulin with trypsin, chymotrypsin, plasmin and papain, *Arch. Biochem. Biophys.*, 221, 261, 1983.
60. **Enghild, J. J., Salvesen, G., Brew, K., and Nagase, H.,** Interaction of human rheumatoid synovial collagenase (matrix metalloproteinase I) and stromelysin (matrix metalloproteinase 3) with human α_2-macroglobulin and chicken ovostatin. Binding kinetics and identification of matrix metalloproteinase cleavage sites, *J. Biol. Chem.*, 264, 8779, 1989.
61. **Maxfield, F. R., Schlessinger, J., Schechter, Y., Pastan, I., and Willingham, M. C.,** Collection of insulin, EGF and α_2-macroglobulin in the same patches on the surface of cultured fibroblasts and common internalization, *Cell,* 14, 805, 1978.
62. **Van Leuven, F., Cassiman, J. J., and Van den Berghe, H.,** Uptake and degradation of α_2-macroglobulin-protease complexes in human cells in culture, *Exp. Cell Res.*, 117, 273, 1978.
63. **Kaplan, J. and Nielsen, M. L.,** Analysis of macrophage surface receptors. I. Binding of α_2-macroglobulin-protease complexes to rabbit alveolar macrophages, *J. Biol. Chem.*, 254, 7323, 1979.
64. **Kaplan, J. and Nielsen, M. L.,** Analysis of macrophage surface receptors. II. Internalization of α_2-macroglobulin-trypsin complexes by rabbit alveolar macrophages, *J. Biol. Chem.*, 254, 7329, 1979.
65. **Van Leuven, F., Cassiman, J. J., and Van den Berghe, H.,** Demonstration of an α_2-macroglobulin receptor in human fibroblasts, absent in tumor-derived cell lines, *J. Biol. Chem.*, 254, 5155, 1979.
66. **Mosher, D. F. and Vaheri, A.,** Binding and degradation of α_2-macroglobulin by cultured fibroblasts, *Biochim. Biophys. Acta,* 627, 113, 1980.
67. **Willingham, M. C. and Pastan, I.,** The receptosome: an intermediate organelle of receptor-mediated endocytosis in cultured fibroblasts, *Cell,* 21, 67, 1980.
68. **Dickson, R. B., Nicolas, J.-C., Willingham, M. C., and Pastan, I.,** Internalization of α_2-macroglobulin in receptosomes. Studies with monovalent microscopic markers, *Exp. Cell Res.*, 132, 488, 1981.
69. **Dickson, R. B., Willingham, M. C., Gallo, M., and Pastan, I.,** Inhibition by bacitracin of high affinity binding of [^{125}I]-α_2-macroglobulin to plasma membranes, *FEBS Lett.*, 126, 265, 1981.
70. **Imber, M. J. and Pizzo, S.,** Clearance and binding of two electrophoretic "fast" forms of human α_2-macroglobulin, *J. Biol. Chem.*, 256, 8134, 1981.
71. **Dickson, R. B., Schlegel, R., Willingham, M. C., and Pastan, I.,** Reversible and irreversible inhibitors of clustering of α_2M in clathrin-coated pits on the surface of fibroblasts, *Exp. Cell Res.*, 140, 215, 1982.
72. **Feldman, S. R., Ney, K. A., Gonias, S. L., and Pizzo, S. V.,** In vitro binding and in vivo clearance of human α_2-macroglobulin after reaction with endopeptidases from four different classes, *Biochem. Biophys. Res. Commun.*, 114, 757, 1983.
73. **Gliemann, J., Larsen, T. R., and Sottrup-Jensen, L.,** Cell association and degradation of α_2-macroglobulin-trypsin complexes in hepatocytes and adipocytes, *Biochim. Biophys. Acta,* 756, 230, 1983.
74. **Gonias, S. L., Balber, A. E., Hubbard, W. J., and Pizzo, S. V.,** Ligand binding, conformational change and plasma elimination of human, mouse and rat α-macroglobulin proteinase inhibitors, *Biochem. J.*, 209, 99, 1983.

75. **Hanover, J. A., Cheng, S.-Y., Willingham, M. C., and Pastan, I.,** α_2-Macroglobulin binding to cultured fibroblasts. Solubilization and partial purification of binding sites, *J. Biol. Chem.*, 258, 370, 1983.
76. **Hanover, J. A., Rudick, J. E., Willingham, M. C., and Pastan, I.,** α_2-macroglobulin binding to cultured fibroblasts: identification by affinity chromatography of high-affinity binding sites, *Arch. Biochem. Biophys.*, 227, 570, 1983.
77. **Feldman, S. R. and Pizzo, S. V.,** Purification and characterization of a "half-molecule" α_2-macroglobulin from the southern grass frog: absence of binding to the mammalian α_2-macroglobulin receptor, *Biochemistry*, 25, 721, 1986.
78. **Marynen, P., Van Leuven, F., Cassiman, J.-J., and Van Den Berghe, H.,** A monoclonal antibody to a neo-antigen on α_2-macroglobulin complexes inhibits receptor-mediated endocytosis, *J. Immunol.*, 127, 1782, 1981.
79. **Marynen, P., Van Leuven, F., Cassiman, J.-J., and Van Den Berghe, H.,** Proteolysis at a lysine residue abolishes the receptor-recognition site of α_2-macroglobulin complexes, *FEBS Lett.*, 137, 241, 1982.
80. **Van Leuven, F., Marynen, P., Cassiman, J.-J., and Van Den Berghe, H.,** Monoclonal antibodies toward the receptor-recognition site on α_2-macroglobulin complexes, *Ann. N.Y. Acad. Sci.*, 421, 434, 1983.
81. **Quigley, J. P., Armstrong, P. B., Gallant, P., Rickles, F. R., and Troll, W.,** An endopeptidase inhibitor, similar to vertebrate α_2-macroglobulin, present in the plasma of *Limulus polyphemus*, *Biol. Bull. (Woods Hole, Mass.)*, 163, 402, 1982.
82. **Quigley, J. P. and Armstrong, P. B.,** An endopeptidase inhibitor, similar to mammalian α_2-macroglobulin, detected in the hemolymph of an invertebrate, *Limulus polyphemus*, *J. Biol. Chem.*, 258, 7903, 1983.
83. **Armstrong, P. B. and Quigley, J. P.,** Proteinase inhibitory activity released from the horseshoe crab blood cell during exocytosis, *Biochim. Biophys. Acta*, 827, 453, 1985.
84. **Armstrong, P. B., Rossner, M. T., and Quigley, J. P.,** An α_2-macroglobulinlike activity in the blood of chelicerate and mandibulate arthropods, *J. Exp. Zool.*, 236, 1, 1985.
85. **Hergenhahn, H.-G. and Söderhäll, K.,** α_2-Macroglobulin-like activity in plasma of the crayfish *Pacifastacus leniusculus*, *Comp. Biochem. Physiol.*, 81B, 833, 1985.
86. **Quigley, J. P. and Armstrong, P. B.,** A homologue of α_2-macroglobulin purified from the hemolymph of the horseshoe crab *Limulus polyphemus*, *J. Biol. Chem.*, 260, 12715, 1985.
87. **Armstrong, P. B. and Quigley, J. P.,** *Limulus* α_2-macroglobulin. First evidence in an invertebrate for a protein containing an internal thiol ester bond, *Biochem. J.*, 248, 703, 1987.
88. **Spycher, S. E., Arya, S., Isenman, D. E., and Painter, H.,** A functional, thioester-containing α_2-macroglobulin homologue isolated from the hemolymph of the American lobster *(Homarus americanus)*, *J. Biol. Chem.*, 262, 14606, 1987.
89. **Armstrong, P. B.,** Adhesion and motility of the blood cells of *Limulus*, in *Blood Cells of Marine Invertebrates: Experimental Systems in Cell Biology and Comparative Physiology*, Cohen, W. D., Ed., Alan R. Liss, New York, 1985, 77.
90. **Armstrong, P. B.,** Blood cells of marine invertebrates: A practical guide. Amebocytes of the American "horseshoe crab," *Limulus polyphemus*, in *Blood Cells of Marine Invertebrates: Experimental Systems in Cell Biology and Comparative Physiology*, Cohen, W. D., Ed., Alan R. Liss, New York, 1985, 253.
91. **Young, N. S., Levin, J., and Prendergast, R. A.,** An invertebrate coagulation system activated by endotoxin: evidence for enzymatic mediation, *J. Clin. Invest.*, 51, 1790, 1972.
92. **Tai, J. Y. and Liu, T.-Y.,** Studies on *Limulus* amoebocyte lysate. Isolation of pro-clotting enzyme, *J. Biol. Chem.*, 252, 2178, 1977.
93. **Nakamura, S. and Levin, J.,** Fractionation of *Limulus* amebocyte lysate. Characterization of activation of the proclotting enzyme by an endotoxin-mediated activation, *Biochim. Biophys. Acta*, 707, 217, 1982.
94. **Nakamura, S. and Levin, J.,** Endotoxin-mediated *Limulus* proclotting enzyme activator and detection of a previously undescribed protease (protease N), *Biochem. Biophys. Res. Commun.*, 108, 1619, 1982.
95. **Torano, A. E., Nakamura, S., and Levin, J.,** Properties of the clotting enzyme responsible for endotoxin-mediate *Limulus* clotting, *Thromb. Res.*, 34, 407, 1984.
96. **Levin, J.,** The role of ameobocytes in the blood coagulation mechanism of the horseshoe crab *Limulus polyphemus*, in *Blood Cells of Marine Invertebrates: Experimental Systems in Cell Biology and Comparative Physiology*, Cohen, W. D., Ed., Alan R. Liss, New York, 1985, 145.
97. **Nakamura, T., Morita, T., and Iwanaga, S.,** Lipopolysaccharide-sensitive serine-protease zymogen (factor C) found in Limulus hemocytes. Isolation and characterization, *Eur. J. Biochem.*, 154, 511, 1986.
98. **Tokunaga, F., Miyata, T., Nakamura, T., Morita, T., Kuma, K., Miyata, T., and Iwanaga, S.,** Lipopolysaccharide-sensitive serine-protease zymogen (factor C) of horseshoe crab hemocytes. Identification and alignment of proteolytic fragments produced during the activation show that it is a novel type of serine protease, *Eur. J. Biochem.*, 167, 405, 1987.
99. **Nakamura, T., Hirai, T., Tokunaga, F., Kawabata, S., and Iwanaga, S.,** Purification and amino acid sequence of Kunitz-type protease inhibitor found in the hemocytes of the horseshoe crab *(Tachypleus tridentatus)*, *J. Biochem.*, 101, 1297, 1987.

100. **Starkey, P. M. and Barrett, A. J.,** Evolution of α_2-macroglobulin. The demonstration in a variety of vertebrate species of a protein resembling human α_2-macroglobulin, *Biochem. J.*, 205, 91, 1982.
101. **Starkey, P. M. and Barrett, A. J.,** Evolution of α_2-macroglobulin. The structure of a protein homologous with human from plaice (*Pleuronectes platessa* L.) plasma, *Biochem. J.*, 205, 105, 1982.
102. **Starkey, P. M., Fletcher, T. C., and Barrett, A. J.,** Evolution of α_2-macroglobulin. The purification and characterization of a protein homologous with human α_2-macroglobulin from plaice (*Pleuronectes platessa* L.) plasma, *Biochem. J.*, 205, 97, 1982.
103. **Armstrong, P. B. and Quigley, J. P.,** unpublished data, 1990.
104. **Kurecki, T., Kress, L. F., and Laskowski, M.,** Purification of human plasma α_2-macroglobulin and a_1-proteinase inhibitor using zinc chelate chromatography, *Anal. Biochem.*, 99, 415, 1979.
105. **Nelles, L. P. and Schnebli, H. P.,** Substructure of the rat a-macroglobulin proteinase inhibitors, *Hoppe-Seyler's Z. Physiol. Chem.*, 363, 677, 1982.
106. **Jones, J. M., Creeth, J. M., and Kekwick, R. A.,** Thiol reduction of human α_2-macroglobulin. The subunit structure, *Biochem. J.*, 127, 187, 1972.
107. **Feldman, S. R., Gonias, S. L., and Pizzo, S. V.,** Model of α_2-macroglobulin structure and function, *Proc. Natl. Acad. Sci. U.S.A.*, 82, 5700, 1985.
108. **Armstrong, P. B., Mangel, W. F., Wall, J. S., Hainfield, J. F., Van Holde, K. E., and Quigley, J. P.,** Structure of α_2-macroglobulin from the arthropod, *Limulus polyphemus*, *J. Biol. Chem.*, 266, 2526, 1991.
109. **Swenson, R. P. and Howard, J. B.,** Amino acid sequence of the tryptic peptide containing the alkylamine-reactive site from human α_2-macroglobulin. Identification of δ-glutamylmethylamide, *J. Biol. Chem.*, 255, 8087, 1980.
110. **Tack, B. F., Harrison, R. A., Janatova, J., Thomas, M. L., and Prahl, J. W.,** Evidence for presence of an internal thiolester bond in third component of human complement, *Proc. Natl. Acad. Sci. U.S.A.*, 77, 5764, 1980.
111. **Larsson, L.-J. and Bjork, I.,** Kinetics of appearance of sulfhydryl groups in α_2-macroglobulin on reaction of the inhibitor with amines, *Biochemistry*, 23, 2802, 1984.
112. **Straight, D. L. and McKee, P. A.,** Characterization of thrombin to α_2-macroglobulin, *J. Biol. Chem.*, 259, 1272, 1984.
113. **Strickland, D. K. and Bhattacharya, P.,** Kinetics of the conformational alterations associated with nucleophilic modification of α_2-macroglobulin, *Biochemistry*, 23, 3115, 1984.
114. **Larsson, L.-J., Olson, S. T., and Bjork, I.,** Kinetics of conformational changes and inactivation of human α_2-macroglobulin on reaction with methylamine, *Biochemistry*, 24, 1585, 1985.
115. **Howard, J. B., Swenson, R., and Eccleston, E.,** The methylamine reactive site and protease inhibition in α_2-macroglobulin, *Ann. N.Y. Acad. Sci.*, 421, 160, 1983.
116. **Howard, J. B., Vermeulen, M., and Swenson, R. B.,** The temperature-sensitive bond in human α_2-macroglobulin is the alkylamine-reactive site, *J. Biol. Chem.*, 255, 3820, 1980.
117. **Feldman, S. R., Gonias, S. L., Ney, K. A., Pratt, C. W., and Pizzo, S. V.,** Identification of "embryonin" as bovine α_2-macroglobulin, *J. Biol. Chem.*, 259, 4458, 1984.
118. **Esnard, F., Gutman, N., Morijahed, A. E., and Gauthier, F.,** Rat plasma α_1-inhibitor$_3$: a member of the α-macroglobulin family, *FEBS Lett.*, 182, 125, 1985.
119. **Harpel, P. C., Hayes, M. B., and Hugli, T. E.,** Heat-induced fragmentation of human α_2-macroglobulin, *J. Biol. Chem.*, 254, 8669, 1979.
120. **Seya, T. and Nagasawa, S.,** Limited proteolysis of the third component of human complement, C3, by heat treatment, *J. Biochem. (Tokyo)*, 89, 659, 1981.
121. **Sottrup-Jensen, L., Borth, W., Hall, M., Quigley, J. P., and Armstrong, P. B.,** Sequence similarity between α_2-macroglobulin from the horseshoe crab, *Limulus polyphemus*, and proteins of the α_2-macroglobulin family from mammals, *Comp. Biochem. Physiol.*, 96B, 621, 1990.
122. **Sepkoski, J. J.,** A kinetic model of Phaneroxoic taxonomic diversity. I. Analysis of marine orders, *Paleobiology*, 4, 223, 1978.
123. **Brasier, M. D.,** The Cambrian radiation event, in *The Origin of Major Invertebrate Groups*, House, M. R., Ed., Academic Press, London, 1979, 103.
124. **Runnegar, B.,** The Cambrian explosion: animals of fossils?, *J. Geol. Soc. Aust.*, 29, 395, 1982.
125. **Kan, C.-C., Solomon, E., Belt, K. T., Chain, A. C., Hiorns, L. R., and Fey, G.,** Nucleotide sequence of cDNA encoding human α_2-macroglobulin and assignment of the chromosomal locus, *Proc. Natl. Acad. Sci. U.S.A.*, 82, 2282, 1985.
126. **Sottrup-Jensen, L., Lønblad, P. B., Jones, C. M., and Stephanik, T. M.,** Primary structure of human α_2-macroglobulin. II. Primary structure of eight CNBr fragments located in the NH_2-terminal half of α_2-macroglobulin accounting for 603 amino acid residues, *J. Biol. Chem.*, 259, 8304, 1984.
127. **Sullivan, J. D. and Watson, S. W.,** Purification and properties of the clotting enzyme from *Limulus* lysate, *Biochem. Biophys. Res. Commun.*, 66, 848, 1975.

128. **Liu, T.-Y., Sied, R. S., Tai, J. Y., Liang, S.-M., Sakmar, T. P., and Robbins, J. B.,** Studies on *Limulus* lysate coagulating system, in *Biomedical Applications of the Horseshoe Crab (Limulidae),* Cohen, E., Ed., Alan R. Liss, New York, 1979, 147.
129. **Nakamura, S., Takagi, T., Iwanaga, S., Niwa, M., and Takahashi, K.,** A clottable protein (coagulogen) of horseshoe crab hemocytes, Structural change of its polypeptide chain during gel formation, *J. Biochem. (Tokyo),* 80, 649, 1976.
130. **Nakamura, S., Iwanaga, S., Harada, T., and Niwa, M.,** A clottable protein (coagulogen) from amoebocyte lysate of Japanese horseshoe crab *(Tachypleus tridentatus).* Its isolation and biochemical properties, *J. Biochem. (Tokyo),* 80, 1011, 1976b.
131. **Nakamura, S., Takagi, T., Iwanaga, S., Niwa, M., and Takagasgi, K.,** Amino acid sequence studies on the fragments produced from horseshoe crab coagulogen during gel formation: homologies with primate fibrinopeptide B, *Biochem. Biophys. Res. Commun.,* 72, 902, 1976.
132. **Nakamura, S., Iwanaga, S., Niwa, M., and Harada, T.,** A clottable protein (coagulogen) from hemocyte lysate of Japanese horseshoe crab *(Tachypleus tridentatus):* isolation, properties and gel formation, *Jpn. J. Med. Sci. Biol.,* 30, 74, 1977.
133. **Tai, J. Y., Seid, R. C., Huhn, R. D., and Liu, T.-Y.,** Studies on *Limulus* amoebocyte lysate. II. Purification of the coagulogen and the mechanism of clotting, *J. Biol. Chem.,* 252, 4773, 1977.
134. **Mosesson, M. W., Wolfenstein-Todel, C., Levin, J., and Bertrand, O.,** Characterization of amebocyte coagulogen from the horseshoe crab *Limulus polyphemus, Thromb. Res.,* 14, 765, 1979.
135. **Takagi, T., Hokama, Y., Morita, T., Iwanaga, S., Nakamura, S., and Niwa, M.,** Amino acid sequence studies on horseshoe crab *(Tachypleus tridentatus)* coagulogen and the mechanism of gel formation, in *Biomedical Applications of the Horseshoe Crab (Limulidae),* Cohen, E., Ed., Alan R. Liss, New York, 1979, 169.
136. **Solum, N. O.,** The coagulogen of *Limulus polyphemus* hemocytes. A comparison of the clotted and non-clotted forms of the molecule, *Thromb. Res.,* 2, 55, 1973.
137. **Shishikura, F., Nakamura, S., Takahashi, K., and Sekiguchi, K.,** Horseshoe crab phylogeny based on amino acid sequences of the fibrinopeptide-like peptide C, *J. Exp. Zool.,* 223, 89, 1982.
138. **Harada, T., Morita, T., Iwanaga, S., Hakamura, S., and Niwa, M.,** A new chromogenic substrate method for assay of bacterial endotoxins using *Limulus* hemocyte lysate, in *Biomedical Applications of the Horseshoe Crab (Limulidae),* Cohen, E., Ed., Alan R. Liss, New York, 1979, 209.
139. **Nakamura, S., Morita, T., Harada-Suzuke, T., Iwanaga, S., Takahashi, K., and Niwa, M.,** A clotting enzyme associated with the hemolymph coagulation system of a horseshoe crab *(Tachypleus tridentatus):* its purification and characterization, *J. Biochem.,* 92, 781, 1982.
140. **Armstrong, P. B., Levin, J., and Quigley, J. P.,** Role of endogenous proteinase inhibitors in the regulation of the blood clotting system of the horseshoe crab, *Limulus polyphemus, Thromb. Haemostasis (Stuttgart),* 52, 117, 1984.
141. **Hergenhahn, H.-G., Aspan, A., and Söderhäll, K.,** Purification and characterization of a high-M_r proteinase inhibitor of pro-phenol oxidase activation from crayfish plasma, *Biochem. J.,* 248, 223, 1987.
142. **O'Connor-McCourt, M. D., and Wakefield, L. M.,** Latent transforming growth factor-β in serum. A specific complex with α_2-macroglobulin, *J. Biol. Chem.,* 262, 14090, 1987.
143. **Huang, S. S., O'Grady, P., and Huang, J. S.,** Human transforming growth factor β-α_2-macroglobulin complex is a latent form of transforming growth factor β, *J. Biol. Chem.,* 263, 1533, 1988.
144. **James, K.,** Interactions between cytokines and α_2-macroglobulin, *Immunol. Today,* 11, 163, 1990.
145. **Debanne, M. T., Bell, R., and Dolovich, J.,** Uptake of proteinase-α-macroglobulin complexes by macrophages, *Biochim. Biophys. Acta,* 411, 295, 1975.
146. **Debanne, M. T., Regoeczi, E., and Dolovich, J.,** Serum protease inhibitors in the blood clearance of subtilisin A, *Br. J. Exp. Pathol.,* 54, 571, 1973.
147. **Katayama, K. and Fujita, T.,** Studies on biotransformation of elastin. III. Effects of elastase-binding proteins in serum on the disappearance of [^{125}I]-labeled elastase from blood, *Biochim. Biophys. Acta,* 336, 165, 1974.
148. **Katayama, K. and Fujita, T.,** Studies on biotransformation of elastase IV. Tissue distribution of ^{125}I-labeled elastase and intracellular distribution in liver after intravenous administration in rats, *Biochim. Biophys. Acta,* 336, 178, 1974.
149. **Ohlsson, K.,** Interaction between alpha$_1$-antitrypsin and alpha$_2$-macroglobulin in the elimination of proteolytic enzymes, *Protides Biol. Fluids,* 23, 43, 1976.
150. **Ohlsson, K. and Laurell, C.-B.,** The disappearance of enzyme-inhibitor complexes from the circulation of man, *Clin. Sci. Mol. Med.,* 51, 87, 1976.
151. **Eddeland, A. and Ohlsson, K.,** The elimination in dogs of trypsin-α-macroglobulin complexes inactivated by the Kazal or the Kunitz inhibitor, *Hoppe-Seyler's Z. Physiol. Chem.,* 359, 379, 1978.
152. **Fuchs, H. E., Shifman, M. A., and Pizzo, S. V.,** In vivo catabolism of α_1-proteinase inhibitor-trypsin, antithrombin III- thrombin and α_2-macroglobulin-methylamine, *Biochim. Biophys. Acta,* 716, 151, 1982.

153. **Lanchantin, G. F., Plesset, M. L., Friedmann, J. A., and Hart, D. W.,** Dissociation of esterolytic and clotting activities of thrombin by trypsin-binding macroglobulin, *Proc. Soc. Exp. Biol. Med.,* 121, 444, 1966.
154. **Starkey, P. M. and Barrett, A. J.,** Human cathepsin B1. Inhibition by α_2-macroglobulin and other serum proteins, *Biochem. J.,* 131, 823, 1973.
155. **Armstrong, P. B., Quigley, J. P., and Rickles, F. R.,** The *Limulus* blood cell secretes α_2-macroglobulin when activated, *Biol. Bull. (Woods Hole, Mass.),* 178, 137, 1990.
156. **Nachman, R. L. and Harpel, P. C.,** Platelet α_2-macroglobulin and α_1 antitrypsin, *J. Biol. Chem.,* 251, 4514, 1976.
157. **Plow, E. F. and Collen, D.,** The presence and release of α_2-antiplasmin from human platelets, *Blood,* 58, 1069, 1981.
158. **Shuster, C. N.,** Observations on the natural history of the American horseshoe crab, *Limulus polyphemus, Woods Hole Oceanogr. Inst. Contr. No. 564 (3rd Rep. Invest. Methods Improving Shellfish Resources Mass.),* 18, 1950.
159. **Shuster, C. N.,** A horseshoe "crab" grows up, *Ward's Natl. Sci. Bull. (Rochester, N.Y.),* 28, 3, 1954.
160. **Ropes, J. W.,** Longevity of the horseshoe crab, *Limulus polyphemus, Trans. Am. Fish Soc.,* 90, 79, 1961.

Chapter 11

THIOESTER-CONTAINING PROTEINS IN THE NORTH AMERICAN LOBSTER *(HOMARUS AMERICANUS)*

Stefan Spycher and Robert H. Painter

TABLE OF CONTENTS

I. INTRODUCTION

Predominant among the agents by which vertebrates promote humoral immunity are the two protein systems known as the immunoglobulins and complement. These protein systems seem to be present in the blood of all vertebrates but with somewhat simplified structure in the lower forms. There is good evidence at the protein level that the immunoglobulins as such are not found in invertebrates, although there are no reports of the unequivocal failure to detect genes encoding immunoglobulin analogues in invertebrates. As one descends the evolutionary tree, the progressive simplification of the variety of immunoglobulins found in more primitive vertebrate species further suggests the genesis and elaboration of the variety of these proteins during vertebrate evolution.

The complement system is an even more complex array of proteins of specialized function occurring in the circulatory fluids of the organism. In recent years there has been an increasing number of reports which indicate that proteins with properties corresponding to those of the key component of the complement system, namely C3, can be found in the blood of lower vertebrates,[1-3] and in a number of cases evidence of complement function has been reported in invertebrate hemolymph[4-8] (see also Chapter 14). While the complement system depends to a great extent for activation on its interaction with antibody complexed with antigen, it can also be triggered by the surface chemistry of viruses, bacteria, and other foreign bodies.[9] Since the consequences of complement activation can all be observed in the absence of antibodies, there is no *a priori* reason to suppose that the absence of antibodies implies the absence of complement. Indeed, there are good grounds for believing that complement may predate immunoglobulins in evolution.

Most investigators have sought evidence of the complement proteins in various species by looking for the characteristic activities of complement, notably lysis of red cells or bacteria, or for evidence of the opsonization of target cells or bacteria for vertebrate or autologous phagocytes. These reports while exciting have not resulted in any clear identification of proteins which may be compared in a chemical way, sequence and molecular weight, for example, with the proteins of the vertebrate complement system.[10]

II. UNIQUE CHEMICAL PROPERTIES OF THIOESTER PROTEINS

An alternative approach is to look for molecules with chemical properties resembling those of certain complement proteins. For example, the thioester bond found in the C3 and C4 molecules lends itself to easy detection by chemical means.[11] The thioester bond is formed between the γ-carboxyl function of a glutamyl residue and the sulfhydryl group of a cysteine residue which is located in the polypeptide chain three residues NH_2-terminal to the glutamyl residue.[12,13] This unusual intrachain linkage has been found only in C3, C4, and α_2-macroglobulin and proteins related to these molecules.[14,15] The presence of the bond can be revealed by its reactivity with nucleophiles, most notably with [^{14}C]methylamine and by the concomitant release of a reactive thiol.[11,16] The thioester bond is an important feature of the molecules in which it occurs, enabling the essential function of covalent binding of the protein to a target. For the complement proteins, this causes opsonization and facilitates cell lysis; for α_2-macroglobulin, it secures the trapped protease and prevents dissociation. In all cases, the fission of the bond permits molecular rearrangements that reveal important new properties of the molecule which further enhance its function, the appearance of new binding and catalytic sites in the case of C3 and C4,[17] and a receptor site in the case of α_2-macroglobulin.[18,19] This is the approach which will be reported in this chapter. Since the thioester bond seems to be an essential constituent of C3 and C4, we chose to look for proteins that contained the thioester bond rather than to seek, in the first instance, for functional evidence of biological activities normally associated with complement activation. In this way, we hoped to identify proteins that could then be tested for functional significance and for which amino acid sequence comparisons might be informative. We would also gain some information on the origins of this unusual structure, the thioester bond.

III. LOBSTER α_2-MACROGLOBULIN

A. BACKGROUND

The animal chosen for this work was the North American lobster, *Homarus americanus*. It was readily obtainable from local markets and a useful volume of hemolymph (20 to 30 ml) could be drawn from each animal. Hemolymph was drawn from animals which had been kept at 4°C for 3 h, and it was centrifuged at 5000 rpm for 5 min at 4°C to sediment the hemocytes. The supernatant plasma was decanted into plastic bottles and brought to 20 m*M* EDTA and pH 7.5.

The initial experiments which we conducted were to establish whether there were proteins which specifically bound [^{14}C]methylamine when incubated with human plasma or lobster hemolymph. As a control, the plasma or hemolymph was pretreated with cold methylamine. In both human and lobster there were a wide variety of proteins labeling with the [^{14}C]methylamine, some labeling specifically and some nonspecifically. However, because of the very strong nonspecific uptake of the [^{14}C] label by albumin and the hemocyanins, no reliable analysis could be made of these gels.

Experiments on fractions separated by ion exchange chromatography indicated clearly the presence of a thioester-containing protein of high molecular weight. This protein bound [^{14}C]methylamine and revealed itself on autoradiography. The protein was purified as described,[20] and was shown to have a subunit molecular weight of approximately 180 kDa, a value which corresponds with that of the human α_2-macroglobulin subunit[21] (Figure 1). Under nonreducing conditions, the molecular weight of the protein was estimated to be about 440 kDa, and by sedimentation equilibrium analysis the molecular weight was calculated to be close to 340 kDa. It was concluded, therefore, that the molecule was a dimeric structure, consisting of two M_r 180 kDa polypeptide monomers linked by disulfide bonds. N-terminal amino acid sequence analysis indicated that the sequence was homogeneous and that the subunits were identical (Table 1). The molecule underwent autolytic fragmentation when boiled in SDS, giving rise to two fragments of M_r 125 and 55 kDa on SDS-PAGE (Figure 1B). Human α_2-macroglobulin behaves similarly under these conditions giving fragments of M_r 120 and 60 kDa.[22]

Conclusive evidence for the presence of an internal thioester was provided by an experiment in which the protein was reacted with thiopropyl-Sepharose in the presence of [^{14}C]methylamine. The methylamine ruptured the thioester bond, labeling the protein and allowing the released thiol group to react with the activated thiol on the thiopropyl-Sepharose.[23] The complex was thoroughly washed and digested with trypsin which left only the [^{14}C]radiolabeled peptide associated with the thiopropyl-Sepharose. This was eluted with β-mercaptoethanol-containing buffer, purified by reversed phase high performance liquid chromatography, and sequenced. The sequence shown in Table 2 is consistent with that expected for an internal thioester. An identical amino acid sequence was recently determined by Hall et al.[24] for the thioester region of the crayfish, *Pacifastacus leniusculus*.

B. SEQUENCE COMPARISONS

Comparison of the amino-terminal sequence of the lobster protein with known amino-terminal sequences of human and rat α_2-macroglobulins and with human pregnancy zone protein revealed striking differences (Table 1). Only 4 residues out of the 22 are identical. Ser-1, Met-9, Ser-14, and Thr-20, together with Pro-6 (which is displaced to position 7 in

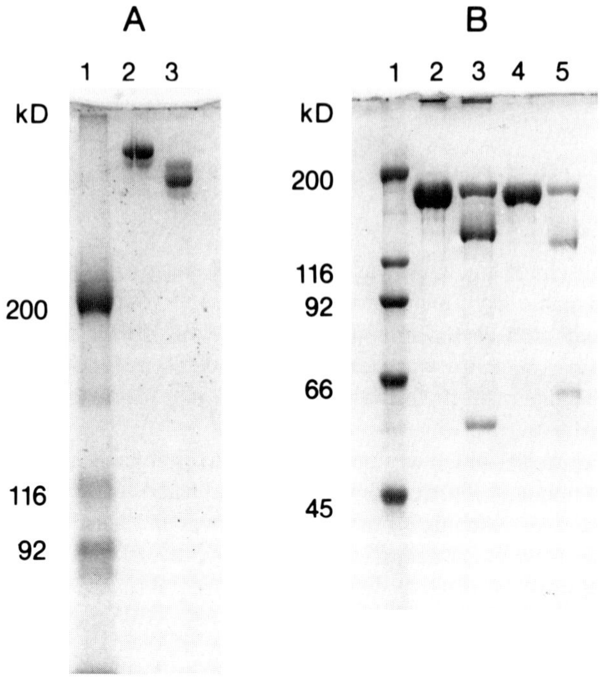

FIGURE 1. SDS gel electrophoresis in 7.5% separating gel of purified lobster α_2-macroglobulin. Lanes A1 and B1, protein molecular weight standards (myosin, 200 kDa, β-galactosidase, 116 kDa; phosphorylase b, 92 kDa; bovine serum albumin, 66 kDa; ovalbumin, 45 kDa); gel A (run under nonreducing conditions); lane A2, purified, unreduced lobster α_2-macroglobulin; lane A3, purified, human α_2-macroglobulin; gel B (run under reducing conditions); lane B2, methylamine treated lobster α_2-macroglobulin subjected to conditions which induce autolytic cleavage in native human α_2-macroglobulin; lane B3, lobster α_2-macroglobulin treated as in B2, but without prior incubation with methylamine; lane B4, methylamine treated human α_2-macroglobulin treated as in B3; lane B5, human α_2-macroglobulin treated as in B2. In each case 6 μg of protein was loaded per lane.

TABLE 1
Amino Terminal Amino Acid Sequence of Lobster α_2-Macroglobulin and Other Related Proteins

1	10	20

Human α_2-macroglobulin[33]
Ser-Val-Ser-Gly-Lys-Pro-Gln-Tyr-Met-Val-Leu-Val-Pro-Ser-Leu-Leu-His-Thr-Glu-Thr-Thr-Glu-...
Rat α_2-macroglobulin[33]
Ser-Ala-Pro-Gly-Lys-Pro-Ile-Tyr-Met-Val-Met-Val-Pro-Ser Leu-Leu
Human pregnancy zone protein[34]
Thr-Glu-Pro-Gln-Tyr-Met-Val-Leu-Val-Pro-Ser-Leu-Leu-His-Thr-Glu-Ala-Val-...
Lobster α_2-macroglobulin[20]
Ser-Tyr-Ile-Ile-Thr-Thr-Pro-Arg-Met-Trp-Val-Ala-Gly-Ser-Pro-Ala-Gln-Val-Arg-Thr-Tyr-Val-...

Note: The cDNA sequence indicates Gln in position 4 of the sequence of rat α_2-macroglobulin.[34]

TABLE 2
Sequence of Thioester-Containing Peptide of Lobster α_2-Macroglobulin Compared with Other Relevant Protein Sequences (Derived from cDNA Sequences)

Human α_2-macroglobulin[36]
Met-Pro-Tyr-Gly-Cys-Gly-Glu-Gln-Asn-Met-Val-Leu-Phe-Ala-Pro-Asn-...
Rat α_2-macroglobulin[35]
Met-Pro-Tyr-Gly-Cys-Gly-Glu-Gln-Asn-Met-Val-Leu-Phe-Ala-Pro-Asn-...
Human C3[37]
Thr-Pro-Ser-Gly-Cys-Gly-Glu-Gln-Asn-Met-Ile-Gly-Met-Thr-Pro-Thr-...
Murine C3[38]
Thr-Pro-Ala-Gly-Cys-Gly-Glu-Gln-Asn-Met-Ile-Gly-Met-Thr-Pro-Thr-...
Human C4[39]
Leu-Pro-Arg-Gly-Cys-Gly-Glu-Gln-Thr-Met-Ile-Tyr-Leu-Ala-Pro-Thr-...
Murine C4[40]
Leu-Pro-Gln-Gly-Cys-Ala-Glu-Gln-Thr-Met-Ile-Tyr-Leu-Ala-Pro-Thr-...
Lobster α_2-macroglobulin[20]
Met-Pro-Tyr-Gly-Cys-Gly-Glu-Gln-Asn-Met-Val-Asn-Phe-Ala-Pro-Asn-...

the lobster) are common to all the α_2-macroglobulins. Elsewhere in the sequence the correspondence of residues is very low. We estimate the overall homology at about 18%, which is an indication that the functional importance of this part of the molecule must be low. Hall et al.[24] reported for the crayfish, *P. leniusculus,* an N-terminal sequence (11 amino acids) with nine residues identical with the lobster sequence. Only residue 3 (Ile to Val exchange) and residue 8 (Arg to Lys exchange) were different.

C. FUNCTIONAL TESTS

To provide evidence for the identity of this protein as α_2-macroglobulin, a number of functional tests were performed. Table 3 shows the ability of the purified protein to inhibit the proteolytic activity towards [^{14}C]radiolabeled casein of a variety of proteolytic enzymes. Release of [^{14}C]radioactivity by the proteolytic enzymes was in each case dramatically reduced. Pretreatment of the α_2-macroglobulin abolished its ability to inhibit the trypsin. In another experiment, a fixed quantity of trypsin was treated with purified lobster α_2-macroglobulin, and the stoichiometry of their reaction was measured (Figure 2). The number of molecules of dimeric lobster α_2-macroglobulin which just protects [^{14}C]casein from hydrolysis was calculated to be 1.3 mol per mol of active trypsin, which suggests a 1:1 equivalence.

Finally, to demonstrate the cagelike trapping of the protease which is characteristic of α_2-macroglobulin, the protection of the esterolytic activity of trypsin to soybean trypsin inhibitor (SBTI) inhibition was demonstrated with lobster α_2-macroglobulin (Figure 3).[25] A fixed amount of trypsin (1 mg, determined spectrophotometrically) was incubated with increasing amounts of lobster α_2-macroglobulin in the presence or absence of SBTI. The amidolytic activity towards the synthetic substrate *N*-benzoyl-DL-arginine-*p*-nitroanilide (BAPNA) was measured spectrophotometrically at 410 nm. The BAPNA amidolytic activity was plotted as a function of the ratio mg α_2-macroglobulin:mg trypsin. As may be seen in Figure 3, in the absence of α_2-macroglobulin SBTI blocks the amidolytic activity of trypsin. However, in the presence of increasing amounts of α_2-macroglobulin, the amidolytic activity of the trypsin approaches the activity of the trypsin measured in the absence of SBTI. The equivalence point of the reaction of the protease with the inhibitor was obtained from the intersection of the two linear portions of the curve; after appropriate corrections based on the active site titration of the trypsin and on the active thioester content of the α_2-macroglobulin, a stoichiometry of 1.26 mol of α_2-macroglobulin per mol of trypsin was calculated.

TABLE 3
α_2-Macroglobulin Inhibition of the Hydrolysis of [^{14}C]Casein by Various Proteolytic Enzymes and the Effect of Pretreatment of α_2-Macroglobulin with Methylamine

Enzyme tested	Addition of α_2-macroglobulin	[^{14}C]casein cpm[a] released
No enzyme	−	570
Chymotrypsin	−	8,538
Chymotrypsin	+	854
Papain	−	7,900
Papain	+	1,286
Pepsin	−	13,821
Pepsin	+	1,144
Thermolysin	−	11,892
Thermolysin	+	1,460
Trypsin	−	18,325
Trypsin	+	1,760
Trypsin	+[b]	16,067

Note: 1 μg of each of the different proteases was incubated with [^{14}C]casein in the absence or presence of 30 μg lobster α_2-macroglobulin or 30 μg methylamine-treated lobster α_2-macroglobulin.

[a] cpm = counts per minute.
[b] α-Macroglobulin pretreated with methylamine.

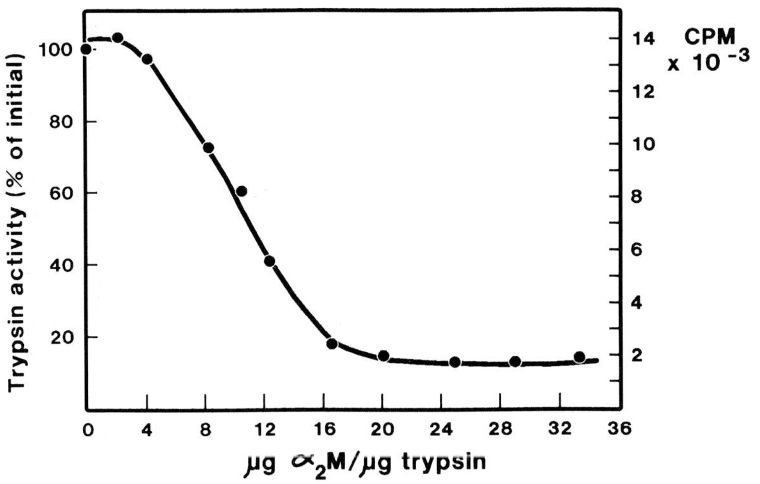

FIGURE 2. [^{14}C]casein assay of α_2-macroglobulin as described in Section III.C. Briefly, a fixed amount of trypsin was preincubated with varying amounts of α_2-macroglobulin and then with a fixed quantity of [^{14}C]casein for 10 min at room temperature. The extent of caseinolysis was measured by the amount of [^{14}C] remaining soluble after precipitation with 10% trichloracetic acid.

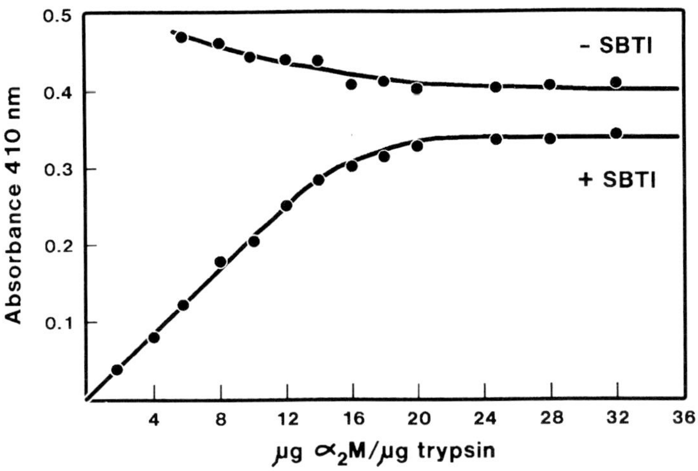

FIGURE 3. BAPNA assay of lobster α_2-macroglobulin as described in Section III.C. Briefly, 1 µg of trypsin was incubated with the α_2-macroglobulin for 10 min at room temperature then with 2 µg of SBTI for 5 min. The samples were then incubated with BAPNA and the esterolysis measured by the absorbance at 410 nm. Parallel experiments were carried out in the absence of SBTI.

D. DISCUSSION

Thus the molecule has physical properties, such as molecular weight and subunit structure, which closely resemble those of human α_2-macroglobulin, it reacts with proteolytic enzymes to inhibit their proteolytic activity towards large molecular weight substrates and blocks the reaction of these enzymes with high molecular weight inhibitors such as SBTI. Its structure differs from that of the human protein, which is a tetramer, in being a dimer but other primitive species, such as those of frog and plaice,[26,27] are also dimeric. Of special interest is the fact that, whereas the dimeric molecules found in these lower vertebrates are not covalently linked, the dimeric lobster molecule is linked by disulfide bonds. The molecule purified from *Limulus polyphemus* is reported to be trimeric[28] (see also Chapter 10).

No tests were conducted for the expression of receptor sites such as those found in human α_2-macroglobulin following reaction with proteases. Such sites are not normally found in dimeric α_2-macroglobulins. Preliminary estimates of the concentration of α_2-macroglobulin in hemolymph indicate a level of 0.5 mg/ml.

Ratcliffe[29] and Johansson and Soderhall[30] have proposed that an important aspect of invertebrate defense rests in the system of proteases which give rise to active phenol oxidase. The control of such a system is thought to be brought about by protease inhibitors. Hergenhahn et al.[31] have purified a protease inhibitor from crayfish plasma which inhibits the prophenol-oxidase-activating enzyme. This enzyme differs from α_2-macroglobulin. Indeed it is not known whether α_2-macroglobulin can inhibit the prophenol-oxidase-activating system; however, given the important roles that α_2-macroglobulin inhibitable proteolytic enzymes play in the defense system of vertebrates, it is an intriguing possibility that its abundance in lobster hemolymph may be important in regulating the defense mechanisms of the lobster.

IV. OTHER THIOESTER-CONTAINING LOBSTER PROTEINS

A. INTRODUCTION

Having established the existence of thioester-containing α_2-macroglobulin in the lobster in this way we tried to detect other thioester-containing proteins in lobster hemolymph, in particular, proteins resembling C3 and C4. The method used was to activate the proteins of hemolymph with methylamine in the presence of thiopropyl-Sepharose, causing them to form a disulfide bond with the thiopropyl-Sepharose. Other proteins were aggressively washed from the matrix and the bound proteins specifically eluted with β-mercaptoethanol. The eluted proteins were run on SDS-PAGE, and the polypeptide chains identified by molecular weight and by reference to standard proteins. The recovery of lobster α_2-macroglobulin served as an internal control on the effectiveness of the reaction procedure in hemolymph, and the ability to detect all the thioester-containing proteins in human plasma was verified by conducting the experiment in parallel on human serum.

B. EXPERIMENTAL: REACTION OF THE PROTEINS OF HEMOLYMPH OR PLASMA WITH THIOPROPYL-SEPHAROSE

Lobster hemolymph was drawn from a chilled lobster and centrifuged to remove hemocytes as described before. The hemolymph was treated with $^1/_{200}$ volume of a saturated solution of phenylmethylsulfonyl fluoride (PMSF) in absolute ethanol; $^1/_{100}$ volume of 2 *M* Tris, pH 7.5; and $^1/_{10}$ volume of 0.2 *M* EDTA, pH 7.0. The pH of the mixture was adjusted with 1 *M* NaOH to pH 8.2. Two 2-ml aliquots of this preparation were each mixed with 1 g of wet thiopropyl-Sepharose (synthesized according to Axen et al.[23]). To one aliquot, methylamine was added to a final concentration of 200 m*M*, and both aliquots were flushed with nitrogen and incubated with shaking for 15 h. The samples were washed six times with a solution of 20 m*M* Tris, 100 m*M* NaCl, 1 m*M* EDTA, pH 7.5; five times with 6 *M* urea, 1 *M* NaCl, 20 m*M* Tris, pH 7.5; and five times with 6 *M* guanidine hydrochloride, 20 m *M* Tris, pH 7.5. Any proteins that remained bound to the thiopropyl-Sepharose were eluted by loading it into a 10-ml syringe used as a column, washing with 20 m*M* ammonium bicarbonate, and eluting by adding 10 μl/ml β-mercaptoethanol to the bicarbonate solution. The protein eluate (2.5 ml) was concentrated in an Amicon 30 Ultrafilter to about 400 μl. The samples were treated with four times SDS sample buffer at 37°C for 30 min and run on a 7.5% SDS separation gel with a 3% stacking gel. Human plasma samples were treated in an identical fashion.

C. RESULTS

Figure 4 shows the result of this experiment. In the reaction between human plasma and thiopropyl-Sepharose, the effect of methylamine is striking. In the absence of methylamine, the eluted protein consists largely of albumin (molecular weight about 65 kDa); however, when 200 m*M* methylamine is present in the reaction mixture, a number of extra bands are seen on the gel, corresponding to the polypeptide chains of the known thioester-containing proteins in human plasma. These can be identified by reference to the purified proteins run in adjacent lanes and to the molecular weight standards. Beginning at the top of the gel, the highest molecular weight polypeptide at 180 kDa corresponds to the α_2-macroglobulin subunit. Next at 105 kDa, the α chain of C3 is followed by a faint band at 95 kDa, the α chain of C4. This appears to have been largely degraded to C4d at about 47 kDa. The β chains of C3 and C4 occur together at about 78 kDa and the γ chain of C4 is evident at 33 kDa, but faint because it is lost through degradation to C4d.

Figure 4B shows the results of the reaction between lobster hemolymph and thiopropyl-Sepharose. In the absence of methylamine, there are major and minor bands visible at about 80 kDa, which probably represent subunits of hemocyanin, and these bands also occur when methylamine is present. The only additional band in the presence of methylamine is that at 180 kDa, which corresponds to that of purified α_2-macroglobulin from lobster. One concludes from this that if C3- and/or C4-like proteins exist in hemolymph, they do not have thioester bonds or that the form of the thioester is much less susceptible to nucleophilic attack by methylamine than in the human. Since the thioester is fundamentally important to the function

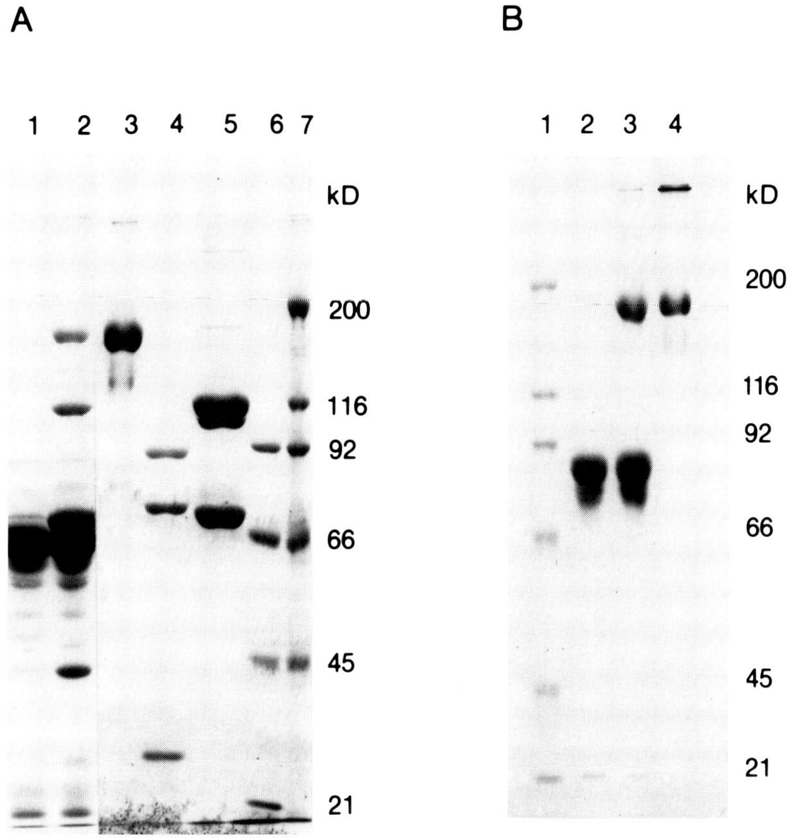

FIGURE 4. SDS-PAGE gels of eluates from thiopropyl-Sepharose after reacting with: (A) human plasma, or (B) lobster hemolymph, in the presence or absence of methylamine. (A) Human plasma eluates from thiopropyl-Sepharose: lane 1, without addition of methylamine; lane 2, with 200 m*M* methylamine. Purified human thioester-containing proteins were run as standards as follows: lane 3, α_2-macroglobulin; lane 4, C4; lane 5, C3; molecular weight standards in lane 6 and lane 7. (B) Lobster hemolymph eluates from thiopropyl-Sepharose: lane 2, without methylamine; lane 3, with 200 m*M* methylamine; lane 4, purified lobster α_2-macroglobulin; lane 1, molecular weight standards.

of C3 and C4 in terms of attachment and molecular conformation, it seems unlikely, therefore, that proteins corresponding in function to human C3 and C4 exist in this invertebrate.

D. DISCUSSION

These findings do not necessarily rule out the existence of such molecules, for as in the case of the human C4 allotypes, these molecules may differ in their sensitivity to specific nucleophiles, and the hypothesis still needs further testing with other classes of nucleophiles.

An alternative and more conclusive test would be to examine cDNA libraries prepared from various organs or cells of the lobster. Mixed oligonucleotides, which correspond to the various possible coding sequences for the α_2-macroglobulin, C3 and C4 thioester bond region depicted in Table 2 (Cys-Gly-Glu-Gln-Asn-Thr-Met), could be used to screen for clones containing this sequence. The α_2-macroglobulin sequence, which would be identified with an oligonucleotide mixture corresponding to the α_2-macroglobulin-specific N-terminal amino acid sequence of the thioester bond region shown in Table 2 (Met-Pro-Tyr-Gly-Cys-Gly), would serve as a positive control on the methodology, so that the failure to detect cDNA clones with sequences other than those corresponding to α_2-macroglobulin should confirm the absence of other proteins with this structure in this invertebrate. On the other hand, cDNA sequences of other clones reacting with thioester region specific oligonucleotide probes should provide interesting comparisons with sequences of vertebrate complement proteins. Shortly before submitting this manuscript, Enghild et al.[32] presented data supporting their hypothesis that *Limulus polyphemus* α_2-macroglobulin is both a proteinase inhibitor and part of a lytic system. The question whether lobster α_2-macroglobulin might have similar properties remains to be investigated.

V. SUMMARY

The thioester bond in proteins such as α_2-macroglobulin, C3, and C4 allows their detection in vertebrate sera by virture of their reactivity toward nucleophiles, such as methylamine, and the concomitant release of an active thiol. The methylamine binds to the protein as a γ-glutamyl methylamide and may be used to radiolabel the protein. The free thiol can form disulfides with other thiols and can be used to attach the protein to a solid support such as thiopropyl-Sepharose and facilitate its separation. When these techniques were applied to the proteins in the hemolymph of the North American lobster *(Homarus americanus)*, a protein closely similar to human α_2-macroglobulin was identified. This protein has a molecular weight of 342 kDa and a subunit molecular weight of 180 kDa. The molecule is composed of two disulfide-linked monomers. The purified protein inhibits proteolytic enzymes, but protects the esterolytic activity of trypsin toward low molecular weight substrates from inactivation by the soybean trypsin inhibitor. The N-terminal amino acid sequence of the protein revealed an overall identity of 18% when compared with the human protein, though the sequence of the thioester-containing peptide was highly conserved with respect to human α_2-macroglobulin and to other proteins having a thioester bond. Comparison of the reactivity of lobster hemolymph and human plasma toward thiopropyl-Sepharose following activation with methylamine showed that, although polypeptide chains corresponding to those of C3, C4, and α_2-macroglobulin were readily apparent when the experiment was conducted with human plasma, only the subunit chains of α_2-macroglobulin were seen with lobster hemolymph. These preliminary results suggest that thioester-containing complement proteins may not occur in this species.

ACKNOWLEDGMENTS

We would like to acknowledge the technical support of Sudha Arya and Dan Ek as well as the advice and expertise of David Isenman.

REFERENCES

1. **Nonaka, M., Fujii, T., Kaidoh, T., Natsuume-Sakai, S., Nonaka, M., Yamaguchi, N., and Takahashi, M.,** Purification of a major serum protein of rainbow trout *(Salmo gairdneri)* homologous to the third component of mammalian complement, *J. Biol. Chem.*, 259, 6327, 1984.
2. **Nonaka, M., Fujii, T., Kaidoh, T., Natsuume-Sakai, S., Nonaka, M., Yamaguchi, N., and Takahashi, M.,** Purification of a lamprey complement protein homologous to the third component of the mammalian complement system, *J. Immunol.*, 133, 3242, 1984.
3. **Sekizawa, A., Fujii, T., and Katagiri, C.,** Isolation and characterization of the third component of complement in the serum of the clawed frog, *Xenopus laevis*, *J. Immunol.*, 133, 1436, 1984.
4. **Day, N. K. B., Gewurz, H., Johannsen, R., Finstad, J., and Good, R. A.,** Complement and complement-like activity in lower vertebrates and invertebrates, *J. Exp. Med.*, 132, 941, 1970.

5. **Anderson, R. S., Day, N. K. B., and Good, R. A.,** Specific hemagglutinin and a modulator of complement in cockroach hemolymph, *Infect. Immun.,* 5, 55, 1972.
6. **Aston, W. P., Chadwick, J. S., and Henderson, M. J.,** Effect of cobra venom factor on *in vivo* immune response in *Galleria mellonella* to *Pseudomonas aeruginosa, J. Invertebr. Pathol.,* 27, 171, 1976.
7. **Bertheussen, K.,** Complement-like activity in sea urchin coelomic fluid, *Dev. Comp. Immunol.,* 7, 21, 1983.
8. **Phillips, D., Menger, M., Chadwick, J. S., and Aston, W. P.,** A cobra venom factor (CVF)-induced C3 convertase activity in the hemolymph of *Galleria mellonella, Dev. Comp. Immunol.,* 11, 37, 1987.
9. **Cooper, N. R.,** The classical complement pathway: activation and regulation of the first complement component, *Adv. Immunol.,* 37, 151, 1985.
10. **Koch, C. and Nielson, H. E.,** Activation of vertebrate complement by *Helix Pomatia* haemolymph, *Dev. Comp. Immunol.,* 8, 15, 1984.
11. **Tack, B. F.,** The beta-Cys-gamma-Glu thiolester bond in human C3, C4 and alpha$_2$-macroglobulin, *Springer Semin. Immunopathol.,* 6, 259, 1983.
12. **Tack, B. F., Harrison, R. A., Janatova, J., Thomas, M. L., and Prahl, J. W.,** Evidence for the presence of an internal thiolester bond in third component of human complement, *Proc. Natl. Acad. Sci. U.S.A.,* 77, 5764, 1980.
13. **Harrison, R. A., Thomas, M. L., and Tack, B. F.,** Sequence determination of the thiolester site of the fourth component of human complement, *Proc. Natl. Acad. Sci. U.S.A.,* 78, 7388, 1981.
14. **Sottrup-Jensen, L., Petersen, T. E., and Magnusson, S.,** A thiolester in alpha$_2$-macroglobulin cleaved during proteinase complex formation, *FEBS Lett.,* 121, 275, 1980.
15. **Sottrup-Jensen, L., Hansen, H. F., Mortensen, S. B., Petersen, T. E., and Magnusson, S.,** Sequence of the active thiolester in human alpha$_2$-macroglobulin, *FEBS Lett.,* 123, 145, 1981.
16. **Swensen, R. P. and Howard, J. B.,** Characterization of alkylamine-sensitive site in alpha$_2$-macroglobulin, *Proc. Natl. Acad. Sci. U.S.A.,* 76, 4313, 1979.
17. **Lambris, J. D. and Muller-Eberhard, H. J.,** The multifunctional role of C3: structural analysis of its interactions with physiological ligands, *Mol. Immunol.,* 23, 1237, 1986.
18. **Hanover, J. A., Willingham, M. C., and Pastan, I.,** Receptor-mediated endocytosis of alpha$_2$-macroglobulin: solubilization and partial purification of the fibroblast alpha$_2$-macroglobulin receptor, *Ann. N.Y. Acad. Sci.,* 421, 410, 1983.
19. **Kaplan, J. and Keough, E. A.,** Studies on the physiology of macrophage receptors for alpha$_2$-macroglobulin X protease complexes, *Ann. N.Y. Acad. Sci.,* 421, 442, 1983.
20. **Spycher, S. E., Arya, S., Isenman, D. E., and Painter, R. H.,** A functional, thioester-containing alpha$_2$-macroglobulin homologue isolated from the hemolymph of the American lobster *Homarus americanus, J. Biol. Chem.,* 262, 14606, 1987.
21. **Hall, P. K. and Roberts, R. C.,** Physical and chemical properties of human plasma alpha$_2$-macroglobulin, *Biochem. J.,* 173, 27, 1978.
22. **Harpel, P. C., Hayes, M. B., and Hugli, T. E.,** Heat induced fragmentation of human alpha$_2$-macroglobulin, *J. Biol. Chem.,* 254, 8669, 1979.
23. **Axen, R., Drevin, H., and Carlsson, J.,** Preparation of modified agarose gels containing thiol groups, *Acta Chem. Scand.,* 29, 471, 1975.
24. **Hall, M., Soderhall, K., and Sottrup-Jenson, L.,** Amino acid sequence around the thiolester of alpha$_2$-macroglobulin from plasma of the crayfish, *Pacifastacus leniusculus, FEBS Lett.,* 254, 111, 1989.
25. **Ganrot, P. O.,** Determination of alpha$_2$-macroglobulin as trypsin-protein esterase, *Clin. Chim. Acta,* 14, 493, 1966.
26. **Starkey, P. M. and Barrett, A. J.,** Evolution of alpha$_2$-macroglobulin: the demonstration of a variety of vertebrate species of a protein resembling human alpha$_2$-macroglobulin, *Biochem. J.,* 205, 91, 1982.
27. **Feldman, S. R. and Pizzo, S. V.,** Purification and characterization of a "half-molecule" alpha$_2$-macroglobulin from the southern grass frog: absence of binding to the mammalian alpha$_2$-macroglobulin receptor, *Biochemistry,* 25, 721, 1986.
28. **Quigley, J. P. and Armstrong, P. B.,** A homologue of alpha$_2$-macroglobulin purified from the hemolymph of the Horseshoe crab, *Limulus polyphemus, J. Biol. Chem.,* 260, 12715, 1985.
29. **Ratcliffe, N. A.,** Invertebrate immunity — a primary for the nonspecialist, *Immunol. Lett.,* 10, 253, 1985.
30. **Johansson, M. W. and Soderhall, K.,** Exocytosis of the prophenol oxidase activating system from crayfish hemocytes, *J. Comp. Physiol., B.* 156, 175, 1985.
31. **Hergenhahn, H. G., Aspan, A., and Soderhall, K.,** Purification and characterization of a high-M_r proteinase inhibitor of prophenol oxidase activation from crayfish plasma, *Biochem. J.,* 248, 223, 1987.
32. **Enghild, J. J., Thogersen, I. B., Salvesen, G., Fey, G. H., Figler, N. L., Gonias, S. L., and Pizzo, S. V.,** Alpha-macroglobulin from *Limulus polyphemus* exhibits proteinase inhibitory activity and participates in a hemolytic system, *Biochemistry,* 29, 10070, 1990.
33. **Sottrup-Jensen, L., Stepanik, T. M., Wierzbicki, D. M., Jones, C. M., Lønblad, P. B., Kristensen, T., Mortensen, S. B., Petersen, T. E., and Magnusson, S.,** The primary structure of human alpha$_2$-macroglobulin and localization of a Factor XIIIa cross-linking site, *Ann. N.Y. Acad. Sci.,* 421, 41, 1983.

34. **Sottrup-Jensen, L., Folkersen, J., Kristensen, T., and Tack, B. F.,** Partial primary structure of human pregnancy zone protein: extensive sequence homology with human alpha$_2$-macroglobulin, *Proc. Natl. Acad. Sci. U.S.A.*, 81, 7353, 1984.
35. **Gehring, M. R., Shiels, B. R., Northemann, W., deBruijn, M. H. L., Kan, C. C., Chain, A. C., Noonan, D. J., and Fey, G. H.,** Sequence of rat liver alpha$_2$-macroglobulin and acute phase control of its messenger RNA, *J. Biol. Chem.*, 262, 446, 1987.
36. **Kan, C. C., Solomon, E., Belt, K. T., Chain, A. C., Hiorns, L. R., and Fey, G.,** Nucleotide sequence of cDNA encoding human alpha$_2$-macroglobulin and assignment of the chromosomal locus, *Proc. Natl. Acad. Sci. U.S.A.*, 82, 2282, 1985.
37. **De Bruijn, M. H. L. and Fey, G. H.,** Human complement component C3: cDNA coding sequence and derived primary structure, *Proc. Natl. Acad. Sci. U.S.A.*, 82, 708, 1985.
38. **Fey, G. H., Lundwall, A., Wetsel, R. A., Tack, B. F., de Bruijn, M. H. L., and Domdey, H.,** Nucleotide sequence of complementary DNA and derived amino acid sequence of murine complement protein C3, *Philos. Trans. R. Soc. London Ser. B*, 306, 333, 1984.
39. **Belt, K. T., Carroll, M. C., and Porter, R. R.,** Structural basis of the multiple forms of human complement component C4, *Cell*, 36, 907, 1984.
40. **Sepich, D. S., Noonan, D. J., and Ogata, R. T.,** Complete cDNA sequence of the fourth component of murine complement, *Proc. Natl. Acad. Sci. U.S.A.*, 82, 5895, 1985.

Chapter 12

ANTIBACTERIAL PROTEINS IN INSECTS OTHER THAN LEPIDOPTERA AND DIPTERA AND IN SOME OTHER INVERTEBRATES

Peter Götz and Tina Trenczek

TABLE OF CONTENTS

I. INTRODUCTION

In 1918, Glaser[1] immunized insects *(Melanoplus femur-rubrum)* by bacterial injection *(Bacillus poncei),* and found that the immunized grasshoppers were able to overcome subsequent challenge injections with lethal doses of bacteria. Many other insects have since been treated in a similar or related way, but with different results. Some insects behave like Glaser's grasshoppers and gain antibacterial activity; others do not. From the excellent investigations by Hans Boman's group in Stockholm and research in other laboratories, the chemical nature of such inducible antibacterial factors has been elicited for some insects, mainly from the lepidopteran and dipteran orders. A survey of the information in these groups will be given in Chapter 13.

On the basis of these results from Lepidoptera and Diptera, one would like to know more about the occurence of inducible antibacterial proteins in other insects. For this reason, we have collected existing information on this topic from the literature, and performed a series of experiments with insects, which have not yet been tested for inducible antibacterial activity. The presentation of data on antibacterial proteins in insects other than Lepidoptera and Diptera follows the taxonomic order, and includes a general discussion of antibacterial proteins and their possible roles in the immune system in insects and in other invertebrates.

II. MATERIAL AND METHODS

Antibacterial activity can be either bacteriostatic, bacteriolytic, or bactericidal. Screening for antibacterial factors in the hemolymph or other tissues of an insect can be performed using photometry (suited to detect reduction of bacterial growth or lysis), viable counts (for detection of bacteriostatic or bactericidal activity), and inhibition zone assay with dead (lysis) or live bacteria (growth inhibition, or killing) as described in textbooks of microbiology (antibiotica screening). The inhibition zone assay is useful for screening many small samples at once, and also allows quantitative analyses.[2,3] However, this method provides no information about the number and nature of the antibacterial factors involved, potency of single factors, and putative combinatory effects. For such studies, isolation of the factors is necessary.

To test whether antibacterial activity is inducible, time-response studies yield only preliminary information. True induction (i.e., *de novo* synthesis of proteins depending on new transcription) can be distinguished only from activation of preexisting precursor molecules or release of stored compounds by the application of actinomycin D and/or cycloheximid, prior to bacterial injection. To estimate the biological significance of inducible antibacterial activity, the occurrence of protective immunity should be tested. After immunization with sublethal doses, successfully immunized insects withstand a subsequent injection with normally lethal doses (challenge injection).

To determine the chemical nature of antibacterial factors, biochemical separation methods (gel filtration, chromatography, electrophoresis, electrofocusing), in combination with chemical analysis, have to be applied. An elegant and quick way to see whether the antibacterial activity is linked to certain protein bands is the combination of PAGE and inhibition test by overlaying the gel with nutrient agar containing the test bacteria.[3] However, information about the protein itself can be obtained only with immunological methods, analysis of the amino acid composition, and finally amino acid sequencing of the molecule. In our study, the main aim was to screen insects of various orders for the occurrence of inducible antibacterial factors. Some information was collected about the chemical nature of the antibacterial factor and the biological significance of the immunization phenomenon. In general, the following test program was applied:

1. Screening for bacteriostatic or bactericidal hemolymph factors with *Escherichia coli* plate test before and after bacterial (*Enterobacter cloacae* β12) injection of test insects
2. Screening for bacteriolytic, lysozyme-like factors in hemolymph of normal and preinjected insects with freeze-dried *Micrococcus luteus* test plates
3. Dose-reponse studies
4. Tests for protective immunity by challenge injection with live bacteria into preinjected (''immunized'') insects
6. Investigation of the protein nature of the antibacterial factors with PAGE, and subsequent overlaying of the gels with *Escherichia coli* in nutrient agar

For immunization, insects were injected into the abdominal hemocoel with live, log-phase *Enterobacter cloacae* β12 (2×10^8 bacteria per ml) that were washed twice and diluted in the appropriate physiologic saline. Unless stated otherwise, injection volumes were 10 μl, and doses were given in number of bacterial cells per animal. Composition of the different salines used was selected according to the Biological Data Book:[4] for *Dysdercus intermedius* and Phasmidae: 280 mg NaCl, 148 mg KCl (amounts given per 100 ml), pH 6.8; for *Acheta domesticus:* 475 mg NaCl, 80 mg KCl, 10 mg $MgSO_4 \times 7H_2O$, 12 mg $CaCl_2 \times 2\ H_2O$, 5 mg Na_2HPO_4, 6 mg KH_2PO_4, 100 mg glucose, 3.4 g saccharose, pH 7.5; for cockroaches: 600 mg NaCl, 80 mg KCl, 5 mg NaH_2PO_4, 6 mg KH_2PO_4, 100 mg glucose, pH 7.5.

Hemolymph was collected in ice-cooled tubes that contained a few crystals of phenylthiourea (PTU) to prevent melanization. Hemolymph from Orthoptera, Phasmida, Blattaria, and Hemiptera was immediately centrifuged (10,000 *g*, 5 min) to remove hemocytes. This treatment prevented total gelation of the hemolymph. The supernatant was tested for antibacterial activity.

To test for other antibacterial factors (bacteriostatic or bacteriocidal), inhibition zone assays were performed according to Faye and Wyatt[5] with minor modifications. Test plates were prepared with either 5 or 7 ml nutrient agar (Standard I, Merck) and seeded with 2×10^4/ml and 2×10^5/ml viable log-phase bacteria. *Escherichia coli* D31 and D22 as well as *M. luteus* 11 (all from H. G. Boman, University of Stockholm) were used as test bacteria. *E. coli* D22 is an envA mutant with defective cell division being sensitive especially for bacteriostatic factors.[6]

It is well known that standard curves for inhibition zone tests differ between different batches of agar plates and also vary when different hemolymph preparations or hemolymph from different species are used as sources for antibacterial factors. In our laboratory, a pool of immune hemolymph from *H. cecropia* was used to standardize each new batch of test plates. The antibacterial activity of undiluted immune hemolymph of this pool (stored in aliquots at −20°C) was arbitrarily taken as 100 relative units (RU) of antibacterial activity. When using plates seeded with viable *M. luteus,* we chose chicken egg white lysozyme (HEL) as a standard. As independent means for controlling alterations in the immune hemolymph pool, the antibioticum gentamycin (Serva) was used as standard (Figure 1). This

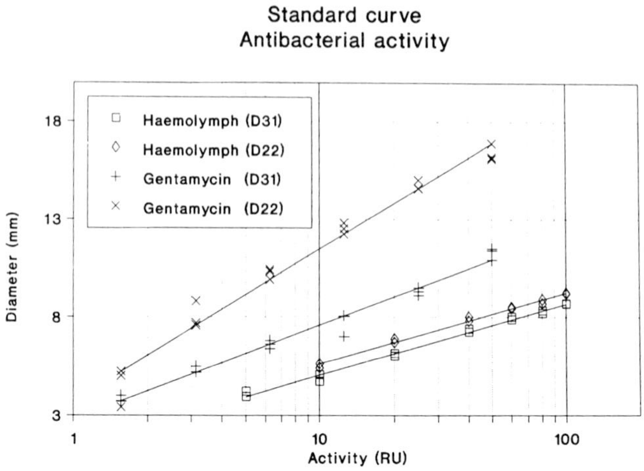

FIGURE 1. Calibration of antibacterial activity against *E. coli*. Pooled hemolymph from immunized *Hyalophora cecropia* pupae or gentamycin at different dilutions in saline was added to agar plates with *E. coli* D31 or *E. coli* D22, as described in Section II for inhibition zone assay. Activity of undiluted immune hemolymph was set as 100 RU. Zone diameters were measured after incubation at 37°C for 24 h. Regression coefficients for hemolymph/*E. coli* D31 is r = 0.97, for hemolymph/*E. coli* D22 r = 0.98, for gentamycin/*E. coli* D31 r = 0.95, and for gentamycin/*E. coli* D22 r = 1.0.

enabled us to confidently compare the plate batches and to detect putative alterations in the standard immune hemolymph pool (e.g., decreasing antibacterial activity by proteolytic degradation). All values given in the tables are means of the indicated number of replicates ± standard errors.

Lysozyme-like activity was quantified by lytic zone assay as described by Mohrig and Messner,[2] using freeze-dried *M. luteus* ([f], Sigma) in concentration of 5 or 1 mg/ml agar. Different dilutions of hen egg white lysozyme (HEL; Sigma) were used as standards. Activity is given in equivalents to μg/ml HEL (Figure 2).

SDS-PAGE was performed with either 15% homogeneous or 7 to 20% gradient acrylamide slab gels according to Laemmli,[7] but using a double concentrated running buffer. Electrophoresis was carried out at constant current (10 mA/gel, overnight). Gels were stained in either Coomassie blue R 250 or with silver nitrate as published elsewhere.[8] Molecular weight markers were from Pharmacia in the range from 14.4 to 94 kDa (LMW) and 3 to 17.1 kDa (myoglobin). Acidic PAGE with bacterial overlay (*E. coli* D31) was done according to Hultmark et al.[3]

Sources of insects tested — *Acheta domesticus, Bacula spp.* and *B. extradentatum*, and *Tenebrio molitor:* from pet shops, Berlin. *Apis mellifera:* Swedish University of Agricultural Sciences, Uppsala. *Blaberus giganteus, Carausius morosus,* and *Gromphadorhina sp:* Institute of Zoology, Free University of Berlin, laboratory culture. *Dipetalogaster maximus:* Institute of Zoology, University of Freiburg, courtesy of Dr. Schaub, Freiburg. *Dysdercus intermedius:* Institute of Zoology, University of Tübingen, courtesy of Dr. F. Dittmann. *Aeshna cyanea, Leptinotarsa decemlineata,* and *Pyrrhocoris apterus:* field collected, Berlin. *Leucophaea maderae:* Institute of Zoology, University of Munich, courtesy of Prof. H. Bohn. *Necrophorus vespilloides:* courtesy of Dr. J. Müller, Bielefeld. *Oryctes nasicornis:* Botanical Garden, Berlin. *Pimpla (Coccygonimus) turionella:* courtesy of Dr. Bogenschütz, Freiburg.

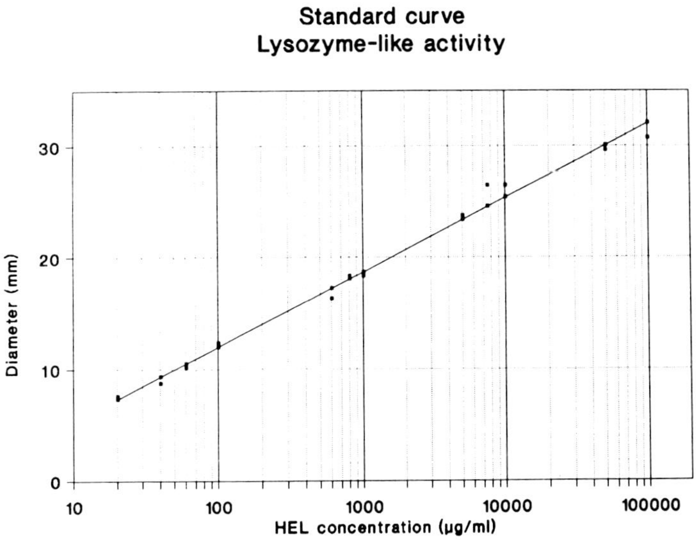

FIGURE 2. Calibration of lysozyme-like activity against hen egg white lysozyme (HEL) at different dilutions in saline was added to agar plates with freeze-dried *M. luteus,* as described in Section II for lytic zone assay. Zone diameters were measured after incubation at 37°C for 24 h. Regression coefficient was calculated as r = 0.96.

III. RESULTS

A. ODONATA

1. *Aeshna cyanea*

Riemer[9] was unable to demonstrate inducible antibacterial activity against *E. coli* in the hemolymph of larvae of the dragonfly *A. cyanea*. Lysozyme-like activity was detected in larvae preinjected with *Enterobacter cloacae* (Table 1). Untreated larvae were not tested by Riemer,[9] but Mohrig and Messner[2] reported lysozyme activity (5 to 20 µg/ml) for *A. cyanea.*

B. PHASMIDA

2. *Carausius morosus*

Hemolymph from larvae of the stick insect *C. morosus* investigated 4 d after injections of either saline or *E. cloacae* β12 (10^3 to 4 × 10^6 bacteria per animal) did not show any antibacterial activity against *Escherichia coli* D31 or live *M. luteus*.[9] Larvae and adults of other stick insects displayed low lysozyme-like activity, if untreated, and an increase of the lysozyme titer 24 h after injection of *Enterobacter cloacae* (Table 2).

C. ORTHOPTERA

Grasshoppers and crickets have been repeatedly investigated in regard to antibacterial activity. In fact, the ability of insects to produce antibacterial activity upon injection of bacteria was discovered by Glaser[1] using grasshoppers *(Melanoplus femur-rubrum)* as test animals. This first investigation was followed by a series of additional studies with different orthopterans.

1. *Locusta migratoria*

Larvae and adults of *L. migratoria* can be protected against lethal doses of *Bacillus thuringiensis* by preinjection with doses of the same bacterium.[10] The protection becomes efficient 6 to 12 h after immunization and persists for more than 2 weeks. During the first

TABLE 1
Antibacterial Activity in Hemolymph of *Aeshna cyanea*[9]

Stage	n	Injected material	Dose (bacteria per animal)	Time p.i. (h)	Antibacterial activity against *E. coli* D31 (RU)	Antibacterial activity against *M. luteus* (1) (μg/ml HEL)
Larvae	2	None		—	0	n.d.
	3	*E. cloacae*	3×10^4	24	0	n.d.
	4		5×10^5	48	0	18 ± 11
	4		5×10^5	96	0	21 ± 3

Note: Larvae of *A. cyanea* were injected with different doses of log-phase *E. cloacae* β12 and hemolymph collected at different times post-injection (p.i.) for inhibition/lysis with *M. luteus* and *E. coli* D31, respectively; n.d., no data.

TABLE 2
Antibacterial Activity in Hemolymph of *Phasmidae*

Species/stage	n	Injection (bacteria/insect)	Time p.i. (h)	Antibacterial activity *E. coli* D31 (RU)	Antibacterial activity *E. coli* D22 (RU)	Antibacterial activity *M. luteus* (f) (μg/ml HEL)
Carausius morosus (Larvae)	3	*E. cloacae* 10^3—4×10^6	96	0	0	0[a]
Bakula sp. (Larvae)	3	None	24	0	0	5 ± 3
	3	*E. cloacae* 4×10^4	24	0	0	19 ± 11
Bakulum sipiloida (Adult)	3	*E. cloacae* 4×10^4	24	0	0	16 ± 1
Bakulum extradentatum (Larvae)	5	None	24	0	0	9 ± 2
	5	Saline	24	0	0	52 ± 23
	5	*E. cloacae* 4×10^4	24	0	0	36 ± 24

[a] Tested against live *M. luteus* 11.

3 h after injection, the lysozyme titer in the hemolymph increases drastically (20-fold) and returns to the normal level before the activity against *B. thuringiensis* becomes evident. Parallel to the *in vivo* protection, antibacterial activity can be demonstrated in the plasma as reduction in the growth rate of bacterial cultures (*B. thuringiensis*, *Escherichia coli*, and *Pseudomonas aeruginosa*).

The antibacterial activity is induced by bacteria (*B. thuringiensis, P. aeruginosa,* and *E. coli*) as well as by nonliving material, such as ovalbumin or iron saccharate. Injection of saline alone did not induce antibacterial activity. The immunizing effect is not specific: hemolymph from *L. migratoria* injected, e.g., with *B. thuringiensis* or iron saccharate showed to a large extent similar antibacterial activity against all three bacteria tested. Biochemical investigation revealed that this antibacterial activity is related to a compound of low molecular weight (1.5 kDa), which is heat stable and resistant to proteolytic and glucolytic enzymes.[11] The antibacterial compound is not identical with lysozyme and has no similarities with the antibacterial proteins known from Lepidoptera and Diptera.

The lysozyme of *L. migratoria* belongs to the chicken type of lysozyme (HEL) and has a molecular weight of 16.5 kDa.[12] It is located in granules of coagulocytes and granular cells, as shown by application of antilysozyme antibodies. The storage of lysozyme in these

TABLE 3
Antibacterial Activity in Hemolymph of *Acheta domesticus*

Stage	n	Injection (bacteria/insect)	Time p.i. (h)	Antibacterial activity against *E. coli D31* (RU)	*E. coli* D22 (RU)	*M. luteus* (f) (μg/ml HEL)
Adults[13]	8	*E. cloacae*	ca. 100	0	n.d.	n.d.
Nymphs	5	None	24	0	0	148 ± 65
	5	Saline	24	0	0	118 ± 55
	5	*E. cloacae* 2×10^4	24	0	0	694 ± 410

Note: n.d., no data.

TABLE 4
Antibacterial Activity in Hemolymph of *Gryllus bimaculatus*[14]

Stage	n	Injected material	Dose (bacteria/insect)	Time p.i. (d)	Antibacterial activity against *M. luteus* (1) (μg/ml HEL)
Adult	50[a]	Saline		3	87 ± 3
(females)	50	Bouillon		3	64 ± 3
	50	*B. thuringiensis*[b]	5×10^5	3	163 ± 13[c]
	28		4×10^7	3	170 ± 20[c]
	26	*E. coli*[b]	4×10^6	3	127 ± 18
	25	*E. coli*[d]	1×10^7	3	171 ± 26[c]
	25	*S. marcescens*[d]	1×10^5	3	189 ± 12[c]

[a] Number of insects per test group.
[b] Heat-killed bacteria.
[c] Significant difference ($p < 0.05$) to saline injection.
[d] Viable bacteria.

hemocytes explains the rapid increase of the hemolymph titer upon injection of bacteria or foreign materials. It ensures the presence of lysozyme after wounding, which causes exocytosis of coagulocytes in connection with clotting. In the case of nodule formation and encapsulation, where granular cells are involved,[12] exocytosis of granules again triggers lysozyme activity.

2. *Acheta domesticus*

In our screening experiments, *A. domesticus,* injected with *Enterobacter cloacae* β12, did not develop antibacterial activity against *Escherichia coli* D22. Lysozyme was present in untreated nymphs and adults, and the titer increased after bacterial inoculation (Table 3).

3. *Gryllus bimaculatus*

Mack,[14] Weber,[15] and Schneider[16] studied in detail the defense reaction of *G. bimaculatus* against bacteria. Saline as well as bacterial injections with sublethal doses reduced the mortality of lethal doses of *B. thuringiensis* DSM 350, *E. coli* K12, and *Serratia marcescens* to 30%. No antibacterial activity against *E. coli* could be induced by injections of heat-killed or live bacteria of the same species. Only the lysozyme titer was, 3 d post-injection (p.i.), significantly higher (Table 4). As in *Acheta,* lysozyme is constitutively expressed in untreated animals, and the differences found between the hemolymph titers of different developmental stages and sexes are highly significant (*t* test, multiple range test).[14]

In further investigations, lysozyme-like activity (expressed in HEL equivalents) was found in the esophagus (123 μg/mg), proventriculus and midgut (0.3 to 4.6 μg/mg), flight muscle (5.7 μg/mg), ovaries (4.8 μg/mg), fat body (4.6 μg/mg), and hemopoietic organ (35 μg/mg).[14] Unfortunately, the samples were not tested for chitinase activity. This related enzyme might be responsible for the high lytic activity found in the salivary glands (267 mg/mg) directly after the imaginal molt. The lysozyme-like activity in fat body and hemopoietic organ varies after injections of different bacteria, but for conclusive interpretation, supplementary studies on release of stored lysozyme-like activity would be necessary.

Three new protein bands were detectable on SDS-PA-gels (M.W. 30, 17, and 10 kDa) 48 h after bacterial provocation.[15] Three different lysozymes were characterized with molecular weights of 15.4 (I), 15.0 (II), and 11.1 (III) kDa and different pH optima (I, pH 5 and II, pH 8).[16] All three forms are detectable in the same individual cricket, indicating gene duplication for this protein.[16] For *Gryllus campestris,* Mohrig and Messner[2] measured a lysozyme titer of 100 to 200 μg/ml hemolymph.

D. BLATTARIA

Injections of *Enterobacter cloacae* β12 (6×10^3 to 6×10^6 bacteria per animal) into larvae of *Blaberus giganteus* did not induce measurable antibacterial activity against *Escherichia coli* D31 within 2 d (n = 5 per dose tested).[9] Similar results were obtained with *B. craniifer,*[13] *Periplaneta americana,*[17] *Gromphadorhina* spp., and *Leucophaea maderae.*[21] Lysozyme-like activity (against *M. luteus*) was detected in *P. americana* by Mohrig and Messner,[2] who reported lysozyme titers of 10 to 200 μg/ml HEL equivalents. No activity against *M. luteus* was found in *B. giganteus,*[9] *L. maderae,* and *Gromphadorhina* spp.[21]

E. HEMIPTERA

1. *Oncopeltus fasciatus*

Gingrich investigated the acquired humoral response in adult specimens of the large milkweed bug *O. fasciatus.*[18] In this hemipteran, antibacterial activity can be induced by a single injection of heat-killed *Pseudomonas aeruginosa.* The biological significance of this immunization could be demonstrated *in vivo* as enhanced resistance against challenge injections with lethal doses of live cells of the same bacterium. Mortality (determined 60 h after the challenge injection) decreased significantly 12 h after immunization and remained low for 4 d. Parallel to this protective immunity, antibacterial activity against *P. aeruginosa* and *E. coli,* but not against *Proteus vulgaris, Bacillus subtilis,* or *Micrococcus pyogenes* var. *aureus,* was shown *in vitro* by lytic zone assay on agar plates (Figure 3). A similar but less pronounced antibacterial activity could be provoked by injection of 2% bovine albumin, 1% suspension of human erythrocytes, and 0.9% saline, and by bare needle injury. Some attempts were made by Gingrich[18] toward a chemical characterization of the antibacterial factor; it was found to be resistent to exposure to heat (75°C), 1 *N* NaOH and 1 *N* HCl for 1 h. The factor remained associated with proteins in electrophoretically fractionated serum, but was found in the supernatant after trichloroacetic acid (TCA) precipitation.[18]

2. *Pyrrhocoris apterus*

Stimulated by the successful immunization of some insects, mainly larvae of the greater wax moth, *Galleria mellonella,* and adults of *O. fasciatus,* Kamp[19] tested another hemipteran *P. apterus* for this ability. Using heat-killed *Pseudomonas aeruginosa,* he was able to induce antibacterial activity in *G. mellonella,* but not in *Pyrrhocoris apterus.* Injection of 1×10^6 heat-killed cells of *Pseudomonas aeruginosa* in larvae of *G. mellonella* provoked reduced mortality against viable bacteria after only 6 h. The protective activity reached its peak after 24 to 48 h and had disappeared on 6 d post-immunization. In *Pyrrhocoris apterus,* no differences in resistance to challenge injections were noted between immunized and non-

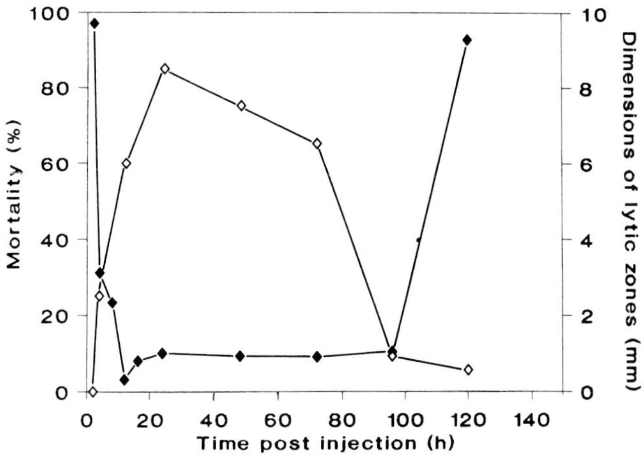

FIGURE 3. Effect of immunization on mortality and bacteriolytic activity in adult *Oncopeltus fasciatus*. Insects were injected with 4 × 10^6 heat-killed cells of *P. aeruginosa* and mortality (♦——♦) registered after challenge injection with 300 live cells of the same bacterium. Bacteriolytic activity (◇——◇) was determined by lytic zone assay with *P. aeruginosa* test plates. (Adapted from Gingrich, R. E., *J. Insect Physiol.*, 10, 179, 1964.)

TABLE 5
Antibacterial Activity in Hemolymph of *Pyrrhocoris apterus*

Stage	n	Injection (bacteria/insect)	Time p.i. (h)	Antibacterial activity: *E. coli* D31 (RU)	*E. coli* D22 (RU)	*M. luteus* (f) (μg/ml HEL)
Adults	5	None	24	0	0	4 ± 2
	5	Saline	24	0	0	4 ± 2
	6	*E. cloacae* 2 × 10^4	24	0	0	18 ± 2

immunized animals. Application of actinomycin C (to inhibit DNA transcription) 75 to 90 min post-immunization did not prevent the appearance of antibacterial activity.

Our own investigations on *P. apterus* gave similar results. No antibacterial activity against *E. coli* D31 and D22 was provoked by injections of *Enterobacter cloacae* β12 (2 × 10^4 per animal) and the lysozyme titer raised only weakly (Table 5).

3. *Rhodnius prolixus*

Azambuja et al. found inducible antibacterial activity in *R. prolixus*.[20] Adult insects were injected with 1 μl of culture medium, containing 1 × 10^6 cells of *Streptococcus mutans*. Development of antibacterial activity was assayed by the inhibitory effect of immune hemolymph on bacterial growth cultures (photometric killing assay) and reduction of viable counts in bacterial suspensions (Figure 4). The antibacterial activity induced by small inocula of *S. mutans* was also effective against *Arthrobacter luteus*, a Gram-negative bacterium, but not against two strains of *Escherichia coli* (Gram-negative) and *Streptococcus faecalis* (Gram-positive).

The antibacterial hemolymph factor was partly isolated by gel filtration of immune hemolymph. It is heat-stable, sensitive to trypsin treatment, and has a molecular weight of 7 kDa. The activity is probably of bacteristatic nature, since it acts only on growing bacteria.

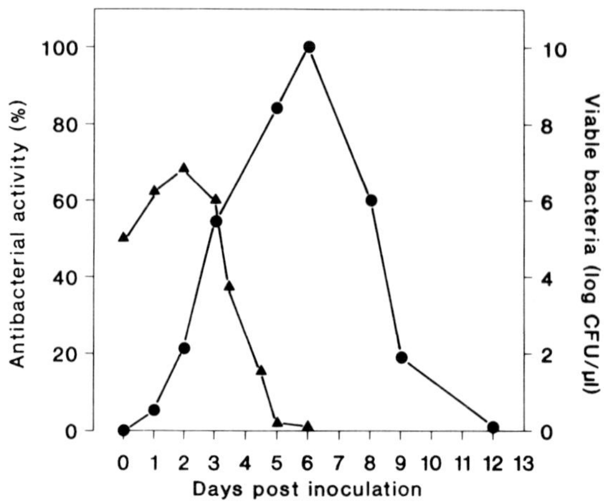

FIGURE 4. Development of antibacterial activity (as growth reduction in %; ●——●) and bacterial clearance in hemolymph from adult *Rhodnius prolixus* at different times after immunization (with 1 × 10^6 cells of *S. mutans*). CFU = colony forming units (▲——▲). (Adapted from Azambuja, P., Freitas, C. C., and Garcia, E. S., *J. Insect Physiol.*, 32, 807, 1986.)

The inducible antibacterial protein from *R. prolixus* develops its bactericidal effect probably by disrupting the cell wall or outer membrane of the bacteria. This was concluded from the observation that labeled DNA was released from bacteria incubated with immune hemolymph.[20] Surprisingly no lysozyme-like activity (tested against *M. luteus*) could be detected. In immunized insects protective activity was demonstrated on day 3; it reached its maximum around day 6 and gradually disappeared until day 11 (Figure 5). Mortality of controls remained near 100% throughout the experiment.

4. *Dipetalogaster maximus*

Larvae of *D. maximus* express antibacterial activity against *E. coli* D31 after immunization with *Enterobacter cloacae* β12. This reaction is dose dependent (Table 6). Concomitantly, the normally low lysozyme titer becomes elevated after bacterial injection but not after injection of saline (Table 7).[9] The proteinaceous nature of the antibacterial activity could be verified by acidic PAGE with bacterial overlay (*Escherichia coli* D31). The position of the protein band was between cecropins A and B of *Hyalophora cecropia* taken as a standard on the same gels.

5. *Dysdercus intermedius*

The cotton bug, *D. intermedius,* is very sensitive to injections of bacteria as well as saline (Table 8). In three independent pilot experiments, adults were injected with *Enterobacter cloacae* β12 (4 × 10^4 bacteria per animal) or saline and hemolymph investigated 24 h later.[21] As in *Dipetalogaster,* we observed considerable differences in the antibacterial activity of different individuals. Occasionally, some activity against *Escherichia coli* was found in hemolymph of even untreated animals. In contrast, lysozyme titers were extremely low.

Hemolymph proteins from normal and immunized adults have been investigated by native and SDS-PAGE. Hemolymph samples from both saline and bacteria-treated bugs showed new bands with molecular weights of about 67, 16, 14, and 10 kDa, which were not detectable in the hemolymph of untreated animals (Figure 6). The 16-kDa protein stains

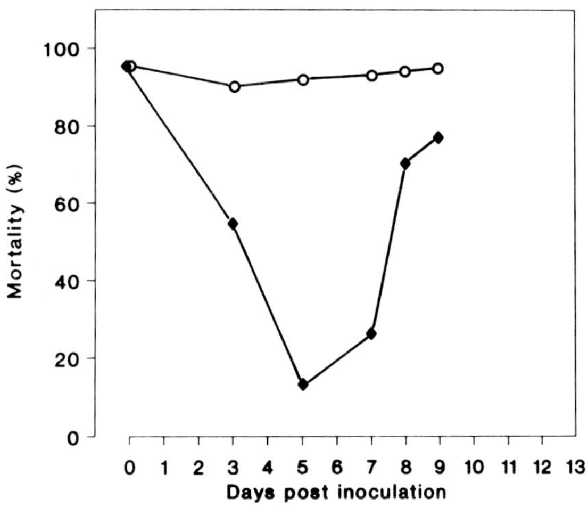

FIGURE 5. Mortality of normal (○——○) and immunized (♦——♦) adult *Rhodnius prolixus* after challenge injection. Insects were immunized with 1 μl of culture medium, containing 1 × 10^6 cells of *S. mutans* or 1 μl of medium without bacteria (control). At different days thereafter, challenge injections with 10^9 cells of *S. mutans* were applied and mortality determined 24 h later. Each point represents the mean ± SE of four groups of 10 to 15 insects. (Adapted from Azambuja, P., Freitas, C. C., and Garcia, E. S., *J. Insect Physiol.*, 32, 807, 1986.)

TABLE 6
Time Course and Dose Response of Antibacterial Activity (in RU) against *E. coli* D31 in Hemolymph of Normal and Immunized *Dipetalogaster maximus*[9]

Injection (bacteria/insect)	Antibacterial activity in RU[a] against *E. coli*				
	day 1	day 2	day 3	day 4	day 6
Saline	n.d.	72 ± 202 (n = 7)	0 (n = 3)	0 (n = 7)	n.d.
3 × 10^4	n.d.	31 ± 0 (n = 5)	122 ± 57 (n = 4)	75 ± 35 (n = 4)	94 ± 67 (n = 5)
7 × 10^4	0 (n = 5)	5 ± 4 (n = 10)	104 ± 31 (n = 11)	63 ± 43 (n = 8)	n.d.
2 × 10^6	16 ± 10 (n = 5)	64 ± 31 (n = 6)	72 ± 30 (n = 6)	116 ± 34 (n = 4)	n.d.
8 × 10^6	14 ± 6 (n = 5)	27 ± 2 (n = 3)	67 ± 16 (n = 4)	150 ± 71 (n = 4)	n.d.

Note: n.d., no data; antibacterial activity of untreated animals was zero.

[a] At different times p.i.

better with Coomassie blue than with silver nitrate. On gels after native PAGE, the active band against *E. coli* D31 migrates about half the distance of cecropins A and B from *H. cecropia*.[21]

F. COLEOPTERA

1. *Anthonomus grandis*

Inducible antibacterial activity in Coleoptera was first reported by Ourth an[illegible]lley[22]

TABLE 7
Time Course and Dose Response of Antibacterial Activity (μg/ml HEL) against *M. luteus* Hemolymph of Normal and Immunized *Dipetalogaster maximus*

Injection (bacteria/insect)	Antibacterial activity in μg/ml HEL[a] against *M. luteus* day 1	day 2	day 3	day 4	day 6
Saline	n.d.	27 ± 6 (n = 3)	17 ± 1 (n = 3)	18 ± 3 (n = 3)	n.d.
3×10^4	n.d.	n.d.	48 ± 17 (n = 4)	31 ± 8 (n = 4)	78 ± 26 (n = 5)
7×10^4	31 ± 5 (n = 5)	55 ± 4 (n = 9)	85 ± 18 (n = 9)	67 ± 16 (n = 8)	n.d.
7×10^5	33 ± 2 (n = 5)	69 ± 13 (n = 5)	106 ± 0 (n = 5)	112 ± 12 (n = 4)	n.d.
8×10^6	37 ± 6 (n = 5)	70 ± 6 (n = 3)	70 ± 21 (n = 4)	150 ± 23 (n = 3)	n.d.

Note: n.d., no data. Lysozyme titer of untreated animals was ca. 20 μg/ml HEL equivalents.

[a] At different times p.i.

TABLE 8
Antibacterial Activity in Hemolymph of *Dysdercus intermedius*[21]

Stage	n	Injection (bacteria/insect)	Time p.i. (h)	Antibacterial activity against *E. coli* D31 (RU)	*E. coli* D22 (RU)	*M. luteus* (f) (μg/ml HEL)
Adult	6	None	24	34 ± 14	24 ± 15	0
(female)	6	Saline	24	1433 ± 331	1397 ± 453	8 ± 2
	6	*E. cloacae* 2×10^4	24	1738 ± 313	986 ± 220	13 ± 4

from adults of the cotton boll weevil, *A. grandis*. The insects were inoculated with 1×10^5 live or heat-killed cells of *Serratia marcescens* and then tested for phagocytosis, lysozyme activity, and bactericidal activity against live *S. marcescens* and *Micrococcus luteus*.

Maximal phagocytosis was observed 16 h after injection of live *S. marcescens* when 16.3% of the plasmatocytes contained engulfed bacteria. After this time, the insects were not able to withstand the increasing bacteremia and became moribund. Immunization was therefore performed with heat-killed *S. marcescens* (Figure 7). Inoculated as well as uninoculated beetles showed lysozyme-like activity. This activity against heat-killed *M. luteus* increased about 30 h after inoculation with *S. marcescens* and reached its peak value (6.9 μg/ml) 48 h p.i. The lysozyme titer dropped below its normal value 4 d after inoculation. A lysozyme-independent bactericidal activity was demonstrated by growth inhibition zone assay with live *S. marcescens*. This activity became detectable 12 h after bacterial injection; it reached its peak at 24 h p.i. and disappeared at 36 h p.i.

2. *Tenebrio molitor*

Larvae, pupae, and adults of *T. molitor* express bactericidal activity in their hemolymph after inoculation with the nonpathogenic bacterium *Enterobacter cloacae* β12.[9] Detailed studies showed that the titer of antibacterial activity against *Escherichia coli* D31 as well as against live *M. luteus* is dose- and time dependent, and reaches maximal values within

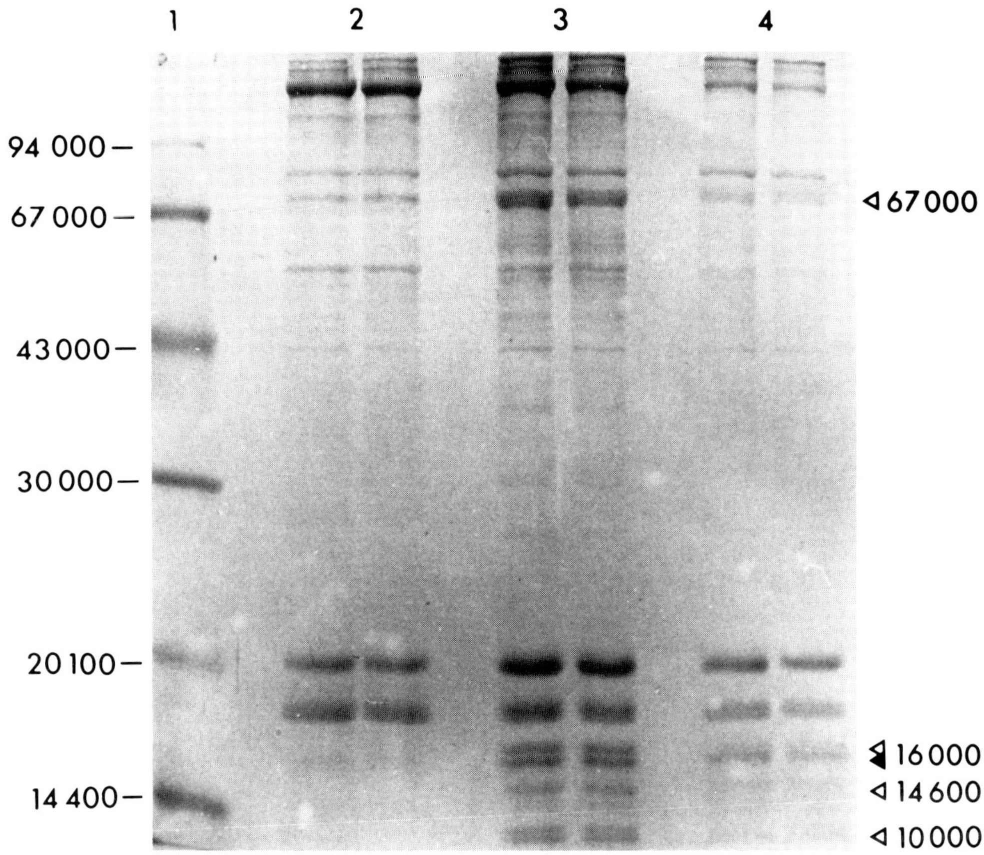

FIGURE 6. SDS-PAGE of *Dysdercus intermedius* hemolymph proteins. Samples were run in a 15% polyacrylamide gel using a modified Laemmli system and stained with Coomassie Blue R250, as described in Section II. Lane 1: reference proteins. Lanes 2-4: hemolymph samples (corresponding to 1 μl original volume) from untreated animals (2), insects 24 h after saline injection (3), and 24 h after injections with *E. cloacae* β12 (4). New protein bands, present only in injected insects, are indicated by <. The protein, which stains better with Coomassie blue than with silver, is marked by ◀.

4 d (76 ± 26 RU and 286 ± 28 g/ml HEL; Figure 8). It is interesting that doses below 10^3 bacteria per insect did not induce higher lysozyme-like activity than saline injection (Figure 8B) and that even 10^4 bacteria per animal are necessary to provoke activity against *E. coli* D31 (Figure 8A). Furthermore, significant differences were found in the activity against *E. coli* D31 between larvae, pupae, and adults (data not shown). The activity against *E. coli* can be attributed to a protein band with similar electrophoretic mobility as cecropin A of *H. cecropia*.[9] Increased lysozyme-like activity was also produced by previous injection with India ink.[2]

3. *Eleodes* spp.

Enterobacter cloacae strain 4A (1 × 10^6 log-phase cells in water) was used by Spies et al.[23] to immunize adult dark beetles, *Eleodes* spp. Immune hemolymph in dilution of 1:420 still inhibited growth of *Escherichia coli* HB 101 in wells of microtiter plates starting from 100 bacterial cell per well in LB broth. Acid PAGE of such immune hemolymph revealed three inhibition zones in overlaying agar gels with suspended *E. coli* D31. The mobility of these bands was similar to that of cecropin- and attacin-like proteins from *Manduca sexta*.

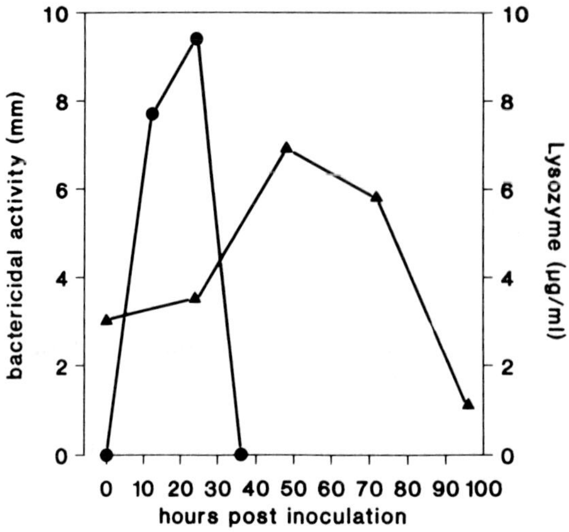

FIGURE 7. Antibacterial activity in hemolymph of *Anthonomus grandis* after inoculation with heat-killed *S. marcescens*. Antibacterial activity (●——●) is expressed as diameter of inhibition zones on agar plates seeded with live *S. marcescens* (referred to as "bactericidal activity") and as lysozyme equivalents ("lysozyme µg/ml"; ▲——▲) estimated with heat-killed *M. luteus* test plates. (Adapted from Ourth, D. D. and Smalley, D. L., *J. Invertebr. Pathol.*, 36, 104, 1980.

4. *Necrophorus vespilloides*

Even though the effect of various doses of *Enterobacter cloacae* β12 (4×10^3 to 10^7 bacteria per animal) was investigated at different times post-infection (2 to 6 d, n = 5 per time and dose), no antibacterial activity against *Escherichia coli* D31, but some activity against *Micrococcus luteus* (up to 14 µg/ml) could be detected in the hemolymph of adult *N. vespilloides*. All pupae treated with 6×10^4 or 1×10^6 *Enterobacter cloacae* β12 per animal died only 1 d after bacterial injection.[9]

5. *Other Coleoptera*

Reinicke,[13] who injected different doses of *E. cloacae* β12 (10^2 to 10^7 bacteria per animal) into larvae of *Leptinotarsa decemlineata* and *Oryctes nasicornis,* could not provoke antibacterial activity against *Escherichia coli* D31. Lysozyme titers of a number of field-collected beetles have been reported by Mohrig and Messner,[2] who noted the following titers (expressed as HEL equivalents): larvae of *Rhagium mordax* 10 to 25 µg/ml, adults of *R. mordax* 25 to 100 µg/ml, larvae of *Elater* spp. 10 to 50 µg/ml, adults of *Carabus auratus* 10 to 25 µg/ml ml, larvae of *Carabus granulatus* 25 to 100 µg/ml, adults of *Dytiscus dimidiatus* 25 to 50 µg/ml, adults of *Meloe violaceus* 50 to 100 µg/ml, and larvae of *Melolontha melolontha* 250 to 500 µg/ml.

G. HYMENOPTERA

1. *Apis mellifera*

Agglutinating substances against *Bacillus larvae* and *B. subtilis* were already demonstrated by Gilliam and Jeter.[24] However, the first proof of the inducibility of antibacterial factors was given by Craig et al.[25] A necessity for the induction of antibacterial activity against *Escherichia coli* D22 seems to be the invasion of microorganisms into the hemocoel. Infections set by feeding *Nosema apis* spores, raising heavy intracellular infections in the gut epithelium, did not result in any detectable activity (Table 9).[26,29] The occurrence of

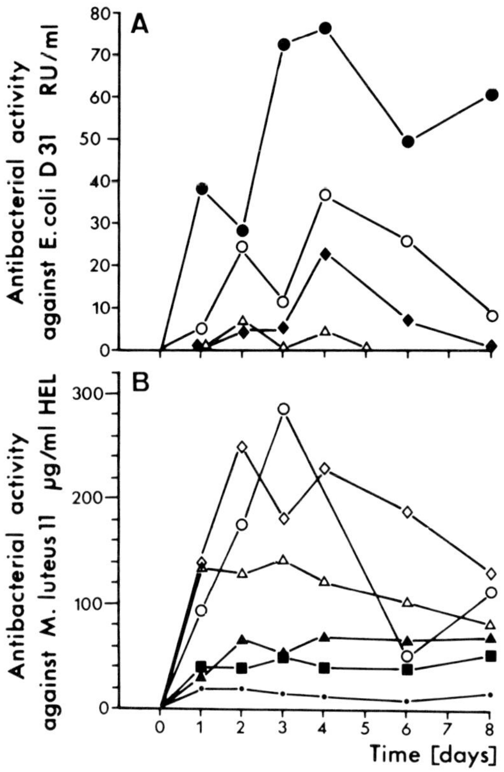

FIGURE 8. (A) Induction of activity against *E. coli* D31 in the hemolymph of *Tenebrio molitor* larvae after injection of *E. cloacae* β12. The different doses used were 6 × 10^4 (△——△), 1 × 10^5 (◆——◆), 1 × 10^6 (○——○), and 2 × 10^6 (●——●) bacteria per animal. Each point represents the mean of 4 to 13 individuals. (B) Induction of lysozyme-like activity in the hemolymph of *T. molitor* larvae. The different doses used for immunization were 1 × 10^3 (■——■), 4 × 10^4 (▲——▲), 6 × 10^4 (△——△), 1 × 10^6 (○——○), and 8 × 10^5 (◇——◇) bacteria per animal. Activity after saline injection is marked with (•——•). Each point represents the mean of 4 to 11 animals.

TABLE 9
Antibacterial Activity in Hemolymph of *A. mellifera*[29]

Stages	n	Injection (bacteria/insect)	Time p.i. (h)	Antibacterial activity against *E. coli* D22 (RU/µl)	Antibacterial activity against *M. luteus* (f) (µg/ml HEL)
Workers	5	None	24	0	22 ± 3
	5	Saline	24	0	n.d.
		E. cloacae[a]			
	5	1 × 10^4	24	320 ± 134	n.d.
	5	2 × 10^4	24	741 ± 155	130 ± 23
	12	*Nosema*[a]	24	0	n.d.

Note: n.d., no data.

[a] Doses are given in bacteria per animal and spores per animal, respectively.

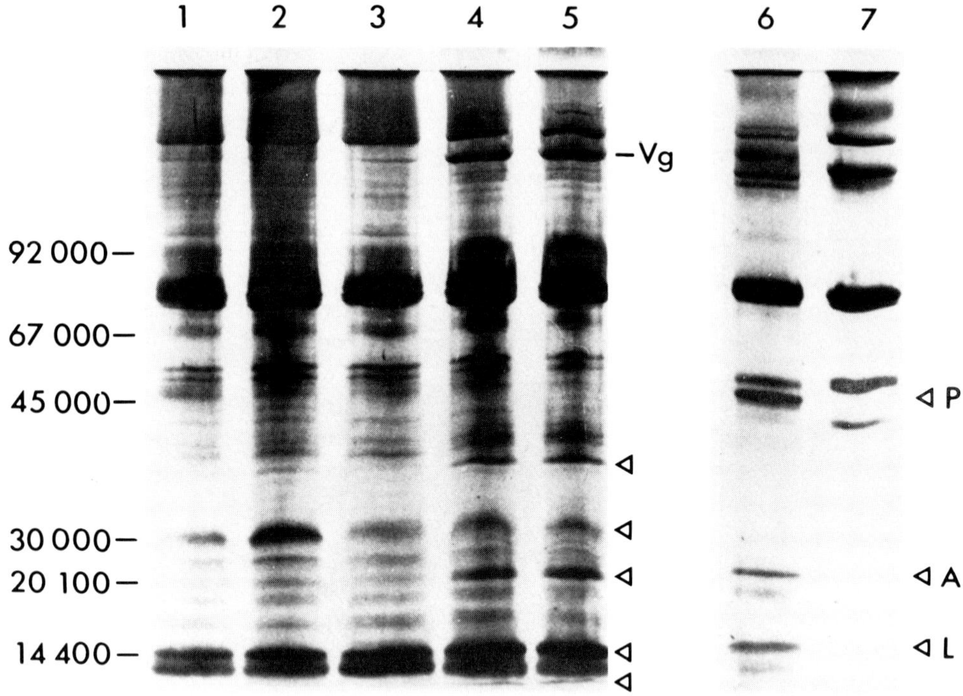

FIGURE 9. SDS-PAGE of *Apis mellifera* hemolymph proteins. Samples were separated on 10 to 15% polyacrylamide gel using a modified Laemmli system and stained with silver nitrate, as described in Section II. Molecular weights are indicated on the left, and protein bands induced or influenced by the treatment of the honeybees are marked on the right (<Vg indicates vitellogenin). Lane 1: untreated bees. Lane 2: bees injected with saline. Lane 3: bees fed with *Nosema apis* spores. Lane 4, 5: bees injected with *E. cloacae* β12. As reference, hemolymph samples from immunized (lane 6) and untreated (lane 7) *Hyalophora cecropia* pupae were separated on the same gel. "Immune proteins" of *H. cecropia* are indicated by their capital letter: P4 (P), attacin (A), and lysozyme (L).

two molecules with the mass of 2108 and 3878 was correlated to the appearance of antibacterial activity. Even though the mass of 3878 is close to that of cecropin D of *H. cecropia* with 3793, immunoblot analysis failed to confirm this hypothesis. The nature of the molecule with the mass of 2108 was later described as apidaecin.[27] Injection of saline alone did not result in detectable antibacterial activity, even though new proteins occurred in the hemolymph (Figure 9).[26,28]

Detailed studies on the immunity in the honeybee led to the discovery of a new class of antibacterial proteins, the apidaecins.[27] Casteels and co-workers[27] could detect antibacterial activity as early as 8 h p.i. of *E. coli*. Apidaecin is inducible in adults at least to a titer of 50 nmol/ml (100 μg/ml) and to a lower extent in last instar larvae where the inactive precursor molecule also appears in the hemolymph. Furthermore, they were able to demonstrate by reverse phase HPLC that India ink or latex beads induced the same nine factors as bacteria did. One of these factors, the apidaecin, has been purified and the amino acid sequence determined. It consists of three closely related proline rich peptides of 18 amino acids, molecular weight of 2.1 kDa, and is probably of nonhelical structure. The mode of action seems to be bacteriostatic. Apidaecins are highly effective against Gram-negative bacteria, such as *Salmonella, Shigella, Rhizobium, Agrobacterium, Erwinia,* and *Pseudomonas,* but show little or no activity against the Gram-positive bacteria *Corynebacterium* spp., *Serratia marcescens, Bacillus alvei, B. subtilis,* and *B. thuringiensis* var. *tenebrionis*. Chemically synthesized apidaecins displayed similar bacteriostatic activity as the natural apidaecin. No activity could be demonstrated against phospholipid layers indicating probable inactivity against eukaryotic cells.[27]

2. *Pimpla turionella*

Antibacterial activity (against *E. coli* and *Micrococcus luteus*) in the hemolymph of the parasitoid hymenopteran *P. turionella* could be detected only in some individuals.[9] The inconsistent results may partly be caused by the difficulty in collecting enough hemolymph from this insect.

H. ANTIBACTERIAL ACTIVITY IN HEMOLYMPH (BODY FLUID) OF OTHER INVERTEBRATES BESIDES INSECTS

1. Crustacea

Acton et al.[30] have demonstrated an inducible antibacterial activity in the hemolymph of the American lobster, *Homarus americanus*. The preexisting bactericidal activity increased two- to fourfold after injection of 10^9 formalin-killed cells of a Gram-negative bacterium, previously isolated from the spiny lobster *Panulirus argus*. The titer remained elevated throughout the observation time (11 d). Even higher bactericidal activity, which also increased after bacterial inoculation, was found in the hepatopancreas of the American lobster.[31]

2. Myriapoda

At least two antibacterial substances have been found in the hemolymph of myriapods, a lysozyme and one substance, which is different from lysozyme.[32] Growth of *Enterobacter cloacae* was inhibited after inoculation of diplopods and chilopods, independent of the bacterial strain used for inoculation. In two species (*Chicobolus* sp. and *Rhapidostreptus virgator*), activity against living *M. luteus* increased after inoculation with *E. cloacae* β12. Activity against *Escherichia coli* was detected only in hemolymph of the milliped *Trianostreptus triodus*.[33]

3. Annelida

A *Glycera*-antibacterial factor (GAF) was described by Chain and Anderson,[34,35] who studied the bactericidal activity of coelomic fluid from the polychaete *Glycera dibranchiata* against the Gram-negative bacterium *Serratia marcescens*, using growth as well as plating assays. The bactericidal action consists of a rapid initial binding of GAF to the bacterial surface, which is followed by the killing action. Three to four molecules of GAF were calculated to be sufficient to disrupt the bacterial membrane. The factor is a heat-labile protein with a molecular weight of >100 kDa.

Earthworms (Lumbricidae) are favorite animals for immunological investigations probably because of their easily availability, their powerful cellular defense reactions, and also their economic importance. Lectins, prophenoloxidase, and cytotoxic and antibacterial activity have been demonstrated in coelomic fluid of these annelids.[36,37] Growth inhibition of Gram-positive and Gram-negative bacterial pathogens was studied in the coelomic fluid of *Eisenia foetida andrei*. The antibacterial activity was found to be mediated by at least three different proteins (M.W. 20, 40, and 45 kDa). In addition, Lassègues et al.[38] demonstrated increased antibacterial activity after primary injection with LD_{50} doses of the pathogenic bacteria *Aeromonas hydrophila* (Gram-negative) and *Bacillus megaterium* (Gram-positive). Such inoculations totally protected the annelids against a 200- to 2000-fold injection dose over a period of 4 to 11 d.

4. Mollusca

Reaction of mollusks to naturally or experimentally introduced microorganisms are primarily of cellular nature. In general, foreign organisms are efficiently phagocytosed by granular cells. Lectins with agglutinating and opsonizing activity are involved in the clearing reaction. The response of the defense system of the fresh-water snail *Lymnaea stagnalis* was enhanced through previous contact with heat-killed *Escherichia coli* or *Streptococcus saprophyticus.*[39] Subsequently injected bacteria are eliminated significantly faster in preinjected than in untreated snails. The elevated resistance is not specific to the bacterial strain used for primary injection. Humoral antibacterial activity seems to play a minor role in the immune reactions of mollusks. However, McDade and Tripp[40] detected lysozyme in the hemolymph of the oyster *Crassostrea virginica*. Hemolymph of untreated *C. virginica* showed not only antibacterial activity against *M. luteus, B. subtilis,* and *B. megaterium,* but also against *E. coli* and some other Gram-negative bacteria.[41]

5. Echinodermata

Lysozyme activity is present in high concentration in coelomocytes of the echinoderm *Holothuria polii.*[42] This bacteriolytic activity was heat-stable in acidic buffer but heat-labile in alkaline medium. The coelomic fluid itself contained only a low lysozyme concentration as demonstrated by spectrophotometry and plate test with ghosts from *M. luteus*.

IV. DISCUSSION

A. LYSOZYME

Lysozyme (EC 3.2.1.17) was discovered as antibacterial agent in mammals in 1922 by Fleming, and was later reported from the silkmoth, *Bombyx mori.*[43,44] Mohrig and Messner,[2,45] in their classical study, demonstrated the occurrence of lysozyme in a great number of insects from different orders; in other arthropods; and in annelids, mollusks, and nematodes. With agar plate inhibition zone tests (used before as antibiotica assay) they were able to visualize lysozyme occurrence not only in hemolymph but also in intestine and other tissues of invertebrates.

Lysozyme from *Galleria mellonella* was purified and completely characterized by Powning and Davidson.[46] The occurrence and nature of lysozyme in difference insect species have been reviewed by Jollès et al.,[47] and were further discussed by Boman and Hultmark.[48] Four different types of lysozymes have been described.[49] The following properties are known for lysozyme: it is a basic protein of low molecular weight, heat stable (100°C) at low pH and heat labile in alkaline media. In general, lysozyme is demonstrated using live or dead (heat-killed or freeze-dried) bacteria or ghosts of *M. luteus* (formerly referred to as *M. lysodeikticus*). Clear zones on agar plates with dead bacteria (or ghosts) indicate lytic activity, whereas inhibition zones on plates with colony-forming, live bacteria signal bacteriostatic or bacteriocidal activity but are no proof for lysis.[50] The lytic activity against *M. luteus* is also influenced by the ionic strength of the medium.[51]

The detection of lysozyme might be hampered by the activity of chitinase (EC 13.2.1.4, also a β-1,4-acetylhexosamidase) or other cell wall degrading enzymes, such as exoacetylglucoseamidases or endopeptidases.[52,53] False positive results might also be caused by other antibacterial proteins such as cecropins or defensins with activity against Gram-positive bacteria, especially when inhibition zone assays with live *M. luteus* are used. However, strong inhibition or lysis of *M. luteus*, in combination with absence of activity against *E. coli*, can be considered as a good proof for the presence of lysozyme. In addition, the previously mentioned characteristics of lysozyme (molecular weight, pH optimum, heat resistance at low pH, etc.) should also be verified.

As summarized in Table 10, activity against *M. luteus* was found in the large majority of insects investigated and, whenever tested, significantly increased in response to previous bacterial injections. Interestingly, the only insects where lysozyme was not constantly detectable are several cockroaches and one hemipteran species. These insects carry symbionts in specialized cells or even organs (mycetocytes and mycetoms, respectively) which can be sensitive to lysozyme, as Malke demonstrated in 1964.[56] Malke succeeded in producing cockroaches free from symbionts (aposymbiotic) by lysozyme application. The Gram-positive symbionts of *Periplaneta americana* showed cell wall deficiencies, which might come from the lysozyme activity in this host. Occurrence of Gram-positive bacteria as symbionts seems to depend on a moderated influence of lysozyme, which keeps bacterial development under control. One might therefore expect that the lysozyme titer in insects with Gram-positive bacteria as symbionts is generally lower, as compared with other insects. Unfortunately, lysozyme titers reported in literature cannot be easily compared with one another. From Table 7, we can see that activity against *M. luteus* in hemolymph after bacterial injection alters, depending on dose and time. Even under natural conditions, the actual lysozyme titer of an insect might be influenced by previous injury or infection. Furthermore, lysozyme activity might be different in various organs of an insect, as it is reported from *Euscelis communis,* where lysozyme is present in the hemolymph, but was not detected in the mycetocytes.[57,58]

B. OTHER ANTIBACTERIAL PROTEINS

Nonlysozyme-like antibacterial activity is usually tested with *Escherichia coli* as test bacterium, and its concentration is calculated from diameter of inhibition zones. Although the *E. coli* K12 strain has a remarkable sensitivity to cecropin- and attacin-like proteins, it is possible to miss antibacterial activity by using this assay only. For example, Hoffmann[10] discovered an inducible bacteriostatic activity in *L. migratoria* by adding immune hemolymph to growth cultures of *Bacillus thuringiensis.* The choice of the test bacteria is also critical, as demonstrated by the results from Azambuja with *Rhodnius prolixus.*[20] This insect does not express antibacterial activity against *E. coli*, but does so against another Gram-negative bacteria, *Arthrobacter luteus.* Therefore, negative results with the *E. coli* test system must not be taken as proof for missing antibacterial activity. In addition, the choice of the bacterial strains used for immunization also influences the possibility of detecting antibacterial activity. Different bacterial strains may have different provoking capabilities. To induce similar levels of antibacterial activity in *G. mellonella* or *H. cecropia* more cells of *E. coli* are needed than of *Enterobacter cloacae* β12.[63]

Searching for antibacterial activity with *Escherichia coli*, the plate test remained without success in most of the hemimetabolous insects tested (Table 10). In some Hemiptera and the majority of holometabolous insects, such activity could be provoked by preinjection of the insects with *E. cloacae.* The activity against *Escherichia coli* in the hemolymph of preinjected (''immunized'') *Dipetalogaster maximus* and *Dysdercus intermedius* (Hemiptera), *Tenebrio molitor* (Coleoptera) or the honeybee, *Apis mellifera* (Hymenoptera), was as high as it is known from Lepidoptera and Diptera. The protein nature of the antibacterial substances has been proved by PAGE. Their possible relationship with known compounds (e.g., cecropins or diptericins) can be decided only by tests with specific antibodies or by amino acid sequence analysis. The antibacterial compounds from representatives of Hemiptera and Coleoptera have not yet been characterized but corresponding investigations are under way. The chemical nature of the antibacterial proteins from honeybees has already been elucidated.[26,27] They represent a new family of inducible peptide antibiotics, called apidaecin.

C. ACQUIRED IMMUNITY IN INSECTS

In a number of insects, injection of bacteria induces appearance of new or increase in previously present antibacterial activity. This change in antibacterial activity can be caused by *de novo* synthesis as shown in Lepidoptera or Diptera. In other cases, the increased activity is the result of activation and/or release of stored compounds. These effects are not identical to acquired immunity known in vertebrates. In vertebrates, increased reactivity after contact with provoking material is accompanied by specificity and memory, two characteristics which are not known from antibacterial reactions of invertebrates. The antibacterial

TABLE 10a
Constitutive (c) and Inducible (i) Antibacterial Activity in Insects (Except Lepidoptera and Diptera)

		Antibacterial activity against						
		E. coli		*M. luteus*		Other bacteria		
Species	**Stage**	**c**	**i**	**c**	**i**	**c**	**i**	**Ref.**
Odonata								
Aeshna cyanea	L			+				2
	L	–	–	+	+			9
Phasmidia								
Carausius morosus	L	–	–	+	+			9
Bakula sp.	L	–	–	+	+			x
B. extradentatum	L	–	–	+	+			x
B. spiloida	A	–	–		+			x
Orthoptera								
Acheta domesticus	L	–	–					12
A. domesticus	L	–	–	+	+			x
Decticus verrucivorus	A			+				2
Gryllus bimaculatus	L,A	–	–	+	+	–	–	14—16
G. campestris	A			+				2
Locusta migratoria	A	–	+	+	+	–	+	10—12
Melanoplus bivitattatus							+	54
M. femur-rubrum							+	1
Blattaria								
Periplaneta americana						Tetrahymena		55
	A			+				2
				+[a]				53
		–	–					17
						Cobra venom		59—62
Blaberus giganteus	L	–	–	–	–			9
Blabera fusca	A			+				2
Leucophaea maderae	A	–	–	–	–			x
Gromaphadorina sp.	L	–	–	–	–			x

Note: A, adult; L, larva; P, pupa; x, see results in this publication.

[a] Chitinase not excluded.

substances known so far have very little specificity; their activity is not directed exclusively against the bacterium, which was used for inoculation. Not all bacteria have the capacity to induce increased antibacterial activity, but those who do elicit identical effects: release or *de novo* synthesis of certain antibacterial compounds.

There is also little or no indication for the occurrence of an immunological memory in insects. A well-known example, which is often interpreted as proof for specificity and immunological memory, comes from cockroaches inoculated with bee and snake venom. Karp and Rheins[59-62] reported that in the cockroach, *P. americana,* injection of inactivated honeybee toxin (phospholipase A_2) results in a protective response. The resistance lasts over about 14 d; it is specific for the toxin used and can be transferred to naive animals by the injection of cell-free hemolymph. Repeated injection elicits a secondary response which is more rapid and powerful. The secondary response can be induced only by the specific toxin (not by western cottonmouth moccasin venom) and needs lower doses than the generation of the primary response. The proteins released or synthetized in response to honeybee toxin form specific precipitates when moving against the toxin in Ouchterlony diffusion tests (see also Chapter 15).

TABLE 10b
Constitutive (c) and Inducible (i) Antibacterial Activity in Insects (Except Lepidoptera and Diptera)

		Antibacterial activity against						
		E. coli		*M. luteus*		Other bacteria		
Species	Stage	c	i	c	i	c	i	Ref.
Hemiptera								
Onocopeltus fasciatus	A	−	+			−	+	18
Pyrrhocoris apterus	A					−	−	19
	A	−	−	+	+			x
Rhodnius prolixus	A	−	−	−	−	−	+	20
Dipetalogaster maximus	L	−	+	+	+			9
Dysdercus intermedius	A	+	+	+	+			21
Palomena viridissima	A			+				2
Coleoptera								
Carabus auratus	A			+				2
C. granulatus	A			+				2
Dysticus dimidiatus	A			+				2
Elater spp.	L			+				2
Meloe violaceus	A			+				2
Melolontha melolontha	L			+				2
Rhagium mordax	L,A			+				2
R. sycophanta	L			+				2
Tenebrio molitor	L,A			+	+			2
	L,P,A	−	+	+	+			9
Anthonomus grandis	A			+	+	−	+	22
Eleodes spp.	A	−	+					23
Necrophorus vespilloides	A	−	−	−	+			9
Oryctes nasicornis	L	−	−					13
Leptinotarsa decemlineata	L	−	−					13
Hymenoptera								
Apis mellifera	L,P,A			+				2
	A	−	+					25,26
	P,A	−	+			−	+	27
	A	−	+	+	+			29
Pimpla turionella	A	−	+	+	+			9
Vespula germanica	A			+				2

Note: x, see results in this publication.

V. SUMMARY

The biological significance of occurrence and inducibility of antibacterial proteins in insects is still a mystery. No one can realistically evaluate the survival value of these substances for those insects that possess the coding genes and control mechanisms for their induction and expression. Some of these insect-borne antibiotics, such as lysozyme, are common in nearly all insect species investigated so far. The few exceptions, where lysozyme and other antibacterial substances are not present in usual concentrations, are insects with bacterial symbionts, a fact used to underline the idea of a regulatory role of such substances for coexistence with microorganisms. Intensive research activity of the last few years has led to the identification and characterization of new and interesting groups of peptides with antibacterial activity. Their occurrence is not restricted to insects. Recently we have learned that cecropin-like substances are also found in vertebrate organisms; and defensins, primarily known from vertebrates, have been discovered in Diptera. As demonstrated in this review, inducible antibacterial proteins, e.g., from Hemiptera and Coleoptera, still await identification. To better understand the defense mechanisms of insects, further studies, particularly those focusing on the interactions between cellular and humoral reactions, have to be done.

REFERENCES

1. **Glaser, R. W.,** On the existence of immunity principles in insects, *Psyche,* 25, 39, 1918.
2. **Mohrig, W. and Messner, B.,** Immunreaktionen bei Insekten. I. Lysozym als grundlegender antibakterieller Faktor im humoralen Abwehrmechanismus der Insekten, *Biol. Zentralbl.,* 87, 439, 1968.
3. **Hultmark, D., Steiner, H., Rasmuson, T., and Boman, H. G.,** Insect Immunity. Purification and properties of three inducible bactericidal proteins from hemolymph of immunized pupae of *Hyalophora cecropia, Eur. J. Biochem.,* 106, 7, 1980.
4. **Altman, P. L. and Dittmer, D. S., Eds.,** *Biology Data Book,* Vol. 3, 2nd ed., Federation of American Societies of Experimental Biology, Bethesda, MD, 1974.
5. **Faye, I. and Wyatt, G. R.,** The synthesis of antibacterial proteins in isolated fat body from *Cecropia* silkmoth pupae, *Experientia,* 36, 1325, 1980.
6. **Sidén, I. and Boman, H. G.,** *Escherichia coli* mutants with an altered sensitivity to cecropin D, *J. Bacteriology,* 154, 170, 1983.
7. **Laemmli, U. K.,** Cleavage of structural proteins during the assembly of the head of bacteriophage T4, *Nature (London),* 227, 680, 1970.
8. **Trenczek, T. and Faye, I.,** Synthesis of immune proteins in primary cultures of fat body from *Hyalophora cecropia, Insect Biochem.,* 18, 299, 1988.
9. **Riemer, M.,** Vergleichende Untersuchungen zur Immunisierbarkeit verschiedener Insekten, diploma thesis, Free University Berlin, 1988.
10. **Hoffmann, D.,** Introduction of antibacterial activity in the blood of the migratory locust *Locusta migratoria* L., *J. Insect Physiol.,* 26, 539, 1980.
11. **Lambert, J. and Hoffmann, D.,** Mise en évidence d'un facteur actif contre des bacteries gram négatives dans le sang de *Locusta migratoria, C. R. Séances Acad. Sci. Paris Ser. III,* 300, 425, 1985.
12. **Zachary, D. and Hoffmann, D.,** Lysozyme is stored in the granules of certain hemocyte types in *Locusta, J. Insect Physiol.,* 30, 405, 1984.
13. **Reinicke, G.,** Natürliche und experimentell induzierbare Immunität bei Larven der Schmeißfliege *Calliphora erythrocephala* (Insecta:Diptera), diploma thesis, Free University Berlin, 1982.
14. **Mack, I.,** Versuche zur Bestimmung des Titers von antibakteriellen–lysozymähnlichen — Substanzen in der Hämolymphe und der Nachweis dieser Stoffe in Organen von *Gryllus bimaculatus* (de Geer), diploma thesis, Bonn, 1982.
15. **Weber, M.,** Untersuchungen zum Proteinspektrum der Hämolymphe von *Gryllus bimaculatus* (de Geer) nach bakterieller Infektion, diploma thesis, Bonn, 1982.
16. **Schneider, P. M.,** Purification and properties of three lysozymes from hemolymph of the cricket *Gryllus bimaculatus* (De Geer), *Insect Biochem.,* 15, 511, 1985.
17. **Boman, H. G. and Steiner, H.,** Humoral Immunity in *Cecropia* pupae, *Curr. Top. Microbiol. Immunol.,* 94/95, 75, 1981.
18. **Gingrich, R. E.,** Acquired humoral immune response of the large milkweed bug, *Oncopeltus fasciatus* (Dallas), to injected materials, *J. Insect Physiol.,* 10, 179, 1964.
19. **Kamp, H.,** Untersuchungen zur humoralen Immunität bei *Pyrrhocoris apterus* und *Galleria mellonella, Z. Vgl. Physiol.,* 58, 441, 1968.
20. **De Azambuja, P., Freitas, C. C., and Garcia, E. S.,** Evidence and partial characterization of an inducible antibacterial factor in the hemolymph of *Rhodnius prolixus, J. Insect Physiol.,* 32, 807, 1986.
21. **Trenczek, T. and Preik, H.,** unpublished data, 1990.
22. **Ourth, D. D. and Smalley, D. L.,** Phagocytic and humoral immunity of the adult cotton boll weevil, *Anthonomus grandis* (Coleoptera:Curculionidae), to *Serratia marcescens, J. Invertebr. Pathol.,* 36, 104, 1980.
23. **Spies, A. G., Karlinsey, J. E., and Spence, K.,** The immune proteins of the darking beetle, *Eleodes* (Coleoptera:Tenebrionidae), *J. Invertebr. Pathol.,* 47, 234, 1986.
24. **Gilliam, M. and Jeter, W. S.,** Synthesis of agglutinating substances in adult honeybees against *Bacillus larva, J. Invertebr. Pathol.,* 16, 69, 1970.
25. **Craig, A. G., Trenczek, T., Engström, Å., and Bennich, H.,** Analysis of crude insect hemolymph by plasma desorption mass spectrometry to probe an immune response, *Adv. Mass Spectrom.,* 11B, 1354, 1988.

26. **Craig, A. G., Trenczek, T., Fries, I., and Bennich, H.,** Mass spectrometric identification of peptides present in immunized and parasitized hemolymph from honeybees without purification, *Biochem. Biophys. Res. Commun.*, 165, 637, 1989.
27. **Casteels, P., Ampe, Ch., Jacobs, F., Vaeck, M., and Tempst, P.,** Apidaecins: antibacterial peptides from honeybees, *EMBO J.*, 8, 2387, 1989.
28. **Trenczek, T.,** Vitellogenin-Vorkommen bei Drohnen der Honigbiene *(Apis meliffera* L.), disseration, Tübingen, 1984.
29. **Trenczek, T. and Fries, I.,** unpublished data, 1988.
30. **Acton, R. T., Weinheimer, P. F., and Evans, E. E.,** A bactericidal system in the lobster *Homarus americanus, J. Invertebr. Pathol.*, 13, 463, 1969.
31. **Mori, K. and Stewart, J. E.,** Natural and induced bactericidal activities of the hepatopancreas of the American lobster, *Homarus americanus, J. Invertebr. Pathol.*, 32, 171, 1978.
32. **Xylander, W. E. R. and Nevermann, L.,** Antibacterial activity in the hemolymph of myriapods (Arthropoda), *J. Invertebr. Pathol.*, 56, 206, 1990.
33. **Van der Walt, A., McClain, E., Puren, A., and Savege, N.,** Phylogeny of arthropod immunity. An inducible humoral response in the Kalahari millipede, *Triaenostreptus triodus* (Attems), *Naturwisenschaften,* 77, 189, 1990.
34. **Chain, B. M. and Anderson, R. S.,** Antibacterial activity of the coelomic fluid of the polychaete, *Glycera dibranchiata.* I. The kinetics of the bactericidal reaction, *Biol. Bull.* (Woods Hole), 164, 28, 1983.
35. **Chain, B. M. and Anderson, R. S.,** Antibacterial activity of the coelomic fluid of the polychaete, *Glycera dibranchiata.* II. Partial purification and biochemical characterization of the active factor, *Biol. Bull.* (Woods Hole), 164, 41, 1983.
36. **Valembois, P., Roch, Ph., and Lassègues, M.,** Antibacterial molecules in annelides, in *Immunity in Invertebrates,* Brehélin, M., Ed., Springer, Berlin, 1986, 74.
37. **Lassalle, F., Lassègues, M., and Roch, Ph.,** Protein analysis of earthworm coelomic fluid. IV. Evidence, activity induction and purification of *Eisenia fetida andrei* lysozyme (Annelidae), *Comp. Biochem. Physiol.*, 91B, 187, 1988.
38. **Lassègues, M., Roch, Ph., and Valembois, P.,** Antibacterial activity of *Eisenia fetida andrei* coelomic fluid: evidence, induction, and animal protection, *J. Invertebr. Pathol.*, 53, 1, 1989.
39. **Van der Knaap, W. P. W. and Meuleman, E. A.,** Interaction between the immune system of lymnaeid snails and trematode parasites, *Symp. Zool. Soc. London,* 56, 179, 1986.
40. **McDade, J. E. and Tripp, M. R.,** Lysozyme in the hemolymph of the oyster, *Crassostrea virginica, J. Invertebr. Pathol.*, 9, 531, 1967.
41. **Rodrick, G. E. and Cheng, Th. C.,** Kinetic properties of lysozyme from the hemolymph of *Crassostrea virginica, J. Invertebr. Pathol.*, 24, 41, 1974.
42. **Canicatti, C. and Roch, P.,** Studies on *Holothuria polii* (Echinodermata) antibacterial proteins. I. Evidence for and activity of a coelomocyte lysozyme, *Experientia,* 45, 756, 1989.
43. **Flemming, A.,** On a remarkable bacteriolytic element found in tissues and secretions, *Proc. R. Soc.*, 93B, 306, 1922.
44. **Cappellato, M. and Naprozzi, E. A.,** Fattori di immunita aspecifica nell emolinfa di *Bombyx mori, Boll. Ist Sicroterap. Milano,* 39, 1960, 40.
45. **Mohrig, W. and Messner, B.,** Immunreaktionen bei Insekten. II. Lysozym als antibakterielles Agens im Darmtrakt von Insekten, *Biol. Zentralbl.*, 87, 705, 1968.
46. **Powning, R. F. and Davidson, W. J.,** Studies on insect bacteriolytic enzymes. I. Lysozyme in hemolymph of *Galleria mellonella* and *Bombyx mori, Comp. Biochem. Physiol.*, 45B, 669, 1973.
47. **Jollès, J., Schoentgen, F., Croizier, G., Croizier, L., and Jollès, P.,** Insect lysozymes from three different species of lepidoptera. Their structural relatedness to the C (chicken) type lysozyme, *J. Mol. Evol.*, 14, 267, 1979.
48. **Boman, H. G. and Hultmark, D.,** Cell-free immunity in insects, *Annu. Rev. Microbiol.*, 41, 103, 1987.
49. **Jollès, P. and Jollès, J.,** What's new in lysozyme research?, *J. Mol. Cell Biochem.*, 63, 65, 1984.
50. **Goetz, P. and Boman, H. G.,** Insect immunity, in *Comprehensive Insect Physiology, Biochemistry and Pharmacology,* Vol. 3, Kerkut, G. A. and Gilbert, L. A., Eds., Pergamon Press, Oxford, New York, 1985, 453.
51. **Gupta, D. K., von Figura, K., and Hasilik, A.,** Conditions for a reliable application of the lysoplate method in the determination of lysozyme, *Clin. Chim. Acta,* 165, 73, 1987.
52. **Powning, R. F. and Izykiewicz, H.,** Lysozyme-like action of enzymes from the cockroach *Periplaneta americana* and from some other sources, *J. Insect Physiol.*, 13, 1293, 1967.
53. **Bernier, I., Landureau, J. Ch., Grellet, P., and Jollès, P.,** Characterization of chitinases from hemolymph and cell cultures of cockroach *(Periplaneta americana), Comp. Biochem. Physiol.*, 47B, 41, 1974.
54. **Stephens, J. M.,** Bactericidal activity of hemolymph of some normal insects, *J. Insect Pathol.*, 5, 61, 1963.

55. **Seaman, G. R. and Robert, N. L.,** Immunological response of male cockroaches to injection of *Tetrahymena pyriformis, Science,* 161, 1359, 1968.
56. **Malke, H.,** Production of aposymbiotic cockroaches by means of lysozyme, *Nature (London),* 204, 1223, 1964.
57. **Houk, E. J. and Griffiths, G. W.,** Intracellular symbiotes of the homoptera, *Annu. Rev. Entomol.,* 25, 161, 1980.
58. **Douglas, A. E.,** Mycetocyte symbiosis in insects, *Biol. Rev. (Cambridge),* 64, 409, 1989.
59. **Karp, R. D. and Rheins, L. A.,** Induction of specific humoral immunity to soluble proteins in the American cockroach *Periplaneta americana.* II. Nature of the secondary response, *Dev. Comp. Immunol.,* 4, 629, 1980.
60. **Rheins, L. A., Karp, R. D., and Butz, A.,** Induction of specific humoral immunity to soluble proteins in the American cockroach *Periplaneta americana.* I. Nature of the primary response, *Dev. Comp. Immunol.,* 4, 447, 1980.
61. **Rheins, L. A. and Karp, R. D.,** The humoral immune response in the American cockroach, *Periplaneta americana:* reactivity to a defined antigen from honeybee venom, Phospholipase A_2, *Dev. Comp. Immunol.,* 8, 791, 1984.
62. **Karp, R. D.,** Preliminary characterization of the inducible humoral factor in the American cockroach *(Periplaneta americana), Dev. Comp. Immunol.,* 9, 569, 1985.
63. **Götz, P. and Trenczek, T.,** personal observation.

Chapter 13

ANTIBACTERIAL IMMUNITY IN LEPIDOPTERA

June S. Chadwick and W. Peter Aston

TABLE OF CONTENTS

I. INTRODUCTION

Much of the early work on insect immunity, as well as the more current biochemical and molecular approaches have been concentrated on lepidopterans. A satisfactory approach to the subject requires integration of knowledge relating to bacteriology, insect physiology, and the field of mammalian immunology where sophisticated technological advances can, on occasion, be applied to research in insect immunity. Until recently, investigations have not embodied all these approaches, but in the last decade, a number of excellent reviews dealing with insect immunity have been published.[1-8] These reviews indicate the complexity of the immune response in insects and the fact that variation from one species to another occurs, even within the Lepidoptera; thus, in some respects, this argues against the proposal made by Klein[9] that "invertebrates can thrive without a sophisticated immune system".

Marchalonis,[10] in a summation of some current problems concerning invertebrate immunology, again pointed out the long-standing controversy on whether or not insects can mount an immune response that is comparable to that of vertebrates. He stated that, though some investigators may feel this issue is still not satisfactorily resolved, it is quite apparent that insects possess a variety of induced molecules capable of inactivating bacteria. In this chapter, we will attempt to review the current state of the art in respect to antibacterial activity in Lepidoptera, as well as focus on some work that has recently been carried out in our laboratories. All animals need immune systems to provide protection against invading microorganisms or parasites. To provide a better understanding of immune phenomena in insects, it is convenient to divide, arbitrarily, the induced immune response into three overlapping phases:

Induction — In this stage, the inducer, naturally or artificially, breaches the structural barrier(s) and enters into contact with appropriate cellular or soluble components of the immune system. There is discrimination between self and nonself, and stimulation of the next two phases.

Mediation — This involves the complex events between the recognition of the nonself inducer and the generation of active effector cells and substances.

Effector Phase — It includes containment, destruction, and elimination of the inducer by the generated effectors. Adverse physiologic reactions are counteracted and repair mechanisms are activated to bring about a return to the normal state. Naturally occurring effector cells and cell-free substances may be present and act at any phase. Regulation of the inducible responses may also occur at any stage.

To facilitate our own research, we have formulated a model (Figure 1) focused on the immune response of the greater wax moth, *Galleria mellonella,* which could be adapted to investigate immune phenomena in other insects.

II. GENERAL CHARACTERISTICS OF THE IMMUNE RESPONSE

Although there may be slight variations from one lepidopteran species to another, the features outlined in Table 1 are generally typical of the immune response. The lack of specificity as well as the lack of memory mentioned in Table 1 indicate that immunity in the Lepidoptera is not truly adaptive. The only reported evidence for adaptive immunity in any insect order is that found in the American cockroach by Duwel-Eby and Karp[11] (see also Chapter 15); this was not an antibacterial immunity but a response to venoms. These authors[11] described the response as antibody-like, and the substance involved as a protein with a molecular weight of 102 kDa. They defined immunologic competence as did Hildemann et al.[12] by three functional components: (1) selective or specific recognition, (2) cytotoxic or antagonistic reaction following sensitization, and (3) inducible memory or selectively altered reactivity on second contact. It remains to be seen whether a lepidopteran

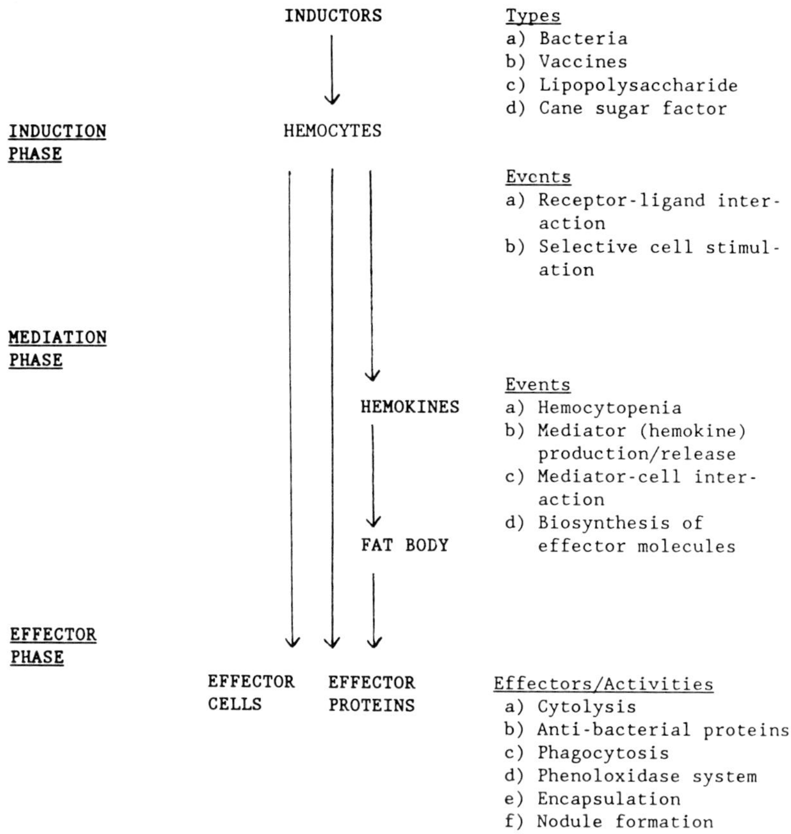

FIGURE 1. A model for investigating the immune response in *Galleria mellonella* and immune phenomena in other insects.

TABLE 1
General Features of Induced Protective Immunity in Lepidoptera

Develops in hours
Persists no more than 72 to 96 h
Nonspecific
Memory not involved
In vitro antibacterial activity present in cell-free hemolymph
Inducible antibacterial proteins of low molecular weight
Cellular involvement present
Decreased total hemocyte count

insect may yet answer these criteria. Table 2 provides a profile of the activities that are associated with a typical course of the immune response in larvae of the greater wax moth, following vaccination.

III. EXPERIMENTAL INSECTS

An analysis of the literature shows that the three lepidopterans most frequently employed for immunologic studies are larvae of *Galleria mellonella* used in our laboratory,[4,13] pupae of *Hyalophora cecropia* or *Samia cynthia* investigated by Boman and his group,[5,6,14] and more recently larvae of *Manduca sexta* by Dunn.[8,15,16] To a lesser extent, pupae of *Antheraea pernyi*[17] and *Bombyx mori*[18-20] have been used on some occasions by various investigators.

TABLE 2
Typical Development and Duration of Protective Immunity and Other *In Vitro* Biological Activities in *Galleria mellonella* Larvae after Administration at Time Zero of an Immunizing Preparation

Time	Activity[a]					
(h)	Protective	Bactericidal	Lysozyme	Hemolytic	Cytotoxic	Phospholipase
0	−	−	+	+ −	+ −	+
2	+ −	−	+	+	+ + +	+
3	ND	ND	+	+	+ +	ND
8	+	+	+ +	+	ND	ND
12	+ + +	+ + +	+ + +	+ +	+ −	+ + +
24	+ + + +	+ + + +	+ + + +	+ + + +	+ −	ND
30	+ + +	+ + +	+ + + +	+ + +	+ −	+ +
48	+	+	+ + + +	+ +	+ −	+ +
60	+ −	+ −	+ + +	+	+ −	+ −

Note: ND, not done; −, not detectable; + −, very weak; + + + +, very strong.

[a] The development of *in vivo* protective immunity and the determinations of antibacterial and lysozyme activities are described by Pryce et al.[35] The profile of the hemolytic activity is described by Phipps et al.,[53] and the development of the cytotoxic and phospholipase C activities are described by Chiu et al.[51,85]

Whether the antibacterial effector substances produced might have varied had different stages of insects been used in individual investigations is currently unclear. It has been reported[21] that Russell from the laboratory of Dunn has shown that during metamorphosis in *M. sexta,* bacteria from the midgut may enter the body cavity when the structure of the insect's digestive tract is being re-formed. The insect then releases large quantities of antibacterial peptides to sterilize the gut. These peptides differ from those made in response to invasive infections. In this recent report, there is speculation that antibacterial peptides may well be different from one stage of development to another in a given insect. If this proves to be so, considerable rethinking of the interpretation of the activities of effector substances may be necessary.

IV. INDUCTORS OF ANTIBACTERIAL IMMUNITY

A less well-studied phase of the immune response, resulting in antibacterial immunity, is the inductive phase. Conventionally, immune responses have been generated in Lepidoptera following injection with agents, such as killed or attenuated bacteria,[4,13,22] nonpathogenic bacteria,[5,23,24] or bacterial products, such as lipopolysaccharide (LPS)[4,25] or peptidoglycan.[15,16] Glinski and Jarosz[26] unsuccessfully attempted to induce immunity in *G. mellonella* by the oral route. In our laboratory until recently,[27,28] we primarily used agents such as vaccines or LPS. We have always investigated immunity in relation to protection against a pathogenic bacterium having a low LD_{50}, i.e., highly virulent by injection, and this is a prerequisite feature to any aspect of our investigations.[4] The end result achieved in relation to the inducing agent will vary with the pathogen, that is, immunity against *Pseudomonas aeruginosa* may be easily induced using any of the inducers just mentioned.[4,22] Immunity against *Proteus* species, with low LD_{50} like that of *Pseudomonas,* is best achieved with a homologous preparation,[22,29] whereas immunity against highly virulent *Serratia marcescens* cannot be achieved with any of the recognized inducers.[22,30,31] Protection against pathogens is optimal for about 20 h and all but disappears after 72 h. Phagocytosis of most bacteria is not prominent. In every circumstance where an immune response is evoked by the agents mentioned earlier, changes described in Table 1 and humoral antibacterial factors are elicited. In this area in the last decade, some major advances in insect immunity have been made (see Section VI).

TABLE 3
Representative LD_{50}s of *Pseudomonas aeruginosa* and *Proteus mirabilis* in *Galleria mellonella* 24 h after Injection with Vaccine[a] or Cane Sugar Factor (CSF)

Treatment	LD_{50} of *P. aeruginosa*	LD_{50} of *P. mirabilis*
None	1.7	4
Vaccine	>5000	>2000
CSF	>3000	>500

[a] Homologous formalized vaccine of test pathogen.

Dunn and Dai[32] stated that insects respond to the presence of bacteria in their body cavity by a combination of events: hemocyte-mediated sequestration, induced synthesis of bacteria-regulated hemolymph proteins, and subsequent killing of the bacteria; in *M. sexta,* peptidoglycan is the signal. The bacteria have to be processed to generate elicitors *in vivo*. These authors suggest that lysozyme, always present in the hemolymph, may do the processing. Kanost et al.[33] characterized the response in *M. sexta* to peptidoglycan from *Micrococcus luteus*. By 48 h after ingestion, lysozyme activity was tenfold greater than in untreated insects. Other antibacterial peptides were also induced by small fragments of peptidoglycan. The response in *Manduca* is believed to be quite distinct from that of other lepidopteran species, as these insects neither respond solely to wounding nor injection of soluble proteins or nonbacterial particles.

Until recently, agents that have classically evoked the response were bacterially derived. It is well known that hemolymph of *Galleria* melanizes very rapidly, which presents a problem with *in vitro* procedures, such as electrophoresis or antibacterial assays. Morton,[27] who was investigating immunity of *Galleria* to *Proteus mirabilis,* found that cane sugar factor (CSF), which Ashida[34] used as an antimelanizing agent, was effective in preventing melanization of wax moth hemolymph. Antimelanizing agents are tested to ensure they are neither antibacterial nor inducers of immunity. Interestingly, an initial screening of the characteristics of CSF indicated that it might induce protective immunity, and Morton[27] showed that it had a positive effect on all parameters of immunity to *P. mirabilis*. Pryce[28] in our laboratory then investigated the comparative effects of CSF to vaccine on all the important parameters of immune responses to *Pseudomonas*. An effective immunizing agent for *Galleria* induces protective immunity in which the LD_{50} of a challenge organism is significantly increased in 24 h.

Table 3 shows typical results obtained when larvae of *Galleria* were injected with CSF 24 h prior to challenge with *Pseudomonas or Proteus* species. As was the case when we had induced immunity against *Proteus* with vaccine or LPS, protection against *Proteus* was not as marked as against *Pseudomonas*. Even so, considerable immunizing efficiency was due to CSF, the LD_{50} being about 250 times higher. In both instances, vaccines seemed to induce the best protection. Like vaccine also, CSF induced immunity that was weakening within 36 h after its injection. In addition, CSF compared favorably to vaccine in eliciting all the characteristic parameters of immunity in *Galleria:* (1) antibacterial activity in the hemolymph, (2) decrease in total hemocytes in hemolymph, (3) increase in lysozyme levels, and (4) change in the electrophoretic profile during the period of protective immunity. In preliminary analysis of CSF,[35] it was found to have a small protein component (less than 5%) and about 90% glucose. Therefore, there was roughly 5% of unknown material. The

TABLE 4
Lepidopteran Hemokines

Hemokine	Insect source	Activity	Ref.
Plasmatocyte-depletion factor	*Galleria mellonella*	Decrease in circulating plasmatocytes	84
Hemokinin	*Samia cynthia, Anthergea polyphemus, Hyalophora cecropia*	Activation of hemocytes	153
Encapsulation promoting factor (EPF)	*Heliothis virescens, H. zea*	Stimulation of hemocytes to start or join encapsulation	154
Phagocytosis stimulating factor	*G. mellonella*	Stimulation of phagocytes	155
*Gal*lysin 2	*G. mellonella*	TNF-like	51,52

possibility that any glucose polymer might be immunogenic was more or less abrogated by the fact that the injection of either dextran or levan (both glucose polymers) had insignificant immunogenic effect.[35]

The discovery that CSF was immunogenic has been advantageous as we now have a simple immunizing substance, which should make it easier to conduct studies to determine how the immune system gets turned on in lepidopteran insects. Recognition of substances as foreign is evidenced by the fact that Lepidoptera differentiate self from nonself, but the question posed by Vasta[36] is valid: which are the effector molecules that bind to the nonself substance and trigger the production and release of effector molecules? To learn more of this, CSF might be a useful product. The activities of digests of CSF could be compared to various known structural components of bacteria. A study of the effects of CSF on defined hemocyte populations may provide evidence as to which are reactive trigger cells. Moreover, a study attempting to determine the connection of CSF activity with the phenoloxidase system and its role in the induction of immunity might well lead to evidence to support or refute the theory that the phenoloxidase system is linked to the immune system; this has been suggested by a number of investigators.[37-41]

V. MEDIATORS

To account for the generation of wide spectrum of effector mechanisms in insects, the events following induction are probably equally diverse and are likely to include stimulation of susceptible hemocyte subpopulations, selective localization of cells (hemocytopenia), biosynthesis of induced proteins, secretion of soluble mediators, and interaction between mediators and immune cells.

Recent models for the antibacterial responses in insects implicate a rapid response to the invaders by the hemocytes.[7,42-44] Responses of hemocytes may be a direct interaction with the bacteria[7,44,45] and/or the release of constitutive or induced soluble factors that communicate with other hemocytes, other tissues, or other soluble factors.[44,45] Beck et al.[46] use the term cytokine to describe invertebrate molecules produced by cells involved in host defense and used to communicate with similar or different cells. This group of investigators recently reviewed the evidence for invertebrate cytokines, including a primitive IL-1 from the echinoderm, *Asterias forbesi,* and eight North American species of tunicates.[46-48] IL-2-like and tumor necrosis factor (TNF)-like activities have also been isolated from echinoderms.[46] We propose that the term hemokine would be a more appropriate term for proteins with cytokine-like functions that are produced by hemocytes. Table 4 lists some lepidopteran hemokines.

The invertebrate IL-1 and TNF have similar activities and structural properties to their vertebrate analogues.[46] Since IL-1 and TNF share many functional activities,[49,50] it is feasible to speculate that these cytokines possessed a similar activity in a single molecule in the distant past.[46] Habicht's group[48] suggest that IL-1 is an ancient, biologically important molecule that has been conserved throughout the evolution of host defense mechanisms. As yet no IL-1 activity has been reported in lepidopteran insects. The IL-1 activities reported in the tunicates have no identifiable TNF-like activities determined by using the standard assays for vertebrate TNF.[47] A conclusion that tunicates do not possess TNF-like molecules must await further experimentation. The TNF-like activity found in echinoderms[46] was active in the murine L-cell cytotoxicity assay for TNF and was associated with a macromolecule (40 to 50 kDa), approximately pI 4.0

Recent evidence from our laboratory suggests that a lepidopteran analogue of TNF is present in the hemolymph of wax moth larvae.[51,52] We have shown that a hemolymph-derived cytotoxic activity exists in low levels in the hemolymph of untreated *G. mellonella,* and significantly higher levels of activity can be transiently induced shortly following an injection of LPS derived from *Salmonella typhimurium* or *P. aeruginosa*.[51] This is the second of two *Galleria*-derived lysins that we have identified,[51,53] and the name *Gal*lysin 2 is proposed to describe its activity and differentiate it from the other lysin, *Gal*lysin 1, that is described in Section VI.

*Gal*lysin 2 is highly cytotoxic for the human tumor cell lines, K562 and Raji; and one murine tumor cell line, L929. Optimal cytotoxic activity is detected in cell-free hemolymph at about 2 h postinoculation of the LPS. After this time, there is a rapid decline in the activity to preinoculation levels, which are then maintained throughout the peak period of the *in vivo* induced protective response, i.e., 18 to 24 h after an immunizing dose of the LPS (Table 2). The LPS-induced increase in *Gal*lysin 2 is dose dependent over a range of 0.02 to 0.20 μg LPS per larva, but larger doses of the LPS did not increase the cytotoxicity further. It is interesting that 0.02 μg LPS per larva, which does not afford any protection to the insect, induces the same level of *Gal*lysin 2 as a routine immunizing dose of 2.5 μg per larva.[54] In contrast to the induction of antibacterial effector activity, which depends on the carbohydrate portion of the LPS,[55] the lipid A structure is required for *Gal*lysin 2 induction. The functional role of *Gal*lysin 2 has not yet been determined. Its early appearance following an immunizing dose of LPS and several features that liken it to TNF suggest that *Gal*lysin 2 may be acting as a hemokine.

The TNF-like properties of *Gal*lysin 2 include

1. The cytotoxic activity is inhibited by polyclonal and monoclonal antibodies specific for human TNF; antibodies with other specificities are not inhibitory.
2. Cell-free hemolymph containing *Gal*lysin 2 reacts with polyclonal antibodies to human TNF in an ELISA.
3. LPS induces both *Gal*lysin 2 and TNF and, like *Gal*lysin 2, the murine TNF induction requires the lipid A portion of the LPS molecule.[56-58]
4. *Gal*lysin 2 and vertebrate TNF are cytolytic for tumor cell lines.[59]
5. There is a similarity in the kinetics of the *in vivo* production of *Gal*lysin 2 and TNF; optimal induction occurs at 2 h postinjection.[56,57]

TNF can have a regulatory effect on the vertebrate immune system.[49,50] The *Gal*lysin 2 hemokine may have a comparable role in the mediation phase of the induced antibacterial response in *G. mellonella*. A possible source for *Gal*lysin 2 would be the macrophage-like cells (plasmatocytes) described in the wax moth larvae by Chain and Anderson.[60] Chiu[52] has developed a model to suggest how, through the expression of TNF-like activities, the *Gal*lysin 2 hemokine might mediate phenomena that are induced by inoculation of *G. mellonella* with LPS (Figure 2). This model also presents a scheme whereby other hemokines

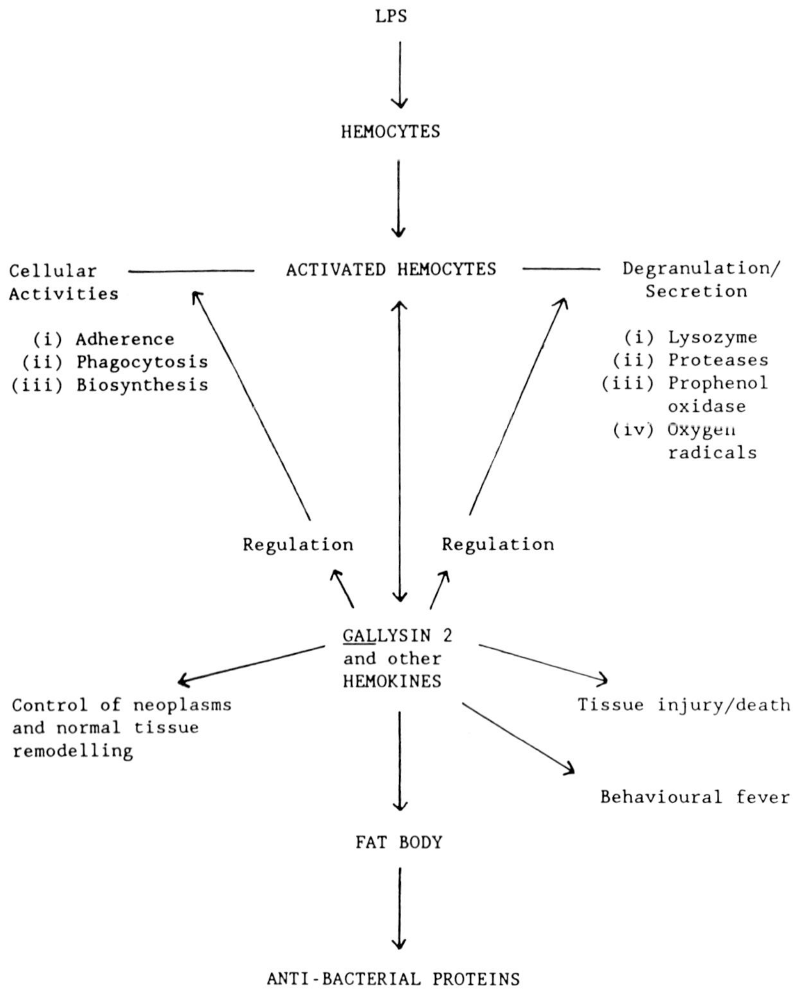

FIGURE 2. A model suggesting the pluripotent effects of *Gal*lysin 2 and other hemokines in the insect immune system.

might regulate the immune response and contribute to its antibacterial, inflammatory, tissue remodeling, and surveillance functions.

Hemocytes, probably plasmatocytes, and insect hemokines such as *Gal*lysin 2 may play important roles in normal tissue homeostasis. Neoplasms have been reported in various insects,[61-65] and endogenous low levels of *Gal*lysin 2 could act as an antitumor agent, like TNF, to maintain homeostasis. The vertebrate analogues of plasmatocytes and *Gal*lysin 2 (e.g., macrophages and TNF) are thought to play important roles in maintaining tissue integrity during aging, injury, and development of cancer.[66] It is tempting to speculate that *Gal*lysin 2 might assist plasmatocytes in their reported roles of phagocytosing cellular and tissue debris,[67] in wound healing, and in the formation of new tissues.[68,69] Both TNF and IL-1 induce fevers that are believed to enhance the defense of vertebrates during infections.[70] Behavioural fevers that have been reported in insects in response to LPS or parasites are thought to benefit the hosts by enhancing their immune responses.[71,72] These examples indicate that hemokines probably play central and crucial roles in the mediation phase of the antibacterial response in insects.

DeVerno et al.[73] showed that cell-free hemolymph from wax moth larvae that had been injected 3 h previously with an immunizing dose of LPS could transfer to normal larvae protection against *Pseudomonas aeruginosa*. Whole hemolymph and hemocytes, collected during the initial stages of the immune response, were also capable of conferring protection. These results suggest that a soluble mediator, a hemokine, produced by hemocytes early in the response was able to stimulate other cells to produce antibacterial factors.

The fat body, the major metabolic organ of insects, is the biosynthetic site for many hemolymph components,[74,75] and it has been shown to be a source of inducible antibacterial activity.[16,76-80] Faye and Wyatt[76] demonstrated that the fat body of immune diapausing pupae of *Hyalophora cecropia* released a soluble form of antibacterial activity. With this in mind and as a consequence of evidence that hemocytes of *G. mellonella* play an early role in the generation of inducible humoral immunity against bacteria, DeVerno et al.[73] proposed that hemocytes of immune insects release a hemokine that can stimulate fat body cells to synthesize antibacterial proteins. It was subsequently shown that *in vitro* mixtures of fat body and cell-free hemolymph from nonimmunized larvae generated a proteinaceous antibacterial activity; this production was enhanced by including in the reaction mixture LPS and/or hemocytes from nonimmunized larvae.[78] The explanation for these results was that LPS, an inducer of immunity, caused the release of a hemocyte factor(s), which acted in conjunction with, or directly on, the fat body resulting in an enhanced production of antibacterial factors.

Dickinson et al.[80] have demonstrated that several tissues of *M. sexta* larvae can be induced by bacterial cell wall peptidoglycan to express high levels of mRNA for antibacterial proteins. However, it was also shown that the main producer of antibacterial proteins, such as lysozyme and cecropin-like peptides, was the fat body. Earlier work by this group supported the hypothesis that the regulation of antibacterial protein synthesis by the fat body required a signal from the hemocytes.[16] Whether the signal is derived exogenously, e.g., a peptidoglycan fragment,[16,44,80] or an endogenous product of stimulated hemocytes,[78] e.g., a hemokine, has not yet been resolved.

Hemocytopenia is a characteristic event of the lepidopteran immune responses.[73,81-83] Following the injection of an immunizing dose of dead bacteria, there is a rapid and selective depletion of plasmatocytes as the total hemocyte count drops. Chain and Anderson[84] identified a hemokine, named plasmatocyte-depletion factor (Table 4), that is released from *G. mellonella* hemocytes, subsequent to their interaction with the bacterial vaccine. Geng and Dunn[83] reported that plasmatocyte depletion in *M. sexta* is not a consequence of wounding, and that an antibacterial response may be elicited with peptidoglycan without effecting plasmatocyte depletion. Although it was shown that evoking a plasmatocyte depletion did not correlate with phagocytosis of particles, there appeared to be a correlation with nodule formation. The actual fate of plasmatocytes is not known but it is speculated that the depletion is the result of nodule formation.

The enhanced synthesis of lysozyme in immunized insects is dose dependent[83] from a threshold number of 10^3 to 10^4 bacteria to a maximum number of 10^7 organisms. This same range of doses produces a dose-dependent depletion of plasmatocytes, which leads to the suggestion that there exists a functional relationship between antibacterial protein synthesis and nodule formation. As plasmatocyte depletion responds to increases in the number of bacteria entering the hemocoel, there is an increase in the production of lysozyme levels to provide a reserve defense in the compromised insect.

Phospholipase C (PLC) activity, determined by a colorimetric assay using *p*-nitrophenylphosphorylcholine as the substrate, is present in low levels in the hemolymph of unimmunized *G. mellonella*.[85] Immunization of the larvae with LPS causes about a 25-fold increase in the PLC activity, and a maximum level is reached over a time period that coincides with the maximum *in vivo* induced protective response and *in vitro* antibacterial activity (Table 2). There is a gradual decline in PLC activity to near normal levels by about 72 h postinjection of the LPS. Over the same 72-h period, there was no increase in PLC level following an injection of endotoxin-free water. The coincidence of maximum levels of induced PLC and antibacterial activity suggests a role for PLC in the effector stage of the protective immune response. However, it has not been determined whether PLC has the role of an effector or a mediator. Chiu et al.[85] have suggested that during the depletion of hemocytes there is cell damage that leads to a release of cellular PLC into the hemolymph. The released PLC could cleave immunologically important membrane-bound material from cell surfaces. The molecules might have an antibacterial effect, or they may act on other hemocytes to enhance the immune system.

The activation events of the phenoloxidase system and the coagulation system that lead to antibacterial effector mechanisms have been reviewed recently by Lackie.[7] Also well described in a recent review are the early events of the synthesis of cecropins in *H. cecropia*.[42] Dickinson et al.[80] have confirmed the earlier partial evidence that *de novo* synthesis of RNA is required for the induction of bactericidins in lepidopteran insects. It was shown that bactericidin transcripts could be detected in the fat body as early as 1 h postinjection of peptidoglycan; the mRNA reached a maximum level between 8 to 12 h and then remained constant up to at least 20 h before starting to decline. This elegant study also provided evidence that the primary control of bactericidin synthesis was probably at the level of gene transcription, and that although the fat body was the main site of bactericidin synthesis, peptidoglycan could induce bactericidin mRNA expression in a variety of other tissues of *M. sexta*.

Although the cecropin family of molecules has been described as having no effect on eukaryote membranes,[42,86] recent observations have provided evidence to the contrary. In addition to cytocidal effects on yeasts and protozoans,[87-89] cecropins have been shown to exhibit a variety of regulatory effects on the differentiation, growth, and function of murine hemopoietic cells.[17] In light of these regulatory functions, it is possible that cecropins may not function solely as effector proteins in the lepidopteran immune response, but could also exert some mediator activity.

VI. EFFECTORS

The end products of activation of the mediation phase are summarized in Figure 1 and include (1) increased LD_{50}, (2) antibacterial activity in cell-free hemolymph, (3) production of low molecular weight antibacterial proteins, and (4) nodule formation. This review will focus on antibacterial activity in cell-free hemolymph and the substances involved therein.

Molecular characterization of antibacterial factors was initiated by the Swedish group led by Boman, and several descriptive reports have been published.[5,90,91] These investigators have largely used the pupae of *H. cecropia* and *S. cynthia,* and nonpathogenic living bacteria as inducing agents. Their research has concentrated on three groups of proteins: cecropins, attacins, and lysozyme. Cecropins are a family of strongly basic proteins with molecular weight of 4 kDa. Up to seven proteins have been classified as members of the cecropin family and designated A through F as well as a factor G.[92] The three major cecropins are designated A, B, and D.[6,93] They are described as having potent antibacterial activity against certain Gram-negative bacteria; the greatest activity always appears to be demonstrable against *Escherichia coli* strain D21, which is nonpathogenic for *Galleria* and most other lepidopterans. The structure of cecropins has been confirmed by synthesis, cecropin A by Andreu et al.[94] and cecropin D by Fink et al.[95] Cecropin D[96] was synthesized by solid-phase methods and was indistinguishable from natural cecropin D. When analogues were prepared, it was suggested that a strongly basic NH_2-terminal segment was a prerequisite for antibacterial activity. Conformational analyses and circular dichroism measurements of natural and synthetic cecropins[95,96] reveal a basic helical amphipathic amino terminal segment that was connected to a hydrophobic helical carboxy terminal region by a flexible nonhelical hinge region.

Recently, precursor forms of cecropins A and B have been synthesized by solid-phase methods, and these consist of 62 to 64 residues.[97] The three genes responsible for cecropin production are believed to have arisen through gene duplication from a common ancestor.[95] Cecropin-like substances with similar activities may occur in other insect orders. An inducible antibacterial factor occurring in the hemolymph of *Rhodnius prolixus* was suggested as being similar to cecropins by De Azambuja et al.[98] Interestingly, Lee et al.[99] have proposed that insect cecropins may have a relation to mammalian cecropins. A high proline content in such substances is presumed to be necessary for bacteriolytic activity of cecropins and other peptide antibiotics.[100] Lee et al.[99] have isolated antibacterial peptides from pig intestine and identified them as mammalian cecropins; these peptides displayed activity against both *E. coli* and *Bacillus megaterium,* and Lee et al.[100] propose their importance as possible defense agents in the intestine. Hybrid peptides containing sequences for antibacterial peptide, cecropin A, and from bee venom toxin, mellittin, had high activity against *Staphylococcus aureus.*[91]

Attacins have larger molecular weights than cecropins (20 to 23 kDa) and show good antibacterial immunity only against *E. coli* strain D21 and two other Gram-negative bacteria, *Acinetobacter calcoaceticus* and *Pseudomonas maltophilia,* isolated from silkworm gut.[101,102] The activity was found to be against the outer membrane of *E. coli* and is believed to facilitate the activity of cecropins and lysozyme.[102] Sequences have been discovered for the attacins, and it is believed the attacin genes have arisen through gene duplication and have given rise to six attacins.[101] Substances found in honeybees, and which were originally thought to resemble attacins, were unique in having a high proline content.[103] These were called apadecins and were postulated to have bacteriostatic rather than bactericidal activity. In the Diptera, a substance called Diptericin A presumed to be different from all other antibacterial substances in Diptera[104] was found to have activity against *E. coli* similar to that exhibited by the attacins (see Chapter 12).

Defensins may represent another important link between the vertebrate and invertebrate immune system. They were first identified as microbiocidal cationic peptides in rabbit alveolar macrophages,[105] but were later isolated from human and rabbit neutrophils.[106] These peptides are composed of 29 to 34 amino acid residues and are rich in cysteine. Defensins have not been reported in Lepidoptera but have been reported in two dipteran insects.[107,108]

Lysozyme is a naturally occurring protein found in the hemolymph of most lepidopterans. It increases during, but persists subsequent to, an immune response. By itself, lysozyme acts only on Gram-positive bacteria or on *E. coli* that are defective in cell division.[4] Lysozyme tends to increase at the same rate as the development of protective immunity, but persists well beyond the duration of protective immunity or bactericidal activity.[109] Glinski and Jarosz[26] reported that, after feeding viable *P. aeruginosa* cells to *Galleria,* a negligibly elevated level of lysozyme activity appeared. Jarosz et al.[110] had also reported that in the same insect any slight disturbance in body integrity activated the defense system, thus markedly increasing hemolymph lysozyme activity. This supports the findings of Trenczek and Faye[79] that injury to *H. cecropia* pupae would result in the synthesis of some immune proteins by the fat body. Young[111] has suggested that the conservation of certain stress proteins throughout the animal kingdom may be related to broad resistance to pathogens. While insects were not specifically mentioned in this review, it is tempting to speculate that such proteins could exist within the Lepidoptera. Though there is fairly good evidence that lysozyme is not the prime substance involved in antibacterial immunity, it undoubtedly has a role to play. Very likely, other antibacterial proteins that are currently being reported in the literature are probably related to one of these three groups. For example, Japanese workers headed by Shiba[112] referred to ''lepidopterans'' as substances which arise in *B. mori* and which have considerable similarities to cecropins. In 1990, another group of investigators in Japan headed by Morishima[19] reported the isolation and structure of cecropins in the silkworm. These cecropins were purified by two-step chromatography and separated into two major and two minor forms by reverse-phase HPLC. The two major cecropins had structures similar to cecropins of *Hyalophora.* An unusual finding was the occurrence of δ-hydroxylysine in three of the four cecropins. Greatest antibacterial activity was exhibited against *E. coli* and the least against staphylococci.

Recently, Anderson et al.[113] compared the humoral response to injection of bacteria in the cabbage looper, *Trichoplusia ni,* to that in *H. cecropia.* The induced proteins were found to be related to cecropins, attacins, and lysozyme in *Hyalophora.* The antibacterial response was elicited effectively by bacteria, while the injection of *Autographa californica* virus had no greater inducing effect than an injection of Ringer's solution.

Because of the varying antibacterial activity of the different inducible humoral factors, it is apparent that no individual factor is capable of protecting the insect host from all kinds of infections. In view of this, Boman and Hultmark[42] proposed that attacins, cecropins, and lysozyme may act synergistically to constitute a multicomponent humoral defense mechanism. Various investigators including Faye and Wyatt[76] and Dunn[16] as well as DeVerno[78] in our laboratory, have presented evidence that these antibacterial substances are synthesized by the fat body.

Hemolymph-derived activities that are capable of lysing vertebrate erythrocytes are found in many classes of invertebrates, including echinoderms[114,115] mollusks,[116-118] annelids,[119-122] and arthropods.[123] There is substantial variation in the activities and biochemical properties of these lysins, and rarely has there been an association with the animal's immune response system.[45,124] The recognition that invertebrate lysins and vertebrate complement can lyse similar target erythrocytes has led to the speculation that both systems are phylogenetically related.[116,125-127]

Although no direct analogy has been demonstrated, there is convincing evidence to suggest that some invertebrates, including insects, have complement-like factors[53,128-133] (see Chapter 14). Evidence that could support any notion that the vertebrate complement system and the invertebrate complement-like factors share a common ancestry is indirect and at best tentative.

There exists in the cell-free hemolymph of *G. mellonella* a factor or factors that can react with the cobra venom factor to produce an enzymic activity capable of cleaving the α-chain of the third component of mammalian complement, C3.[133] Investigating the hemolytic potential of this system, we discovered a hemolysin that was unrelated to the C3-cleaving activity.[53,134] The lysin is present in normal and immune hemolymph of *G. mellonella* larvae and is active against sheep, human, guinea pig, and rabbit erythrocytes. The low level of hemolytic activity that exists in normal larvae is increased by vaccination with formalized *P. aeruginosa,* and the increase and subsequent decline parallels the induced *in vivo* antibacterial activity (Table 2). The active material has been purified and is a protein with an apparent molecular weight of 70 to 75 kDa.[53] Although the hemolytically active material was originally referred to as HAM, *Gal*lysin 1 is now used to describe this *Galleria*-derived lysin.

*Gal*lysin 1 is distinct from other inducible lytic activities found in cell-free hemolymph of insects, e.g., lysozyme, attacins, cecropins, phospholipases, and *Gal*lysin 2.[4,7,42,44,45,51,53,85,135] *Gal*lysin 1 is also distinct from the other reported invertebrate hemolysins, most of which are labile at 56°C and/or calcium dependent. Stable at 56°C, *Gal*lysin 1 functions effectively in the presence of EDTA and has a molecular weight higher than the other isolated invertebrate hemolysins.[45,120]

P. aeruginosa, E. coli, and *Proteus mirabilis* are killed by cell-free hemolymph from vaccinated *G. mellonella* larvae as shown in Table 5. Purified *Gal*lysin 1 alone and purified *G. mellonella* lysozyme alone have no antibacterial activity for these organisms,[134] but together they have effective antibacterial activity (Table 5). Substitution of insect-derived

TABLE 5
Antibacterial Activity of *Gal*lysin 1

Test sample	Killing *in vitro* of bacterial targets[a] (%)		
	P. aeruginosa	*E. coli*	*P. mirabilis*
Hemolymph	>90	>90	40—50
*Gal*lysin 1	<5	<5	<5
G. mellonella lysozyme	<5	<5	<5
Egg white lysozyme	<5	ND[b]	ND[b]
G. mellonella lysozyme plus *Gal*lysin 1	60—65	60—65	30—40
Egg white lysozyme plus *Gal*lysin 1	40—45	ND[b]	ND[b]

[a] An overnight bacterial culture of 50 μl of a 10^{-5} dilution was mixed with a test sample of 100 μl incubated at 28°C for 40 min, then brought to 1.0 ml with ice-cold saline, and plated (0.1 ml) on tryptic soy broth agar. After incubation overnight at 37°C the colonies were counted. Antibacterial activity was expressed as the percent killed with respect to saline controls (0% killing). Lysozyme was made to a final concentration of 250 units/ml.

[b] ND, not done.

lysozyme by an avian egg white lysozyme was also effective when incubated with the bacteria and *Gal*lysin 1 at 28°C; a more potent effect was seen at 37°C. Presumably the relative effects of *Gal*lysin 1 and lysozyme in this interdependent killing are the lysis of the peptidoglycan by the lysozyme and a membrane attack by *Gal*lysin 1. The involvement of other hemolymph factors in this antibacterial activity is unlikely, since *Gal*lysin, purified to a single protein,[136] is effective in the presence of lysozyme from a nonhemolymph source. The concomitant increase in the activities of *Gal*lysin 1 and lysozyme to their maximum levels, at the peak of the *in vivo* protective response, in vaccinated larvae suggests that they play an important role in the insects' defense armory. The mechanism whereby *Gal*lysin 1 attacks bacterial membranes is as yet unknown, as is its relationship, if any, to the well-described action of attacins and cecropins. In our initial report of this hemolytic activity, we suggested that the functional activity and molecular size of *Gal*lysin 1 are related to the functional activities and molecular sizes of the cytolysins of mammalian complement and cytotoxic T-lymphocyte systems.[53,137] This is still a tentative hypothesis, but it does indicate that links may exist between the vertebrate complement system and the invertebrate defense mechanisms.

As mentioned earlier, the concept of certain nonspecific antibacterial factors in invertebrates being phylogenetically related to one or more of the opsonic, phlogistic, lytic, or control components of the vertebrate complement system has been proposed by many investigators. The recent findings of Leonard et al.[138] and Enghild et al.[139] provide some evidence to support this concept.

The coelomic fluid of *Asterias forbesi* possesses a hemolytic activity that is antibacterial.[138] The data suggest that coelomic fluid contains components which might constitute a primitive alternative pathway in which a heat-stable protein acts as a substrate for a heat-labile serine protease. This is reminiscent of some recent work in our own laboratory that demonstrated the generation of a cobra venom factor (CVF)-induced C3 convertase in the cell-free hemolymph of *G. mellonella*.[133] This C3-convertase activity was detected by the enzymic cleavage of the α chain of the third component of the mammalian complement system, C3, in a manner analogous to the CVF-induced alternative pathway C3-convertase of vertebrate complement, i.e., CVF Bb.[140] Since CVF was previously shown to inhibit both the development of the *in vivo* immune response in vaccinated *G. mellonella* and the *in vitro* antibacterial response, we postulate that a natural substrate for the insect derived C3-convertase enzyme is probably a hemolymph factor(s) that is required for the development and expression of immunity. The hemolymph factors that react with CVF and the natural substrate of the induced enzyme are most likely related, functionally, to Factor B and C3 of the mammalian alternative pathway. It is interesting to note that one of early indications that invertebrate complement-like activities existed was a report that *A. forbesi* coelomic fluid reacted with CVF to generate a cleaved product of vertebrate C3 that had activity similar to the α chain derived anaphylatoxin.[141]

The pivotal component in both the classical and alternative complement pathways is C3.[140] This protein belongs to a group of molecules, consisting of the α_2-macroglobulin family of proteinase inhibitors and the complement components C3, C4, and C5 that probably arose from a common ancestor.[139,142] These proteins have several stretches of amino acid sequence similarity and many of them, including α_2-macroglobulin, C3, and C4 have an internal β-cysteinyl-γ-glutamyl thiol ester.[142] Enghild et al.[139] have shown that hemolymph from *Limulus polyphemus* exhibits complement-like and proteinase activities (see also Chapter 10). Furthermore, both of these activities are associated with a thiol ester containing α-macroglobulin that is referred to as limac, a name which describes the *Li*mulus α_2-*ma*croglobulin *c*omplement-like protein. Limac appears to be a component of a complement-like lytic system. An α_2-macroglobulin with an internal thiol ester bond has been isolated from another marine invertebrate, *Homerus americus*[143] (see also Chapter 11). There are, as yet, no reports of α_2-macroglobulin analogues in insects. The α_2-macroglobulin is a highly conserved protein[142] that is able to bind and modify the action of a wide range of physiologically important molecules, including several cytokines of the immune system.[66] The concept of an insect α_2-macroglobulin analogue modifying the course and activity of the induced immune response, and possibly the functions of hemokines, is plausible, particularly in the light of evidence for the existence of C3 analogues.

While there has been considerable research focus and progress on the structure of the effector substances and their genetic aspects, the end result of the effector phase, which represents yet another parameter of immunity, i.e., mechanisms involved in destruction of the pathogen, is not well understood.

Damage to bacteria results very rapidly when they are exposed to hemolymph from insects injected with homologous *Pseudomonas* or *Proteus* vaccine. A similar result was found when bacteria were exposed to hemolymph from insects injected with CSF. Figures 3a and 3b represent typical electron micrographs depicting the type of damage occurring when *Pseudomonas* or *Proteus* are exposed to hemolymph from insects immunized by CSF. A normal cell of *P. mirabilis* is shown in Figure 3a, and a bacterium of the same species with extensive damage after 15-min exposure to immune hemolymph is depicted in Figure 3b. Destruction of *Pseudomonas* was more rapid and complete in 7 min.[144] These electron micrographs at best indicate only gross damage though a membrane attack is apparent. Steiner et al.[145] constructed liposomes to investigate the activity of some cecropins. They demonstrated that liposomes were disrupted by cecropins and proposed that bacterial lysis may be the result of similar membrane attack by cecropins on the phospholipid layer. Such observations were confirmed by Jaynes[88] who demonstrated in electron micrographs that large pores were found on membranes of bacteria after treatment with cecropins. We have initiated some studies utilizing enzyme assays, indicating specific enzyme leakage to help us pinpoint target sites and determine progressive damage to various cell structures.

VII. INHIBITION OF EFFECTOR ACTIVITY

The bacterial species *Serratia marcescens* served as a model in our laboratory in order to focus on a situation where some indicators of the immune response seem to occur, but not all parameters are met, and where protective immunity does not occur.[31] As experimental strains, we used a highly virulent strain of *Serratia,* designated *S. marcescens* Rh (LD_{50} 1.0); a strain of intermediate virulence, *S. marcescens* Br (LD_{50} 2×10^5); and a nonpathogenic strain presumed to be *Serratia* but though closely related, later positively identified as *Enterobacter cloacae*. We were unable to achieve protective immunity against *Serratia* strain Rh, although newly synthesized basic proteins similar to those found in the immune response to *Pseudomonas* were induced with *Serratia* vaccine. Interestingly, it was possible to raise the LD_{50} of *Serratia* strain Br by two logs after immunization. Evans[31] in our laboratory investigated the interrelation of the virulence of *Serratia* to the lack of immune response against it, that is the failure to achieve protection against highly virulent strains of *Serratia*. Protease production has been correlated with virulence in human *Serratia* infections,[146] and similarly some years ago Grimont and Grimont[147] and Lysenko[148] presented evidence that enzymes of *Serratia* are the toxic factors for insects. Evans[31] isolated and purified a protease from Rh strain with the characteristics shown in Table 6. This protease is unique because most *Serratia* do not produce a single polypeptide protease and also because it is reactivated by ferric ions. *Serratia* strain Br did not produce protease. As basic proteins are considered antibacterial and hence protective, it appeared feasible that perhaps immunity was not achieved because protease rendered these proteins inactive. Flyg and Xanthopoulos[149] had reported that proteases of *S. marcescens* nonspecifically cleaved hemolymph proteins of *Cecropia,* including "the immune proteins", presumably the cecropins. In addition to having a low LD_{50}, 10 μl of the isolated protease of this strain Rh was toxic to normal larvae but not to vaccinated larvae. Despite their resistance to the protease of this strain, vaccinated larvae were not protected against the whole bacteria. This suggests a complex virulence pattern for *Serratia*.

Evans[31] further demonstrated that hemolymph from *Serratia* vaccinated *Galleria* did not maintain this pattern if exposed to protease *in vitro*. Not only did this protease exhibit activity against the basic proteins synthesized on induction by vaccine, but also some of the proteins occurring in normal hemolymph were degraded. On the basis of these findings, one might speculate that if a bacterium possesses enzymes that can destroy these immune proteins, then protective immunity will not occur and it will not be possible to demonstrate antibacterial immunity. In other words, an organism which possesses a virulence factor which can circumvent these proteins will flourish.

Following this hypothesis a bit further, Evans[31] isolated a mutant of her *Serratia* Rh strain, which was protease deficient. Apart from protease production the organism appeared identical to the parent strain and had an LD_{50} of 2.3 for normal insects also like that of the wild type. However, immunized insects fared no better against this organism than they did against the wild strain. In this instance, there was no protease to destroy immune proteins; thus the explanation for lack of protection appears to be more complex than mere degradation of these proteins.

There are various ways in which pathogenic bacteria, such as *Serratia,* might circumvent the induced, potentially antibacterial proteins, and some possibilities are suggested in Table 7. With respect to *Serratia,* mechanism #3 would have been expected to be a possible explanation, but the results of Evans[31] suggest this may not be so. Another possibility is that the outer membrane of this strain of *Serratia* may be so constructed that the proteins cannot enter. Again, the organism may not possess suitable receptors for activity of antibacterial factors, or, otherwise the communication system is silent for want of a receptor.

VIII. CONCLUSIONS

This brief consideration of antibacterial immunity in Lepidoptera indicates that the defense system of Lepidoptera has many facets. Although much has been accomplished at the biochemical and molecular levels, there is more we still need to investigate. Further attention must be paid to mechanisms of recognition as well as to the potential significance of cooperation among effectors, particularly how newly reported substances, such as *Gal-*

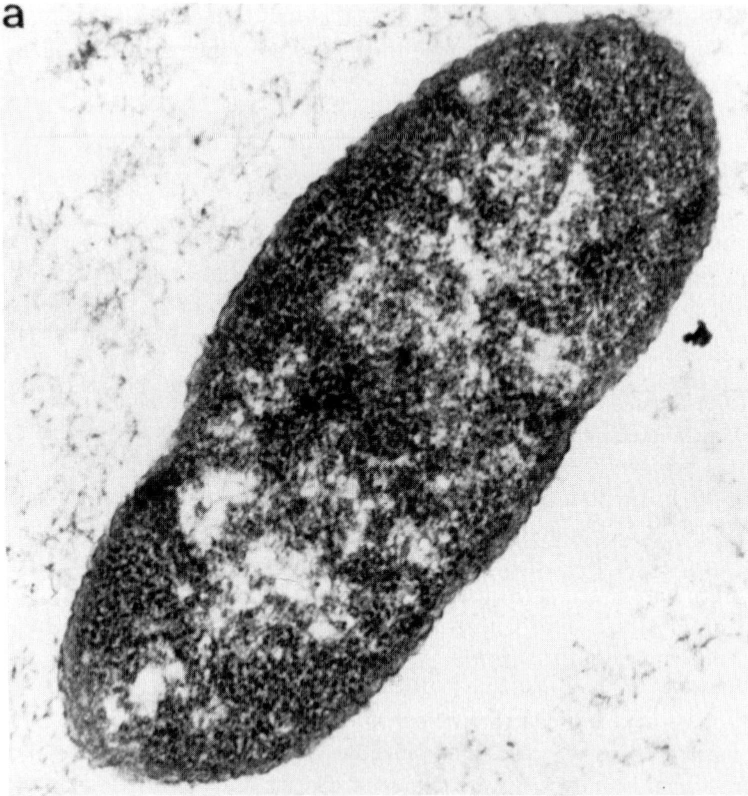

FIGURE 3. Transmission electron micrographs of *Proteus mirabilis,* strain 11-1A. (a) Untreated cell (× 57,724). (b) Cell exposed for 15 min to hemolymph from larvae of *Galleria mellonella* injected 18 h previously with cane sugar factor. Arrowheads indicate damaged areas. Note the loss of cell contents (× 43,750).

TABLE 6
Some Characteristics of a Protease Isolated from *Serratia marcescens* Strain Rh

A single subunit 56,000 Da protein by SDS-PAGE
Isoelectric point, 6.00
Optimum activity at pH 9.5
Completely inactivated by EDTA
Reactivated by ferric ions
Sensitive to temperature above 50°C
Otherwise stable and nonautodegradative

TABLE 7
Possible Mechanisms by Which a Bacterial Strain May Circumvent Activity of Induced Products with Potential Antibacterial Activity

Toxigenicity of strain
  Endotoxins
  Exotoxins
Activation of host factors to cause self-damage
Inhibition of antibacterial factors
Structural features of bacterial strain leading to insusceptibility to factor(s)
Recognition repertoire of insect defense system does not accommodate structural features of bacteria

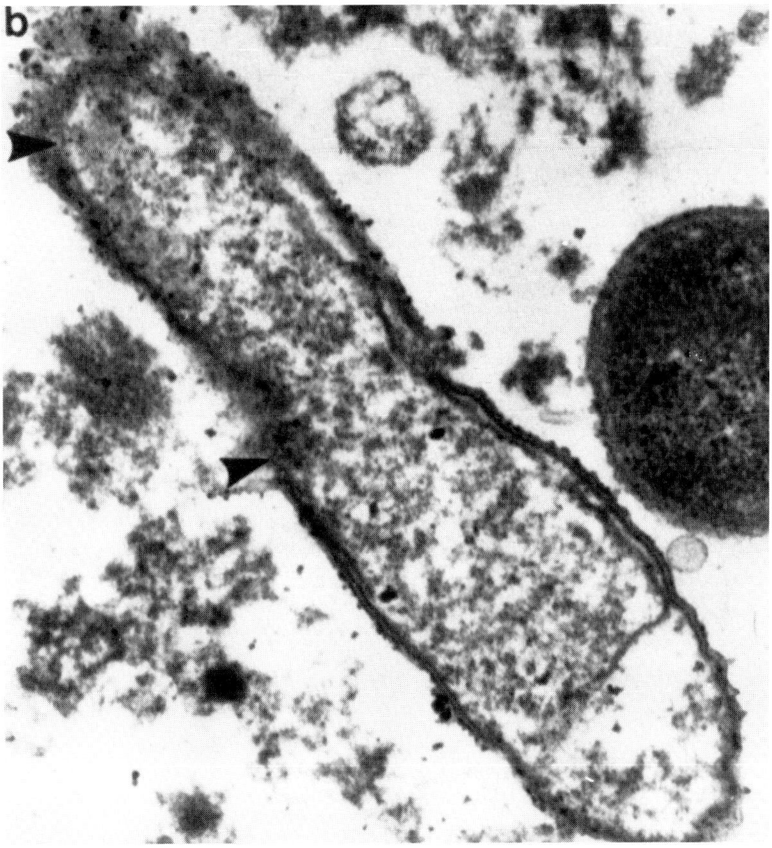

FIGURE 3b.

lysins 1 and 2 might interact with earlier established substances, such as cecropins and attacins.

Most immunologists agree that invertebrates exist in a milieu of pathogenic organisms without possessing a sophisticated immune system. The antibacterial substances discovered in the Lepidoptera serve as good examples of the contention of Reinisch and Litman[150] that efficient defense mechanisms do exist in invertebrates and that possibly some of these are predictive of more complex immunologic functions in higher vertebrates. These authors go so far as to speculate that active substances in invertebrates actually mimic the immunologic activity of antibody molecules.

While care must be taken not to draw analogies from the mammalian system, it is probably fair to say that as in the cellular immune system of mammals, there are a number of inhibitors, promoters, and suppressors and cells with these various functions which must be examined in conjunction with the effector substances. It is only then that we may understand why we get different responses to different bacterial species and why different responses to pathogenic and nonpathogenic bacteria might be expected. Moreover, there are some recent investigators who have evidence to show that inducible antibacterial factors are not solely responsible for defense. Researchers such as Matha[151] suggest that it would be valuable to investigate induced hydrolases with respect to cellular and humoral aspects of the defense mechanism, believing proteins such as cecropins and attacins represent only a part of insect humoral immunity. Clearly, there are many remaining avenues providing stimulating challenges. An increased attention to such areas would not only contribute to better understanding of immune mechanisms in insects at the molecular level but also, as indicated by Cooper,[152] produce information for comparative immunology, which is sadly lacking in interest. Moreover, such information might well point to the possible use of insects as new animal models in many aspects of research, particularly as candidates for testing evolutionary models, a position taken by Law and Wells,[75] because of the large variations among insect species.

IX. SUMMARY

The immune response in Lepidoptera is generally short-lived and nonspecific. It is characterized by an increase in LD_{50} against pathogenic bacteria, antibacterial activity in cell-free hemolymph, production of low molecular weight proteins, and nodule formation. Antibacterial immunity is complex, both cellular and humoral factors appearing to be involved. Demonstrable antibacterial immunity against various Gram-negative bacteria occurs after induction by avirulent bacteria, or attenuated virulent bacteria or their products. Recently, CSF, a nonbacterial product, has been shown to induce antibacterial immunity and the accompanying parameters of immunity in *Galleria mellonella.* The events that occur between the induction of an immune response and the expression of antibacterial activity are probably mediated, in most instances, by soluble cell products referred to as invertebrate cytokines or hemokines.

Basic proteins referred to as cecropins or attacins are probably responsible for most antibacterial activity. These substances have been well characterized in cecropia moths and have been cloned and synthesized; they have been reported to be related to similar antibacterial peptides occurring in mammals. In addition to antibacterial immunity in *G. mellonella,* two substances referred to as *Gal*lysin 1 and *Gal*lysin 2 both inducible by LPS occur at particular stages of the immune response in *G. mellonella. Gal*lysin 1 is a hemolytic factor which in conjunction with lysozyme is antibacterial against Gram-negative pathogens and while present in small quantities in unimmunized larvae, increases following the profile of the immune response curve. *Gal*lysin 2 occurs early in the response and is cytoxic for two human tumor cell lines and may be a lepidopteran analogue of TNF.

The response in *G. mellonella* is not the same to all Gram-negative pathogens. Although similar low molecular weight proteins are induced against *S. marcescens,* no antibacterial activity occurs. Possible reasons for this may be related to virulence factors of the bacterium.

ACKNOWLEDGMENTS

We thank Ms. Marilyn Menger and Dr. Myron Szewczuk, Department of Microbiology and Immunology, Queen's University at Kingston, Ontario for their assistance in the preparation of the reference section. The research in our laboratories is supported by the Natural Sciences and Engineering Research Council of Canada. Recent graduate students involved in our investigations reported in this chapter are Pamela DeVerno, David Morton, Michael Pryce, Elizabeth Evans, David Phipps, and Jonathan Chiu.

REFERENCES

1. **Lackie, A. M.,** Invertebrate immunity, *Parasitology,* 80, 393, 1980.
2. **Ratcliffe, N. A., White, K. N., Rowley, A. F., and Walters, J. B.,** Cellular defense systems of the Arthropoda, in *The reticuloendothelial systems; phylogeny and ontogeny.* Vol. 3, Cohen, N. and Sigel, M. M., Eds., Plenum Press, New York, 1982, 167.

3. **Cooper, E. L.,** Overview of humoral factors in invertebrates, *Dev. Comp. Immunol.,* 9, 577, 1985.
4. **Chadwick, J. S. and Dunphy, G. B.,** Antibacterial and antiviral factors in arthropod hemolymph, in *Hemocytic and Humoral Immunity in Arthropods,* Gupta, A. P., Ed., John Wiley & Sons, New York, 1986, 287.
5. **Boman, H. G. and Hultmark, H. G.,** Cell free immunity in insects, *Trends Biochem. Sci.,* 6, 306, 1981.
6. **Boman, H. G., Faye, I., Hofsten, P. V., Kockum, K., Lee, J-Y., Xanthopoulos, K. G., Bennich, H., Engstrom, A., Merrifield, B. R., and Andreu, D.,** Antibacterial immune proteins in insects — a review of some current perspectives, in *Immunity in Invertebrates,* Brehelin, M. Ed., Springer-Verlag, Berlin, 1986, 63.
7. **Lackie, A. M.,** Immune mechanisms in insects, *Parasitol. Today,* 4, 98, 1988.
8. **Dunn, P. E.,** Antibacterial defensive responses of the tobacco hornworm *Manduca sexta, J. Cell. Biochem.,* Suppl. 10 Part C, 65, 1986.
9. **Klein, J.,** Are invertebrates capable of anticipatory immune responses?, *Scand. J. Immunol.,* 29, 499, 1989.
10. **Marchalonis, J. J.,** Ancient problems and new beginnings, in *Defense molecules,* Marchalonis, J. J. and Reinisch, C. L., Eds., Alan R. Liss, New York, 1990, 295.
11. **Duwel-Eby, L. E. and Karp, R. D.,** The inducible humoral immune response to soluble proteins in the American cockroach, in *Defense Molecules,* Marchalonis, J. J. and Reinisch, C. L. Eds., Alan R. Liss, New York, 1990, 63.
12. **Hildemann, W. H., Clark, E. A., and Raison, R. L.,** Phylogeny of immunocompetence, in *Comprehensive immunogenetics,* Elsevier, New York, 1981, 302.
13. **Chadwick, J. S. and Aston, W. P.,** Effector mechanisms involved in the protective response in *Galleria mellonella* towards bacterial pathogens, *Proc. Int. Colloq. Invertebr. Pathol.,* 1976, 204.
14. **Boman, H. G., Nilsson-Faye, I., Paul, K., and Rasmuson, T.,** Insect Immunity. I. Characteristics of an inducible cell-free antibacterial reaction in hemolymph of *Samia cynthia* pupae, *Infect. Immun.,* 10, 136, 1974.
15. **Dunn, P. E. and Drake, D. R.,** Fate of bacteria injected into naive and immunized larvae of the tobacco hornworm *Manduca sexta, J. Invertebr. Pathol.,* 41, 77, 1983.
16. **Dunn, P. E., Dai, W., Kanost, M. R., and Geng, C.,** Soluble peptidoglycan fragments stimulate antibacterial protein synthesis by fat body from larvae of *Manduca sexta, Dev. Comp. Immunol.,* 9, 559, 1985.
17. **Okai, Y. and Qu, X.,** Effects of a purified cecropin D from a Chinese silk moth on growth, function and differentiation of murine hemopoietic cells, *Immunol. Lett.,* 20, 127, 1989.
18. **Yoshida, H. and Ashida, M.,** Microbial activation of two serine enzymes and prophenoloxidase in the plasma fraction of hemolymph of silkworm, *Bombyx mori, Insect. Biochem.,* 16, 539, 1986.
19. **Morishima, I., Suginaka, S., Ueno, T., and Hirano, H.,** Isolation and structure of cecropins, inducible antibacterial peptides, from the silkworm, *Bombyx mori, Comp. Biochem. Physiol.,* 95B, 551, 1990.
20. **Yoshida, S., Yamashita, M., Yonehara, S., and Eguchi, M.,** Properties of fungal protease inhibitors from the integument and haemolymph of the silkworm and effect of an inhibitor on the fungal growth, *Comp. Biochem. Physiol.,* 95B, 559, 1990.
21. **Anon.,** *ASM News,* 56, 460, 1990.
22. **Chadwick, J. S.,** Induction and effector mechanisms of insect immunity, *Comp. Pathobiol.,* 3, 85, 1977.
23. **Boman, H. G., Faye, I., Pye, A., and Rasmuson, T.,** The inducible immunity system of giant silk moths, *Comp. Pathobiol.,* 4, 145, 1978.
24. **Boman, H. G., Boman, A., and Pigon, A.,** Immune and injury responses in *Cecropia* pupae — RNA isolation and comparison of protein synthesis *in vivo* and *in vitro, Insect. Biochem.,* 11, 33, 1981.
25. **Chadwick, J. S. and Vilk, E.,** Endotoxins from several bacterial species as immunizing agents against *Pseudomonas aeruginosa* in *Galleria mellonella, J. Invertebr. Pathol.,* 13, 410, 1969.
26. **Glinski, Z. and Jarosz, J.,** Efforts to induce defence responses in the greater wax moth larvae by oral feeding of insect pathogenic bacteria, *Comp. Biochem. Physiol.,* 85, 673, 1986.
27. **Morton, D. B.,** An investigation of some parameters of the immune response of *Galleria mellonella* to *Proteus mirabilis,* M.Sc. thesis, Queen's University, Kingston, Ontario, Canada, 1987, 1.
28. **Pryce, M. J.,** Cane sugar factor as a putative inducing agent of immunity in *Galleria mellonella:* a partial characterization, M.Sc. thesis, Queen's University, Kingston, Ontario Canada, 1987, 1.
29. **Chadwick, J. S.,** Serological responses of insects, *Fed. Proc.,* 26, 1675, 1967.
30. **Chadwick, J. S.,** Antibacterial immunity in *Lepidoptera, Symposium VIII S3, Proc. 18th Int. Congr. Entomol.,* Vancouver, Canada, 1988, 253.
31. **Evans, E. B.,** An investigation of the insect pathogenic properties of *Serratia marcescens* in relation to the immune response of *Galleria mellonella,* M.Sc. thesis, Queen's University, Kingston, Ontario, Canada, 1987, 1.
32. **Dunn, P. E. and Dai, W.,** Bacterial peptidoglycan: a signal for initiation of the bacteria-regulated synthesis and secretion of lysozyme in *Manduca sexta,* in *Defense Molecules,* Marchalonis, J. J. and Reinisch, C. L., Eds., Alan R. Liss, New York, 1990, 33.

33. **Kanost, M., Dai, W., and Dunn, P.,** Peptidoglycan fragments elicit antibacterial protein synthesis in larvae of *Manduca sexta, Arch. Insect Biochem. Physiol.,* 8, 147, 1988.
34. **Ashida, M.,** A cane sugar factor suppressing activation of prophenoloxidase in hemolymph of the silkworm, *Bombyx mori, Insect Biochem.,* 11, 57, 1981.
35. **Pryce, M. J., Aston, W. P., and Chadwick, J. S.,** Cane sugar factor as an inducing agent of immunity in *Galleria mellonella, Dev. Comp. Immunol.,* 14, 369, 1990.
36. **Vasta, G. R.,** Invertebrate lectins, C-reactive proteins and serum amyloid-structural relationships and evolution, in *Defense Molecules,* Marchalonis, J. J. and Reinisch, C. L., Eds., Alan R. Liss, New York, 1990, 183.
37. **Nappi, A. J.,** The role of melanization in the immune reaction of larvae of *Drosophila algonquin* and *Pseudocoila bochei, Parasitology,* 66, 23, 1973.
38. **Soderhall, K.,** Prophenoloxidase activating system and melanization — a recognition mechanism of arthropods? A review, *Dev. Comp. Immunol.,* 6, 601, 1982.
39. **Leonard, C., Ratcliffe, N. A., and Rowley, A. F.,** The role of prophenoloxidase activation in non-self recognition and phagocytosis by insect blood cells, *J. Insect. Physiol.,* 31, 789, 1985.
40. **Brookman, J. L., Ratcliffe, N. A., and Rowley, A. F.,** Optimization of a monolayer phagocytosis assay and its application for studying the role of the prophenol oxidase system in the wax moth *Galleria mellonella, J. Insect Physiol.,* 34, 337, 1988.
41. **Dunphy, G. B. and Chadwick, J. S.,** Effects of selected carbohydrates on the prophenoloxidase cascade system to the adhesion of strains of *Pseudomonas aeruginosa* and *Proteus mirabilis* to hemocytes of nonimmune larvae of *Galleria mellonella, Can. J. Microbiol.,* 35, 524, 1989.
42. **Boman, H. G. and Hultmark, D.,** Cell-free immunity in insects, *Annu. Rev. Microbiol.,* 41, 103, 1987.
43. **Chadwick, J. S.,** Hemolymph changes with infection or induced immunity in insects and ticks, in *Invertebrate Immunity,* Maramorosch, K. and Shope, R. E., Eds., Academic Press, New York, 1975, 241.
44. **Dunn, P. E.,** Biochemical aspects of insect immunology, *Annu. Rev. Entomol.,* 31, 321, 1986.
45. **Ratcliffe, N. A., Rowley, A. F., Fitzgerald, S. W., and Rhodes, C. P.,** Invertebrate immunity: basic concepts and recent advances, *Int. Rev. Cytol.,* 97, 183, 1985.
46. **Beck, G., O'Brien, R. F., and Habicht, G. S.,** Invertebrate cytokines: the phylogenetic emergence of interleukin-1, *BioEssays,* 11, 62, 1989.
47. **Beck, G., Vasta, G. R., Marchalonis, J. J., and Habicht, G. S.,** Characterization of interleukin-1 activity in tunicates, *Comp. Biochem. Physiol.,* 92B, 93, 1989.
48. **Beck, G. and Habicht, G. S.,** Isolation and characterization of a primitive interleukin-1-like protein from an invertebrate, *Asterias forbesi, Proc. Natl. Acad. Sci. U.S.A.,* 83, 7429, 1986.
49. **Mantovani, A. and Dejana, E.,** Cytokines as communication signals between leukocytes and endothelial cells, *Immunol. Today,* 10, 370, 1989.
50. **Le, J. and Vilcek, J.,** Biology of disease — tumor necrosis factor and interleukin 1: cytokines with multiple overlapping biological activities, *Lab. Invest.,* 56, 234, 1987.
51. **Chiu, J. K. C., Aston, W. P., and Chadwick, J. S.,** Induction of a cytotoxic factor in the hemolymph of *Galleria mellonella* by a subimmunizing dose of lipopolysaccharide (LPS), *Soc. Invertebr. Pathol. (Annu. Meet.),* 22, (Abstr.), 97, 1989.
52. **Chiu, J. K. C.,** A novel hemolymph-derived cytotoxic activity in *Galleria mellonella,* M.Sc. thesis, Queen's University, Kingston, Ontario, Canada, 1990, 1.
53. **Phipps, D. J., Chadwick, J. S., Leeder, R. G., and Aston, W. P.,** The hemolytic activity of *Galleria mellonella* hemolymph, *Dev. Comp. Immunol.,* 13, 103, 1989.
54. **De Verno, P. J.,** The functional inter-relationships of cellular and humoral factors in the immune response of *Galleria mellonella,* M.Sc. thesis, Queen's University, Kingston, Ontario, Canada, 1983, 1.
55. **Chadwick, J. S.,** An assessment of the ability of individual moieties of *Pseudomonas aeruginosa* endotoxins to induce immunity in larvae of *Galleria mellonella, J. Invertebr. Pathol.,* 17, 299, 1971.
56. **Gifford, G. E. and Flick, D. A.,** Natural production and release of tumour necrosis factor, in *Tumour Necrosis Factor and Related Cytotoxins,* Ciba Foundation Symposium, John Wiley & Sons, New York, 1987, 3.
57. **Henricson, B. E., Benjamin, W. R., and Vogel, S. N.,** Differential cytokine induction by dose of lipopolysaccharide and monophosphoryl lipid A that result in equivalent early endotoxin tolerance, *Infect. Immun.,* 58, 2429, 1990.
58. **Flad, H. D., Loppnow, H., Feist, W., Wang, M-H., Brade, H., Kusumoto, S., Rietschel, E. Th., and Ulmer, A. J.,** Interleukin 1 and tumour necrosis factor: studies on the induction by lipopolysaccharide partial structures, *Lymph. Res.,* 8, 235, 1989.
59. **Beutler, B. and Cerami, A.,** The biology of cachectin/tumour necrosis factor — a primary mediator of the host response, *Annu. Rev. Immunol.,* 7, 625, 1989.
60. **Chain, B. M. and Anderson, R. S.,** Selective depletion of the plasmatocytes in *Galleria mellonella* following injection of bacteria, *J. Insect. Physiol.,* 28, 377, 1982.
61. **Gateff, E.,** Malignant neoplasms of genetic origin in *Drosophila melanogaster, Science,* 200, 1448, 1978.

62. **Gateff, E.**, Gonial cell neoplasm of genetic origin affecting both sexes of *Drosophila melanogaster*, in *Progress in Clinical and Biological Research*, Vol. 85B, Burger, M. M. and Weber, R., Eds., Alan R. Liss, New York, 1982, 621.
63. **Gateff, E.**, Genetics of oncogenesis in *Drosophila melanogaster*, *J. Cancer Res. Clin. Oncol.*, 111, 44, 1986.
64. **Matz, G. and Bergoin, M.**, Experimental tumours in *Leucophaea maderae* (Dictyoptera): a viral etiology, *J. Invertebr. Pathol.*, 43, 424, 1984.
65. **Sato, C., Kawamoto, F., and Kumada, N.**, A new virus isolated from the cockroach *Periplaneta fuliginosa* (Serville), *Microbiol. Immunol.*, 23, 207, 1979.
66. **James, K.**, Interactions between cytokines and alpha-2-macroglobulin, *Immunol. Today*, 11, 163, 1990.
67. **Arnold, J. W.**, Hemocytes of insects, in *The Physiology of Insecta*, Vol. 5, Rockstein, M., Ed., Academic Press, New York, 1974, 201.
68. **Wigglesworth, V. B.**, The circulatory system and associated tissues, in *The Principles of Insect Physiology*, Chapman and Hall, London, 1972, 411.
69. **Ratcliffe, N. A. and Rowley, A. F.**, A comparative synopsis of the structure and function of the blood cells of insects and other invertebrates, *Dev. Comp. Immunol.*, 3, 189, 1979.
70. **Polla, B. S.**, A role for heat shock proteins in inflammation?, *Immunol. Today*, 9, 134, 1988.
71. **Bronstein, S. M. and Conner, W. E.**, Endotoxin-induced behavioural fever in the Madagascar cockroach, *Gromphadorhina portentosa*, *J. Insect Physiol.*, 30, 327, 1984.
72. **Louis, C., Jourdan, M., and Cabanac, M.**, Behavioural fever and therapy in the orthoptera *Gryllus bimaculatus* during infection by an intracellular pathogenic procaryotic *Rickettsiella grylli*, *Am. J. Physiol.*, 250, 991, 1986.
73. **DeVerno, P. J., Aston, W. P., and Chadwick, J. S.**, Transfer of immunity against *Pseudomonas aeruginosa* P11-1 in *Galleria mellonella* larvae, *Devel. Comp. Immunol.*, 7, 423, 1983.
74. **Wyatt, G. R.**, Regulation of protein and carbohydrate metabolism in insect fat body, *Verh. Dtsch. Zool. Ges.*, 1974, 209, 1975.
75. **Law, J. H. and Wells, M. A.**, Insects as biochemical models, *J. Biol. Chem.*, 264, 16335, 1989.
76. **Faye, I. and Wyatt, G. R.**, The synthesis and uptake of hemolymph storage proteins by the fat body of the greater wax moth, *Galleria mellonella*, *Experientia*, 36, 1325, 1980.
77. **Abu-Hakimu, R. and Faye, I.**, An ultrastructural and autoradiographic study of the immune response in *Hyalophora cecropia* pupae, *Cell Tissue Res.*, 217, 311, 1981.
78. **DeVerno, P. J., Chadwick, J. S., Aston, W. P., and Dunphy, G. B.**, The *in vitro* generation of an antibacterial activity from the fat body and hemolymph of non-immunized larvae of *Galleria mellonella*, *Dev. Comp. Immunol.*, 8, 537, 1984.
79. **Trenczek, T. and Faye, I.**, Synthesis of immune proteins in primary cultures of fat body from *Hyalophora cecropia*, *Insect Biochem.*, 18, 299, 1988.
80. **Dickinson, L., Russell, V., and Dunn, P. E.**, A family of bacteria-regulated, cecropin D-like peptides from *Manduca sexta*, *J. Biol. Chem.*, 263, 19424, 1988.
81. **Stephens, J. M.**, Bactericidal activity of hemolymph of some normal insects, *J. Insect. Pathol.*, 5, 61, 1963.
82. **Gagen, S. J. and Ratcliffe, N. A.**, Studies on the *in vivo* cellular reactions and fate of injected bacteria in *Galleria mellonella* and Pieris brassicae larvae, *J. Inertebr. Pathol.*, 28, 17, 1976.
83. **Geng, C. and Dunn, P. E.**, Plasmatocyte depletion in larvae of *Manduca sexta* following injection of bacteria, *Dev. Comp. Immunol.*, 13, 17, 1989.
84. **Chain, B. M. and Anderson, R. S.**, Inflammation in insects: the release of a plasmatocyte depletion factor following interaction between bacteria and hemocytes, *J. Insect Physiol.*, 29, 1, 1983.
85. **Chiu, J. K. C., Aston, W. P., and Chadwick, J. S.**, A suggested role for phospholipase-C activity in the immune response of *Galleria mellonella*, *J. Invertebr. Pathol.*, in press, 1990.
86. **Steiner, H., Hultmark, D., Engstrom, A., Bennich, H., and Boman, H. G.**, Sequence and specificity of two antibacterial proteins involved in insect immunity, *Nature (London)*, 292, 246, 1981.
87. **Jaynes, J. M., Burton, C. A., Barr, S. B., Jeffers, G. W., Julian, G. R., White, K. L., Enright, F. M., Klei, T. R., and Laine, R. A.**, *In vitro* cytocidal effect of novel lytic peptides on *Plasmodium falciparum* and *Trypanosoma cruzi*, *FASEB J.*, 2, 2878, 1988.
88. **Jaynes, J.**, Peptides to the rescue, *New Scientist*, 16, 42, 1989.
89. **Gwadz, R. W., Kaslow, D., Lee, J., Maloy, W. L., Zasloff, M., and Miller, L. H.**, Effects of magainins and cecropins on the sporogonic development of malaria parasites in mosquitoes, *Infect. Immun.*, 57, 2628, 1989.
90. **Boman, H. G. and Steiner H.**, Humoral immunity in *Cecropia* pupae, *Curr. Top. Microbiol. Immunol.*, 94, 95, 75, 1981.
91. **Boman, H. G., Wade, D., Boman, I. A., Wahllin, B., and Merrifield, R. B.**, Antibacterial and antimalarial properties of peptides that are cecropin-melittin hybrids, *Fed. Eur. Biochem. Soc.*, 259, 103, 1989.

92. **Hultmark, D., Engstrom, A., Bennich, H., Kapur, R., and Boman, H. G.,** Insect immunity: isolation and structure of cecropin D and four minor antibacterial components from *Cecropia* pupae, *Eur. J. Biochem.*, 127, 207, 1982.
93. **Boman, H. G., Faye, I., V Hofsten, P., Kockum, K., Lee, J-Y, Xanthopoulos, K. G., Bennich, H., Engstrom, A., Merrifield, B. R., and Andreu, D.,** Proceedings in life sciences: immunity in invertebrates: cells, molecules, and defense reactions, in *Antibacterial Immune Proteins in Insects — a Review of Some Current Perspectives,* Brehelin, M., Ed., Springer-Verlag, Berlin, 1986, 63.
94. **Andreu, D., Merryfield, R. B., Steiner, H., and Boman, H. G.,** Solid-phase synthesis of cecropin A and related peptides, *Proc. Nat. Acad. Sci. U.S.A.*, 27, 6475, 1983.
95. **Fink, J., Merrifield, R. B., Boman, A., and Boman, H. G.,** The chemical synthesis of cecropin D and an analog with enhanced antibacterial activity, *J. Biol. Chem.*, 264, 6260, 1989.
96. **Fink, J., Boman, A., Boman, H., and Merrifield, R. B.,** Design, synthesis and antibacterial activity of cecropin-like model peptides, *Int. J. Peptide Protein Res.*, 33, 412, 1989.
97. **Boman, H., Boman, I., Andreu, D., Li, Z., Merrifield, R., Schlenstedt, G., and Zimmermann, R.,** Chemical synthesis and enzymic processing of precursor forms of cecropins A and B, *J. Biol. Chem.*, 264, 5852, 1989.
98. **De Azambuja, P., Freitas, C. C., and Garcia, E. S.,** Evidence and partial characterization of an inducible antibacterial factor in the hemolymph of *Rhodnius prolixus, Insect Physiol.*, 32, 807, 1986.
99. **Lee, J., Boman, A., Chuanxin, S., Andersson, M., Jornvall, H., Mutt, V., and Boman, H. G.,** Antibacterial peptides from pig intestine: isolation of a mammalian cecropin, *Proc. Natl. Acad. Sci. U.S.A.*, 86, 9159, 1989.
100. **Lee, S., Mihara, H., Aoyagi, H., Kato, T., Izumiya, N., and Yamasaki, N.,** Relationship between antimicrobial activity and amphiphilic property of basic model peptides, *Biochim. Biophys. Acta,* 862, 211, 1986.
101. **Hultmark, D., Engstrom, A., Andersson, K., Steiner, H., Bennich, H., and Boman, H. G.,** Insect immunity. Attacins, a family of antibacterial proteins from *Hyalophora cecropia, EMBO J.*, 2, 571, 1983.
102. **Engstrom, P., Carlsson, A., Engstrom, A., Tao, Z-J., and Bennich, H.,** The antibacterial effect of attacins from the silk moth *Hyalophora cecropia* is directed against the outer membrane of *Escherichia coli, EMBO J.*, 3, 3347, 1984.
103. **Casteels, P., Ampe, C., Riviere, L., Van Damme, J., Elicone, C., Fleming, M., Jacobs, F., and Tempst, P.,** Isolation and characterization of abaecin, a major antibacterial response peptide in the honeybee *(Apis mellifera), Eur. J. Biochem.*, 187, 381, 1990.
104. **Dimarcq, J-L., Keppi, E., Dunbar, B., Lambert, J., Reichhart, J-M., Hoffmann, D., Rankine, S. M., Fothergill, J. E., and Hoffmann, J. A.,** Insect immunity. Purification and characterization of a family of novel inducible antibacterial proteins from immunized larvae of the dipteran *Phormia terranovae* and complete amino-acid sequence of the predominant member, diptericin A, *Eur. J. Biochem.*, 171, 17, 1988.
105. **Selsted, M. E., Brown, D. M., Delange, R. J., and Lehrer, R. I.,** Primary structures of MCP-1 and MCP-2, natural peptide antibiotics of rabbit lung macrophages, *J. Biol. Chem.*, 258, 14485, 1983.
106. **Selsted, M. E., Brown, D. M., Delange, R. J., Harwig, S. S. L., and Lehrer, R. I.,** Primary structures of six antimicrobial peptides of rabbit peritoneal neutrophils, *J. Biol. Chem.*, 260, 4579, 1985.
107. **Matsuyama, K. and Natori, S.,** Purification of three antibacterial proteins from the culture medium of NIH-Sape-4, an embryonic cell line of *Sacrophaga peregrina, J. Biol. Chem.*, 263, 17112, 1988.
108. **Lambert, J., Keppi, E., Dimarcq, J., Wicker, C., Reichhart, J., Dunbar, B., Lepage, P., Dorsselaer, A., Hoffmann, J., Fothergill, J., and Hoffmann, D.,** Insect immunity. Isolation from immune blood of the dipteran *Phormia terranovae* of two insect antibacterial peptides with sequence homology to rabbit lung macrophage bactericidal peptides, *Proc. Natl. Acad. Sci. U.S.A.*, 86, 262, 1989.
109. **Chadwick, J. S.,** Relation of lysozyme concentration to acquired immunity against *Pseudomonas aeruginosa,* in *Galleria mellonella, J. Invertebr. Pathol.*, 15, 455, 1970.
110. **Jarosz, J., Glinski, Z., and Bartyzel, M.,** Increased activity of blood lysozyme in the greater wax moth larvae affected perorally by bacterial invaders with chitinase added, *Comp. Biochem. Physiol.*, 87, 189, 1987.
111. **Young, R. A.,** Stress proteins and immunology, *Annu. Rev. Immunol.*, 8, 401, 1990.
112. **Shiba, T., Teshima, T., Nakai, T., and Kitazawa, M.,** Structure-activity relationship of lepidopteran self-defense peptide of *Bombyx mori, Tetrahedron,* 44, 787, 1988.
113. **Anderson, D., Gunne, H., Hellers, M., Johansson, H., and Steiner, H.,** Immune responses in *Trichoplusia ni* challenged with bacteria or baculoviruses, *Insect Biochem.*, 20, 537, 1990.
114. **Canicatti, C.,** Hemolysins: pore-forming proteins in invertebrates, *Experientia,* 46, 239, 1990.
115. **Canicatti, C. and D'Ancona, G.,** Cellular aspects of *Holothuria polii* immune response, *J. Invert. Pathol.*, 53, 2, 152, 1989.
116. **Bertheussen, K.,** Complement-like activity in Sea Urchin coelomic fluid, *Dev. Comp. Immunol.*, 7, 21, 1983.

117. **Anderson, R. S.,** Inducible hemolytic activity in *Mercenaria mercenaria* hemolymph, *Dev. Comp. Immunol.*, 5, 575, 1981.
118. **Leippe, M. and Renwrantz, L.,** Release of cytotoxic and agglutinating molecules by *Mytilus* hemocytes, *Dev. Comp. Immunol.*, 12, 297, 1988.
119. **Roch, P., Valembois, P., Davant, N., and Lasegues, M.,** Protein analysis of earthworm coelomic fluid. II. Isolation and biochemical characterization of the *Eisenia fetida andrei* factor (EFAF), *Comp. Biochem. Physiol.*, 69B, 829, 1981.
120. **Roch, P., Canicatti, C., and Valembois, P.,** Interactions between earthworm hemolysins and sheep red blood cell membranes, *Biochim. Biophys. Acta*, 983, 193, 1989.
121. **Tuckova, L., Rejnek, J., Siam, P., and Ondrejova, R.,** Lytic activities in coelomic fluid of *Eisenia foetida* and *Lumbricus terrestris*, *Dev. Comp. Immunol.*, 10, 181, 1986.
122. **Kauschke, E. and Mohrig, W.,** Comparative analysis of hemolytic and hemagglutinating activities in the coelomic fluid of *Eisenia foetida* and *Lumbricus terrestris* (annelida, lumbricidae), *Dev. Comp. Immunol.*, 11, 331, 1987.
123. **Weinheimer, P. F., Evans, E. E., Stroud, R. M., Acton, R. T., and Painter, B.,** Comparative immunology: natural hemolytic system of the spiny lobster, *Panulirus argus*, *Proc. Soc. Exp. Biol. Med.*, 130, 322, 1969.
124. **Valembois, P., Roch, P., Lassegues, M., and Cassand, P.,** Antibacterial activity of the hemolytic system from the earthworm *Eisenia fetida andrei*, *J. Invert. Pathol.*, 40, 21, 1982.
125. **Day, N. K., Gewurz, H., Johannsen, R., Finstad, J., and Good, R. A.,** Complement and complement-like activity in lower vertebrates and invertebrates, *J. Exp. Med.*, 132, 941, 1970.
126. **Bertheussen, K.,** Receptors for complement on echinoid phagocytes. II. Purified human complement mediates echinoid phagocytosis, *Dev. Comp. Immunol.*, 6, 635, 1982.
127. **Canicatti, C.,** Membrane damage by coelomic fluid from *Holothuria polii* (Echinodermata), *Experientia*, 43, 611, 1987.
128. **Day, N. K. B., Geiger, H., Finstad, J., and Good, R. A.,** A starfish hemolymph factor which activates vertebrate complement in the presence of cobra venom factor, *J. Immunol.*, 109, 164, 1972.
129. **Anderson, R. S.,** Phagocytosis by invertebrate cells *in vitro:* biochemical events and other characteristics compared with vertebrate phagocytic systems, in *Invertebrate immunity*, Maramorosch, K. and Shope, R. E., Academic Press, New York, 1975, 153.
130. **Bertheussen, K.,** Complement and lysins in invertebrates, *Dev. Comp. Immunol.*, Suppl. 3, 173, 1984.
131. **D'Cruz, O. J. M. and Day, N. K.,** Structural and functional similarities between the major hemolymph protein of fall armyworm and cat C4 binding protein of the complement system, *Dev. Comp. Immunol.*, 9, 541, 1985.
132. **Aston, W. P., Chadwick, J. S., and Henderson, M. J.,** Effect of Cobra venom factor on the *in vivo* immune response in *Galleria mellonella* to *Pseudomonas aeruginosa*, *J. Invertebr. Pathol.*, 27, 171, 1976.
133. **Phipps, D., Menger, M., Chadwick, J. S., and Aston, W. P.,** A cobra venom factor CVF induced C3 convertase activity in the hemolymph of *Galleria mellonella*, *Dev. Comp. Immunol.*, 11, 37, 1987.
134. **Phipps, D. J., Chadwick, J. S., and Aston, W. P.,** A novel antibacterial activity in the hemolymph of *Galleria mellonella*, *Annu. Meet. Can. Soc. Microbiologists*, (Abstr.), 1, 1989.
135. **Powning, R. F. and Davidson, W. J.,** Studies on insect bacteriolytic enzymes. I. Lysozyme in hemolymph of *Galleria mellonella* and *Bombyx mori*, *Comp. Biochem. Physiol.*, 45, 669, 1973.
136. **Phipps, D. J., Chadwick, J. S., and Aston, W. P.,** unpublished data, 1990.
137. **Podack, E. R.,** Molecular mechanisms of cytolysis by complement and by cytolytic lymphocytes, *J. Cell. Biochem.*, 30, 133, 1986.
138. **Leonard, L. A., Strandberg, J. D., and Winkelstein, J. A.,** Complement-like activity in the sea star, *Asterias forbesi*, *Dev. Comp. Immunol.*, 14, 19, 1990.
139. **Enghild, J. J., Thogersen, I. B., Salbesen, G., Fey, G. H., Figler, N. L., Gonias, S. L., and Pizzo, S. V.,** Alpha-macroglobulin from *Limulus polyphemus* exhibits proteinase inhibitory activity and participates in a hemolytic system, *Biochemistry*, 29, 10070, 1990.
140. **Volanakis, J. E.,** Participation of C3 and its ligands in complement activation, *Curr. Top. Microbiol. Immunol.*, 153, 1, 1989.
141. **Day, N., Good, R. A., and Muller-Eberhard, H. J.,** "C3 proactivator" in starfish hemolymph, *Fed. Proc.*, 31, 788, 1972.
142. **Sottrup-Jensen, L.,** Alpha-macroglobulins: structure, shape and mechanism of proteinase complex formation, *J. Biol. Chem.*, 264, 11539, 1989.
143. **Spycher, S. E., Arya, S., Isenman, D. E., and Painter, R. H.,** A functional, thioester-containing alpha-2-macroglobulin homologue isolated from the hemolymph of the American lobster *(Homerus americans)*, *J. Biol. Chem.*, 262, 14606, 1987.
144. **Chadwick, J. S., DeVerno, P. J., Chung, K. L., and Aston, W. P.,** Effects of hemolymph from immune and non-immune larvae of *Galleria mellonella* on the ultra-structure of *Pseudomonas aeruginosa*, *Dev. Comp. Immunol.*, 6, 433, 1982.

145. **Steiner, H., Andreu, D., and Merrifield, R. B.,** Binding and action of cecropin and analogues of antibacterial peptides from insects, *Biochim. Biophys. Acta,* 939, 260, 1988.
146. **Matsumoto, K., Maeda, H., Takata, K., Kamata, R., and Okamura, R.,** Purification and characterization of four proteases from a clinical isolate of *Serratia marcescens* kum 3958, *J. Bacteriol.,* 157, 225, 1984.
147. **Grimont, P. A. D. and Grimont, F.,** The genus *Serratia, Annu. Rev. Microbiol.,* 32, 221, 1978.
148. **Lysenko, O.,** Chitinase of *Serratia marcescens* and its toxicity for insects, *J. Invertebr. Pathol.,* 27, 385, 1976.
149. **Flyg, C. and Xanthopoulos, K. G.,** Insect pathogenic properties of *Serratia marcescens.* Passive and active resistance to insect immunity studied with protease deficient and phage-resistant mutants, *J. Gen. Microbiol.,* 129, 453, 1983.
150. **Reinisch, C. and Litman, G.,** Evolutionary immunobiology, *Immunol. Today,* 10, 278, 1989.
151. **Matha, V.,** Changes in *Galleria mellonella* larvae hemolymph hydrolases induced by immunization with nonpathogenic bacteria, *Microbios Lett.,* 36, 43, 1987.
152. **Cooper, E. L.,** New animal models in immunobiology research, in *Defense Molecules,* Marchalonis, J. J. and Reinisch, C. L., Eds., Alan R. Liss, New York, 1990, 255.
153. **Cherbas, L.,** The induction of an injury reaction in cultured hemocytes from saturniid pupae, *J. Insect. Physiol.,* 19, 2011, 1973.
154. **Ratner, S. and Vinson, S. B.,** Phagocytosis and encapsulation: cellular immune responses in arthropoda, *Am. Zool.,* 23, 185, 1983.
155. **Mohrig, W. and Schittek, D.,** Phagocytosis stimulating mediators in insects, *Acta Biol. Med. Ger.,* 38, 953, 1979.

Chapter 14

IDENTIFICATION AND CHARACTERIZATION OF INSECT HEMOLYMPH PROTEINS INTERACTING WITH THE MAMMALIAN COMPLEMENT CASCADE

Osmond J. D'Cruz

TABLE OF CONTENTS

I. INTRODUCTION

The function of the immune system is to preserve the biological identity of the self and to eliminate all the possible sources of nonself. The task of eliminating nonself or altered self antigens without attacking self can be conducted properly only if the system has the ability to discriminate precisely between self and nonself. In vertebrates, this essential feature of the immune system has evolved to a sophisticated level of complexity with an array of cytolytic effector cells and molecules.[1] Generally, there are three pathways of immune-mediated cytolysis: one is triggered by antibodies and complement (C) proteins; the other pathway is mediated by specific killer cells; and the third pathway involves antibodies and cytotoxic cells.

The C system is the effector arm of the humoral immune system. The human C cascade comprises a group of at least 20 well-characterized proteins of blood plasma,[2] and can be activated by either the classical or alternative pathways. Both pathways have homologous components, use related activation and regulatory mechanisms, and result in similar inflammatory mediators. Although the primary function of the C system is to destroy foreign invaders, there is now substantial evidence that they can also act against the body's own cells, to cause their premature destruction or to fundamentally alter their biological function.[3,4] In this context, an understanding of the phylogenetic origin of C proteins and its interactions with cellular targets is of particular importance to understanding the origin of transplantation rejections and C-fixing autoimmune disorders.

Phylogenetic data indicate that the alternative C pathway (ACP) is older and that the classical pathway (CP) "evolved" from it.[5] Both pathways produce a divalent cation-dependent C3 convertase enzyme, C3bBb (alternative) or C4b2a (classical), that cleaves C3 into C3b and C3a. C3b of the alternative pathway enzyme is structurally very similar to C4b of the classical convertase.[6] In addition to sequence homology, C3 and C4 both possess a thioester bond, which enables covalent binding to cell surface, and have virtually identical mechanisms of activation and degradation. Similarly, factors B and C2, the proenzyme precursors of Bb and C2a, respectively, are homologous. Their genes are closely linked within the major histocompatibility complex in humans, and are derived from a gene duplication. The negative regulators, factor H (which binds to C3b), and C4-binding protein (which binds to C4b) are also homologous and genetically linked.

The ACP is nonadaptive and responds to simple polymeric structures, such as yeast and bacterial cell walls, without the requirement for C-fixing antibodies.[7] Phylogenetic studies have looked for generation of hemolytic activity in response to polysaccharides or cobra venom factor. Such ACP-like activity has been detected not only in the sera of the most primitive vertebrates, such as hagfish and lamprey, but also in the hemolymph of some invertebrates, namely, starfish, horseshoe crab, and sipunculid worm.[8,9]

The most primitive species for which the classical pathway has been demonstrated is the nurse shark, where components functionally related to C1, C2, C3, C4, C8, and C9 of the C system exist,[10] and the shark C1 can be activated by sheep red cells sensitized with shark serum. Molecules resembling C3, an essential component of the classical and alternative C pathways, have been isolated from lamprey.[5] In lamprey, this is associated with ACP-like system that can be activated by yeast polysaccharides. IgM-like molecules, with antibody function, occur in the serum of even the most primitive vertebrates,[11] but are absent in invertebrates.[12]

Insects are known to lack lymphocytes and immunoglobulins, and therefore have developed a relatively simple sequence of inducible antimicrobial peptides that are distinctively different from those of mammalian origin.[13,14] In this chapter, I will focus on the ability of insect hemolymph (HL) proteins to interact with the mammalian C system.[15,16] I report the biochemical nature of these C-interacting proteins, their relationship to larval-pupal transition, and their homology with mammalian C regulators.

II. INTERACTION OF LARVAL *SPODOPTERA* HEMOLYMPH WITH THE CLASSICAL COMPLEMENT PATHWAY

Hemolymph of larval fall armyworm, *Spodoptera frugiperda,* has inhibitors for the classical and alternative pathways of the mammalian C system.[15]

The inhibitory property of *Spodoptera* HL for the classical pathway (CP) of C was detected using a two-step hemolytic assay.[17] In the first step, dilutions of inhibitor fractions were preincubated with $EAC1^{gp}4^{hu}$ intermediate cells. In the second step, the cells were incubated with limiting amounts of guinea pig C2 for EAC142 formation. The hemolytic sites were then developed by the addition of a 1:20 dilution of guinea pig serum in EDTA as the source of C3 to C9. The number of hemolytic sites generated was determined using $EAC1^{gp}4^{hu}$ as the control.

For the isolation of the CP inhibitor by the limited C2 assay described previously, cell-free larval HL was precipitated with 50% $(NH_4)_2SO_4$ saturation, and after dialysis was passed through a DEAE-Sephacel column. The CP, recovered in the flow through fraction was gel-filtered twice on Bio-Gel 1.5 *m*.

The purified CP inhibitor, which was identified as the major HL protein, showed a single band of apparent M_r 520 kDa (Figure 1A, lane B). This molecular weight is characteristic of insect larval serum protein (LSP).[18,19] On 10% SDS-polyacrylamide gel electrophoresis (PAGE) under nonreducing conditions, two nonidentical subunits with M_r 110 and 80 kDa designated heavy (H) and light (L), respectively, were observed (Figure 1B, lane A). On partial reduction, the molecular weight of the heavy subunit was reduced to 90 kDa (Figure 1B, lane B). Complete reduction resulted in identical subunits of M_r 80 kDa (Figure 1C, lane B). These observations suggested that the CP inhibitor is a hexamer, composed of two nonidentical subunits. The conversion of the heavy to lighter subunit may be due to limited cleavage of the heavy subunit upon complete reduction. During the purification steps, two major HL proteins designated LSP-I and LSP-II were identifiable. The EAC142 lysis inhibitory activity was associated with LSP-I. (Figure 1D, lane B). LSP-II on complete reduction revealed nonidentical subunits of 165,000 and 80,000 Da, respectively (Figure 1D, lane A).

Electron microscopy of *Spodoptera* CP inhibitor (LSP-I) using both rotary shadowing and traditional methods showed a pure population of cylindrical particles approximately 10 nm in diameter and about 11.5 nm in length (Figure 2A, B). The presence of several visible subunits in each particle agrees with the SDS-PAGE analysis and together with the lack of detectable cysteine residues suggests that the subunits are associated noncovalently, also a characteristic feature of the insect LSP[18,19] and mammalian C4-binding protein (C4bp).[20]

Preincubation of $EAC1^{gp}4^{hu}$ with purified *Spodoptera* LSP-I (50 to 200 μg) followed by addition of limiting amounts of C2 brought about a dose-dependent inhibition of lysis of EAC142 cells by guinea pig C3 to C9. No differences in inhibitory properties were observed when LSP-I was simultaneously incubated with C2. These results suggested that *Spodoptera* LSP-I interacts at the C4b step and inhibits the formation of classical C3 convertase (EAC142), a property also characteristic of mammalian C4-binding protein.[21,22]

The developmental profile of LSP-I was followed functionally as well as by electrophoretic analysis of native larval HL at various stages of larval development. As shown in Figure 3, the increase in CP inhibitor activity of larval HL was concomitant with the accumulation of LSP-I.

These findings suggested that *Spodoptera* LSP-I interacts with a component in the mammalian C cascade at the C4 step and is thus characteristic of mammalian C4bp. This is the first report that an insect protein exists, which resembles a mammalian C protein in function, as well as in some of the structural properties.

The human C4bp is known to modulate the classical C3 convertase in three ways.[21,22]

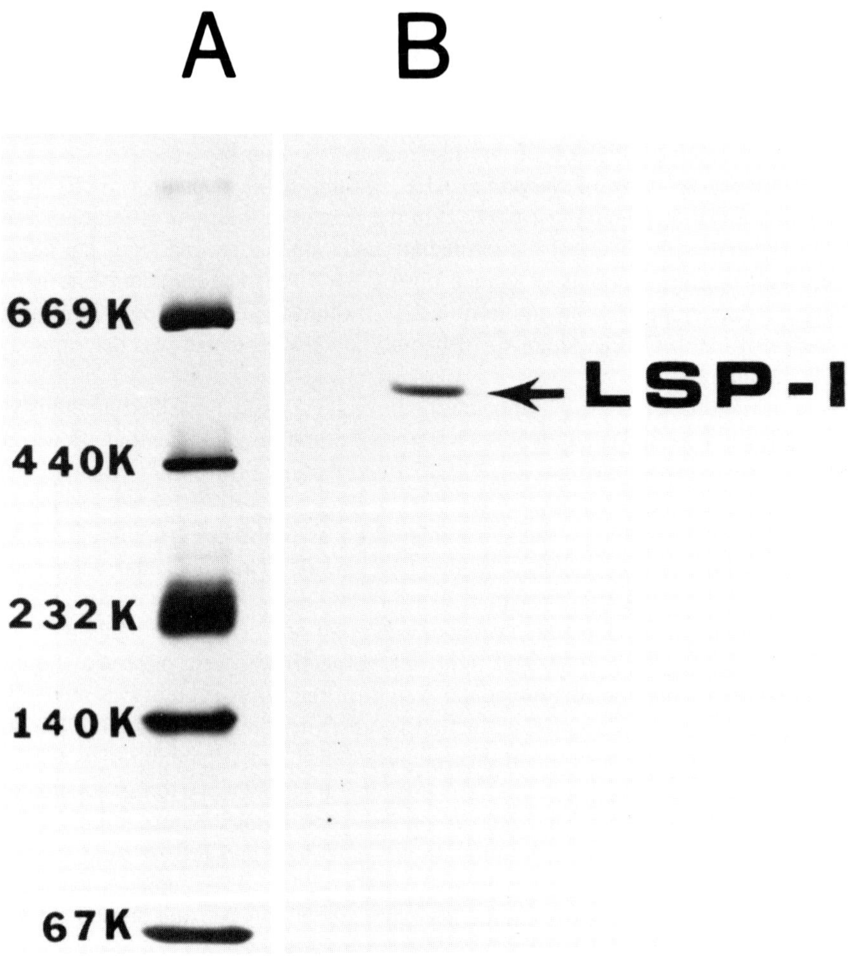

FIGURE 1A. A 4 to 30% gradient PAGE of purified *Spodoptera* LSP-I. Lane A, marker proteins of known molecular weight; lane B, active fraction from Bio-Gel A 1.5 *m*. The electrophoresis was conducted at 125 V for 15 h. Running buffer was Tris-boric acid-EDTA pH 8.5. The gel was stained with Coomassie brilliant blue R-250. Standard globular proteins used as markers were thyroglobulin (669 kDa), ferritin (440 kDa), catalase (232 kDa), lactate dehydrogenase (140 kDa), and bovine serum albumin (67 kDa).

At least one of these reaction mechanisms is detectable in *Spodoptera* LSP-I. The exact molecular mechanism of the interaction of LSP-I with cell-bound C4b is unknown at present. However, our earlier finding that chymotrypsin digests of both *Spodoptera* LSP-I and mammalian C4bp generate a common cleavage fragment of 55 kDa, and since a comparable fragment (48 kDa) from human C4bp (540 kDa) retains the cofactor activity for proteolysis of cell-bound C4b by Factor I, it is likely that this reaction has been conserved during phylogeny.[23] That sera from some lower vertebrates can cleave cell-bound C4b also supports this view.[24] Furthermore, it is also of interest that appearance of insect LSP during the larval-pupal transition is stimulated by a steroid hormone (ecdysone).[25,26] Mammalian C4bp also appears to be under the control of a steroid hormone (testosterone).[27]

It is provocative that insect larval serum proteins, which account for 40 to 60% of the total HL protein during the larval-pupal transition at a time when 80 to 90% of larval tissue undergoes histolysis,[28] are also inhibitors of lysis of mammalian cells by mammalian C. One might postulate that the major LSP of insects *in vivo* could play a significant role in modulating histolytic events during the larval-pupal-adult metamorphosis.

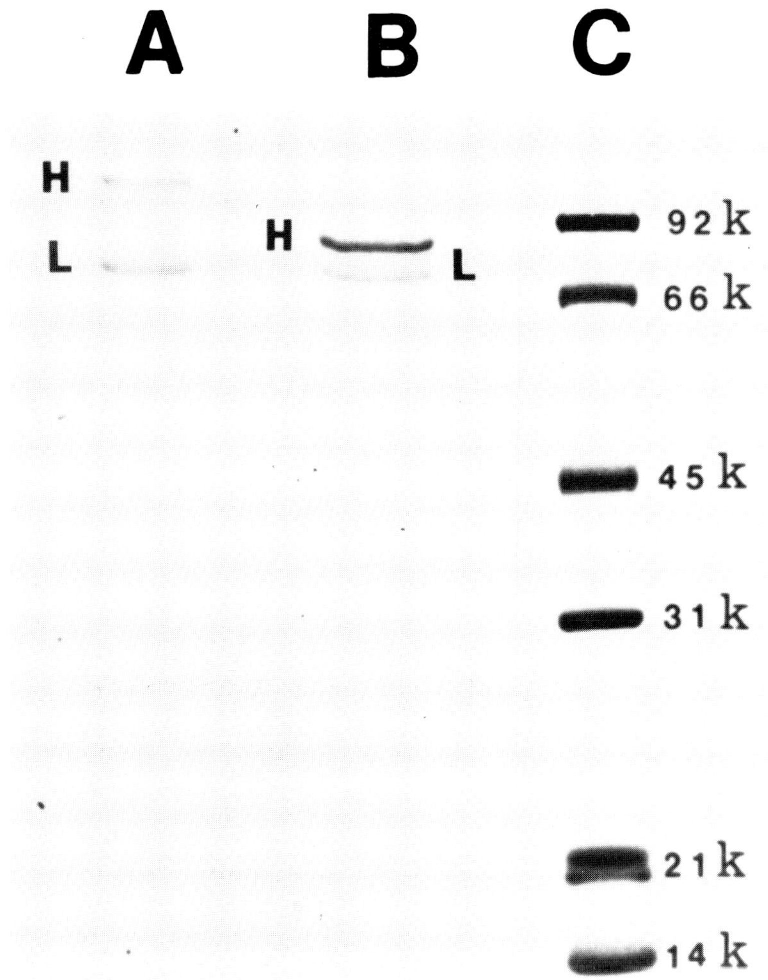

FIGURE 1B. Subunit pattern of *Spodoptera* LSP-I on 10% SDS-PAGE. Lane A, nonreduced LSP-I; lane B, partially reduced LSP-I. Lane C, molecular weight standards: phosphorylase b 92.5 kDa, BSA 66 kDa, ovalbumin 45 kDa, carbonic anhydrase 31 kDa, soybean trypsin inhibitor 21 kDa, and lysozyme 14 kDa. Running buffer was 0.25 *M* Tris-HCl, pH 8.3, 0.192 *M* glycine, 0.1% SDS. Electrophoresis was performed at 100 V for 8 h and the gel was stained with Coomassie blue.

III. INTERACTION OF LARVAL *SPODOPTERA* HEMOLYMPH WITH THE ALTERNATIVE COMPLEMENT PATHWAY

The inhibitory property of *Spodoptera* HL for the alternative C pathway (ACP) was detected on the basis of its ability to inhibit the lysis of rabbit red blood cells (RaRBC) by human C.[15] Briefly, dilution of HL fractions were preincubated with 1:3 dilutions of human serum chelated with EGTA, followed by the addition of RaRBC. The degree of hemolysis of cells at 37°C was monitored by recording the turbidity of the reaction mixture at 700 nm.[29]

The differential binding property of ACP and CP inhibitors of *Spodoptera* HL to bind DEAE-Sephacel was used to separate these two C inhibitors from HL.[15] The ACP inhibitor bound to DEAE-Sephacel and was eluted with a salt gradient (100 to 300 m*M* NaCl). The bound inhibitor was further purified by sequential gel-filtration chromatography on Bio-Gel 1.5 and 0.5 *m* and finally on Sephacryl S-200.

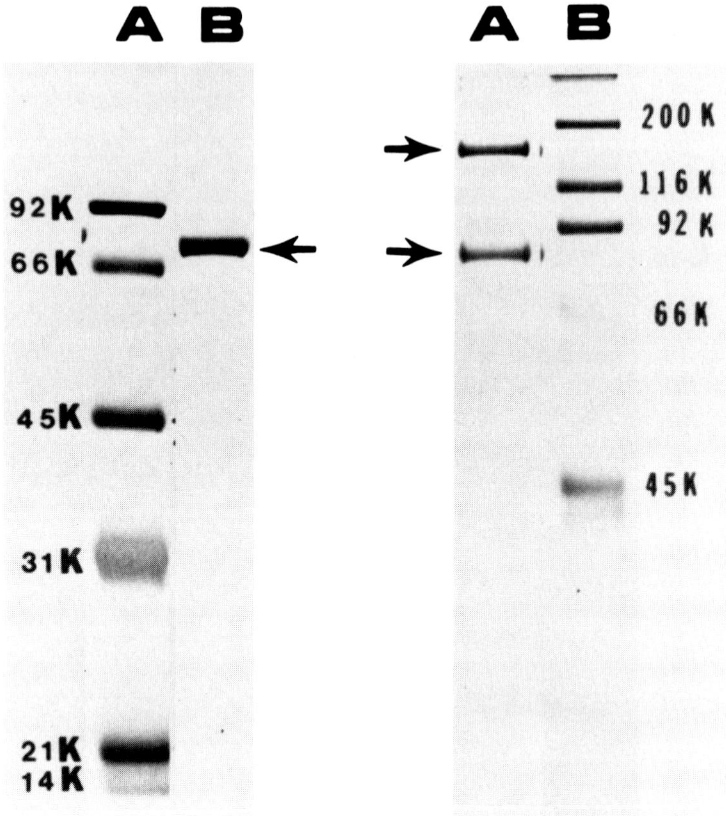

FIGURE 1C & D. (C) Lane B, 10% SDS-PAGE of completely reduced LSP-I; (D) lane A, 10% SDS-PAGE of completely reduced LSP-II.

The purified ACP inhibitor showed a single band of apparent M_r 110 kDa on 10% SDS-PAGE under nonreducing conditions (Figure 4). The amino acid composition of ACP inhibitor revealed the virtual absence of methionine and low cysteine content.

Preincubation of normal human serum (NHS) with purified ACP (0.5 to 5 μg) followed by the addition of RaRBC showed a dose-dependent inhibition of lysis and also resulted in characteristic agglutination of RaRBC.

Three probable mechanisms can be expected by which the insect HL protein inhibits mammalian ACP. First, the inhibition could result from the blocking of either the C3b receptor or acceptor sites on the RaRBC. Second, the inhibitor could compete with C3 for the C3 convertase. Third, the inhibitor could react with an early product of C3b degradation. If the inhibitor protein can react selectively with a degraded product of C3b, it could provide a sensitive and specific indicator for human C3b degradation.

IV. INTERACTION OF LARVAL *BOMBYX* HEMOLYMPH WITH HUMAN COMPLEMENT

Hemolymph of larval silkworm, *Bombyx mori,* revealed the presence of an activator and an inhibitor for the human ACP.[30]

A kinetic C assay was used to assess the xenogeneic interaction between larval *Bombyx* HL and the human C.[29] In this system, the lysis of RaRBC in EGTA (1 m*M*) chelated NHS, was tested in the presence of various stages of fifth instar larval HL. Continuous spectrophotometric readings of absorbance were recorded following the addition of RaRBC to the

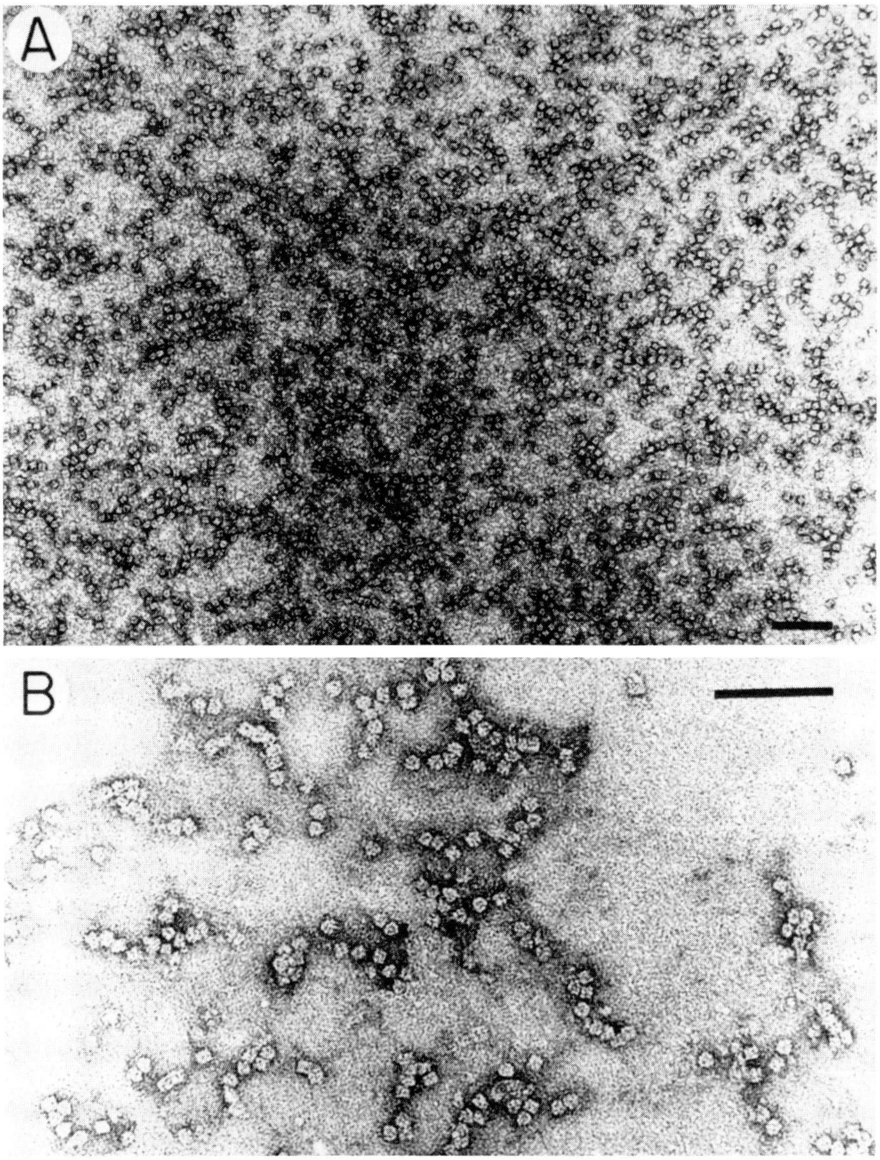

FIGURE 2. Electron micrographs of LSP-I. LSP-I (50 μg/ml) in 0.1 *M* NH_4 acetate 0.05 *M* NH_4HCO_3, pH 7.4 was absorbed to thin carbon films and negatively stained with 1% uranyl formate. The grids were imaged at 80 kV in a Zeiss EM-10CR electron microscope (× 50,000). (Bar = 100 nm.)

reaction mixture set at 37°C. The time required for the lag period (t_0) and for 50% of the cells to lyse ($t_{1/2}$) was calculated from each lytic curve.

Figure 5 represents a typical recorder tracing of lysis of RaRBC by a 1:3 dilution of NHS chelated with 1 m*M* EGTA (curve a). The lytic curve is characterized by a long lag phase (t_0, 510 s) and a relatively linear phase of decline in absorbance ($t_{1/2}$, 240 s). The minimum time for RaRBC lysis in undiluted NHS was 40 s for t_0 and 12 s for $t_{1/2}$. Addition of early fifth instar larval HL to NHS (1:3) reduced t_0 (120 s) but delayed $t_{1/2}$ (315 s, curve b). Addition of late fifth instar larval HL to NHS (1:3) reduced both t_0 (105 s) and $t_{1/2}$ (70 s, curve c). Addition of heated HL from early and late larval stages brought about a maximum reduction of t_0 and $t_{1/2}$ and was comparable to lysis of RaRBC in undiluted NHS (curve d).

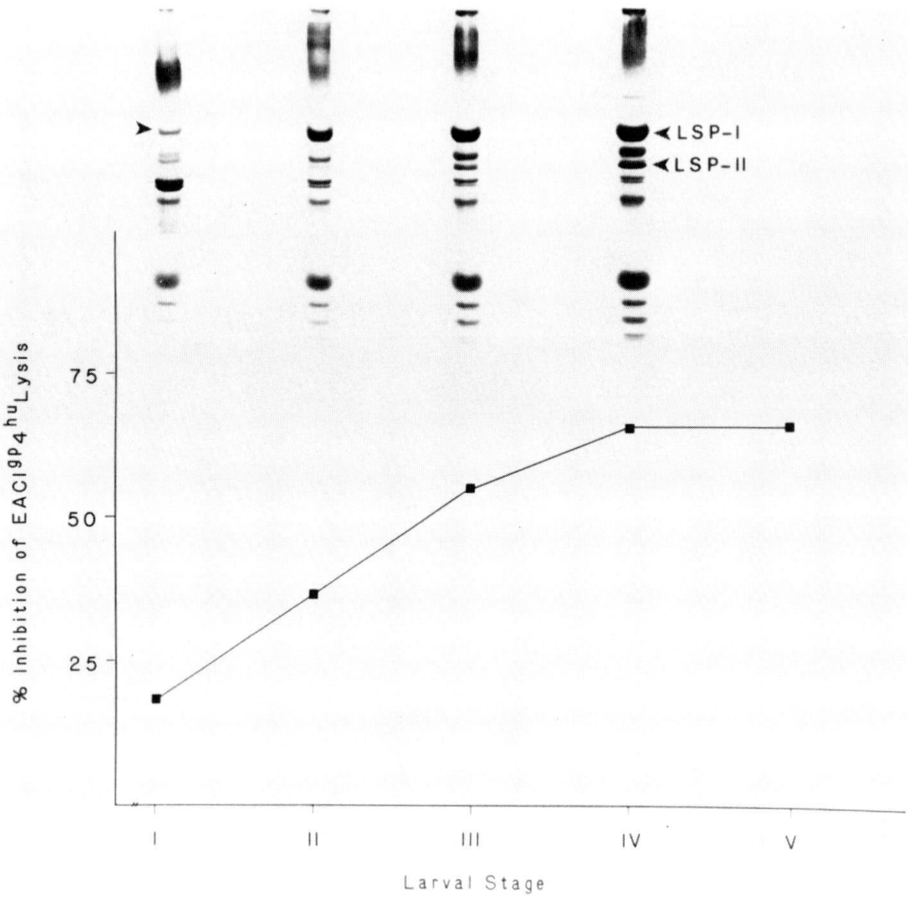

FIGURE 3. Classical C pathway inhibitory activity and HL protein profile during larval development: 50 μl of pooled cell-free HL from each stage (4-d interval) was tested for the inhibition of EAC14 cell lysis by the limited C2 assay; 5 μl of pooled HL from each stage was also electrophoresed on a 4 to 30% gradient gel. Running buffer: Tris-boric acid-EDTA, pH 8.5. The electrophoresis was carried out at 125 V for 15 h. The gel was stained with Coomassie brilliant blue R-250.

By contrast, addition of dialyzed HL from all larval stages completely inhibited the lysis of RaRBC by NHS (curve e). These variations in hemolytic curves, obtained in the presence of HL, suggested that larval *Bombyx* HL can activate as well as inhibit the lysis of RaRBC by human C. The lysis-inhibiting property was more pronounced in the early larval HL, while the lysis-enhancing property was more pronounced in the late fifth instar larval HL.

When we compared the RaRBC lysis curves obtained for 90 sera from healthy adults in the presence and absence of larval HL, consistent differences were noted. Despite the individual variations in t_0 and $t_{1/2}$ of NHS (1:3) for RaRBC lysis (represented by solid triangles, t_0 250 to 500 s, $t_{1/2}$ 95 to 410 s), the presence of heated (●) HL consistently reduced both t_0 and $t_{1/2}$ dramatically (60 s) of each of the 90 NHS (Figure 6, cluster a). Again, consistent differences were observed when HL from early and late fifth instar larval stages was incubated with each of the normal human sera (■). Early larval HL reduced t_0 but delayed $t_{1/2}$ (Figure 6, cluster c), whereas late larval HL samples reduced both t_0 and $t_{1/2}$ of NHS (Figure 6, cluster b). Mid-larval HL showed intermediate values of t_0 and $t_{1/2}$ between those of early and late larval HL. These observations indicated that the ability of larval HL to activate or inhibit human C-mediated lysis of RaRBC was larval stage dependent. Addition of HL during the hemolytic reaction revealed that the two HL components interacted

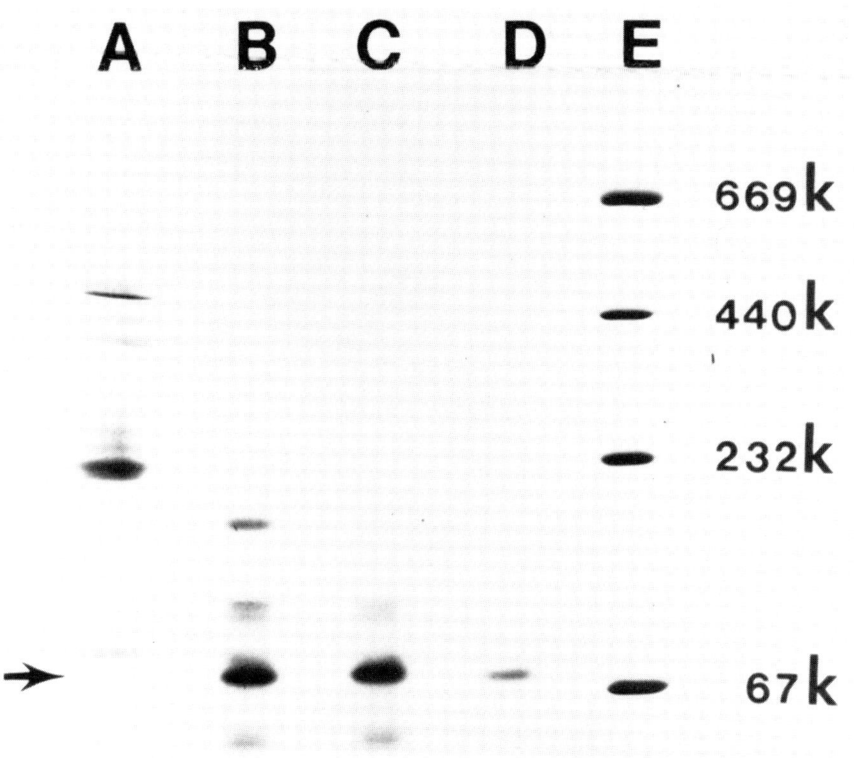

FIGURE 4. A 4 to 30% gradient PAGE of the ACP inhibitor purified from larval *Spodoptera* HL. Lane A, DEAE-Sephacel bound active fraction; lane B, Bio-Gel 1.5 *m* active fraction; lane C, Bio-Gel 0.5 *m* active fraction; lane D, Sephacryl S-200 active fraction. Lane E represents marker proteins of known molecular weight.

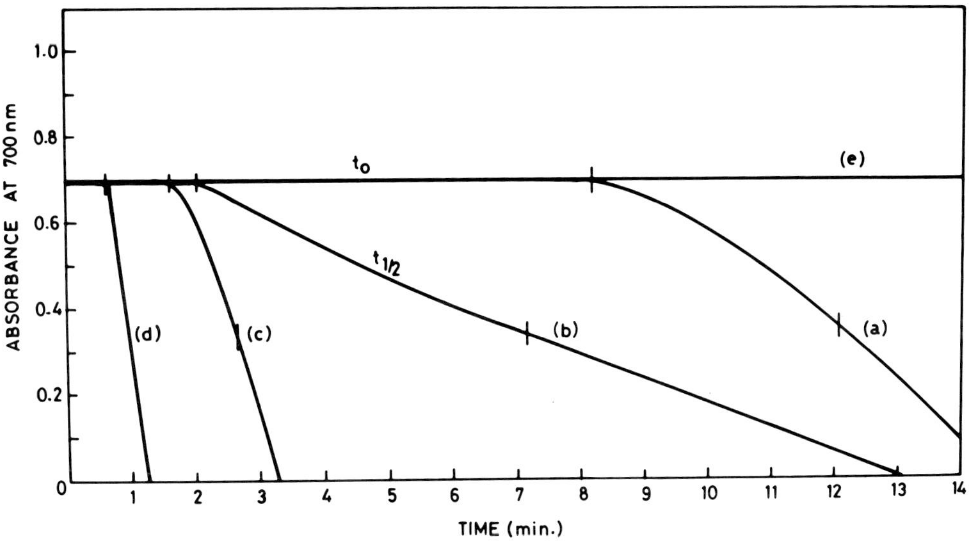

FIGURE 5. Recorder tracings of the kinetic hemolytic reaction using a dilution of NHS (1:3) and RaRBC as target cells. Curve a, NHS control (1:3); curve b, NHS (1:3) preincubated with 100 μl of fifth instar early (3 d) larval HL; curve c, NHS (1:3) preincubated with 100 μl of fifth instar late (9 d) larval HL; curve d, NHS (1:3) preincubated with 100 μl of heated larval HL or undiluted NHS; curve e, NHS (1:3) preincubated with dialyzed HL from early and late larval stages.

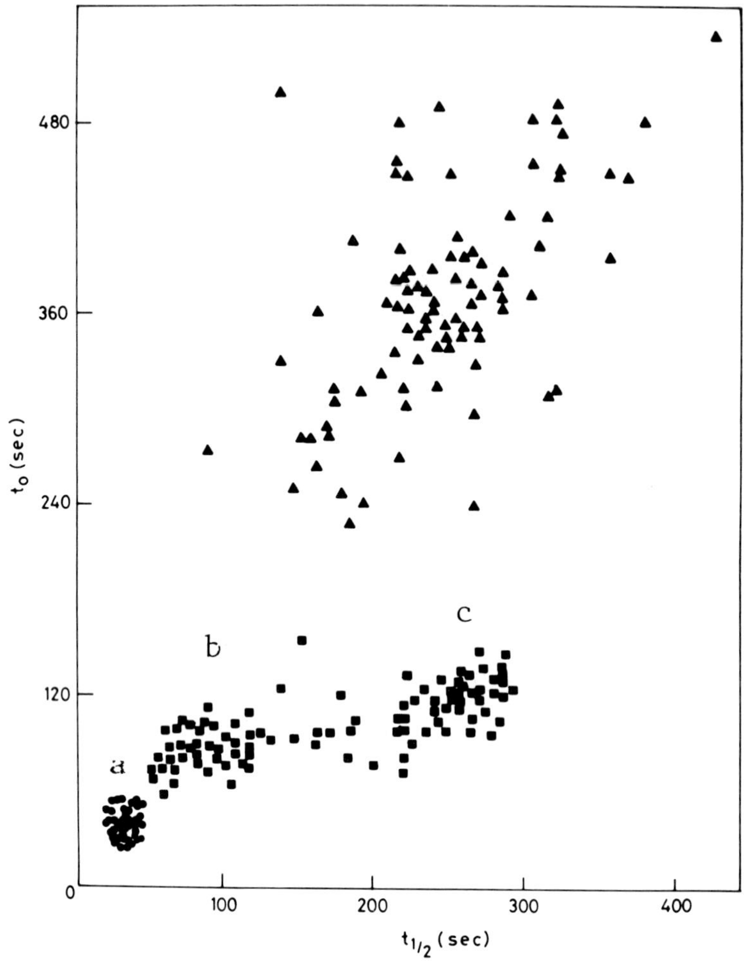

FIGURE 6. Relationship between t_0 and $t_{1/2}$ obtained for 90 NHS samples in the absence (▲) and presence of fresh HL (■) and heated HL (●). NHS (1:3) standard preincubated with 100 μl of HL from early and late larval stages (cluster c and b); corresponding NHS (1:3) preincubated with 100 μl HL (56°C/30 min) from fifth instar early and late larval stages (cluster a).

at two different steps in the lytic reaction, and the variations in the hemolytic curves observed was dependent upon the availability of the activator in free form in the HL.

To substantiate this view, HL from early fifth instar larvae (1 to 5 d) and HL during the larval-pupal transition (6 to 9 d) were each tested with NHS. As shown in Figure 7A, early larval HL preferentially reduced t_0 rather than $t_{1/2}$, whereas late larval HL reduced both t_0 and $t_{1/2}$ maximally. This effect of late larval HL can also be demonstrated by heat treatment of early larval HL (Figure 7B). Upon heat treatment of larval HL, the C-inhibiting property was abolished and the C-activating property was enhanced. Taken together, these studies suggested that although HL from both early and late larval stages has C-activating as well as C-inhibiting effects on NHS, the C-activating property of HL is enhanced during the larval-pupal transition, thereby masking the effect of inhibitor. The lysis-promoting property of late larval HL, was not due to the disappearance of the inhibitor, but due to the complete release of the C activator in free form during the larval-pupal transition.

The ACP inhibitor was fractionated from early fifth instar *Bombyx* HL by precipitation with $(NH_4)_2SO_4$, anion exchange chromatography on DEAE-Sephacel, and gel-filtration

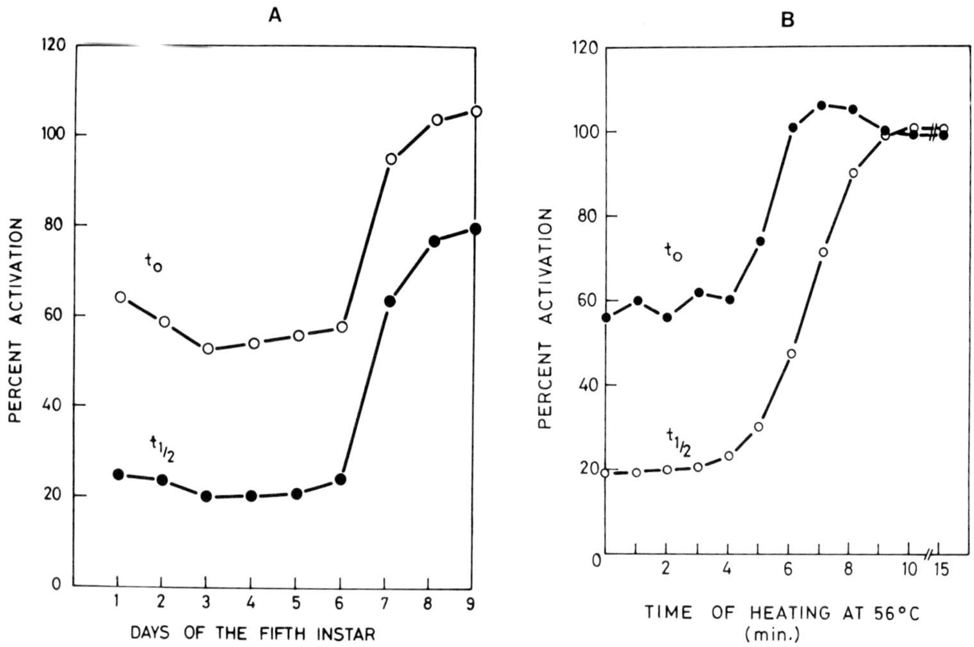

FIGURE 7. (A) C-activating property of HL obtained from different stages of larval development. Fifth instar larvae were bled every day from the first day of last molting stage until pupation; 100 μl of cell-free HL was preincubated with NHS (1:3) and the variations obtained for t_0 and $t_{1/2}$ were calculated from the recorder tracings for each sample. The percent activation in t_0 and $t_{1/2}$ was derived from t_0 control/t_0 test and $t_{1/2}$ control/$t_{1/2}$ test indices. (B) Heat-activation of HL complement activator; 1 ml of HL was incubated at 56°C; every 60 s intervals the tubes were chilled, centrifuged, and 100 μl of the supernatant was incubated with NHS (1:3). The percent activation in t_0 and $t_{1/2}$ was calculated with reference to the t_0 and $t_{1/2}$ indices obtained for undiluted serum.

chromatography on Bio-Gel 0.5 *m* and Sephacryl-200. The semipurified active fraction revealed a prominent band of M_r 120 kDa under nonreducing (Figure 8, lane 2) and reducing (Figure 8, lane 3) conditions on 10% SDS-PAGE. Functionally, the ACP inhibitors isolated from *Bombyx* and *Spodoptera* HL were identical.

The C activator, a low molecular weight peptide (1.4 kDa), was purified from heat-treated fifth instar larval HL by molecular sieve chromatography on Sephadex G-75, followed by Bio-Gel P-10 and P-2 and differential dialysis. A preliminary analysis of the amino acid content of the C-activating peptide revealed unusually high content of lysine.

When *Bombyx* HL was absorbed with several bacterial pathogens of the silkworm, (isolated from diseased larvae) the C-activating peptide of HL was specifically depleted by protein A bearing *Staphylococcus aureus*. Incubation of protein A positive ^{14}C-labeled *Staphylococcus aureus* with larval *Bombyx* HL also promoted their binding and internalization by *Bombyx* hemocytes *in vitro*. These findings suggested that the complement activator peptide of the HL may act as an opsonin *in vivo*.

The differential ability of larval HL to affect the rate of lysis of RaRBC as well as the heat activation kinetics is also indicative of the presence of a third component acting as a modulator for the HL peptide during larval development. It is likely that during the larval-pupal transition, the lysis-promoting peptide could be released by limiting proteolysis of the modulator protein.

The specific induction/release of the cell-lysis-enhancing HL peptide during the larval-pupal transition and its modulation by a regulator protein in HL, suggests a causal association of C-interacting *Bombyx* HL proteins with events leading to histolysis of specific larval tissues that take place during the metamorphosis of silkworm.

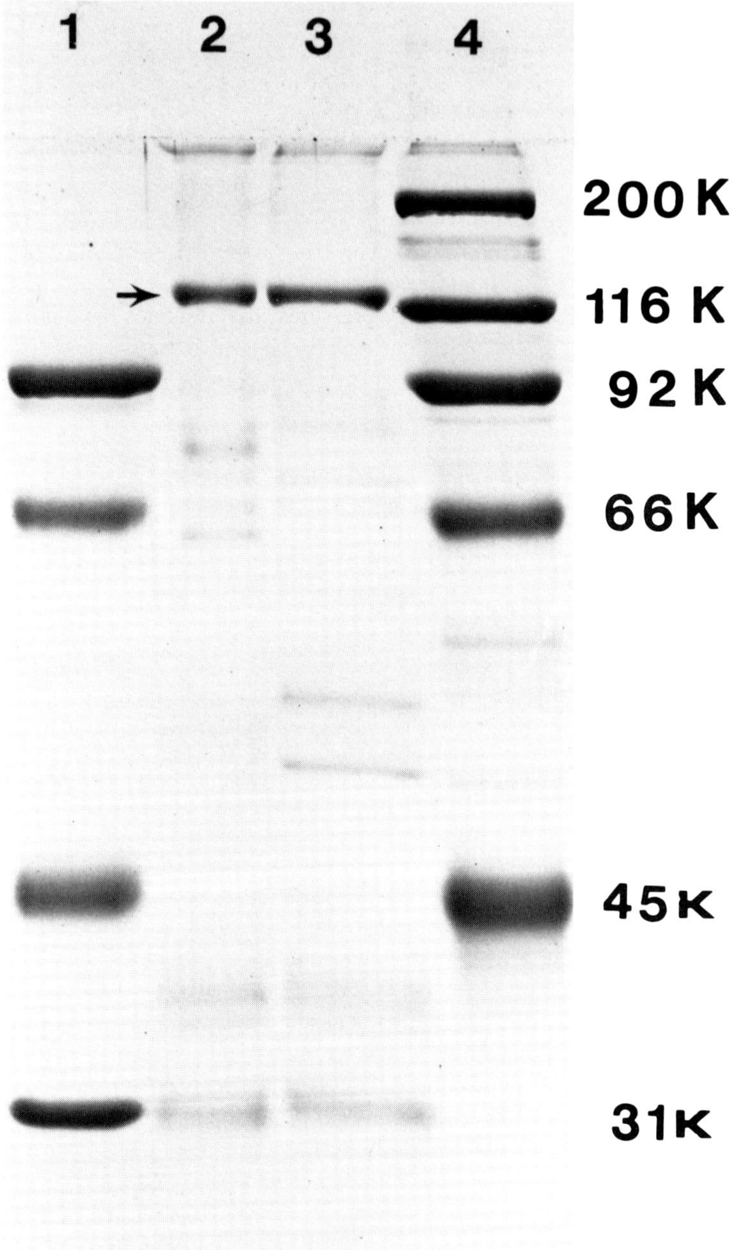

FIGURE 8. A 10% SDS-PAGE of partially purified ACP inhibitor fractionated from larval *Bombyx* HL. Lane 1, LMW marker proteins; lane 2, nonreduced active fraction; lane 3, reduced fraction; lane 4, HMW marker proteins.

V. SUMMARY

From the context of phylogeny, the presence of specific inhibitors of the classical and alternative C pathway of mammalian C in insect HL raises several speculations in view of the ability of the C system to distinguish self from nonself, based on cell surface differences. The fact that insect HL proteins can selectively interact at the C3 and C4 regulatory steps in the mammalian C cascade, might suggest that similar reactions may also be operative *in vivo*; however, the detection of these reactions might require appropriate assays. Since C3 and C4 have considerable sequence and structural homologies which is also consistent with gene duplication, it is highly probable that these key components of the vertebrate C may also exist in insect HL. These molecules could be the natural substrates for either inhibitor proteins. Alternatively, the presence of specific inhibitors for mammalian C in insect HL in the absence of a detectable C-like reaction suggests that the regulatory proteins of the vertebrate C perhaps existed prior to the appearance of a C-type reaction. Some of these inhibitors might have had a common origin with other important metabolic reactions. Thus, C3, which is central to the C system in vertebrates, may have been the original precursor molecule in insects, while the evolutionary pressures in vertebrates have provided several pathways mediated by C3 to assemble the cytolytic C3 and C5 convertases on the cellular target and corresponding acceptor proteins to regulate their amplification. The variable ability of insect HL proteins to either potentiate or inhibit these reactions at the C3 convertase levels suggests that C system may have originally evolved as a dual system for development and defense. Indeed, recent studies have shown that the terminal C pathway inhibitors (clusterin and vitronectin) of human blood plasma can also mediate cell recognition and/or cell activation by the receptors of the cytoadhesin/integrin protein family. Therefore, with the availability of purified C components and monoclonal antibodies to C proteins, investigations on the molecular mechanisms of insect HL proteins interacting with mammalian C relating to insect development promise to open up new horizons in the area of insect immunity.

ACKNOWLEDGMENT

I am grateful to Dr. Alfred F. Esser, University of Florida, Gainesville for the electron micrographs of purified LSP-I.

REFERENCES

1. **Vitetta, E., Fernandez-Botran, R., Myers, C. D., and Sanders, V. M.,** Cellular interactions in the humoral immune response, *Adv. Immunol.*, 45, 1, 1989.
2. **Muller-Eberhard, H. J.,** Molecular organization and function of the complement system, *Annu. Rev. Biochem.*, 57, 321, 1988.
3. **D'Cruz, O. J., Haas, G. G., Jr., and Lambert, H.,** Evaluation of antisperm complement-dependent immune mediators in human ovarian follicular fluid, *J. Immunol.*, 144, 3841, 1990.
4. **D'Cruz, O. J., Haas, G. G., Jr., Wang, B., and DeBault, L. E.,** Activation of human complement by IgG antisperm antibody and the demonstration of C3 and C5b-9 mediated immune injury to human sperm, *J. Immunol.*, 146, 611, 1991.
5. **Nonaka, M., Fuji, T., Kaidoh, T., Natsuume-Sakai, S., Yamaguchi, N., and Takahashi, M.,** Purification of a lamprey complement protein homologous to the third component of the mammalian complement system, *J. Immunol.*, 133, 3242, 1984.
6. **Volanakis, J. E.,** Participation of C3 and its ligands in complement activation, *Curr. Top. Microbiol. Immunol.*, 153, 1, 1989.
7. **Atkinson, J. P. and Farries, T.,** Separation of self from non-self in the complement system, *Immunol. Today*, 8, 212, 1987.
8. **Day, N. K. B., Gewruz, H., Johannsen, R., Finstad, J., and Good, R. A.,** Complement and complement-like activity in lower vertebrates and invertebrates, *J. Exp. Med.*, 132, 941, 1970.
9. **Day, N. K. B., Geiger, H., Finstad, J., and Good, R. A.,** A starfish hemolymph factor which activates vertebrate complement in the presence of cobra venom factor, *J. Immunol.*, 109, 164, 1972.
10. **Jensen, J. A., Festa, E., Smith, D. S., and Cayer, M.,** The complement system of the nurse shark: hemolytic and comparative characteristics, *Science*, 214, 566, 1981.

11. **Linthicum, D. S. and Hildemann, W. H.,** Immunological responses of pacific hagfish. III. Serum antibodies to cellular antigens, *J. Immunol.,* 105, 912, 1970.
12. **D'Cruz, O. J., Day, N. K., and Good, R. A.,** The phenomena of immunity in invertebrates, in *Iwanami Series in Immunology,* Benacerraf, B. and Ishizaka, K., Eds., Shoten Publ., Tokyo, 1985, chapter 1.
13. **Steiner, H., Hultmark, D., Engstrom, A., Bennich, H., and Boman, H. G.,** Sequence and specificity of two antibacterial proteins involved in insect immunity, *Nature (London),* 292, 246, 1981.
14. **Van Hofsten, P., Faye, I., Kockum, K., Lee, J.-Y., Xanthopoulous, K. G., Boman, I. A., Boman, H. G., Engstrom, A., Andreu, D., and Merrifield, R. B.,** Molecular cloning, cDNA sequencing, and chemical synthesis of cecropin B from *Hyalophora cecropia, Proc. Natl. Acad. Sci. U.S.A.,* 82, 2240, 1985.
15. **D'Cruz, O. J. M. and Day, N. K.,** Purification of a human alternative complement pathway inhibitor from hemolymph of larval fall armyworm *(Spodoptera frugiperda), Biochem. Biophys. Res. Commun.,* 1229, 1426, 1984.
16. **D'Cruz, O. J. M. and Day, N. K.,** Structural and functional similarities between the major hemolymph protein of fall armyworm and C4 binding protein of the complement system, *Dev. Comp. Immunol.,* 9, 541, 1985.
17. **Gee, A.,** Molecular titration of components of the classical complement pathway, *Methods Enzymol.,* 93, 339, 1983.
18. **Wyatt, G. R. and Pan, M. L.,** Insect plasma proteins, *Annu. Rev. Biochem.,* 47, 779, 1978.
19. **Roberts, D. B. and Brock, H. W.,** The major serum proteins of Dipteran larvae, *Experientia,* 37, 103, 1981.
20. **Dahlback, B., Smith, C. A., and Muller-Eberhard, J. H.,** Visualization of human C4 binding protein and its complexes with vitamin K-dependent protein S and complement protein C4b, *Proc. Natl. Acad. Sci. U.S.A.,* 80, 3461, 1983.
21. **Fujita, T., Gigli, I., and Nussenzweig, V.,** Human C4-binding protein II role of proteolysis of C4b by C3b inactivator, *J. Exp. Med.,* 148, 1044, 1978.
22. **Gigli, I., Fujita, E., and Nussenzweig, V.,** Modulation of the classical pathway C3 convertase by plasma proteins C4 binding protein and C3b inactivator, *Proc. Natl. Acad. Sci. U.S.A.,* 76, 6596, 1979.
23. **Nagasawa, S., Mizuguchi, K., Ichihara, C., and Koyama, J.,** Limited chymotryptic cleavage of human C4-binding protein: isolation of a carbohydrate-containing core domain and an active fragment, *J. Biochem.,* 92, 1329, 1982.
24. **Kaidoh, T. and Gigli, I.,** Phylogeny of C4b-C3b cleaving activity: similar fragmentation patterns of human C4b and C3b produced by lower animals, *J. Immunol.,* 139, 194, 1987.
25. **Lepesant, J. A., Lepesant, J. K., and Garen, A.,** Ecdysone-inducible functions of larval fat bodies in *Drosophila, Proc. Natl. Acad. Sci. U.S.A.,* 75, 5570, 1978.
26. **Jowett, T. and Postlethwait, J. H.,** Hormonal regulation of synthesis of yolk proteins and a larval serum protein (LSP2) in Drosophila, *Nature (London),* 292, 633, 1981.
27. **Ferriera, A., Carrington, P. W., and Nussenzweig, V.,** Testosterone control of serum levels of C4-binding protein in mice, *J. Immunol.,* 121, 1213, 1978.
28. **Sridhara, S.,** Macromolecular changes during insect metamorphosis, in *Metamorphosis,* Gilbert, L. I. and Ferieden, E., Eds., Plenum Press, New York, 1981, chap. 6.
29. **Polhill, R. B., Jr., Pruitt, K. M., and Johnston, R. B., Jr.,** Kinetic assessment of alternative complement pathway activity in a hemolytic system. Experimental and mathematical analysis, *J. Immunol.,* 121, 363, 1978.
30. **D'Cruz, O. J. M.,** Interaction of larval silkworm hemolymph with human complement, *Fed. Proc.,* 42, 1237, 1983.

Chapter 15

ADAPTIVE HUMORAL IMMUNITY IN THE AMERICAN COCKROACH

Laura E. Duwel-Eby, Lenore M. Faulhaber, and Richard D. Karp

TABLE OF CONTENTS

I. INTRODUCTION

In higher vertebrates, the immune response to the invasion of microorganisms and other foreign substances is specific. There are basically two kinds of immune responses: anticipatory and nonanticipatory.[1] Anticipatory immune responses are based on the ability of an organism to produce enough specific receptors capable of binding any distinct antigen (Ag) the organism might encounter. The binding of such an Ag will lead to the induction of a specific immune reaction with the ability to differentiate even very slight biochemical differences between molecules. In addition, anticipatory immune responses demonstrate immunologic memory, that is, a primed animal recognizes that it has had a previous exposure to an Ag, and reacts in an accelerated fashion to a repeat exposure. Anticipatory immune responses are therefore Ag-specific, mediated by Ag-specific receptors, highly discriminatory between self and nonself, and characterized by anamnestic reactivity upon reexposure to Ag. In contrast, nonanticipatory immune responses do not anticipate the encounter with a particular immunogen and thus, can only respond to all Ags indiscriminately with little specificity.

Vertebrates have developed highly sophisticated anticipatory and nonanticipatory immune responses.[1] In mammals, anticipatory responses are mediated by antibodies (Ab) or lymphocytes (i.e., transplantation reactions), whereas nonanticipatory reactions are carried out by macrophages, natural killer cells, eosinophils, coagulation proteins, lymphokines, and many other elements. It is common knowledge that invertebrates, such as insects and other arthropods, are capable of nonanticipatory responses; however, one of the raging controversies in immunology is whether invertebrates are also capable of specific anticipatory responses to Ag. It has been argued that: (1) invertebrates lack the complexity and cell numbers needed to sustain an inducible anticipatory response and (2) invertebrates generate larger populations and have shorter life spans, and therefore, can simply out-reproduce environmental insults.[2] It is of interest to note that many invertebrates have life spans equal to or greater than that of the laboratory mouse and thus, face the same selective pressures for survival to achieve reproductive maturity. Evidence is quickly accumulating which indicates that invertebrates have evolved an impressive array of cellular and humoral defense strategies to maintain their integrity.[3-7] It is now of critical importance to elucidate these immune mechanisms and to determine whether invertebrates are truly capable of anticipatory immune responses similar to vertebrates or whether the invertebrate mechanisms, although protective, are fundamentally different.

By far, the majority of current studies concerned with defining the nature of invertebrate immunity have utilized insect systems. Most of the inducible humoral factors described to date have been generated by the injection of bacterial vaccines in one form or another. The existence of antibacterial activity in insects was first reported by Glaser[8] in so-called stimulated grasshoppers. Soon after, Metalnikov,[9] and others through the years had shown that low doses of injected bacteria led to the protection of these animals upon subsequent challenge with lethal doses of the microbes. However, this immunity was also found to be generated by using bacterial products or components. For instance, Chadwick and Vilk[10] described studies in which *Galleria mellonella* larvae were immunized with vaccines derived from *Pseudomonas aeruginosa* endotoxins, resulting in a significant decrease in the percent mortality after challenge with LD_{50} doses of the active toxin (see also Chapters 12 and 13). In addition, Dunn et al.[11] have shown that soluble peptidoglycan fragments from bacterial cell walls can stimulate effective antibacterial responses in *Manduca sexta*.

The pioneering work of Boman and co-workers[12] had a tremendous influence on the burgeoning field of insect immunology when they reported the existence of two families of inducible antibacterial peptides, which were named cecropins and attacins. The cecropins, which are basic proteins with M_r 4 kDa are capable of lysing various species of Gram-negative bacteria. The attacins, with M_r 20 to 23 kDa have both basic and acidic forms, and also show a fair degree of nonspecificity. Both cecropins and attacins display peak antibacterial activity within 72 h postinjection of bacteria, and then rapidly decline in activity. In addition, the lepidopteran response to bacteria fails to demonstrate immunologic memory upon reexposure to the original microbial pathogen. Since that time, other inducible antibacterial peptides, such as the diptericins and defensins, have also been discovered.[13-15] Thus, it appears that holometabolous insects have developed the capacity to quickly meet bacterial challenges, even though nonspecifically, by the production of various small, active peptides.

Kamon and Shulov[16] reported on one of the first definitive studies, demonstrating a nonbacterial inducible humoral response in an insect system. They were able to induce acquired immunity in adult locusts by employing repeated injections of sublethal doses of scorpion venom and comparing the resulting mortality with that of nonimmunized control insects. This response was found to provide only short-term, nonspecific protection that peaked at 24 h postinjection.[17] In similar studies, Bettini[18] immunized groups of houseflies with a low dose of crude spider venom to assay for inducible immunity, while Chadwick and Vilk[10] in the study described before used greater wax moth larvae to test for immunity against injected pseudomonad endotoxin. Both approaches resulted in the generation of short-term protection. However, a secondary response could not be induced upon repeat exposure to the immunizing Ag.

In contrast, results in our lab have reported the generation of a humoral (cell-free) protective factor in the American cockroach, *Periplaneta americana*, which is induced by the injection of inactivated protein toxins, such as snake phospholipase A_2 (SPA2),[19] and other nontoxic soluble proteins, such as bovine gamma globulin (BGG)[20] and creatine phosphokinase (CPK).[21] The response has been shown to be long-term in nature, as well as displaying both specificity and memory,[22-23] which are hallmark criteria for the definition of true adaptive immune reactivity. The molecule mediating this effect is apparently a large protein present in the circulating hemolymph and is capable of precipitating its corresponding Ag in Ouchterlony gels.[24] Thus, this inducible humoral factor (IHF) from the roach may be a worthy candidate as an invertebrate Ab-like molecule. In order to establish any structural similarity and evolutionary relatedness between the roach IHF and vertebrate Ab, current studies in our lab are focused on identifying the mediating molecule responsible for this inducible protection. The results from our studies on the American cockroach should provide valuable information regarding the definition of invertebrate immunity and the evolution of immune reactivity in general.

II. THE BIOLOGY OF THE PRIMARY IMMUNE RESPONSE IN THE AMERICAN COCKROACH

Initial studies to investigate the immunocompetency of the American cockroach by Rheins et al.[22] used soluble protein venoms from the honeybee (*Apis mellifera*) and the western cottonmouth moccasin (*Agkistrodon piscivorus*) as Ag. These studies utilized a functional approach to evaluate the roach's responsiveness to lethal toxic proteins. Adult male roaches were immunized with a single injection of honeybee toxoid (HBTd) and assayed for immunity by scoring for survival after a lethal challenge dose of the active toxin. The results demonstrated (Figure 1) that a protective response began to develop within 3 d postimmunization. The response continued to gain in strength as reflected by a 62.9% survival rate at 1 week and then peaked at 80% after 14 d of rest postprimary injection. The rate of protection began to decline after 3 weeks and did not significantly differ from controls by the fourth week. Less than 10% of the control animals that received Burns Tracey saline (BTS), instead of toxoid, survived the challenge injection of HBT. Thus, the primary response

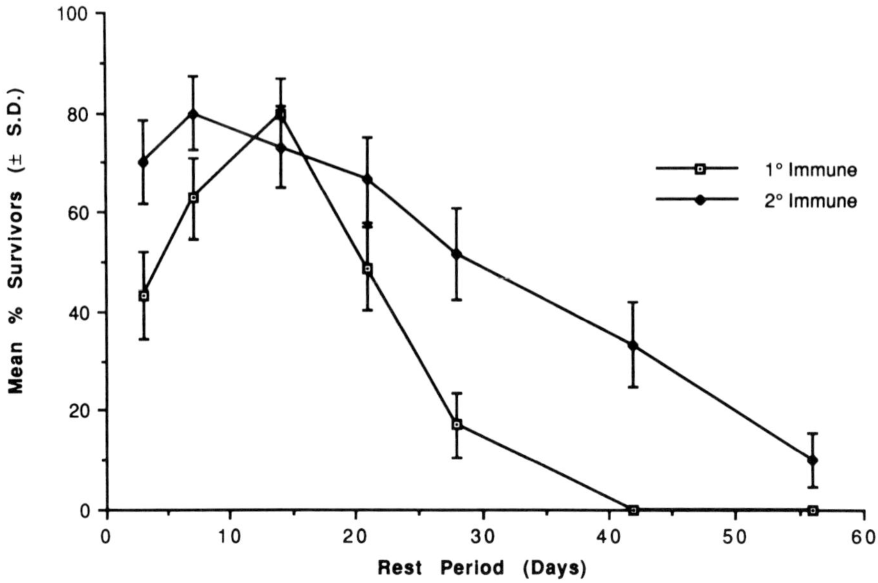

FIGURE 1. The primary and secondary response of the adult male cockroach to honeybee toxoid. (From Karp, R. D., original.)

of the American cockroach demonstrated long-term protection similar to that seen for vertebrate immune responses, in contrast to the short-term humoral factors generated in other insect systems.

Antigen specificity is a characteristic of vertebrate anticipatory immune responsiveness that permits vertebrates to distinguish fine biochemical differences at the molecular level. Thus, in order for the roach IHF to be considered as an Ab-like molecule, it would be critical that the response demonstrate specificity for the immunizing Ag. To evaluate the selectivity of the roach immune response to Ag, a heterologous venom was used for challenge. Cockroaches were immunized with either HBTd or CMVd and then rested for 2 weeks postinjection to allow for the induction of a maximum protective response.[22] Animals receiving heterologous toxin challenges demonstrated poor protection, since only 10% of the roaches immunized against HBTd survived challenge with CMV, and just 5% of the CMVd-primed animals were protected against HBT.

In contrast, animals injected with HBTd and challenged with HBT demonstrated a 84.2% survival rate, while a slightly lower but significant rate of 65% protection was observed for CMVd immunized animals that were challenged with CMV. None of the control roaches receiving sterile saline as the primary injection survived challenge with either toxin. Many arthropods generate humoral factors that inactivate viruses, bacteria, and protozoans, but generally with little specificity.[25] Thus, the roach IHF, like vertebrate immunoglobulins, was capable of selectively responding to the venoms in an Ag-specific manner. In addition, passive transfer experiments demonstrated that the mediating factor was humoral in nature, since protection could be transferred to naive recipient roaches via cell-free immune hemolymph.[22] Differential immunity, however, was not observed upon transfer of circulatory hemocytes from immune animals to naive recipients.

III. NATURE OF THE SECONDARY IMMUNE RESPONSE

The two key features of an adaptive immune system are specificity and memory. In vertebrates, repeated exposure to an Ag will trigger an anamnestic response that appears more quickly, persists for longer periods of time, and is characterized by a general enhancement of activity. The cellular basis of memory is the expansion of populations of Ag-specific lymphocytes during the primary response so that there is an increased frequency of cells capable of responding to that particular Ag.[26] To demonstrate the presence of immunologic memory in the roach, groups of adult males were given 1° and 2° injections of HBTd, spaced 2 weeks apart.[23] Roaches were then challenged with active HBT at various time periods postsecondary immunization. A secondary response was readily induced in the roaches, since hyperimmune animals rested 3 d, and then challenged, displayed a significant increase ($p = 0.005$) in the protection rate (Figure 1). A 70% survival rate was observed for roaches receiving 1° and 2° HBTd injections, compared with a 43.3% survival rate seen in animals receiving only a primary dose of Ag. In addition, there was a quicker attainment of peak activity. The peak response of 80% occurred after only 7 d, a week earlier than observed in the primary response. The secondary response was also found to be more prolonged and more persistent, since a high rate of protection continued for 2 to 3 weeks, and did not begin to decline until the fourth week of the response.

Thus, a second injection of toxoid into previously immunized animals induced a more intense and prolonged response, confirming that roaches primed with certain Ags are capable of eliciting immunologic memory. The secondary response also demonstrated specificity for the immunizing Ag, since HBTd primed animals were unable to generate an anamnestic response when a heterologous Ag, such as CMVd, was administered as the second injection. By all criteria, a typical anamnestic response has been demonstrated for the humoral immune response in the American cockroach. The induction of immunologic memory in an invertebrate system provides new evidence that invertebrates are capable of both anticipatory and nonanticipatory immune responses, and that origins of vertebrate immunity may, in fact, be found among invertebrates.

IV. PRECIPITIN ACTIVITY OF THE INDUCIBLE HUMORAL FACTOR

The fundamental basis of Ab-mediated immunity depends on the reactions of Ag with Ab molecules.[27] *In vivo*, mammalian IgG Ab molecules act as antitoxins to neutralize toxic proteins by combining with the toxin, blocking their reactions, and preventing the onset of pathological symptoms. When Ag and Ab molecules react, they form large polymeric macromolecules, with solubility properties that result in their precipitation. The precipitation of Ag-Ab complexes can be readily visualized by allowing the Ag and Ab proteins to mix in an agar gel. Rheins and Karp[24] utilized this method to determined if the roach IHF was analogous to an Ab-like molecule. To test immune hemolymph for precipitin activity, a series of Ouchterlony gel double diffusion tests were performed. Undiluted HBT was placed in the center well of a 1% Noble agar Ouchterlony gel, and twofold serial dilutions of anti-HBTd hyperimmune hemolymph were placed in the surrounding wells. After 48 h of incubation, at least one very strong homologous precipitin band was observed between the hemolymph and Ag. As with the *in vivo* response, this reaction was specific, since there was no evidence of precipitin band formation when HBT was incubated with saline-injected control hemolymph or with heterologous hemolymph from CMVd immunized animals. Treatment of anti-HBTd immune hemolymph with trypsin completely abrogated the formation of precipitin bands, confirming that the IHF was protein in nature. In addition, trypsin-treated immune hemolymph used for passive transfer experiments lost its ability to confer protection to naive recipient roaches. The results from these experiments indicated that the roach IHF is a circulating hemolymph protein capable of acting as a precipitating Ab-like molecule *in vitro*.

V. THE EFFECT OF GENDER AND AGE ON THE RESPONSE

The generation of long-term Ag-specific 1° and 2° responses to protein toxins provides evidence that the American cockroach is capable of generating an adaptive immune response similar to those observed in vertebrate species. In light of this, other biological parameters of immunity were investigated to define additional similarities between vertebrate and invertebrate humoral systems. In general, the females of most vertebrate species generate superior responses, as compared with males.[28-29] It has been theorized that since the female is responsible for the development and nurturing of offspring, she must be better prepared to maintain her immunologic integrity in order to ensure the success of the species as a whole. To test this law of nature, adult female roaches were immunized with HBTd, rested for various periods of time, and then challenged with a lethal dose of HBT.[30] The results showed that a protective response was generated in both sexes within 3 d postimmunization. A comparison between male and female primary responses showed that male roaches rested 7 d following the 1° immunization had a 62.9% survival rate, whereas females had a significantly greater protection rate of 92.0% at the same time period. Female roaches not only reached peak activity a week sooner (day 7 vs. 14), but also they reached higher levels (92 vs. 80%) and maintained the response at these higher levels for a longer period of time (approximately twice as long as the male response). These data support the general phenomenon found among the vertebrates in which the female of the species appears to react and defend itself better immunologically than males.

Another typical trait associated with vertebrate immunity is that both humoral and cell-mediated immunity are not optimal in the very young and decline in an age-related fashion after reaching peak responsiveness approximately at the time of sexual maturity.[31] To study the ontogenetic development of immunity in the American cockroach, animals representing distinct developmental stages were tested for their ability to generate a protective immune response to HBT.[32] Immature nymphs (animals weighing between 200 to 500 mg) displayed a significant lag in reactivity, as compared with young adults. This delay may have been due to the nymph's inability to process the Ag as efficiently as the adult. Despite this lag, the nymphs were still capable of achieving adult levels of immunity against the HBT by 2 weeks postimmunization. On the other hand, old adult roaches (animals at least 5 months into adulthood) displayed a significant decline in reactivity during early phases of the response and, in general, demonstrated a less vigorous response. Both nymphs and old adult roaches generated good secondary responsiveness, which indicated that once cells were primed, they reacted just as well as those of the adult. The results from these experiments indicated that the developmental stage of the roach was directly correlated to the degree of immunocompetency possessed by the animal. As with vertebrates, young roaches appeared to lack the developmental capacity to respond as quickly to antigenic challenge as mature adults, whereas older adults began to lapse into immunosenescence.

VI. THE ANTIBACTERIAL RESPONSE OF THE AMERICAN COCKROACH

It was only natural that we would become interested in determining if a paurometabolous insect, such as the cockroach, also generates potent antibacterial responses. The injection of bacteria into the hemocoel of the American cockroach has led to some very interesting preliminary results. At this point, we feel that the response may be composed of both the lepidopteran-type antibacterial response as well as the true immune response elicited by soluble protein antigens. Figure 2 represents the results from an experiment in which adult male cockroaches were immunized with either pyrogen-free BTS (control saline) or glutaraldehyde-fixed *Pseudomonas aeruginosa* cells. Animals were then challenged with a LD_{100}

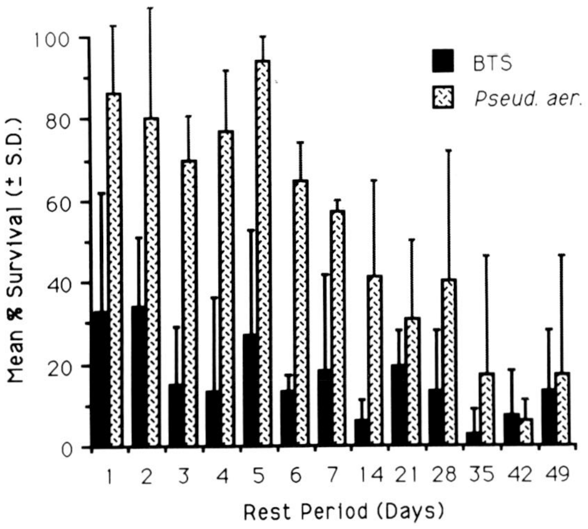

FIGURE 2. Kinetics of the antibacterial response. Adult male cockroaches were immunized with either Burns Tracey saline (BTS) or glutaraldehyde fixed *P. aeruginosa* cells (*Pseud. aer.*). Animals were challenged with viable *P. aeruginosa* cells at the times indicated after immunization. The percent survival was determined 48 h after challenge. (From Faulhaber, L. M. and Karp, R. D., in preparation.)

dose of viable cells at various days postimmunization. The average percent survival from three trials per time point was then plotted. It would appear that the response is diphasic in nature. The first phase of the response appeared to peak within the first 24 h after immunization and began to decline by 72 h after immunization. The second phase became apparent by day four postimmunization, and peaked by day five. By day 21 postimmunization, there was borderline significance between the control group and the group immunized with fixed *P. aeruginosa* cells.

To test the specificity of the response, unrelated bacteria were used for immunization of animals prior to challenge with viable *P. aeruginosa*. The Gram-negative bacterium *Serratia marcescens* and the Gram-positive bacterium *Streptococcus lactis* were fixed with glutaraldehyde and used for immunization at doses comparable to that used for *P. aeruginosa*. The results shown in Figure 3 compare the response induced by immunization with BTS and *P. aeruginosa* (as seen in Figure 2), with the response induced by immunization with *S. marcescens* and *S. lactis*. The first phase of the response appeared to be nonspecific in nature, since protection against challenge is provided by both the use of Gram-positive and Gram-negative bacteria for immunization. It should be noted that in addition to the data presented, other Gram-positive and Gram-negative bacteria have been tested and similar results have been observed. The second phase of the response displayed a higher degree of specificity. No significant protection was provided by Gram-positive bacteria, and although immunity was induced by immunizing with a Gram-negative bacterium, the percent survival was consistently lower than for animals immunized with *P. aeruginosa*, the same bacterium used for challenge.

Animals used in the studies just discussed were less than 1 month into adulthood. It is beginning to appear that younger adults are capable of producing only a relatively nonspecific response to bacteria, whereas the response of older animals may be much more specific. Cockroaches can live as long as 3 years; thus, animals less than 1 month or even 2 months into adulthood are considered to be young adults. The majority of the previous studies in the area of antibacterial responses in insects have been done on the juvenile stages of the

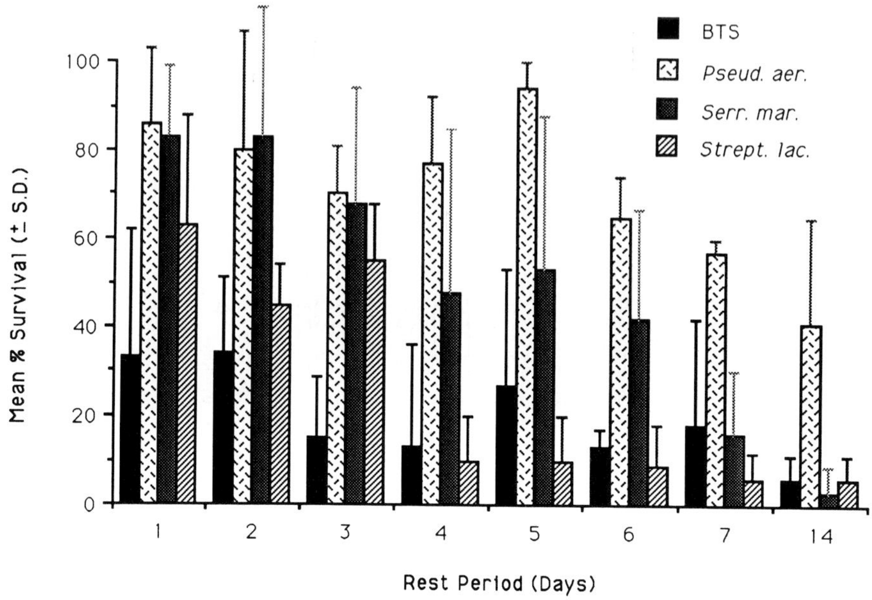

FIGURE 3. Kinetics of the specificity of the antibacterial response. Adult male cockroaches were immunized with Burns Tracey saline (BTS), or glutaraldehyde-fixed cells of either *P. aeruginosa (Pseud. aer.)*, *Serratia marcescens (Serr. mar.)*, or *Streptococcus lactis (Strept. lac.)*. Animals were challenged with viable *P. aeruginosa* cells at the times indicated after immunization and percent survival determined 48 h postchallenge. (From Faulhaber, L. M. and Karp, R. D., in preparation.)

species, and these responses have displayed a low degree of specificity. Thus, our results in the roach may make sense since this animal may need several months of development as an adult before it has the capacity to generate a specific response.

The ability of the roach to distinguish between various types of bacteria at later time points in the response leads to the hypothesis that the induction of a humoral antibacterial factor has occurred during this period. As seen in Figure 2, survival was greatest when animals were immunized 5 d prior to challenge. Polyacrylamide gel electrophoresis (PAGE) was used to separate hemolymph proteins from animals immunized at this time point. A second group of animals was challenged with BTS or *P. aeruginosa* 5 d after immunization with BTS or *P. aeruginosa*, respectively. The survivors from the *P. aeruginosa* group and animals from the BTS-control group were bled 48 h after challenge, and the hemolymph proteins were analyzed. Both nonreducing PAGE and SDS-PAGE analysis indicated an increased protein band in animals immunized with, or immunized and challenged with *P. aeruginosa* as compared with controls. As seen in the NR-PAGE gel (Figure 4), there was an increase in the band designated NR-2b. The corresponding increased band in the reduced proteins had a molecular weight of approximately 91 kDa (Figure 5). PAGE analysis of hemolymph from animals immunized 2 d prior to bleeding (the middle of the first phase of the response) showed similar results (data not shown) and interestingly no apparent increase in lower molecular weight proteins. The fact that the first phase of the response was so similar to the response attributed to attacin production in the giant silk moth, i.e., nonspecific protection peaking within 72 h after injection of bacteria, leads one to suspect that this low molecular weight hemolymph protein might exist in the American cockroach. However, if the increased 91-kDa protein has antibacterial properties and its production begins within 48 h postimmunization, it may be partially responsible for the increased survival. It is very likely that cellular responses may also be involved and hemocyte stimulation may be responsible for the low specificity represented by the first phase of the response, thus leaving humoral factors to play a reduced role in defense at these early times after immunization.

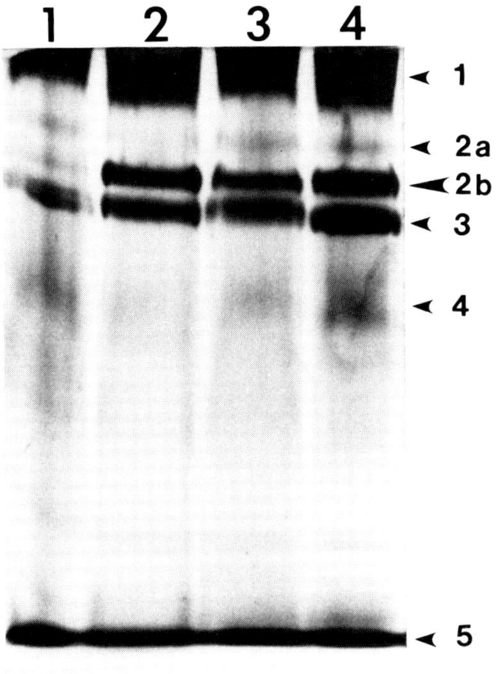

FIGURE 4. A 6% NR-PAGE analysis of hemolymph samples from adult male cockroaches. Animals were either immunized with Burns Tracey saline (BTS) (lane 1), immunized and challenged with BTS (lane 2), immunized with glutaraldehyde-fixed *P. aeruginosa* cells (lane 3), or immunized with fixed cells and challenged with viable bacteria (lane 4). (From Faulhaber, L. M. and Karp, R. D., in preparation.)

VII. SDS-PAGE ANALYSIS OF THE INDUCIBLE HUMORAL FACTOR TO SOLUBLE PROTEINS

Since we had demonstrated that the protective response to HBT in the roach was humoral in nature, we next investigated whether immunization resulted in the production of unique or enhanced proteins in the hemolymph after injection of soluble protein Ag. Initial characterization of the roach hemolymph proteins using SDS-PAGE had shown that a number of hemolymph proteins were newly induced or enhanced in response to injection of HBTd, compared with saline injected controls.[21] Two unique proteins with reduced M_r 220 and 162 kDa were observed in immune samples. Densitometric scans of the gels further revealed an increase in quantity of several proteins with M_r 115, 102, 95, and 45 kDa. The 115-kDa protein was later determined to be gender-specific (observed only in female roaches) and was therefore not considered to be a common mediator of the roach humoral immune response. This left the 220-, 162-, 102-, 95-, and 45-kDa bands as possibly being associated with the roach immune response to HBTd.

HBT is actually a complex mixture of several components, including phospholipase A_2, melittin, hyaluronidase, and several other low molecular weight polypeptides and hormones. HBT was thus too complex to be used for subsequent isolation and characterization of the IHF. It was determined that the PA2 protein component of the venom could totally mimic the toxicity and immunogenicity of HBT in the roach.[19] When inactivated PA2 was used as the immunizing Ag, the only observable differences between immune and control hemolymph

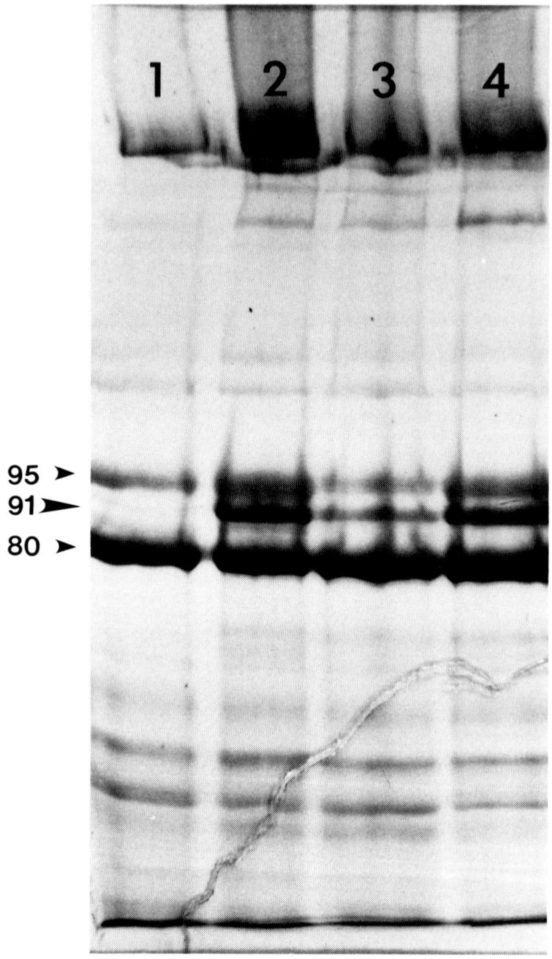

FIGURE 5. A 6% SDS-PAGE analysis of hemolymph samples from adult males cockroaches. Animals were either immunized with Burns Tracey (BTS) (lane 1), immunized with and challenged with BTS (lane 2), immunized with glutaraldehyde-fixed *P. aeruginosa* cells (lane 3), or immunized with fixed cells and challenged with viable bacteria (lane 4). (From Faulhaber, L. M. and Karp, R. D., in preparation.)

from adult female roaches were increases in the 115- and 102-kDa bands. Similar results were obtained using PA2 from snake venom or unrelated Ags such as BGG[20] and CPK.[21] Since the 115-kDa band was gender specific, the only protein that was consistently enhanced by all Ags tested was the 102-kDa band.

We have previously shown that the roach IHF is capable of precipitating Ag in Ouchterlony gels.[24] Thus, extensive absorption studies were undertaken to identify the hemolymph proteins that were involved in the precipitation reaction. In these studies, anti-HBTd hemolymph was incubated with equal amounts of either HBTd, PA2d, or several specificity control Ags, such as BGG, ovalbumin, and cytochrome c. SDS-PAGE analysis of the resulting precipitates revealed that the 102-kDa band was consistently precipitated when HBTd or PA2d were incubated with the anti-HBTd immune hemolymph.[21] In contrast, the unrelated Ags failed to produce any detectable precipitates when incubated with the immune samples. From these results, the 102-kD band seems to be strongly associated with the roach immune response to soluble protein Ag.

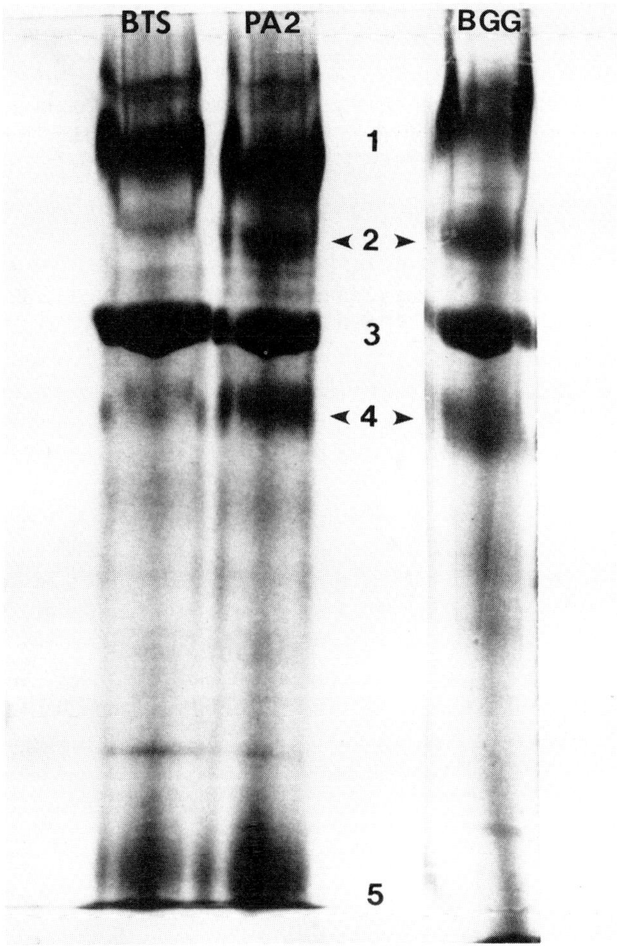

FIGURE 6. A 6% NR-PAGE analysis of hemolymph samples from female roaches demonstrating enhancement of bands 2 and 4 in response to immunization with SPA2d or BGG. BTS, hemolymph from saline injected controls; PA2, female anti-SPA2d hemolymph; BGG, female anti-BGG hemolymph. (From Duwel-Eby, L. E. and Karp, R. D., in preparation.)

VIII. NR-PAGE ANALYSIS OF THE INDUCIBLE HUMORAL FACTOR (IHF)

In vivo, the roach encounters Ag with hemolymph proteins that are in their nondenatured configuration. To determine which native proteins might serve as primary mediators of the roach immune response, NR-PAGE was performed to examine any changes in the protein profile in response to immunization. The 102-kDa band, although present in increased quantities after immunization, may not be the sole mediator of the roach immune reactivity. For this reason, current studies are focused on examination of the hemolymph proteins in their native state to identify all potential immune mediators. Roach hemolymph was resolved into five major protein bands on 6% NR-PAGE gels. Comparison of immune and control hemolymph, using SPA2d or BGG as the immunizing Ag, has revealed a consistent enhancement in NR-PAGE bands 2 and 4 (Figure 6). Interestingly, SDS-PAGE analysis of the individual NR-PAGE bands has revealed that both bands 2 and 4 contain species of the 102-kDa protein previously shown to be enhanced in immune hemolymph samples.

IX. IDENTIFICATION OF *IN VITRO* BINDING PROTEINS

Results from NR-PAGE analysis indicate that native bands 2 and 4 are primary candidates as mediators of the roach's immune reactivity. However, these protein bands can only be considered true immune mediators if they are shown to be capable of binding Ag. Initial attempts to confirm that these bands played an active role in the roach's immune response utilized affinity chromatography to identify binding proteins *in vitro*. In these experiments, whole hemolymph from animals immunized with SPA2d or anti-BTS (control) hemolymph was incubated at room temperature with the immunizing Ag, SPA2. To isolate those proteins capable of binding Ag, the hemolymph samples were then subjected to affinity chromatography in which rabbit Ab against the SPA2d was covalently linked to the column. The rabbit Ab should have bound to the SPA2, which was already bound to the roach factor proteins, creating an Ab-Ag-IHF complex on the column. The roach immunoregulatory proteins were then recovered from the column by decreasing the pH of the starting buffer. The starting buffer (containing the unbound proteins) and the low pH wash (containing the eluted immune complexes) were analyzed by SDS-PAGE and NR-PAGE electrophoresis. Analysis of the washes by NR-PAGE supported our speculations that bands 2 and 4 were acting as Ag-binding proteins. Comparison of immune and control starting buffer and low pH washes revealed that NR-PAGE bands 2 and 4 were retained by the column and were eluted by the decrease in pH in the immune samples.[20] Analysis of control hemolymph indicated that some nonspecific uptake of Ag occurred in control samples since band 2 appeared to have bound some SPA2 and thus was retained by the column. In addition, SDS-PAGE gels confirmed that the column was indeed binding the SPA2/IHF complexes since the Ag was exclusively recovered in the low pH wash. The results from the affinity chromatography experiments support the idea that NR-PAGE bands 2 and 4 contained immunoregulatory proteins.

In a second *in vitro* approach to establish bands 2 and 4 as immune mediators, cell-free hemolymph was first fractionated on a CM Affi-Gel Blue column. Unbound proteins were collected in the starting buffer (buffer A), whereas bound proteins were eluted using the same buffer containing 1.4 *M* salt. The CM Affi-Gel Blue concentrated NR-PAGE bands 2 and 4 in the salt wash, placing them in nearly equal concentration with bands 1 and 3, the major proteins in crude unfractionated hemolymph (Figure 7). The buffer A and salt washes were then run on 6% NR-PAGE gels and electroblotted onto nitrocellulose prior to incubation with [^{125}I]radiolabeled Ag for 2 h at room temperature. Binding of Ag by any of the hemolymph proteins was then visualized by autoradiography. Preliminary results using SPA2d as Ag have demonstrated uptake of Ag by NR-PAGE bands 2 and 4 in immune hemolymph (Figure 8). In contrast, there was a total absence of binding for any proteins in the buffer A wash. In the control experiment, using hemolymph from roaches injected with sterile saline, slight binding was observed for only band 2 in the salt wash (Figure 9). It is interesting to note that SDS-PAGE analysis of the buffer A and salt washes has revealed that the 102-kDa band was exclusively found in the salt wash (Figure 10).

X. DEMONSTRATION OF BIOLOGICAL ACTIVITY *IN VIVO*

The results from *in vitro* affinity column chromatography and Western blot binding experiments raised the possibility that several proteins present in NR-PAGE bands 2 and 4 may be involved in mediating the roach immune response to soluble protein Ag. The significant level of uptake of Ag by band 2 in control hemolymph indicated that there could be naturally occurring nonspecific mechanisms protecting the roach from foreign Ag, which act in conjunction with more specific defense strategies that are induced by toxic foreign substances, such as SPA2. In order for the *in vitro* findings to be considered an accurate reflection of the events occurring *in vivo*, it was necessary to demonstrate biological activity

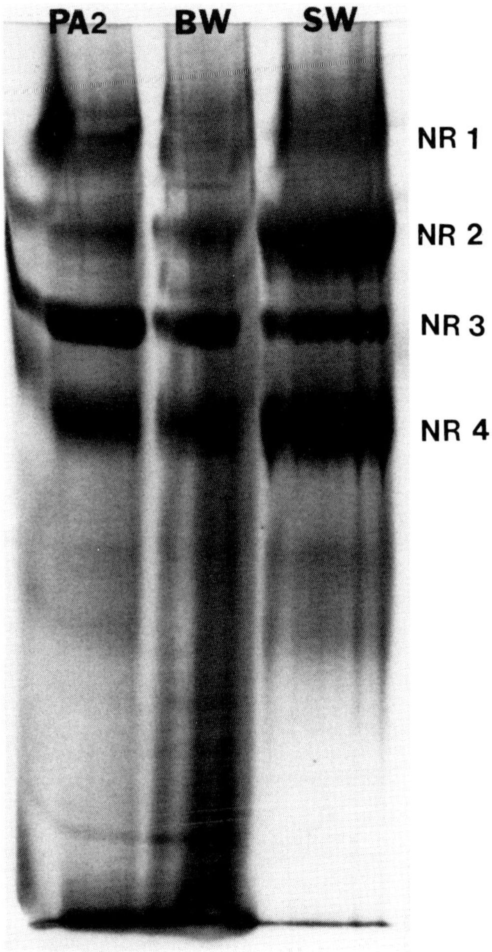

FIGURE 7. A 6% NR-PAGE analysis of CM Affi Gel blue column washes from an immune hemolymph sample. PA2, unfractionated female anti-SPA2d hemolymph; BW, buffer A(starting buffer) wash containing unbound proteins; SW, salt wash containing bound proteins eluted with 1.4 *M* NaCl. (From Duwel-Eby, L. E. and Karp, R. D., in preparation.)

for our proposed IHF proteins. We have previously shown that passive transfer of whole immune hemolymph conferred protection on naive recipients.[22] Therefore, this same approach is being used to test the ability of individual protein bands eluted from NR-PAGE gels to transfer protective activity to naive recipients. To evaluate the ability of excised proteins to transfer immunity, animals were once again immunized with the inactivated SPA2d. After fractionation over a CM Affi-Gel Blue column, the salt wash was collected and electrophoresed on 6% NR-PAGE gels. The individual salt wash NR-PAGE bands were then excised and the protein electroeluted from the gel prior to injection into naive recipient roaches. To assay for immunity, the recipients were challenged with a lethal dose of the immunizing Ag 1 h after injection of the NR-PAGE bands and scored for percent survivors 48 h posttransfer. A preliminary trial of this approach has shown that passive transfer of either bands 2 or 4 from the salt wash of immune hemolymph conferred a survival rate of 60% in recipients, in sharp contrast to little (20%) or no protection (0%) for bands

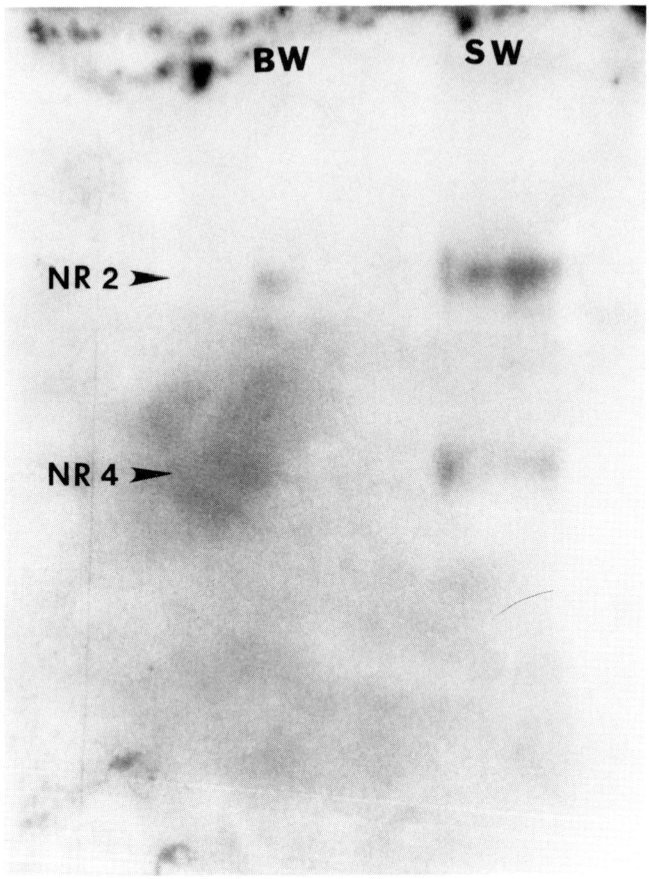

FIGURE 8. Autoradiograph demonstrating binding of Ag by bands 2 and 4 in immune hemolymph. Buffer A and salt washes from roaches immunized with SPA2d were run on 6% NR-PAGE gels, electroblotted onto nitrocellulose, and then incubated with [^{125}I] labeled SPA2. Binding of Ag was then visualized by autoradiography. BW, buffer A-eluted proteins from immune hemolymph; SW, salt wash-eluted proteins from immune hemolymph. (From Duwel-Eby, L. E. and Karp, R. D., in preparation.)

1 and 3, respectively. Control animals receiving whole immune hemolymph demonstrated a protection rate of 80%, whereas none of the recipients injected with the hemolymph from saline-injected donors survived. Thus, both the *in vitro* and *in vivo* experiments have indicated that NR-PAGE bands 2 and 4 appear to be directly associated with the American cockroach's humoral immune response. It is essential that further trials of these approaches be conducted using both native and reduced proteins to allow us to identify which proteins directly mediate the roach's humoral immune reactivity, so that further comparative studies can be conducted.

XI. DISCUSSION

One of the more controversial subjects in comparative immunology today is whether or not any invertebrate species has evolved an antibody-like response. One problem is that immunologists automatically take that to mean that such a response has to be mediated by an immunoglobulin molecule. Since such a molecule has yet to be discovered in any invertebrate species, this logic dictates that adaptive humoral immunity therefore does not

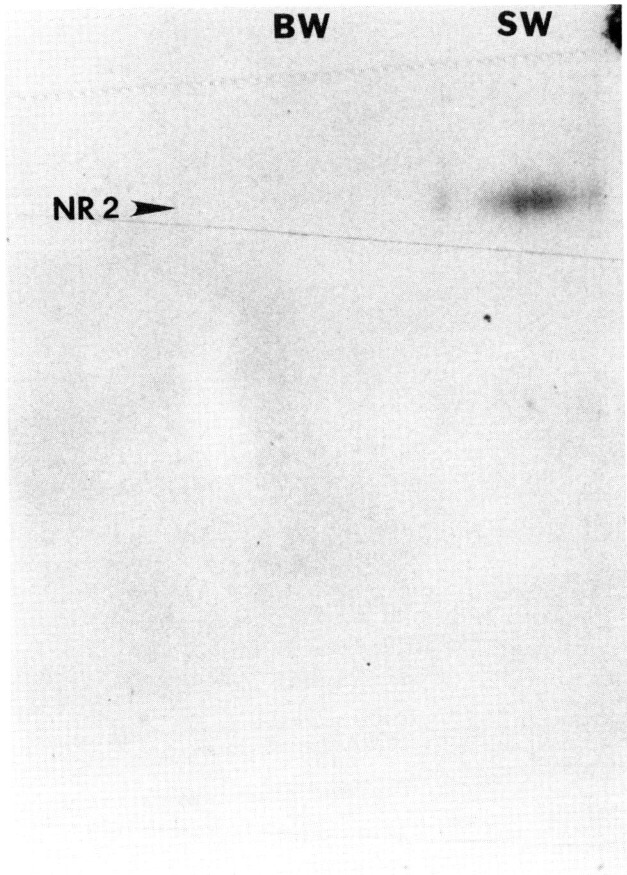

FIGURE 9. Autoradiograph of control hemolymph from roaches injected with sterile saline demonstrating binding of Ag by band 2. The eluates from buffer A and salt washes from control samples were run on 6% NR-PAGE, electroblotted onto nitrocellulose, and then incubated with [^{125}I] labeled SPA2. Binding of Ag was then visualized by autoradiography. BW, buffer A proteins from control hemolymph; SW, salt wash proteins from control samples. (From Duwel-Eby, L. E. and Karp, R. D., in preparation.)

exist in these ''lower'' life-forms. However, we must keep in mind that we are dealing with evolution and that immunoglobulins surely did not appear on the scene fully evolved. Instead, any molecule that complex must first have gone through some precursor stages. Basically, it is really important that such a response possesses specificity and memory. The effector molecule mediating such activity may or may not be related to immunoglobulins, but that does not make a response displaying specificity and memory any less valid as representing true adaptive immunity in the animal.

Our findings in studies on the American cockroach indicate that a humoral response exists in this advanced invertebrate that biologically mimics, in almost every detail, the type of immune reactivity found in mammals. As surprising as this may seem to some, it begins to make sense when one considers the life history of this animal. Unlike their holometabolous cousins, paurometabolous insects maintain the same morphological form during their existence and can have relatively long life spans. This means that they are faced with the same problems for survival as the laboratory mouse: how does one exist in a sea of toxic substances and microbial pathogens that threaten its existence and reproductive potential? Adaptive

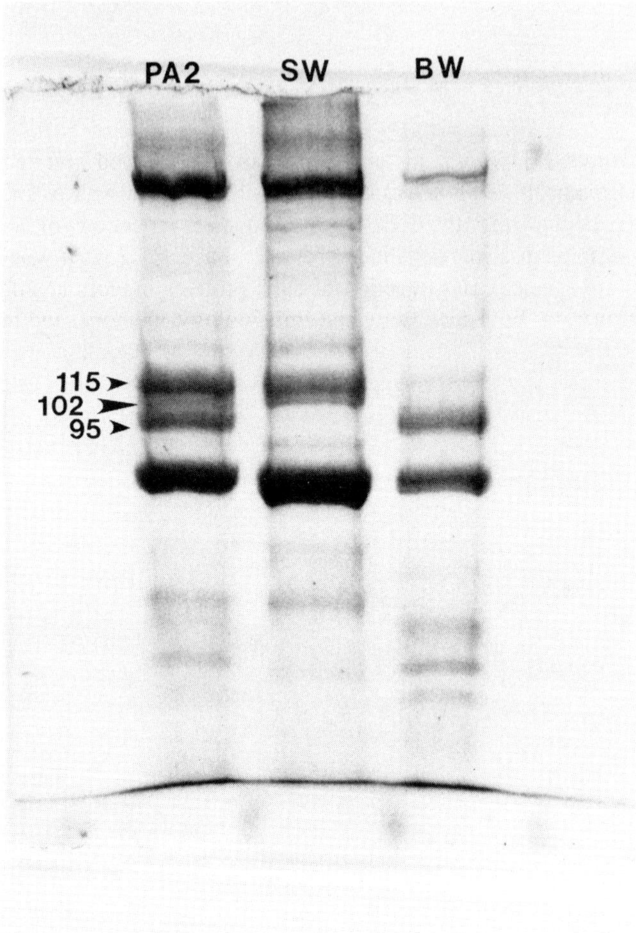

FIGURE 10. A 7.5% SDS-PAGE analysis of the buffer A and salt wash-eluted proteins from immune hemolymph after fractionation over a CM Affi-Gel Blue column. PA2, unfractionated female anti-SPA2d hemolymph; SW, salt wash-eluted proteins from female anti-SPA2d hemolymph; BW, buffer A-eluted proteins from female anti-SPA2d samples. (From Duwel-Eby, L. E. and Karp, R. D., in preparation.)

immunity has been a very effective answer to this problem for higher animals, but apparently this concept did not originate with them. Long-lived invertebrates, such as the cockroach, also needed a way to cope with repeated exposures to noxious and/or threatening substances over their life span. As such mechanisms evolved in these animals, quality of life must have improved for these organisms, as it did for vertebrates. More importantly, it allowed them to reproduce for longer periods of time, ensuring the survival of the species.

Time will tell if the inducible humoral factor in the cockroach has any relationship at all to members of the immunoglobulin superfamily found in vertebrates. It may be that insects have evolved an immune molecule that, although physically dissimilar to vertebrate immunoglobulin, nevertheless, effectively carries out the concept of adaptive immunity in these animals. However, it cannot be ignored that the studies represented here, in both holometabolous and paurometabolous insects, have brought us another step closer to defining the nature of invertebrate immunity, and will also add to the growing knowledge concerning the evolution and origins of anticipatory immune reactivity in general.

XII. SUMMARY

The use of insect models for the study of invertebrate immune mechanisms has become increasingly attractive to investigators such that insects have moved to the forefront of productive research on invertebrate adaptive immunity. We have been studying the American cockroach, *Periplaneta americana*, as an example of a long-lived invertebrate that must deal with repeated exposures to noxious chemical substances as well as infectious microorganisms. We have found that the cockroach responds to injections of soluble proteins, both toxic and nontoxic, in a manner that is very reminiscent of what would be expected to occur in vertebrate species. The injection of such proteins induces an adaptive humoral response that demonstrates both specificity and immunologic memory, and thus qualifies as a true immune response. Every biological parameter tested indicates that the roach humoral response mimics the characteristics of vertebrate immunity. For instance, both younger and older animals have diminished capacity to respond; females have a stronger response than males; the mediating factor is protein in nature; and the mediating factor will bind and specifically precipitate the original inducing antigen. SDS-PAGE gel analysis indicates that a 102-kDa protein band is consistently enhanced by immunization of roaches with numerous protein substances. Nonreducing PAGE (NR-PAGE) gel analysis indicates that two major protein bands are enhanced following immunization and that a major component of each of these bands is the 102-kDa-reduced protein. Proteins eluted from the enhanced NR-PAGE bands will passively transfer biological activity against the antigen to naive animals. These same native protein bands also demonstrate *in vitro* binding capacity. Thus, the use of immunoprecipitation promises to be a valuable approach for isolating the mediating factor.

We are just starting to explore the response to bacteria in the American cockroach and finding some unexpected results. The antibacterial responses that have been so well characterized in holometabolous insects reveal that the typical protective response that is generated is mediated by various inducible peptides that arise very quickly, do not last very long, and are nonspecific in nature. The response to *Pseudomonas aeruginosa* in the cockroach appears to be biphasic in nature. The first part of the response is very quick, as in holometabolous insects, and is relatively nondiscriminatory. However, the second part of the response, which peaks at 5 d postimmunization, is more persistent and begins to show selective specificity. Thus, the cockroach appears to have evolved mechanisms for coping with a hostile environment that differ from those of short-lived holometabolous insects.

ACKNOWLEDGMENTS

The work presented in this chapter that has emanated from our lab has been supported by NSF Research Grants PCM 8316140, DCB 8702382, NIH Research Grant GM 39398, and the University of Cincinnati Foundation Fund for Comparative Immunology.

REFERENCES

1. **Klein, J.,** *Immunology — The Science of Self-Nonself,* John Wiley & Sons, New York, 1982.
2. **Klein, J.,** Are invertebrates capable of anticipatory immune responses?, *Scand. J. Immunol.,* 29, 499, 1989.
3. **Karp, R. D. and Duwel-Eby, L. E.,** Adaptive immune responses in insects, in *Phylogenesis of Immune Functions,* Warr, G. and Cohen, N., Eds., CRC Press, Boca Raton, FL, 1991, 2.
4. **Ratcliffe, N. A., Rowley, A. F., Fitzgerald, S. W., and Rhodes, C. P.,** Invertebrate immunity: basic concepts and recent advances, *Int. Rev. Cytol.,* 97, 183, 1985.
5. **Dunn, P. E.,** Biochemical aspects of insect immunology, *Annu. Rev. Entomol.,* 31, 321, 1986.
6. **Cooper, E. L.,** Ed., Inducible responses in invertebrates, in: Proceedings from 4th ISDCI Cong., Special Issue, *Dev. Comp. Immunol.,* 13, 445, 1989.

7. **Karp, R. D.,** Cell-mediated immunity in insects, *Bioscience,* 40, 732, 1990.
8. **Glaser, R. W.,** On the existence of immunity principles in insects, *Psyche,* 25, 38, 1918.
9. **Metalnikov, S.,** Immunité naturelle ou acquise des chenelles de *Galleria mellonella, C. R. Soc. Biol. Paris,* 83, 817, 1920.
10. **Chadwick, J. S. and Vilk, E.,** Endotoxins from several bacteria species as immunizing agents against *Pseudomonas aeruginosa* in *Galleria mellonella, J. Invertebr. Pathol.,* 13, 410, 1969.
11. **Dunn, P. E., Dai, W., Kanost, M. R., and Geng, C.,** Soluble peptidoglycan fragments stimulate antibacterial protein synthesis by fat body from larvae of *Manduca sexta, Dev. Comp. Immunol.,* 9, 559, 1985.
12. **Boman, H. G. and Hultmark, D.,** Cell-free immunity in insects, *Annu. Rev. Microbiol.,* 41, 103, 1987.
13. **Dimarcq, J. L., Keppi, E., Dunbar, B., Lambert, J., Reichhart, J. M., Hoffmann, D., Rankine, S. M., Fothergill, J. E., and Hoffmann, J. A.,** Insect immunity. Purification and characterization of a family of novel inducible antibacterial proteins from immunized larvae of the dipteran *Phormia terranovae* and complete amino acid sequence of the predominant member, dipteran A, *Eur. J. Biochem.,* 171, 17, 1988.
14. **Reichhart, J. M., Essrich, M., Dimarcq, J. L., Hoffmann, D., Hoffmann, J. A., and Laguex, M.,** Insect immunity. Isolation of cDNA clones corresponding to diptericin, an inducible antibacterial peptide from *Phormia terranovae* (Diptera). Transcriptional profiles during immunization, *Eur. J. Biochem.,* 182, 423, 1989.
15. **Lambert, J., Keppi, E., Dimarcq, J. L., Wicker, C., Reichhart, J. M., Dunbar, B., Lepage, P., Van Dorsselaer, A., Hoffmann, J. A., Fothergill, J., and Hoffmann, D.,** Insect immunity: isolation from immune blood of the dipteran *Phormia terranovae* of two insect antibacterial peptides with sequence homology to rabbit lung macrophage bactericidal peptides, *Proc. Natl. Acad. Sci. U.S.A.,* 86, 262, 1989.
16. **Kamon, E. and Shulov, A.,** Immune response of locusts to venom of the scorpion, *J. Invertebr. Pathol.,* 7, 192, 1965.
17. **Kamon, E.,** Specificity of the immune response of locusts to scorpion venom, *J. Invertebr. Pathol.,* 7, 199, 1965.
18. **Bettini, S.,** Acquired immune response of the housefly, *Musca domestica* (Linnaeus), to injected venom of the spider *Lactrodectus mactans tredecimagluttatus* (Rossi), *J. Invertebr. Pathol.,* 7, 378, 1965.
19. **Rheins, L. A. and Karp, R. D.,** The humoral immune response in the American cockroach, *Periplaneta americana*: reactivity to a defined antigen from honey bee venom, phospholipase A_2, *Dev. Comp. Immunol.,* 8, 791, 1984.
20. **Duwel-Eby, L. E. and Karp, R. D.,** The inducible humoral immune response to soluble proteins in the American cockroach, in *Defense Molecules — UCLA Symposium on Molecular and Cellular Biology,* Marchalonis J. and Reinisch C., Eds., Academic Press, New York, 121, 1990, 63.
21. **George, J. F., Karp, R. D., Rellahan, B. L., and Lessard, J. L.,** Alterations of the protein composition in the haemolymph of American cockroaches immunized with soluble proteins, *Immunology,* 62, 505, 1987.
22. **Rheins, L. A., Karp, R. D., and Butz, A.,** Induction of specific humoral immunity to soluble proteins in the American cockroach (*Periplaneta americana*). I. Nature of the primary response, *Dev. Comp. Immunol.,* 4, 447, 1980.
23. **Karp, R. D. and Rheins, L. A.,** Induction of specific humoral immunity to soluble proteins in the American cockroach (*Periplaneta americana*). II. Nature of the secondary response, *Dev. Comp. Immunol.,* 4, 629, 1980.
24. **Rheins, L. A. and Karp, R. D.,** An inducible humoral factor in the American cockroach (*Periplaneta americana*): precipitin activity that is sensitive to a proteolytic enzyme, *J. Invertebr. Pathol.,* 40, 190, 1982.
25. **Hildemann, W. H., Clark, E. A., and Raison, R. L.,** Phylogeny of immunocompetence, in *Comprehensive Immunogenetics,* Elsevier, New York, 1981, 302.
26. **Roitt, I., Brostoff, J., and Male, D.,** Cell cooperation in the antibody response, in *Immunology,* 2nd ed., Gower Medical Pub., London, England, 1989, 8.1.
27. **Atlas, R. M.,** The immune response, in *Microbiology — Fundamentals and Applications,* Macmillan, New York, 1984, 500.
28. **Lahita, R. G.,** *Basic and Clinical Immunology,* 4th ed., Stites, D. P., Stobo, J. D., Fudenberg, H., and Wells, J. V., Eds., Lange, Lost Altos, CA, 1982.
29. **Belisle, E. H. and Strausser, H. R.,** Sex-related immunocompetence of Balb/c mice. II. Study of immunologic responsiveness of young, adult and aged mice, *Dev. Comp. Immunol.,* 5, 661, 1981.
30. **Rheins, L. A. and Karp, R. D.,** Effect of gender on the inducible humoral immune response to honey bee venom in the American cockroach *(Periplaneta americana), Dev. Comp. Immunol.,* 9, 41, 1985.
31. **Makinodan, T. and Kay, M. M.,** Age influence on the immune system, *Adv. Immunol.,* 29, 287, 1980.
32. **Rheins, L. A. and Karp, R. D.,** Ontogeny of the invertebrate humoral immune response: studies on various developmental stages of the American cockroach, *Dev. Comp. Immunol.,* 9, 395, 1985.

Chapter 16

ANTIGENIC STIMULATION OF HEMAGGLUTININ PRODUCTION IN INSECTS

Hiroto Komano

TABLE OF CONTENTS

I. INTRODUCTION

Many insects are known to have humoral lectins,[1,2] which are generally categorized into two classes or groups on the basis of their modes of lectin production. One group is induced upon infection of the insect by pathogens, such as bacteria. The second class of lectins is always present in the hemolymph, is not inducible, and does not have clear functions. The inducible lectins appear in response to invasion by foreign substances and seem to function in the removal of such substances. Because lectin induction in insects has not been studied widely, this discussion will be restricted to studies of lectin induction in the hemolymph of *Sarcophaga peregrina* (flesh fly).

We have found that a galactose-binding lectin (named *Sarcophaga* lectin) is induced in the hemolymph of third instar larvae of *S. peregrina* upon injury. In addition, it is also induced in the embryo and pupal stages during normal development. A number of results strongly suggest that this lectin has a dual role in this fly: as a defense protein in response to injury and as part of the normal morphogenetic development of the organism (Figure 1). Questions regarding the mechanism of induction of this lectin fall into two categories. Is the induction mechanism after the stimulus of injury analogous with mammalian antigenic stimulation? Is the induction, which occurs upon injury, similar in mechanism to the induction of this lectin during normal development?

The first part of this chapter is devoted to descriptions of the various stages of lectin induction and the organ which synthesizes this lectin; the second part describes the regulation of lectin induction at the transcriptional level; and the third part describes humoral factors that are involved in lectin induction. Finally, it closes with a discussion of this and other insect lectins and a summary of the information presented here.

II. INDUCTION OF *SARCOPHAGA* LECTIN IN *SARCOPHAGA PEREGRINA*

A. SARCOPHAGA LECTIN IS INDUCED UPON INJURY[3,4,5]

When the body wall of third instar larvae of *S. peregrina* is injured with a hypodermic needle, the hemagglutinating activity in the hemolymph increases markedly (Table 1). Interestingly, similar increases in hemagglutinating activity can be detected also in pupal hemolymph. This agglutinin is a galactose-binding lectin (named *Sarcophaga* lectin). As summarized in Table 2, it has a molecular weight of 190 kDa and consists of four α subunits and two β subunits, with molecular weights of 32 and 30 kDa, respectively.

The induction of this lectin is enhanced by injecting third instar larvae with sheep red blood cells[6] (Figure 2). Such enhancement was also detected after injection of other foreign substances, such as *Escherichia coli*. Thus, this lectin appears to be synthesized in response to invasion by foreign substances. Similar enhancement of a lectin in hemolymph was observed in *Antheraea perni* (oak silk moth),[7] but not in *Periplaneta americana* (American cockroach)[8] or *Bombyx mori* (silk moth).[9] It also has been reported that the lectin of *Hyalphora cecropia* is not inducible.[10] It is not clear whether these two types of lectins have the same function, but it seems that the inducible lectins function as defense proteins against the attack of pathogens or similar invading substances.

Sarcophaga lectin has been shown to play a role in removing sheep red blood cells, which were introduced into the abdominal cavity,[11] although the mechanism of this process is not clear. Several antibacterial substances also have been shown to be induced upon injury.[12-15] (See also the Chapter 9). This lectin appears to function as a defense protein in cooperation with antibacterial compounds for removing foreign substances, which invade through the damaged body wall.

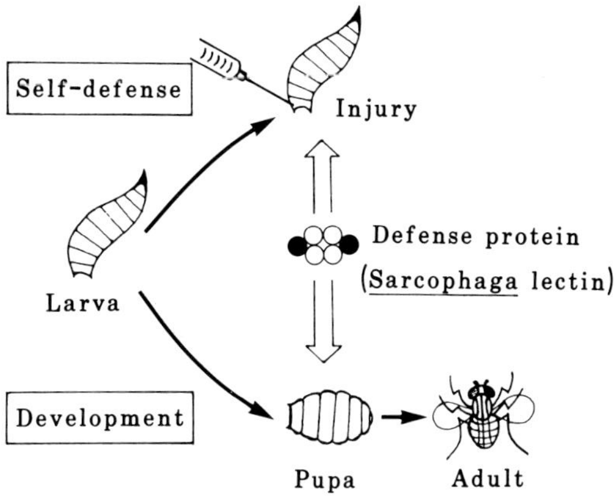

FIGURE 1. Dual function of *Sarcophaga* lectin. *Sarcophaga* lectin is induced both upon injury of larvae and during development. Upon injury, *Sarcophaga* lectin functions as a defense protein for removing foreign substances invading through the damaged body wall, and during development, *Sarcophaga* lectin is essential for morphogenesis.

TABLE 1
Hemagglutinating Activity in the Hemolymph of Larvae and Pupae

Hemolymph	Activity ($titer^{-1}$)	Specific activity ($titer^{-1}$/mg protein/ml)
Normal larvae	2	0.01
Injured larvae[a]	64	0.46
Pupae[b]	16	0.46

Note: Sheep red blood cells were used as indicator cells. Hemagglutinating activity was determined using serial twofold dilutions of hemolymph. Activity is expressed as the reciprocal of the titer.

[a] Hemolymph was collected 48 h after injuring the body wall.
[b] Hemolymph was collected 48 h after puparium formation.

From Komano, H., Mizuno, D., and Natori, S., *J. Biol. Chem.*, 255, 2919, 1980. With permission.

TABLE 2
Characterization of *Sarcophaga* Lectin

Molecular weight	190 kDa
Subunit structure	$\alpha_4\ \beta_2$ (α, 32 kDa; β, 30 kDa)
Hapten sugar	Galactose
Site of synthesis	Fat body

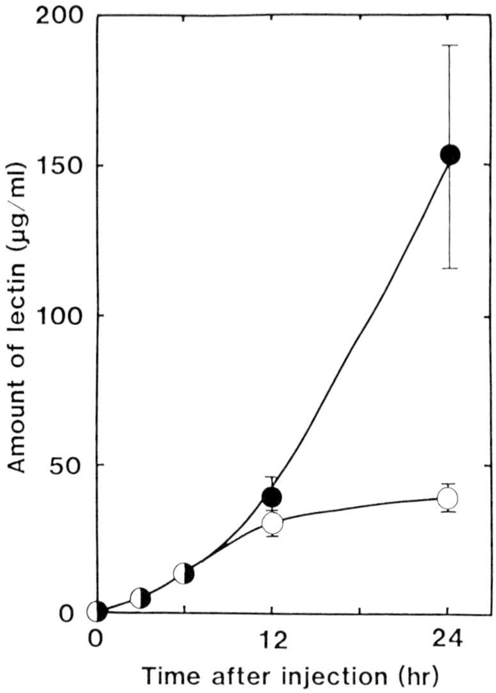

FIGURE 2. Change in the amount of *Sarcophaga* lectin in the hemolymph after administration of foreign substances to third instar larvae. Sheep red blood cells suspended in insect saline or insect saline alone were injected into larvae. Hemolymph was collected at the indicated times and the amount of *Sarcophaga* lectin measured by radioimmunoassay. ●, Injection with sheep red blood cells; ○, injection with insect saline. (From Takahashi, H., Komano, H., and Natori, S., *J. Insect Physiol.*, 32, 771, 1986. With permission.)

B. *SARCOPHAGA* LECTIN IS INDUCED DURING DEVELOPMENT[4,6]

Very interestingly, the *Sarcophaga* lectin has been found to be induced during development of the pupae and embryo. Figure 3 shows the amount of *Sarcophaga* lectin measured by radioimmunoassay during the pupal stage. The same lectin also was detected in the embryo. Because there are several reports that lectins exist in the embryo and pupae of many insects,[1] the question arises as to whether lectins are essential during development. Our work shows that *Sarcophaga* lectin is synthesized by imaginal disk cells in the presence of β-ecdysone in an *in vitro* culture system, and is necessary for development of the imaginal disk into adult structures.[15a] Thus, we believe that the *Sarcophaga* lectin has a dual function: as a defense protein against injury or pathogenic invasion and as part of the morphogenetic development of the organism. It also seems that the induction of *Sarcophaga* lectin during development may be regulated by β-ecdysone.

C. SITE OF SYNTHESIS OF *SARCOPHAGA* LECTIN[4,16]

It has been clearly shown that *Sarcophaga* lectin is synthesized in the fat body, although there are several reports that the site of insect lectin synthesis is in the hemocytes or other tissues[17,18] (see also Chapter 2). The fat body is the largest and most well-developed organ in the larvae and synthesizes most hemolymph proteins. As shown in Figure 4, the *Sarcophaga* lectin gene was expressed exclusively in the fat body and not in other tissues,

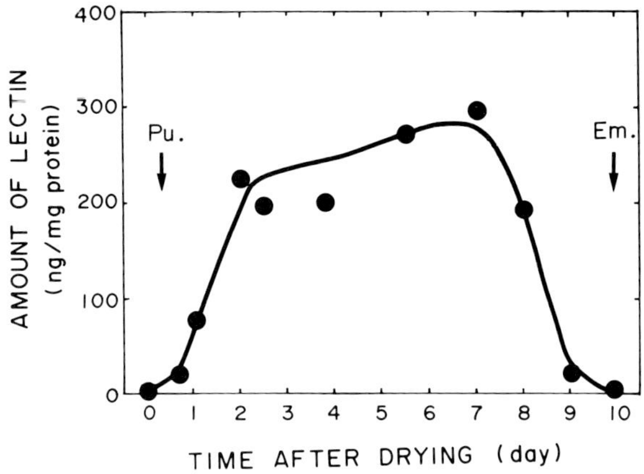

FIGURE 3. Change in the amount of *Sarcophaga* lectin in pupae. Points are the average of duplicate measurements. Arrows indicate the time of pupation (Pu.) and emergence of adults (Em.), respectively. (From Komano, H., Nozawa, R., Mizuno, D., and Natori, S., *J. Biol. Chem.*, 258, 2143, 1983. With permission.)

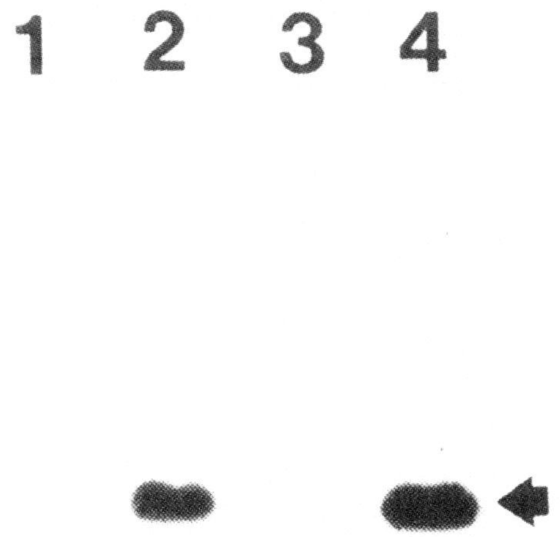

FIGURE 4. Fat body-specific expression of the *Sarcophaga* lectin gene. RNA was extracted from pooled tissues and analyzed by RNA blot hybridization. RNA extracted from: lane 1, whole body of normal larvae; lane 2, whole body of injured larvae; lane 3, carcass of injured larvae; lane 4, fat body of injured larvae. Arrow indicates the position of *Sarcophaga* lectin mRNA. (From Shiraishi, S. and Natori, S., *FEBS Lett.*, 232, 163, 1988. With permission.)

indicating that the fat body is the only organ that synthesizes *Sarcophaga* lectin in response to body injury. Figure 5 shows the amount of lectin in the fat body and in the hemolymph in the first 80 h after injury. These results show that *Sarcophaga* lectin is synthesized transiently in the fat body after injury, and then secreted into the hemolymph. Other defense proteins also are synthesized in this organ.[19]

The analysis of proteins synthesized *in vitro*, directed by poly(A) RNA from the fat

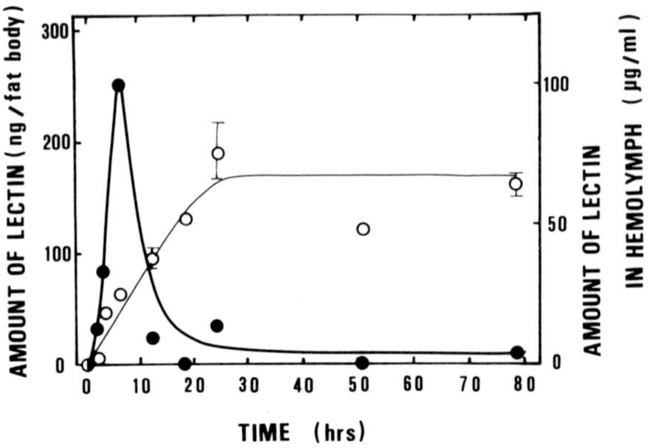

FIGURE 5. Change in the amount of *Sarcophaga* lectin in the fat body and hemolymph with time after injury of the body wall. The amounts of lectin in homogenates of fat bodies and in the hemolymph were measured by radioimmunoassay. ●, Fat body; ○, hemolymph. (From Kamano, H., Nozawa, R., Mizuno, D., and Natori, S., *J. Biol. Chem.*, 258, 2143, 1983. With permission.)

body, show that poly(A) RNA contains only mRNA for the α subunit of *Sarcophaga* lectin (Figure 6).[20] In addition the primary structure of α- and β subunits has been found to be very similar.[21] Therefore, it seems that the β subunit is derived from the α subunit.

III. INDUCTION OF *SARCOPHAGA* LECTIN IS REGULATED AT THE TRANSCRIPTIONAL LEVEL[6]

We examined expression of the *Sarcophaga* lectin gene at various stages during *Sarcophaga* development by RNA blot hybridization. As shown in Figure 7, no significant expression of the *Sarcophaga* lectin gene was observed in third instar larvae, unless the body wall had been injured. During normal development, the *Sarcophaga* lectin gene was expressed in the embryo and in the early pupal stage. These results are consistent with the data obtained by radioimmunoassay of *Sarcophaga* lectin.

As mentioned previously, the induction of *Sarcophaga* lectin is enhanced by injection of foreign substances. To examine whether the enhanced synthesis of *Sarcophaga* lectin was due to activation of transcription of the lectin gene, we compared the expression of the *Sarcophaga* lectin gene in the larvae after injection of insect saline alone and with sheep red blood cells. Figure 8 shows that the expression of the *Sarcophaga* lectin gene in the fat body was enhanced by injection with either insect saline alone or with sheep red blood cells. However, a clear difference was observed in the persistence of this mRNA. Namely, the life of the mRNA synthesized in response to sheep red blood cells appeared to be much longer than that of lectin mRNA synthesized in response to insect saline alone. The higher level of mRNA was maintained for more than 24 h after injection of the sheep red blood cells. It is not clear whether this was due to a difference in the stability of the mRNA or to the active transcription of the *Sarcophaga* lectin gene lasting longer when a foreign substance, such as sheep red blood cells, is injected into larvae. These results are consistent with those showing changes in the level of *Sarcophaga* lectin in Figure 2.

Thus, it is clear that both during development and in an emergency, such as injury, the *Sarcophaga* lectin gene is activated and *Sarcophaga* lectin is synthesized *de novo*. An *in vitro* transcription system for the *Sarcophaga* lectin gene has been established,[22] and results from this system should elucidate whether or not the transcription factors required for activation of the *Sarcophaga* lectin gene during injury and during development are identical.

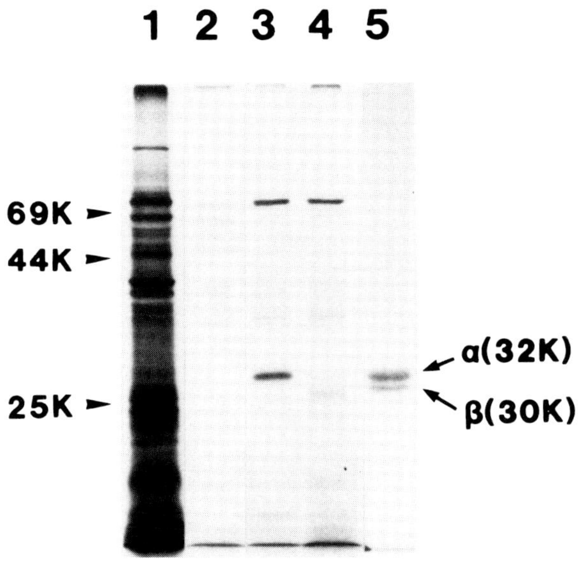

FIGURE 6. Analysis of protein synthesized *in vitro* directed by poly(A)$^+$RNA. Poly(A)$^+$RNA was translated *in vitro* using a rabbit reticulocyte lysate containing [^{35}S]methionine, and the immunoprecipitate formed with antibody against *Sarcophaga* lectin was analyzed by SDS-polyacrylamide gel electrophoresis followed by fluorography. Lane 1, total protein synthesized in the translation mixture; lane 2, precipitate formed with normal serum; lane 3, immunoprecipitate formed with antiserum; lane 4, the same as lane 3 but with cold *Sarcophaga* lectin added to the reaction mixture before addition of antiserum; lane 5, *Sarcophaga* lectin stained with Coomassie brilliant blue. Arrows show the positions of the α subunit and the β subunit of *Sarcophaga* lectin. (From Takahashi, H., Komano, H., Kawaguchi, N., Kitamura, N., Nakanishi, S., and Natori, S., *J. Biol. Chem.*, 260, 12228, 1985. With permission.)

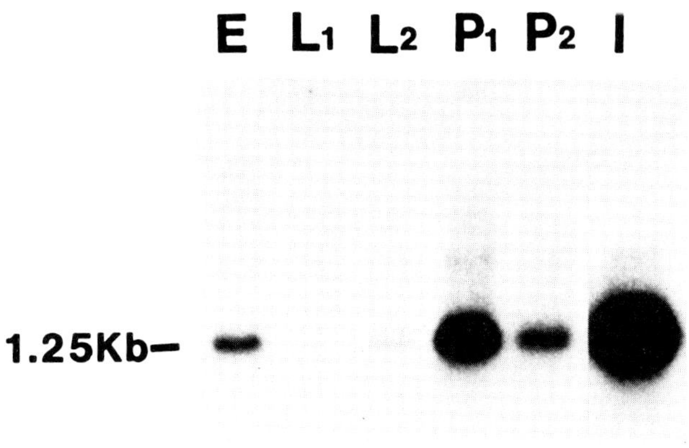

FIGURE 7. Expression of the *Sarcophaga* lectin gene at developmental stages and after injury. RNA was extracted from embryos, larvae, and pupae. The amount of mRNA was examined by RNA blot hybridization using cDNA from *Sarcophaga* lectin as a probe. The bar of 1.25 kilobase pairs shows the position of *Sarcophaga* lectin mRNA. Lane E, embryo; lane L_1, early stage of third instar larvae; lane L_2, late stage of third instar larvae; lane P_1, early stage of pupae; lane P_2 late stage of pupae; lane I, injured third instar larvae. (From Takahashi, H., Komano, H., and Natori, S., *J. Insect Physiol.*, 32, 771, 1986. With permission.)

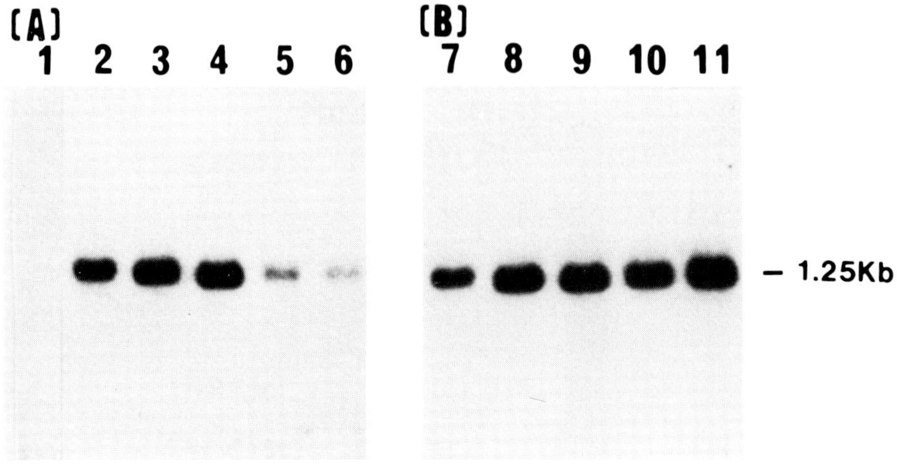

FIGURE 8. Expression of the *Sarcophaga* lectin gene after administration of foreign substances to third instar larvae. Insect saline alone (A) or sheep red blood cells suspended in insect saline were injected into third instar larvae. Fat bodies were excised at the indicated times, RNA was extracted, and the amount of *Sarcophaga* lectin mRNA was examined by RNA blot hybridization. Lane 1, RNA prepared from normal third instar larvae. RNAs were prepared 1.5 h (lanes 2 and 7), 3 h (lanes 3 and 8), 6 h (lanes 4 and 9), 12 h (lanes 5 and 10), and 24 h (lanes 6 and 11) after injection. (From Takahashi, H., Komano, H., and Natori, S., *J. Insect Physiol.*, 32, 771, 1986. With permission.)

IV. HUMORAL MEDIATOR-DEPENDENT ACTIVATION OF THE *SARCOPHAGA* LECTIN GENE[16,23]

In this section, the mechanism of transmission of the signal indicating that injury has occurred to the fat body where the *Sarcophaga* lectin is selectively activated, is described. There are at least two possible scenarios explaining signal transduction. Insects have an open vascular system, and their tissues are surrounded by hemolymph. Thus, one possibility is that a mediator that interacts directly with the fat body cells is secreted from the damaged body wall into the hemolymph. A second possibility is that some substance secreted from the damaged body wall stimulates another tissue to secrete a mediator molecule that interacts with the fat body cells. Clear results regarding this question were obtained in ligation experiments.

The middle of the bodies of third instar larvae were ligated with cotton thread, and the posterior regions of the larvae were then pricked with a hypodermic needle. After 3 h, the anterior and posterior regions were separated at the point of ligation, and the contents of each were examined for *Sarcophaga* lectin mRNA by RNA blot hybridization. As summarized in Figure 9, no appreciable amount of *Sarcophaga* lectin mRNA was detected in either the posterior or anterior regions of the fat body of unligated and undamaged larvae. After injury of the posterior body wall, both the posterior and anterior fat body were found to synthesize almost the same amount of *Sarcophaga* lectin mRNA. A significant amount of *Sarcophaga* lectin mRNA also was detected in ligated larvae that had not been injured in the body wall, probably due to body injury caused by ligation. In these larvae, much more *Sarcophaga* lectin mRNA was synthesized in the anterior region of the fat body than in the posterior region, suggesting the participation of some tissue in the anterior region of the body in activation of the *Sarcophaga* lectin gene. Probably a substance secreted from the body wall when it is damaged by ligation is transmitted to the tissue in the anterior region of the body and promotes the secretion of a mediator molecule that stimulates fat body cells to synthesize *Sarcophaga* lectin mRNA. This notion is supported by the finding that pricking the body wall in the posterior region after ligation did not result in any obvious increase in lectin mRNA in the posterior region.

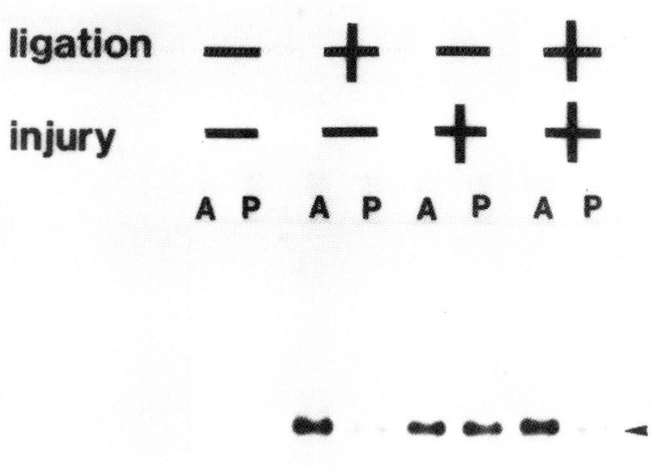

FIGURE 9. Effect of ligation on expression of the *Sarcophaga* lectin gene. The middle of the bodies of third instar larvae were ligated with cotton thread, and the posterior body wall was pricked with a hypodermic needle when appropriate. Posterior and anterior regions were separated with fine scissors and pooled separately. RNA was extracted from the two regions, which contain the posterior and anterior regions of the fat body, respectively, and analyzed by RNA blot hybridization. (+) Ligation or injury; (−) no ligation or injury; (A) anterior and (P) posterior regions of the fat body. Arrowhead indicates the position of *Sarcophaga* lectin mRNA. (From Shiraishi, S. and Natori, S., *FEBS Lett.*, 232, 163, 1988. With permission.)

We also investigated the effect of hemolymph prepared from injured larvae in activating *Sarcophaga* lectin gene in fat body cells. For this, larvae were ligated in the middle of the body, and then the anterior half was excised. The posterior region of the body did not synthesize *Sarcophaga* lectin mRNA, as described before. The effect of injecting hemolymph through the ligated wound into this body sac containing the posterior region of the fat body on the expression of the *Sarcophaga* lectin gene, was examined. No appreciable mRNA synthesis was detected in the fat body after simply inserting a hypodermic needle into the body cavity, but after injection of hemolymph from injured larvae, the *Sarcophaga* lectin gene was clearly activated (Figure 10).

These results strongly suggest that injuring the body wall induces some tissue in the anterior region of the larvae to produce a mediator molecule that is essential for selective activation of the *Sarcophaga* lectin gene. Many endocrine organs, including the brain, ring gland, and corpus allatum, are located in the anterior region of the body, and one of these organs probably releases the mediator molecule. Figure 11 shows a model of the mechanism of activation of the *Sarcophaga* lectin gene upon injury. Several antibacterial substances also are induced in response to body injury (see also Chapters 12 and 13). Therefore, the same mediator molecule may be responsible for activation of several acute-phase protein genes. From the analysis of the mediator molecule using a system for *in vitro* culture of the fat body that mimics the situation *in vivo*, it was suggested that the mediator molecule is a heat-stable, low molecular weight substance.

The mechanism of enhancement of *Sarcophaga* lectin production by injection of foreign substances remains to be elucidated. Hemocytes might be involved in this process, just as mammalian macrophages play an important role in the recognition of foreign antigens. Since

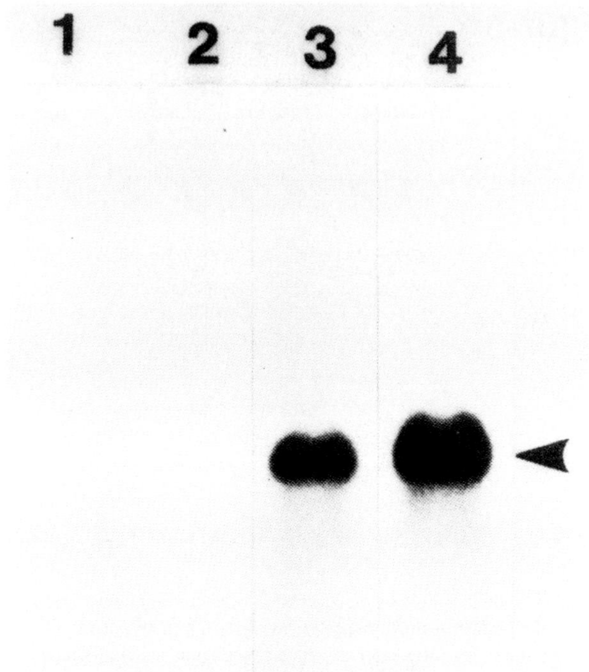

FIGURE 10. Expression of the *Sarcophaga* lectin gene in the posterior region of the fat body after injection of hemolymph. Larvae were ligated in the middle of the body, and the anterior half was cut off with fine scissors. Then 2-μl samples of hemolymph were injected through the ligate wound into the posterior body sac. After incubation of this sac for 2 h, RNA was extracted and analyzed by RNA blot hybridization. Posterior body sac (lane 1) of normal larvae (control); (lane 2) posterior body sac into which a hypodermic needle had been inserted via the wound (mock injection); (lane 3) after injection of hemolymph from normal larvae; (lane 4) after injection of hemolymph from injured larvae. Arrowhead indicates the position of *Sarcophaga* lectin mRNA. (From Shiraishi, S. and Natori, S., *FEBS Lett.*, 232, 163, 1988. With permission.)

it is known that the *Sarcophaga* lectin gene also is activated transiently in the early embryonic and pupal stages of *Sarcophaga* development, another interesting problem is whether the same mediator molecule is involved in the expression of this gene during the development of *Sarcophaga*.

V. DISCUSSION

It is uncertain whether other insect lectins are induced by the same mechanism as *Sarcophaga*. It has been clearly demonstrated that the site of synthesis of *Sarcophaga* lectin is the fat body. However, there are several reports that hemocytes synthesize humoral lectins (see also Chapter 2). Various insects might have their own system of producing lectins, or various kinds of lectins in the hemolymph may be produced in different tissues or cells. The mechanism of production of *Sarcophaga* lectin after injury is unique in comparison with the vertebrate immunoresponses that have been elucidated. As illustrated in Figure 11, the signal that body injury has occurred is first transmitted to a tissue present in the anterior region of the body. From there mediator molecules that interact directly with fat body cells are secreted into the hemolymph. The tissue in the anterior region of the body, which is

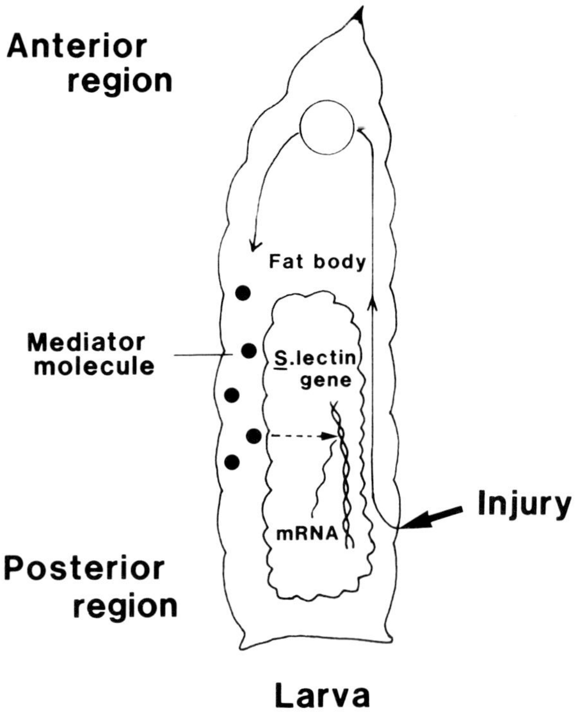

FIGURE 11. A model of the mechanism of activation of the *Sarcophaga* lectin gene upon injury. The stimulus of body injury is first transmitted to a certain tissue present in the anterior region of the body, and from there a mediator molecule that interacts directly with fat body cells is secreted into the hemolymph. On interaction with this mediator molecule, the fat body begins to synthesize mRNA for *Sarcophaga* lectin.

involved in this process, must still be elucidated. This finding is noteworthy because it is possible that there is a similar, as yet unknown, signal in the mammalian defense system, which signals the brain, nervous system, or other tissues, and/or works with the immune system in mediating macrophage and lymphocyte recognition of foreign substances.

Production of *Sarcophaga* lectin is regulated at the transcriptional level and is enhanced by injection of foreign substances. Exactly how the enhancement occurs is still unclear, but this is an important question that must be resolved in order to understand the insect immune system. It has been reported that in Schneider cells, the gene for cecropin, which is an antibacterial peptide in *Drosophila*, is induced by lipopolysaccharide.[24] It is possible that components of pathogen, such as lipopolysaccharide, may directly act on the fat body and induce the lectin. On the other hand, it also has been reported that the release of agglutinin from grasshopper fat body cultures was not enhanced by addition of microbial cell wall components.[18] Hemocytes might be involved in this process like mammalian macrophages in the initial recognition of foreign antigens.

Another important problem is whether the induction of *Sarcophaga* lectin during development requires the same mediator as the one that appears to be involved in the response to injury. Since it has been shown in an *in vitro* culture system that imaginal disks produce *Sarcophaga* lectin in response to β-ecdysone, there seem to be several activation mechanisms of the *Sarcophaga* lectin gene.

It is not yet clear which molecules participate in the antigenic stimulation of insect lectin production. It is hoped that the information presented here regarding the *Sarcophaga* lectin will help lead to a clear understanding of this and related problems in the insect immune system.

VI. SUMMARY

This chapter has described research regarding the production of *Sarcophaga* lectin in the flesh fly, *Sarcophaga peregrina*, and its induction mechanism. *Sarcophaga* lectin is a galactose-binding lectin, which is induced upon injury and during normal development in *S. peregrina*. This lectin is synthesized in the fat body, and its induction is regulated at the transcriptional level. Its induction by bodily injury is enhanced by injection of foreign substances. The mechanism of stimulus transmission from the time of injury until lectin production is as follows: bodily injury stimulates transmission of a message or impulse to a tissue in the anterior region of the body, which secretes a mediator molecule into the hemolymph. After this mediator molecule interacts with the fat body, it begins to synthesize *Sarcophaga* lectin mRNA.

It is not yet clear which molecules participate in the antigenic stimulation of *Sarcophaga* lectin or whether the mediator molecules involved in the induction of *Sarcophaga* lectin during development are the same as those which are involved in *Sarcophaga* lectin production in response to injury. Several reports on lectin induction in other insects describe the site of lectin synthesis to be hemocytes or tissues other than the fat body, and some other lectins appear to be uninducible. Nevertheless, it is hoped that information regarding *Sarcophaga* lectin induction will help lead to a clear understanding of the mechanism of insect lectin production.

REFERENCES

1. **Yeaton, R. W.,** Invertebrate lectins. I. Occurrence, *Dev. Comp. Immunol.*, 5, 391, 1981.
2. **Yeaton, R. W.,** Invertebrate lectins. II. Diversity of specificity, biological synthesis and function in recognition, *Dev. Comp. Immunol.*, 5, 535, 1981.
3. **Komano, H., Mizuno, D., and Natori, S.,** Purification of lectin induced in the hemolymph of *Sarcophaga peregrina* larvae on injury, *J. Biol. Chem.*, 255, 2919, 1980.
4. **Komano, H., Nozawa, R., Mizuno, D., and Natori, S.,** Measurement of *Sarcophaga peregrina* lectin under various physiological conditions by radioimmunoassay, *J. Biol. Chem.*, 258, 2143, 1983.
5. **Kubo, T., Komano, H., Okada, M., and Natori, S.,** Identification of hemagglutinating protein and bactericidal activity in the hemolymph of adult *Sarcophaga peregrina* on injury of the body wall, *Dev. Comp. Immunol.*, 8, 283, 1984.
6. **Takahashi, H., Komano, H., and Natori, S.,** Expression of the lectin gene in the *Sarcophaga peregrina* during normal development and under conditions where the defense mechanism is activated, *J. Insect Physiol.*, 32, 771, 1986.
7. **Qu, X.-M., Zhang, C.-F., Komano, H., and Natori, S.,** Purification of a lectin from the hemolymph of chinese oak silk moth (*Antheraea perni*) pupae, *J. Biochem.*, 101, 545, 1987.
8. **Kubo, T. and Natori, S.,** Purification and some properties of a lectin from the hemolymph of *Periplaneta americana* (American cockroach), *Eur. J. Biochem.*, 168, 75, 1987.
9. **Suzuki, T. and Natori, S.,** Identification of a protein having hemagglutinating activity in the hemolymph of the silkworm, *Bombyx mori, J. Biochem.*, 93, 583, 1983.
10. **Castro, V. M., Boman, H. G., and Hammarstorm, S.,** Isolation and characterization of a group of isolectins with galactose/*N*-acetylgalactosamine specificity from the giant silk moth *Hyalophora cecropia, Insect Biochem.*, 17, 513, 1987.
11. **Komano, H. and Natori, S.,** Participation of *Sarcophaga peregrina* humoral lectin in the lysis of sheep red blood cells injected into the abdominal cavity of larvae, *Dev. Comp. Immunol.*, 9, 31, 1985.
12. **Okada, M. and Natori, S.,** Primary structure of Sarotoxin I., an antibacterial protein in the hemolymph of *Sarcophaga peregrina* (flesh-fly) larvae, *J. Biol. Chem.*, 260, 7174, 1985.

13. **Ando, K. and Natori, S.,** Purification of Sarcotoxin II, antibacterial proteins of *Sarcophaga peregrina* (flesh-fly) larvae, *Biochemistry,* 26, 226, 1987.
14. **Baba, K., Okada, M., Kawano, T., Komano, H., and Natori, S.,** Purification of Sarcotoxin III, a new antibacterial protein of *Sarcophaga peregrina, J. Biochem.,* 102, 69, 1987.
15. **Matsuyama, K. and Natori, S.,** Molecular cloning of cDNA for sapecin gene during the development of *Sarcophaga peregrina, J. Biol. Chem.,* 263, 1717, 1988.
15a. **Kawaguchi, N., Komano, H., and Natori, S.,** Involvement of *Sarcophaga* lectin in the development of imaginal discs of *Sarcophaga peregrina* in an autorcrine manner, *Dev. Biol.,* 144, 86—93, 1991.
16. **Shiraishi, S. and Natori, S.,** Humoral mediator-dependent activation of the *Sarcophaga* lectin gene, *FEBS Lett.,* 232, 163, 1988.
17. **Amirante, G. A.,** Production of heteroagglutins in haemocytes of *Leucophaea maderae* L., *Experientia,* 124, 526, 1976.
18. **Stiles, B., Bradley, R. S., Stuart, G. S., and Harpner, K. D.,** Site of synthesis of the haemolymph agglutinin of *Melanoplus differentialis, J. Insect Physiol.,* 34, 1077, 1988.
19. **Takahashi, H., Komano, H., Kawaguchi, N., Obinata, M., and Natori, S.,** Activation of the secretion of specific proteins from fat body following injury to the body wall of *Sarcophaga peregrina* larvae, *Insect Biochem.,* 14, 713, 1984.
20. **Takahashi, H., Komano, H., Kawaguchi, N., Kitamura, N., Nakanishi, S., and Natori, S.,** Cloning and sequencing of cDNA of *Sarcophaga peregrina* humoral lectin induced on injury of the body wall, *J. Biol. Chem.,* 260, 12228, 1985.
21. **Komano, H., Mizuno, D., and Natori, S.,** A possible mechanism of induction of insect lectin, *J. Biol. Chem.,* 256, 7087, 1981.
22. **Kobayashi, A., Hirai, H., Kubo, T., Ueno, K., Nakanishi, Y., and Natori, S.,** Cloning and *in vitro* transcription of the *Sarcophaga* lectin gene, *Biochim. Biophys. Acta,* 1009, 244, 1989.
23. **Shiraishi, A. and Natori, S.,** Humoral factor activating the *Sarcophaga* lectin gene in cultured fat body, *Insect Biochem.,* 19, 261, 1989.
24. **Samakovlis, C., Kimbrell, D. A., Kylsten, P., Engstrom, A., and Hultmark, D.,** The immune response in *Drosophila*: pattern of cecropine expression and biological activity, *EMBO J.,* 9, 2969, 1990.

Chapter 17

WHY ARE INSECT CELLS IMMUNE TO HIV?

Alagarsamy Srinivasan, Claude E. Monken, and Ayyavoo Velpandi

TABLE OF CONTENTS

I. INTRODUCTION

The acquired immunodeficiency syndrome (AIDS) was first recognized in 1981 with the occurrence of *Pneumocystis carinii* pneumonia and Karposi's sarcoma in previously healthy homosexual males and intravenous drug users.[1] These same clinical manifestations later appeared in both hemophiliacs and blood transfusion recipients; they were then associated with an immune-deficient condition which led to the recognition of an acquired T-cell defect as the basis of the syndrome. Within 3 years of the first report of AIDS, the causative agent of the disease was isolated and named the human immunodeficiency virus (HIV).[2] The disease has already been reported in more than 142 countries worldwide.[3]

Besides opportunistic infections, which are common in HIV-infected individuals, a higher incidence of cancer and neurological syndromes has also been reported.[4-6] Epidemiological studies revealed that HIV is transmitted through sexual intercourse, intravenous drug use, and transfusion of blood and blood products. Speculations have been raised that there is a potential for the transmission of HIV by arthropods in Africa and other regions.[7-11] However, there is substantial evidence against the transmission of HIV by blood-sucking insects in an African context.[12] In the U.S., AIDS is found most frequently in the two traditional high-risk groups: male homosexuals and intravenous drug users. If insects were a factor in the transmission of HIV, the incidence of AIDS in children would be higher because children are bitten more frequently by mosquitoes than are adults. As it is, the incidence of AIDS increases rapidly with sexual activity.

II. COMPARISON OF IMMUNE RESPONSES TO INFECTION IN HIGHER VERTEBRATES AND INSECTS

Infection with retroviruses elicits both humoral and cellular immune responses, as observed in other viral infections.[13] Virally infected cells, expressing virus-specific proteins on the cell surface, can serve as targets for immune defenses. Humoral response is elicited by viral structural proteins, envelope glycoproteins, and regulatory proteins. Antibodies against the envelope proteins can neutralize viral particles by binding to them directly. In addition, antibodies can also bind to cells expressing viral proteins, making them susceptible to complement-mediated or cell-mediated lysis.[13]

Cellular immunity can be conferred by several mechanisms in vertebrates. Cytotoxic T-cells can lyse virus-infected cells expressing viral antigens. Helper T-cells and T-cells mediating delayed-type sensitivity upon infection with an agent start producing specific lymphokines which can amplify the phagocytic activity of macrophages. Natural killer cells can also be increased by the production of lymphokines such as γ interferons. The immune response of the host can be impaired by a decreased production of specific lymphokines or by infection of the macrophages and lymphocytes.

In contrast to the high degree of specificity observed in the immune response of higher vertebrates, insects exhibit a generalized response to infection. Although insects possess neither lymphocytes nor immunoglobulins, they exhibit both humoral and cellular immunity against invading pathogens.[14] Of the different types of blood cells (prohemocytes, plasmatocytes, granulocytes, coagulocytes, spherulocytes, and oenocytoids), only the granulocytes and the plasmatocytes participate in cellular immunity in all arthropods.[15] Insects effectively remove bacteria and unicellular fungi by phagocytosis; both plasmatocytes and granulocytes are involved in this process. Larger aggregates of foreign cells or parasites are also encapsulated by the same two immunocytes. Nodule formation occurs as a response to a large number of foreign cells.[16] Humoral immunity in insects is provided by 15 to 25 different proteins, including attacins, cecropins, and lysozymes, with antiviral and antibacterial activity.[14] (See also Chapters 2, 4, 12, and 13.)

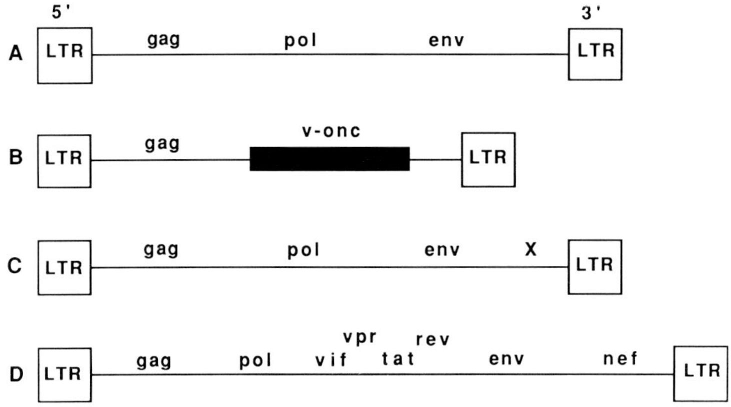

FIGURE 1. Schematic diagram of the genetic structure of retroviruses. (A) Replication-competent virus; (B) acute transforming virus; (C) T-cell lymphotropic virus; (D) human immunodeficiency virus. Abbreviations: *env*, envelope gene; *gag*, group-specific antigen gene; LTR, long terminal repeat sequences; *nef*, negative regulatory factor gene; *pol*, polymerase gene; *rev*, regulator of virion protein expression gene; *tat*, transactivator gene; *vif*, viral infectivity gene; *v-onc*, viral oncogene; *vpr*, virus protein R gene; X, open reading frame encoding regulator proteins.

HIV replication has not yet been attempted in insect hemocytes. The insect cells predominantly used to study HIV are cell lines of the moth, *Spodoptera frugiperda*, and of the fruit fly, *Drosophila melanogaster*. (The *S. frugiperda* cell lines were derived from a primary explant culture of pupal immature ovaries,[17] and the *D. melanogaster* cell lines originated from late embryonic-stage tissue.[18]) In view of the suggestion that the insect granulocyte is a hybrid of B- and T-lymphocytes and is also functionally similar to the macrophage,[15] the study of HIV replication in insect immunocytes seems warranted, since the main targets of HIV infection are the T-cells and macrophages.

III. INSECTS AS POSSIBLE VECTORS FOR TRANSMISSION OF HIV

The observation of HIV nucleic acid sequences in insects[7] and the possibility of transmission of HIV by mouthparts of several species of insects prompted a series of studies regarding insects as potential vectors for transmission. There are two ways in which a bloodsucking insect can spread the virus: biologically and mechanically.[19] Biological transmission occurs when an insect ingests blood infected with a virus which then replicates in its tissues (especially in the salivary glands) and gets transmitted to different hosts. Mechanical transmission occurs simply by transfer of virus between hosts through contaminated mouthparts. There is evidence for mechanical transmission of hepatitis B virus by the common bedbugs, *Cimex lectularius*[20] and *Cimex hemipterus*.[21] Since the known routes of transmission of HIV are the same for the hepatitis B virus, a similar mode of transmission of HIV by bedbugs was suspected.[22] Such a possibility is also supported by the mechanical transmission by insects of a number of retroviruses, including Friend leukemia virus, equine infectious anemia virus, and bovine leukemia virus.[23-29] However, it should be pointed out that the possibility of mechanical transmission depends on the extent of viremia in the infected individual. Experiments with bovine leukemia virus indicate that in a donor animal with low virus titer, transmission by flies does not occur.[29]

IV. STRUCTURAL AND FUNCTIONAL ORGANIZATION OF HIV

In light of the demonstration of the mechanical transmission of retroviruses, the genetic organization of HIV is discussed in the context of other retroviruses. Figure 1 shows the schematic diagram of the genetic organization of animal and human retroviruses. Retroviruses can be divided into two categories: replication competent and replication defective. Replication-competent viruses, represented by line A, contain the *gag* (group antigen), *pol* (polymerase), and *env* (envelope) genes, which encode the viral structural proteins, reverse transcriptase (RT), and envelope glycoproteins, respectively. The *gag, pol* and *env* sequences are essential for viral replication and are bound by long terminal repeat (LTR) sequences. Replication-defective viruses, which have also been termed acute transforming viruses, are represented by line B. They are derived from replication-competent viruses but have lost one or more genes and thus depend on proteins supplied by competent viruses for their replication. Human retroviruses and other members of the lentivirus family of retroviruses have many features in common with replication-competent retroviruses. They possess several open reading frames, in addition to the essential *gag, pol,* and *env* genes, which encode proteins that regulate both viral and cellular gene expression.[30,31] Line C represents both human T-cell lymphotropic virus and bovine leukemia virus, and line D represents human immunodeficiency virus. HIV shares transactivation properties with human T-cell lymphotropic viruses and with bovine leukemia virus. The HIV genome is, however, more complex than all other retroviruses, having seven additional open reading frames capable of encoding proteins.[32,33] The designated genes and their translational products identified in the HIV genome are given in Table 1. The functions of some of the genes are not yet understood.

In addition to the coding sequences, an important domain of the viral genome is the LTR region. The 5′ LTR is noncoding and contains regulatory sequences that govern the expression of the viral gene from the integrated proviral DNA.[30,34] The HIV-LTR functions in a variety of cell types.[30,35] The functional domains present in the LTR include the TATA box located upstream from the transcriptional start site (+1) and polyadenylation signal sequences located upstream from the boundary of the R and U5 regions. In addition to the enhancer sequences, LTR contains several sequences that have the potential to bind cellular factors which include AP-1, USF, Sp1, and LBP.[30]

V. ACCURATE SYNTHESIS AND PROCESSING OF HIV-ENCODED PROTEINS IN INSECT CELL LINES

The requirement of many eukaryotic proteins for proper posttranslational modification necessitated the development of eukaryotic expression vectors to augment those prokaryotic vectors already in use. Employing cell lines derived from the moth *Spodoptera frugiperda*, baculovirus vectors have been successfully used to produce a number of proteins,[36] including various HIV gene products (Table 2). Initially, the HIV gene was inserted into a plasmid vector downstream from the promoter for the baculovirus polyhedrin gene. The chimeric gene in the recombinant plasmid is then transfected with viral (AcNPV) DNA into *S. frugiperda* cell lines. Homologous recombination between the plasmid and the viral DNA occurs in the regions flanking the chimeric gene and allows its insertion into the genome of AcNPV. Jeang et al.[37] expressed the first 72 amino acids (coding exon 1) of *tat* protein in SF-9 cells by using a baculovirus vector *Autographa californica*. While SF-9 cells synthesized the 72 amino acid *tat* protein, they failed to support the synthesis of the complete *tat* protein, containing 101 amino acids. The underlying mechanism for such differential synthesis of *tat* protein is not known. The 72 amino acid *tat* protein synthesized in insect cells was able to activate HIV-LTR-directed expression. The expression of the *gag* protein in insect cell lines showed proper synthesis and posttranslational processing resembling that which occurs

TABLE 1
Open Reading Frames Identified in HIV Genome and Their Translational Products

Gene designation	Translational product	Function
	Structural Genes	
gag (group antigen)		
Precursor	p55	
MA	p17	Matrix
CA	p24	Capsid
NA	p7	Nucleocapsid
pol (polymerase)		
PR	p10	Protease
IN	p32	Endonuclease
RT	p66, p51	Reverse transcriptase
env (envelope)		
Precursor	gp160	
SU	gp120	Surface envelope
TM	gp41	Transmembrane envelope
	Regulatory Genes	
tat (transactivator)	p14	Transactivation of all HIV proteins
rev (regulator of expression of virion)	p19	Regulation of expression of virion proteins
nef (negative factor)	p27	Reduces virus expression
	Other Genes	
vif (virion infectivity factor)	p23	Required for infectivity as cell-free virus
vpr (virus protein R)	p16	Role in efficient virus replication
vpu (virus protein U)	p16	Role in virion release
tev (shares sequences with *tat*, *env*, and *rev* Proteins)	p28	Unknown

TABLE 2
Synthesis of HIV Gene Products in Insect Cells[a] through Baculovirus Recombinant Vectors

HIV gene used for expression	Modifications	Ref.
gag	Phosphorylated Myristylated	38,39
env	Glycosylated	40,41, 53—55
tat		37
rev		56

[a] Insect cell line SF-9 (*S. frugiperda*).

in authentic HIV replication.[38,39] Metabolic labeling experiments of the baculovirus-expressed *gag* products have demonstrated that p17 is myristylated at its amino terminus and that p24 is phosphorylated. Unprocessed precursor *gag* p55 in the absence of specific viral protease appeared to form retroviral core-like particles within the cytoplasm of *S. frugiperda* cell lines.

The *env* gene of HIV encodes a precursor glycoprotein of 160 kDa, which is cleaved

TABLE 3
Insects and Cell Lines used for HIV Expression Studies

Insects used	Cell lines used	Survival studies	HIV expression studies	Ref.
	Bedbug			
C. lectularius		+		9
C. hemipterus		+	+	22
	Mosquitoes			
A. aegypti		+		9
T. amboinensis		+	+	43
	SF-9 (*S. frugiperda*		+	37,46
	MEI 218 (*D. melanogaster*)		+	46
	S-2 (*D. melanogaster*)		+	48,49
	SF-8 (*S. frugiperda*)		+	54
	RML 19 (*D. variabilis*)		+	46

to produce mature virion glycoproteins gp120 and gp41. It has been shown that insect cell lines recognize the same glycosylation signal on polypeptides as do mammalian cells, although the pattern of glycosylation appears to be different. Though Hu et al.[40] showed the processing of HIV *env* gene products in an *S. frugiperda* cell line, Rusche et al.[41] reported that the *env* precursor did not appear to be cleaved, despite it appearance on the surface of these same cells. Goats immunized with *env* protein synthesized in *S. frugiperda* cell lines have high titers of antibodies that neutralize virus infection, as measured by viral gene expression or cell cytolysis. The sera from AIDS, AIDS-related complex, and asymptomatic HIV-infected individuals react positively with *env* proteins made in insect cell lines. Table 2 summarizes the list of HIV genes expressed in insect cell lines and their posttranslational modifications.

The *rev* gene of HIV encodes a protein which posttranscriptionally regulates the expression of the HIV structural proteins by relieving the nuclear sequestration of mRNAs containing viral *gag* or *env* coding regions. Recently, Ivey-Hoyle and Rosenberg[42] examined the expression of the HIV-I gene in *Drosophila* embryo-derived S-2 cells. By transfecting a pair of vectors, one containing the HIV *env* gene, the other containing the HIV *rev* gene, and both driven by the *Drosophila* inducible metallothionein promoter, they showed that gp160 is expressed in S-2 cells and its expression is dependent on the coexpression of the *rev* gene product. This result demonstrates that the mechanism of *rev* gene action on *env* mRNA, which occurs in mammalian systems, is conserved in the *Drosophila* cell line.

VI. SURVIVAL OF HIV IN INSECTS

In an effort to analyze the mechanical transmission of HIV by insects, studies were carried out with the common bedbug, *C. lectularius*, and the mosquito, *Aedes aegypti* (Table 3). Lyons et al.[9] allowed *C. lectularius* and *A. aegypti* to engorge on defibrinated blood to which tissue culture supernatant from HIV-infected cells had been added. The insects were homogenized and added to the permissive HIV-negative H9 cell line 60 min after feeding. Samples derived from the bedbug showed positive results by RT assay, indicating the active replication of HIV. Further studies showed that HIV survives for 4 h in *C. lectularius* and for 1 to 2 h in *C. hemipterus*.[22] On the other hand, samples derived from mosquitoes remained negative by RT assay, indicating the lack of survival of HIV. Webb et al.[43] conducted similar experiments and showed that HIV survived for 2 h in intrathoracically inoculated

Toxorhynchites amboinensis. In contrast, it has also been shown that HIV can persist in orally fed *A. aegypti* for at least 48 h.[19] Direct inoculation of bedbugs with high doses of HIV showed the presence of infectious virus for as long as 8 d.[43] The results obtained from these studies suggest that the reaction of the insect immune system to HIV infection is slow.

VII. DOES HIV REPLICATE IN INSECTS?

Considering the survival of HIV for longer periods of time in insects (previous section), it is important to know whether insects support active replication of HIV; the biological transmission of HIV depends on this parameter. Studies with *A. aegypti* indicated that HIV did not replicate in this strain. Similar negative data were also reported by Webb et al.,[43] using the bedbug (*C. hemipterus*). HIV is known to bind CD4 antigen for its entry into T-cells and monocytes;[44,45] its binding to cell lines established from mosquitoes *A. aegypti* and *Culex pipiens* was also demonstrated.[7] Despite its binding to these cells, replication of HIV fails to occur.

Srinivasan et al.[46] utilized cell lines developed from ticks (*Dermacentor variabilis*) (RML19), moths (*S. frugiperda*) (SF9), and fruit flies (*Drosophila melanogaster*) (MEI 218) to study HIV production. Biologically infectious HIV proviral DNA[47] was transfected into the three cell lines by the calcium phosphate co-precipitation method.[48] Culture supernatants from the transfected cells were tested for the release of HIV particles by RT assay. All of the transfected cultures showed only background RT activity, as observed in the untransfected culture, indicating that these insect cell lines did not support the synthesis and assembly of HIV particles upon transfection of proviral DNA.

HIV-LTR-directed expression was also analyzed by transfection of a chimeric plasmid, containing the HIV-LTR linked to the bacterial chloramphenicol acetyl transferase (CAT) gene. This assay showed a barely detectable level of CAT activity, indicating that the HIV promoter was not active in arthropod cells.[47] McDonald et al.[49] also observed no CAT activity in *D. melanogaster* S-2 cells upon transfection with HIV-LTR-CAT plasmids.

VIII. REPRESSION OF HIV EXPRESSION IN INSECTS: POSSIBLE MECHANISMS

HIV, like other lentiviruses, can produce a latent infection in different target tissues.[5] The mechanism by which an actively replicating virus becomes a nonreplicating latent virus is not known. The interplay of HIV encoded proteins with host cell proteins may contribute to the latent virus infection. McDonald et al.[49] analyzed expression of the HIV-LTR linked to the CAT gene in *Drosophila* S-2 cells. The results, based on CAT assay, showed that the HIV-LTR is barely active in these cells. HIV-LTR-directed expression can, however, be minimally activated by co-transfection of the HIV trans-activator gene *tat*. Analysis of the transfected cells revealed that HIV-LTR-initiated CAT-RNA is relatively abundant. Further, ribonuclease protection studies pointed out that transcription was correctly initiated within the HIV LTR. The differential level of CAT message and the level of CAT activity indicated that HIV-1 expression is effectively repressed in *Drosophila* S-2 cells predominantly at the posttranscriptional level. This observation is consistent with the results that HIV replication is blocked in insect cell lines.

Besides utilizing CD4 antigen as the receptor for its entry into cells, HIV uses alternate receptors for entry into non-$CD4^+$ cells.[50,51] The utilization of this pathway is evident in the glial cell lines. Studies conducted by Maddon et al.[52] revealed that successful HIV infection of a cell requires not only a suitable cell surface receptor for the virus but also an intracellular environment for viral gene expression and replication. It was reported that an established mouse cell line containing CD4 antigen did not show replication of HIV despite its binding to the receptor on the cell surface. Further, the level of transcriptional activation exhibited by *tat* is also less in mouse cells than the level in human T-cells. Such a situation may exist in the insect cell due to the lack of appropriate cellular factors that act in concert with *tat* product. Additionally, the intracellular environment of insect cells also may not favor the stability of some of the crucial HIV proteins. The *tat* protein of HIV belongs in this category because the expression of complete *tat* protein was not successful in insect cell lines, although other HIV-encoded proteins were synthesized and modified properly.[53-56]

IX. SUMMARY

The human immunodeficiency virus is transmitted through the exchange of bodily fluids such as semen and blood; whether HIV can be transmitted by bloodsucking insects is therefore a relevant and important question. Insects may transmit viruses by two means: biological and mechanical. To transmit HIV by the biological mode, the virus must replicate in the insect. Through the use of baculovirus vectors, several functionally active HIV proteins have been synthesized in *S. frugiperda* cell lines. However, biologically infectious HIV proviral DNA, transfected into *D. variabilis, S. frugiperda,* and *D. melanogaster* cell lines, failed to produce viral particles. Both the HIV regulatory sequences governing expression of the viral genes, as well as the important transactivator protein *tat*, were barely active in the insect cell lines. This lack of replication assures that the transmission of HIV by insects through the biological mode is not possible.

The mechanical transmission of several animal retroviruses has been documented. Under experimental conditions, HIV was shown to persist for up to 2 d in mosquitoes and up to 8 d in bedbugs. The persistence of HIV in bedbugs and mosquitoes presents a situation where the mechanical transmission of HIV is still possible through interrupted feeding by the residual blood present in the contaminated mouthparts. Several studies concluded that, based on the low level of HIV viremia,[57] the lack of replication of HIV in insects, and the extremely small amounts of blood on mouthparts, it is unlikely that insects can transmit HIV.

ACKNOWLEDGMENTS

We acknowledge NIH for support (AI29306-01, AI25380, and CA10815) and the Wistar Institute editorial staff for the preparation of this review.

REFERENCES

1. **Curran, J. W., Morgan, W. M., Hardy, A. M., Jaffe, H. W., Darrow, W. W., and Dowdle, W. R.,** The epidemiology of AIDS: current status and future prospects, *Science*, 229, 1352, 1985.
2. **Curran, J. W., Jaffe, H. W., Hardy, A. M., Morgan, W. M., Selik, R. M., and Dondero, T. J.,** Epidemiology of HIV infection and AIDS in the United States, *Science*, 239, 610, 1988.
3. **Schochetman, G., Epstein, J. S., and Zuck, T. F.,** Serodiagnosis of infection with the AIDS virus and other human retroviruses, *Annu. Rev. Microbiol.*, 43, 629, 1989.
4. **Srinivasan, A. and Reddy, E. P.,** Viruses and cancer of the immune system *Curr. Opinion Immunol.*, 2, 714, 1990.
5. **Fauci, A. S.,** The human immunodeficiency virus: infectivity and mechanisms of pathogenesis, *Science*, 239, 617, 1988.
6. **Wong-Staal, F. and Gallo, R. C.,** Human T-lymphotropic retroviruses, *Nature (London)*, 317, 395, 1985.

7. **Becker, J. L., Hazan, O., Nugeyre, M. T., Rey, F., Spire, B., Barre-Simoussi, F., Georges, A., Teulires, L., and Chermann, J. C.**, Infection of insect cell lines by HIV, agent of AIDS, and evidence for HIV proviral DNA in insects from Central Africa, *C. R. Acad. Sci. III*, 303, 1986.
8. **Cook, L. F.**, Hepatitis B and AIDS in Africa, *Med. J. Aust.*, 142, 661, 1985.
9. **Lyons, S. F., Jupp, P. G., and Schoub, B. D.**, Survival of HIV in the common bedbug, *Lancet*, 2, 45, 1986.
10. **Zuckerman, A. J.**, AIDS and insects, *Br. Med. J.*, 292, 1094, 1986.
11. **Miike, L.**, AIDS-related issues. Do insects transmit AIDS? in *Office of Technology Assessment Staff Paper*, U.S. Government Printing Office, Washington, D.C., 1987.
12. **Piot, P. and Schofield, C. J.**, No evidence of arthropod transmission of AIDS, *Parasitol. Today*, 2, 294, 1986.
13. **Sharpe, A. H. and Fields, B. N.**, Pathogenesis of viral interactions, *N. Engl. J. Med.*, 312, 486, 1985.
14. **Boman, H. G. and Hultmark, D.**, Cell-free immunity in insects, *Annu. Rev. Microbiol.*, 41, 103, 1987.
15. **Gupta, A. P.**, Arthropod immunocytes, in *Hemocytic and Humoral Immunity in Arthropods*, Gupta, A. P., Ed., John Wiley & Sons, New York, 1986, chap. 1.
16. **Han, S. S. and Gupta, A. P.**, Arthropod immune system. II. Encapsulation of implanted nerve cord and "plain gut" surgical suture by granulocytes of *Blattella germanica* (L.) (Dictyoptera:Blattellidae), *Zool. Sci.*, 6, 303, 1989.
17. **Vaughn, J. L., Goodwin, R. H., Tompkins, G. J., and McCawley, P.** The establishment of two cell lines from the insect *Spodoptera frugiperda*. (Lepidoptera:Noctuidae), *In Vitro*, 13, 213, 1977.
18. **Schneider, I.**, Cell lines derived from later embryonic stages of *Drosophila melanogaster*, *J. Embryol. Exp. Morphol.*, 27, 353, 1972.
19. **Booth, W.**, AIDS and insects, *Science*, 237, 355, 1987.
20. **Jupp, P. G., McElligott, S. E., and Lecatsas, G.**, The mechanical transmission of hepatitis B virus by the common bedbug (*Cimex lectularius* L) in South Africa, *S. Afr. Med. J.*, 63, 77, 1983.
21. **Ogston, C. W., Wittenstein, F. S., London, W. T., and Millman, I.**, Persistence of hepatitis B surface antigen in the bedbug *Cimex hemipterus* (Fabr.), *J. Infect. Dis.*, 140, 411, 1979.
22. **Jupp, P. G. and Lyons, S. F.**, Experimental assessment of bedbugs (*Cimex lectularius* and *Cimex hemipterus*) and mosquitoes (*Aedis aegypti formosus*) as vectors of human immunodeficiency virus, *AIDS*, 1, 171, 1987.
23. **Rehacek, J., Fisher, R. G., and Inecke, D. H.**, Oncogenic viruses in vertebrates transmitted by hematophagous arthropods, *Adv. Virus Res.*, 20, 159, 1976.
24. **Fisher, R. G., Luecke, D. H., and Rehacek, J.**, Friend leukemia virus (FLV) activity in certain arthropods. III. Transmission studies, *Neoplasma*, 20, 255, 1973.
25. **Foil, L. D., Meek, C. L., Adams, W. V., and Issel, C. J.**, Mechanical transmission of equine infectious anemia virus by deer flies (*Chrysops flavidus*) and stable flies (*Stomoxys calcitrans*), *Am. J. Vet. Res.*, 44, 155, 1983.
26. **Issel, C. J., Adams, W. V., Miik, L., and Ochoa, R.**, Transmission of equine infectious anemia from horses without clinical signs of disease, *J. Am. Vet. Med Assoc.*, 180, 272, 1982.
27. **Bech-Nielsen, S., Piper, C. E., and Ferrer, J. F.**, Natural mode of transmission of the bovine leukemia virus: role of blood-sucking insects, *Am. J. Vet. Res.*, 39, 1098, 1978.
28. **Buxton, B. A., Schultz, R. D., and Collins, W. E.**, Role of insects in the transmission of bovine leukosis virus: potential for transmission by mosquitoes, *Am. J. Vet. Res.*, 43, 1458, 1982.
29. **Foil, L. D., Seger, C. L., French, D. D., Issel, C. J., McManus, J. M., Ohrberg, C. L., and Ransey, R. T.**, Mechanical transmission of bovine leukemia virus by horse flies (Diptera:Tabanidae), *J. Med. Entomol.*, 25, 374, 1988.
30. **Haseltine, W. A.**, Development of antiviral drugs for the treatment of AIDS: strategies and prospects, *J. AIDS*, 2, 311, 1989.
31. **Cann, A. J. and Karn, J.**, Molecular biology of HIV: new insights into the virus life-cycle, *AIDS*, 3 (Suppl. 1), S19, 1989.
32. **Gallo, R. C., Wong-Staal, F., Montagnier, L., Haseltine, W., and Yoshida, M. C.**, HIV/HTLV gene nomenclature, *Nature (London)*, 333, 504, 1988.
33. **Laurence, J.**, Update: HIV-1 gene nomenclature, *AIDS Res. Hum. Retroviruses*, 4, 7, 1988.
34. **Chen, I. S. Y.**, Regulation of AIDS virus expression, *Cell*, 47, 1, 1986.
35. **Cullen, B.**, Transactivation of human immunodeficiency virus occurs via a bimodal mechanism, *Cell*, 46, 973, 1986.
36. **Kang, C. Y.**, Baculovirus vectors for expression of foreign genes, *Adv. Virus Res.*, 35, 177, 1988.
37. **Jeang, K.-T., Shank, P. R., Rabson, A. B., and Kumar, A.**, Synthesis of functional human immunodeficiency virus *tat* protein in baculovirus as determined by a cell-cell fusion assay, *J. Virol.*, 62, 3874, 1988.
38. **Overton, H. A., Fujii, Y., Price, I. R., and Jones, I. M.**, The protease and *gag* gene products of the human immunodeficiency virus: authentic cleavage and post-translational modification in an insect cell expression system, *Virology*, 170, 107, 1989.

39. **Madisen, L., Travis, B., Hu, S.-L., and Purchio, A. F.,** Expression of the human immunodeficiency virus *gag* gene in insect cells, *Virology,* 158, 248, 1987.
40. **Hu, S.-L., Kosowski, S. G., and Schaaf, K. F.,** Expression of envelope glycoproteins of human immunodeficiency virus by an insect virus vector, *J. Virol.,* 61, 3617, 1987.
41. **Rusche, J. R., Lynn, D. L., Robert-Guroff, M., Langlois, A. J., Lyerly, H. K., Carson, H., Krohn, K., Ranki, A., Gallo, R. C., Bolognesi, D. P., Putney, S. D., and Matthews, T. J.,** Humoral immune response to the entire human immunodeficiency virus envelope glycoprotein made in insect cells, *Proc. Natl. Acad. Sci. U.S.A.,* 84, 6924, 1987.
42. **Ivey-Hoyle, M. and Rosenberg, M.,** Rev-dependent expression of human immunodeficiency virus type 1 gp160 in *Drosophila melanogaster cells, Mol. Cell. Biol.,* 10, 6152, 1990.
43. **Webb, P. A., Happ, C. M., Maupin, G. O., Johnson, B. J., Ou, C.-Y., and Monath, T. P.,** Potential for insect transmission of HIV: experimental exposure of *Cimex lectularius* and *Toxorhynchites amboinensis* to human immunodeficiency virus, *J. Infect. Dis.,* 160, 970, 1989.
44. **Dalgleish, A. G., Beverly, P. C. L., Clapham, P. R., Crawford, C. H., Greaves, M. F. and Weiss, R. A.,** The CD4 (T4) antigen is a component of the receptor for the AIDS retrovirus, *Nature (London),* 312, 763, 1984.
45. **Klatzman, D., Champagne, E., Chamaret, S., Gruest, J., Guetard, D., Hercend, T., Gluckman, J. C., and Montagnier, L.,** T4 molecule behaves as the receptor for human retrovirus LAV, *Nature (London),* 312, 767, 1984.
46. **Srinivasan, A., York, D., and Gohan, C. A.,** Lack of HIV replication in arthropod cells, *Lancet,* 1, 1094, 1987.
47. **Srinivasan, A., Anand, R., York, D., Ranganathan, P., Feorino, P., Schochetman, G., Curran, J., Luciw, P., and Sanchez-Pescador, R.,** Molecular characterization of human immunodeficiency virus from Zaire: nucleotide sequence analysis identifies conserved and variable domains in the envelope gene, *Gene,* 52, 71, 1987.
48. **DiNocera, P. P. and David, I. B.,** Transient expression of genes introduced into cultured cells of *Drosophila, Proc. Natl. Acad. Sci. U.S.A.,* 80, 7095, 1983.
49. **McDonald, J. F., Josephs, S. F., Wong-Staal, F., and Strand, D. J.,** HIV-1 expression is post-transcriptionally repressed in *Drosophila* cells, *AIDS Res. Hum. Retroviruses,* 5, 79, 1989.
50. **Fenyo, E. M., Albert, J., and Asjo, B.,** Replicative capacity, cytoplastic effect and cell tropism of HIV, *AIDS,* 3 (Suppl. 1), S5, 1989.
51. **Srinivasan, A., Dorsett, D., York, D., Bohan, C., and Anand, R.,** Human immunodeficiency virus replication in human brain cells, *Arch. Virol.,* 99, 135, 1988.
52. **Maddon, P. J., Dalgleish, A. G., McDougal, J. S., Clapham, P. R., Weiss, R. A., and Axel, R.,** The T4 gene encodes the AIDS virus receptor and is expressed in the immune system and the brain, *Cell,* 47, 333, 1986.
53. **McQuade, T. J., Pitts, T. W., and Tarpley, W. G.,** A rapid solution immunoassay to quantify binding of the human immunodeficiency virus envelope glycoprotein to soluble CD4, *Biochem. Biophys. Res. Commun.,* 163, 172, 1989.
54. **Wells, D. E. and Compans, R. W.,** Expression and characterization of a functional human immunodeficiency virus envelope glycoprotein in insect cells, *Virology,* 176, 575, 1990.
55. **Moore, J. P., McKeating, J. A., Jones, I. M., Stephens, P. E., Clements, G., Thomson, S., and Weiss, R. A.,** Characterization of recombinant gp120 and gp160 from HIV-1: binding to monoclonal antibodies and soluble CD4, *AIDS,* 4, 307, 1990.
56. **Daefler, S., Klotman, M. E., and Wong-Staal, F.,** Trans-activating *rev* protein of the human immunodeficiency virus 1 interacts directly and specifically with its target RNA, *Proc. Natl. Acad. Sci. U.S.A.,* 87, 4571, 1990.
57. **Ho, D. D., Moudgil, T., and Alam, M.,** Quantitation of human immunodeficiency virus type 1 in the blood of infected persons, *N. Engl. J. Med.,* 321, 1621, 1989.

Biomedical Applications and Techniques

Chapter 18

APPLICATIONS OF LAL IN PHARMACEUTICALS AND MEDICAL DEVICES

Ronald N. Berzofsky and Karen Zink McCullough

TABLE OF CONTENTS

I. INTRODUCTION

A. HISTORICAL DEVELOPMENT

The use of *Limulus* amebocyte lysate (LAL) to detect and control the presence of pyrogenic substances in pharmaceuticals and medical devices is a relatively recent development. However, the association of pyrogenic reactions with the intravenous injection of fluids is over 200 years old. This historical association has been previously reviewed[1,2] and falls outside the scope of this chapter. However, it is instructive to appreciate the major concepts, which comprise our current understanding and philosophy and which serve as a rationale for the need to control the levels of pyrogens in pharmaceuticals and medical devices.

To understand better the historical developments, it is necessary to define three words that are currently used interchangeably when referring to pyrogen testing. The word *pyrogen*, originally used by Billroth[3] in 1862, applies to any material which induces a fever. *Pyrogen* describes an *in vivo* characteristic. A solution that contains a fever-producing substance is said to be pyrogenic. The word *endotoxin* refers to a specific pyrogen associated with the membrane of Gram-negative bacteria. Endotoxins are the most prevalent pyrogen found in aqueous solutions. The word *lipopolysaccharide* (LPS) describes the biochemical structure of endotoxin. Endotoxins contain a fatty acid portion termed *lipid A* and a long chain of repeating sugars or polysaccharides (see also Chapter 19). LPSs are purified endotoxins and are used as standards for *in vivo* pyrogen assays or *in vitro* endotoxin assays.

Early observations in the mid-1700s and 1800s suggested that intravenous injection of fluids was always associated with the onset of fever.[4,5] Although one might conclude that the "act" of intravenous injection intrinsically triggered a pyrogenic response independent of the material injected, Panum[6] postulated the existence of a heat-stable pyrogenic material present in the injected fluid as the causative agent. Subsequent investigations on the nature of this fever-inducing material suggested that it was of bacterial origin[3,7-9] and exclusively associated with Gram-negative bacteria.[1] Pfeiffer is reported to have proposed the term "endotoxin" to describe the membrane-associated toxin of *Vibrio*.[2]

The concern over the presence of endotoxin in pharmaceuticals was raised by the studies of Wechselmann,[10] Muller,[11] Samelson,[12] and Bendix and Bergmann.[13] Hort and Penfold[14-16] expanded the findings that a Gram-negative bacterial substance was responsible for pyrogenicity. These workers are credited also with designing the first rabbit pyrogen assay.[2]

Subsequently, confirmatory observations by Seibert,[17,18] Seibert and Mendel,[19] and Rademaker[20,21] reinforced the concept that bacterial contamination was present in all pyrogenic injectables, and that through careful technique to avoid bacterial contamination, nonpyrogenic injectables could be produced. Their work led to a collaborative study, which standardized the rabbit pyrogen test, and allowed pyrogen testing to attain pharmacopeial recognition in 1942.

The rabbit pyrogen test remained the exclusive official pyrogen assay, referenced in the U.S. Pharmacopeia (USP) for over 25 years, and is still referenced extensively in the majority of USP drug monographs. However, events occurring in the late 1950s and early 1960s were to lead to a significant alternative to the rabbit pyrogen test for determining pyrogenicity in pharmaceuticals and medical devices.

B. BIOCHEMISTRY

The biochemical basis for the use of *Limulus* amebocyte lysate in the detection of endotoxin lies entirely in the coagulation reaction inherent in *Limulus* blood. The ability of *Limulus* blood to form a gelatinous clot was initially described by Howell,[22] Loeb,[23,24] and Blanchard.[25] These early investigations showed that coagulation was induced by foreign substances and that the circulating amebocyte (= granulocytes) was involved in the reaction.

The association between endotoxin and coagulation began in 1956, when Bang[27] reported that infecting the horseshoe crab, *Limulus polyphemus*, with *Vibrio*, a Gram-negative bacteria, caused a fatal intravascular coagulation. By 1964, Levin and Bang[28] had demonstrated that extracts of the circulating amebocytes would gel in the presence of Gram-negative bacterial endotoxin.

Limulus blood has always held a certain fascination because of its bluish appearance. The blue color is due to hemocyanin, a copper-based oxygen acceptor. The hemocyanin, while produced by the cyanocytes, is extra-corporeal and remains in the plasma after centrifugation of whole blood. The amebocyte is the only cell present in *Limulus* blood.[23,28,29] It represents a homogeneous population of nucleated, disk-shaped granular cells.[30] The amebocyte contains all the components of the entire coagulation system. Cell-free hemolymph does not gel in the presence of endotoxin and does not enhance the endotoxin-mediated reaction associated with amebocyte extracts.[28,31] Lysates from washed amebocytes have been prepared by lysis in hypotonic solutions,[31,32] by physical disruption using mechanical shear,[33,34] or by freeze-thaw cycles.[31,33] The sensitivity of amebocyte lysate to endotoxin can be increased by subsequent extraction with organic solvents.[35]

A simplified description of the *Limulus* coagulation reaction sequence was presented by Young et al.[36] They fractionated amebocyte lysate into two active fractions using gel filtration chromatography. Fraction I, the column void, contained a heat-labile component and represented material >75 kDa. Fraction II represented a heat-stable component of lower molecular weight. A mixture of both fractions was necessary for the regeneration of the reactivity to endotoxin seen in unfractionated lysate. Incubation of endotoxin with fraction I material prior to the addition of material from Fraction II greatly accelerated the time course of coagulation. This supported the two-step mechanism originally proposed by Levin and Bang.[31] Namely, proenzyme (Fraction I) in the lysis is activated by the presence of endotoxin. This active enzyme subsequently catalyzed the conversion of a clotting protein (Fraction II) into an insoluble gel. The rate-limiting reaction is that of endotoxin and the proclotting enzyme, with the concentration of active clotting enzyme formed being proportional to the initial concentration of endotoxin.

The coagulation reaction is not as simple as originally proposed. It exists as a multicomponent cascade, which is initiated by endotoxin and terminates with gelation. The majority of the biochemical analysis has been performed by Japanese workers using lysate from another species of horseshoe crab, *Tachypleus tridentatus*,[37,38] but their observations are thought to apply to *Limulus* as well. The first coagulation factor, Factor C, directly interacts with endotoxin to form Factor $\overline{C}$. An anti-LPS factor, found in both *Limulus* and *Tachypleus*, inhibits the interaction of Factor C and endotoxin.[39] Factor $\overline{C}$ possesses enzymatic activity, which activates Factor B. Factor $\overline{B}$, the active form of Factor B, is then responsible for activating the proclotting enzyme. Factors B and C have not been directly demonstrated in *L. polyphemus*; however Nakamura and Levin[40,41] have described an activator of the proclotting enzyme and protease N in *Limulus*. The exact relationship of these *Limulus* proteins to Factor B and C of *Tachypleus* is unknown at present.

The clotting protein, coagulogen, is the last component of the coagulation cascade. It comprises almost half of total protein present in amebocyte lysate.[36] The active clotting enzyme is a serine protease[42] and hydrolyzes a single peptide bond within the coagulogen[42,43] to form a shorter peptide, coagulin. Once generated, the coagulin self-associates, forming a three-dimensional lattice structure, and eventually gels.

The catalytic nature of each activated component in the coagulation cascade serves to amplify the next step in turn. This amplification most probably results in the extreme sensitivity of *Limulus* amebocyte lysate to endotoxin. All commercial lysates are capable of detecting picogram (pg) quantities (10^{-12}) of endotoxin.

In addition to the endotoxin-mediated cascade which activates the proclotting enzyme, another protein component, Factor G, appears to represent an alternative coagulation pathway.[38] Factor G is present in both *Tachypleus* and *Limulus*[38] and is activated by β-1,3-D-glucans. Factor G directly activates the proclotting enzyme, which in turn causes the gelation of coagulogen.

C. METHODOLOGY

1. Gel-Clot Assay

The most commonly employed LAL method is the gel-clot assay. This assay utilizes the entire endotoxin-mediated cascade in addition to the clotting protein to produce a gelatinous clot after incubation with endotoxin. Basically, equal volumes of sample and LAL (typically 0.1 ml each) are combined in a 10 × 75 mm glass tube. After an incubation period of 60 min at 37°C, the tubes are inverted 180°. A positive result is indicated by a clot, which withstands the inversion. By titrating the lysate with an endotoxin of known potency, the minimum concentration of endotoxin required to yield a positive clot can be determined. This minimum endotoxin concentration, or end point, is referred to as the lysate sensitivity.

The gel-clot assay can be used as a purely qualitative limits test to rank samples as either positive or negative, i.e., greater than or less than the lysate sensitivity. However, by titrating positive samples one can obtain a semiquantitative measure of the endotoxin concentration in unknowns by multiplying the last positive sample dilution by the lysate sensitivity.

Several variations of the gel-clot method have been developed, which use much smaller quantities of LAL and involve novel end point detection methods; most employ microscope slides or coverslips, and score positives by an increase in viscosity,[44] pattern after evaporation,[45] staining with dye,[46,47] or ability of the reaction mix to migrate within a capillary tube.[48] Other modifications include performing the assay in microtiter plates, followed by the addition of dye,[49] or performing the assay in capillary tubes, followed by testing each tube for its ability to draw up dye.[50]

Harris and Feinstein[51] described the LAL-bead assay. The assay uses a hydrophobic polyolefin bead to bind endotoxin in the test sample. After a gentle wash procedure, each bead is incubated in the presence of LAL and the reaction tubes are observed for classical gelation.

The gel-clot assays suffer from the disadvantage that they are at best semiquantitative and require preparing multiple dilutions of samples to determine a positive/negative end point. Various alternative methods have been developed to yield more quantitative results.

2. Turbidimetric End Point and Kinetic Assays

During the process of clot formation, the reaction mixture becomes increasingly more turbid and contains a larger proportion of insoluble clotting protein.[31] The turbidimetric assays measure this increase in turbidity as a function of endotoxin concentration. Endotoxin concentrations in unknowns are determined by comparing the resultant turbidity to a standard curve. Since turbidity is a precursor to gelation, these assays are more sensitive to endotoxin concentration. Turbidimeric assays have been described, which measure turbidity at the end of a incubation period, i.e., turbidimetric end point,[52,53] or measure turbidity continually during the reaction and utilize reaction rates, i.e, turbidimetric kinetic assay.[54] The latter has the advantage that it can quantitate endotoxin over a greater concentration range. A variation of the end point turbidimetric technique has been reported by Dubczak et al.[55] based on nephelometry.

3. Colorimetric Assay

Another quantitative method stemming from the gel-clot reaction is the colorimetric assay.[56] This assay utilizes the Lowry protein assay to quantitate the amount of insoluble clotting protein generated as a result of the endotoxin present. As with the other quantitative techniques, endotoxin concentrations in unknowns are determined by comparing optical density to a standard curve.

4. Chromogenic End Point and Kinetic Assays

All the methods previously described utilize the natural clotting protein as the substrate for the endotoxin-mediated coagulation cascade, and employ various methods to detect and quantify the gelation reaction. The chromogenic assays still utilize the initial portions of the enzyme cascade, but substitute a synthetic chromogenic peptide as substrate for the activated clotting enzyme in place of the clotting protein.[57-59] The chromogenic substrates are hydrolyzed by the clotting enzyme releasing the terminal chromogenic moiety and generating a yellow color.

The chromogenic end point assay is normally performed in a two-step assay, where LAL and sample are incubated, followed by the addition of substrate. The resultant absorbance is determined spectrophotometrically after the reaction is stopped with acetic acid. Like the other quantitative procedures, endotoxin in unknowns is determined from a standard curve. The assays are rapid, less than 15 min,[58] and the sensitivity to endotoxin can be increased by extending the LAL/sample incubation.

The chromogenic kinetic assay utilizes a single co-lyophilized LAL/substrate reagent, which is incubated with the sample and monitored spectrophotometrically for the appearance of hydrolyzed substrate. Endotoxin in unknowns is interpolated from standard curves which are linear over a five-log concentration range.

II. REGULATORY ASPECTS

A. REGULATORY HISTORY

The development of regulatory policies for LAL use has been a cooperative effort between national and international regulatory agencies, including the U.S. Food and Drug Administration (FDA), the U.S. Pharmacopeia Committee of Revision (USP), the European Pharmacopoeia Commission (EP), the Society of Japanese Pharmacopoeia (JP), lysate reagent manufacturers, and the parenteral drug and medical device industry. Throughout development of the methodology, regulatory agencies have taken their lead from data presented by LAL researchers, issuing or modifying documents in response to scientific advances made by industry.

In 1970, Cooper et al.[60] published data on the use of LAL for the detection of endotoxin in short-lived radiopharmaceuticals. This was the first published report on the use of LAL as an end-product release test for a parenteral products and prompted the rest of the industry to consider the LAL test as an alternative to the USP rabbit pyrogen test. The first regulatory acknowledgment of LAL as a useful tool for the detection of endotoxin came in 1973, when FDA announced that LAL reagent was to be considered a biological product, subject to license.[61] Although many large pharmaceutical companies had, by this time, been experimenting with LAL, there were insufficient data and experience to support replacement of the USP rabbit pyrogen test for the final release of product. FDA stated that until standards for purity and potency (sensitivity) of the reagent could be established, the test was to be used only as an in-process monitoring tool. Later that year, FDA announced proposed manufacturing standards for LAL reagent, taking the first step in regulating the quality of the product.[62]

By the time the next regulatory notice appeared in the Federal Register on November 4, 1977,[63] the pharmaceutical industry had gained considerable experience with the test. The largest body of data initially came from those involved in testing large volume parenterals (LVPs) and medical devices,[56,64,65] but data from other segments of the parenteral industry, including antibiotics,[66,67] vaccines,[68] antitumor agents,[69] and other SVPs[70,71] began to appear in the literature. In the November 4 notice, FDA outlined a conditional approval for the use of LAL as a final release test for licensed biological products and medical devices, provided manufacturers of these products submitted separate amendments to each product license including methodology, test criteria, and other supporting data. In addition, device manufacturers had to show that the LAL test was at least equivalent to the rabbit test, thereby equating the presence of endotoxin *in vitro* with pyrogenicity *in vivo*.

Equivalency on a product-by-product basis was labor intensive and costly for many parenteral manufacturers, but by 1980, many reports on rabbit/LAL comparisons had already been published.[72-78] The data presented in the later studies defined two terms used in determining rabbit/LAL equivalency. The average pyrogenic dose (APD) is that amount of endotoxin that causes a total temperature rise in a USP eight rabbit pyrogen test of ≥3.70°C. The threshold pyrogenic dose (TPD) is defined as the lower 95% confidence limit of the APD, and was experimentally determined to be 1.0 ng/kg with a number standard endotoxin preparations, including Difco's *Escherichia coli* 055:B5, and EC-2, the FDA *E. coli* Reference Standard Endotoxin at the time. The HIMA study, as reported by Dabbah and collaborators,[76] went so far as to suggest that if a laboratory could demonstrate a LAL test failure rate of ≥50% at an endotoxin level of 0.1 ng/ml* of *E. coli* 055:B5 endotoxin; the test could be considered qualified as equivalent to the USP rabbit pyrogen test. Despite the fact that FDA had not yet approved the use of LAL for end-product release of injectable drug product other than licensed biologicals, the pharmaceutical industry continued to collect data on the use of the test with a variety of products.[79-82] Industry researchers and reagent manufacturers also experimented with modifications of the established gel-clot assay with methods (such as a bead assay,[51] a slide micro-method,[44-46] a capillary micro-method)[48] and quantitative procedures (such as the chromogenic substrate method,[58,83] a nephelometric method,[55] a turbidimetric assay,[52] and a colorimetric procedure[56]). In response to continued research and data acquisition by the parenteral industry, the U.S. Pharmacopeia published its first chapter on the bacterial endotoxins test (BET) in USP XX.[84] The focus of this chapter was on the development of a single, consistent gel-clot method, including suggested procedures for reagent label claim verification, product validation, and routine testing. The USP chapter did not, however, suggest endotoxin limits for the release testing of compendial articles, and did not offer the LAL test a replacement for the USP rabbit pyrogen test.

FDA published its first "draft guideline for validation of the Limulus amebocyte lysate test as an end-product pyrogen test for human and veterinary injectable drugs other than licensed biologicals" in January of 1980.[85] A guideline, as issued by FDA, does not constitute a legal requirement, but outlines procedures that are acceptable to the agency. Deviations from the procedure suggested in the guideline are acceptable, but may require additional supporting validation data. Significant sections of the 1980 guideline included:

1. Validation procedures for drugs including initial laboratory qualifications, and inhibition/enhancement screens, which had to be performed for each drug product (see Section II.B as follows)
2. The imposition of an endotoxin limit of 0.25 EU/ml** of drug product
3. Concurrent rabbit/LAL testing of at least three batches of each drug product to determine the absence of nonendotoxin pyrogenic substances

Industry welcomed the publication of the two regulatory documents as acknowledgment by FDA and USP of the increased speed, sensitivity, and cost-effectiveness of LAL over

* An endotoxin pass/fail limit of 0.1 ng/ml for LAL assays is equivalent to the rabbit threshold pyrogenic dose of 1 ng/kg when the test solution is administered at 10 ml/kg.

** EU, Endotoxin Unit. An Endotoxin Unit is defined as 0.2 ng FDA Reference Standard Endotoxin EC-2.[86]

the USP rabbit pyrogen test. There was, however, some concern over a number of proposals in the 1980 FDA guideline:

1. Industry argued that the production of fever in an animal was a function of the total amount of endotoxin injected, not the amount of endotoxin per milliliter of the drug. It was suggested that the endotoxin limit should be dose dependent (i.e., endotoxin limit per dose) rather than the proposed guideline limit of 0.25 EU/ml, which had applied to all drug products, regardless of drug concentration or dose.
2. Industry presented data to support an increase in the endotoxin limit from 2.5 EU/kg (the 1980 limit of 0.25 EU/ml × 10 ml/kg USP rabbit dose for many products) to 5.0 EU/kg, which represented an experimentally determined acceptable release limit with a significant safety factor over the USP rabbit pyrogen test.[76-78]
3. Industry petitioned that FDA drop the requirement for concurrent rabbit/LAL testing, arguing that such testing would prove valuable only for large numbers of samples.

FDA and USP responded to continuing research of the parenteral industry and concerns about certain provisions of the proposed regulatory documents. USP revised the BET in 1982 and 1983.[87,88] The changes proposed by USP were intended to further define parameters for the gel-clot test method, but did not address the question of endotoxin limits, which was still a major concern of the pharmaceutical industry. In 1983, FDA published a revised guideline, "draft guideline for validation of the *Limulus* amebocyte lysate test as an end-product endotoxin test for human and animal parenteral drugs, biological products, and medical devices".[89] The revised guideline consolidated the guidelines that had been previously issued separately for drugs, biologicals, and medical devices into one comprehensive document. Major revisions to the guideline included the elimination of concurrent rabbit/LAL testing, the decision to set endotoxin limits based on the maximum human or rabbit dose of the product (whichever was higher), and the adjustment of the endotoxin limit for all drugs (except intrathecals*) from 2.5 to 5.0 EU/kg. In 1984 USP again revised the BET chapter to better conform to the procedures described in the FDA 1983 guideline.[90]

Having set a new dose-dependent endotoxin limit and realizing that many concentrated drug products, by their nature, interfere with the formation of a firm gel in the clot test, FDA and USP proposed that a maximum valid dilution factor (MVD) be determined for each product tested. The MVD is the calculated limit for the dilution of an incompatible drug product to overcome test interference, yet still provide a valid test:

$$\text{MVD} = \frac{\text{K} \times \text{C}}{\lambda \times \text{M}}$$

where K = endotoxin limit (5.0 EU/kg, or 0.2 EU/kg for intrathecals), C = drug concentration (in units or mg/ml), λ = lysate sensitivity (EU/ml), and M = maximum rabbit or human dose (mg or ml/kg).

The terms in the MVD calculation are the endotoxin limit (K), the concentration of drug product (C), the lysate sensitivity (λ), and the drug dose (M). For any given drug product presentation, K, C, and M remain constant. The MVD, therefore, is inversely affected by the numerical value of the lysate sensitivity. Raising the numerical value of λ lowers the MVD and lowering the numerical value raises the MVD. The tester, therefore, can increase or decrease the MVD as necessary to help overcome test interference or to facilitate testing, by changing the lysate sensitivity.

In 1983, USP replaced the rabbit pyrogen test with the BET in individual monographs

* The endotoxin limit for intrathecals was set as 0.2 EU/kg.

for 29 radiopharmaceutical products and five parenteral waters — water for injection, bacteriostatic water for injection, sterile water for inhalation, sterile water for injection, and sterile water for irrigation.[88] USP has since proposed replacing the rabbit pyrogen test with the BET in 230 individual monographs.[91,92]

Since 1983, the major advances in LAL technology have included the marketing and routine use of quantitative LAL methods, including turbidimetric end point, turbidimetric kinetic, chromogenic end point, and chromogenic kinetic assays. The BET was modified in USP XXI and subsequent supplements to be more consistent with the validation criteria outlined in the FDA guideline.[91,92] Gel-clot is still the primary method described in the BET chapter, although the monograph states that turbidimetric (including turbidimetric kinetic) and colorimetric (i.e., chromogenic end point and kinetic) assays may be used if ''shown to comply with the requirements for alternative methods''. In other words, if a company manufacturing a product labeled ''nonpyrogenic, USP'' chooses to use a LAL method other than gel-clot, they must prove that the new method is at least equivalent to the USP BET method. In December 1987, FDA released a ''final'' guideline for LAL use.[93] This guideline contains minor changes and refinements to the gel-clot method, but more importantly, recognizes the quantitative methods by including detailed descriptions of suggested methods for the validation and use of these new assays.

Although the gel-clot method still remains the official BET method, proposed revisions to the monograph in USP XXII offer descriptions of the quantitative methods and significant changes to the gel-clot method to conform with the provisions of the FDA guideline.[94]

The current European Pharmacopoeia (EP) Chapter on ''Bacterial Endotoxins'',[95] like its U.S. counterpart, focuses attention on the gel-clot method. The two chapters are similar in content, describing procedures for validation, standard evaluation, and routine testing. There are, however, two important differences in the methods described in the two compendia. First, the European Pharmacopoeia does not require that the analyst use a licensed lysate for endotoxin determination. The EP is not a licensing agency, and therefore cannot control the certification of lysates used in all of the countries covered by EP regulations. Furthermore, the EP could not require the use of an FDA licensed lysate since the U.S. is not a member nation of EP. The second major difference between the USP and EP is the endotoxin standard used in the validation and control of the LAL test. The current USP Reference Standard Endotoxin (RSE) is lot F, a lyophilized preparation derived from *Escherichia coli* strain 113:H10:K, and is equivalent to the FDA RSE, EC-5. The European Pharmacopoeial Biological Reference Preparation (BRP) is a liquid endotoxin preparation derived from a different Gram-negative bacterium, *Salmonella abortus equi* and is standardized against the current World Health Organization (WHO) standard, a lyophilized preparation formulated from the same endotoxin stock used to prepare USP RSE, lot F.

Currently, in addition to the European Pharmacopoeia, many countries are subject to additional or modified specifications imposed by their own national pharmacopoeias. The Japanese Pharmacopoeia references the gel-clot method and has produced its own lyophilized endotoxin standard from *Escherichia coli* strain UKT-B.[96] In the future, the European Community (EC) intends to regulate drugs by issuing common technical specifications and regulatory requirements through the EC Commission. Any drug product meeting the EC specifications will be guaranteed free circulation among member nations. Those products being exported to the U.S., however, must still meet FDA/USP specifications.

Although the strains, histories, extraction procedures, and purification methods for the pharmacopeial standards are well documented, the limited supply and high cost of these standards make it impractical to use them routinely in the lab. The USP, EP, and FDA methods permit laboratories to use more cost-effective Control Standard Endotoxins (CSEs) packaged in LAL test kits, provided the CSE is well characterized and is calibrated using the pharmacopoeial reference preparation. However, many CSEs come from bacterial strains other than the reference standards, and thus may vary considerably in potency.

The major component of bacterial endotoxins is lipopolysaccharide, or LPS. Endotoxin preparations may contain fragments of other cellular components as impurities, but the two terms, endotoxin and LPS will be used interchangeably in this discussion. LPS is composed of two major parts, the hydrophobic lipid A portion and the hydrophilic polysaccharide portion (commonly called the "O" region). The structure of the lipid A portion is fairly well conserved, but the nature (length and chemical composition) of the polysaccharide side chain varies between genera, species, and even strains of Gram-negative bacteria. The lipid A portion of the molecule has been shown to be responsible for numerous *in vivo* and *in vitro* effects of endotoxin, including the ability to produce fever (pyrogenicity), macrophage activation, complement activation, and LAL gelation, whereas the diverse polysaccharide side chains are responsible for the unique antigenic properties of different endotoxins (for a review, see Pearson,[2] and also Chapter 19).

A number of factors affect the reactivity of various endotoxins with LAL reagent:

1. Theoretically, those molecules with short polysaccharide side chains are more potent than those with long side chains, because they contain a higher proportion (by weight) of the toxic lipid A portion. Similarly, the purity of the endotoxin preparation affects the potency. The less pure the preparation, the lower the proportion of lipid A. The potency of different endotoxins, therefore, should not be strictly compared on a weight basis.
2. The potencies of different lots or batches of endotoxin preparation from the same strain of bacterium may vary depending on culture conditions, and extraction procedures, which may affect the purity of the final product.
3. The formulation of lysates from different reagent manufacturers, or even normal lot-to-lot variations in lysate from the same manufacturer, may cause them to react differently with a given lot of endotoxin. Thus the potency of any control endotoxin is lysate dependent and must be determined for each lot of lysate/CSE combination in use in the lab.

Obviously, each of these factors affecting endotoxin potency contributes to the variability of the test, especially when comparing data from laboratories using different standards. To alleviate some of the problem, the standards are all assigned a potency based on LAL activity, rather than on weight. The USP, FDA, and Japanese standards are labeled in Endotoxin Units (EUs), whereas the WHO and EP standards are labeled in International Units (IUs). The potencies of all Control Standard Endotoxins (CSEs) used routinely in the lab are determined relative to the USP, Japanese, or EP standards, and are reported as EU/ng or IU/ng. Although the potency of the Japanese standard is expressed in Endotoxin Units, equivalent to the USP standard, and the potency of the EP and WHO standards is assigned such that one IU = one EU, in practice, variation in endotoxin preparations or lysate lots frequently changes this 1:1 ratio, requiring that lysate manufacturers label their products to be used internationally in EUs (USP), EUs (Japanese), and IUs. Differing EU:IU ratios or differing EU (USP):EU (Japanese) ratios may also affect the determined amount of endotoxin in a product to be sold internationally, depending on which standard is used in the release specification for the product.

B. VALIDATION

There are a number of test-related variables, which may affect the performance of the LAL assay when testing pharmaceutical products:

1. Many pharmaceutical products interfere with the performance of the test.[66,97-102] Guilfoyle and Munson[97] reported that of 587 pharmaceuticals tested, 78% interfered with

the LAL assay in some manner. Interference may be defined as the inability to recover an endotoxin spike within acceptable limits. Inhibition is characterized by the apparent recovery of less than a known endotoxin spike. Inhibition may be caused by the drug itself (e.g., β-lactam antibiotics), excipients in the drug formulation (e.g., preservatives or buffers), or the chemistry of the drug product (e.g., high or low pH, chelators of divalent cations, mercaptans). Enhancement is characterized by the apparent recovery of more than a known endotoxin spike. It, too, may be caused by the nature of the product or by levels of endotoxin that may already be in the product, such as the natural contamination of material of plant or animal origin.

2. LAL reagent formulations from different lysate manufacturers may exhibit different interference pattern as when used to test the same product.[103] Because of lysate/product compatibility problems, some laboratories will find that they must investigate the use of more than one lysate reagent to successfully validate their company product line.[104]

In view of these challenges facing the user of LAL reagent, government regulatory agencies have outlined procedures to be followed for the validation of the LAL method in the testing of pharmaceutical products. Complete descriptions of these methods may be found in the FDA guideline,[93] USP XXII bacterial endotoxins test,[105] the European Pharmacopoeia[95] the Japanese Pharmacopoeia,[96] and individual lysate manufacturer instructions.

1. Initial Quality Control Procedure

Before any official LAL tests are performed in the laboratory, two qualification steps must be performed. First, the label claim sensitivity of each lot of lysate reagent to be used must be verified. For the gel-clot method, verification is accomplished by making serial twofold dilutions of either the appropriate Reference Standard Endotoxin (RSE) or another usually provided in the reagent test kit, the Control Standard Endotoxin (CSE), to concentrations bracketing the label claim sensitivity of the reagent, and testing them in replicate. The label claim is verified if the geometric mean of the observed end points is within one twofold dilution of the labeled sensitivity of the reagent.

For the quantitative methods currently in use, a CSE or RSE standard is diluted to 4 to 6 different concentrations extending over the labeled linear range of the assay. Replicate tests are made of each endotoxin concentration, and linear regression/correlation analysis is performed on the resulting standard curve. The labeled linear range of the reagent is confirmed if the absolute value of the coefficient of correlation, r, is $\geq$0.980.

Before performing any official tests, each analyst must be qualified. The procedure is the same as for the label claim verification — if the analyst confirms the label claim sensitivity or the labeled linear range of the reagent, he/she is qualified to perform tests routinely.

2. Inhibition/Enhancement Testing

Inteference of the LAL test by pharmaceutical products has been well documented previously. FDA and USP methods require that the degree of interference of each drug formulation or each group of devices with a common chemical formulation be determined before performing any LAL tests. For medical devices, the test solution is an eluate or extract, prepared for each device by either rinsing or soaking the product in nonpyrogenic water; while for drug products the test solution is a dilution of the product, not to exceed the MVD (Section II.A). The gel-clot procedure requires that the device eluate/extract or a dilution of drug product be spiked with a known amount of endotoxin. Serial twofold dilutions of this spiked solution is prepared with additional unspiked device eluate/extract or the same concentration of endotoxin-free drug product, so that the test solution remains constant, but the endotoxin concentration is varied. The test solution shows no interference if the geometric mean of the observed end points is within one twofold dilution of the labeled sensitivity of the reagent. If interference is observed by either not recovering the spike at all (inhibition) or recovering more endotoxin than the spike (enhancement), regulatory documents allow the testing of another dilution of drug product, again not to exceed the MVD.

Most interference in drug product can be overcome by simple dilution in nonpyrogenic water to reduce the inhibitory/enhancing substance to a noninterfering level. Guilfoyle and Munson[97] reported that 72% of pharmaceuticals found to be inhibitory when tested undiluted could be successfully validated at a dilution within the MVD, whereas only 6% of products tested were still inhibitory after dilution. Since devices themselves have no dose, there is no provision in the regulatory methods for dilution of the extract to reach a noninterfering level. Therefore, if interference of the gel-clot method is observed with an eluate/extract from a medical device, the FDA guideline states that an alternate technique must be used to measure endotoxin content.

In addition to dilution, other test modifications or adjustments may be used to overcome interference. The LAL assay has an optimum pH range between 6.0 to 7.5; thus, some products may require pH adjustment prior to LAL testing. Although most LAL reagents contain buffers as part of their formulations to allow for maximum pH flexibility, products which contain strong acids, bases, or buffers may cause the pH of the LAL/sample mixture to be outside this optimum range. Other products may affect the aggregation state of endotoxin and cause inhibition. The use of dispersing agents may be helpful in this regard. The use of Pyrosperse™, an endotoxin dispersing agent, has been reported to overcome product inhibition in several, otherwise LAL incompatible pharmaceutical preparations.[97,106] Weiss[107] has suggested another philosophy in overcoming product inhibition, namely, the removal of the inhibitory substance. The fact that endotoxins are relatively large molecules (M_r >1,000 kDa) while the interfering substances are usually relatively small (M_r <10 kDa) suggests that endotoxins can be isolated by molecular exclusion using an ultrafilter. The product is filtered through a low molecular weight exclusion filter and any endotoxin present is trapped, while the interfering substances pass through the filter. After washing the filter to remove residual interfering substances, the endotoxin is eluted from the filter in a non-inhibitory fluid and assayed using LAL. This method is useful when the interfering substance has a molecular weight significantly lower than endotoxin. Pfeiffer and Weiss[108] have reported success with this approach in the testing of amino acid solutions and antibiotics.

For the validation of the quantitative methods, turbidimetric end point and kinetic assays and the chromogenic end point assay, the device eluate/extract or a dilution of drug product (not to exceed the MVD) is spiked with CSE or RSE at a concentration of four λ, where λ is equal to the lowest concentration used to generate the water standard curve. The test solution shows no interference if the four λ spike can be recovered to within $+/-25\%$. For the validation of the chromogenic kinetic method, the device eluate/extract or a dilution of drug product (not to exceed the MVD) is spiked with CSE or RSE at a concentration of 0.5 or 5.0 EU/ml, depending on the endotoxin release limit and the testing dilution. The test solution shows no interference if the endotoxin spike can be recovered within $+/-50\%$. As with the label claim verification, the standard curve must meet the requirements for linearity for a valid test. Again, because dilution of a device eluate/extract to reach a noninterfering level is not allowed in the guideline, the observation of inhibition/enhancement requires that an alternate technique be used to measure or detect endotoxin.

In the case of some drug products or medical devices with extreme interference problems, a standard curve may be prepared in an endotoxin-free eluate/extract or a dilution on endotoxin-free product, not to exceed to MVD. This is commonly called a "product standard curve". In addition to meeting the requirements for linearity, FDA requires that the product standard curve meet a slope specification, as inhibitory products will tend to depress the slope and enhancing products will tend to raise the slope.

Once validation criteria are met on three lots of a product, raw material, or sample from a given point in the manufacturing process, the test may be used routinely to monitor endotoxin levels in raw materials, in-process samples, and finished, packaged drug product.

III. SPECIFIC APPLICATIONS

A. RAW MATERIAL AND IN-PROCESS TESTING

The total endotoxin content (pyroburden) in an aseptically processed drug or medical device is the sum of the individual pyroburdens of each factor, which contributes to the manufacturing of that product.[109] Raw materials, packaging components, environment, and manufacturing process itself, therefore, all must be viewed by the drug or device manufacturer as potential endotoxin contributors. Periodic screening of selected raw materials and crucial processing steps will help the manufacturer identify endotoxin contributors or spot rising trends in endotoxin levels, helping ensure the lowest possible pyroburden in the finished product. Prior to the introduction of the LAL test, the testing of such samples by the USP rabbit pyrogen test was prohibitively costly, time-consuming, and labor intensive.

Raw materials used in aseptic processing come from many sources. Those of plant or animal origin, such as heparin (from beef lung or porcine mucosa) or some carbohydrates, may have been exposed in nature to Gram-negative bacteria, and thus may contain detectable levels of endotoxin. Bulk chemicals, such as organic solvents, may be free of endotoxin, owing to detoxification of the lipid A portion of the LPS molecule by extreme chemical environments, such as high or low pH, or extreme temperature fluctuations during processing.[110,111] Water is perhaps the most important raw material used in the processing of a parenteral product. Water is basic to every phase of the operation from the rinsing and cleaning of vessels and packaging components to its use as a solvent in the actual manufacturing process. City water usually contains high levels of endotoxin, because methods used to reduce bacterial levels at the water treatment facility do not necessarily remove endotoxin. Ion exchange columns and resin beds used at the pharmaceutical manufacturing site for softening or other water pretreatments are prime candidates for endotoxin contributors, by offering good environments for colonization by Gram-negative bacteria.[79,112,113] It cannot be assumed that routine column or bed preparation and regeneration will necessarily eliminate endotoxin contamination or viable Gram-negative bacteria. Validation procedures designed to assess the endotoxin removal efficiency of such preparation and regeneration methods must be developed, and intermittent checks on the endotoxin content of the column/resin bed effluent should be performed to assure that endotoxin levels are kept to a minimum.

The efficiency of proven pyrogen removal procedures for the generation of water for injection (WFI), such as distillation, ultrafiltration, or reverse osmosis may be easily ascertained by quantitating endotoxin in the water before and after treatment using either gel-clot or quantitative LAL assays. Once the water is found to be nonpyrogenic, the LAL test should be used to assess the WFI storage system ability to maintain the water in a nonpyrogenic state, by periodic testing of samples taken from storage tanks or circulating loops. Plotting endotoxin concentration over a period of time allows the manufacturer to spot trends in rising endotoxin concentration and to perform filter changes or system flushes as necessary before the endotoxin levels become critical.

The biotechnology industry uses LAL extensively to screen culture media, enzyme preparations, and fermentation products for endotoxin content.[114] The monitoring of raw materials and processes are especially important to this new branch of the pharmaceutical industry for a number of reasons:

1. Mammalian cell systems used in the production of proteins may be susceptible to contamination by endotoxin found in growth media and water, potentially lowering the yield of product.[115]

2. Gram-negative cell systems used in the production of some recombinant products may actually contribute endotoxin to the final product.
3. Since biotechnology products are usually produced in relatively small batches, failure of a lot due to endotoxin contamination could prove extremely costly.

Pyroburden in a product is not static, but may fluctuate during processing of a drug or device. The introduction of water and some raw materials may increase the endotoxin level in the product, while extremes in pH, solvent content, or temperature (such as in the extrusion or molding of plastics or glass for medical devices) may actually decrease endotoxin levels. Monitoring of processing steps with LAL will identify critical or malfunctioning points and may detect extremely low levels or slowly rising levels of endotoxin, allowing the manufacturer to spot trends and implement process changes to correct the problem before endotoxin limits are exceeded. At least two instances of the use of LAL to correct processing problems have been reported in the literature.[82] In one case, one of four filling heads used in the aseptic fill of a peritoneal dialysis solution was identified as an endotoxin contributor. In the other case, a diatomaceous earth filter, designed to clarify product, was found to be contributing endotoxin to an albumin product. In both instances, the use of the rapid, cost-effective LAL test allowed the manufacturer to test large numbers of samples, identify the problem, and correct the situation before entire batches of the product had to be reworked or destroyed owing to endotoxin contamination.

A manufacturing step designed to depyrogenate a product (e.g., ultrafiltration) is a prime candidate for frequent LAL monitoring to ensure that the equipment is working properly. As with the water systems, processing steps designed to depyrogenate a product can be monitored by determining the pyroburden of the product upstream and downstream of the validated depyrogenating step using either gel-clot or a LAL quantitative method.

In response to current good manufacturing practices[116] and the Food and Drug Administration "guideline on sterile drug products produced by aseptic processing,"[117] many pharmaceutical companies are implementing extensive raw material and in-process testing programs using LAL. For most companies, the gel-clot LAL test has been the test of choice for release testing of their products. However, gel-clot, when used as a limits test, is only semiquantitative. A positive test indicates that there is an amount of endotoxin in the sample that is equal to or greater than the label claim sensitivity of the reagent. Likewise, a negative result indicates only that the level of endotoxin in the sample is less than the label claim sensitivity of the reagent. While this information is useful, it cannot be used to trend endotoxin levels in products or raw materials. The gel-clot test can be used as an assay by making extensive dilutions of the product, testing each dilution, and multiplying the label claim sensitivity by the reciprocal of the end point dilution factor in the series. Using the test in this manner can provide the manufacturer with information that can be used in trend analysis, but the procedure, with its extensive dilution series, can be extremely labor intensive and time-consuming.

A number of methods have been described in the literature for the quantitation of endotoxin using LAL, but the most widely used are the turbidimetric kinetic[54,118] and the chromogenic end point and kinetic assays.[58,119,120] These quantitative methods are ideally suited for the routine screening of raw materials and in-process samples for endotoxin content. Current instrumentation can allow for greater sample through-put by analyzing up to 96 samples at once. Depending on the sensitivity required by the laboratory, the reaction time for these methods may be as little as 16 min, as opposed to 1 h for the gel-clot test, so that a large number of samples can be analyzed in a relatively short time. Time savings is especially important for a point of use or in-process test, because in many cases production personnel will submit a sample to the quality assurance (QC) lab and wait for a determination before proceeding. Data management is another advantage of the quantitative systems. Each of the instruments currently in use can be interfaced with a laboratory or personal computer, for which the software can automatically issue reports, chart trends, or perform any number of data management tasks.

B. DEPYROGENATION

Depyrogenation may be defined as the elimination of all substances capable of producing a fever. The pyrogenic substance of most interest to the pharmaceutical industry is endotoxin (with lipopolysaccharide [LPS] as a major component), which is found in the cell walls of both viable and nonviable Gram-negative bacteria. Elimination of these substances may be accomplished either by inactivation or removal with the method of choice, depending largely in the chemical composition of the material to be depyrogenated. Methods of depyrogenation by removal include ultrafiltration, affinity chromatography, dilution, through adsorption to a specific medium, such as asbestos, charcoal or charged filter media, hydrophobic attraction, or reverse osmosis. These methods are widely used in production situations, where the product may be heat or chemically labile. Depyrogenation by inactivation of the toxic lipid A portion of the LPS molecule may be accomplished by a number of methods, including long exposure to high temperatures, acid/base hydrolysis, oxidation, alkylation, moist heat, and ionizing radiation. These methods are perhaps best suited to those materials which are resistant to heat and chemicals, such as glass packaging components and metal processing equipment and utensils.[2,121,122] No matter what the method of depyrogenation, a known amount of endotoxin can be added to the material before processing and assayed using LAL after processing to determine the efficiency of the procedure.

Endotoxin inactivation by dry heat is widely described in the literature and is the method of choice for the processing of packaging components and processing equipment, including glass vials, ampoules, syringes, metal machine parts, tools, and sampling utensils. A number of studies have been published describing the kinetics of LPS destruction.[123-129] These data support the use of a Control Standard Endotoxin (CSE) challenge as an important complement to physical and chemical measurements in the monitoring of a dry-heat depyrogenation process. The USP chapter "Sterilization and Sterility Assurance of Compendial Articles"[130] recognize the value of such an *in vitro* biological challenge by recommending that an endotoxin challenge be included in the validation of any dry-heat process designed to render glassware or packaging components nonpyrogenic. The chapter states that the process should demonstrate at least a three-log cycle reduction in the difference between the recoverable input endotoxin ($\geq$1000 EU/challenge unit) and any residual endotoxin present after processing. To facilitate the monitoring of such processes, laboratories are preparing carrier devices or "endotoxin indicators" analogous in function to biological indicators used for sterilization studies.[131] Since different surfaces and materials have different heat penetration/distribution characteristics, the indicator device should be the same as, or as chemically similar as possible in size, shape, and chemical composition to the material being depyrogenated.

A Control Standard Endotoxin is used to spike the indicator units for depyrogenation studies. The endotoxin is "fixed" onto the unit by a drying process, usually exposure to ambient laminar flow, lyophilization, or vacuum drying. Adsorption of endotoxin to glass is a well-documented phenomenon, but the degree of adsorption depends on a number of factors, including endotoxin concentration in the original inoculum (the more potent the inoculum, the higher the recovery), the glass type and shape, the method used to fix the endotoxin onto the glass surface, and the recovery procedure.[132,133] Quantitation of endotoxin spikes at the time of inoculation ("initial spike") and after fixation onto the glass surface ("recoverable endotoxin") is necessary to determine the loss due to adsorption. USP suggests a recoverable endotoxin level of at least 1000 EU/units.

After depyrogenation, a small amount of nonpyrogenic water is added to the processed endotoxin indicator. The "reconstituted" unit is then subjected to a recovery process to leach any residual endotoxin from the vessel walls. Recovery typically is accomplished by sonication, vortexing, or a combination of both. The resulting solution is assayed using LAL to determine the amount of "residual endotoxin" remaining after depyrogenation. According to USP, the endotoxin challenge portion of the depyrogenation validation is successful if the initial amount of recoverable endotoxin is $\geq$1000 EU/unit and there is a determined reduction in endotoxin level of at least three logs.

LAL testing allows the laboratory to perform endotoxin challenges as an *in vitro* biological method of validating processes and equipment designed to depyrogenate. The gel-clot test may be used as a limits test, and the results reported as <x EU/challenge unit, where x is a lysate sensitivity at least three logs lower than the recoverable endotoxin. The gel-clot test may also be performed as an assay by testing serial dilutions of the residual endotoxin solution until an end point is reached. Multiplication of the reciprocal of the dilution factor by the lysate sensitivity gives a quantitative determination of the residual endotoxin. The extensive dilutions required to use the gel-clot method as an assay are very labor intensive, time-consuming, costly, and prone to technical error. The use of a validated quantitative LAL method (such as turbidimetric kinetic, chromogenic end point, or kinetic assays) offers the lab a quick, reproducible, sensitive, quantitative method of determining recoverable and residual endotoxin levels in indicator units. Because of this increased testing efficiency, more samples can be tested, allowing the pharmaceutical manufacturer to monitor processes more closely. Many companies are routinely including endotoxin indicators in routine production runs, much as biological indicators are included in each sterilization run, to monitor these critical processes.

The concept of an endotoxin challenge to validate depyrogenation methods can be extended beyond the application of dry-heat processing of glass and metal. Endotoxin challenges are routinely used to validate and monitor any of the other endotoxin removal/inactivation methods listed before. In these cases, the carrier might be a liquid (either water or product), a rubber stopper, or a medical device, but the idea is the same — quantitation of endotoxin with LAL before and after the depyrogenation procedure results in data on the efficiency of the method.

IV. SUMMARY

Throughout the development of intravenous pharmaceuticals and implantable medical devices, pyrogenic reactions have always been closely associated with the introduction of foreign material into the body. The principal pyrogen in aqueous solution, one of the biggest concerns to pharmaceutical and medical device manufacturers, is a component of the cell wall of Gram-negative bacteria called endotoxin. The observation that endotoxin caused gelation in extracts of *Limulus* amebocytes eventually was expanded to the development of an *in vitro* assay for endotoxin.

Limulus amebocyte lysate (LAL) contains a multicomponent enzyme cascade, which is initiated by endotoxin and terminates in coagulation. A variety of methods have been developed in an attempt to make the assay of endotoxin more simple, faster, more sensitive, or quantitative. Four methods are currently commercially available, namely, the gel-clot, turbidimetric kinetic, the chromogenic end point, and the chromogenic kinetic assays.

LAL is recognized worldwide as an alternative to *in vivo* pyrogen assays. The U.S. Pharmacopeia, FDA, European Pharmacopoeia, and Japanese Pharmacopoeia have guidelines for the validation and use of LAL in pharmaceutical and medical device manufacturing. These agencies have established limits for the amount of endotoxin allowable in products and endotoxin standards for use in the standardization of reagents.

The use of LAL has proved invaluable in controlling the level of endotoxin in a finished product. The endotoxin contribution of raw materials and packaging material can be monitored. In-process testing at critical production steps can identify additional sources of endotoxin contamination, and depyrogenation processes can be validated by assessing the degradation of endotoxin challenges. The speed, economics, and sensitivity of the LAL assays allow such an in-depth approach to the control of endotoxin in pharmaceuticals and medical devices.

REFERENCES

1. **Westphal, O., Westphal, U., and Sommer, T.,** The history of pyrogen research, in *Microbiology*, Schlessinger, D., Ed., American Society for Microbiology, Washington, D. C., 1977, 221.
2. **Pearson, F. C., III,** *Pyrogens, Endotoxins, LAL Testing, and Depyrogenation*, Marcel Dekker, New York, 1985.
3. **Billroth, T.,** Biobachtungstudien über das wundfieber und accidentelli wund-krankheiten, *Arch. Klin. Chir.*, 2, 578, 1862.
4. **von Haller, A.,** *Elements Physiologiae Corporis Humani*, Vol. 3, 1757.
5. **Gaspard, B.,** Mèmoire physiologique sier les maladies purulentes et putrides, sur la vaccine etc, *J. Physiol. (Paris)*, 2, 1, 1822.
6. **Panum, P. L.,** Das putride gift, die bakterien, die putride infektion oder intoxikation und die septikämie, *Arch. Pathol. Anat. Pysiol. Klin. Med.*, 60, 301, 1874.
7. **Burdon-Sanderson, J.,** On the process of fever. Part III. Pyrexia, *Practitioner*, 417, 1876.
8. **Jona, J. L.,** A contribution to the experimental study of fever, *J. Hyg.*, 15, 169, 1916.
9. **Centanni, E.,** La immunità stomogene (terza immunità), in *Trattato di Immunologia*, Societa Endrice Libraria, Rome, 1921, 149.
10. **Wechselmann, W.,** Neuere Erfahrungen über intravenöse Salvarsan-injektionen ohne Reaktionserscheinungen, *Muench. Med. Wochenschr.*, 58, 1510, 1911.
11. **Müller, P. T.** Über den Bakteriengehalt des in Apotheken erhältlichin destillerten Wassers, *Muench. Med. Wochenschr.*, 58, 2739, 1911.
12. **Samelson, S.,** Über das sogenannte Kochsalzfieber, *Monatsschr. Kinderheilkd.*, 11, 125, 1913.
13. **Bendix, B. and Bergmann, J.,** Über das sogenannte Kochsalzfieber, *Monatsschr. Kinderheilkd.*, 11, 387, 1913.
14. **Hort, E. C. and Penfold, W. J.,** The relation of salvarsan fever to other forms of infection fever, *Proc. R. Soc. Med. (Part III)*, 5 (Path. Sect.), 131, 1912.
15. **Hort, E. C. and Penfold, W. J.,** Microorganisms and their relation to fever, *J. Hyg.*, 12, 361, 1912.
16. **Hort, E. C. and Penfold, W. J.,** A critical study of experimental fever, *Proc. R. Soc. London, Ser. B*, 85, 174, 1912.
17. **Seibert, F. B.,** Fever-producing substance found in some distilled waters, *Am. J. Physiol.*, 67, 90, 1923.
18. **Seibert, F. B.,** The cause of many febrile reactions following intravenous injections, *Am. J. Physiol.*, 71, 621, 1925.
19. **Siebert, F. B. and Mendel, L. B.,** Protein fevers, with special reference to casein, *Am. J. Physiol.*, 67, 105, 1923.
20. **Rademaker, L.,** The cause and elimination of reactions after intravenous infusions, *Ann. Surg.*, 92, 195, 1930.
21. **Rademaker, L.,** Reactions after intravenous infusion. A further report on their elimination, *Surg. Gynecol. Obstet.*, 56, 956, 1933.
22. **Howell, W. H.,** Observations upon the chemical composition and coagulation of the blood of *Limulus polyphemus, Callinectis hastatus*, and *Cucumaria, Johns Hopkins Univ. Cir.*, 5, 4, 1885.
23. **Loeb, L.,** On the blood lymph cells and inflammatory processes of *Limulus*, *J. Med. Res.*, 7, 145, 1902.
24. **Loeb, L.,** Über die zweite Gerinnung des Blutes von Limulus, *Biochem. Z.*, 16, 157, 1909.
25. **Loeb, L. and Blanchard, K. C.,** The effect of various salts on the outgrowth from experimental amebocyte tissue near isoelectric point and with the addition of acid or alkali, *Am. J. Physiol.*, 60, 277, 1922.
26. **Loeb, L.,** Amoebocyte tissue and amoeboid movement, *Protoplasma*, 4, 596, 1928.
27. **Bang, F.,** A bacterial disease of *Limulus polyphemus*, *Bull. Johns Hopkins Hosp.*, 98, 325, 1956.
28. **Levin, J. and Bang, F. B.,** The role of endotoxin in the extracellular coagulation of *Limulus* blood, *Bull. Johns Hopkins Hosp.*, 115, 265, 1964.

29. **Dumont, J. J., Anderson, E., and Winner, G.,** Some cytologic characteristics of the hemocytes of *Limulus* during clotting, *J. Morphol.*, 119, 181, 1966.
30. **Armstrong, P. B.,** Motility of the *Limulus* amebocyte, in *Biomedical Applications of the Horseshoe Crab (Limulidae),* Cohen, E., Ed., Alan R. Liss, New York, 1979, 73.
31. **Levin, J. and Bang, F. B.,** Clottable protein in *Limulus*: its localization and kinetics of its coagulation by endotoxin, *Thromb. Diath. Haemorrh.*, 19, 186, 1968.
32. **Jorgensen, J. H. and Smith, R. F.,** Preparation, sensitivity, and specificity of *Limulus* lysate for endotoxin assay, *Appl. Microbiol.*, 26, 43, 1973.
33. **Takagi, T., Hokama, Y., Morita, S., Iwanaga, S., Nakamura, S., and Niwa, M.,** Amino acid sequence studies on horseshoe crab (*Tachypleus tridentatus*) coagulogen and the mechanism of gel formation, in *Biomedical Applications of the Horseshoe Crab (Limulidae),* Cohen, E., Ed., Alan R. Liss, New York, 1979, 169.
34. **Yin, E. T., Galanos, C., Kinsky, L. S., Bradshaw, R. A., Wessler, S., Luderitz, O., and Sarmiento, M. E.,** Picogram-sensitive assay for endotoxin: gelation of *Limulus polyphemus* blood cell lysate induced by purified lipopolysacaharides and Lipid A from Gram-negative bacteria, *Biochim. Biophys. Acta,* 261, 284, 1972.
35. **Sullivan, J. D. and Watson, S. W.,** Factors affecting the sensitivity of *Limulus* lysate, *Appl. Microbiol.*, 28, 1023, 1974.
36. **Young, N. S., Levin, J., and Prendergast, R. A.,** An invertebrate coagulation system activated by endotoxin: evidence for enzymatic mediation, *J. Clin. Invest.*, 51, 1790, 1972.
37. **Nakamura, T., Morita, T., and Iwanaga, S.,** Biochemical characterization of protein components associated with hemolymph coagulation system of *Limulus (Tachypleus tridentatus), Thromb. Haemost.*, 50, 405, 1983.
38. **Morita, T., Nakamura, T., Miyata, T., and Iwanaga, S.,** Biochemical characterization of *Limulus* clotting factors and inhibitors which interact with bacterial endotoxin, in *Bacterial Endotoxin: Structure, Biomedical Significance, and Detection with the Limulus Amebocyte Lysate Test,* tenCate, J. W., Buller, H. R., Sturk, A., and Levin, J., Eds., Alan R. Liss, New York, 1985, 53.
39. **Tanaka, S., Nakamura, T., Morita, T., and Iwanaga, S.,** *Limulus* anti-LPS factor: an anticoagulant which inhibits the endotoxin-mediated activation of *Limulus* coagulation system, *Biochem. Biophys. Res. Commun.*, 195, 717, 1982.
40. **Nakamura, S. and Levin, J.,** Fractionation of *Limulus* amebocyte lysate, *Biochim. Biophys. Acta,* 707, 217, 1982.
41. **Nakamura, S. and Levin, J.,** Endotoxin — mediated *Limulus* proclotting enzyme activator and detection of a previously undescribed protease (protease N), *Biochem. Biophys. Res. Commun.*, 108, 1619, 1982.
42. **Tai, J. Y. and Kui, T.,** Studies on *Limulus* amebocyte lysate. II. Purification of the coagulogen and the mechanism of clotting, *J. Biol. Chem.*, 252, 4773, 1977.
43. **Nakamura, S., Takagi, T., Iwanaga, S., Niwa, M., and Takahashi, K.,** Amino acid sequence studies on the fragments produced from horseshoe crab coagulogen during gel formation. Homologies with primate fibrinopeptide B, *Biochem. Biophys. Res. Commun.*, 72, 902, 1976.
44. **Frauch, P.,** Slide test as a micro-method of a modified *Limulus* endotoxin test, *J. Pharm. Sci.*, 63, 808, 1974.
45. **Goto, H., Wantanabe, N., and Nakamura, S.,** Studies on a simple *Limulus* test, a slide method, *Jpn. J. Exp. Med.*, 47, 523, 1977.
46. **Okuguchi, S.,** Improvement of the micromethod for the *Limulus* lysate test, *Microbiol. Immunol.*, 22, 113, 1978.
47. **Melvaer, K. L. and Fystro, D.,** Modified micromethod of the *Limulus* amebocyte lysate assay for endotoxin, *Appl. Environ. Microbiol.*, 43, 493, 1982.
48. **Flowers, D. J.,** A microtechnique for endotoxin assay by using *Limulus* lysate, *Med. Lab. Sci.*, 36, 171, 1979.
49. **Prior, R. B. and Spagna, V. A.,** Adaptation of a microdilution procedure to the *Limulus* lysate assay for endotoxin, *J. Clin. Microbiol.*, 10, 394, 1979.
50. **Gardi, A. and Arpagaus, G. R.,** Improved microtechnique for endotoxin assay by the *Limulus* amebocyte lysate test, *Anal. Biochem.*, 109, 382, 1980.
51. **Harris, N. S. and Feinstein, R.,** The LAL bead assay for endotoxin, in *Biomedical Applications of the Horseshoe Crab (Limulidae),* Cohen, E., Ed., Alan R. Liss, New York, 1979, 265.
52. **Teller, J. D. and Kelly, K. M.,** A turbidimetric *Limulus* amebocyte assay for the quantitative determination of Gram negative bacteria, in *Biomedical Applications of the Horseshoe Crab (Limulidae),* Cohen, E., Ed., Alan R. Liss, New York, 1979, 423.
53. **Valois, F. W.,** Quantitative method for determining less than a pg/ml of LPS, in *Biomedical Applications of the Horseshoe Crab (Limulidae),* Cohen, E., Ed., Alan R. Liss, New York, 1979, 415.
54. **Novitsky, T. J., Ryther, S. S., Case, M. J., and Watson, S. W.,** Automated LAL testing of parenteral drugs in the Abbott MS-2, *J. Parenter. Sci. Technol.*, 36, 11, 1982.

55. **Dubczak, J. A., Cotter, R., and Dastoli, F. R.,** Quantitative determination of endotoxin by nephelometry, in *Biomedical Applications of the Horseshoe Crab (Limulidae),* Cohen, E., Ed., Alan R. Liss, New York, 1979, 403.
56. **Nandan, R. and Brown, D. R.,** An improved *in vitro* pyrogen test: to detect picograms of endotoxin contamination in intravenous fluids using *Limulus* amebocyte lysate, *J. Lab. Clin. Med.,* 89, 910, 1977.
57. **Fujita, Y. and Nakamura, C.,** Preparation and application of a new endotoxin determination kit, Pyrodick, using a chromogenic substrate, in *Endotoxins and Their Detection with the Limulus Amebocyte Lysate Test,* Watson, S. W., Levin, J., and Novitsky, T. J., Eds., Alan R. Liss, New York, 1982, 173.
58. **Friberger P., Knos, M., and Mellstam, L.,** A quantitative endotoxin assay utilizing LAL and a chromogenic substrate, in *Endotoxins and Their Detection with the Limulus Amebocyte Lysate Test,* Watson, S. W., Levin, J., and Novitsky, T. J., Eds., Alan R. Liss, New York, 1982, 195.
59. **Scully, M. G., Newman, Y. M., Clark, S. E., and Kakkar, V. V.,** Evaluation of a chromogenic method for endotoxin measurement, *Thromb. Res.,* 20, 263, 1980.
60. **Cooper, J. F., Levin, J., and Wagner, H. N.,** New rapid *in vitro* test for pyrogen in short-lived radiopharmaceuticals, *J. Nucl. Med.,* 11, 310, 1970.
61. United States Food and Drug Administration, Status of biological substances used for detecting bacterial endotoxins, *Fed. Reg.,* 38, 1404, 1973.
62. United States Food and Drug Administration, *Limulus* amebocyte lysate: additional standards, *Fed. Reg.,* 38, 26130, 1973.
63. United States Food and Drug Administration, Licensing of *Limulus* amebocyte lysate: use as an alternative for rabbit pyrogen test, *Fed. Reg.,* 42, 57789, 1977.
64. **Jorgensen, J. H. and Smith, R. F.,** Rapid detection of contaminated intravenous fluids using the *Limulus in-vitro* endotoxin assay, *Appl. Microbiol.,* 26, 521, 1973.
65. **Weary, M. and Baker, B.,** Utilization of the *Limulus* amebocyte lysate test for pyrogen testing large volume parenterals, administration sets, and medical devices, *Bull. Parenter. Drug Assoc.,* 31, 127, 1977.
66. **McCullough, K. Z. and Scolnick, S. A.,** Effect of semisynthetic penicillins on the *Limulus* lysate test, *Antimicrob. Agents Chemother.,* 9, 856, 1976.
67. **Newsome, P.,** Penicillins and the *Limulus* amebocyte lysate test for endotoxin, *J. Pharm. Pharmacol.,* 29, 704, 1977.
68. **Rastogi, S. C., Hochstein, H. D., and Seligman, E. B.,** Statistical determination of endotoxin content in influenza virus vaccine by the *Limulus* amebocyte lysate test, *J. Clin. Microbiol.,* 6, 144, 1977.
69. **Siegel, S. E., Nachum, R., Leimbrock, S., and Karon, M.,** Detection of bacterial endotoxin in antitumor agents, *Cancer Treat. Rep.,* 60, 9, 1976.
70. **Sullivan, J. D. and Watson, S. W.,** Inhibitory effect of heparin on the *Limulus* test for endotoxin, *J. Clin. Microbiol.,* 2, 151, 1975.
71. **Murata, H., Kobayaski, M., Iio, M., Yamada, H., and Chiba, K.,** Sensitivity of the *Limulus* test and inhibitory factors in the radiopharmaceuticals, *J. Nucl. Med.,* 17, 1088, 1976.
72. **Cooper, J. F., Levin, J., and Wagner, H., Jr.,** Quantitative comparison of *in vitro* and *in vivo* methods for the detection of endotoxin, *J. Lab. Clin. Med.,* 78, 138, 1971.
73. **Eibert, J., Jr.,** Pyrogen testing: horseshoe crabs vs. rabbits, *Bull. Parenter. Drug Assoc.,* 26, 253, 1972.
74. **van Noordwijk, J. and DeJong, Y. C.,** Comparison of the *Limulus* test for endotoxins with the rabbit test for pyrogens of the European Pharmacopoeia, *J. Biol. Stand.,* 4, 131, 1976.
75. **Tomasulo, P. A., Levin, J., Murphy, P. A., and Winkelstein, J. A.,** Biological activities to titrated endotoxins: correlations of the *Limulus* lysate assay with rabbit pyrogen and complement-activation assays for endotoxin, *J. Lab. Clin. Med.,* 89, 308, 1977.
76. **Dabbah, R., Ferry, E., Jr., Gunther, D. A., Hahn, R., Mazur, P., Neely, M., Nicolas, P., Pierce, J. S., Slade, J., Watson, S., Weary, M., and Sanford, R. L.,** Pyrogenicity of *E. coli* 055:B5 endotoxin by the USP rabbit test — a HIMA collaborative study, *J. Parenter. Drug Assoc.* 34, 212, 1980.
77. Parenteral Drug Association, A response to *Fed. Reg.,* 45, January 18, 1980, interindustry communication, 1980.
78. **Tsuji, K., Steindle, K. A., and Harrison, S. J.,** *Limulus* amebocyte assay for detection and quantitation of endotoxin in a small volume parenteral product, *Appl. Environ. Microbiol.,* 40, 533, 1980.
79. **Cooper, J. R.,** Acceptance of the *Limulus* test as an alternative pyrogen test for radiopharmaceutical and intrathecal drugs, in *Biomedical Applications of the Horseshoe Crab (Limulidae),* Cohen, E., Ed., Alan R. Liss, New York, 1979, 345.
80. **Harrison, S. J., Tsuji, K., and Enzinger, R. M.,** Application of LAL for detection of endotoxin in antibiotic preparations, in *Biomedical Applications of the Horseshoe Crab (Limulidae),* Cohen, E., Ed., Alan R. Liss, New York, 1979, 353.
81. **Guzman, R. and Harng-Shen Kuo,** Pyrogen testing of an intravenous fat emulsion, in *Biomedical Applications of the Horseshoe Crab (Limulidae),* Cohen, E., Ed., Alan R. Liss, New York, 1979, 333.
82. **Mascoli, C. C. and Weary, M. E.,** *Limulus* amebocyte lysate (LAL) test for detecting pyrogens in parenteral injectable products and medical devices: advantages to manufacturers and regulatory officials, *J. Parenter. Drug Assoc.,* 33, 81, 1979.

83. **Harada, T., Morita, T., Iwanaga, S., Nakamura, S., and Niwa, M.**, A new chromogenic substrate method for assay of bacterial endotoxins using *Limulus* hemocyte lysate, in *Biomedical Applications of the Horseshoe Crab (Limulidae)*, Cohen, E., Ed., Alan R. Liss, New York, 1979, 209.
84. Committee of Revision, Bacterial endotoxin test, in *The United States Pharmacopeia, XX,* Board of Trustees, United States Pharmacopeial Convention, Rockville, 1979, 888.
85. United States Food and Drug Administration, Draft guideline for validation of the *Limulus* amebocyte lysate test as an end-product pyrogen test for human and veterinary injectable drugs other than licensed biologicals, 1980.
86. **Hochstein, D. H.**, The LAL test versus the rabbit pyrogen test for endotoxin detection, *Pharm. Technol.*, August, 37, 1981.
87. Committee of Revision, Bacterial endotoxin test, in *The United States Pharmacopeia, XX, Addendum A to Suppl. 3,* Board of Trustees, United States Pharmacopeial Convention, Rockville, 1982, 299.
88. Committee of Revision, *The United States Pharmacopeia, XX, Addendum A to Suppl. 4,* Board of Trustees, United States Pharmacopeial Convention, Rockville, 1983.
89. United States Food and Drug Administration, Draft guideline for validation of the *Limulus* amebocyte lysate test as an end-product endotoxin test for human and animal parenteral drugs, biological products, and medical devices, 1983.
90. Committee of Revision, Bacterial endotoxin test, in *The United States Pharmacopeia, XX, Suppl. 5,* Board of Trustees, United States Pharmacopeial Convention, Rockville, 1984, 1100.
91. Committee of Revision, Bacterial endotoxin test, in *The United States Pharmacopeia, XXI,* Board of Trustees, United States Pharmacopeial Convention, Rockville, 1984, 1165.
92. Committee of Revision, Bacterial endotoxin test, in *The United States Pharmacopeia, XXI, Suppl. 1,* Board of Trustees, United States Pharmacopeial Convention, Rockville, 1984, 1768.
93. United States Food and Drug Administration, Guideline on validation of the *Limulus* amebocyte lysate test as an end-product endotoxin test for human and animal parenteral drugs, biological products, and medical devices, 1987.
94. Committee of Revision, In-process revision, *Pharm. Forum.*, 16, 679, 1990.
95. Council of Europe, Bacterial endotoxins, in *European Pharmacopoeia,* 2nd ed., Part II, 11th fascicule, Maisonneuve S. A., France, 1987, V.2.1.9.1.
96. Subcommittee on the Japanese Pharmacopoeia, Bacterial endotoxins test, in *Supplement to the Pharmacopoeia of Japan,* 11th ed., The Society of Japanese Pharmacopoeia, Yakuji Nippo, Tokyo, 1988, 1588.
97. **Guilfoyle, D. E. and Munson, T. E.**, Procedures for improving detection of endotoxin in products found incompatible for direct analysis with *Limulus* amebocyte lysate, in *Endotoxins and Their Detection With the Limulus Lysate Test,* Watson, S. W., Levin, J., and Novitsky, T. J., Eds., Alan R. Liss, New York, 1982, 79.
98. **McCullough, K. Z.**, The use of LAL as an alternative to the CFR rabbit pyrogen test for disodium ticarcillin, in *Endotoxins and Their Detection With the Limulus Lysate Test,* Watson, S. W., Levin, J., and Novitsky, T. J., Eds., Alan R. Liss, New York, 1982, 91.
99. **Case, M. J., Ryther, S. S., and Novitsky, T. J.**, Detection of endotoxin in antibiotic solutions with *Limulus* amebocyte lysate, *Antimicrob. Agents Chemother.*, 23, 649, 1983.
100. **Twohy, C. W., Duran, A. P., and Munson, T. E.**, Endotoxin contamination of parenteral drugs and radiopharmaceuticals as determined by the *Limulus* amebocyte lysate method, *J. Parenter. Sci. Technol.*, 38, 190, 1984.
101. **Remillard, J. F., Gould, M. C., Roslansky, P. F., and Novitsky, T. J.**, Quantitation of endotoxin in products using the LAL kinetic turbidimetric assay, in *Detection of Bacterial Endotoxins with the Limulus Amebocyte Lysate Test,* Watson, S. W., Levin, J., and Novitsky, T. J., Eds., Alan R. Liss, New York, 1987, 197.
102. **Cooper, J. F.**, Resolving LAL test interferences, *J. Parenter. Sci. Technol.*, 44, 13, 1990.
103. **Twohy, C. W., Nierman, J. L., and Duran, A. P.**, Comparison of *Limulus* amebocyte lysates from different manufacturers, *J. Parenter. Sci. Technol.*, 37, 93, 1983.
104. **McCullough, K. Z.**, Variability in the LAL test, *J. Parenter. Sci. Technol.*, 44, 19, 1990.
105. Committee of Revision, Bacterial endotoxin test, in *The United States Pharmacopeia, XXII,* Board of Trustees, United States Pharmacopeial Convention, Rockville, 1989, 1493.
106. **Kannegieter, E. M. and Baggerman, C.**, A new method to reduce electrolyte inhibition of the detection of endotoxins in large volume parenterals, *J. Parenter. Sci. Technol.*, 38, 17, 1984.
107. **Weiss, A. R.**, A new ultrafiltration unit for the removal of substances interfering with the LAL-endotoxin-test, in *Detection of Bacterial Endotoxins with the Limulus Amebocyte Lysate Test,* Watson, S. W., Levin, J., and Novitsky, T. J., Eds., Alan R. Liss, New York, 1987, 235.
108. **Pfeiffer, M. and Weiss, A. R.**, Removal of LAL-test interfering low molecular weight substances by ultrafiltration, in *Detection of Bacterial Endotoxins With the Limulus Amebocyte Lysate Test,* Watson, S. W., Levin, J., and Novitsky, T. J., Eds., Alan R. Liss, New York, 1987, 251.

109. **McCullough, K. Z.,** Process control: in-process and raw material testing using LAL, *Pharm. Technol.,* 12, 40, 1988.
110. **McCullough, K. Z. and Novitsky, T. J.,** Detoxification of endotoxin by acid and base, in *Depyrogenation,* Parenteral Drug Association Technical Report No. 7, Philadelphia, 1985, 78.
111. **Weary, M.,** Depyrogenation, in *Pyrogens: Endotoxins, LAL Testing and Depyrogenation,* Pearson, F. C., III, Ed., Marcel Dekker, New York, 1985, 201.
112. **Novitsky, T. J.,** Monitoring and validation of high purity water systems with the *Limulus* amebocyte lysate test for pyrogens, *Pharm. Eng.,* 4, 21, 1984.
113. **Dawson, M. E., Novitsky, T. J., and Gould, J. J.,** Microbes, endotoxins and water, *Pharm. Eng.,* 8, 9, 1988.
114. Parenteral Drug Association Quality Assurance/Quality Control Task Group, Current practices in endotoxin and pyrogen testing in biotechnology, *J. Parent. Sci. Technol.,* 44, 39, 1990.
115. **Gould, M. C.,** Endotoxin in vertebrate cell culture: its measurement and significance, in *Uses and Standardization of Vertebrate Cell Cultures,* Monograph No. 5, Patterson, M. K., Ed., Tissue Culture Association, Rockville, 1983, 125.
116. United States Food and Drug Administration, Current good manufacturing practice for finished pharmaceuticals, in *Code of Federal Regulations, Food and Drugs,* Title 21, part 211, Office of the Federal Register, National Archives and Records Administration, Washington, D.C., 1987, 77.
117. United States Food and Drug Administration, Guideline on sterile drug products produced by aseptic processing, 1987.
118. **Albaugh, B. R. and Chandler, C. B.,** Automated methodology for the *Limulus* amebocyte lysate (LAL) assay using the microplate reader, in *Endotoxins and Their Detection With the Limulus Lysate Test,* Watson, S. W., Levin, J., and Novitsky, T. J., Eds., Alan R. Liss, New York, 1982, 183.
119. **Oishi, H., Takaoka, A., Hatayama, Y., Matsuo, T., and Sakata, Y.,** Automated *Limulus* amebocyte lysate (LAL) test for endotoxin analysis using a new Toxinometer ET-201, *J. Parenter. Sci. Technol.,* 39, 194, 1985.
120. **Weary, M., Dubczak, J., Wiggins, J., Bohon, J., Brandon, A., and Pearson, F. C., III.,** Validating an LAL chromogenic substrate pyrogen test for large volume parenterals, in *Detection of Bacterial Endotoxin With the Limulus Amebocyte Lysate Test,* Watson, S. W., Levin, J., and Novitsky, T. J., Alan R. Liss, New York, 1987, 307.
121. Parenteral Drug Association, *Depyrogenation,* Technical Report No. 7, Parenteral Drug Association, Philadelphia, 1985.
122. **Weary, M. and Pearson, F. C., III.,** A manufacturer's guide to depyrogenation, *Biopharmacology,* 4, 22, 1988.
123. **Robertson, J. H., Gleason, D., and Tsuji, K.,** Dry heat destruction of lipopolysaccharide: design and construction of dry-heat destruction apparatus, *Appl. Environ. Microbiol.,* 36, 705, 1978.
124. **Tsuji, K. and Harrison, S. J.,** Dry heat destruction of lipopolysaccharide: dry heat destruction kinetics, *Appl. Environ. Microbiol.,* 36, 710, 1978.
125. **Tsuji, K. and Lewis, A. R.,** Dry heat destruction of lipopolysaccharide: mathematical approach to process evaluation, *Appl. Environ. Microbiol.,* 36, 715, 1978.
126. **Tsuji, K. and Harrison, S. J.,** *Limulus* amebocyte lysate — a means to monitor inactivation of lipopolysaccharide, in *Biomedical Applications of the Horseshoe Crab (Limulidae),* Cohen, E., Ed., Alan R. Liss, New York, 1979, 367.
127. **Akers, M. J., Ketron, K. M., and Thompson, B. R.,** F value requirements for the destruction of endotoxin in the validation of dry-heat sterilization/depyrogenation cycles, *J. Parenter. Sci. Technol.,* 36, 23, 1982.
128. **Ludwig, J. D. and Avis, K. E.,** Validation of heating cell for precisely controlled studies on the thermal destruction of endotoxin in glass, *J. Parenter. Sci. Technol.,* 42, 9, 1988.
129. **Ludwig, J. D. and Avis, K. E.,** Dry heat inactivation of endotoxin on the surface of glass, *J. Parenter. Sci. Technol.,* 44, 4, 1990.
130. Committee of Revision, Sterilization and sterility assurance of compendeal articles, in *The United States Pharmacopeia, XXI,* Board of Trustees, United States Pharmacopeial Convention, Rockville, 1984, 1347.
131. LAL User's Group Endotoxin Indicators Committee, Preparation and use of endotoxin indicators for depyrogenation process studies, inter-group communication, 1988.
132. **Novitsky, T. J., Schmidt-Gengenback, J., and Remillard, J. F.,** Factors affecting recovery of endotoxin adsorbed to container surfaces, *J. Parenter. Sci. Technol.,* 40, 284, 1986.
133. **Jensch, U. E., Gail, L., and Klavehn, M.,** Fixing and removing of bacterial endotoxin from glass surfaces for validation of dry heat sterilization, in *Detection of Bacterial Endotoxins with the Limulus Amebocyte Lysate Test,* Watson, S. W., Levin, J., and Novitsky, T. J., Eds., Alan R. Liss, New York, 1987, 283.

Chapter 19

ENDOTOXIN DETERMINATION WITH *LIMULUS* LYSATE IN CLINICAL DIAGNOSIS AND HYGIENIC CONTROL

Petter Friberger

TABLE OF CONTENTS

I. INTRODUCTION

Endotoxins from Gram-negative bacteria can activate an enzyme system in the blood cells (amebocytes = granulocytes) from the horseshoe crab (*Limulus polyphemus*). This reaction cascade can be monitored in several ways, and extremely small amounts of endotoxin can be measured with various techniques. As endotoxin is found practically everywhere, there is great risk for serious errors due to contamination of the sample during handling. Endotoxin determination is of interest in different types of samples, which nearly all affect the reaction conditions. This makes it necessary to use a large variety of techniques. When performed in a careful manner, such assays have proved to be of great benefit in several application areas, e.g., pharmacy, clinical diagnosis, and hygienic control.

II. ENDOTOXIN BIOCHEMISTRY

Endotoxins are lipopolysaccharides from the outer membrane of Gram-negative bacteria. Gram-positive bacteria, viruses, or fungi do not contain endotoxins. Endotoxins constitute a class of substances, which have the same basic structure (Figure 1). When tested in the *Limulus* amebocyte lysate (LAL) system, various purified endotoxin preparations behave differently (Table 1). This is probably owing to the differences in preparation (extraction, treatment, purification, storage, and additives). Further studies are necessary to determine if there are also significant species differences. Until then, at least we will have to use only well-characterized preparations for the standardization of our experiments.

Lipid A (Figure 2) consists of a disaccharide, two phosphate groups (one causes an activity reduction to 1 to 20%), and six fatty acids (five cause an activity reduction to about 10% and four or fewer give a total loss of activity).

Endotoxin preparations for standardization are available from various agencies, e.g., U.S. Pharmacopoeial Convention Inc., Rockville, U.S. and WHO, National Institute for Biological Standards and Control, U.K.

III. EFFECTS OF ENDOTOXIN

Endotoxin is an important virulent factor, but not the sole toxic factor in the pathogenesis of Gram-negative bacteria. It is probably not a toxin in a classical sense, as it is the host's answer to endotoxin which creates toxic effects. Both pathogenic and saprophytic bacteria may possess endotoxin. However, other factors, such as adhesion to endothelial cells, are also important for bacterial pathogenesis.

Little is known about the factors that control the resorption or the secretion of endotoxin, and significant species differences are expected. The state of dispersion of the endotoxin preparation, used in various experiments, markedly affects the responses produced. The basic mechanisms by which endotoxin functions involve both cellular and humoral systems, and the complex interrelationship of these systems makes it difficult to define and distinguish between direct and indirect effects of endotoxin. Bacteria commonly found in septic complications are listed in Table 2. Except from circulating bacteria, endotoxin may originate from a septic focus and through resorption from the intestine, which contains more than 2 g of endotoxin (probably partially bound to the bile salts) resulting from a disease state (e.g., obstructive jaundice) and irradiation. Resorption can contribute significantly to the endotoxin content in the blood during a disease. It is possible that other microorganisms can trigger this resorption, explaining connections between such microorganisms and Gram-negative sepsis. The concentration of endotoxin in portal blood should then be significantly higher than in systemic blood. Regularly, however, no endotoxin is found in the portal vein.[2] Introduction of endotoxin via parenteral drugs, medical devices like dialysis filters, catheters, and surgical gloves has also been considered.

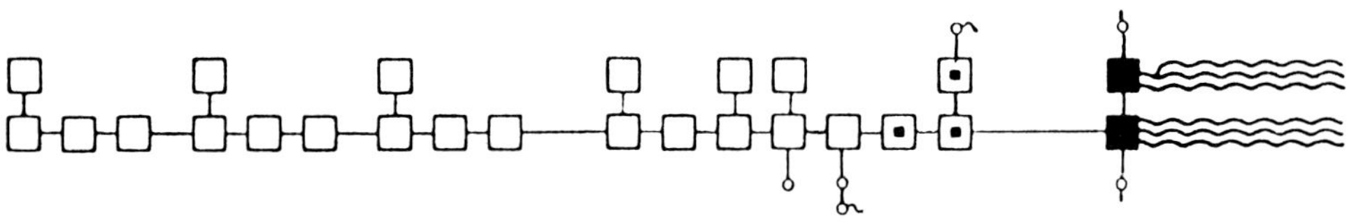

FIGURE 1. The endotoxin molecule consists of an antigenic polysaccharide, a rather constant polysaccharide core and lipid A. It usually exists as an aggregate containing at least 15 to 20 molecules (M_r 15 to 20 kDa each). (From C. Galanos, 1979.)

TABLE 1
The Effect of 44 Different Gram-Negative Bacterial Endotoxins

	Relative activity		Relative activity
E. coli		*S. abortus equi*	78
0111:B4	100		
	85		135
	90	*S. enteritidis*	86
		Wild type (795)	
055:B5	65	*S. typhimurium*	65
	123	Wild type (LT2)	142
	114	Re-mutant (G30-C21)	56
026:B6	51	*S. typhosa*	179
		0901	
	162	*Se. marcescens*	62
1027:B8	156		69
	124		163
11100:K235	45	*Sh. dysenteriai*	7
K235	20	*Sh. flexneri*	21
K12, lipid A	210	*P. aeruginosa*	7
0113	102	14-x	14
0128:B12	76	FD, Type 1	31
J5	19	*Y. enterocolitica*	35
EC4	2	*V. choloerae inaba*	121
		569B	
S. minnesota			
Wild type	140	*K. pneumoniae*	23
	40		
R595, re-mutant	17		
	254		
R595, lipid A	15		
R60 (Ra)	103		
R345 (Rb)	22		
R5 (Rc)	295		
R7 (Rd)	242		

Note: The relative activity — COATEST® kit endotoxin (120 EU/ng) set as 100 — is given. E. = *Escherichia*; K. = *Klebisella*; P. = *Pseudomonas;* S. = *Salmonella*; Se. = *Serratia*; Sh. = *Shigella*; V. = *Vibrio*; Y. = *Yersinia*.

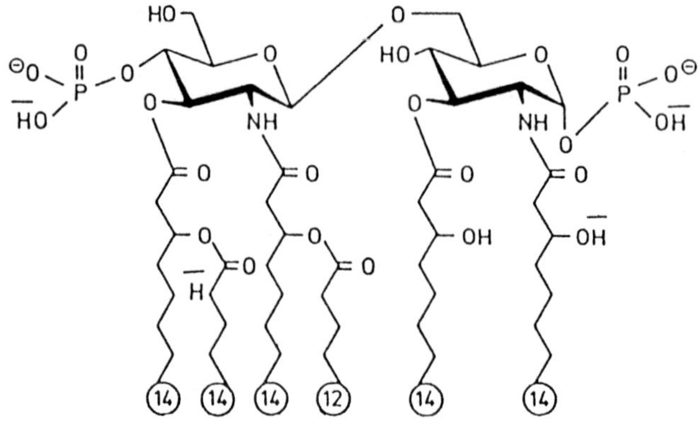

FIGURE 2. Basic structure of lipid A (M_r 3 kDa) consists of a disaccharide, two phosphate groups, and six fatty acid residues. At H̄ lipid A from *E. coli, P. mirabilis,* and *S. minnesota* have different substituents. (From E. Th. Rietschel.)

TABLE 2
Gram-Negative Bacteria Most Frequently Causing Sepsis

Bacteria	Total number of cases %
E. coli	33
Klebsiella	13
Pseudomonas	11
Proteus providencia	7
Enterobacter	6
Serratia	3
Anaerobes	9
Other	6
Polymicrobial	12

In normal human blood, only a fraction of the theoretical endotoxin activity from bacteria is released. Some of the activity may, however, originate from living bacteria. During Gram-negative infections, different patterns of endotoxin concentration in blood are to be expected. Three principal types are demonstrated in Figure 3. In acute fulminating meningococcemia, where in the nonreacting individual the bacteria rapidly multiply in the bloodstream until they disintegrate at a later stage, endotoxin is rapidly released in copious quantities (the spike in pattern 3, Figure 3). Endotoxin in the circulation probably forms a complex with high density lipoproteins (HDL) and/or glycoproteins in the plasma and remains in this complex until it is cleared from circulation. Data about the half-life of endotoxin activity in the blood vary between 30 min and 10 h. The ''smooth type'' endotoxins are taken up in the liver by macrophages and granulocytes, and the ''rough type'', by hepatocytes and Kupffer cells. Other pathways of endotoxin elimination are macrophages and granulocytes in other organs, such as the lungs and the adrenal glands. Some endotoxin persists in the liver for over 5 weeks before being secreted in the bile in a chemically altered, virtually inactive form. It has been postulated that the toxic effects of endotoxin are inactivated by the chemical modification of its ''toxic site'', lipid A, probably in combination with the proceeding KDO part. Some excretion, in a degraded form, also takes place in the urine. Endotoxin can create both a tolerance (stimulation of the immune response) and hyperreactivity (Schwartzman reaction) at a second stimulation, depending on doses and time lag.

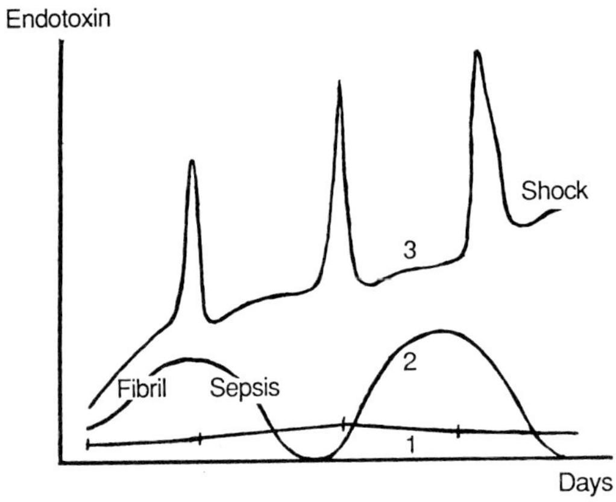

FIGURE 3. Three different patterns of endotoxin concentration in blood plasma. (1) Pattern expected in chronic, often asymtomatic, sepsis in, e.g., elderly or patients with cirrhosis of the liver. (2) Acute sepsis. (3) Acute sepsis with endotoxin spikes — often severe condition which leads to septic shock.

The reticuloendothelial system (RES) is recognized as a collection of mobile and fixed phagocytic cells present in circulation, but also in lymph nodes, lung, liver, spleen, and bone marrow. Mononuclear phagocytic cells, monocytes, and macrophages as well as granulocytes (polymorphonuclear neutrophilic leukocytes) all contribute. Products such as peroxides and free hydroxyl radicals, platelet-activating factors (PAF), proteases, complement proteins, thromboplastin, fibronectin, and numerous bioactive peptides important in wound repair, inflammation, circulatory shock, trauma, and immunologic surveillance (immunostimulants) are secreted by the RES when triggered by endotoxin of foreign cells (Figure 4). Among the newly found proteins/peptides (cytokines) released from the cells of the RES, such as interleukins, cachectin (tumor necrosis factor, TNF), and colony-stimulating factor (CSF), interferons are important in the pathogenesis.

The oxygen-centered radicals can induce lysis of cell membranes, inactivate various enzymes, fragment and mutate nucleic acids, degrade macromolecules (like collagen) of the extra cellular matrix, and inactivate important serine protease inhibitors (like α-antitrypsin).

Endothelial cell permeability is influenced directly by endotoxin, but the cells are also affected through free radicals and by interleukin-1 (IL-1). These cells then release tissue thromboplastin (Figure 5). Endothelial thrombomodulin contributes to the activation of protein C, leading to inactivation of coagulation Factor V and Factor VIII. The fibrinolytic system is activated through the release of tissue plasminogen activator (t-PA) from the endothelial cells, and in the presence of fibrin there is a rapid production of plasmin. From the same endothelial cells, the main inhibitor of t-PA, plasminogen activator inhibitor (PAI), can also be released by similar stimuli.

The complement and free radicals from activated granulocytes lead to lung microvascular injury. This leads finally to organ failure because of acidosis due to fibrin deposition circulatory breakdown and to edema. Tissue kallikreins activate collagenases and gelatinases, which then degrade connective tissues. Such disastrous events have to be met by treatment using a broad spectrum of therapeutic agents at the same time as circulation is maintained. Protease inhibitors as well as inhibitors of lysosomal enzymes, antioxidants, and radical scavengers are then required.

During hypotension, prerenal insufficiency, which is a reversible condition causing low urinary volumes in combination with elevated osmolality and low sodium concentration,

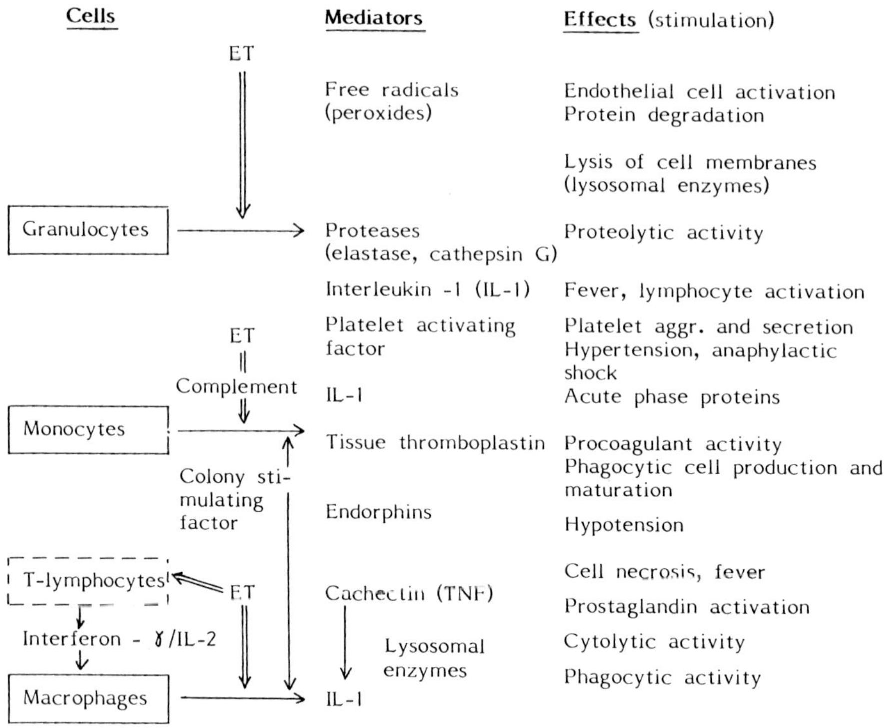

FIGURE 4. Reactions of the reticuloendothelial system (RES). Phagocytic cells are activated by endotoxin (ET) and release various mediators (also humoral factors contribute); a number of effects have been registered. The exact pathways are still obscure.

may appear. Blood urea concentration may also rise during this stage. Acute tubular necrosis caused by endotoxin is a less reversible condition. These conditions are naturally very much affected by the underlying disease.

Microcirculatory disorders due to intravascular coagulation may damage the liver. Liver insufficiency in shock is serious because it involves almost 90% of the clearing cells of the RES.

Recently it has been proved that endotoxin influences myocardial function in a similar way as the other organs, namely, by increasing the coronary capillary pressure and creating myocardial edema which can lead to dysfunction.

Just because so many metabolic and defense systems are triggered by endotoxin and progress in parallel for hours and days, it has been impossible to clearly define how they are interrelated and in which order they act. There may be many more effects of endotoxin, which have been judged to come from other sources, because endotoxin has certainly been an interfering factor in many experiments *in vitro* as well as in animals and humans. A careful control of contamination with endotoxin is indeed of utmost importance in all biomedical research.

Once the diagnosis of Gram-negative sepsis has been considered clinically, the next step is to identify the source of infection — the urinary tract, the lower respiratory tract, the biliary tract or elsewhere within the abdomen (Table 3) or, in a hospitalized patient, an intravascular device — to guide subsequent management. Identifying the initial source of infection may allow surgical extirpation, drainage of an abscess (see Section V.D). New therapeutic measures such as substitution with protease inhibitors may be effective in delaying the more serious shock condition. It has been anticipated that specific antibodies to cachectin and related cytokines may offer new therapeutic directives to treat a broad spectrum of diseases.

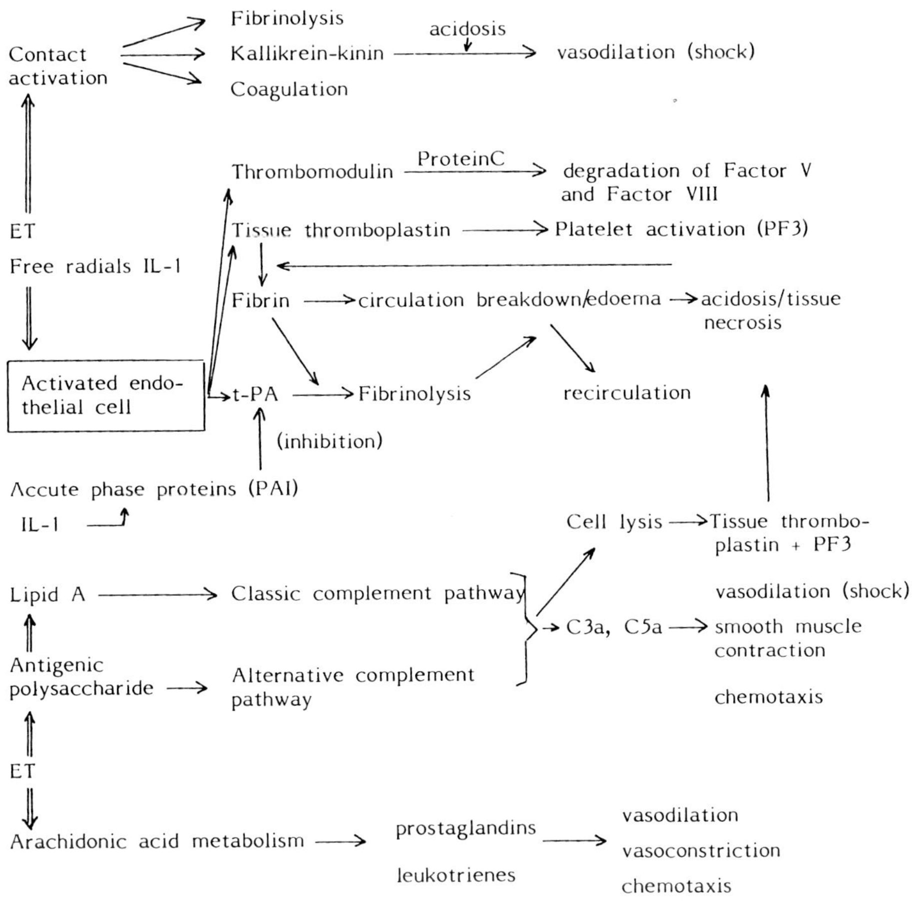

FIGURE 5. The humoral pathways of endotoxin action contribute to shock, DIC, and circulatory breakdown in lung, liver, kidney and heart, often leading to tissue necrosis and even death. Some of the abbreviations are explained in the text.

TABLE 3
Source of Bacteremia

Primary infection	Approximative proportion %
Urinary tract	40
Respiratory tract	10
Gastrointestinal or intraperitoneal	10
Skin and surgical wound	5
Biliary tract	5
Miscellaneous	30

Note: Overall data, however, vary significantly between receptions involved.

IV. METHODS OF ENDOTOXIN DETERMINATION

It has long been considered necessary to document that products meant for parenteral administration are free from contaminants which are hazardous to the health of the treated patient. To accomplish this, it is important to have a sensitive and reliable procedure for the determination of endotoxin. By using only bacteriological tests, the risk for misinter-

TABLE 4
Data from the QA Laboratories of Travenol U.S. Plants Demonstrates the Difference in Sensitivity between the LAL and Rabbit Tests

Ranges with the LAL test (EU/ml)	LAL test number of samples	Number of rabbit test failures
<0.25	679	0
0.25—1.0	816	16
>1.0	136	15

pretation is obvious. The presence of endotoxin can under certain circumstances be an early warning of bacterial contamination because bacteria and spores might be hidden, and thus survive various sterilization treatments. The presence of endotoxins without significant bacterial growth may also be caused by bacteriophages, antiseptics, and antibiotics. Regrowth may also occur at a later stage.

Endotoxins produce fever due to the induction of the release of cytokines, which acts on the thermoregulatory center in the hypothalamus. The rabbit pyrogen test has been used since the early 1940s. In this test, the sample is injected into a rabbit ear vein, and the body temperature of the animal is followed. An increase in the temperature of about 0.5°C (mean value) is considered as a positive result. A colony of four or eight rabbits is used each time. The introduction and the acceptance of the *Limulus* amebocyte lysate (LAL) test have been relatively slow. This is a common feature when regulatory authorities are involved. Besides, it took a long time before it was realized that quite a few products, containing significant amounts of endotoxin, passed the rabbit pyrogen test in many laboratories less experienced in animal experiments than the quality assurance (QA) departments of the large pharmaceutical industries (Table 4). The cases where the rabbit failed to detect contamination have steadily increased. In the U.S., the Food and Drug Administration (FDA) has lately considered it impossible to give proper guidelines to guarantee a certain sensitivity in the rabbit test. Therefore, the LAL test has been suggested as the primary test for endotoxin, and guidelines how to perform the test are given in the U.S. Pharmacopoeia. The formation of a gel after 1 h incubation means that endotoxin is present at a level above the sensitivity level of the LAL (Figure 6). The LAL test is thought to be much safer than the old rabbit pyrogen test (Table 5). By measuring the time to the formation of a precipitate in the gel, it is possible to make a quantitative determination of endotoxin (turbidimetric procedure).

The enzyme that clots or precipitates the LAL also splits certain chromogenic substrates. This detection of the reaction in the LAL is even more sensitive and less influenced by disturbances than the turbidimetric and gelation procedures (Table 6).

Other types of endotoxin assays — functional, biochemical, chemical, and immunochemical — have been used but they are without any doubt more difficult to perform and/or less sensitive; more important, for a heterogeneous group of compounds such as the endotoxins, they are far less documented.

V. THE LAL ASSAY

The activation of the LAL reagent by endotoxin has been extensively studied, and the assumed mechanism for the lysate from the Japanese horseshoe crab is described in Figure 7 (see also Chapter 18).

To date, the most frequently used LAL method is by the gelation procedure, which is a ''pass or fail'' type of method (Figure 6). In this method, equal volumes of LAL reagent and sample are mixed in a test tube and incubated at 37°C for 1 h. The outcome (gel =

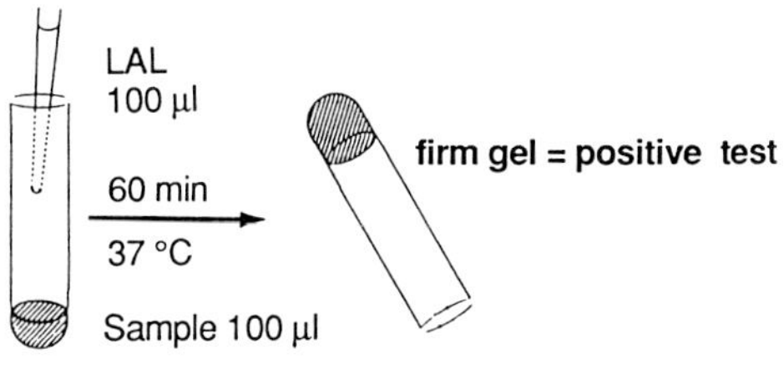

FIGURE 6. Clot method. This method utilizes the formation of a firm gel as the mark for failing the test. This gel is fragile and it can be seriously disturbed by minor shakings of the incubator when forming.

TABLE 5
Improvements Obtained Using the LAL Test Instead of the U.S. Pyrogen Rabbit Test

The LAL test is
Simpler
More rapid
More sensitive
Less expensive (less than half price)
More reproducible
More versatile
Less influenced by other compounds
More specific

TABLE 6
Approximate Sensitivity of Endotoxin Methods

Rabbit test	>2.0 EU/ml
LAL clot test	0.1 EU/ml
Chromogenic LAL test	0.01 EU/ml

fail) is tested by turning the tube upside down. To get a fair quantitative estimate, serial dilutions of a standard (Table 7) and of the sample can be used. Four to eight parallel determinations are demanded by the Pharmacopoeia.

There have been several suggestions as to how this LAL method should be performed in a direct quantitative manner. At an early stage of method development, one of the inventors of the LAL tests, Jack Levin, used a clotting time procedure. The time to the formation of a clot is inversely proportional to the amount of endotoxin. As the LAL reagent has been made more sophisticated and sensitive, it has also become sensitive to slight shaking — the gel formation is disturbed and usually irreversibly destroyed when only touched. Then it is not possible to check whether the gel has been formed by tilting the tube as performed in a classical blood clotting time procedure. A turbidimetric approach has therefore been utilized.

There are two principal effects that can be measured, either separately or in combination, namely, time to the formation of the clotting enzyme (lag phase = clotting time) or activity of the enzyme formed (slope of the progress curve). Both seem to be related to the amount of endotoxin (Figure 8). These parameters can be measured directly on the LAL + sample reaction mixture (formation of turbidity in the clot — measured here with a photometer at 380 nm) or determined by using a chromogenic substrate and a photometer to measure the

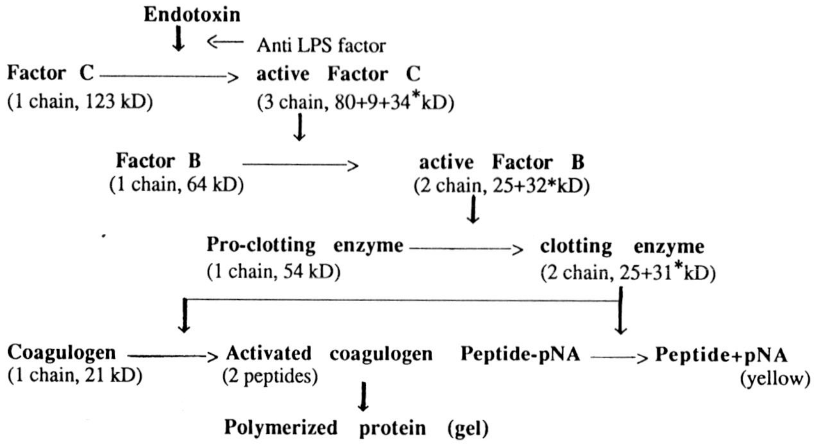

FIGURE 7. The cells of the hemolymph of the horseshoe crab are washed and lysed by using ultrasound. Then inhibitory factors are eliminated by chloroform or surfactants. The chain of reactions triggered by endotoxin induces a certain lag (5 to 8 min) in the activation of the clotting enzyme as well as a remarkably high sensitivity of the method.[1] *, Signifies chain with active site.

TABLE 7
Semiquantitative Determination of Endotoxin with a LAL Gelation Procedure

	Endotoxin potency (EU/ml)						
Replicate	**2**	**1**	**0.5**	**0.25**	**0.125**	**0.062**	**0.031**
1	+	+	+	−	−	−	−
2	+	+	+	+	+	−	−
3	+	+	+	+	−	−	−
4	+	+	+	+	+	−	−

Note: The endotoxin standard is diluted to 2 EU/ml and then further diluted 1:2 six times; + means a firm gel; statistical evaluation gives a lysate sensitivity = 0.21 EU/ml (geometric mean).

activity of the proteolytic enzyme that activates the coagulogen. If the chromogenic substrate is included in the LAL reagent from the start (before or at mixing with the sample), it is possible to measure both lag time (to the appearance of proteolytic activity) and the rate of activation of the final protease, as well as maximal activity of this enzyme. If the chromogenic substrate is added at a point in time when a certain amount of enzyme is activated (the most frequently used procedure), the result depends on both the lag time and the rate of enzyme activation. Figure 8 shows an example where the turbidimetric technique is used. If a synthetic chromogenic peptide substrate is included and the absorbance measured at 405 nm, it will look similar. The absorption will probably appear earlier, and the plateau may come earlier because of substrate depletion. The concentrations of the clotting protein and of the synthetic substrate both will naturally influence the kinetics. These two substrates have been found to be competitive.

It is also obvious from Figure 8 that the range that can be used is narrower (e.g., 5 to 50 picogram [pg]/ml at 60 min), using a fixed incubation time compared with the so-called "kinetic procedure" where a large number of readings are made for each sample combined with automated data treatment.[3] The disadvantage with a narrow range has also been considered in the chromogenic (two-stage procedure) technique (Figure 9). By using a microplate

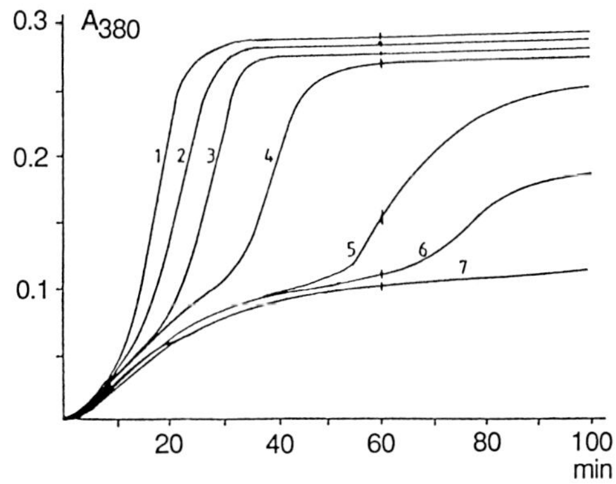

FIGURE 8. Progress curves obtained with the turbidimetric technique; 100 pg is approximately equivalent to 1 EU of endotoxin. (1) 4000 pg/ml, (2) 1000 pg/ml, (3) 250 pg/ml, (4) 63 pg/ml, (5) 16 pg/ml, (6) 8 pg/ml, and (7) 0 pg/ml. (From Ditter, B., Becker, K. P., Urbaschek, R., and Urbaschek, B., *Arzneim.-Forsch.*, 33, 681, 1983. With permission.)

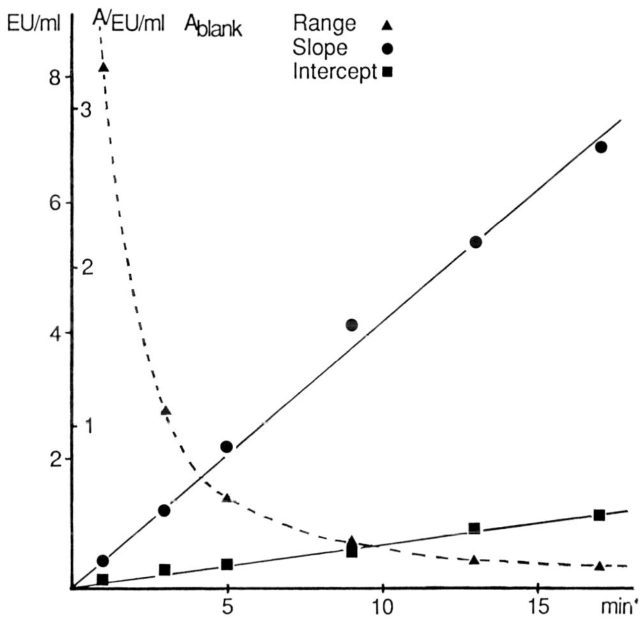

FIGURE 9. Using a microplate method (principle see Figure 10) with 15 min activation at 37°C (LAL of mean sensitivity) the time of hydrolysis can be varied to obtain different measurement ranges (EU/ml, A_{405}, <1.4, χ) from 0—0.3 to 0—8 EU/ml. The slope of the standard curve (A/EU/ml, ●) increases linearly with time as does the intercept (blank, ■); r is in most cases ≥0.99 and at the smallest ranges still ≥0.97. (From *Pharmeuropa,* 1, 5—8, 1989.)

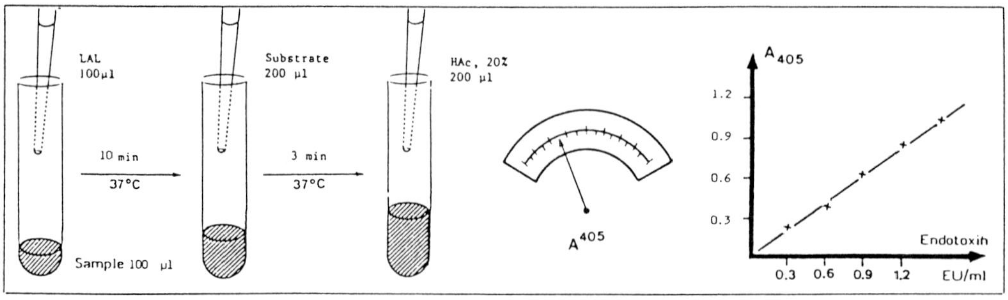

FIGURE 10. In the first stage of the chromogenic LAL test, endotoxin in the sample activates LAL. In the second stage, the activity is determined with a chromogenic peptide substrate. The absorbance is read at 405 nm in a photometer, and the amount of endotoxin is determined by using a standard curve or by using standards in the same run. Two ranges, 0.1 to 1.0 and 0.01 to 0.1 EU/ml, the latter demanding 25 and 5 min incubation, respectively, have been suggested.

reader, it is possible to expand the activity range by doing repeated readings. Then it is possible to make sure that the substrate is not depleted even at high activity and that sufficient reading time is used at low activity.

Using a single-stage chromogenic technique, it is possible to use a kinetic technique similar to that presented by Ditter et al.[3] (cf. Figure 8). The technique must then be slightly modified to fit the new, probably more sensitive detection. The "kinetic technique" demands some equipment not generally available (automated photometer readers and computors), which in the very near future will be expected to appear in some research laboratories and possibly in the quality assurance laboratories of the larger pharmaceutical companies.

For clinical purposes, nearly all suggested procedures are of the "pass or fail" type or close to it, probably because so far it has been difficult to correlate endotoxin concentration to the clinical condition, even if the presence of endotoxin gives a high specificity. What has been utilized to date is the knowledge about the existence of endotoxin activity over a certain level and, occasionally, whether it increases or decreases. These studies (see Section V.D) have proved that they can be effective as diagnostic and prognostic tools.

For the control or parenterals, raw materials, production conditions, and final products, the critical endotoxin concentration is well defined, and a narrow range (one order of magnitude) has proved to be sufficient in most cases. However, as pointed out before, some efforts have been made to broaden or to modify the measuring range.

For hygienic control, as a quick alternative to culturing techniques, an endotoxin method with a range covering several orders of magnitude may be of significant benefit. However, until the LAL method has been accepted in routine hygienic control, a less demanding (regarding equipment) methodology probably has to be used.

The chromogenic LAL test (Figure 10) available today is built on the gelation procedure. Because the formation of a firm gel (Figure 7) is not needed in the chromogenic LAL assay, the lysate production is simplified. This is a proper way to avoid the main obstacle of the gelatin method. Having a quantitative procedure and an excellent endotoxin standard, it is also easy to control the production of the LAL reagent. All LAL preparations made for the gel procedure can be used in the chromogenic assay.[4]

The test reagents are made optimal for the assay of endotoxin in pure water. As the activation of LAL is influenced by most compounds, inhibition can be expected for most types of samples to be tested. Inhibitory effects caused by the samples can be largely or totally avoided (less than 30% inhibition is recommended) by dilution with endotoxin-free water and heat treatment (denaturation of proteins). To control this, the product has to be used for standardization of the assay. This has to be checked for each new batch of kits because the various LAL preparations may be influenced differently by the various sample constituents, all depending on the amount and quality of the LAL constituents participating in the activation of the final proteolytic activity (Figure 7). The various LAL factors may vary in concentration as well as in rate-limiting potency. There is no reason to postulate that inhibitory effects by the various sample constituents should be directly additive, as they may influence different stages of the LAL activation (Figure 7).

Another reason why a low inhibition is preferred is that the LAL assay based on comparison of environmental endotoxin and purified endotoxin (standard) can respond differently to these endotoxins, regarding inhibition of the LAL activation reaction. The LAL test in much material has proved to be surprisingly specific (see also Section V.B).

A. THE ADVANTAGES OF A QUANTITATIVE METHOD

A pass or fail method may be sufficient for the approval of final products or constituents, but for all types of investigative work, it is necessary to have a quantitative procedure that is also significantly more sensitive (0.001 EU/ml or 0.1 pg/ml is the practical limit in the chromogenic procedure) than what is usually demanded for passing the limit of acceptance at a control. It is definitely a sound strategy to choose a quantitative and safe approach to control all products. By using a chromogenic method, it is possible to study inhibitory effects and to develop suitable strategies to minimize them. A 1:10 dilution with endotoxin-free water is usually sufficient. It is, however, important that the linear relationship between endotoxin and the response of the assay, as obtained in water, remains. A quantitative method also makes it possible to control, and thereby avoid contamination during sampling and handling of the samples. The stability of the LAL has to be under control. The consequence of what has been stated before is that products presently tested with the clot procedure should be worth retesting with the chromogenic procedure in order to find out more exactly how the product inhibits the assay, to control the risk of contamination at sampling and performing the assay, to investigate the risk of contamination in the production, and to check the quality and stability of the LAL, endotoxin, and water reagents. The LAL producers usually use quantitative procedures in their own investigations.

By using a quantitative method, it is possible to trace the source of contamination in parenteral drugs whether it comes from the various constituents or if there is a growth of Gram-negative bacteria. Cases have been reported where endotoxin has been detected in samples earlier than bacteria. This may be the result of the speed of the endotoxin test and of its sensitivity, and also of the difficulties in finding proper conditions for bacterial growth. The savings that can be achieved by a careful control of the production are definitely of significance in the pharmaceutical industry as well as in most kinds of production where a high hygienic standard is necessary (see Section D.6). Air and water pollution in the production plant may be controlled as well.

B. SPECIFICITY OF THE LAL ASSAY

From what has been reported before, it is clear that the LAL test is unusually well documented (Table 4). A large number of combinations of endotoxins from different species present in or added to many types of parenteral drugs have been tested. Only a few compounds at very high concentrations, other than endotoxin, can activate the LAL reagent (Table 8).

Falsely low results can be obtained by aggregation and losses of endotoxin, and by inhibitory effects on the LAL activation as well as by compounds, which can influence gelation or precipitation. Misinterpretation of these reasons can, however, usually be avoided by using control experiments. Here a quantitative procedure is of great help.

Users of the LAL method in the control of parenteral drugs must show that they can perform the assay and that the assay is applicable to their products. Measures have to be taken to overcome inhibition and to make a suitable choice of measuring range (Table 9).

TABLE 8
Substances That Have Been Suspected of Causing False Positive Results.

Proteolytic enzymes — can be denatured by heat
Cellulose from certain filters — does not activate all types of LAL
CMPS — a carboxymethylated β-1,3-glucan — gives only low enzymatic activity
Peptidoglycan is about 10^3 to 10^5-fold less potent than endotoxin
Lipoteichoic acids from Gram-positive streptococcal or lactobacillus cell wall — have the same low potency
Polyribonucleoic acids
Palmitoyldextran — a phosphate derivative and polysaccharide

Note: At relative concentrations of 3 to 6 orders of magnitude above endotoxin these substances can give a positive LAL test. It is, however, unclear to what extent the effect is just due to endotoxin contamination.

TABLE 9
Some Strategies Useful for Sample Preparation and Method Modification

Choose a "suitable dilution" of the sample.
A "necessary dilution" can be titrated.
Biological samples should usually be heat denaturated.
Ultrafiltration devices are available.
The incubation times have to be modified.

TABLE 10
The Most Convenient Way to Avoid Interference by the Sample Constituents on the LAL Test is Dilution with Endotoxin-Free Water

		Suitable dilution of the sample to obtain 0.7/0.07 EU/ml	
Total dose (ml)	**Threshold conc (EU/ml)**	**0.1—1.0 EU/ml range**	**0.01—0.1 EU/ml range**
1	350	1:500	1:5000
10	35	1:50	1:500
100	3.5	1:5	1:50
1000	0.35	—	1:5

Note: For the clot procedure, the "maximum valid dilution" (MVD) has to be considered. In the chromogenic assay, a "suitable dilution" determines the threshold dose; 5 EU/kg body weight is administered within a single 1-h period. As the threshold dose has been defined using purified endotoxin standards and environmental endotoxin has a similar effect in the LAL test but ten times less effect regarding pyrogenicity, there seems to be a good safety margin. This dose will elicit a minimum pyrogenic response in 50% of the rabbits injected.

The COATEST® kit method is valid within a range of little more than one order of magnitude. Therefore, the sample usually has to be diluted to reach this range. The basic method, developed for water samples, is used in the 0.1 to 1.0 EU/ml range (1 EU ~ 0.1 ng) in the sample added to the LAL reagent. By increasing the incubation time, it is possible to increase the sensitivity more than ten times. Once the critical concentrations are determined, either by calculation of the "threshold pyrogenic dose" or by any other criteria, the sample can be diluted with endotoxin-free water to an extent so that this critical concentration becomes 0.7 or 0.07 EU/ml, respectively, or less in the sample in order to reach a suitable level on the scale used (Table 10).

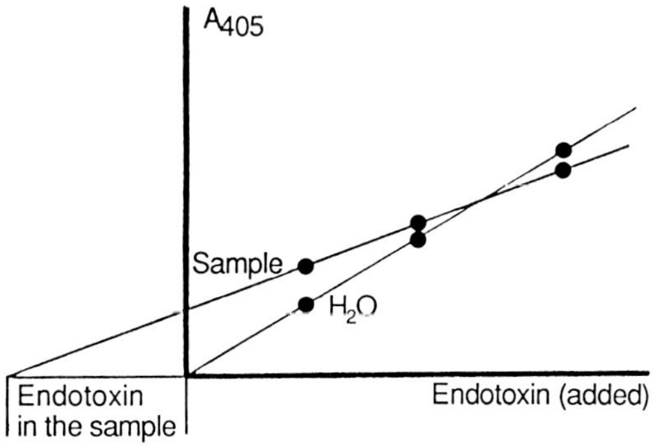

FIGURE 11. Internal standardization of a sample, which may also show inhibitory effects compared to water (less than 30%). For the plasma system, the H_2O curve is exchanged for the "pooled normal plasma" curve.

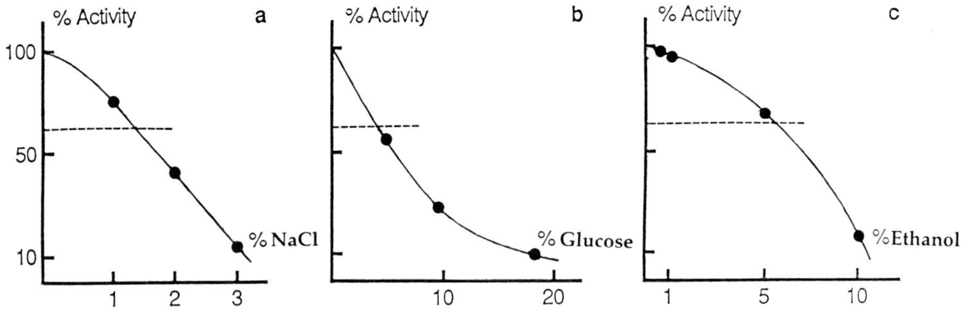

FIGURE 12. Effect of different concentrations of sodium chloride (a), glucose (b), and ethanol (c) on the chromogenic LAL assay. Samples containing these constituents should be diluted with endotoxin-free water to contain 1.3% NaCl, 4% glucose, or 6% ethanol or less. These levels can change slightly between LAL batches (From *Pharmeuropa,* 1, 5—8, 1989.)

A standard curve is established by adding endotoxin standard to a sample of an endotoxin-free batch of the product to be tested. If a linear correlation (e.g., r ≥0.98) is obtained, this curve or standards determined at the same time as the sample are used for calculation. Internal standardization is possible by using at least two standard dilutions (Figure 11).

The LAL reagent is influenced by most types of samples because this rather complex mixture of reagents has a narrow optimum regarding several reaction conditions (Figure 7). Therefore, inhibitory effects are usually expected. Stimulatory effects can usually be explained by elimination of binding to surfaces or to the formation of a more reactive aggregate.

The sensitivity can be extended to about 0.001 EU/ml, but this is only needed in a very few cases. It is possible to widen the range by using two different dilutions of the sample instead of duplicates of one single dilution. A five times difference (dilute 1 + 4) is suggested. Before establishing the procedure, it is necessary to prove that inhibitory effects are less than 30% and that a similar possible potentiation is less than 30%. If the method is modified to obtain an absorbance <1.5, a larger potentiation is acceptable. The inhibitory effects are studied by making standard curves using both endotoxin-free water and an endotoxin-free batch of the sample in question diluted to the predetermined level. This level can be built on "critical concentration" (see as follows) or on a test of inhibition using at least three different dilutions of the sample (Figure 12). Such a technique is chosen by those who wish to make the test as sensitive as possible, also making it possible to control the effect of

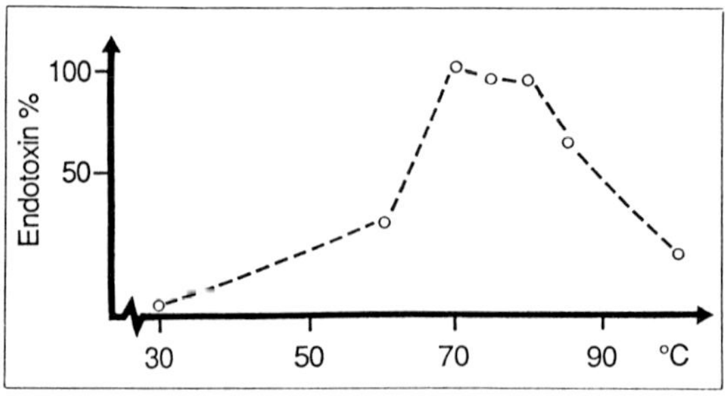

FIGURE 13. Plasma is diluted 1:10 with water and heated to a certain temperature for 5 min. At temperatures below 70°C inhibitors against the LAL enzymes are still active. Above 80°C some endotoxin is lost. After being heated to 70 to 80°C for 5 min the sample can be chilled and analyzed with the LAL test without centrifugation.

changes in handling the product in the production. The linearity of the standard curves also has to be established for each product in order to make it possible to utilize only two standards in the routine use of the assay.

In cases where the "threshold concentration" is very low because the product is used at a high volume or the inhibitory effect of the sample demands an unusually high dilution, it is necessary to increase the sensitivity of the method. This can be done primarily by increasing the first incubation time (activation) up to a maximum of 30 min at 37°C after which usually little increase in activity is obtained, and secondarily by increasing the substrate hydrolysis time. This time should probably also be limited to 30 min because of practical difficulties with high blanks (possibly due to minute contamination), which appear at long incubation times.

Effects due to proteins and other compounds present in biological samples are effectively eliminated by dilution and heat treatment, which denature such compounds. It is then important to avoid precipitation in which endotoxin can disappear. In samples containing or consisting of biological material, e.g., plasma fractions, protease inhibitors may influence the enzymes activated in the LAL. These inhibitors are conveniently destroyed by heating a 1:10 dilution to 70 to 80°C for 5 min in a water bath. The sample should then be mixed several times during heating (Figure 13). Purified albumin is, however, an exception because heat treatment of this product may result in a considerable loss of endotoxin activity. Fetal calf serum contains so little inhibitors that a dilution (e.g., 1:100) is preferred and also often needed because of high endotoxin content.

If these measures are not sufficient, it is also possible to remove low molecular weight (5 kDa or less) inhibitory substances by using a specially designed ultrafiltration unit, Ultrasart D20, from Sartorius GmbH, Germany (Figure 14). It is, however, important to ensure that endotoxin is not bound to the membrane.

Other types of treatment of the sample to avoid difficulties are as follows: $MgCl_2$ is added to samples containing metal-binding compounds. If acid is unsuitable for stopping the reaction with the chromogenic substrate, it is possible to use 10% SDS. When a fat emulsion is being tested, an organic solvent may be used to stop the reaction and at the same time to make the sample transparent.[1,5]

There are still unsolved questions regarding the cogency of endotoxin quantification. Do endotoxins from different species behave equivalently *in vitro* test systems and in the patient, except that the individuals respond differently? We certainly know very little about the quantitative aspects and must be very careful in our method design and in the evaluation of the results. The species differences between the Gram-negative bacteria, the occasion of

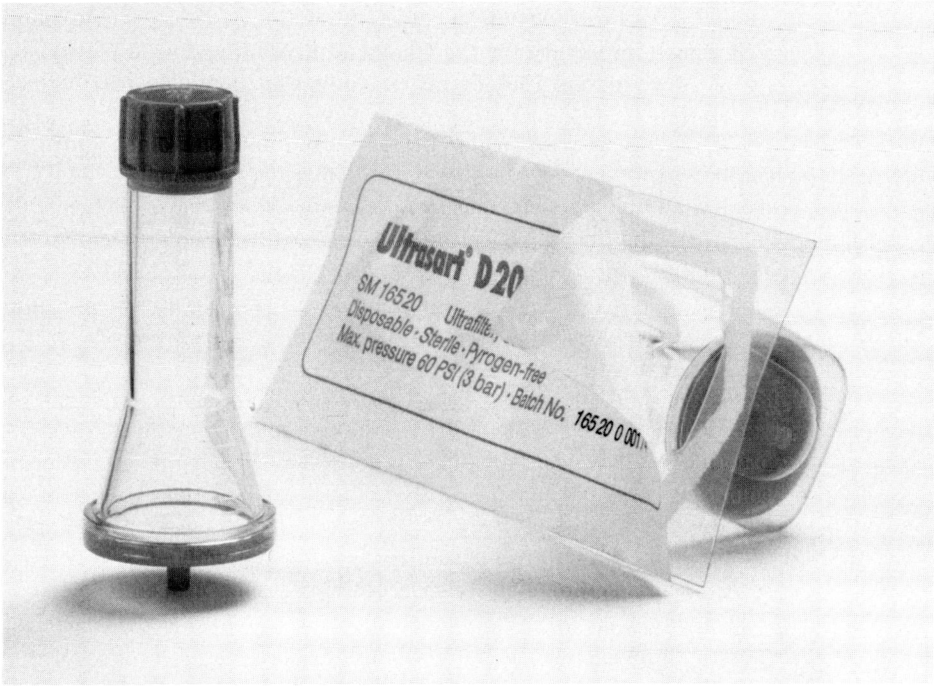

FIGURE 14. Ultrasart D20 filter is an endotoxin-free plastic device. The top is connected to a pressure tank, and the solution containing small molecules penetrates the filter placed in the bottom.

sampling, and the preparation of the samples can all seriously influence the results. Therefore, all sample handling has to be carefully standardized and control experiments have to be included.

It is often necessary to stop the bacterial growth after sampling by heating the sample. The effect of dissociation or aggregation should be controlled by standardization of procedures for sample preparation. Often, care must be taken that endotoxin is not lost on surfaces of the test tubes or in precipitates. Adsorption effects may be more critical for the extracted standard endotoxins than for the average sample unless the sample is very much diluted.

Diluted and heat-treated plasma gives practically the same effects as water, whether the endotoxin is added before or after heat treatment. It has been reported, however, that standard curves can vary between individuals (see Section D.1 following).

In performing sample manipulations, it is important to avoid contamination. Losses of endotoxin, especially purified endotoxin standards (e.g., due to adsorption to surfaces), can easily lead to significant misinterpretations. Addition of albumin, free of endotoxin, to a concentration of 0.1% has been found efficient in most cases, but cannot be considered completely to solve problems due to surfaces.

C. CHOICE OF METHODS

The basic chromogenic LAL test (Figure 10) is usually the best choice when adapting the technique to a new type of sample. It is designed to be safe and well controlled. For routine purposes, however, factors such as those in Table 11 may be worth considering. There are obviously possible method modifications. Some are briefly described here.

1. Single Assays

After reconstitution 100 μl of LAL can be dispensed into test tubes and immediately frozen. These tubes can be used one by one with or without standards over a 2-week period.

TABLE 11
Factors Important for the Choice of Procedure for Routine Endotoxin Determination

Equipment available
Number of tests in each series
Reproducibility needed
Ease of performance
Cost of reagents
Preference of technician and/or responsible person

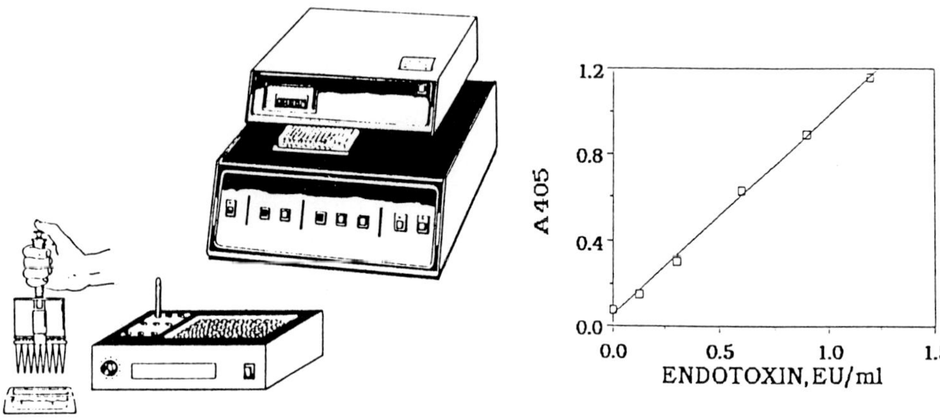

FIGURE 15. Reagents are added to the microplate with samples in the incubator. The absorbance is then read in a micro-ELISA reader. In each series 96 tests can be made. A standard curve is included.

The LAL should then be used immediately after thawing. This technique is suitable when simple screening of a single or a few samples occurs and when the reproducibiliy is not critical.

2. Multiple Assays

For large series of assays, microplate techniques are preferred (Figure 15). This is conveniently performed by scaling down the volumes to half, using multichannel pipettes or hand dispensers and an ELISA plate reader. Here errors can be introduced by mishandling the pipettes and by inefficient mixing and heating. Changing the incubation time from 10 to 13 min can compensate for the less efficient temperature regulation. In this test modification, 96 assays can be performed in less than 30 min. The smaller volume and comparatively fewer standards reduce the cost per assay. The procedure can be performed at room temperature by expanding the incubation time to 60 min and by choosing a suitable substrate reaction time between 3 and 10 min, depending on temperature and selected range (Figure 16). If the plate is read at several incubation times, it is possible to obtain data at a much wider range (see also Section 4).

3. A Simpler Procedure

If a simple single-stage procedure is preferred, the substrate is added to the LAL followed by the sample. A 20-min incubation (time depending on sensitivity needed) at 37°C is used before acid is added to stop the reaction.

4. A Wider Range

See a suggested procedure in Figure 17. It is possible to obtain a wider sensitivity range by using the sample at two dilutions (inhibition must be considered). Dilutions with fivefold

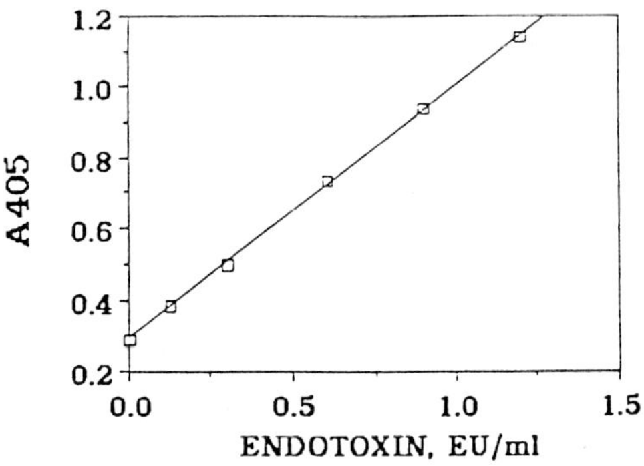

FIGURE 16. Standard curve obtained at room temperature (20°C) by using 60-min activation and 6-min color development.

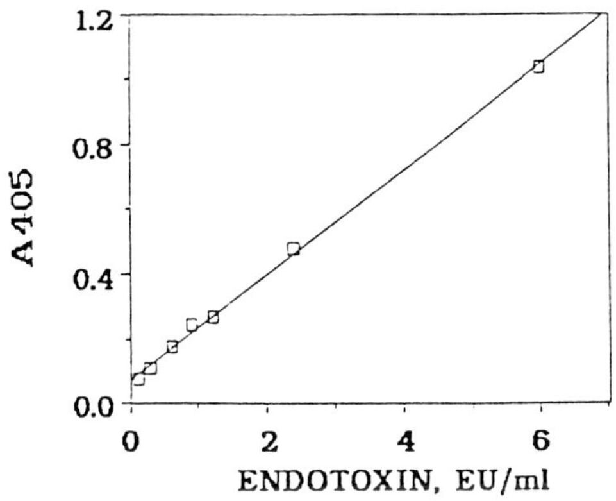

FIGURE 17. Wide range standard curve obtained by using 12-min activation and 2-min color development in a microplate.

differences may be a good choice. Deviation from a linear regression then has to be considered. The results can be estimated even by ocular inspection of the sample color. Another alternative is to use 15-min activation in a microplate technique and read the color after different hydrolysis times (Figure 9). A single-stage procedure and different incubation times can also be utilized. In such a case, standards within a wide range will have to be used.

5. Reduced Cost

A micromethod, using one fifth of all reagents, 15-min incubation in the first stage, and 4 min for color development, has also been suggested.[6] The reaction is terminated with 500 μl of 10% acetic acid to obtain a sufficient final volume to be read in a semimicro cuvette. This method has not been checked in our quality control, and some kit batches may behave unsatisfactorily in this method variation. Using a microplate technique, the incubation can be shorter and the reaction stopped by using only 200 μl or even less of 10% acetic acid.

6. Increased Sensitivity

An increased sensitivity or a strategy for analyzing an intensively yellow sample is also possible. The sensitivity of the assay can be increased by using diazotization or Schiff's base to potentiate the color.[7] This technique can also be used when the sample itself is strongly yellow.

7. Pass-or-Fail Technique

By using a pass-or-fail technique when there is a wide variation between amounts of endotoxin in the sample, leading to approval or rejection of the sample, it can be performed by using ocular inspection of the sample color. By using such a technique, it may be possible to control, e.g., certain perishables directly in the store without having to send samples to a laboratory (see also Section D.6). By such a test, both time and waste can be saved.

These are examples of how the measuring technique can be varied to fit the need of the user. Such variations can, however, create problems if the user has no experience with the basic technique.

D. USE IN THE CLINIC

1. Endotoxin Determination in Blood

Because of some practical difficulties in obtaining endotoxin-free platelet rich plasma for the preparation of standard curves in plasma, other media such as water and albumin solutions (0.5% or less) have been used. These curves gave consistently about 10% lower slopes than platelet-rich plasma, probably because of less efficiency in avoiding losses on surfaces in a purer system. The same results are obtained whether the endotoxin is added to the plasma before or after heat treatment.

Others have reported significant (3 to 10 times) stimulation by plasma in their assays. This is not expected, as interferences usually are negative because the reaction conditions are easily rendered less optimal, and most substances, which are supposed to interfere, are expected to be denaturated after the heat treatment. Losses in precipitates due to heat treatment can appear under certain circumstances, e.g., in Tris buffer. Such losses may create differences in standard curves between individual plasmas (from different individuals or at different sampling occasions) owing to differences in plasma constituents. In an investigation comparing 12 plasmas from patients and healthy persons we obtained no significant differences between individual plasmas and water.

It has been reported, however, that standard curves made by using plasma from certain individuals are different (usually lower) than those obtained with pooled normal plasma. Even healthy individuals gave different standard curves. Two types (Figure 18) or combinations of them may theoretically be found. Such effects can then lead to a significant underestimation of the severity of the disease in some patients. Bound endotoxin can, however, be recovered by adding 1% Tween 80 and sonicating the heat-treated plasma.[29] The possibility of adding surfactants before heat treatment has not been investigated yet. Possible additions to improve the solubility of the endotoxin or to decrease the losses in, e.g., plasma, are surfactants (Saporin, 1%); Pyrosperse®; trimethylamin; and SDS (0.1% and diluted 100 times before assay), combined with sonication.

The individual potency to eliminate endotoxin from the blood plasma can be determined by adding endotoxin *in vitro* to the patient's plasma, incubating it for a fixed-time interval (e.g., 30 min), and determining the residual endotoxin activity. The difference in activity between unincubated and incubated endotoxin activity is the "elimination capacity". Qualitative differences may also exist (cf. Figure 18). The amount of endotoxin to be used as well as the incubation time has still to be evaluated. The endotoxin neutralizing capacity is normally below 0.1 EU/ml (n >100) or practically undetectable. In a population of 100 patients, it varies between 0.2 to 8 EU/ml.[30] This type of assay may even give more relevant data than the ordinary endotoxin test, but a combination is probably preferred. Hopefully, available endotoxin standards can be used in such experiments to mimic the effect of environmental endotoxins. If the difference between individual plasmas is consistent and

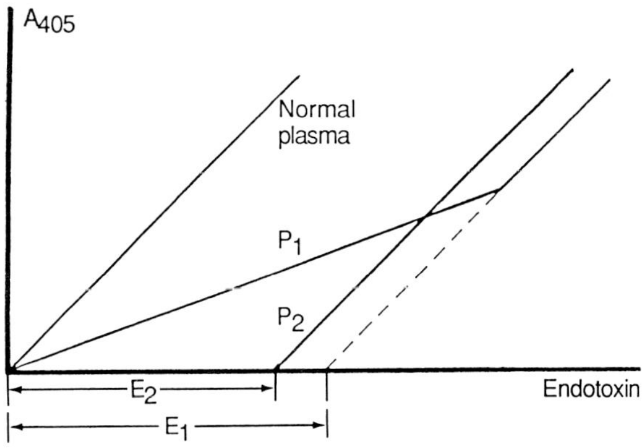

FIGURE 18. Endotoxin standard curve obtained in pooled normal plasma (practically equivalent to water) and standard curves performed in plasmas from individual patients have been reported to be different. Two types of effects are possible. In plasmas like P_1, there is a decreased sensitivity due to some plasma component in which concentration is taken as limited. In plasmas like P_2, it seems that some of the endotoxin is totally lost when added to the plasma. A mixture of the two types is also theoretically possible. It is quite obvious how the endotoxin neutralizing capacity (E_2) of P_2 is calculated, but regarding P_1, the effect (E_1) may be more difficult to derive.

not a method artifact, an internal standardization technique is necessary. By adding known amounts of endotoxin standard to the sample, it is possible to determine the endotoxin concentration by internal standardization (Figure 11). Such data also reveal if there is an inhibiting capacity present (fast-acting partial inhibiting effect).

A procedure suggested by Bode[29] utilizes the addition of 0.25 EU endotoxin standard per ml of plasma to an aliquot of the sample. Next both the sample and the "spiked" sample are analyzed, and the difference is a measure of the slope of the standard curve. Then it is possible to determine the amount of endotoxin present from the point where the standard curve cuts the X axis by setting $A_x = 0$ (cf. Figure 11), and use the absolute figure or subtract the absorbance of the sample from the spiked sample and calculate $X = 0.25 \cdot \frac{\text{A sample}}{\text{A } 0.25}$ EU/ml. The blank (caused by both the reagents and the sample) is determined without using any significant incubation time before the acid stop. If the adjusted reading for the sample is zero or negative, there is no endotoxin in the sample, but it may bind endotoxin if added. In such cases, it may even be advisable to use two different spikes (e.g., 0.25 and 0.50 EU/ml) to be sure to obtain a reliable value of the slope.

Another approach is that each individual sample is compared with a standard curve obtained in pooled normal plasma so that the "absolute" effect on the LAL system is the factor to be correlated to the clinical condition. That means that a plasma with a low slope on the standard curve will be underestimated regarding absolute endotoxin concentration, and one with high slope will be overestimated. This gives a combined measure regarding slope (sensitivity) and endotoxin concentration. Naturally the two measures: (1) endotoxin activity and (2) slope of the standard curve must also be investigated separately before the final method is settled. These still hypothetical procedures of evaluation and quantification will have to be investigated in practical experiments for final evaluation. Finding differences between individual plasmas, one may also suspect other biological samples to show individual properties.

TABLE 12
Risk Factors for Gram-Negative Bacteremia and Endotoxic Shock

Fatal underlying disease, infections, cancer
Newly born, especially premature
Advanced age (>60 years)
Cirrhosis, especially with ascites or biliary tract disease
Diabetes mellitus
Granulocytopenia
Multiple trauma or severe burns
Immunosuppressive therapy
Previous local infection, especially intraabdominal, urinary
Invasive device, receipt of large volumes of parenteral fluid
Hemodialysis
Irradiation of the abdomen
Major surgery, especially abdominal and gynecologic
Inappropriate antimicrobial therapy
Pancreatitis
Peritonitis
Meningitis
Diverticulitis
Ulcerative colitis

Clinical conditions concomitant with high endotoxin concentrations in blood, include sepsis in connection with surgery and multitrauma, malignant tumor, leukemia, decompensated liver cirrhosis, and systemic lupus erythematosus with uremia (Table 12).

Sepsis induces an activation of the complement, coagulation, fibrinolysis, and kallikrein-kinin systems. It has been demonstrated that a significant decrease/consumption in the plasma proteins antithrombin, plasminogen, and prekallikrein are early warnings of disseminated intravascular coagulation (DIC) and septic shock.[2,3] In surgical and multitrauma patients subjected to intensive care, it is important to follow, e.g., antithrombin concentration in plasma.[3] If it stays low, it is a warning for septic complications. If endotoxin assays in blood confirm this finding, a blood sample for culturing should be taken and treatment with a broad-spectrum antibiotic started immediately. Further clinical and laboratory signs are given in Table 13.

In a recent review, it was concluded that there were still some doubts about the reliability of endotoxin determination in blood.[8] In a more recently performed investigation,[9] endotoxin was determined in plasma and the results compared with bacterial culture. The conclusion was that the endotoxin method, utilizing a chromogenic substrate, was sensitive enough and that the specificity was excellent (99%).

In a prospective study,[10] 483 consecutive febrile patients were controlled using blood cultures and endotoxin determinations. The results obtained (Table 14) show that the endotoxin assay is a more accurate predictor of sepsis than the traditional bacterial culture test. As Gram-negative bacteremia without septicemia is common, a test for endotoxin is necessary to identify patients at risk for septicemia and septic shock. Septic shock in the absence of bacteremia is also common.

The same group has also recently reported an investigation on endotoxemia/bacteremia at urosepsis. The study involved 80 consecutive fibrile patients (>38°C) with a Gram-negative infection of the urinary tract (>15 leukocytes and $>10^5$ cfu [colony forming units] per ml of urine). Both tests had a sensitivity of 79%, but the endotoxin test had a specificity (94%), which was superior to the blood culture test (specificity 56%).

Endotoxin levels as high as or above 10 EU/ml are frequently found in systemic meningococcal disease. Such levels are always accompanied by severe septic shock, multiorgan failure, extensive coagulopathy, and complement activation.

TABLE 13
Clinical Signs and Laboratory Abnormalities Suggesting Bacteremic Infection in Different Stages

Fever, chills or shaking rigors, diaphoresis
Hypotension, circulatory shock
Hyperventilation, respiratory failure
Clouded mentation
Nonspecific gastrointestinal symptoms, pain, vomiting, diarrhea
Oliguria
Cutaneous lesions, mucosal bleeding, pustules, ecthyma
Leukocytosis, vacuolated neutrophils
Disseminated intravascular coagulation, thrombocytopenia, FDP
Hypocomplementemia
Liver function abnormalities
Respiratory alkalosis to acidosis with lactemia
Renal failure

TABLE 14
Results Obtained in a Prospective Study of 483 Patients Using a Bacterial Culturing and a Chromogenic Endotoxin Assay

Test	Sensitivity (%)	Specificity (%)	Pos. predictor value (%)	Neg. predictor value (%)
Culture	89	82	16	19
Endotoxin	78	96	48	99

Note: Found were 109 positive cultures and 31 endotoxin levels >0.05 EU/ml.

For the determination of endotoxin in blood, some features have to be considered. First, the method has to be very sensitive to make it possible to determine endotoxin in the 0.05 to 1 EU/ml range without obtaining errors due to contamination. A correct timing of sampling in relation to the development of the disease is also of great importance because the endotoxin concentration in the circulation varies considerably (Figure 3). The use of a chromogenic LAL method seems, however, to allow a surprisingly high specificity, even in clinical samples.

It is not yet clear what type of sample would be the best to analyze to obtain the most reliable diagnosis of endotoxemia. Platelet-rich plasma has been used most frequently but more recent data obtained from analyzing patients' plasmas indicate that platelet-poor plasma,[11] whole blood,[12] and blood serum[13] can be expected to give approximately the same results. Serum is suitable because it is often available with less risk of contamination. Regarding whole blood, however, there is an obvious risk of hemolysis. Working with plasma, one will avoid possible losses of endotoxin on fibrin, but it is has not been proved yet that these losses take place.

There are mainly three factors in plasma samples that have to be eliminated, namely, plasma inhibitors acting on the LAL enzymes, substances eliminating endotoxin, and proteases activated in septic patients' plasma, which can hydrolyze the chromogenic substrate. Therefore, the blood has to be sampled on ice, and then all three interfering activities have to be eliminated by dilution and heat treatment. Other procedures also exist. Chloroform extraction is now mostly abandoned, but alkaline and acid treatment are still used with apparently good results. Acidification, neutralization, and centrifugation are more laborious, however, than 5-min incubation at 70 to 80°C (Figure 13). The decrease in sensitivity due to a ten times dilution of the sample is no problem with the present reagents. When this technique of sample preparation is used, no precipitate is formed, and there is very little,

TABLE 15
Endotoxin Determination in Urine Using the COATEST® Kit

No. of patients	Sensitivity (%)	Specificity (%)
324	100	97
664	89	99

TABLE 16
Method for Endotoxin Detection in Urine[19]

Urine sample diluted 1:10 in 0.1% Tween 80 (room temp)	50 μl
LAL reagent (kept on ice or fresh at room temp)	50 μl
Incubate at room temperature for 5 min	
Substrate + buffer (1 + 1) mix and observe for a period not exceeding 1 ½ min	100 μl

Note: Positive test, strong distinctive yellow color.

From Saubolle, M. A., *Bacterial Endotoxins: Structure, Biomedical Significance, and Detection with Limulus Lysate Test,* Alan R. Liss, New York, 1985, 369. With permission.

if any, cross-reactivity by β-1,3-glucans.[14] The same good results are not found if another quantitative procedure, utilizing the lysate from *Tachypleus tridentatus*, is used.[15] The reason for this discrepancy may depend on differences in the enzyme system involved. Furthermore, this latter procedure is reported to lead to a much higher endotoxin-neutralizing effect.[16]

Every effort must be made to identify the source of bacteremia and, if possible, to remove it by such procedures as surgically draining an abscess, decompressing an obstructed ureter or common bile duct, closing an intestinal perforation, removing an infected intravascular catheter, discontinuing an intravenous or intraarterial infusion that may contain contaminants, or removing a contaminated ventilator or hemodialysis machine. Occasionally, the liquid used in hemodialysis contains endotoxin, and the filter is often ineffective to high amounts of endotoxin. Absorption on solid-phase bound *Polymyxins* or antibodies has been suggested. Plasmapheresis has been used with some success. Such measures are crucial to survival and in many instances are far superior to antimicrobial therapy in importance.

2. Endotoxin Determination in Urine

As many as 10% of elderly men and 20% of elderly women have chronic asymtomatic bacteriuria. As urine can contain large amounts of Gram-negative bacteria, there are no difficulties in detecting them with the chromogenic LAL test. The limit for clinical bacteriuria has been set at 10^5 cells per ml. In recent LAL tests performed with the gel or turbidimetric methods in 2056 patients, 296 of 303 patients with $\geq 10^5$ Gram-negative bacteria per ml showed positive LAL tests.[17] In 94% of the cases, the LAL procedure used could distinguish between urines above and below 10^5 Gram-negative bacteria per ml. Contamination from the outer urinary tract is rarely as high as 10^5 cells/ml. If that should happen, it is not expected to be reproduced a second time. A proper sampling is of great importance and has been successful in two recent investigations by Nachum[17] (Table 15).

The test used has been designed as a pass-or-fail type of test. It is extremely simple and can be performed in a small laboratory or in a doctor's office to identify bacteriuric patients in open health care (Table 16). Pregnant women with risk of perinatal morbidity and mortality, with pyelonephritis appearing as a common complication in untreated patients can easily be identified. In other categories of patients, other cut-off levels or other ranges may be of interest, and the assay has to be designed accordingly. The effect of inhibitors in urine

TABLE 17
Method for Endotoxin Detection in Cerebrospinal Fluid

Cerebrospinal fluid (incubate at 37°C)	100 μl
LAL reagent (room temperature immediately after reconstitution)	100 μl
Incubate at 37°C for 10 min	
Substrate + buffer (1 + 1) kept at 37°C	200 μl
Incubate at 37°C for 3 min	
Acetic acid 20%	200 μl

Note: Read A at 405 nm. Positive test, color more intense than the blank.

has not yet been extensively studied. Such inhibitors may originate from the kidney (naturally occurring), from the bacteria, or from the blood, e.g., nephrotic syndrome. They can certainly be eliminated by heat treatment in a similar fashion as with blood plasma.

3. Endotoxin Determination in Cerebrospinal Fluid

This application of the LAL test has been rather successful because plasma protein inhibitors are usually absent in the cerebrospinal fluid. The number of Gram-negative bacteria commonly found in the cerebrospinal fluid during meningitis (10^3 to 10^8 cells/ml) easily exceeds the threshold sensitivity of the traditional LAL test. Eight different studies totaling 1297 patients have been reported: 396 were diagnosed to have bacterial meningitis and 324 of them had the disease; 299 of these patients had a positive LAL test. This means a sensitivity of 92% and probably a still higher specificity. In spite of these very positive results, the test does not yet seem to be approved by the FDA.

The chromogenic LAL test has been evaluated in two investigations. Thomas et al.[18] found an endotoxin level in cerebrospinal fluid below 0.2 EU/ml in patients without meningitis. Saubolle[19] investigated 146 patients, and the method used gave an overall sensitivity of 93% with a specificity of 100% (Table 17). The chromogenic assay was found to be more sensitive than direct gram stain, culture, and antigen detection. The assay is fast and allows an appropriate treatment at an early stage of the disease. Interfering plasma compounds, which can leak from the blood compartment during a diseased state, can be destroyed by heat in a way analogous to plasma. In a retrospective review of data, it was shown that out of 1504 tests on cerebrospinal fluid all with positive Haemophilius or Neisseria gave a positive LAL test.[20] The overall sensitivity and specificity of the LAL test were 97 and 99%, respectively.

4. Endotoxin Determination in Urethral Exudates and Cervical Secretions

The LAL test has been used in several studies for the diagnosis of gonorrhea in men and women. By diluting the sample at least 1:50, and in most cases ten times more, it has been possible to avoid interference by endotoxins from other Gram-negative bacteria.

Summarizing data from 11 recent studies, the overall result was 98% sensitivity and 88% specificity.[17] More recent data with a chromogenic procedure combined with a new improved sampling technique used for testing 200 patients showed a sensitivity of 95% and specificity of 97%.[21] By using this technique, it is possible for a private physician to make a presumptive diagnosis of gonorrhea within 15 min without the need of a microscope and staining supplies. In contrast to bacteriologic techniques, the endotoxin assay does not give rise to false negatives due to antibiotic treatment.

5. Other Samples Which May Be Tested with a Chromogenic LAL Test

There is practically no limit to what the chromogenic LAL test can be used for. Mentioned in combination with the gel technique, are synovial fluid (septic arthritis), peritonitis fluid, corneal ulcer material, amniotic fluid, middle ear fluid, saliva, sputum, plaque from the teeth, tear fluid, and various rinse solutions.

Delayed recovery of chronic otitis media with effusions in children was correlated to the finding of endotoxin (chromogenic LAL assay) and lysosomal enzymes in mucoid (20 ± 41 pg/ml) and purulent (81 ± 83 pg/ml) middle ear effusions.

Smalley et al.[22] found that the chromogenic LAL test detected 66 (100% sensitivity, 88% specificity) culture-proven, Gram-negative peritonitis cases among 185 continuous ambulatory peritonal dialysis patients with clinical evidence of infectious peritonitis. The endotoxin content in the positive samples was greater or equal to 0.8 EU/ml. This test with an analysis time of 13 min will make an early initiation of appropriate therapy possible. The assay is not influenced by antibiotic therapy and can therefore be used for monitoring treatment efficacy (Clayman et al.)[23] A successful peritonitis treatment has also been reported to be accompanied by a substantial decrease in plasma endotoxin.

Complete gastrointestinal decontamination has been used with success in bone marrow transplantation. This has been controlled by using the chromogenic LAL method. The endotoxin concentration has then usually been decreased from about 1 mg/g feces to about 10 to 50 ng/g.

Amniotic fluid from 76 patients was tested with the LAL gel test and a 69% sensitivity was found.[24] Less than 5% false positives were found.

Endotoxins from Gram-negative bacteria play an important role in the development of symptoms and disease in workers in environments, such as cotton industries, sewage plants, and compost plants. Dust is sampled with cellulose acetate 0.8 μm filters (Millipore) and air with Cambridge absolute filter using a personal sampler or a stationary sampler in the breathing zone area, with a flow rate of 10 to 20 l/min and a total sampling volume of 100 to 800 l. The filter is then rinsed in 1% saporin for 20 min. Hot 1% SDS has also been found efficient.

Endotoxin background in the air is about 0.1 EU/m^3. At 2000 EU/m^3 acute bronchoconstriction and chronic inflammation are expected; at 5000 EU/m^3 chest tightness in connection with byssinosis is common and fever is obtained when the air contains more than 10,000 EU/m^3. Then large numbers of blood platelets migrate to the lung tissues followed by neutrophils.

In dialysis water (88 samples), the endotoxin concentration was found to be about 20 EU/ml. Out of 17 hemodialysis patients 16 showed elevated (0.2 to 5 EU/ml) concentrations in their serum (ref. level 0 to 0.2 EU/ml, n = 99) immediately before hemodialysis. In 13 of these patients, the endotoxin concentration was increased during and immediately after hemodialysis.

6. Use for Hygienic Control

The LAL test is suitable as a quick test for the control of hygienic conditions in all types of processing or production where Gram-negative bacteria appear. This is especially valid for all stages of food processing, water control, air pollution control, and hygienic control in factories and medical facilities (Table 18). The presence of endotoxin in food is rather common and the lack of effect of milligram amounts taken orally is understandable, when one realizes that the amount of endotoxin in the intestine is often larger than 2 g. Little is yet known about the effect when larger amounts are consumed. Gram-negative bacteria and endotoxins are, however, important parameters in the study of hygienic conditions of products and in their production and storage.

A reason why it is suitable to measure just endotoxin in foodstuff is that the growth of Gram-negative bacteria is significant at the low temperature at which the food is usually stored. Endotoxin can also be determined in boiled, irradiated, and otherwise treated food where all bacteria have been killed. In this respect, the endotoxin test is rather unique. Many people are exposed to endotoxin in their professions, cotton products manufacturing, chicken and pig farms, car wash, paper production, waste water treatment, leading to byssinasis.

TABLE 18
Cases Where an Endotoxin Determination Is Suitable in Hygienic Control

Meat, fish, egg, and dairy production
Production plants and machines
Domestic and imported raw material
Evaluation of new techniques of food handling
Shelf life of food in storehouses and shops
Restaurants and similar facilities
Fresh- and wastewater
Air in farms and factories
During air conditioning and humidification
Health care facilities and veterinary dispensaries
In development of medical care devices and facilities

Bacterial culture and endotoxin determination are complementary, each providing important information about the hygienic state of the material tested.

Water—The chromogenic LAL test (Figure 10) is attractive for a bacterial test of water, because it is very fast and nearly all the bacteria found in high purity water as well as in the marine environment are Gram-negative, purified laboratory water containing anywhere from 0.01 to 10 EU/ml (1 EU $\approx$ 0.1 ng). Distillation or filtration with a positively charged filter can then be used to render the water endotoxin free ($<$0.01 EU/ml). The quality of the water used in, e.g., the production of electronic chips has been successfully controlled with the chromogenic LAL method.

Milk—The amount of endotoxin found in milk is in the order of 10 to 100 EU/ml, and in milk from cattle with mastitis of various kinds, more than 100 EU/ml is regularly found. The correlation between bacterial counts and endotoxin in ultrahigh-temperature-treated milk is usually very good, 10 EU of endotoxin corresponding to approximately 10^4 cfu (in raw milk $2 \cdot 10^3$ cfu). After pasteurization, the expected amount of endotoxin is, however, only recovered to about 50% (cf. Figure 13). Gram-negative saprophytic organisms are mainly responsible for proteolytic and lipolytic degradation of milk and milk products, and these processes result in sensoric failures in cell-culturing procedures. Working with a comparatively narrow range such as 10 to 100 EU/ml, a chromogenic LAL test is very suitable to follow the production in terms of cumulative contamination by Gram-negative bacteria. As endotoxins may exist as whole cells, cell fragments, or smaller particles, and thus may behave differently in the LAL test, a proper standardization of the sample treatment is crucial.

Meat—The endotoxin concentration has been shown to increase with onset of meat spoilage.[26] Factors enabling an estimate of aerobic mesophilic count from the endotoxin concentration was made using a LAL method capable of detecting 10^2 to 10^3 cfu/g. Normal contamination with Gram-negative bacteria of pork is $5 \cdot 10^3$ to 10^6 cfu/g; in beef it is 10^2 to 10^5 cfu/g; and in minced meat, $5 \cdot 10^6$ to 10^7 cfu/g. The Gram-negative *Pseudomonas* (at 0 to 5°C), *Acinetobacter* (room temperature), and *Moraxella* dominate. Endotoxin content in ground beef and fish, as determined with the LAL test, can be used as a rapid approximation of viable plate counts. Only at high endotoxin levels may an additional test of microbial quality be necessary. The production of animal food (dog, cat, mink, etc.) from second selection raw material like cadavers or other material from slaughterhouses, where a contamination or resorption from the intestines is possible, should be controlled. Also, meat rejected because of bacterial contamination has been processed and sold as approved goods. These are typical cases, disclosed by journalistic investigations, where a quantitative endotoxin test could have revealed the truth and even given guidance to a correct handling. Officials of public health could use a quantitative LAL test to clarify many types of cheating as exemplified before.

Miscellaneous—In all types of biotechnical industry, a control of Gram-negative contamination of, e.g., yeast fermentation is important. When, e.g., *Escherichia coli* is used for gene technological purposes, all endotoxins must be eliminated. Such work has to be monitored and controlled with sensitive quantitative assays. At endotoxin levels of 2000 EU/m^3 or more in the air, pathological reaction can be expected (see Section 5.D). This has to be considered in, e.g., farming and industries processing flax and cotton. Clinically, endotoxin inhalations may cause fever, bronchoconstriction, and chronic irritation with mucos gland hyperplasia. Many commercially available liquid culture media (e.g., fetal calf serum) have been found to contain more than 250 EU/l endotoxin. They can be purified by using Polymyxin B Sepharose gel with a capacity to bind more than 2 EU/ml.[27] Besides these areas of applications, there are numerous occasions where a quick and simple test for hygienic control is very much appreciated. It can be used in connection with control of any kind of health care facilities and instruments. The value of this is to protect both goods and employees. In the control of warehouses, shops, and restaurants as well as in the control of containers for transport and storage, a simple and quick test is very useful. Whenever a bacterial culturing is performed, an endotoxin test is a faster and usually better choice.

E. GUIDELINES FOR METHOD ADAPTATION

The chromogenic LAL method developed for water samples is used in the 0.1 to 1.2 EU/ml (1 EU = 0.1 ng) range. The sample has to be diluted to an extent so that this range is reached. Often several different dilutions have to be analyzed. Steps of about five times dilution would make it possible to have at least one dilution within the expected range.

Before establishing the procedure, it is necessary to prove that inhibitory effects are less than 30%. This is performed by making standard curves using both endotoxin-free water and an endotoxin-free sample of the preparation to be analyzed, diluted to the predetermined level.

In cases where a high sensitivity is needed, e.g., because of a large dilution to overcome inhibition, this can be obtained by increasing the first incubation time (activation) up to approximately 30 min and/or increasing the substrate hydrolysis time. This time should probably also be limited to 30 min because of practical difficulties with high blanks, which often appear with long incubation times.

In samples containing or consisting of biological matter, especially blood and meat, protease inhibitors may influence the enzymes activated in the LAL. These inhibitors are conveniently destroyed by heating a 1:10 water dilution to 70 to 80°C for 5 min in a water bath (Figure 13). The sample should be mixed several times during heating. If these measures are not sufficient, it is possible to remove low molecular weight (5 kDa or less) substances by using a specially designed ultrafiltration unit (Ultrasart D20) from Sartorius GmbH, Göttingen, Germany.

Meat—Samples are taken from the surface layer of meat or fish, ground, heat-treated, and centrifuged. Bacteria in the sample may be killed by heating to 100°C for 5 to 10 min. If 1% SDS is used, a more efficient solubilization of the endotoxin may be obtained but interfering substances can also appear. SDS may also interfere with the LAL assay, if not removed or diluted extensively (100 times). Usually the supernatant is added to the LAL reagent. Another alternative is to cut the meat in thin slices to avoid contamination; 5 g is added to 100 ml endotoxin-free water. An ultrasonic mixer is used for extraction. A sample is taken and diluted ten times and then further diluted 1:5 several times with endotoxin-free water. The diluted samples are heat-treated at 75°C for 7 min. After cooling, the samples are analyzed using the water method. If the sample contains substances, which bind metal ions, $MgCl_2$ can be added. The substrate reaction can be stopped by using SDS when acetic acid is not suitable. If the sample is a fat emulsion, an organic solvent can be used to stop the color development.

Milk—Before assay, milk is prepared in a similar way to plasma. It has been suggested that 20 ml of milk (2 g of cheese, butter, or dry milk is added to 18 ml of endotoxin-free water) is heated in boiling water for 15 min to kill all bacteria. After cooling, 1.0 ml of endotoxin-free 15% lactic acid is added to the milk. The sample is centrifuged for 15 min at 1000 *g*. A 5 ml amount of the supernatant is neutralized with 0.4 ml 1 mol/l Tris and centrifuged for 15 min at 5000 *g* before assay.[28] As the endotoxin concentration in milk is high, a considerable dilution 10 to 100 times is necessary. This can be done according to the previous description by directly using a more diluted Tris buffer. An alternative method of pretreatment is just to dilute the milk, e.g., ten times with 0.02 mol/l Tris buffer, pH 7.5, before assay. The color development is later stopped by adding acetic acid containing 0.1 mol/l $CaCl_2$, and the precipitate is removed by using 0.45 μm membrane filters (Schleicher & Schnell GmbH, Dassel, Germany) before measurement. Single control of a final product is at present the most common. By using a quantitative assay it is now possible to control many stages in the production and thus make it possible to find sources of contamination. This creates an opportunity for the present assay to be a great cost saver in many types of food refinement and production, and also leads to an improved security in health.

F. FUTURE POSSIBILITIES

Most of what is described in this chapter belongs in fact to the headline of "future possibilities" because the chromogenic LAL test, in most applications, only recently has appeared in the research community. It is now time to use it in practical routine, not only to avoid the dangerous effects of endotoxin, but also to learn how to deal with contamination of various kinds. Being a sensitive quantitative method, it will be possible to avoid all types of interferences mainly by predilution of the sample. Contamination and losses due to adsorption to surfaces can now be handled more satisfactorily. New automated or semiautomated equipment is also available, both to conduct quick single determinations and to perform the assay on large series of samples.

VI. SUMMARY

Endotoxins constitute a class of substances with the same basic structure. They are of great interest because they are a definite threat to life if they affect us in an adverse way. The sometimes rather obscure mechanisms of endotoxin are discussed. The LAL test is suitable to determine small amounts at the picogram level of endotoxin. The assay is, however, sensitive to contamination, to losses on surfaces, and to most substances, which at high concentrations interfere with the reaction. Each type of sample, therefore, demands its own method variation. Guidelines on how to design suitable methods for the determination of endotoxin in various samples (e.g., pharmaceuticals; clinical samples like blood, urine, cerebrospinal fluid, etc.; and food stuff) are given in this chapter. In many cases dilution and heat treatment (70 to 80°C) have been successful in eliminating inhibitory effects and interfering activities. Relevant data collected during more than 10 years of work with the LAL assay are described and utilized to offer the reader advice in handling endotoxin assays.

ACKNOWLEDGMENT

In addition to published data, this text deals with ideas resulting from discussions with numerous friends and colleagues whom I acknowledge here collectively.

REFERENCES

1. **Friberger, P., Sörskog, L., Nilsson, K., and Knös, M.,** The use of a quantitative assay in endotoxin testing, in *Detection of Bacterial Endotoxins with the Limulus Amebocyte Lysate Test,* Watson, S. W., Levin, J. P., and Novitsky, T. J., Eds., Alan R. Liss, New York, 1987, 149.
2. **van Deventer, S. J. H., Knepper, A., Landsman, J., Lawson, J., tenCate, J. W., Sturk, A., and Pauw, W.,** Endotoxin in portal blood, *Hepatogastroenterology,* 35, 223, 1988.
3. **Ditter, B., Becker, K. P., Urbaschek, R., and Urbaschek, B.,** Quantitativeer Endotoxin-Nachweis Automatisierter, Kinetischer Limulus-Amöbozyten-Lysat-Microtiten-Test mit Messung probenabhängigen Interferenzen, *Arzneim.-Forsch.,* 33, 681, 1983.
4. **Friberger, P., Knös, M., and Mellstam, L.,** A quantitative endotoxin assay utilizing Limulus lysate and a chromogenic peptide substrate, in *Endotoxin and Their Detection with the Limulus Amoebocyte Lysate Test,* Watson, S. W., Levin, J. P., and Novitsky, T. J., Eds., Alan R. Liss, New York, 1982, 195.
5. **Friberger, P.,** The design of a reliable endotoxin test, in *Bacterial Endotoxins: Structure, Biomedical Significance and Detection with the Limulus Amoebocyte Lysate Test,* tenCate, J. W., Büller, H. R., Sturk, A., Levin, J., Eds., Alan R. Liss, New York, 1985, 405.
6. **Martens H. J. M.,** Quantitative photometric determination of endotoxin in water and parenterals, *(Dutch) Ziekenhuisfarmacie,* 1, 89, 1985.
7. **Friedman, R. B., Kwaan, H. C., and Szczecinsky, M.,** Improved sensitivity of chromogenic substrate assays for urokinase and plasmin, *Thromb. Res.,* 12, 37, 1977.
8. **Elin, R. J. and Hosseini, J.,** Clinical utility of the Limulus Amebocyte Lysate (LAL) test, in *Bacterial Endotoxins; Structure, Biomedical Significance and Detection with the Limulus Amebocyte Lysate Test,* tenCate, J. W., Büller, H. R., Sturk, A., Levin, J., Eds., Alan R. Liss, New York, 1985, 307.
9. **Büller, H. R., tenCate, J. W., Sturk, A., and Thomas, L. L. M.,** Validity of the endotoxin assay in postsurgical patients, in *Bacterial Endotoxins, Structure, Biomedical Significance and Detection with the Limulus Amebocyte Lysate Test,* tenCate, J. W., Büller, H. R., Sturk, A., Levin, J., Eds. Alan R. Liss, New York 1985, 405.
10. **van Deventer, S. J. H., Pauw, W., tenCate, J. W., Jansen, M. E., Büller, H. R., and Sturk, A.,** Clinical evaluation of febrile patients of an optimized endotoxin assay in blood, in *Detection of Bacterial Endotoxins with the Limulus Lysate Test,* Watson, S. W., Levin, J., and Novitsky, T. J., Eds., Alan R. Liss, New York, 1987, 489.
11. **Tachiyama, G., Sakon, M., Kambayashi, J., Ohshiro, T., and Mori, T.,** Chromogenic assay of endotoxin in platelet poor or rich plasma, *Thromb. Res.,* 41, 309, 1986.
12. **Sakon, M., Tachiyama, G., Kambayashi, J., Ohshiro, T., and Mori, T.,** Studies on blood sample preparation for clinical endotoxin assay, *Thromb. Res.,* 43, 361, 1986.
13. **Eriksson, Ö., Friberger, P., Sjölin, J., Kulander, L., and Friman, G.,** Endotoxinbestämning en ny metod att påvisa gramnegativ septikemi, *Läkartidningen,* 85, 3390, 1988.
14. **Friberger, P.,** Chromogenic peptide substrates. Their use for the assay of factors in the fibrinolytic and the plasma kallikrein-kinin system, *Scand. J. Clin. Lab. Invest.,* 42 (Suppl. 162), 87, 1982.
15. **Obayashi, T., Tamura, H., Tanaka, S., Ohki, M., Takahashi, S., Arai, M., Masuda, M., and Kawai, T.,** A new chromogenic endotoxin-specific assay using recombined limulus coagulation enzymes and its clinical applications, *Clin. Chim. Acta,* 149, 55, 1985.
16. **Yajima, Y., Fukuda, I., Otsuki, M., Suzuki, H., and Goto, Y.,** Stability of endotoxin detected in human plasma against endotoxin-inactivating factor (EIF): quantitative analysis of EIF using chromogenic endotoxin assay, *Tohoken J. Exp. Med.,* 150, 317, 1986.
17. **Nachum, R.,** Usefulness of LAL assay for rapid detection of Gram-negative bacteriuria, in *Detection of Bacterial Endotoxins with the Limulus Amebocyte Lysate Test,* Watson, S. W., Levin, J., and Novistky, T. J., Eds., Alan R. Liss, New York, 1987, 431.
18. **Thomas, L. L. M., tenCate, H., Sturk, A., Joop, K., and tenCate, J. W.,** Quantitative chromogenic endotoxin determination in cerebrospinal fluid, *Clin. Chim. Acta,* 127, 137, 1983.
19. **Saubolle, M. A.,** Chromogenic limulus amebocyte lysate assays as an aid in the diagnosis of meningitis, in *Bacterial Endotoxins: Structure, Biomedical, Significance, and Detection with Limulus Lysate Test,* tenCate, J. W., Büller, H. R., Sturk, A., and Levin, J., Eds., Alan R. Liss, New York, 1985, 369.
20. **Saubolle, M. A. and Jorgensen, J. H.,** Use of the limulus amebocyte lysate test as a cost-effective screen for Gram-negative agents of meningitis, *Diagn. Microbiol. Infect. Dis.,* 7, 177, 1987.
21. **Prior, R. B. and Spanga, V. A.,** Rapid evaluation of gonococcal and non-gonococcal methritis in men with limulus amebocyte lysate and a chromogenic substrate, *J. Clin. Microbiol.,* 17, 485, 1983.
22. **Smalley, D. L., Baddour, L. M., and Kraus, Jr., A. P.,** Rapid detection of Gram-negative bacterial peritoritis by the limulus amalbocyte lysate assay, *J. Clin. Microbiol.,* 24, 882, 1987.
23. **Clayman, M. D., Raymond, A., Colon, D., Moffitt, C., Wolf, C., and Neilson, E. G.,** The limulus embocyte lysate assay. A rapid and sensitive method for diagnosing early Gram-negative peritonitis in patients undergoing continuous ambulatory peritoreal dialysis, *Arch. Intern. Med.,* 147, 337, 1987.

24. **Romero, R., Kadan, N., Hobbins, J. C., and Duff, G. W.,** Infection and labor: the detection of endotoxin in amniotic fluid, *Am. J. Obstet. Gynecol.,* 157, 815, 1987.
25. **Fallowfield, H. J. R. and Patterson, J. T.,** Potential value of the limulus lysate assay for the measurement of meat spoilage, *J. Food Technol.,* 20, 467, 1985.
26. **Mølvig, J. and Baek, L.,** Removal of endotoxin from culture media by a polymyxin B Sepharose column, *Scand. J. Immunol.,* 26, 611, 1987.
27. **Mottar, J.,** A colorimetric endotoxin assay for the determination of the bacteriological quality of milk, *Neth. Milk Dairy,* 41, 137, 1987.
28. **Svensson, A. and Hahn-Hägerdal, B.,** Comparison of a gelation and a chromogenic limulus (LAL) assay for the detection of Gram-negative bacteria, and the application of the latter assay to milk, *J. Dairy Res.,* 267, 1987.
29. **Bode, C.,** personal communications.
30. **Becker, K.,** personal communications.

Taxonomic Index

TAXONOMIC INDEX

A

B

C

D

N

O

P

R

S

Subject Index

SUBJECT INDEX

A

C

D

E

F

G

H

I

J

K

L

N

O

P

Q

R

S

T

U

V

W

X

Y

Z